T0155342

NEUROMETHODS

Series Editor
Wolfgang Walz
University of Saskatchewan
Saskatoon, SK, Canada

For further volumes:
http://www.springer.com/series/7657

Lateralized Brain Functions

Methods in Human and Non-Human Species

Edited by

Lesley J. Rogers

School of Science and Technology, University of New England, Armidale, Australia

Giorgio Vallortigara

Centre for Mind/Brain Sciences, University of Trento, Trento, Trento, Italy

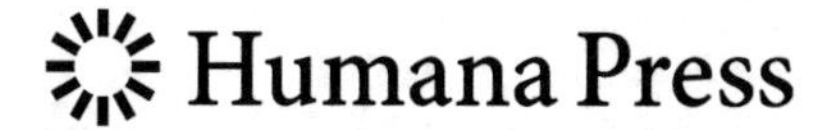

Editors
Lesley J. Rogers
School of Science and Technology
University of New England
Armidale, Australia

Giorgio Vallortigara
Centre for Mind/Brain Sciences
University of Trento
Trento, Trento, Italy

ISSN 0893-2336 ISSN 1940-6045 (electronic)
Neuromethods
ISBN 978-1-4939-8283-7 ISBN 978-1-4939-6725-4 (eBook)
DOI 10.1007/978-1-4939-6725-4

Softcover reprint of the hardcover 1st edition 2017

Printed on acid-free paper

This Humana Press imprint is published by Springer Nature
The registered company is Springer Science+Business Media LLC
The registered company address is: 233 Springer Street, New York, NY 10013, U.S.A.

Preface

Brief History of the Field and the Role of Animal Models

Beginning with the first discoveries of lateralized brain function and behavior in two avian species and rats, made in the 1970s and early 1980s (1–3), investigation of lateralization in nonhuman species has become a burgeoning field of research, expanding to include a wide number of vertebrate species (4–9) and, more recently, even to include invertebrate species (10, 11).

The availability of animal models for studying lateralization has produced two important changes. Firstly, it has promoted a resurgence of interest in a research area that previously had been confined to human neuropsychology and neurology and had shown signs of declining interest because it appeared to be incapable of tackling the most basic issues, namely the causes of lateralization of the brain and its biological function (12). The entering of animal biology into the study of lateralization has meant not only the availability of new system models with better accessibility to neural structures and functions but also different ways of considering the development of lateralization and its origins and biological significance in evolutionary terms (13). In regard to the latter, for example, theoretical tools from game theory and population genetics have prompted unexpected advances in our understanding of the evolution of brain asymmetry (14–18).

A second important change, which ultimately provided the impetus for this book, relied on the availability of animal models that allowed scientists to develop and use new techniques for studying brain and behavioral lateralization. In addition, it is recognized that these animal models are potentially able to enhance understanding of individual differences in behavior, including aberrant or unusual behavioral and neural conditions in humans, as well as in nonhuman species (19–21). Many different experimental approaches have been applied, each adapted to reveal lateralization in different species or to do so more precisely in species already known to be lateralized. This has led to a broad range of techniques with varying success in revealing lateralization of both brain function and brain structure.

Research in the neuroscience of lateralization in humans has, of course, also progressed in recent years, especially since neuroimaging and neurostimulation techniques have become a source of new data on lateralization. Interestingly, however, the development of techniques to be used with animal model systems has also had an influence on research of lateralization in humans. An example is provided by the ethological methods developed to investigate lateralization in everyday behavior in nonhuman species, methods that are now used widely in humans as well (22, 23).

Now that the field has been established it seems timely to publish in one volume the various methods used to measure lateralization in the different species. Although functional lateralization has been investigated in humans for many decades and in nonhuman species increasingly so over the last two or three decades, the methods used have not previously been gathered together in one volume. Therefore, we invited leaders in the field to write chapters on the methods they have used to investigate functional lateralization in a range of different species, including humans. We hope that by collecting these contributions together in one volume we are able to assist newcomers to the field. Also by presenting the methods adopted to investigate lateralization in a broad range of species, we hope to stimulate new research and provide a basis from which hypotheses can be tested and compared across species.

New methods developed to test different species need to take into account species differences in sensory and motor systems. We expect this line of study to continue as researchers explore the evolution of brain and behavioral lateralization (24). In fact, study of a wide range of vertebrate species has led to the realization that a basic pattern of asymmetry of function is common across vertebrate species (25). The right hemisphere is specialized to attend to novel stimuli, including predators, to control social behavior, recognize faces, and process global information using spatial cues (26). The left hemisphere is specialized for focused attention needed to perform learned tasks, to follow rules, and to categorize stimuli (9). This division of function is present in humans and other vertebrate species.

Important to understanding lateralization is knowledge of how it develops. For this direction of research a few key species have been the focus of detailed study, including domestic chicks (e.g., 27–29), pigeons (30), and zebrafish (31–34). These species were chosen as models for studying the development of lateralization because their stages of development are known rather precisely and because sensory inputs can be manipulated with the aim of determining the role of sensory experience in the development of lateralization. In all three of these species, there is clear evidence that exposing the developing embryo to light is essential for the development of visual lateralization. For example, light exposure of avian embryos during the final stages of development before hatching stimulates the right eye but not the left because the embryo's head is twisted to the side so that the left eye is occluded by the embryo's body (8). This difference in stimulation of the left and right eyes at a time when connections between eye and brain first become functional is essential in establishing certain visual lateralities, a fact known because these functions are not lateralized if the embryos are incubated in the dark during this critical stage of development (see Chap. 19).

To discover the causation of each type of functional lateralization is another approach taken by researchers of brain lateralization. This has led to detailed examination of the structural differences between the left and right hemispheres, examined in both humans and other species, as well as hemispheric differences in connections between neurons and in neurotransmitter concentrations.

Of course, detailed knowledge of the function of lateralization is essential. What are the advantages of having a lateralized brain and, on the contrary, what disadvantages might be apparent? In this direction of investigation, it is important to consider two distinct types of lateralization: individual and population lateralization (24). Individual lateralization is that present in individual members of a species but favoring the left in some individuals and the right side in others, resulting in no directional bias within the group or population. Population lateralization, also referred to as directional lateralization, is present when the majority of individuals are lateralized in the same direction. Handedness in humans is an example of the latter, as also is the lateralization of visual processing in birds that we have discussed above.

In this volume we have included some chapters with methods that are simple to apply and others that require more sophisticated techniques, often newly available (e.g., Chap. 11 on optogenetics). Each set of methods can lead to discovery of different levels or types of lateralization and, contrary to one approach superseding another, all methods can be complementary in advancing understanding.

Chapters 1–6 address measuring lateralization by scoring behavior induced by inputs to one or the other side of the brain and in a range of species. These chapters include some of the classic methods developed by experimental psychologists to deal with hemispheric specialization, such as tachistoscopic viewing and dichotic listening (Chap. 1 by Ocklenburg), which have been improved to the highest level of technical precision and sophistication, and the study of split-brain patients of the clinical neuropsychological tradition (Chap. 2 by Fabri, Foschi, Pierpaoli, and Polonara). The roots of all these methods can be traced back to the early sensory physiology and psychology of the nineteenth century. In fact, it was Gustav Fechner, the founder of psychophysics, who made the first inquiry about the possible outcome of disconnecting the two cerebral hemispheres: "The two cerebral hemispheres, while beginning with the same moods, predispositions, knowledge, and memories, indeed the same consciousness generally, will [when divided through the middle] thereafter develop differently according to the external relations into which each will enter" (Gustav Theodor Fechner, 1860, in Zangwill 1974 (35)).

New methods have come from ethology and include measurements of eye preferences (particularly in animals with laterally placed eyes and complete decussation at the optic chiasma) and ear preferences (Chap. 3 by Rogers), as well as preferences in nostril use and olfactory stimulation (Chap. 4 by Siniscalchi). Behavioral methods have been developed to study lateralization in invertebrates (Chap. 6 by Frasnelli); thanks to the limited number of neurons in these model species, the combination of behavioral analyses together with the sophisticated molecular and genetic techniques available for some invertebrate species (e.g., the fruit fly) research on invertebrates promises the possibility of important breakthroughs in the study of brain asymmetries in the years to come. In fact, the study of hand and limb preferences, which has been traditionally the province of an allegedly unique phenomenon, i.e., human handedness, has been deeply challenged by the mounting evidence from studies on nonhuman primates and other species, both mammals and birds, showing a variety of asymmetries in limb usage (36). Here the availability of precise techniques of recording the use of the limbs in natural and seminatural conditions may prove to be crucial in comparing strength and direction of handedness in different species (Chap. 5 by Forrester).

Chapters 7–11 cover neurobiological methods used to reveal lateralization. Again, these include both well-established techniques such as lesion studies (Chap. 7 by Manns), electrophysiology and pharmacology (Chap. 8 by McCabe), tract tracing (Chap. 9 by Stöckens and Güntürkün), and early gene expression (Chap. 10 by Patton, Uysal, Kellog, and Shimizu), as well as the new optogenetic methods that allow selective activation or blocking of specific circuits or synapses in the left or right side of the brain (Chap. 11 by El-Gaby, Kohl, and Paulsen).

Chapters 12–15 address imaging techniques, electroencephalographic techniques, and transcranial stimulation to reveal lateralization. The mixture of human and nonhuman animal research here is apparent, with noninvasive techniques such as transcranial magnetic stimulation (Chap. 12 by Cattaneo) and electroencephalographic stimulation (Chap. 13 by Mazza and Pagno) mainly used for research in humans, fMRI imaging used to compare human and nonhuman primates species (Chap. 14 by Hopkins and Phillips), and other more recent methods with much finer spatial and temporal resolution, such as two-photon microscopy, confined at present to system models such as insects (Chap. 15 by Paoli, Andrione, and Haase).

For a long time the genetic basis of lateralization was limited to the study of inheritance of handedness in humans (37). New genetic techniques in studying lateralization in humans (Chap. 16 by Paracchini and Scerri), zebrafish (Chap. 17 by Duboué and Halpern), and *C. elegans* (Chap. 18 by Vidal and Hobert) are covered in Chaps. 16–18 describing powerful techniques to address the role of genes in the establishment and development of brain asymmetry.

Last, but by no means least, are Chaps. 19–21 covering methods used to study the development of lateralization and to do so by manipulation of sensory exposure (Chap. 19 by Chiandetti), hormone levels (Chap. 20 by Beking, Geuze, and Groothuis), and model systems for the study of lateralized development (Chap. 21 by Blackiston and Levin).

Overall we believe that this collection of papers can provide a state-of-the-art collection of methods currently in use for investigating brain and behavioral asymmetries, thereby nurturing the next generation of scientists in this field. The latter scientists will certainly make further progress, both in methods and theory, in the years to come. Our hope is that this book will develop the field to such an extent that it will lead to a need for a second edition covering a new collection of methods.

Armidale, Australia — *Lesley J. Rogers*
Trento, Italy — *Giorgio Vallortigara*

References

1. Nottebohm F (1971) Neural lateralization of vocal control in a passerine bird. I. Song. J Exp Zool 177:229–261.
2. Rogers LJ, Anson JM (1979) Lateralisation of function in the chicken forebrain. Pharmacol Biochem Behav 10:679–686.
3. Denenberg VH (1981) Hemispheric laterality in animals and the effects of early experience. Behav Brain Sci 4:1–49.
4. Bisazza A, Rogers LJ, Vallortigara G (1998) The origins of cerebral asymmetry: a review of evidence of behavioural and brain lateralization in fishes, amphibians, and reptiles. Neurosci Biobehav Rev 22:411–426.
5. Vallortigara G, Rogers LJ, Bisazza A (1999) Possible evolutionary origins of cognitive brain lateralization. Brain Res Rev 30:164–175.
6. Vallortigara G (2000) Comparative neuropsychology of the dual brain: a stroll through left and right animals' perceptual worlds. Brain Lang 73:189–219.
7. Vallortigara G, Chiandetti C, Sovrano VA (2011) Brain asymmetry (animal). Wiley Interdiscip Rev Cogn Sci 2:146–157. doi:10.1002/wcs.100.
8. Rogers LJ, Andrew RJ (eds) (2002) Comparative vertebrate lateralization. Cambridge University Press, Cambridge.
9. Rogers LJ, Vallortigara G, Andrew RJ (2013). Divided brains: the biology and behaviour of brain asymmetries. Cambridge University Press, Cambridge.
10. Frasnelli E, Vallortigara G, Rogers LJ (2012) Left-right asymmetries of behaviour and nervous system in invertebrates. Neurosci Biobehav Rev 36:1273–1291.
11. Frasnelli E, Haase A, Rigosi E, Anfora G, Rogers LJ, Vallortigara G (2014). The bee as a model to investigate brain and behavioural asymmetries. Insects 5:120–138. doi:10.3390/insects5010120.
12. Efron R (1990) The decline and fall of hemispheric specialization. Lawrence Erlbaum Associates, Hillsdale.
13. Rogers LJ, Vallortigara G (2015) When and why did brains break symmetry? Symmetry 7:2181–2194.
14. Ghirlanda S, Frasnelli E, Vallortigara G (2009) Intraspecific competition and coordination in the evolution of lateralization. Philos Trans R Soc Lond B Biol Sci 364:861–866.
15. Ghirlanda S, Vallortigara G (2004) The evolution of brain lateralization: a game theoretical analysis of population structure. Proc Biol Sci 271:853–857.
16. Raymond M, Pontier D, Dufour AB, Moller A (1996) Frequency-dependent maintenance of left handedness in humans. Proc Biol Sci 263:1627–1633. doi:10.1098/rspb.1996.0238.
17. Vallortigara G (2006) The evolutionary psychology of left and right: costs and benefits of

lateralization. Dev Psychobiol 48:418–427. doi:10.1002/dev.20166.
18. Abrams DM, Panaggio MJ (2012) A model balancing cooperation and competition can explain our right-handed world and the dominance of left-handed athletes. J R Soc Interface 9:2718–2722. doi:10.1098/rsif.2012.0211.
19. Concha ML, Bianco IH, Wilson SW (2012) Encoding asymmetry within neural circuits. Nat Rev Neurosci 13:832–843.
20. Duboc V, Dufourcq P, Blader P, Roussigné M (2015) Asymmetry of the brain: development and implications. Ann Rev Genet 49:647–672.
21. Branson NJ, Rogers LJ (2006) Relationship between paw preference strength and noise phobia in *Canis familiaris.* J Comp Psychol 120(3):176–183.
22. Forrester GS, Pegler R, Thomas MA, Mareschal D (2014) Handedness as a marker of cerebral lateralization in children with and without autism. Behav Brain Res 268:14–21.
23. Marzoli D, Tommasi L (2009) Side biases in humans (*Homo sapiens*): three ecological studies on hemispheric asymmetries. Naturwissenschaften 96(9):1099–1106.
24. Vallortigara G, Rogers LJ (2005) Survival with an asymmetrical brain: advantages and disadvantages of cerebral lateralization. Behav Brain Sci 28:575–589.
25. MacNeilage PF, Rogers LJ, Vallortigara G (2009) Origins of the left and right brain. Sci Am 301:60–67.
26. Rosa Salva O, Regolin L, Mascalzoni E, Vallortigara G (2012) Cerebral and behavioural asymmetry in animal social recognition. Comp Cogn Behav Rev 7:110–138. doi:10.3819/ccbr.2012.70006.
27. Andrew RJ (1991) The nature of behavioral lateralization in the chick. In Andrew RJ (ed) Neural and behavioral plasticity: the use of the chick as a model, Oxford University Pres, Oxford. pp. 536–554.
28. Vallortigara G, Cozzutti C, Tommasi L, Rogers LJ (2001) How birds use their eyes: opposite left-right specialisation for the lateral and frontal visual hemifield in the domestic chick. Curr Biol 11:29–33.
29. Rogers LJ (1996) Behavioral, structural and neurochemical asymmetries in the avian brain: a model system for studying visual development and processing. Neurosci Biobehav Rev 20:487–503.
30. Güntürkün O (2006) Avian cerebral asymmetries: the view from the inside. Cortex 42:104–106.
31. Barth KA, Miklosi A, Watkins J, Bianco IH, Wilson SW, Andrew RJ (2005) fsi zebrafish show concordant reversal of viscera, neuroanatomy and subset of behavioural responses. Curr Biol 15:844–850.
32. Concha ML, Burdine RD, Russell C, Schier AF, Wilson SW (2000) A nodal signaling pathway regulates the laterality of neuroanatomical asymmetries in the zebrafish forebrain. Neuron 28:399–409.
33. Gamse JT, Thisse C, Thisse B, Halpern ME (2003) The parapineal mediates left-right asymmetry in the zebrafish diencephalons. Development 130:1059–1068.
34. Sovrano VA, Andrew RJ (2006) Eye use during viewing a reflection: behavioural lateralisation in zebrafish larvae. Behav Brain Res 167(2):226–231.
35. Zangwill OL (1974) Consciousness and the cerebral hemispheres. In: Dimond SJ, Beaumont JG (eds) Hemisphere function in the human brain. Wiley, New York pp. 264–278.
36. Versace E, Vallortigara G (2015) Forelimb preferences in human beings and other species: multiple models for testing hypotheses on lateralization. Front Psychol 6:233. doi:10.3389/fpsyg.2015.00233.
37. McManus IC (2002) Right hand, left hand. Weidenfeld and Nicolson, London.

Contents

Contributors

MARA ANDRIONE • *Centre for Mind/Brain Sciences, University of Trento, Rovereto, Italy*
TESS BEKING • *Clinical and Developmental Neuropsychology, University of Groningen, Groningen, The Netherlands; Behavioral Biology, Groningen Institute for Evolutionary Life Sciences, University of Groningen, Groningen, The Netherlands*
DOUGLAS J. BLACKISTON • *Biology Department, Center for Regenerative and Developmental Biology, Tufts University, Medford, MA, USA*
LUIGI CATTANEO • *Dipartimento di Neuroscienze, Biomedicina e Movimento, Sezione di Fisiologia e Psicologia, University of Verona, Verona, Italy*
CINZIA CHIANDETTI • *Department of Life Sciences, University of Trieste, Trieste, Italy*
ERIK R. DUBOUÉ • *Department of Embryology, Carnegie Institution for Science, Baltimore, MD, USA*
MOHAMADY EL-GABY • *MRC Brain Network Dynamics Unit, University of Oxford, Oxford, UK; Department of Physiology, Development and Neuroscience, University of Cambridge, Cambridge, UK*
MARA FABRI • *Dipartimento di Medicina sperimentale e clinica, Sezione di Neuroscienze e Biologia cellulare, Università Politecnica delle Marche, Ancona, Italy*
GILLIAN S. FORRESTER • *Department of Psychological Science – Birkbeck, University of London, London, UK*
NICOLETTA FOSCHI • *Centro Epilessia, Clinica di Neurologia, Azienda Ospedaliera-Universitaria Umberto I, Ancona, Italy*
ELISA FRASNELLI • *Department of Psychology, College of Life and Environmental Sciences, University of Exeter, Exeter, UK*
REINT H. GEUZE • *Clinical and Developmental Neuropsychology, University of Groningen, Groningen, The Netherlands*
TON G.G. GROOTHUIS • *Behavioral Biology, Groningen Institute for Evolutionary Life Sciences, University of Groningen, Groningen, The Netherlands*
ONUR GÜNTÜRKÜN • *Department of Psychology, Institute of Cognitive Neuroscience and Biopsychology, Ruhr-University Bochum, Bochum, Germany*
ALBRECHT HAASE • *Centre for Mind/Brain Sciences, University of Trento, Rovereto, Italy; Department of Physics, University of Trento, Trento, Italy*
MARNIE E. HALPERN • *Department of Embryology, Carnegie Institution for Science, Baltimore, MD, USA*
OLIVER HOBERT • *Department of Biological Sciences, Howard Hughes Medical Institute, Columbia University, New York, NY, USA*
WILLIAM D. HOPKINS • *Division of Developmental and Cognitive Neuroscience, Yerkes National Primate Research Center, Atlanta, GA, USA*
LEILANI S. KELLOGG • *Department of Psychology, University of South Florida, Tampa, FL, USA*
MICHAEL M. KOHL • *Department of Physiology, Anatomy and Genetics, University of Oxford, Oxford, UK*

Michael Levin • *Biology Department, Center for Regenerative and Developmental Biology, Tufts University, Medford, MA, USA*
Martina Manns • *Department of Biopsychology, Institute of Cognitive Neuroscience, Ruhr-University Bochum, Bochum, Germany*
Veronica Mazza • *Centre for Mind/Brain Sciences (CIMeC), University of Trento, Rovereto, Italy; IRCSS San Giovanni di Dio, Fatebenefratelli, Brescia, Italy*
Brian McCabe • *Sub-Department of Animal Behavior, University of Cambridge, Cambridge, UK*
Sebastian Ocklenburg • *Institute of Cognitive Neuroscience and Biopsychology, Ruhr-University Bochum, Bochum, Germany*
Silvia Pagano • *Centre for Mind/Brain Sciences (CIMeC), University of Trento, Rovereto, Italy*
Marco Paoli • *Centre for Mind/Brain Science, University of Trento, Rovereto, Italy*
Silvia Paracchini • *School of Medicine, University of St. Andrews, North Haugh, Scotland*
Tadd B. Patton • *Department of Psychological Sciences, Augusta University, Augusta, GA, USA*
Ole Paulsen • *Department of Physiology, Development and Neuroscience, University of Cambridge, Cambridge, UK*
Kimberley A. Phillips • *Department of Psychology, Trinity University, San Antonio, TX, USA; Southwest National Primate Research Center, Texas Biomedical Research Institute, San Antonio, TX, USA*
Chiara Pierpaoli • *Dipartimento di Medicina sperimentale e clinica, Sezione di Neuroscienze e Biologia cellulare, Università Politecnica delle Marche, Ancona, Italy*
Gabriele Polonara • *Dipartimento di Scienze Cliniche Specialistiche ed Odontostomatologiche, Sezione di Scienze Radiologiche, Università Politecnica delle Marche, Ancona, Italy*
Lesley J. Rogers • *School of Science and Technology, University of New England, Armidale, NSW, Australia*
Tom Scerri • *Walter and Eliza Hall, Institute of Medical Research, Parkville, VIC, Australia*
Toru Shimizu • *Department of Psychology, University of South Florida, Tampa, FL, USA*
Marcello Siniscalchi • *Department of Veterinary Medicine – Section of Behavioral Sciences and Animal Bioethics, University of Bari "Aldo Moro,", Bari, Italy*
Felix Ströckens • *Department of Psychology, Institute of Cognitive Neuroscience and Biopsychology, Ruhr-University Bochum, Bochum, Germany*
Ahmet K. Uysal • *Department of Psychology, University of South Florida, Tampa, FL, USA*
Giorgio Vallortigara • *Centre for Mind/Brains Sciences, University of Trento, Rovereto, Italy*
Berta Vidal • *Department of Biological Sciences, Howard Hughes Medical Institute, Columbia University, New York, NY, USA*

The original version of this book was revised. An erratum to this book can be found at DOI 10.1007/978-1-4939-6725-4_22

Part I

Behavioral Methods

Chapter 1

Tachistoscopic Viewing and Dichotic Listening

Sebastian Ocklenburg

Abstract

While advanced neuroimaging methods such as fMRI provide a reliable way to determine individual lateralization of function, these methods are costly and not readily available to every scientist interested in investigating functional hemispheric asymmetries in humans. Behavioral methods of testing humans provide cheaper and easily administered alternatives to fMRI scans and are still widely used in lateralization research today. In the following chapter, two key methods will be reviewed: divided visual field paradigms based on tachistoscopic viewing and the dichotic listening task.

Key words Dichotic listening, Auditory system, Divided visual field paradigm, Tachistoscopic viewing, Visual system, Laterality, Hemispheric asymmetries

1 Tachistoscopic Viewing

1.1 Introduction

There are several methodological options for the researcher to determine, in humans, whether an individual is left- or right-dominant for a specific cognitive function. Apart from invasive procedures such as the Wada test [1], electrophysiological methods like EEG [2] and advanced neuroimaging methods such as PET [3] or fMRI [4] provide reliable ways of determining individual lateralization of cognitive functions. Unfortunately, these methods have in common that they require more or less costly equipment such as MRI scanners or EEG systems, which are not readily available to every scientist interested in investigating functional hemispheric asymmetries in humans. Moreover, some of these methods are not suitable for specific groups of patients (e.g. patients with a pacemaker cannot be tested in a MRI scanner) or may cause discomfort in some participants (e.g. a fMRI scan can be perceived as very unpleasant by claustrophobic individuals due to the narrowness of the scanner tube). By contrast, behavioral measures of hemispheric asymmetries have no contraindications and are cheap and readily available to any researcher. Divided visual field paradigms based on tachistoscopic viewing comprise one of the major groups of behavioral tests used in laterality research (see Table 1 for key findings).

Lesley J. Rogers and Giorgio Vallortigara (eds.), *Lateralized Brain Functions: Methods in Human and Non-Human Species*, Neuromethods, vol. 122, DOI 10.1007/978-1-4939-6725-4_1, © Springer Science+Business Media LLC 2017

Table 1
Key papers for divided visual field paradigms based on tachistoscopic viewing

Study	Finding
[5]	Description of the first tachistoscope
[6]	Used the divided visual field technique to assess interhemispheric transfer time
[7]	Showed that a split-brain patient could name stimuli presented in the right but not in the left visual half-field, indicating left hemispheric dominance for language production and highlighting the role of the corpus callosum in cognition
[8]	Showed that female sex hormones modulate performance in divided visual field tasks for lexical decision, figural comparison, and face discrimination
[9]	Overview about the findings obtained with the divided visual field paradigm and other techniques in split brain patients
[10]	Comprehensive review of the methodology of divided visual field paradigms
[11]	Found positive correlations between divided visual field lateralization and brain activation asymmetries during picture naming and word naming in the MRI scanner, proving that divided visual field paradigms can be a reliable predictor of brain activation asymmetries

Fig. 1 A tachistoscope (Faculty of Psychology, Ruhr-University Bochum)

The term "tachistoscopic" (from Greek "*tachistos*": very rapid and "*skopein*": to view) refers to presentation of visual stimuli for a precisely controlled period of time [12, 13]. Typically, this period of time is very short, e.g. in the millisecond range. Historically, tachistoscopic presentation of visual stimuli was performed using a family of scientific instruments called tachistoscopes [14] (see Fig. 1). Today it is usually performed using a standard PC and monitor.

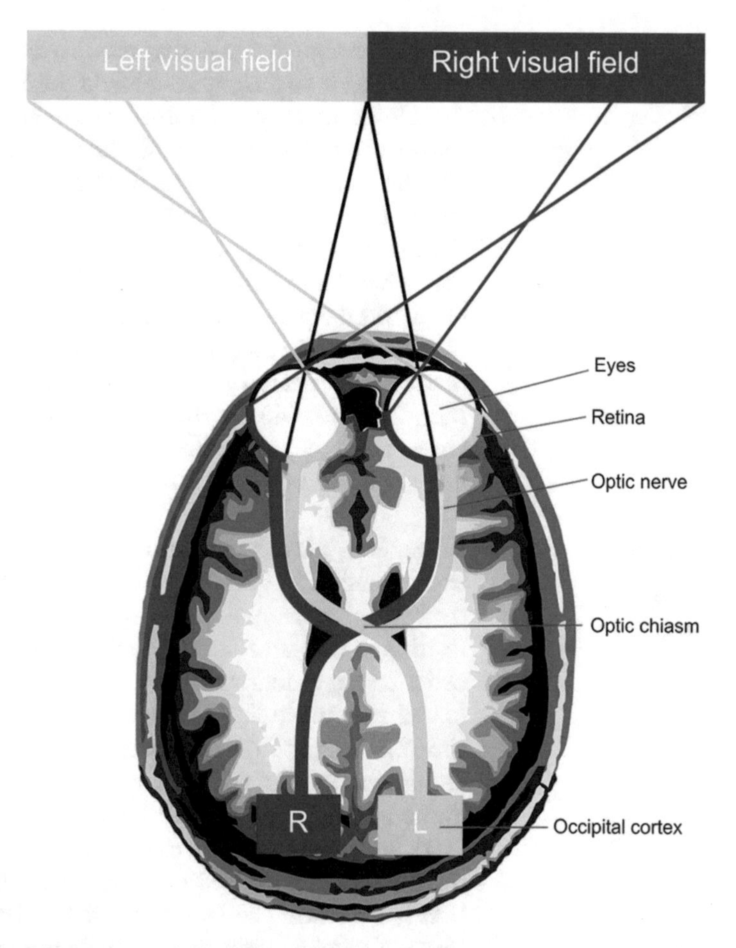

Fig. 2 Neuroanatomical organization of the visual system in humans. For central fixation, information in the left visual field is processed by the right hemisphere and information in the right visual field by the left hemisphere

How can tachistoscopic presentation of stimuli be helpful when investigating lateralization? Due to the structure of the mammalian visual system (see Fig. 2), a stimulus that is presented in the left visual field is initially processed by the right hemisphere and vice versa. Nasal portions of left and right visual input are fed into contralateral hemispheres; distal portions are relayed into ipsilateral hemisphere. Thus, by presenting visual information in only one visual field, this information is initially only processed by the contralateral hemisphere. By comparing performance on left versus right visual field trials, the experimenter can statistically evaluate performance differences for a specific set of stimuli between the left and the right hemisphere. Typically, hemispheric asymmetries are reflected by a processing advantage (e.g. faster reaction times or higher accuracy) in the contralateral visual field (e.g. a left visual field advantage for tasks that require discrimination of complex figures or faces, or a right visual field advantage for verbal tasks [8]).

Unfortunately, stimulus processing in only one hemisphere cannot be retained for long after onset of stimulus presentation. On the one hand, subjects might perform saccades, that is, rapid lateral eye movements that change the position of the foveae relative to the stimuli, possibly resulting in bilateral stimulation. On the other hand, information transfer via the corpus callosum is likely to take place soon after initial information processing, possibly reducing hemispheric asymmetries due to bilateral stimulus processing. Therefore tachistoscopic stimulus presentation (less than 200 ms, ideally shorter) is a must when using the divided visual field technique to investigate hemispheric asymmetries in healthy humans.

1.2 Methods

1.2.1 Prerequisites to Use the Task

In general, participants should have unimpaired or corrected-to-normal vision in both eyes, should be able to keep their head relatively motionless over prolonged periods of time while fixating the fixation cross, and should be able to press response buttons in order to log reactions to the stimuli. Also, participants should not have any neurological damage in the visual system. For this reason, the task might not be optimal for certain clinical groups, and administration in younger children might also be complicated. Some authors recommend testing only right-handed participants [10], at least when investigating differences between left- and right-handers is not part of the study design. This recommendation is based on the fact that left-handers can show different patterns of lateralization than right-handers in a number of cognitive systems, e.g. for language [15]. Thus, including this group could introduce unwanted variance in the data. Also, participants' sex and age should be carefully balanced, as it has been shown that both can possibly influence performance in divided visual field paradigms [16, 17]. Moreover, many psychiatric disorders such as schizophrenia [18] have been found to have an impact on performance in divided visual field paradigms. Hence, as long as the impact of pathological processes on lateralization is not within the focus of the study, it is advised to test neurologically and psychologically healthy participants.

1.2.2 Set-Up

Compared to many other methods to assess individual lateralization, implementation of divided visual field tasks is comparably easy and cost-efficient. The basic set-up requires only a standard PC or other type of personal computer, a monitor, a reaction device, and a chin rest. For stimulus presentation, a large number of different software tools can be used, ranging from specialized stimulus presentation software such as E-Prime 2 (Psychology Software Tools, Inc., Sharpsburg, USA; http://www.pstnet.com), SuperLab 5 (Cedrus Corporation, San Pedrom, USA; http://www.superlab.com) or

Presentation (Neurobehavioral Systems, Inc., Berkeley, USA; https://www.neurobs.com) to programming languages such as Matlab (The MathWorks, Inc., Natick, USA; http://de.mathworks.com/products/matlab/).

With regard the monitor, most research labs using visual paradigms have traditionally used CRT (cathode-ray tube) monitors due to superior temporal and spatial acuity which allows more reliable and precise onsets and offsets of visually displayed stimuli compared to modern LCD (liquid crystal display) monitors (see [19] for an in-depth investigation of this issue). However, since most hardware producers have ceased production of CRT monitors, and since increasingly accurate LCD monitors have become available, more and more researchers have taken interest in using this kind of monitor. Importantly, there seem to be large differences in suitability for vision research between different producers and product types. For researchers interested in acquiring a new monitor for setting up divided visual field paradigms, Ghodrati et al. [20] provide a detailed review of the suitability of several models of LCD monitors for vision research.

With regard to the reaction device, it has been argued that devices connected to the computer via a serial port (such as most specialized response boxes) are preferable to standard interface such as USB (Universal Serial Bus) keyboards or mouse devices, as USB can introduce a slight "lag" between the actual reaction time and the time the reaction is recognized by the computer. Moreover, internal wiring might cause huge differences in reaction timing between different mouse models. Plant et al. [21] reported inter-individual differences between different types of computer mouse devices ranging up to 61 ms in timing error. Thus, devices measuring reaction time have to be carefully chosen and tested before the first participants can be run. As mentioned above, divided visual field paradigms rely on presentation of visual information to only one visual field, which results in this information being processed only by the contralateral hemisphere. Proper administration of such paradigms critically depends on immobilizing the head, and instructing participants to keep body and head still during the whole experiment. Typically, immobilization of the head is realized using a chin rest (see Fig. 3) or a combined chin and forehead rest.

The distance between chin rest and monitor depends on the size and dimensions of the monitor as well as on the size of the stimuli and their distance from the fixation cross (more on this topic in the following sections). In previous divided visual field studies by the author of this chapter, the chin rest was placed at a viewing distance of 57 cm from the monitor [22]. This distance is commonly used because in this case a stimulus with the length of 1° of visual angle is 1 cm long, as the stimulus size on the monitor is the tangent of the stimulus size in visual angle multiplied by the distance between participant and monitor.

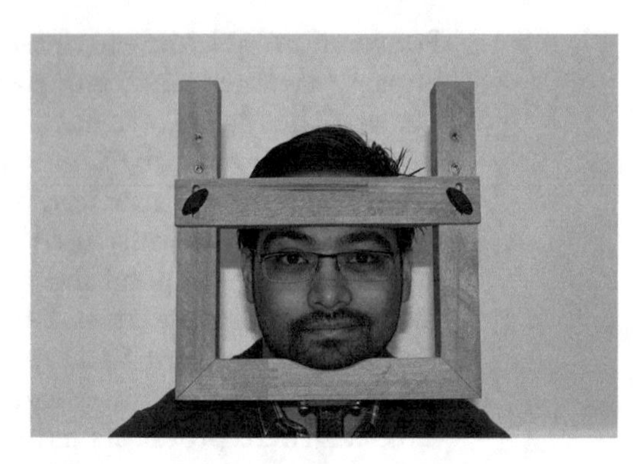

Fig. 3 A participant demonstrating the use of a chin rest

Another factor that has to be taken into account is the testing room. As analysis of left–right differences is central to divided visual field paradigms, great care should be taken to avoid any lateralized distraction in the testing room, such as one-sided light sources or decorative elements. In general, a very plain room that does not contain visual distractions is preferable. Also, dimming the light during the task may aid in avoiding further distraction.

1.2.3 Fixation Control

In order for a divided visual field paradigm to work it is essential that stimuli supposed to be processed by the left hemisphere are only perceived in the right visual field and vice versa. Experimentally, this is obtained by presenting the stimuli at a specific distance from the center of the screen. This distance differs depending on the distance of the participant's eyes from the screen (more on this topic below). For this principle to work, it is essential that the participant fixates the center of the screen throughout the experimental trial without moving his/her eyes left- or rightward. To ensure midpoint fixation, each trial in a divided visual field paradigm starts with the presentation of a fixation cross, with the participant being instructed to fixate this cross throughout the whole experiment. The fixation cross should have high visibility (e.g. black on white background or white on black background). While many published papers do not contain information about the size of the fixation cross, a height and width of 1° of visual angle seems to be a common size [23]. The duration for which the fixation cross is presented on its own before a stimulus is presented varies greatly between studies but in general seems to range from 1000 ms [24] to 2000 ms [22].

While the fixation cross is a critical part of divided visual field paradigms, some authors have provided evidence that simply instructing participants to fixate the cross might not be optimal to control for central fixation, as participants might fixate to the left or the right of the cross on some trials [25]. Several other measures have been suggested to ensure proper fixation control [10]. For example, Bourne and Hole [26] presented a single upper-case consonant instead of a fixation cross for 750 ms before stimulus presentation and asked the participants to verbally report this letter. Trials in which the letter was reported incorrectly were not included in the subsequent analysis, assuming that the participant had not properly fixated the center of the screen in this trial. While this method arguably provides a better way to control fixation than simple instructions, it increases cognitive demand and essentially creates a dual-task paradigm, which might not be desirable for the researcher. To avoid this issue, different methods to directly control the participant's eye movements have been applied. For example, Marzi and Berlucchi [27] ensured lateralization of visual information to one hemisphere by observing the participant's eyes from behind a mirror that was positioned over the fixation point. All trials in which fixation was not maintained throughout stimulus presentation were excluded from later analysis. Since this approach depends to a high degree on the experimenter's subjective assessment, more objective eye movement measurements are preferable. For example, Meyer and Federmeier [28] used a commercially available eye-tracking system (Applied Science Laboratories Model 504 High-Speed Eye-Tracking System) to track eye-movements. If participants moved their eyes from the central fixation cross during stimulus presentation, the trial was excluded from later analysis. The overall number of trials excluded by this procedure was about 12%.

Additional options to control for eye movements are available when the participant is tested within an EEG setting. If dedicated eye electrodes are used, automated artifact rejection procedures can be used to exclude trials in which horizontal or vertical eye movements occurred as determined by on amplitude changes in the eye electrodes. For example, Lange et al. [29] excluded all trials containing horizontal eye movements with an EOG (electro-oculogram) criterion of 15 μV and vertical eye movements with an EOG criterion of 50 μV. If no dedicated eye electrodes are used, another possibility to control for saccadic eye movements is calculating an independent component analysis (ICA) on the EEG data and afterwards identifying (and removing) the component(s) reflecting saccadic eye movements. By manually adding trigger points/markers for the typical box-shaped peaks of the saccade component and exporting the time points to compare them with the data from the divided visual field paradigm, the researcher could exclude post hoc all trials in which saccades occurred (Ocklenburg et al., unpublished data).

1.2.4 Stimulus Presentation in the Left or Right Visual Half Field

Although protocols differ between studies, all divided visual field paradigms have in common that at some point after initial presentation of the fixation cross a stimulus is presented unilaterally, in either the left or the right visual field. In order to ensure that unilateral stimulus presentation leads to (initial) unihemispheric processing in the contralateral hemisphere, three important factors need to be controlled for: duration of stimulus presentation, size of the stimulus, and its distance from the fixation cross. With regard to duration of stimulus presentation, one should generally aspire to present the stimulus as briefly as possible to prevent saccadic eye movements towards the stimulus that would interfere with unilateral stimulus presentation. After onset of a target (which in this case would be the laterally presented stimulus), it takes on average 200 ms for the eyes to start moving towards the target [30]. Thus, it is not advisable to use stimulus durations longer than 200 ms in divided visual field paradigms. Ideally, stimulus durations of 150 ms or less are chosen, as some saccades can have latencies shorter than 200 ms [10].

Stimulus size and distance from the fixation cross are factors that are somewhat interdependent. It is important to note that the division between left and right visual field is not clear-cut, and that the fovea is likely to be bilaterally represented and that a so-called bilateral strip between 0.5° and 3° of the fovea projects to both hemispheres [31]. Thus, in order to ensure unilateral stimulus presentation to the left or right hemisphere, the inside edge of a stimulus should be presented at less than 2°, but better 2.5–3° away from the central fixation cross [10]. For example, Saban-Bezalel and Mashal [32] presented the inside edges of their stimuli 2.8° away from the central fixation cross, while Bourne and Hole [26] placed the inside edges of their stimuli 4° and Marzi and Berlucchi [27] 5° from the center of the screen. Stimulus size in divided visual field paradigms is limited by the fact that peripheral vision is less accurate with greater eccentricity [33]. Thus, stimuli should be kept as small as possible without compromising visibility. Most authors keep stimulus size below 10° of visual angle [34].

Horizontally presented words are probably the most commonly used stimulus type in divided visual field tasks. Typically, two main types of cognitive tasks can be performed [35], either lexical decisions (participants have to decide whether a string of letters represents a word or a meaningless non-word) or semantic decisions (participants have to identify whether a word belongs to a certain category or not). More complicated verbal stimuli have also been used, e.g. Gold et al. [36] conducted a divided visual field paradigm with two word metaphors. In addition to verbal stimuli, several authors used facial stimuli [26, 34, 37, 38], emotional stimuli [39–42] or complex visual figures [8, 43], but in principle any type of visual stimulus can be used.

After stimulus presentation, most experiments involve a blank screen or the fixation cross that is presented during the response period. However, it has been argued that presenting a pattern mask might be advantageous as it prevents afterimages that could confound stimulus duration. For instance, Michałowski and Króliczak [44] presented stimuli for 200 ms and then replaced them with black and white high-contrast pattern masks, which were presented for another 200 ms.

After the end of the stimulus presentation segment, divided visual field paradigms typically contain a segment in which the participant is supposed to react to the stimuli in some way, e.g. by pressing a button to indicate whether a stimulus belonged to a certain category. It is important to balance use of left and right hand for responding, e.g. by asking the participant to switch the hand that is used to respond after a certain number of trials, in order to prevent systematic effects of motor preferences on the results. Also, researchers should be careful about the spatial configuration of the response buttons, as it has been shown that response times are shorter when stimulus location and response button location match as compared to when they do not match (the so-called "Simon effect") [45]. Thus, using the left and right arrow keys on a standard keyboard (or any type of button configuration in which one button is on the left and the other one on the right) might not be optimal. Instead, it is advisable to use buttons that are above and below each other. Since subjects have to fixate the fixation cross and cannot gaze on the response buttons, most divided visual field paradigms rely on only a few, typically two, response alternatives. This is done because more response alternatives might introduce a memory component or lead to unwanted head or eye movements when subjects check the response keys. Typical types of responses are categorization (e.g. word/non-word or normal/altered faces) as used by Hausmann and Güntürkün [8], or Go/No-go [46]. Some older studies [47] also used oral responses (e.g. participants verbally reported what they had seen), but this approach has largely been discontinued because accurate measurement of reaction time is more difficult here than for overt motor responses such as button presses.

The number of trials in a divided visual field paradigm typically depends on the task used (e.g. a Go/No-go task typically needs more trials than a categorization task, since the Go and No-go conditions have different probabilities). The number of trials should, however, in general be rather large to allow for reliable observation. Based on earlier studies [48, 49], Hunter and Brysbaert [50] concluded that divided visual field tasks should at least contain 150 observations. In general, equal numbers of left and right visual field trials should be presented, and presentation of these two trial types should be randomized in order to prevent anticipatory eye movements. To further prevent any other anticipatory effects, it is advisable to jitter the inter-stimulus interval.

Divided visual field paradigms can be quite challenging for participants due to rapid presentation of stimuli and the unusual requirement to keep the head still and fixate the fixation cross for a prolonged period time. It is therefore advisable to start every experimental session with 10–20 training trials under supervision of the experimenter. Also, since keeping head and body still during all experimental trials can be quite difficult for some participants, an adequate number of breaks in between experimental blocks in which the participants are allowed to relax their head and body is advised. While the author is not aware of any scientific investigation of break timing during divided visual field paradigms, a break about every 10 min, with participants themselves deciding when to carry on, appears to be advisable based on personal experience.

1.2.5 Dependent Variables and Statistical Analysis

Typically, two dependent variables are analyzed in divided visual field paradigms, that is, accuracy or error rate (in %), and median or mean reaction time (in ms). Should there be indication for a speed-accuracy tradeoff (e.g. some participants were very slow yet accurate, while others were fast and made many errors), it could also be advisable to additionally calculate so-called inverse efficiency scores (reaction time divided by percentage of correct responses), a measure integrating the two parameters [51]. These values then can be analyzed using repeated measures analysis of variance (ANOVAs), with visual field (left and right) as within-subjects factor and all additional factors of interest as additional within-subject (e.g. for experimental conditions) or between-subjects factors (e.g. when testing patients vs. controls). If the results of a divided visual field paradigms need to be correlated with other variables, it may also make sense to calculate a so-called laterality quotient (LQ), e.g. by using the following formula $LQ = [R - L]/[R + L] \times 100$, with R indicating the value of the dependent variable for right-visual field presentation and L indicating the value of the dependent variable for left-visual field presentation.

1.3 Notes

The obvious advantage of divided visual field paradigms is that they offer a cheap, fast and noninvasive method of determining the dominant hemisphere for a specific task. However, that does not mean they are easy to implement. Great care needs to be taken to avoid measurement errors. This is illustrated by a meta-analysis of divided visual field studies by Voyer [49], who assessed reliability and found that, on average, it was 0.56 for verbal tasks, implying moderate reliability, but only 0.28 for nonverbal tasks. Importantly, divided visual field tasks will yield reliable data, if they are conducted correctly. This assumption has been supported by a study by Hunter and Brysbaert [50], who compared left-handers' reaction time lateralization in two different divided visual field tasks (word and picture naming) to brain activation asymmetry during a mental word generation task in the fMRI scanner. The authors found significant positive correlations between lateralization in both divided visual field paradigms and the fMRI task (picture naming: $r = 0.77$; word naming: $r = 0.63$), and concluded that divided visual field paradigms can be used as a reliable predictor of language-related brain activation asymmetries when designed carefully. Some years later, the group of Brysbaert continued this research in a larger sample [11], again reporting positive correlations between divided visual field lateralization and brain activation asymmetries (picture naming: $r = 0.65$ and word naming: $r = 0.64$). Detailed analysis of different participant groups showed that divided visual field paradigms can mainly predict language lateralization in the fMRI tasks when participants show a consistent pattern in both divided visual field tasks, but that their predictive power is not as good for subjects with inconsistent patterns.

One important caveat when conducting divided visual field paradigms concerns data interpretability. While the findings of Hunter and Brysbaert [50] and Van der Haegen et al. [11] show positive correlations between reaction time laterality in divided visual field paradigms and language-related brain activation asymmetries, the researcher has to keep in mind that reaction times in this kind of task are a rather indirect and coarse measure of brain activation. While they might provide a general idea of whether the left or right hemisphere is dominant for a certain task, further insight into the relation between performance and brain activation is difficult to support with clear evidence from the literature. For example, it is largely unclear how the degree of lateralization in reaction times or accuracy in these tasks relates to the magnitude of brain activation asymmetries. Aside from the extent of brain activation asymmetries, behavioral performance measures can be modulated by several other variables (e.g. by participants' motivation, general intelligence, cognitive control functions, fatigue, and many others). Thus, one should be careful not to over-interpret the results yielded by divided visual field paradigms.

This issue can be addressed by combining the divided visual field technique with other techniques. For example, it can be combined with electrophysiological techniques like EEG (electroencephalogram, see Chap. 13), allowing for deeper insights into the temporal dynamics of the neuronal process underlying task performance [23, 39, 46, 52–55]. Moreover, they have also been used in the fMRI scanner [56, 57], rendering it possible to relate behavioral findings more directly to brain activation patterns.

One interesting tool for researchers aiming to introduce students to the divided visual field technique is the so-called Lateralizer software package [58] (http://cognitrn.psych.indiana.edu/CogsciSoftware/Lateralizer/index.html) which allows students to program and conduct divided visual field paradigms with different types of stimuli, such as faces, words or hierarchical stimuli. Motz et al. [58] reported positive learning outcomes as compared to traditional lecture-based courses, suggesting that this software package might be an interesting addition to classes about hemispheric asymmetries.

1.4 Findings on the Evolution of Lateralization

In general, divided visual field paradigms are not a commonly used method in comparative neuroscience studies investigating lateralization in nonhuman model species. Some work in this regard has

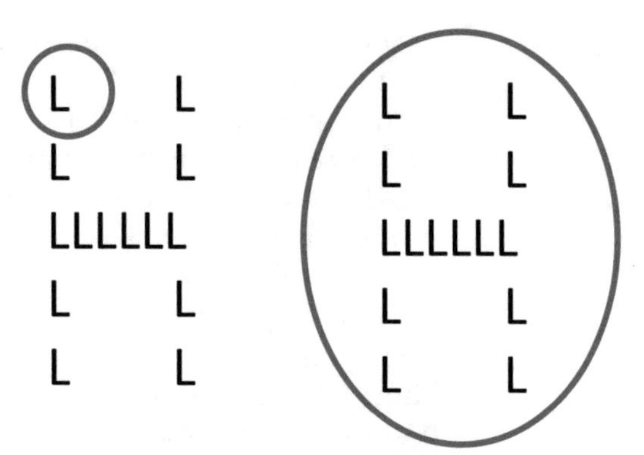

Fig. 4 Examples for hierarchical stimuli used by Hopkins [59]. *Left panel*: Local processing, *Right panel*: Global processing

been done in primates. Hopkins [59] used a divided visual half-field paradigm to investigate lateralization of global and local processing of hierarchical compound stimuli (see Fig. 4) in the chimpanzee (*Pan troglodytes*). For reaction times, he found a right visual field/left hemisphere advantage for local processing but no visual field differences for global processing. Since the results for local processing resembled findings in humans [60], Hopkins [59] concluded that the chimpanzee data suggests homologous lateralization in chimpanzees and humans. Similarly, using a conditional matching-to-sample task in a divided-field paradigm, Dépy et al. [61] found that both humans and baboons showed better distance processing with the left compared to the right hemisphere.

2 Dichotic Listening

2.1 Introduction

In addition to divided visual field paradigms, the dichotic listening task has been the main workhorse paradigm in behavioral laterality research on humans for more than 50 years [62] (see Table 2 for key findings).

Based on a procedure to test attention in air traffic controllers developed by Broadbent [77], Kimura [63, 64] published two landmark papers in which she reported using headphones to acoustically present two different digits simultaneously to the left and right ears of her participants, and asking them to report what they heard (see [78] for a historical perspective of this early dichotic listening research). Kimura [63, 64] showed for the first time that there is an asymmetry of reports for right versus left ear stimuli in most participants, the so-called right-ear advantage (REA). While these early studies used digits as stimuli, most recent dichotic listening studies use a variation of the so-called consonant–vowel (CV) syllables dichotic listening paradigm [65, 66]. In this paradigm, participants wear headphones, and two different

Table 2
Key papers for the dichotic listening task

Study	Finding
[63, 64]	Developed dichotic listening procedure with digits as stimuli and showed right ear advantage for the first time
[65, 66]	Developed the classic consonant–vowel version of the dichotic listening task
[67, 68]	Introduction of the forced-attention version of the dichotic listening task, adding the "forced right" and "forced" left condition to the task to assess the impact of cognitive control on dichotic listening performance
[69]	Study showing that determination of the speech-dominant hemisphere with dichotic listening overlaps to 92% with the results of the Wada test, proving that dichotic listening is a reliable method to assess the language dominant hemisphere
[70]	Meta-analysis showing significantly decreased lateralization in schizophrenia in studies using the consonant–vowel or fused word dichotic listening tasks
[71]	Review article highlighting the role of the corpus callosum for both bottom-up and top-down stimulus process during dichotic listening
[72]	Comprehensive fMRI study with 113 participants investigating the neural correlates of forced-attention dichotic listening
[73, 74]	Introduction of the smartphone-based iDichotic app that allows testing participants with the forced attention version of the dichotic listening paradigm outside traditional laboratory settings
[75]	Large-scale study (1782 participants) showing that sex differences in dichotic listening are age-dependent and small. Male younger adults showed greater asymmetry than female younger adults, but no sex differences were found for children or older adults
[76]	Large-scale study (3680 participants) showing that the right ear advantage for dichotic stimuli increases over 60 years of age

acoustic CV stimuli (e.g. BA and GA) are presented at the same time (see Fig. 5). Participants are asked to indicate which stimulus he or she has heard. Typically, most participants report more syllables presented to the right ear than to the left ear. This is thought to reflect left-hemispheric dominance for speech processing. While sensory input from the ears is transmitted to both auditory cortices, the contralateral projections are stronger and may inhibit ipsilateral projections so that input from the right ear is mostly processed by the left hemisphere, and vice versa [79].

As of 2015, the search term, "dichotic listening" yields more than 2800 publications in the scientific search engine PubMed (http://www.ncbi.nlm.nih.gov/pubmed), probably more than for any other experimental paradigm used to assess lateralization. Its success is also reflected in the fact that Hugdahl [80] published an edited book on dichotic listening, and that several researchers have authored review articles covering the task in general [79,

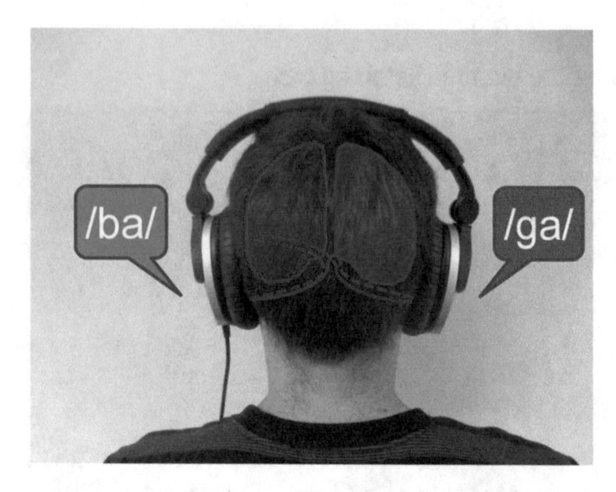

Fig. 5 Schematic representation of the consonant vowel syllable dichotic listening task. Input from the left ear is mainly processed by the right hemisphere and input from the right ear is mainly processed by the left hemisphere

81–83], the role of attention [84], or the role the corpus callosum in dichotic listening [71, 85]. Moreover, several review articles cover alterations of dichotic listening performance following pathological conditions such as Alzheimer's disease [86], depression [87], learning disabilities [88], stroke [89], or schizophrenia [70, 90]. Taken together, the dichotic listening task has proven to be one of the most versatile tasks in laterality research and is a useful addition to the methodological repertoire for every scientist interested in investigating hemispheric asymmetries.

2.2 Methods

2.2.1 Prerequisites to Use the Task

In order to get meaningful results from the Dichotic Listening Task, participants' hearing capabilities should be tested using an audiometer (or audiometric software on a standard PC) prior to testing. Obviously, participants who are deaf or show substantially impaired hearing in one ear need to be excluded. For example, Hahn et al. [91] tested participants at frequencies of 500, 1000, 1500, and 3000 Hz with a MA25 audiometer (MAICO Diagnostic GmbH, Berlin, Germany) and included only participants with interaural differences smaller than 15 dB at any of these frequencies. In addition to the interaural difference criterion, general hearing threshold can be used as an exclusion criterion, e.g. Hahn et al. [92] only included participants with hearing thresholds smaller than 20 dB on both ears. While the exact exclusion criteria differ between studies and depend on the sample that is being tested (e.g. when testing elderly participants, the hearing threshold criterion should be less strict), it is generally advisable to follow a rather conservative approach to prevent artifacts in the data. Another important factor is the linguistic background of the participants in relation to the language in which the stimuli are spoken. Bless et al. [74]

tested participants with different linguistic backgrounds (English, Danish, Norwegian, Hindi, Chinese, Spanish) with Dichotic Listening Tasks with different stimulus languages (English, Norwegian, German, Estonian) and found that participants showed stronger language lateralization when they were tested in their native language. Thus, it is advisable to test only native speakers for the language in which the stimuli were recorded, unless testing differences between native and nonnative speakers in dichotic listening performance is a specific study aim. Also, participants' sex [75], age [76] and handedness [93] should be considered when determining exclusion and inclusion criteria for the test sample, as these variables are known to modulate performance in the Dichotic Listening Task. Moreover, many psychiatric disorders like schizophrenia [90] and neurological diseases like Alzheimer's disease [86] have been found to influence performance in the Dichotic Listening Task, typically by reducing the right ear advantage. Thus, as long as the impact of pathological processes on lateralization is not a focus of the study, it is advisable to test neurologically and psychologically healthy participants.

2.2.2 Set-Up

As for divided visual field paradigms, implementation of dichotic listening tasks is comparably easy and cost-efficient. Since dichotic listening is an auditory task, the environment in which the task is performed should be noise-free, ideally a noise-shielded room. For stimulus presentation, most research labs use a standard PC or other type of personal computer-. The same software tools as for divided visual field paradigms can be used, that is, specialized stimulus presentation software such as E-Prime 2 (Psychology Software Tools, Inc., Sharpsburg, USA; http://www.pstnet.com), SuperLab 5 (Cedrus Corporation, San Pedrom, USA; http://www.superlab.com) or Presentation (Neurobehavioral Systems, Inc., Berkeley, USA; https://www.neurobs.com) or programming languages like Matlab (The MathWorks, Inc., Natick, USA; http://de.mathworks.com/products/matlab/). In addition to classic PC-based testing, a Dichotic Listening Application for Apple iOS Smartphones is available (http://www.dichoticlistening.com/). Bless et al. [73] extensively tested the feasibility of conducting dichotic listening research with this application and concluded that it represents a valid and reliable method for conducting dichotic listening research both in experimentally controlled as well as uncontrolled settings. Therefore, this application opens up interesting opportunities for researchers looking for a way to test participants easily in their own homes or at their bedside in a hospital/clinical setting.

For stimulus presentation, high quality over-the-ear headphones are needed. To the author's best knowledge, there is no published systematic comparison of different types of headphones with regard to usability in dichotic listening tasks. One viable option is DT 770 Pro headphones (Beyerdynamic GmbH, Heilbronn, Germany), as used by Ocklenburg et al. [94], but it is possible to conduct the task with any type of high quality headphones. To minimize possible error sources, headphones without in-built volume control should be used, so that participants cannot accidentally alter the sound level. Since correct application of headphones is essential for successful use of the dichotic listening task, it can be helpful to customize earphones with large "LEFT" and "RIGHT" signs in order to minimize potential left–right confusion by the experimenter. For hygienic reasons, the use of disposable headphone covers is advised. While the earphones could also be disinfected with disinfectant spray, most headphones do not tolerate such treatment very well for a long time. As it is possible that minimal interaural differences between the left and the right headphones might occur and have an impact on test results, it is advisable to conduct half of the trials for each condition with reversed headphones [94].

In order to capture the participants' responses, a standard keyboard can be used on which each stimulus type is represented by one key (e.g. 1 = BA; 2 = DA; 3 = GA; 4 = KA; 5 = PA; 6 = TA). In order to prevent erroneous responses due to the participant forgetting which key represents which syllable, it is advisable to mark the keys, e.g. by taping a small piece of paper showing the respective syllable on the key. Alternatively, a customized response board with a different button for each stimulus type can be used (see Fig. 6). Such a response board has the advantage of minimizing the chance of accidental button presses that may involve buttons not linked to any of the stimuli, or even buttons that may accidentally halt or end the experiment (such as the Windows key on a Windows PC).

In addition to keyboards, touch-screens displaying the different syllables have been used as response devices in dichotic listening studies [73]. Moreover, in several studies, participants were asked to react orally instead of manually, e.g. by speaking the syllable they heard best [95, 96]. In this case, an audio recorder or specific recording software for a PC is needed to record responses.

Fig. 6 A custom response board for the dichotic listening task

2.2.3 Stimuli

While digits were used as stimuli in the classic studies by Kimura [63, 64], most modern dichotic listening studies use verbal stimuli, more specifically consonant–vowel syllables. In the most widely used version of the task, the syllables consist of the six stop-consonants /b/, /d/, /g/, /p/, /t/, /k/, and the vowel /a/ [62]. Since syllables are always presented in pairs, this procedure yields 36 different stimuli (see Table 3).

Thirty of these syllable pairs consist of two different syllables (e.g. /ba/-/da/), while six consist of homonym syllable pairs (e.g. /ba/-/ba/). Stimulus duration is usually around 350–400 ms [95] and should be consistent for all stimuli. Obviously, the stimuli also need to be exactly matched in volume and should be presented at a sound level that is easily perceptible for the participants. Hahn et al. [92] presented stimuli at 80 dB. Another factor that needs to be controlled for is voice onset time, as differences between voice onset times of different consonants are known to affect performance in the dichotic listening task [97]. One parameter that has been varied in several studies is emotional intonation of the stimuli [98, 99]. While the stimuli in a standard CV dichotic listening paradigm are spoken in an emotionally neutral way, some authors used CV syllables that were spoken in different emotional intonations to investigate hemispheric asymmetries in emotion processing. For example, Hahn et al. [91] used the stimulus /da/-/da/ with the two syllables either spoken neutrally or in one of four different emotional intonations (happy, sad, angry, or surprised). Participants had to indicate which emotion they had perceived by pressing response buttons that were labeled with cartoon drawings of emotional faces corresponding to the emotions.

In addition to CV dichotic listening tasks, variants of the paradigm using words as stimuli have been developed, for example the fused rhymed words test developed by Wexler and Halwes [100]. The stimuli in this task were monosyllabic rhymed consonant—vowel—consonant (CVC) words beginning with one of the six stop

Table 3
Thirty-six possible stimulus pairs in the classic CV dichotic listening task

	/ba/	/da/	/ga/	/ka/	/pa/	/ta/
/ba/	**/ba/-/ba/**	/da/-/ba/	/ga/-/ba/	/ka/-/ba/	/pa/-/ba/	/ta/-/ba/
/da/	/ba/-/da/	**/da/-/da/**	/ga/-/da/	/ka/-/da/	/pa/-/da/	/ta/-/da/
/ga/	/ba/-/ga/	/da/-/ga/	**/ga/-/ga/**	/ka/-/ga/	/pa/-/ga/	/ta/-/ga/
/ka/	/ba/-/ka/	/da/-/ka/	/ga/-/ka/	**/ka/-/ka/**	/pa/-/ka/	/ta/-/ka/
/pa/	/ba/-/pa/	/da/-/pa/	/ga/-/pa/	/ka/-/pa/	**/pa/-/pa/**	/ta/-/pa/
/ta/	/ba/-/ta/	/da/-/ta/	/ga/-/ta/	/ka/-/ta/	/pa/-/ta/	**/ta/-/ta/**

Homonym stimulus pairs are given in bold.

consonants b, p, d, t, g, k. The two words in each stimulus pair differed from each other only in the initial consonant. Similar paradigms using words as stimuli have also been used by other authors [101, 102]. Moreover, nonverbal acoustic stimuli have been used in dichotic listening task, for example dichotic chords [103, 104], dichotic rhythms [105], or dichotic tonal patterns [106].

2.2.4 Protocol

The sensory experience of perceiving two different acoustic stimuli at the same time might be somewhat confusing for some participants. For this reason, it is essential to properly instruct participants prior to the test session, e.g. by telling them to always just report the syllable they heard best, and ignore all other syllables. Also, it can be beneficial to show all syllables that can occur throughout the task to the participant before beginning the experiment, e.g. written out on a sheet of paper. After instructing the participant, a few practice trials (e.g. 24) should be run to familiarize the participants with the task. Practice trials should be excluded from later analysis. A trial in a dichotic listening paradigm typically starts with presentation of the stimulus pair (for details see "Stimulus" section). After stimulus presentation, there is typically a silent inter-stimulus interval in which participants have to indicate which syllable they heard best. The length of this interval typically ranges between 2000 ms [107] and 4000 ms [108] for CV paradigms. If more complex stimuli (e.g. words) are used, longer intervals may be more appropriate, e.g. an interval of 5000 ms [109]. In clinical studies involving patient samples, longer intervals might be appropriate, depending on the specific impact of the disorder in question on reaction times in the task.

In the classic CV paradigm, stimulus presentation typically occurs in blocks of 36 stimuli, so that each possible combination of the six syllables is presented once [110]. Some studies do not present the homonym syllable pairs, resulting in blocks of 30 stimuli [111]. Number of blocks differs between studies. For example, Hirnstein et al. [110] presented one block of 36 stimuli, while Foundas et al. [111] presented four blocks of 30 stimuli, one of them being a practice block, resulting in 90 experimental trials. Ocklenburg et al. [107] used four blocks of 36 trials each, resulting in a total of 144 trials. To account for possible differences between responses with the left and the right hand, two of the blocks were answered with the left hand and the other two with the right hand, in counterbalanced order. Moreover, to control for aural differences between the two headphone speakers, headphones were reversed for half of the test runs. In general, four blocks of 36 trials seems to be the maximum of what is commonly used in dichotic listening trials. In fused rhymed words test, Wexler and Halwes [100] used 120 CVC stimuli trials.

One commonly used extension of the classic CV dichotic listening paradigm is the so-called forced attention paradigm, developed by Bryden et al. [67] and Hugdahl and Andersson [68]. In this version of the task, participants are tested in three conditions: a nonforced condition in which participants are instructed to report the syllable they heard best, a forced-right condition in which they are instructed to concentrate only on the syllables presented to the right ear, and a forced-left condition in which participants are instructed to attend only to the syllables presented to the left ear. Systematic comparison of the three conditions allows investigation of the effects of attention on dichotic listening performance [62]. This version of the task has been used in several studies since its invention [112–116] and is also implemented in the iDichotic app [73, 74].

2.2.5 Dependent Variables and Statistical Analysis

Typically, analysis of dichotic listening performance is focused on the number of correctly identified stimuli for each ear. Reaction times are not typically within the focus of analysis. The number of correctly identified stimuli for the left and right ear are typically analyzed using a repeated measures ANOVA with ear (left and right) as within-subjects factor and all additional factors of interest as additional within-subject factors (e.g. for experimental conditions) or between-subjects factors (e.g. when testing patients vs. controls). In addition, several control variables can be analyzed to assess participants' general performance level, e.g. the number of correctly identified homonyms and the overall error rate. If a participant is unable to identify the homonyms correctly or makes too many errors, he or she might not have understood the task instructions correctly or lacked motivation to perform the task properly. Thus, these variables might be helpful for the researcher when deciding on whether or not to exclude a dataset from analysis.

Similar to divided visual field paradigms, the results of a dichotic listening paradigm can also be displayed in a single value [91]. To this end, a so-called laterality quotient (LQ) can be calculated, e.g. by using the following formula $LQ = [R - L]/[R + L] \times 100$, with R indicating the number of right ear reports and L indicating the number of left ear reports.

2.3 Notes

Comparable to divided visual field paradigms, dichotic listening offers a cheap, fast, and noninvasive method to determine the dominant hemisphere for a specific task. In general, dichotic listening seems to have higher reliability than divided visual field tasks. For the CV syllables task, Russell and Voyer [117] reported a test-retest validity of 0.76, while Speaks et al. [118] reported a split-half reliability of 0.90 for 180 trials. Identification of the language-dominant hemisphere with dichotic listening also has a good concordance with the results of the Wada test. Hugdahl et al. [69] tested 13 children and adolescents who were surgically treated for resistant epilepsy and therefore underwent the Wada test with the dichotic listening task before and after surgery. Using discriminant analysis, they found that determining lateralization with the dichotic listening task led a correct classification according to the Wada test results in all but one subject (92 %).

One advantage of the behavioral dichotic listening paradigm is that, like divided visual field paradigms, it can be combined with other techniques such as electroencephalography [119–122], magnetoencephalography [123, 124], positron emission tomography [125–127], or functional magnetic resonance imaging [72, 114, 128–130]. It is also well suited to be used in field studies. For example, Güntürkün et al. [131] used a dichotic listening paradigm to investigate speech perception asymmetries in native whistle-speaking people in a mountainous area of Northeast Turkey. Use of the smartphone-based iDichotic app [73, 74] further extends the appeal of the dichotic listening paradigm for studies outside the laboratory and offers exciting possibilities to test participants in unusual settings.

One thing researchers have to bear in mind when interpreting results of a dichotic listening paradigm is that it is not clear to what extent performance represents solely language dominance or whether other factors confound this effect. In particular, cognitive control functions and attentional biases have been implicated to modulate performance in dichotic listening tasks [132]. Thus, like divided visual field paradigms, dichotic listening provides the researcher with a general idea of whether the left or right hemisphere is dominant for language perception. However, further inferences about the relation between performance and brain activation should only be conducted with great care and possibly the use of neuroscientific techniques beyond behavioral measurements.

2.4 Findings on the Evolution of Lateralization

Dichotic stimulus presentation has been used in many studies in diverse species, for example squirrel monkeys [133], cats [134], budgerigars [135], rabbits [136], or barn owls [137]. However, true dichotic listening tasks in the sense that the animal has to indicate which of several response alternatives it heard so that a REA can be calculated have not been published for nonhuman animals. One line of research that parallels the results of dichotic listening paradigms is the studies of Fitch et al. [138, 139]. These authors showed that rats exhibit a REA for discrimination of monaurally presented tone sequences. They concluded that left-hemisphere dominance for auditory temporal processing in rats might present an evolutionary precursor to left-hemisphere language specialization in humans, which is also in line with comparative studies employing other paradigms [140, 141].

References

1. Baxendale S (2009) The Wada test. Curr Opin Neurol 22:185–189
2. Davidson RJ (1988) EEG measures of cerebral asymmetry: conceptual and methodological issues. Int J Neurosci 39:71–89
3. Rossion B, Dricot L, Devolder A, Bodart JM, Crommelinck M, De Gelder B, Zoontjes R (2000) Hemispheric asymmetries for whole-based and part-based face processing in the human fusiform gyrus. J Cogn Neurosci 12:793–802
4. Westerhausen R, Kompus K, Hugdahl K (2014) Mapping hemispheric symmetries, relative asymmetries, and absolute asymmetries underlying the auditory laterality effect. Neuroimage 84:962–970
5. Volkmann AW (1859) Das Tachistoskop. Berichte über die Verhandlungen der Königlich Sächsischen Gesellschaft der Wissenschaften zu Leipzig. Mathematisch-Physische Classe 11:90–98
6. Poffenberger A (1912) Reaction time to retinal stimulation with special reference to the time lost in conduction through nervous centers. Arch Psychol 23:1–73
7. Gazzaniga MS, Bogen JE, Sperry RW (1962) Some functional effects of sectioning the cerebral commissures in man. Proc Natl Acad Sci U S A 48:1765–1769
8. Hausmann M, Güntürkün O (2000) Steroid fluctuations modify functional cerebral asymmetries: the hypothesis of progesterone-mediated interhemispheric decoupling. Neuropsychologia 38:1362–1374
9. Gazzaniga MS (2005) Forty-five years of split-brain research and still going strong. Nat Rev Neurosci 6:653–659
10. Bourne VJ (2006) The divided visual field paradigm: methodological considerations. Laterality 11:373–393
11. Van der Haegen L, Cai Q, Seurinck R, Brysbaert M (2011) Further fMRI validation of the visual half field technique as an indicator of language laterality: a large-group analysis. Neuropsychologia 49:2879–2888
12. Godnig EC (2003) The tachistoscope: its history and uses. J Behav Optom 14:39–42
13. Correia S (2011) Tachistoscopic presentation. In: Kreutzer J, DeLuca J, Caplan B (eds) Encyclopedia of clinical neuropsychology. Springer, Heidelberg, pp 2461–2463
14. Benschop R (1998) What is a tachistoscope? Historical explorations of an instrument. Sci Context 11:23–50
15. Bryden PJ, Brown SG, Roy EA (2011) Can an observational method of assessing hand preference be used to predict language lateralisation? Laterality 16:707–721
16. Godard O, Fiori N (2012) Sex and hemispheric differences in facial invariants extraction. Laterality 17:202–216
17. Kavé G, Gavrieli R, Mashal N (2014) Stronger left-hemisphere lateralization in older versus younger adults while processing conventional metaphors. Laterality 19:705–717
18. Lam M, Collinson SL, Sim K, Mackay CE, James AC, Crow TJ (2012) Asymmetry of lexico-semantic processing in schizophrenia changes with disease progression. Schizophr Res 134:125–130
19. Elze T, Tanner TG (2012) Temporal properties of liquid crystal displays: implications for vision science experiments. PLoS One 7:e44048
20. Ghodrati M, Morris AP, Price NS (2015) The (un)suitability of modern liquid crystal displays (LCDs) for vision research. Front Psychol 6:303
21. Plant RR, Hammond N, Whitehouse T (2003) How choice of mouse may affect response timing in psychological studies. Behav Res Methods Instrum Comput 35:276–284
22. Ocklenburg S, Arning L, Gerding WM, Epplen JT, Güntürkün O, Beste C (2013) FOXP2 variation modulates functional hemispheric asymmetries for speech perception. Brain Lang 126:279–284
23. Selpien H, Siebert C, Genc E, Beste C, Faustmann PM, Güntürkün O, Ocklenburg S (2015) Left dominance for language perception starts in the extrastriate cortex: an ERP and sLORETA study. Behav Brain Res 291:325–333
24. Walsh A, McDowall J, Grimshaw GM (2010) Hemispheric specialization for emotional word processing is a function of SSRI responsiveness. Brain Cogn 74:332–340
25. Jordan TR, Patching GR, Milner AD (1998) Central fixations are inadequately controlled by instructions alone: implications for studying cerebral asymmetry. Q J Exp Psychol A 51:371–391
26. Bourne VJ, Hole GJ (2006) Lateralized repetition priming for familiar faces: evidence for asymmetric interhemispheric cooperation. Q J Exp Psychol (Hove) 59:1117–1133
27. Marzi CA, Berlucchi G (1977) Right visual field superiority for accuracy of recognition of famous faces in normals. Neuropsychologia 15:751–756
28. Meyer AM, Federmeier KD (2008) The divided visual world paradigm: eye tracking reveals

hemispheric asymmetries in lexical ambiguity resolution. Brain Res 1222:166–183
29. Lange JJ, Wijers AA, Mulder LJ, Mulder G (1999) ERP effects of spatial attention and display search with unilateral and bilateral stimulus displays. Biol Psychol 50:203–233
30. Purves D, Augustine GJ, Fitzpatrick D (2001) Types of eye movements and their functions. In: Purves D, Augustine GJ, Fitzpatrick D, Katz LC, LaMantia AS, McNamara JO, & Williams SM (eds.) Neuroscience, 2nd edn. Sinauer Associates, Sunderland
31. Lindell AK, Nicholls ME (2003) Cortical representation of the fovea: implications for visual half-field research. Cortex 39:111–117
32. Saban-Bezalel R, Mashal N (2015) Hemispheric processing of idioms and irony in adults with and without pervasive developmental disorder. J Autism Dev Disord 45(11):3496–3508
33. Falkenberg HK, Rubin GS, Bex PJ (2007) Acuity, crowding, reading and fixation stability. Vision Res 47:126–135
34. Prete G, Laeng B, Fabri M, Foschi N, Tommasi L (2015) Right hemisphere or valence hypothesis, or both? The processing of hybrid faces in the intact and callosotomized brain. Neuropsychologia 68:94–106
35. Ellis AW (2004) Length, formats, neighbours, hemispheres, and the processing of words presented laterally or at fixation. Brain Lang 88:355–366
36. Gold R, Faust M, Ben-Artzi E (2012) Metaphors and verbal creativity: the role of the right hemisphere. Laterality 17:602–614
37. Tressoldi PE (1987) Visual hemispace differences reflect hemisphere asymmetries. Neuropsychologia 25:625–636
38. Verosky SC, Turk-Browne NB (2012) Representations of facial identity in the left hemisphere require right hemisphere processing. J Cogn Neurosci 24:1006–1017
39. Beaton AA, Fouquet NC, Maycock NC, Platt E, Payne LS, Derrett A (2012) Processing emotion from the eyes: a divided visual field and ERP study. Laterality 17:486–514
40. Jończyk R (2015) Hemispheric asymmetry of emotion words in a non-native mind: a divided visual field study. Laterality 20:326–347
41. Holtgraves T, Felton A (2011) Hemispheric asymmetry in the processing of negative and positive words: a divided field study. Cogn Emot 25:691–699
42. Mashal N, Itkes O (2014) The effects of emotional valence on hemispheric processing of metaphoric word pairs. Laterality 19:511–521
43. Jones B, Anuza T (1982) Effects of sex, handedness, stimulus and visual field on "mental rotation". Cortex 18:501–514
44. Michałowski B, Króliczak G (2015) Sinistrals are rarely "right": evidence from tool-affordance processing in visual half-field paradigms. Front Hum Neurosci 9:166
45. Leuthold H (2011) The Simon effect in cognitive electrophysiology: a short review. Acta Psychol (Amst) 136:203–211
46. Ocklenburg S, Güntürkün O, Beste C (2011) Lateralized neural mechanisms underlying the modulation of response inhibition processes. Neuroimage 55:1771–1778
47. Leehey S, Carey S, Diamond R, Cahn A (1978) Upright and inverted faces: the right hemisphere knows the difference. Cortex 14:411–419
48. Brysbaert M, d'Ydewalle G (1990) Tachistoscopic presentation of verbal stimuli for assessing cerebral dominance: reliability data and some practical recommendations. Neuropsychologia 28:443–455
49. Voyer D (1998) On the reliability and validity of noninvasive laterality measures. Brain Cogn 36:209–236
50. Hunter ZR, Brysbaert M (2008) Visual half-field experiments are a good measure of cerebral language dominance if used properly: evidence from fMRI. Neuropsychologia 46:316–325
51. Bruyer R, Brysbaert M (2011) Combining speed and accuracy in cognitive psychology: is the Inverse Efficiency Score (IES) a better dependent variable than the mean Reaction Time (RT) and the Percentage of Errors (PE)? Psychol Belg 51:5–13
52. Ocklenburg S, Güntürkün O, Beste C (2012) Hemispheric asymmetries and cognitive flexibility: an ERP and sLORETA study. Brain Cogn 78:148–155
53. Ocklenburg S, Ness V, Güntürkün O, Suchan B, Beste C (2013) Response inhibition is modulated by functional cerebral asymmetries for facial expression perception. Front Psychol 4:879
54. Smith ER, Chenery HJ, Angwin AJ, Copland DA (2009) Hemispheric contributions to semantic activation: a divided visual field and event-related potential investigation of time-course. Brain Res 1284:125–144
55. Küper K, Zimmer HD (2015) ERP evidence for hemispheric asymmetries in exemplar-specific explicit memory access. Brain Res 1625:73–83
56. Anders S, Lotze M, Wildgruber D, Erb M, Grodd W, Birbaumer N (2005) Processing of

a simple aversive conditioned stimulus in a divided visual field paradigm: an fMRI study. Exp Brain Res 162:213–219
57. Sung YW, Someya Y, Eriko Y, Choi SH, Cho ZH, Ogawa S (2011) Involvement of low-level visual areas in hemispheric superiority for face processing. Brain Res 1390:118–125
58. Motz BA, James KH, Busey TA (2012) The Lateralizer: a tool for students to explore the divided brain. Adv Physiol Educ 36:220–225
59. Hopkins WD (1997) Hemispheric specialization for local and global processing of hierarchical visual stimuli in chimpanzees (Pan troglodytes). Neuropsychologia 35:343–348
60. Van Kleeck MH (1989) Hemispheric differences in global versus local processing of hierarchical visual stimuli by normal subjects: new data and a meta-analysis of previous studies. Neuropsychologia 27:1165–1178
61. Dépy D, Fagot J, Vauclair J (1998) Comparative assessment of distance processing and hemispheric specialization in humans and baboons (Papio papio). Brain Cogn 38:165–182
62. Hugdahl K (2011) Fifty years of dichotic listening research—still going and going and …. Brain Cogn 76:211–213
63. Kimura D (1961) Some effects of temporal-lobe damage on auditory perception. Can J Psychol 15:156–165
64. Kimura D (1961) Cerebral dominance and the perception of verbal stimuli. Can J Psychol 15:166–170
65. Shankweiler D, Studdert-Kennedy M (1967) Identification of consonants and vowels presented to left and right ears. Q J Exp Psychol 19:59–63
66. Studdert-Kennedy M, Shankweiler D (1970) Hemispheric specialization for speech perception. J Acoust Soc Am 48:579–694
67. Bryden MP, Munhall K, Allard F (1983) Attentional biases and the right-ear effect in dichotic listening. Brain Lang 18:236–248
68. Hugdahl K, Andersson L (1986) The "forced-attention paradigm" in dichotic listening to CV-syllables: a comparison between adults and children. Cortex 22:417–432
69. Hugdahl K, Carlsson G, Uvebrant P, Lundervold AJ (1997) Dichotic-listening performance and intracarotid injections of amobarbital in children and adolescents. Preoperative and postoperative comparisons. Arch Neurol 54:1494–1500
70. Sommer I, Ramsey N, Kahn R, Aleman A, Bouma A (2001) Handedness, language lateralisation and anatomical asymmetry in schizophrenia: meta-analysis. Br J Psychiatry 178:344–351
71. Westerhausen R, Hugdahl K (2008) The corpus callosum in dichotic listening studies of hemispheric asymmetry: a review of clinical and experimental evidence. Neurosci Biobehav Rev 32:1044–1054
72. Kompus K, Specht K, Ersland L, Juvodden HT, van Wageningen H, Hugdahl K, Westerhausen R (2012) A forced-attention dichotic listening fMRI study on 113 subjects. Brain Lang 121:240–247
73. Bless JJ, Westerhausen R, Arciuli J, Kompus K, Gudmundsen M, Hugdahl K (2013) "Right on all Occasions?"—On the feasibility of laterality research using a smartphone dichotic listening application. Front Psychol 4:42
74. Bless JJ, Westerhausen R, von Koss Torkildsen J, Gudmundsen M, Kompus K, Hugdahl K (2015) Laterality across languages: results from a global dichotic listening study using a smartphone application. Laterality 20:434–452
75. Hirnstein M, Westerhausen R, Korsnes MS, Hugdahl K (2013) Sex differences in language asymmetry are age-dependent and small: a large-scale, consonant-vowel dichotic listening study with behavioral and fMRI data. Cortex 49:1910–1921
76. Westerhausen R, Bless J, Kompus K (2015) Behavioral laterality and aging: the free-recall dichotic-listening right-ear advantage increases with age. Dev Neuropsychol 40:313–327
77. Broadbent DE (1954) The role of auditory localization in attention and memory span. J Exp Psychol 47:191–196
78. Kimura D (2011) From ear to brain. Brain Cogn 76:214–217
79. Tervaniemi M, Hugdahl K (2003) Lateralization of auditory-cortex functions. Brain Res Brain Res Rev 43:231–246
80. Hugdahl K (ed) (1988) Handbook of dichotic listening: theory, methods, and research. John Wiley & Sons Ltd, Chichester
81. Bradshaw JL, Burden V, Nettleton NC (1986) Dichotic and dichhaptic techniques. Neuropsychologia 24:79–90
82. Bryden MP (1967) An evaluation of some models of laterality. Effects in dichotic listening. Acta Otolaryngol 63:595–604
83. Bruder GE (1991) Dichotic listening: new developments and applications in clinical research. Ann N Y Acad Sci 620:217–232
84. Hiscock M, Kinsbourne M (2011) Attention and the right-ear advantage: what is the connection? Brain Cogn 76:263–275
85. Musiek FE, Weihing J (2011) Perspectives on dichotic listening and the corpus callosum. Brain Cogn 76:225–232

86. Bouma A, Gootjes L (2011) Effects of attention on dichotic listening in elderly and patients with dementia of the Alzheimer type. Brain Cogn 76:286–293
87. Gadea M, Espert R, Salvador A, Martí-Bonmatí L (2011) The sad, the angry, and the asymmetrical brain: dichotic listening studies of negative affect and depression. Brain Cogn 76:294–299
88. Obrzut JE, Mahoney EB (2011) Use of the dichotic listening technique with learning disabilities. Brain Cogn 76:323–331
89. Niccum N, Speaks C (1991) Interpretation of outcome on dichotic listening tests following stroke. J Clin Exp Neuropsychol 13:614–628
90. Ocklenburg S, Westerhausen R, Hirnstein M, Hugdahl K (2013) Auditory hallucinations and reduced language lateralization in schizophrenia: a meta-analysis of dichotic listening studies. J Int Neuropsychol Soc 19:410–418
91. Hahn C, Neuhaus AH, Pogun S, Dettling M, Kotz SA, Hahn E, Brüne M, Güntürkün O (2011) Smoking reduces language lateralization: a dichotic listening study with control participants and schizophrenia patients. Brain Cogn 76:300–309
92. Hahn C, Pogun S, Güntürkün O (2010) Smoking modulates language lateralization in a sex-specific way. Neuropsychologia 48:3993–4002
93. Demarest L, Demarest J (1981) The interaction of handedness, familial sinistrality and sex on the performance of a dichotic listening task. Int J Neurosci 14:7–13
94. Ocklenburg S, Arning L, Hahn C, Gerding WM, Epplen JT, Güntürkün O, Beste C (2011) Variation in the NMDA receptor 2B subunit gene GRIN2B is associated with differential language lateralization. Behav Brain Res 225:284–289
95. Hugdahl K, Løberg EM, Falkenberg LE, Johnsen E, Kompus K, Kroken RA, Nygård M, Westerhausen R, Alptekin K, Ozgören M (2012) Auditory verbal hallucinations in schizophrenia as aberrant lateralized speech perception: evidence from dichotic listening. Schizophr Res 140:59–64
96. Van der Haegen L, Westerhausen R, Hugdahl K, Brysbaert M (2013) Speech dominance is a better predictor of functional brain asymmetry than handedness: a combined fMRI word generation and behavioral dichotic listening study. Neuropsychologia 51:91–97
97. Arciuli J (2011) Manipulation of voice onset time during dichotic listening. Brain Cogn 76:233–238
98. Castro A, Pearson R (2011) Lateralisation of language and emotion in schizotypal personality: evidence from dichotic listening. Pers Individ Dif 51:726–731
99. Prete G, Marzoli D, Brancucci A, Fabri M, Foschi N, Tommasi L (2014) The processing of chimeric and dichotic emotional stimuli by connected and disconnected cerebral hemispheres. Behav Brain Res 271:354–364
100. Wexler BE, Halwes T (1983) Increasing the power of dichotic methods: the fused rhymed words test. Neuropsychologia 21:59–66
101. Crosson B, Warren RL (1981) Dichotic ear preference for C-V-C words in Wernicke's and Broca's aphasias. Cortex 17:249–258
102. di Stefano M, Salvadori C, Fiaschi E, Viti M (1998) Speech lateralisation in callosal agenesis assessed by the dichotic fused words test. Laterality 3:131–142
103. Hoch L, Tillmann B (2010) Laterality effects for musical structure processing: a dichotic listening study. Neuropsychology 24:661–666
104. Nelson MD, Wilson RH, Kornhass S (2003) Performance of musicians and nonmusicians on dichotic chords, dichotic CVs, and dichotic digits. J Am Acad Audiol 14:536–544
105. Craig JD (1980) A dichotic rhythm task: advantage for the left-handed. Cortex 16:613–620
106. Spreen O, Spellacy FJ, Reid JR (1970) The effect of interstimulus interval and intensity on ear asymmetry for nonverbal stimuli in dichotic listening. Neuropsychologia 8:245–250
107. Ocklenburg S, Ball A, Wolf CC, Genç E, Güntürkün O (2015) Functional cerebral lateralization and interhemispheric interaction in patients with callosal agenesis. Neuropsychology 29:806–815
108. Løberg EM, Jørgensen HA, Hugdahl K (2004) Dichotic listening in schizophrenic patients: effects of previous vs. ongoing auditory hallucinations. Psychiatry Res 128:167–174
109. Bozikas VP, Kosmidis MH, Giannakou M, Kechayas P, Tsotsi S, Kiosseoglou G, Fokas K, Garyfallos G (2014) Controlled shifting of attention in schizophrenia and bipolar disorder through a dichotic listening paradigm. Compr Psychiatry 55:1212–1219
110. Hirnstein M, Hugdahl K, Hausmann M (2014) How brain asymmetry relates to performance—a large-scale dichotic listening study. Front Psychol 4:997
111. Foundas AL, Corey DM, Hurley MM, Heilman KM (2004) Verbal dichotic listening in developmental stuttering: subgroups with atypical auditory processing. Cogn Behav Neurol 17:224–232

112. Conn R, Posey TB (2000) Dichotic listening in college students who report auditory hallucinations. J Abnorm Psychol 109:546–549
113. Hugdahl K, Andersson L, Asbjørnsen A, Dalen K (1990) Dichotic listening, forced attention, and brain asymmetry in right-handed and lefthanded children. J Clin Exp Neuropsychol 12:539–548
114. Jäncke L, Buchanan TW, Lutz K, Shah NJ (2001) Focused and nonfocused attention in verbal and emotional dichotic listening: an FMRI study. Brain Lang 78:349–363
115. Kershner JR (2016) Forced-attention dichotic listening with university students with dyslexia: search for a core deficit. J Learn Disabil 49(3):282–292
116. Obrzut JE, Mondor TA, Uecker A (1993) The influence of attention on the dichotic REA with normal and learning disabled children. Neuropsychologia 31:1411–1416
117. Russell NL, Voyer D (2004) Reliability of laterality effects in a dichotic listening task with words and syllables. Brain Cogn 54:266–267
118. Speaks C, Niccum N, Carney E (1982) Statistical properties of responses to dichotic listening with CV nonsense syllables. J Acoust Soc Am 72:1185–1194
119. Bayazit O, Oniz A, Hahn C, Güntürkün O, Ozgören M (2009) Dichotic listening revisited: trial-by-trial ERP analyses reveal intra- and interhemispheric differences. Neuropsychologia 47:536–545
120. Eichele T, Nordby H, Rimol LM, Hugdahl K (2005) Asymmetry of evoked potential latency to speech sounds predicts the ear advantage in dichotic listening. Brain Res Cogn Brain Res 24:405–412
121. Passow S, Westerhausen R, Hugdahl K, Wartenburger I, Heekeren HR, Lindenberger U, Li SC (2014) Electrophysiological correlates of adult age differences in attentional control of auditory processing. Cereb Cortex 24:249–260
122. Yurgil KA, Golob EJ (2010) Neural activity before and after conscious perception in dichotic listening. Neuropsychologia 48:2952–2958
123. Alho K, Salonen J, Rinne T, Medvedev SV, Hugdahl K, Hämäläinen H (2012) Attention-related modulation of auditory-cortex responses to speech sounds during dichotic listening. Brain Res 1442:47–54
124. Brancucci A, Penna SD, Babiloni C, Vecchio F, Capotosto P, Rossi D, Franciotti R, Torquati K, Pizzella V, Rossini PM, Romani GL (2008) Neuromagnetic functional coupling during dichotic listening of speech sounds. Hum Brain Mapp 29:253–264
125. Beaman CP, Bridges AM, Scott SK (2007) From dichotic listening to the irrelevant sound effect: a behavioural and neuroimaging analysis of the processing of unattended speech. Cortex 43:124–134
126. Hugdahl K, Law I, Kyllingsbaek S, Brønnick K, Gade A, Paulson OB (2000) Effects of attention on dichotic listening: an 15O-PET study. Hum Brain Mapp 10:87–97
127. Lipschutz B, Kolinsky R, Damhaut P, Wikler D, Goldman S (2002) Attention-dependent changes of activation and connectivity in dichotic listening. Neuroimage 17:643–656
128. Dos Santos Sequeira S, Specht K, Moosmann M, Westerhausen R, Hugdahl K (2010) The effects of background noise on dichotic listening to consonant-vowel syllables: an fMRI study. Laterality 15:577–596
129. Van den Noort M, Specht K, Rimol LM, Ersland L, Hugdahl K (2008) A new verbal reports fMRI dichotic listening paradigm for studies of hemispheric asymmetry. Neuroimage 40:902–911
130. Westerhausen R, Passow S, Kompus K (2013) Reactive cognitive-control processes in free-report consonant-vowel dichotic listening. Brain Cogn 83:288–296
131. Güntürkün O, Güntürkün M, Hahn C (2015) Whistled Turkish alters language asymmetries. Curr Biol 25:R706–R708
132. Hugdahl K, Westerhausen R, Alho K, Medvedev S, Laine M, Hämäläinen H (2009) Attention and cognitive control: unfolding the dichotic listening story. Scand J Psychol 50:11–22
133. Don M, Starr A (1972) Lateralization performance of squirrel monkey (Samiri sciureus) to binaural click signals. J Neurophysiol 35:493–500
134. Tollin DJ, Populin LC, Moore JM, Ruhland JL, Yin TC (2005) Sound-localization performance in the cat: the effect of restraining the head. J Neurophysiol 93:1223–1234
135. Welch TE, Dent ML (2011) Lateralization of acoustic signals by dichotically listening budgerigars (Melopsittacus undulatus). J Acoust Soc Am 130:2293–2301
136. Ebert CS Jr, Blanks DA, Patel MR, Coffey CS, Marshall AF, Fitzpatrick DC (2008) Behavioral sensitivity to interaural time differences in the rabbit. Hear Res 235:134–142
137. Moiseff A (1989) Bi-coordinate sound localization by the barn owl. J Comp Physiol A 164:637–644
138. Fitch RH, Brown CP, O'Connor K, Tallal P (1993) Functional lateralization for auditory temporal processing in male and female rats. Behav Neurosci 107:844–850

139. Fitch RH, Brown CP, Tallal P (1993) Left hemisphere specialization for auditory temporal processing in rats. Ann N Y Acad Sci 682:346–347

140. Böye M, Güntürkün O, Vauclair J (2005) Right ear advantage for conspecific calls in adults and subadults, but not infants, California sea lions (Zalophus californianus): hemispheric specialization for communication? Eur J Neurosci 21:1727–1732

141. Ehret G (1987) Left hemisphere advantage in the mouse brain for recognizing ultrasonic communication calls. Nature 325:249–251

Chapter 2

Split-Brain Human Subjects

Mara Fabri, Nicoletta Foschi, Chiara Pierpaoli, and Gabriele Polonara

Abstract

This chapter reviews the neuropsychological and imaging studies, carried out by the author's group and coworkers on split-brain patients in the past 19 years, to investigate the role of the human corpus callosum in the interhemispheric transfer and integration of information. These studies will provide evidence of how the research on split-brain patients may provide a significant contribution to the understanding of lateralized and diffuse brain functions. In particular, by comparing results from total and partial callosotomized patients and with control subjects, many findings have been obtained on the organization and functions of human brain. The studies will be described in a brief overview of other groups' research on similar patients.

Key words Corpus callosum, Interhemispheric connections, fMRI, Neuropsychological testing

Abbreviations

APC	Anterior parietal cortex
BOLD	Blood oxygenation level dependent
CC	Corpus callosum
Cin	Cingulated cortex
CS	Central sulcus
CT	Computerized tomography
CUD	Crossed uncrossed difference
DTI	Diffusion tensor imaging
fMRI	Functional magnetic resonance imaging
IT	Interhemispheric transfer
GAD	Glutamic acid decarboxylase
LVF	Left visual field
MEG	Magnetoencephalography
MRI	Magnetic resonance imaging
PCB	Posterior callosal body
PCG	Postcentral gyrus
PCS	Postcentral sulcus

The original version of this chapter was revised. The erratum to this chapter is available at: DOI 10.1007/978-1-4939-6725-4_22

Lesley J. Rogers and Giorgio Vallortigara (eds.), *Lateralized Brain Functions: Methods in Human and Non-Human Species*, Neuromethods, vol. 122, DOI 10.1007/978-1-4939-6725-4_2, © Springer Science+Business Media LLC 2017

PET	Positron emission tomography
PO	Parietal operculum
PP	Poffenberger paradigm
PPC	Posterior parietal cortex
RF	Receptive field
RHH	Right hemisphere hypothesis
ROI	Region of interest
RT	Reaction time
RTE	Redundant target effect
RVF	Right visual field
SC	Superior colliculus
S-DRT	Same-different recognition test
SI	Primary somatic sensory area
SII	Secondary somatic sensory area
SS	Sylvian sulcus
TFLT	Tactile finger localization test
TNT	Tactile naming test
TPJ	Temporal parietal junction
VF	Visual field
VH	Valence hypothesis

1 Introduction

One of the major and as yet unexplained features of the organization of the central nervous system is that the brain is made up of two hemispheres, each of them receiving afferents mainly from the contralateral, i.e. opposite side of the body [1], and controlling the motor and sensory functions of the contralateral side of the body (see, however, [2] for a possible evolutionary account about the origins of this organization of the vertebrate brain). One consequence is that some functions are lateralized, i.e., the specializations for them are exclusive or prevalent functions of only one hemisphere. On the other hand, each sensory cortical area in each hemisphere contains the representation of the contralateral sensory periphery [3], and discontinuities may occur in processing of stimuli that move from one side of the sensory space to the other.

In normal human brain, both hemispheres are always informed on what is going on and both participate in programming and controlling brain and body activity. It is therefore almost impossible to identify the specific competence of each hemisphere, their difference and their similarities. The hemispheres are connected by the corpus callosum (CC), which provides interhemispheric integration and transfer of information [4].

The CC is topographically organized, as postmortem human studies and data from nonhuman primates have indicated. This organization seems to result in modality-specific regions: the anterior fibers, which connect the frontal lobes, transfer motor information, whereas the posterior fibers, which connect the temporal, parietal, and occipital lobes, are involved in the integration of somatosensory (posterior midbody), auditory (isthmus), and visual (splenium) information [5] (see data and literature in [6, 7]).

After the introduction of the surgical procedure to separate the cerebral hemispheres by dissecting the corpus callosum (callosotomy) and eventually also the anterior and posterior commissures (commissurotomy), a new "experimental model" became available to study the lateralized and diffuse functions of the brain, and the organization of neural circuits subserving different cerebral functions [8]. This model was represented by the so-called split-brain (divided brain) or split-brained patients, who had surgical separation of their hemispheres to different extents.

Split brain patients are subjects in whom the cerebral hemispheres have been surgically disconnected by cutting the corpus callosum and sometimes also the other cerebral commissures (anterior and posterior, more commonly the anterior one). This surgical procedure, performed as a last resort measure to relieve intractable epilepsy by preventing the spread of seizures (electrical discharges) to the whole brain, results in either the partial or complete disconnection between the two hemispheres [9]. The modern procedure typically involves only the anterior 1/3 of the corpus callosum; however, if the epileptic seizures continue, the following 1/3 is resected prior to the remaining 1/3 if the seizures persist. This results in a complete callosotomy in which most of the information transfer between hemispheres is lost.

Due to the functional mapping of the corpus callosum, a partial callosotomy has less detrimental effects because it leaves parts of the corpus callosum intact. There is little functional plasticity observed in partial and complete callosotomies on adults, with the splenium assuming probably most callosal functions [10].

In general, split-brained patients behave in a coordinated, purposeful, and consistent manner, despite the independent, parallel, usually different and occasionally conflicting processing of the same information from the environment by the two disconnected hemispheres. When two hemispheres receive competing stimuli at the same time, the response mode tends to determine which hemisphere controls behavior. Often, split-brained patients are indistinguishable from normal adults. This is due to the compensatory phenomena; split-brained patients progressively acquire a variety of strategies to get around their interhemispheric transfer deficits (see [9]).

Although the first interventions of hemispheric disconnection date back to the 1940s [11], methodical research on split brain patients began in the 1960s in the USA, when the callosotomy was routinely introduced as surgical palliative practice to treat drug-resistant epilepsy.

The first topic studied was the organization and the hemispheric dominance of the known lateralized functions, such as language, reading ability, and speaking ability. Initial research on split brain patients was carried out using neuropsychological methods (see below), analyzed hemispheric specialization, and first introduced the concept of a topographical organization of the corpus callosum [11]. The first studies investigated the language processing abilities of the right hemisphere as well as auditory and emotional reactions. The reported findings showed that the two halves of the brain have numerous functions and specialized skills, and it has been concluded that each hemisphere really has functions of its own. The left hemisphere is thought to be better at writing, speaking, mathematical calculation, reading, and is the primary area for language. The right hemisphere is seen to possess capabilities for problem solving, recognizing faces, symbolic reasoning, art, and spatial relationships [11].

The research with split-brain patients has become, however, more and more uncommon, since callosotomy has been largely replaced by more efficient and less invasive pharmacological treatments for epilepsy in recent years [4, 11]. Nonetheless, the performance of these rare patients in different cognitive tasks can still help us to understand what kind of information is preferentially processed by each hemisphere. In addition, the possibility to compare patients who have undergone different degrees of hemispheric disconnection (totally or partially and, in the latter case, in different regions of the CC) provides us with a unique chance to study hemispheric competences.

1.1 Neuropsychological Techniques

These techniques aimed to and allowed the study of split brain patients by stimulating one hemisphere at the time, in order to evoke a response from the same or opposite hemisphere, and to measure and compare the latency and accuracy in the two conditions. Due to particular organization plans of the different sensory and motor systems, different stimulation strategies have been used.

1.1.1 Naming Test (Tactile or Visual)

Tactile Domain

Some of the tests administered are devised to evaluate the interhemispheric transfer of somatosensory information: viz., the Tactile Naming Test, Same-Different Recognition Test, and intra- and inter-manual Tactile Finger Localization Test. Chance-level performance is calculated based on the number of stimulus alternatives.

The Tactile Naming Test (TNT) evaluates the subject's ability to name tactile stimuli presented only to the right hemisphere (left hand; for a detailed description of the procedure see [6, 12, 13]). A number of common objects, whose nature is unknown to the patient, are presented randomly to each hand and the patient is required to name them correctly. To do this, sensory information from the hand exploring the object must reach the left hemisphere, where usually the primary language areas are located, and which, for this reason, is often named the "spiking hemisphere." The left hand sends information about the explored object to the right hemisphere, and this in turn should reach the language area in the left hemisphere. The usual route is via the corpus callosum. If this way is interrupted, the information cannot reach the language area and the patient is not able to name the object.

The Same-Different Recognition Test (S-DRT) indicates whether the patient can cross-compare tactile stimuli separately presented to the two hemispheres. Patients are required to simultaneously explore manually two common objects held in each hand and to say whether they are the same or different objects. This decision requires interhemispheric transfer of tactile information (see data and literature in [6, 13]).

Finally, the Tactile Finger Localization Test (TFLT) assesses the ability to localize tactile stimuli applied to the proximal or distal phalanx of the fingers of one hand and to transfer the information to the opposite hand [6, 13]. In this test, the examiner lightly stimulates in random succession one of eight points (proximal or distal volar surface of fingers 2–5, a number of stimulations per point) of each hand, and immediately afterwards, the subject is asked to touch the stimulated point using the thumb of the same hand (intramanual task) or the same point in the contralateral hand using the thumb of that hand (intermanual task). The intermanual modality requires the tactile information from one hand to reach the contralateral hemisphere, and then be transferred to the opposite hemisphere in order to match the point on the opposite, not stimulated, hand.

Visual Domain

In this modality the interhemispheric transfer of visual information presented to a single hemisphere (in the contralateral visual field only) is evaluated. Due to the particular organization of the visual central pathways, in order to send the visual information to a single hemisphere before compensatory ocular reflex saccadic movements occur, it is necessary to present the visual stimuli in a *tachistoscopic way* [9], i.e., for a duration of less than 150 ms and in the lateral part of the visual field (at least 5° from the vertical meridian). The subject is told to sit in front of the board and stare at a point in the middle of a screen, and then the light spot will flash in the right or left visual field. When the patients are asked to describe afterward what they saw, they said that only the lights on the right side of the field had lit up. Next, when the lights are flashed on the subjects' left side of their visual field, they usually claim to have not seen any lights at all. The same test can be repeated with a different response modality, i.e., by asking the subjects to point to the lights that lit up. Although subjects can only report seeing the lights flash on the right, they actually point to all the lights in both visual fields. This shows that both brain hemispheres had seen the lights and were equally competent in visual perception. The subjects do not say that they saw the lights when they flashed in the left visual field because the center for speech is located in the brain's left hemisphere. This test supports the idea that the region of the brain associated with speech must be able to communicate with areas of the brain processing the visual information.

Combination of Visual and Tactile Tests

The tactile and visual test can be combined, by presenting subjects with a picture of an object to only their right hemisphere, which subjects are unable to name or describe. If the subjects however can reach under the screen and touch with their left hand various objects, they will be able to pick the one that had been shown in the picture. The subjects will also be able to pick out objects that were related to the picture presented, if that object was not under the screen.

1.1.2 Dichotic Listening

Dichotic listening is used as a behavioral test for hemispheric lateralization of speech sound perception. During a standard dichotic listening test, a participant is presented with two different auditory stimuli simultaneously (usually speech). The different stimuli are directed into different ears over headphones. Participants are asked to pay attention to one or both of the stimuli. Later, they are asked about the content of either the message they were asked to attend to or the message that they were not told to listen to.

An *emotional* version of the dichotic listening task has been developed. In this version individuals listen to the same word in each ear but they hear it in either a surprised, happy, sad, angry, or neutral tone. Participants are then asked to press a button indicating what tone they heard. Usually dichotic listening tests show a right-ear advantage for speech sounds. Right-ear/left-hemisphere advantage is expected, because of evidence from Broca's area and Wernicke's area, which are both located in the left hemisphere. In contrast, the left ear (and therefore the right hemisphere) is often better at processing nonlinguistic and emotional information.

Dichotic listening can also be used as a lateralized speech assessment task. Neuropsychologists have used this test to explore the role of singular neuroanatomical structures in speech perception and language asymmetry. After reviewing many studies, it was concluded [14] that dichotic listening should be considered a test of functional interhemispheric interaction and connectivity, besides being a test of lateralized temporal lobe language function, and the corpus callosum is critically involved in the top-down attentional control of dichotic listening performance, thus having a critical role in auditory laterality.

1.1.3 Reaction Time and Crossed–Uncrossed Difference. The Poffenberger Paradigm

To evaluate motor transfer through the corpus callosum, a simple reaction time (RT) measure can be used, and the crossed–uncrossed difference (CUD), combined in the so-called Poffenberger paradigm (PP). This is a behavioral paradigm in which the simple reaction time (RT) to visual stimuli presented to the hemifield ipsilateral to the responding hand is compared with that to stimuli presented to the contralateral hemifield, a condition requiring an interhemispheric transfer (IT). By using the paradigm originally developed by Poffenberger in 1912, it has been consistently found that in healthy humans it takes on average 4 ms longer to respond to stimuli presented to one visual hemifield with the contralateral than with the ipsilateral hand [15, 16]. This difference, the so-called crossed–uncrossed difference, is believed to reflect the need for callosal IT in the former condition and is considered to depend upon callosal transmission time. In the crossed condition the visual cortex receiving the stimulus and the motor cortex producing the response reside in different hemispheres, and therefore a callosal IT is required. In contrast, in the uncrossed condition, stimulus detection and response production can be carried out within one hemisphere and a commissural pathway is not required. The need for a callosal pathway involving more synapses than the direct intrahemispheric pathway can account for the longer RT in the crossed condition. As mentioned above, convincing evidence supporting a callosal transmission explanation is that the CUD is considerably increased in patients who have undergone callosotomy: healthy subjects have a mean CUD of 3.8 ms with a range between 1.0 and 10.3 ms [17]; those with total callosotomy have a range between 20 and 96 ms [18]. Further evidence that IT is mediated via the corpus callosum comes from functional magnetic resonance imaging (fMRI) studies, which have found direct signs of selectively greater activation in the genu of the CC when subjects perform the crossed conditions of the Poffenberger paradigm [19–21], as compared to the uncrossed condition.

2 Studies Performed with Patients from Ancona

Callosotomy surgery in Ancona started in 1983, as a palliative surgical procedure suitable for some patients with medically intractable epilepsy who were not candidates for focal resective surgery, and was initially performed by the neurosurgeon Isacco Papo [22–24].

Since then, a cohort of split-brain patients has become available for studying in Ancona, and the research on these subjects was initiated. The research performed on these patients will be here listed according to a chronological criterion, to give a taste of the studies on evolution of the brain. In the following paragraphs, however, the research on this population of patients will be described tentatively also according the scientific issues raised, but trying to maintain the chronological order, to let the reader see the research developments.

Since the first intervention, 52 epileptic patients have received the operation. The Ancona surgeons followed the "two-stage" procedure; in most cases the anterior partial callosotomy was sufficient to improve the control of seizure diffusion; in some cases (nine patients), it was necessary to proceed to a second stage callosotomy, leading to a total callosal resection, sometimes involving also the anterior commissure. The first operations could not benefit of a careful radiological control, since magnetic resonance imaging magnetic resonance imaging (MRI) was not yet available, and computerized tomography (CT), acquired at that time only in axial plane, lacked the high degree of spatial resolution achieved in the following years; even later, the first three patients could not be studied with neuroimaging techniques, because of a metallic arterial clip in their brain. Out of the 52 patients, 31 were still followed in 2010, and 20 of them participated to functional and neuropsychological studies (Fig. 1; Tables 1 and 2). The selection was

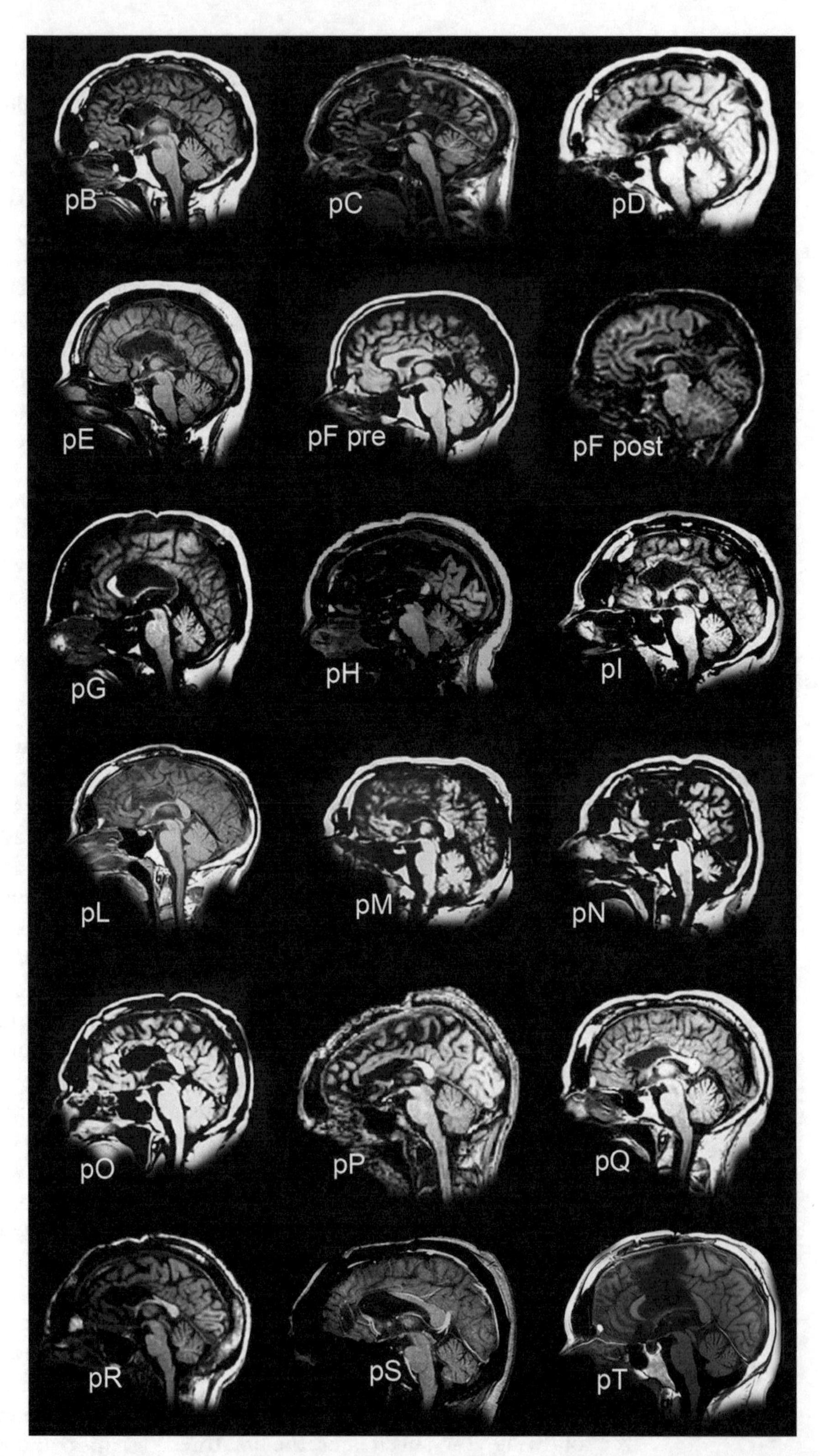

Fig. 1 Midsagittal images of the split-brain patients participating in the studies described. Patients pB–pE have a total callosal resection, as well as pF after the second stage surgery. Patient pG shows a posterior resection. Patients pH–pS display an anterior callosotomy of various extent. Finally, patient pT has a central callosal resection. Patients pA (total callosotomy), pU, and pV (anterior callosotomy) could not be submitted to MRI because of a metallic arterial clip in their brain

Table 1
Data from patients

Code	Gender	I.Q.	Oldfield score[a]	Present age	Age at surgery	Callosal resection	Studies in which participated																										
							s1	s2	s3	s4	s5	s6	s7	s8	s9	s10	s11	s12	s13	s14	s15	s16	s17	s18	s19	s20	s21	s22	s23	s24	s25	s26	s27
pA[b]	M	90	13	Deceased[c]	33	Total	×				×												×	×									
pB	M	81	10	43	22	Total		×					×	×	×	×	×	×	×	×	×		×	×	×	×	×				x	*	
pC	M	83	21	39	19	Total		×	×	×			×					×			×	×			×	×				×	*	x	×
pD	F	93	13	51	31	Total		×					×					×															
pE	F	70	14	36	16	Total		×													×	×	×	×									
pF	F	70	10	44	25	Total		×																	×								
pG	M	88	10	57	42	3/4 post					×	×		×	×						×	×			×			×	×				
pH	F	92	11	69	31	4/5 ant		×																				×	×				
pI	M	70	10	45	19	3/4 ant		×						×	×						×							×	×				
pL	M	70	10	42	21	3/4 ant		×										×			×		×	×									
pM	M	82	45	44	21	3/4 ant		×																									
pN	M	70	12	65	38	3/4 ant		×																									
pO	M	80	14	68	46	2/3 ant		×																									
pP	M	87	10	50	28	3/4 ant		×	×	×								×			×		×	×	×		×			×	x	*	×
pQ	F	70	10	41	20	2/3 ant		×						×	×			×			×												
pR	F	70	12	41	15	2/3 ant																						×	×				
pS	M	80	10	56	36	1/3 ant																					×	×	×				
pT	M		10	31	23	Central																					×	×	×				
pU[b]	F	99	10	55	26	4/5 ant														×			×	×									
pV[b]	F	86	12	62	33	3/4 ant																	×	×									

[a]Oldfield [25]
[b]Patient carrying a metallic clip in the brain
[c]On 2007, at 56 years

Table 2
Studies in which Ancona split brain patients participated, listed in a chronological order

	Authors	Title	Patients involved
s1 [26]	Làdavas E, Cimatti D, Del Pesce M, Tuozzi G. Cognit Emot 7: 95–114, 1993	Emotional evaluation with and without conscious stimulus identification: evidence from a split-brain patient	pA
s2 [12]	Fabri M, Polonara G, Quattrini A, Salvolini U, Del Pesce M, Manzoni T. Eur J Neurosci 11: 3983–3994, 1999	Role of the corpus callosum in the somatosensory activation of the ipsilateral cerebral cortex: an fMRI study of callosotomized patients	pB, pC, pD, pE, pF, pH, pI, pL, pM, pN, pO, pP, pQ
s3 [27]	De Guise E, Del Pesce M, Foschi N, Quattrini A, Papo I, Lassonde M. Brain 122: 1049–1062, 1999	Callosal and cortical contribution to procedural learning	pB, pC, pP
s4 [28]	Arguin M, Lassonde M, Quattrini A, Del Pesce M, Foschi N, Papo I. Neuropsychologia 38: 283–291, 2000	Divided visuospatial attention systems with total and anterior callosotomy	pB, pC, pP
s5 [29]	Aglioti SM, Tassinari G, Fabri M, Del Pesce M, Quattrini A, Manzoni T, Berlucchi G. Eur J Neurosci 13: 195–200, 2001	Taste laterality in the split brain	pA, pG
s6 [13]	Fabri M, Polonara G, Del Pesce M, Quattrini A, Salvolini U, Manzoni T. J Cogn Neurosci 13: 1071–1079, 2001	Posterior corpus callosum and interhemispheric transfer of somatosensory information: an fMRI and neuropsychological study of a partially callosotomized patient	pG
s7 [30]	Fabri M, Polonara G, Quattrini A, Salvolini U. Cerebral Cortex 12: 446–451, 2002	Mechanical noxious stimuli cause bilateral activation of parietal operculum in callosotomized subjects	pB, pC, pD
s8 [31]	Corballis MC, Corballis PM, Fabri M. Neuropsychologia 42: 71–81, 2004	Redundancy gain in simple reaction time following partial and complete callosotomy	pB, pG, pI, pQ
s9 [32]	Hausmann M, Corballis MC, Fabri M. Neuropsychology 17: 602–609, 2003	Line bisection in the split brain	pB, pG, pI, pQ
s10 [33]	Savazzi S, Marzi CA. Neuropsychologia 42, 1608–1618, 2004	The superior colliculus subserves interhemispheric neural summation in both normals and patients with a total section or agenesis of the corpus callosum	pB
s11 [34]	Corballis MC, Barnett KJ, Fabri M, Paggi A, Corballis PM. Neuropsychologia 42, 1852–1857, 2004	Hemispheric integration and differences in perception of a line-motion illusion in the divided brain	pB
s12 [6]	Fabri M, Del Pesce M, Paggi A, Polonara G, Bartolini M, Salvolini U, Manzoni T. Cogn Brain Res 24, 73–80, 2005	Contribution of the posterior corpus callosum to interhemispheric transfer of tactile information	pB, pC, pD, pL, pP, pQ
s13 [35]	Corballis MC, Corballis PM, Fabri M, Paggi A, Manzoni T. Cogn Brain Res 25, 521–530, 2005	Now you see it, now you don't: variable hemineglect in a commissurotomized man	pB

s14 [36]	Hausmann M, Corballis MC, Fabri M, Paggi A, Lewald J. Cogn Brain Res 25, 537–546, 2005	Sound lateralization in subjects with callosotomy, callosal agenesis, or hemispherectomy	pB, pU
s15 [37]	Fabri M, Polonara G, Mascioli G, Paggi A, Salvolini U, Manzoni T. Eur J Neurosci 23, 3139–3148, 2006	Contribution of the corpus callosum to bilateral representation of the trunk midline in the human brain: an fMRI study of callosotomized patients	pB, pC, pE, pG, pI, pL, pP, pQ
s16 [38]	Savazzi S, Fabri M, Rubboli G, Paggi A, Tassinari CA, Marzi CA. Neuropsychologia 45, 2417–2427, 2007	Interhemispheric transfer following callosal resection in humans: role of the superior colliculus	pC, pE, pG
s17 [39]	Ouimet C, Jolicoeur P, Miller J, Ptito A, Paggi A, Foschi N, Ortenzi A, Lassonde M. Neuropsychologia 47: 684–92, 2009	Sensory and motor involvement in the enhanced redundant target effect: a study comparing anterior- and totally split-brain individuals	pA, pB, pE, pL, pP, pU, pV
s18 [40]	Ouimet C, Jolicoeur P, Lassonde M, Ptito A, Paggi A, Foschi N, Ortenzi A, Miller J. Neuropsychologia 48, 3802–3814, 2010	Bimanual crossed–uncrossed difference and asynchrony of normal, anterior-, and totally split-brain individuals	pA, pB, pE, pL, pP, pU, pV
s19 [41]	Pizzini FB, Polonara G, Mascioli G, Beltramello A, Moroni R, Paggi A, Salvolini U, Tassinari G, Fabri M. Brain Res 1312, 10–17, 2010	Diffusion tensor tracking of callosal fibers several years alter callosotomy	pB, pC, pF, pG, pP
s20 [42]	Corballis MC, Birse K, Paggi A, Manzoni T, Pierpaoli C, Fabri M. Neuropsycología 48, 1664–1669, 2010	Mirror-image discrimination and reversal in the disconnected hemispheres	pB, pC
s21 [43]	Miller MB, Sinnott-Armstrong W, Young L, King D, Paggi A, Fabri M, Polonara G, Gazzaniga MS. Neuropsycologia 48, 2215–2220, 2010	Abnormal moral reasoning in complete and partial callosotomy patients	pB, pP, pS, pT
s22 [44]	Fabri M, Polonara G. Neural Plast 2013; doi: 10.1155/2013/251308. 2013	Functional topography of the human corpus callosum: an fMRI mapping study	pG, pH, pI, pR, pS, pT
s23 [45]	Polonara G, Mascioli G, Foschi N, Salvolini U, Pierpaoli C, Manzoni T[§], Fabri M[*], Barbaresi P. J Neuroimaging; doi: 10.1111/jon.12136. 2014	Further evidence for the topography and connectivity of the corpus callosum: an fMRI study of patients with partial callosal resection	pG, pH, pI, pR, pS, pT
s24 [46]	Prete G, D'Ascenzo S, Laeng B, Fabri M, Foschi N, Tommasi L. J Neuropsychol; Dec 11. doi: 10.1111/jnp.12034. 2015	Conscious and unconscious processing of facial expressions: Evidence from two split-brain patients	pC, pP
s25 [47]	Prete G, Marzoli D, Brancucci A, Fabri M, Foschi N, Tommasi L. Behav Brain Res 271:354–364, 2014	The processing of chimeric and dichotic emotional stimuli by connected and disconnected cerebral hemispheres	pB, pP
s26 [48]	Prete G, Fabri M, Foschi N, Brancucci A, Tommasi L. Laterality 26:1–13, 2015	The “consonance effect” and the hemispheres: A study on a split-brain patient	pC
s27 [49]	Prete G, Laeng B, Fabri M, Foschi N, Tommasi L. Neuropsychologia 7. pii: S0028-3932(15)00003-2. 2015.	Right hemisphere or valence hypothesis, or both? The processing of hybrid faces in the intact and callosotomized brain	pC, pP

mainly due to the ability of the patients to cooperate, understand the task to be performed during the test, and to the consent to be submitted to an fMRI investigation.

Initially, neuropsychological studies were carried out, investigating the role of the corpus callosum in the emotional behavior [26], in the learning of a visuomotor skill that involved a motor control from either both hemispheres or a single hemisphere [27], and in the interhemispheric integration of the visuospatial attention system [28].

Shortly later a new topic emerged from research on this group of patients, in whom for the first time the recently introduced functional MRI (fMRI) technique was employed to study the pattern of cortical activation evoked by simple sensory stimulation in split-brain patients. These studies were made possible thanks to a combination of fortuitous events: the presence, at the Ancona Medical School, of researchers with converging interests, and the fMRI equipment newly purchased by the University of Ancona. The visionary researchers were Prof. Tullio Manzoni, the neurophysiologist, interested in the interhemispheric interaction which he also studied in animal models; Prof. Ugo Salvolini, the neuroradiologist, initiated to run fMRI in his division; Dr. Angelo Quattrini, the epileptologist, interested in research other than in clinical.

The first line of research was designed to explore in humans the issue of the bilateral cortical representation of some cutaneous territories, i.e., the hand and midline trunk touch receptors and the hand pain receptors [12, 13, 30, 37]. These topics were the object of previous studies of our group in animal models (cat, and monkey [50]) lead by Prof. Tullio Manzoni.

Later, the lateralization of the cortical representation of gustatory sensitivity was object of studies in control subjects and in split-brain patients [29]. During various imaging studies performed by our group, some BOLD (blood oxygen level dependent) activations were observed within the corpus callosum, with position within the commissure specifically related to the kind of sensory stimulus applied; therefore, the functional topography of the CC was also analyzed in split-brain patients [44, 45], as well as the remaining residual fiber tracts [41].

Publishing papers on the interhemispheric communication studied in callosotomized patients attracted the attention of many researchers in the world, mainly neuropsychologists, who came to Ancona to test our patients for their projects (see Table 2).

In our opinion, the best way to show how the callosotomized patients can help in understanding the organization of brain functions, both lateralized and diffused, is to describe what we studied and how these patients contributed to our research. Since the functional studies on callosotomized patients from the Ancona group are the sole ones performed to date, to our knowledge, they will be described in more detail and with more iconographic material. The other studies, carried out with more classical neuropsychological techniques, are summarized here, and for a better description and discussion the readers are invited to refer to the original papers.

Studies from 1996 to 2015 addressed many issues about brain function. We can recognize some main themes:

1. Interhemispheric transfer of touch-pain stimuli;
2. Interhemispheric transfer of taste stimuli;
3. Functional topography of corpus callosum;
4. Interhemispheric transfer of visual stimuli and motion perception;
5. Interhemispheric transfer and localization of auditory stimuli;
6. Interhemispheric coordination of motor responses;
7. Allocation of attention and learning functions;
8. Hemispheric collaboration in higher brain functions;
9. Emotional hemispheric specialization.

3 Interhemispheric Transfer of Touch and Pain Stimuli (Studies s2, s6, s7, s12, s15)

Over the last few years, fMRI has provided a safe and noninvasive method to study human brain function in vivo. The present report describes the application of fMRI to the study of the role of the corpus callosum in the interhemispheric transfer of somatosensory information.

Hand studies. Human somatosensory evoked potentials, magneto-encephalography (MEG), and positron emission tomography (PET) studies have shown that unilateral tactile stimulation evokes bilateral activation of two main cortical regions involved in the processing of tactile stimuli: one in the posterior parietal cortex (PPC) and one in the parietal operculum (PO; for a review, see [51, 52]). The former region lies in the postcentral gyrus (PCG) around the postcentral sulcus (PCS), posterior and medial to the anterior parietal cortex (APC) forming the posterior bank of the central sulcus (CS) and containing the first somatic sensory area (SI); the PP cortex probably includes cytoarchitectonic areas 2 and 5–7. The second region lies in the cortex forming the upper bank of the Sylvian sulcus (SS) and is commonly identified with the second somatosensory area (SII).

The pathways by which sensory afferents reach ipsilateral PO and PPC were at that time still unknown. A widely held view is that these afferents reach the ipsilateral hemisphere via the CC after relaying in the somatic sensory areas of the contralateral hemisphere. Indeed, latency studies have shown that the ipsilateral activation of both cortical fields is delayed compared with that of the contralateral homotopic regions by a time-lag compatible with interhemispheric transmission time [51]. Although the callosal connections of these cortical regions in humans are unknown, experiments on the callosal connectivity of homologous cortical regions of nonhuman primates and other mammals support the callosal recrossing hypothesis [51].

Bilateral cortical activation of human area SII in PO is also evoked by painful stimulation, as shown by PET, MEG, laser evoked potentials, and, recently, fMRI studies. In normal subjects unilateral painful stimulation generally induced activation of SII in both hemispheres (see data and literature in [30]). In some investigations area SII was activated in parallel with SI, while in others activation occurred also in absence of SI recruitment; moreover, the latencies of activation of contralateral SI and SII and of ipsilateral SII were very similar. Such pain-induced simultaneous activation of contralateral and ipsilateral SII, likely by means of extracallosal pathways, and the fact that patients with lesions involving the PO report altered pain sensitivity suggest that area SII may have a primary role in pain processing.

The area also participates in tactile information processing. Animal studies have suggested its involvement in tactile learning and retention [53] as well as in the integration of afferents from the two body halves (see [54], for a review). For all these reasons it is interesting to explore whether the somatosensory modalities of touch and pain may share a common mechanism of SII activation in the two hemispheres.

Trunk studies. In animals with bilateral symmetry, the midline is the imaginary line dividing the body into two mirror halves. In sensory physiology, midline territories are the cutaneous regions straddling the midsagittal plane and forming a central strip of somatosensory space lacking lateralization, found close to the geometric midline. Sensory midline regions are represented in the first somatic sensory cortex of both hemispheres in cats, macaque monkeys [50].

At the time when our studies were initiated, the pathways by which tactile stimuli from the trunk region reach the ipsilateral PCG were unknown. Neuroanatomical animal studies combining electrophysiological recording and neuronal tracing have shown that in the trunk representation zone both axon terminals and cell bodies of callosal projecting neurons are concentrated in SI, in areas where bilateral receptive field (RF) cells are especially numerous, namely on the borders of cytoarchitectonic areas 3a–3b, 3b–1, and 1–2 in monkeys and in areas 3b and 2 in cats [50]. These findings suggested that the ipsilateral component of bilateral RFs might have a callosal origin, a hypothesis subsequently confirmed also in nonhuman primates (see [50, 51] for a review), although a direct demonstration has not yet been provided in these species.

3.1 Functional MRI Studies

The first functional MRI studies carried out by our group starting 19 years ago investigated the neural mechanism of the bilateral activation of the hand and trunk zones, and the role of the human CC in this process [12, 13, 30, 37, 52, 55]. Particular attention was devoted to the bilateral activation of somatic sensory areas evoked by unilateral tactile stimulation of body regions with different anatomical and functional characteristics (hand and trunk), to assess the role of the commissure in the interhemispheric transfer of somatosensory inputs. As in the neuropsychological studies, the stimulation was administered in such a way as to involve, when possible, only one hemisphere at a time (i.e., unilateral touch or pain stimulation of the hand) (see Table 3).

We first investigated the neural mechanisms underpinning the bilateral activation of somatic sensory cortices elicited by unilateral tactile stimulation of the hand. This activation pattern was then compared with the activation evoked by unilateral painful stimulation of the hand. We then recorded the bilateral activation induced by tactile stimulation of tactile trunk receptors close to the ventral midline, and finally compared this activation pattern with that evoked by hand tactile receptor stimulation. The choice of the hand and trunk regions was based on their anatomo-functional characteristics. The trunk midline is the region where the two halves of the body join and whose cortical representation shifts from one hemisphere to the other (anatomical midline); in contrast, the hands are far removed from one another, but since they are often mobilized together they need to be coordinated (functional midline). The studies described here emphasize the results obtained in a special category of subjects, a group of callosotomized patients, who provide a valuable model to investigate human brain function in living "experimental" preparations. The fMRI studies were organized as follows:

1. cortical somatic sensory areas were investigated in normal volunteers, who received unilateral moving tactile stimuli on their hand and fingers to obtain a "normal activation pattern."
2. to verify the callosal re-crossing hypothesis, the study was also performed in a group of patients subjected to surgical resection of the CC to control medically intractable epilepsy. Subsequently, a patient requiring a second callosotomy accepted to be studied before and after the operation, where the posterior part of the trunk and the splenium of the CC were resected; this patient thus served both as a "control" and as an "experimental" subject;
3. to establish whether the CC contribution depends on the sensory modality, i.e., whether ipsilateral pain-related cortical activation might be mediated by the CC, the pattern of cortical activation evoked by painful mechanical stimulation of the hand was investigated in a group of normal volunteers and in three patients with complete CC resection;
4. to evaluate whether the CC contribution might depend on body regions within the same sensory modality, the activation pattern of cortical somatic sensory areas was investigated in normal volunteers and in a group of callosotomized patients who were tested with unilateral moving tactile stimuli on their ventral trunk surfaces.

Table 3
Functional MRI data and neuropsychological test results of callosotomized patients. Modified with permission from [77].

Subject	Hand touch stimulation		Hand pain stimulation		Trunk touch stimulation		Neuropsychological tests[a]		
	PCG activation	PO activation	PCG activation	PO activation	PCG activation	PO activation	TNT left hand	S-DRT	TFLT intermanual[b]
pQ	Contralateral	Bilateral			Bilateral	Bilateral	100	95	98
pL	Bilateral	Bilateral			Bilateral	Bilateral	93	100	98
pP	Contralateral	Contralateral			Bilateral	Bilateral	100	100	100
pI	Contralateral	Contralateral			Bilateral	Bilateral			
pN	Contralateral	Contralateral							
pM	Contralateral	Contralateral							
pO	Contralateral	Contralateral							
pG	Contralateral	Contralateral			Bilateral	Bilateral	16	24	18
pF	Contralateral	Contralateral							
pH	Contralateral	Contralateral							
pC	Contralateral	Contralateral	Contralateral	Bilateral	Bilateral	Bilateral	60	100	76
pB	Contralateral	Contralateral	Bilateral	Bilateral	Bilateral	Bilateral	15	55	80
pD	Contralateral	Contralateral	Contralateral	Bilateral			14	50	85
pE	Contralateral	Contralateral			Bilateral	Bilateral			

[a]Proportion of correct answers
[b]Mean of right and left hand responses

3.2 Neuropsychological Studies

A group of callosotomized patients underwent specific neuropsychological tests evaluating interhemispheric tactile transfer to compare functional with behavioral data.

3.3 Results from Touch Studies: Hand Stimulation

In all control subjects, cortical activation due to unilateral tactile stimulation of the hand was consistently observed in the following somatic sensory areas (for a detailed description, see [52]):

3.3.1 Control Subjects

- in the contralateral APC, anteriorly in the PCG, just posterior to the CS. This focus likely corresponds to the hand representation zone of SI;
- in the PPC of both hemispheres, posteriorly in the PCG, just anterior to the PCS, a region likely corresponding to cytoarchitectonic areas 2 and 5–7;
- in the PO cortex of both hemispheres, in a cortical region of the upper bank of the SS, which contains SII.

These results confirmed those obtained in previous functional studies (reviewed in [12]).

3.3.2 Callosotomized Subjects

Midsagittal MR slices from the 14 callosotomized subjects showed CC surgical resection to be complete in five patients and partial in the other nine. Accordingly, subjects were assigned to the *total* and the *partial callosotomy* subgroup, respectively. The activation and neuropsychological data of the patients participating in the different studies are described below.

Total Callosotomy

In these five patients, the activation pattern in the hemisphere contralateral to the hand receiving the stimulus was similar to that obtained in control healthy intact-CC subjects, activation foci being consistently observed in the APC, PPC, and PO and, occasionally, in the precentral and medial frontal cortices. Conversely, tactile stimulation of the hand failed to elicit activation foci in the ipsilateral hemisphere, unlike the pattern observed in control subjects (Fig. 2a1 and b1; Table 3). The cortical activation pattern obtained in these subjects during stimulation of one hand was the mirror image of the pattern induced by stimulation of the other hand [12].

Partial Callosotomy

In all nine patients of this subgroup, stimulation of one hand resulted in a pattern of contralateral activation that was similar to the one obtained in control and callosotomized subjects. Contralateral activation foci were consistently observed in the APC, PPC, and PO.

Ipsilateral activation was absent in seven patients, while it was observed in the PO cortex in one subject and in the PO and PPC in the ninth.

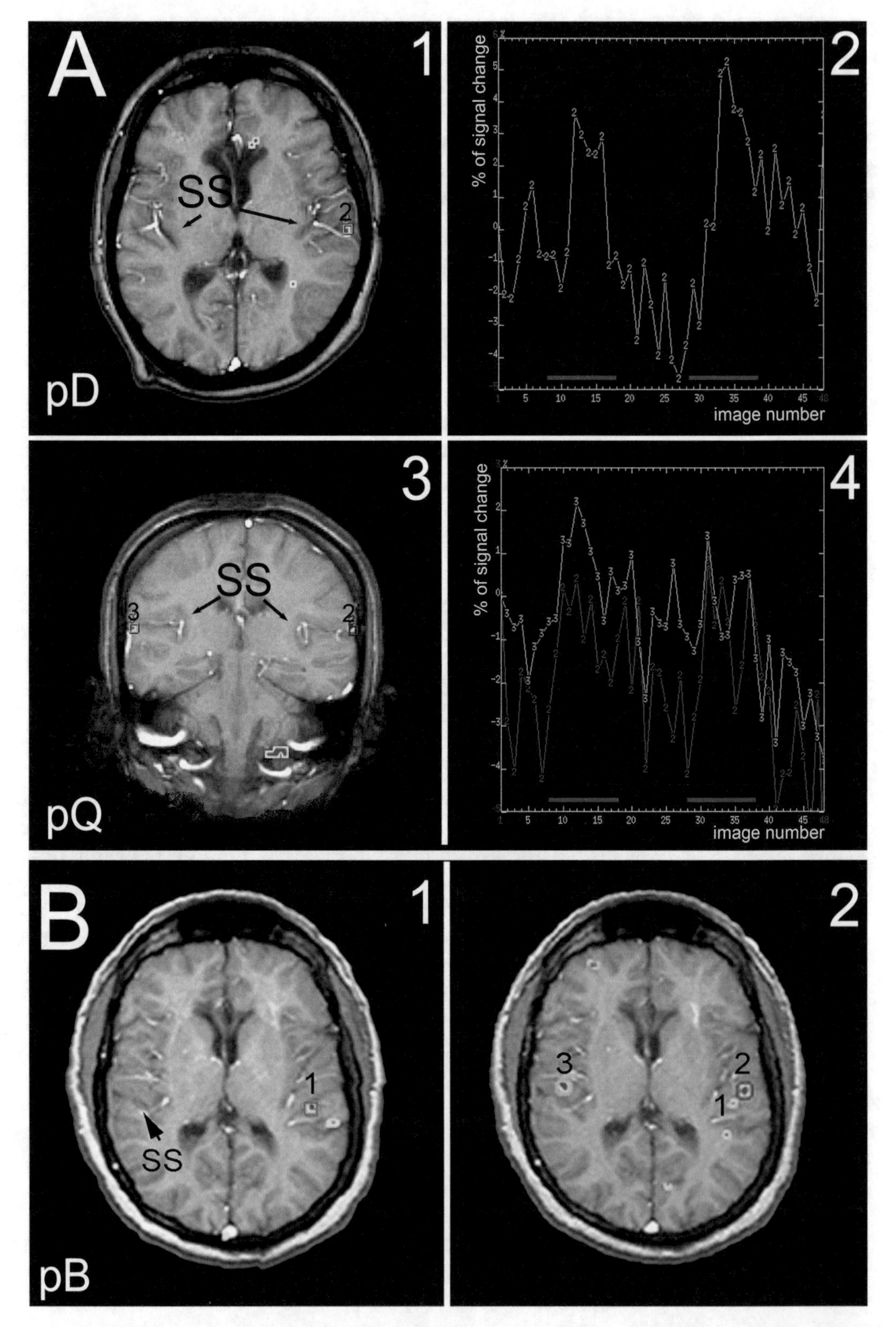

Fig. 2 (**a**) Cortical activation by unilateral tactile stimulation of one hand in two callosotomized patients with total (pD) and partial (pQ) callosal resection, respectively. (**a1**) Axial image showing activation in the contralateral PO (*box 2*) in patient pD. (**a2**) Graph showing the signal increase in ROI 2 of A1. (**a3**) Coronal image showing

Of the seven patients lacking ipsilateral activation, one still retained a small portion of the splenium, one the rostral 2/3 of the CC and two the entire splenium. In the remaining three subjects the posterior region of the CC had been spared. However, since the borders between the different portions of the CC cannot be established with certainty, it is unclear whether in the last three subjects a part of the posterior body of the CC was also extant, besides the splenium. In the two patients in whom unilateral hand tactile stimulation did induce ipsilateral activation in PO or PO and PPC, the spared portion of the CC included the entire splenium and probably also the posterior part of the callosal body (posterior callosal body, PCB). The data from one of these patients, showing activation in the ipsilateral PO cortex, are reported in Figs. 2a3, and 2a4 [12]. Also in these patients, the cortical activation pattern obtained during stimulation of one hand was the mirror image of the pattern induced by stimulation of the other.

During our investigations, a patient requiring a second callosotomy due to drug-refractory epilepsy accepted to be studied before and after surgery. Activation due to unilateral tactile stimulation of the hand was explored with fMRI 1 week before the second operation and 6 months and 1 year afterwards. Before this operation, tactile stimulation of either hand activated the contralateral PPC and areas SI and SII as well as ipsilateral PPC and SII (Fig. 3, top row). In both postoperative sessions, somatosensory activation was detected in contralateral SI, SII, and the PPC, but failed to be recorded in ipsilateral SII and PPC (Fig. 3, bottom row). This case provides direct demonstration of the hypothesis that the posterior third of the body of the CC is crucial for interhemispheric transfer of hand somatosensory information, as strongly suggested by previous data (for a detailed description, see [13]).

3.4 Results from Pain Studies: Hand Stimulation

3.4.1 Control Subjects

Cortical activation due to painful mechanical stimulation of one hand was in line with previous studies and was consistently observed in the following areas [56, 57]:

- in the contralateral APC, anteriorly in the PCG, just posterior to the CS;
- in the PPC of both hemispheres, posteriorly in the PCG, just anterior to the PCS;

Fig. 2 (continued) an activation focus in contralateral (*box 2*) and ipsilateral (*box 3*) PO in patient pQ. (**a4**) Graphs showing the signal increase in ROIs 2 (*graph 2*) and 3 (*graph 3*) of A3. In this and in the adjacent sections no activation was detected in the ipsilateral PO. (**b**) Cortical activation by unilateral painful stimulation of one hand in one callosotomized patient with total callosal resection (pB). (**b1**) Axial image showing activation in the contralateral PO (focus *1*) by tactile stimulation of the right hand. No activation was detected in the ipsilateral PO in this and in adjacent sections. (**b2**) Axial image showing PO activation by painful stimulation of the right hand. Activation is found both in the contralateral (*1* and *2*) and the ipsilateral (*3*) PO. Left hemisphere on the right. *SS* Sylvian sulcus. **a**, modified with permission from [12]; **b**, modified with permission from [30]

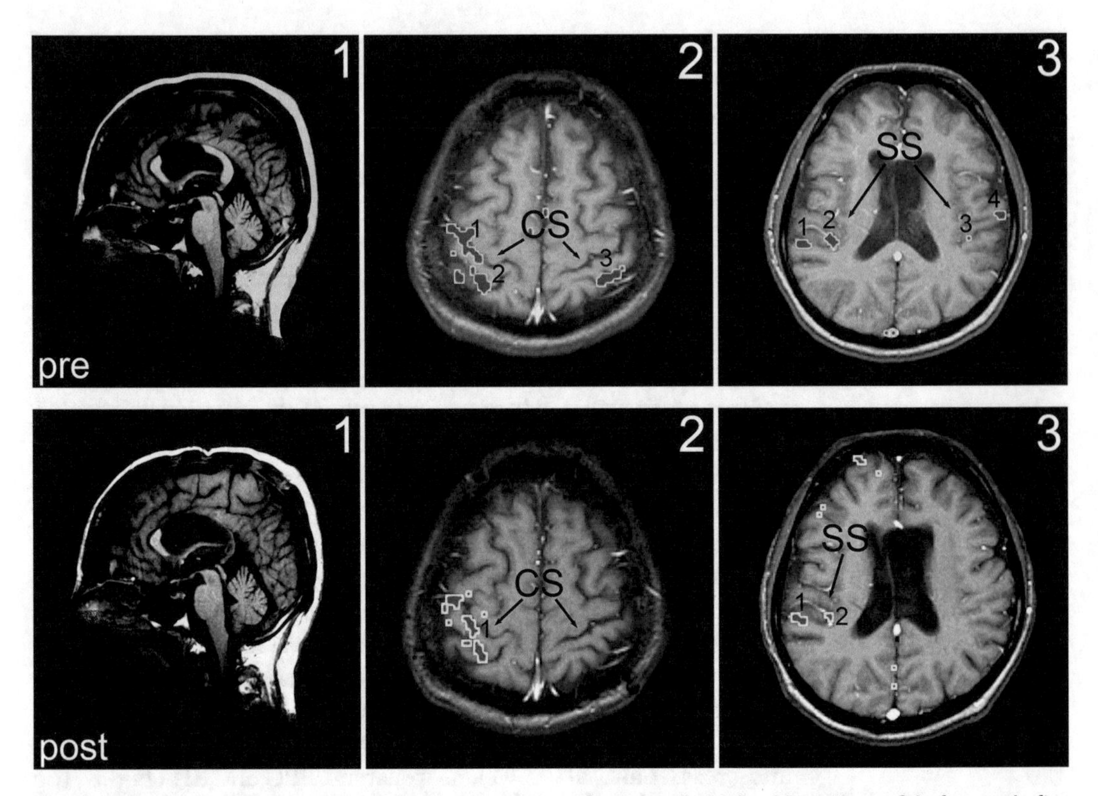

Fig. 3 Cortical activation by unilateral tactile stimulation of one hand obtained in patient pG before and after posterior callosotomy. (*Top row, 1*) MR image of a midsagittal brain slice showing the extent of callosal resection before the second operation. *Red arrow* marks the extant posterior body of the commissure. (*Top row, 2*) Axial image on which the regions activated during tactile stimulation have been superimposed. Main activation foci were in the contralateral APC (*1*) and in the contralateral (*2*) and ipsilateral (3) PPC. (*Top row, 3*) Axial image from a more ventral level, showing activation foci in contralateral (*1, 2*) and ipsilateral (*3, 4*) PO. (*Bottom row, 1*) MR image showing the extent of callosal resection after the second operation. (*Bottom row, 2*) Axial image on which the regions activated during tactile stimulation have been superimposed. Main activation foci were in the contralateral APC (*1*) and PPC (*2*). In this and in the adjacent sections no activation was detected in the ipsilateral PPC. (*Bottom row, 3*) Axial image on which the regions activated during tactile stimulation have been superimposed. The same main activation foci were identified in the contralateral PO (*1, 2*). Left hemisphere on the right. *CS* central sulcus, *SS* Sylvian sulcus. Modified with permission from [13]

- in the PO cortex of both hemispheres, in a cortical region in the upper bank of the SS, which contains SII;
- in the cingulate (Cin) cortex of the contralateral (five cases) and ipsilateral (two cases) hemispheres;
- in the insular cortex of the contralateral (four cases) and ipsilateral (two cases) hemispheres.

3.4.2 Callosotomized Patients

Contralateral activation was observed in the PO in all patients (Fig. 2b2, focus 2), in the APC in two cases and in the PPC in one case; the insular cortex was activated in one patient. The ipsilateral PO was also activated in all patients (Fig. 2b2, focus 3), the insular cortex in 1 and the PP in another, whereas the APC and the cingulate cortex were never activated.

This study indicates that the activation evoked by painful stimulation of the hand in ipsilateral SII is at least partially independent of the CC, suggesting that painful and tactile stimuli might use different pathways to reach the same PO target area [30].

3.5 Results from Touch Studies: Trunk Stimulation

3.5.1 Control Subjects

Cortical activation due to tactile stimulation of the trunk surface was consistently observed:

- in the APC of both hemispheres, anteriorly in the PCG, just posterior to the CS, when the ventromedial surface was stimulated, and solely in the contralateral hemisphere when the lateral surface was stimulated [55]. This focus likely corresponds to the trunk representation zone of SI. These results agree with data from nonhuman primates [50];
- in the PO cortex of both hemispheres, in a cortical region in the upper bank of the SS, which contains SII [55].

These data showed for the first time that in humans, as in nonhuman primates, the trunk regions close to the midline are represented bilaterally in SI.

3.5.2 Callosotomized Subjects

In the patient with an intact splenium and posterior callosal body and in the two with partial anterior resection sparing only the callosal splenium, unilateral stimulation of medial ventral trunk evoked cortical activation in the PCG (Fig. 4a2) and PO of both hemispheres (Fig. 4a3), as in control subjects. The patient with partial posterior callosotomy displayed bilateral cortical activation in area SI when stimulation was applied to the right ventral trunk and contralateral activation when it was applied to the left side. The three patients with complete callosal resection had activation patterns similar to those of the last two groups: in two patients left and right tactile stimulation of the trunk ventral surface close to the midline caused bilateral activation in the PCG (Fig. 4b2) and PO (Fig. 4b3); in the other patient bilateral activation was detected after left side stimulation, and contralateral activation after right side stimulation.

Data from callosotomized patients indicate that the activation evoked in ipsilateral SI by ventral midline stimulation is at least partially independent of the CC (for a detailed description see [37]).

3.6 Results from the Neuropsychological Study

3.6.1 Control Subjects

This group performed well with both hands in the TNT and S-DRT tests (mean 99 % of correct answers), and obtained high scores in the intermanual TFLT task with both hands (mean proportion of correct answers: 95 % and 96 % with the right and left hand, respectively).

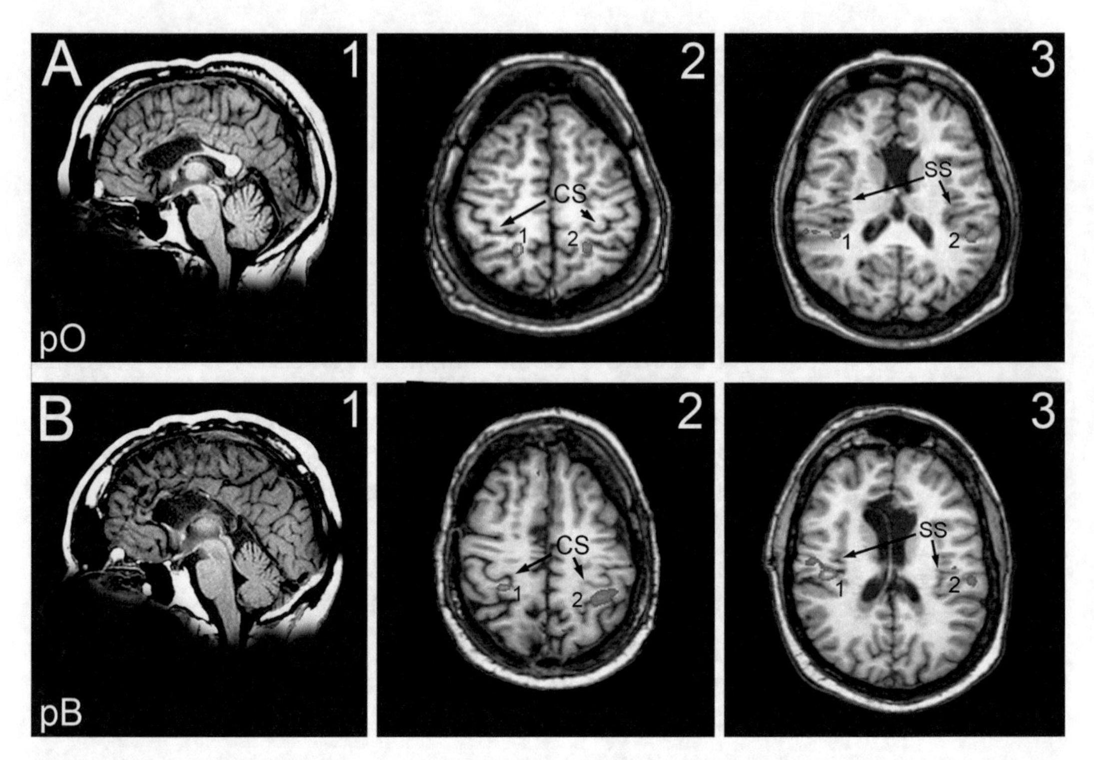

Fig. 4 Cortical activation obtained during tactile stimulation of the left midline trunk surface in two callosotomized patients. (**a1**) MR image of midsagittal brain slice showing the extent of callosal resection in patient pQ. (**a2**) Stimulus-related activation was observed in the PCG of the contralateral hemisphere (focus *1*), in the trunk representation zone of SI, and in a corresponding ipsilateral region (focus *2*). (**a3**) Cortical activation was also evoked in PO, bilaterally in area SII (focus *1*, contralateral, and *2* ipsilateral). (**b1**) MR image of midsagittal brain slice showing complete resection of the corpus callosum in patient pB. (**b2**) Stimulus-related activation was observed in the PCG of the contralateral hemisphere (focus *1*), in the trunk representation zone of SI, and in a corresponding ipsilateral region (focus *2*). **(b3)** Cortical activation was also evoked in PO, bilaterally in area SII (focus *1*, contralateral, and *2* ipsilateral). Left hemisphere on the right. *CS* central sulcus, *SS* Sylvian sulcus. Modified with permission from [37]

3.6.2 Patients with Partial Anterior Callosotomy

All these subjects performed well with both hands in the TNT (mean 93 % and 95 % of correct responses for the right and the left hand, respectively). One had an intact splenium and PCB, while in the other two the splenium and possibly also part of the trunk of the CC were extant. These data indicate that somatosensory information from the left hand reached the speaking hemisphere (very likely the left, all patients having high right-handedness scores), probably along the fibers crossing through the posterior region of the callosal body. These patients also provided 95–100 % of correct answers in the S-DRT and 77–100 % in the intra- and intermanual TFLT tasks with both hands, denoting interhemispheric exchange of tactile information (Table 3).

3.6.3 Patient with Partial Posterior Callosotomy

Before posterior callosotomy, in 100 % of trials, this patient was able to name the familiar objects explored with either hand (TNT), to state whether the object explored with one hand was the same or a different object from that explored with the other hand (SDRT), and to localize with the thumb the points on fingers 2–5 that had been touched by the experimenter in the same (TLFT, intramanual mode) or in the contralateral (TLFT, intermanual mode) hand. The patient's postoperative performance in the TNT was as correct as preoperatively with the right hand, whereas he failed to name correctly the objects explored with the left hand in 89 % and 84 % of trials after 6 months and 1 year, respectively. His performance in the SDRT was also very poor, with correct responses well below chance in both postoperative sessions. In the TFL, performance was as high as preoperatively in the intramanual mode: with both hands he could localize with the thumb the points touched on fingers 2–5 of the same hand without errors 6 months from surgery, and with few errors after 1 year. Conversely, in the intermanual mode he made errors in 82 % of trials in the right-to-left transfer and in 83 % in the opposite direction after 6 months; after 1 year errors were 85 % and 79 %, respectively. This case provides direct demonstration that activation of SII and PPC to stimulation of the ipsilateral hand requires the integrity of the posterior body of the CC.

3.6.4 Patients with Total Callosotomy

Patients with total resections performed well in the TNT with the right hand (93 % of correct responses), but poorly with the left (mean: 30 % of correct responses). Left hand performances were significantly different in patients with total callosotomy compared with those with partial resection, whereas their right hand performances were similar. Performance in the S-DRT was highly variable, with correct responses ranging from 50 % to 100 %. Two of the three patients with complete resection obtained chance level scores (55 % and 50 %, respectively) and failed the TNT with the left hand, while surprisingly the third patient performed well in this test. However, the results from this group were overall significantly poorer than those of the partial callosotomy group. In the TFLT, the three complete callosotomy patients gave 100 % of correct answers with both hands in the intramanual task; in the intermanual task correct responses were 78–93 % with the right hand and 67–90 % with the left. Although these values are above chance (12.5 %), differences from the results of partially callosotomized patients are statistically significant (Table 3).

The results of the fMRI studies described above can be summarized as follows:

1. unilateral tactile stimulation of the hand and fingers evoked bilateral activation of the PPC and area SII in control subjects; in patients with complete or partial callosotomy sparing only the splenium activation was only in contralateral PPC and area SII.

Bilateral activation was also recorded in patients with an extant posterior callosal body and preoperatively in the patient who underwent a second (posterior) callosotomy. Lack of ipsilateral somatosensory cortex activation in callosotomized patients shows that stimulus transmission is mediated by the corpus callosum. The disappearance of ipsilateral SI and SII activation in the patient who was evaluated before and after posterior callosotomy indicates that the fibers carrying activation to ipsilateral SI and SII run in the posterior body of the CC;

2. unilateral painful stimulation of the hand evoked bilateral activation of area SII both in control subjects and in patients with complete callosotomy, suggesting that the two modalities (touch and pain) may use different mechanisms, and that activation of ipsilateral SII by painful stimuli could be elicited by extracallosal pathways, likely subcortical fibers;
3. unilateral tactile stimulation of receptors on trunk regions close to the ventral midline evoked bilateral activation of areas SI and SII in healthy subjects and patients, indicating that afferents from different body regions reach the ipsilateral hemisphere by different ways, likely through a direct subcortical uncrossed pathway.

Comparison of neuropsychological results with fMRI data provides further evidence that the interhemispheric transfer of tactile afferents from hand receptors is likely mediated by the fibers running in the posterior part of the callosal body, confirming that the posterior midbody of the CC is the "tactual" channel.

In the patients participating in the various studies, different kinds of stimulation (i.e., tactile versus painful) and different sites of application of the tactile stimulus (i.e., hand versus trunk midline) resulted in different patterns of bilateral activation of somatic sensory areas. It can be hypothesized that the bilateral cortical representation of peripheral inputs might, in principle, be subserved by different mechanisms, according to the nature and functional significance of the incoming stimulus. Tactile afferents from proximal body regions (e.g. trunk midline) may reach the ipsilateral hemisphere by a direct subcortical uncrossed pathway, independent of the CC (*subcortical*), to ensure anatomical continuity of the cortical representation of the sensory periphery. It has recently been shown that oral structures are represented in SI of both hemispheres and that the ipsilateral representation is at least partially independent of the commissure, since differences in the latency of cortical activation between the hemispheres are not significant [58]. A recent fMRI mapping study showed that other body regions other than the midline territories are represented bilaterally in SI: unilateral tactile stimulation of shoulder, arm, and thigh regions evoked activation foci in PCG of both hemispheres [59], but the neural mechanism of this bilateral representation has not yet been investigated. Alternatively, tactile afferents from distal body regions (e.g. hand) may reach the ipsilateral hemisphere by an indirect cortico-cortical recrossed pathway, dependent on the CC (*callosal*), to ensure functional continuity. Finally, pain afferents from distal body regions (hands) reach the ipsilateral hemisphere by a direct subcortical uncrossed pathway, independent of the CC (*subcortical*), probably to ensure prompt stimulus processing and appropriate reaction to potentially harmful situations. The callosally dependent mechanism, related to a more complex behavior, likely appeared late in mammalian evolution. Indeed, the corpus callosum is first found in the brain of Eutherian mammals [60, 61], possibly "as a more efficient commissural system" [61] connecting increasingly distant sensory cortical areas in the expanding brain, or, more likely, as a consequence of the evolution of a separate motor cortex in Eutherian brains [61, 62] and the emerging ability to perform complex or skilled motor tasks. It is therefore conceivable that the callosal commissure is necessary to connect and integrate effectors of complex behavior (the hands), and not for fusing the left and right sensory periphery, or to alert the nervous system in case of danger, all functions that are found in lower mammals (Prototherians and Metatherians), and that in callosal mammals are accomplished, at least in part, by an alternative commissure and/or by other mechanisms.

The analysis of the hand touch and pain, trunk touch, and neuropsychological studies performed on patients with various degrees of callosal resection prompts the following considerations:

1. activation of ipsilateral PPC and area SII by unilateral tactile stimulation of the hand is mediated by somatic sensory areas in the contralateral hemisphere, through the CC;
2. the callosal fibers carrying activation to the ipsilateral PP cortex, SI, and SII pass through the posterior third of the body of the commissure, anterior to the splenium;
3. activation of ipsilateral area SII by unilateral painful stimulation of the hand is at least partially independent of callosal transmission; nociceptive inputs reach area SII likely via uncrossed ipsilateral subcortical pathways;
4. activation of ipsilateral areas SI and SII by unilateral tactile stimulation of the ventral trunk midline is at least partially independent of the CC;
5. in the last two cases, extracallosal pathways are probably involved: uncrossed subcortical pathways, subcortical commissures, or both.

The contribution of the corpus callosum to the bilateral cortical representation of peripheral sensory inputs is thus critical for tactile afferents from the hands, probably to integrate and coordinate sensory and motor activity of both hands in performing complex tasks. It is not crucial for more primitive functions, i.e., processing tactile afferent stimuli from midline trunk regions or painful stimuli from the hands. In these two cases the corpus callosum could, however, have a general modulatory role: callosal projections may exert a facilitatory action, combining the excitatory signals from the two hemispheres to yield bilateral receptive fields, and/or a synchronizing action, enabling activation of adjacent regions at the same time, or even an inhibitory action, as suggested by the finding that cooling area 3 in monkey SI cortex results in augmented activity and enlargement of the RFs of neurons in the homotopic cortical area of the contralateral hemisphere, and by the presence of callosally projecting GAD-immunoreactive neurons in rat and cat somatic sensory areas [63].

4 Interhemispheric Transfer of Taste Stimuli (Study s5)

4.1 Neuropsychological Studies

Clear evidence on the organization plan is largely lacking for the gustatory system, the lateral organization of which in the human brain has long been the object of considerable controversy. The projection of taste information from each side of the tongue to the gustatory cortex has been reported to be uncrossed in some studies and crossed in other studies (see data and literature in [64]). Previous data from a complete callosotomy patient, typically affected by left hand anomia and left field alexia [65], showed that he was able to verbally identify basic taste stimuli applied to either side of the tongue, though his accuracy and speed of performance were significantly better for left hemitongue stimuli than for right hemitongue stimuli, a difference absent in normal controls. These results are incompatible with an exclusively crossed projection carrying taste information from the tongue to the cortex and instead support the hypothesis of a bilaterally distributed organization of the gustatory pathway. Moreover, on the assumption that the left hemisphere is ultimately responsible for verbally identifying taste stimuli, the left hemitongue advantage could best be attributed to a functional predominance of the ipsilateral over the contralateral component of the gustatory pathway, a predominance masked in normal controls by the equalizing action of the corpus callosum [65]. Since in the callosotomy patient the accuracy and speed of response to right hemitongue stimuli are below the control performance, it seems that the corpus callosum may normally contribute to the transfer, but it is not indispensable for the transfer, since that patient showed an above-chance ability to name taste stimuli from the right hemitongue. Whether taste information from the right hemitongue reaches the left cortex via crossed ascending projections, or by means of extracallosal connections, or by both ways, remains an entirely open question. Two out of the callosotomy Ancona patients with bilateral anatomical integrity of the cortical regions known to be involved in taste information processing participated in a subsequent study. The results of these tests provide a general confirmation of the previous findings [65], and add to the general understanding of the lateral organization of the gustatory pathway in humans. The two patients, one with a complete corpus callosum resection (pA), the other sparing the genu and the rostrum (pG), were tested for discrimination of three basic taste stimuli (sour, bitter, salty) applied to the right or left sides of the tongue. Responses were made by pointing with either hand to written words or images of visual objects corresponding to the stimuli, a language-based discrimination, i.e., a sort of "taste naming test," similar to the tactile and visual ones. In both patients, response accuracy was significantly above chance for both hemitongues but there was a significant advantage for the left side (Table 4). Reaction time was shorter for left stimuli than for right stimuli but the difference was not significant. Eight normal controls matched for age with the patients performed equally well with right and left hemitongue stimuli. Tactile and visual tests showed that the left hemisphere was responsible for language-based responses in the first two patients. The results confirm and extend previous findings in another callosotomy patient, indicating that:

1. taste information from either side of the tongue can reach the left hemisphere in the absence of the corpus callosum;
2. the ipsilateral input from the tongue to the left hemisphere is more potent functionally than the contralateral input and
3. in the normal brain, the corpus callosum, specifically its posterior part including the splenium, appears to equalize the effects of the ipsilateral and contralateral gustatory inputs on the left hemisphere.

Taken together with evidence about lateralized taste deficits following unilateral cortical lesions, the results also suggest that the gustatory pathways from tongue to cortex are bilaterally distributed with an ipsilateral predominance that may be subject to individual variations.

Results from functional studies (not published yet) seem to confirm this hypothesis, suggesting for the taste system an organization plan slightly different from that of the somatic sensory system [66].

Table 4
Percentage correct and mean RT of correct responses for the two hemitongues. Modified with permission from [29]

	Correct responses (%)				
	L hemitongue	**R hemitongue**	**Statistical test used**	**Significance**	
pA					
Mean across all sessions	80.5	54.2	Chi square	$\chi^2 = 14.5$	$P < 0.001$
Pointing to names	79.2	58.3	Chi square	$\chi^2 = 9.2$	$P < 0.01$
Pointing to drawings	83.3	45.8	Chi square	$\chi^2 = 29.0$	$P < 0.001$
Mean RT (s)	6.5	7.0	Unpaired *t*-test	$t_{95} < 1$	n.s.
pG					
Mean across all sessions	80.5	58.3	Chi square	$\chi^2 = 10.5$	$P < 0.01$
Mean RT (s)	12.6	13.2	Unpaired *t*-test	$t_{48} < 1$	n.s.

The probability of success by chance is 33.3 % for each hemitongue. All performances with both hemitongues, except pA's performance with the right hemitongue in the pointing task with tastant-matching drawings, are significantly different from chance at the $P < 0.01$ level at least (chi square tests). All performances with both hemitongues are also significantly inferior to the means of normal controls at the $P < 0.05$ level at least (chi square tests)

5 Functional Topography of Corpus Callosum (Studies s19, s22, s23)

The concept of a topographical map of the corpus callosum (CC) has emerged from human lesion studies and from electrophysiological and anatomical tracing investigations in other mammals. As already pointed out in Sect.1, this organization seems to result in modality-specific regions: the anterior fibers, which connect the frontal lobes, transfer motor information, whereas the posterior fibers, which connect the temporal, parietal, and occipital lobes, are involved in the integration of somatosensory (posterior midbody), auditory (isthmus), and visual (splenium) information [5] (see data and literature in [6, 7]).

All neuropsychological studies here described also confirmed some different functions for the different portions of the corpus callosum, i.e., visual for the splenium, tactile for the posterior body, auditory for the isthmus/splenium, attentive for the anterior part. Over the last few years a rising number of researchers have been reporting fMRI activation in white matter, particularly the CC. During our studies on tactile, gustatory, auditory, and visual sensitivity and of motor activation performed with fMRI, a callosal activation was often observed to be evoked through simple sensory stimulation and motor tasks. We reviewed our published and unpublished fMRI and diffusion tensor imaging data on the cortical representation obtained in 36 normal volunteers and in six patients

Table 5
Characteristics of patients participating in study on the callosal topography and types of stimulation administered. Modified with permission from [45]

Patients							Type of stimulation		
Subject	Age	Gender	Oldfield score	Callosal resection	Years from callosotomy	DTI	Gustatory	Tactile	Visual
							Salty	Hand	
pG	51	M	10 (right)	Partial posterior	7	Yes	L,R	L,R	L,R—C
pH	59	F	11 (right)	Partial anterior	20	Yes		L,R	L,R—C
pI	39	M	10 (right)	Partial anterior	16	Yes	L,R	L,R	L,R—C
pR	32	F	12 (right)	Partial anterior	17	Yes	L,R	L,R	
pS	51	M	10 (right)	Partial anterior	15	Yes	L,R	L,R	
pT	26	M	10 (right)	Partial central	4	Yes	L,R	L,R	L,R

C center
L left
R right

with partial callosotomy (Table 5). Activation foci were consistently detected in discrete CC regions: anterior (taste stimuli), central (motor tasks), central and posterior (tactile stimuli), and splenium (auditory and visual stimuli). Sometimes, taste and tactile stimuli activated splenial fibers, in addition to a more anterior location. This result is compatible with the observation previously reported that patients with only the splenium intact can perform interhemispheric tactile transfer [67], and have above chance performance in taste discrimination task [29, 65] (see literature in [7, 12]). Reconstruction by diffusion tensor imaging (DTI) of callosal fibers connecting activated primary gustatory, motor, somatosensory, auditory, and visual cortices by diffusion tensor tracking showed bundles crossing, respectively, through the genu, anterior and posterior body, and splenium, at sites harboring fMRI foci (Fig. 5) [44, 45].

In this study, the contribution of partially callosotomized patients was particularly important, since the topographical organization described in CC control subjects was confirmed for the callosal portion spared. Taste stimuli evoked bilateral activation of the primary gustatory area in all patients and foci in the anterior CC, when spared. Tactile stimuli to the hand evoked bilateral foci in the primary somatosensory area in patients with an intact posterior callosal body and only contralateral in the other patients. Callosal foci

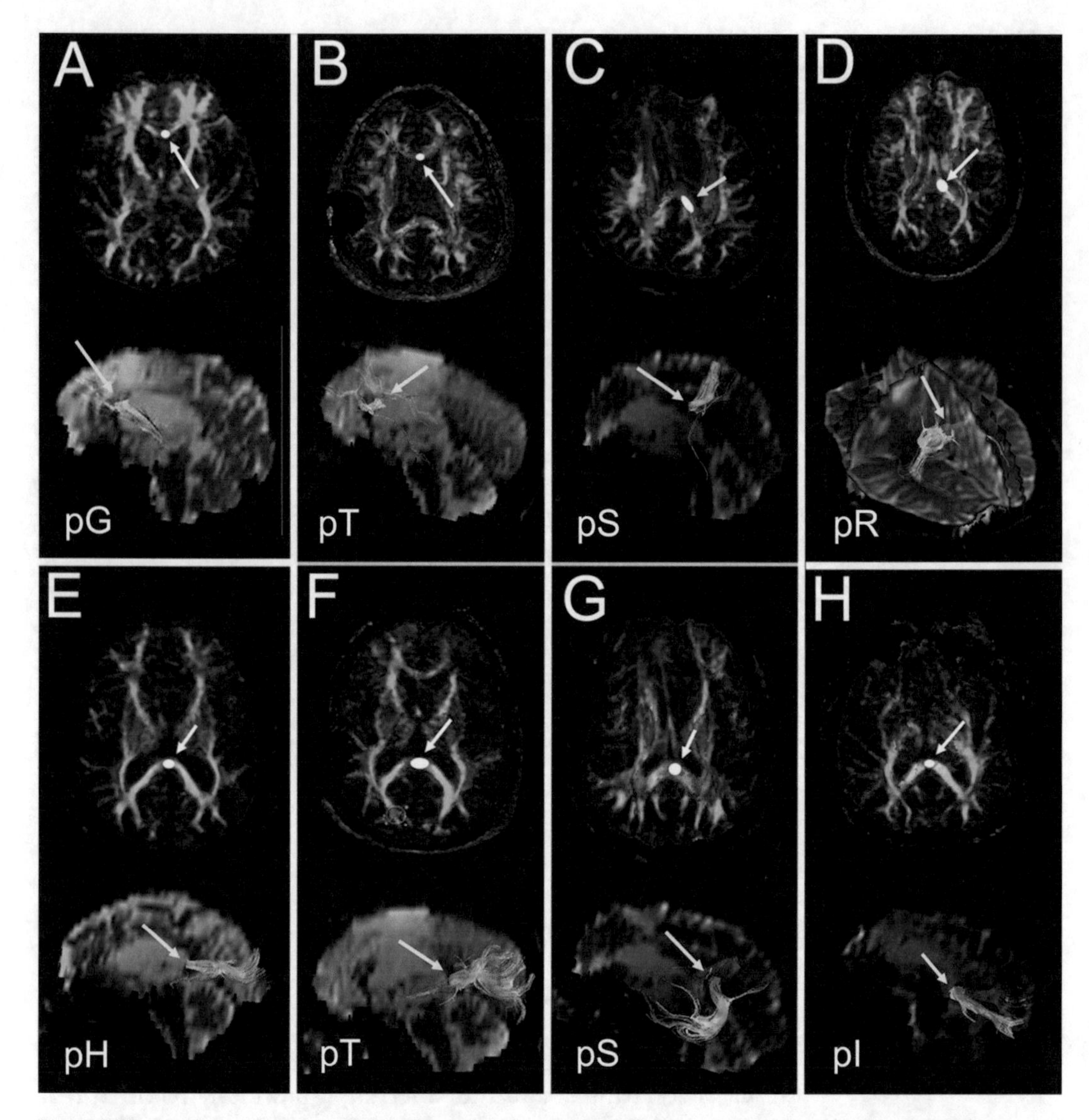

Fig. 5 Diffusion Tensor Imaging (DTI) images showing callosal fibers from Regions of Interest (ROIs) selected in callosal regions where activation foci were evoked by different types of sensory stimuli. In each panel, the top brain figurine represents the structural view with the position of the ROI (*white spot*) and the bottom one the callosal fibers arising from it. (**a**, **b**) ROIs in the genu in patients pG (**a**) and pT (**b**). (**c**, **d**) ROIs placed in the posterior callosal body in patients pS (**c**) and pR (**d**). (**e**, **f**, **g**, **h**) ROIs placed in the splenium in patients pH (**e**), pT (**f**), pS (**g**), and pI (**h**). For axial images, left hemisphere is on the right; for sagittal images, posterior pole is on the right. Modified with permission from [45]

occurred in the CC body, if spared. In patients with an intact splenium central visual stimulation induced bilateral activation of the primary visual area as well as foci in the splenium itself. Present data show that interhemispheric fibers linking sensory areas crossed through the CC at the sites where the different sensory stimuli evoked activation foci, and that topography of callosal foci evoked by sensory stimulation in spared CC portions is consistent with that previously observed in subjects with intact CC (Fig. 6) [44, 45].

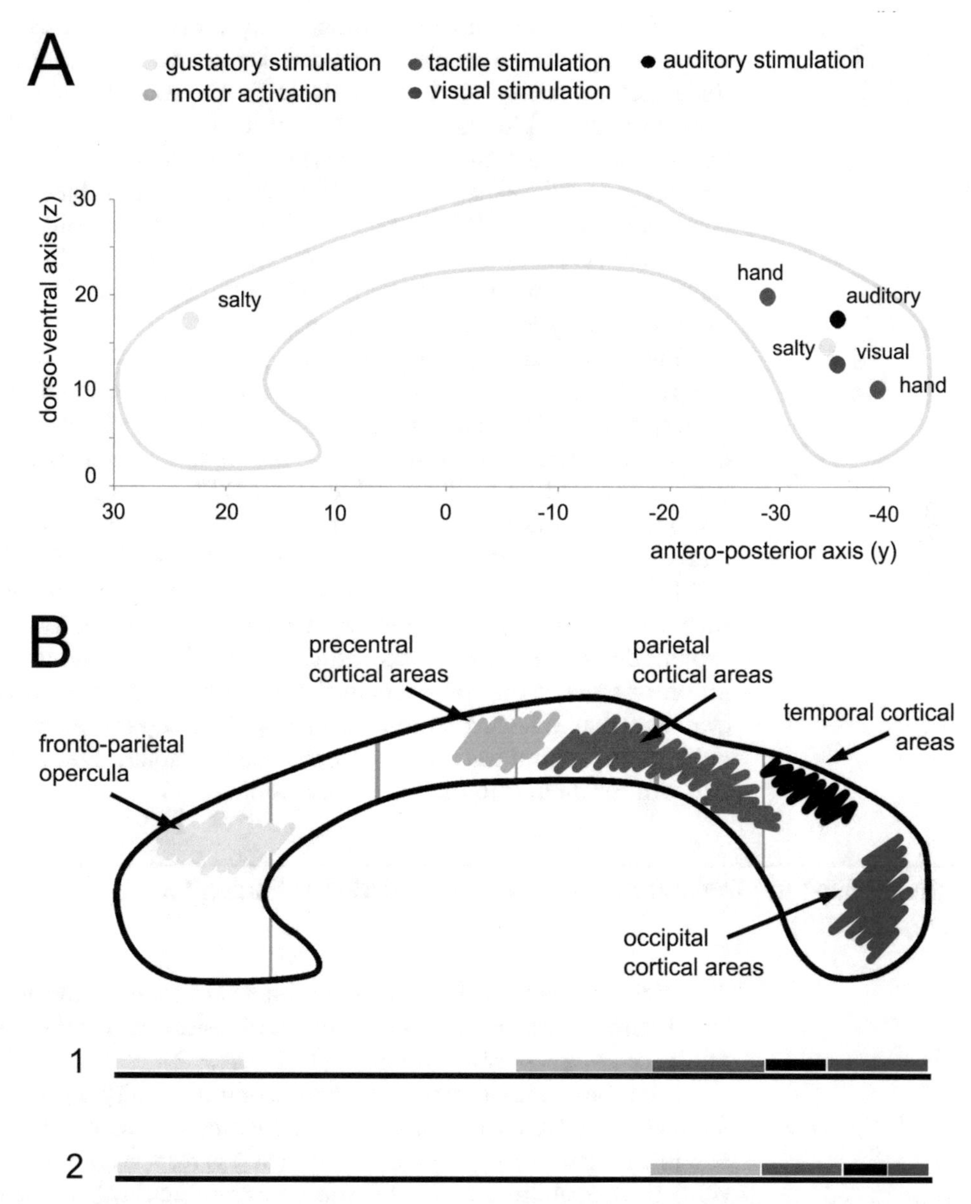

Fig. 6 Summary schematic diagram showing the similar distribution of the callosal activation foci evoked by different stimulus types in partially callosotomized patients (**a**), and (**b**) crossing site of callosal fibers seen in our work and in the studies of Witelson [68] (*gray lines*), Hofer and Frahm [69] (*line 1*), and Chao et al. [70] (*line 2*). (**a**) Each colored dot represents the "mean" value of the y and z Talairach coordinates (reported on the respective Cartesian axes) of the foci evoked by different stimuli. *Yellow*: foci evoked by gustatory stimuli; *green*: foci evoked by hand motor tasks; *red*: foci evoked by tactile stimuli; *black*: foci evoked by auditory stimuli; *blue*: foci evoked by visual stimuli. See the text for a detailed description. (**b**) The crossing sites of interhemispheric fibers connecting the sensory and motor cortical areas activated by the relevant peripheral stimuli are shown. *Vertical gray lines* mark the main CC subdivision according to Witelson [68]. *Line 1* and *line 2* on the bottom show the CC subdivision according to Hofer and Frahm [69] (see their Fig. 3) and Chao et al. [70] (see their Fig. 7): colored tracts mark the CC crossing sites of fibers from frontal opercular cortical areas (*yellow*), motor cortices (*green*), anterior and posterior parietal cortices (*red*), temporal cortices (*black*), and occipital cortices (*blue*). Modified with permission from [44]

Diffusion tensor imaging, other than information of callosal topography, can also provide more detailed in vivo information on the structural preservation of transected white matter tracts than conventional imaging methods. In the study described here [41] it has been shown for the first time that severed callosal fibers can be traced up to 17 years from resection. Five patients subjected to complete or partial callosotomy several years before the study were examined with DTI and compared to a normal control. The study showed for the first time that resected central tracts can be traced in vivo several years after the injury using DTI.

The main finding of the present investigation is the DTI detection of a large amount of resected but still traceable callosal fibers (as well as some intact fibers) several years after callosotomy. Intact fibers are shown to be continuous at the midline, whereas in the case of resected but still traceable fibers DTI does not enable a distinction between proximal stumps and distal degenerating callosal axons with persistent myelin sheaths. Previous histochemical and MR studies did not enable central tracts to be traced in detail such a long time after proximal injury. Even high-resolution MR, though detecting spared fibers at the midline, does not allow resolution of severed callosal fibers in the white matter. Here the first evidence that microstructural changes persist several years after resection of central tracts has been provided, enabling DTI reconstruction of their longitudinal organization.

6 Interhemispheric Transfer of Visual Stimuli and Motion Perception (s9, s10, s11, s16)

This line of research deals with the interhemispheric transfer of visual stimuli and perception of motion illusion in patients with partial or complete resection of the CC (Table 2).

It has been shown that some degree of interhemispheric transfer of visual stimuli occurs also in complete callosotomized patients, and in this case the superior colliculus (SC) is responsible of it. To verify the possibility that the SC subserves interhemispheric neural summation, single or double white visual targets were presented to one or both hemifields in normal participants and in one patient with total callosotomy. Simple reaction time was typically faster with double than single stimuli, a phenomenon known as the redundant target effect (RTE). To explain this effect, two models have been proposed, a probabilistic model and a coactivation one (see literature in [33]). The former postulates that two or more redundant stimuli are processed in separate channels in a horse-race fashion. On each presentation the fastest stimulus wins the race and triggers the response; therefore, multiple stimuli are more likely than single stimuli to yield a fast response. In this model redundant stimuli do not have to converge at some stage of neural processing to produce a response; however, an important prerequisite is that the different channels are independent. At variance, the coactivation model hypothesizes a convergence (neural summation) between the redundant stimuli. This convergence results in a faster response to redundant than to single stimuli. By calculating the so-called race-model violation [33], it is possible to know which model has been adopted. Briefly, Miller's race-inequality test, by using the cumulative frequency distribution (CDF) of RTs, sets an upper limit for probability summation when redundant stimuli are presented. If this limit is violated, then a probabilistic explanation is no longer tenable and the RTE is likely ascribed to a coactivation mechanism, commonly referred to as a "neural" mechanism. It should be pointed out that although the violation of probability models is evidence for neural coactivation, the nonviolation might still be compatible with some moderate amount of neural coactivation not exceeding the upper bound given by the inequality. A larger RTE was found in a patient without a corpus callosum (pB) than in normal subjects, thus confirming previous results. In both groups, the redundancy gain was related to neural coactivation rather than to probability summation. The novel finding was that, when using monochromatic purple stimuli that are invisible to the SC, a similar redundancy gain was observed in both groups; moreover, this redundancy gain was probabilistic rather than neural. Control experiments with monochromatic red stimuli yielded a RTE of the neural type similar to that with white stimuli and this confirmed that the probabilistic RTE found was specific for purple stimuli. In conclusion, visual input to the SC is necessary for interhemispheric neural summation in both normal subjects and in individuals without the corpus callosum, whereas probabilistic summation can occur without a collicular contribution [33].

The contribution of the SC in the interhemispheric transfer of some kind of visual stimuli has been confirmed by the same authors in a subsequent study [38]. It is now common knowledge that after total surgical section of the corpus callosum and of the other forebrain commissures, the IT of simple sensorimotor functions, although severely delayed, is not abolished, and an important question concerns the pathways subserving this residual IT. To answer this question visuomotor IT in split-brain patients was assessed using the Poffenberger paradigm (PP), that is, a behavioral paradigm in which simple RT to visual stimuli presented to the hemifield ipsilateral to the responding hand is compared with that to stimuli presented to the contralateral hemifield, a condition requiring IT. The possibility that the residual IT is mediated by the collicular commissure interconnecting the two sides of the superior colliculus was tested. To this purpose, short-wavelength visual stimuli (purple) were used, which in neurophysiological studies in nonhuman primates have been shown to be undetectable by collicular neurons. It has been shown that, in both totally (pC and pE) and partially callosotomised patients lacking the splenium (pG), IT was considerably longer for purple than for red or achromatic stimuli. This was not the case in healthy participants in whom IT was not affected by color. These data clearly show that the SC plays an important role in IT of sensorimotor information in the absence of the corpus callosum [38].

Similar conclusion on the presence of residual interhemispheric transfer also in absence of the corpus callosum was obtained in another study on a total callosotomy patient from Ancona group (pB) [34], and in a control group of healthy adults. The subjects were shown vertical lines that appeared instantaneously between pairs of rectangles in one or other visual field. When one of the rectangles flashed prior to the presentation of the line, and the line was in the same visual field, all subjects perceived the line as spreading from the flashed rectangle to the other rectangle. Normal subjects and the Ancona callosotomized subject showed a slight but significant right visual-field advantage, perhaps reflecting a left-hemispheric superiority in processing rapid temporal events. The illusion was also induced when the line and the flash were in opposite visual fields in the Ancona callosotomized patient, and about half of the normal subjects. This finding implies that some degree interhemispheric integration can occur even in the absence of the corpus callosum.

Another study strongly indicating a role for the posterior corpus callosum, likely the splenium, in the processing of visual inputs has been performed by examining the line bisection performance in four patients with resection of the corpus callosum (one total, pB, one partial posterior, pG, and two partial anterior, pI and pQ) and in 22 control participants [32]. All subjects were instructed to bisect 17 horizontal black lines 1 mm wide, presented on a white sheet of paper, into two parts of equal length by marking the subjective midpoint of each line with a fine pencil. All participants completed the task with one hand and then repeated it with the other in a balanced order. The control participants showed a leftward bias, especially with the left hand, implying right-hemispheric dominance in spatial attention. Two patients with anterior callosotomy showed similar biases, suggesting that the anterior callosum plays only a small role. The patient with complete callosotomy showed a strong right bias, regardless of hand use. The patient with posterior callosotomy showed the opposite pattern: a strong left bias, regardless of hand use. These data suggest that the posterior corpus callosum normally plays a role in line bisection and that the resection of the posterior corpus callosum produces consistent bias. The direction of the bias depends on which hemisphere assumes the control.

7 Interhemispheric Transfer and Localization of Auditory Stimuli (Studies s14, s26)

The auditory system is largely bilaterally organized but some auditory functions could benefit also from the callosal contribution, thus indicating some degree of lateralization. Ancona split-brain patients participated in some studies, carried out in different periods and by different research groups, investigating the hemispheric difference in the processing of auditory spatial information [36] and in the perception of pleasantness of music chords [48]. Although these studies addressed different topics of auditory function, both confirmed the dominance of the right hemisphere, and seem to suggest a role of the corpus callosum, likely of its posterior pre-splenial portion (isthmus), in equalizing the functions of the two hemispheres.

The question of whether there is a right-hemisphere dominance in the processing of auditory spatial information in the human cortex, as well as the role of the corpus callosum in spatial hearing functions, is still a matter of debate. This issue was approached by investigating two late-callosotomized subjects (one with partial, pU, and one with total callosal resection, pB; see Table 2), in whom sound lateralization was tested using tone bursts. The stimuli were presented to the subject via supra-aural headphones. These stimuli usually evoke an intracranial sound image, along the line joining the ears and shifts of the sound image with respect to the median plane of the head can be easily detected in normal subjects, even if the changes of interaural differences are very small [36]. Besides a significant reduction in their acuity, subjects with total or partial resection of the corpus callosum exhibited a considerable leftward bias of sound lateralization compared to normal controls. This result indicates that the integrity of the corpus callosum is not indispensable for preservation of sound-lateralization ability. On the other hand, transcallosal interhemispheric transfer of auditory information obviously plays a significant role in spatial hearing functions that depend on binaural cues. Moreover, these data are compatible with the general view of a dominance of the right cortical hemisphere in auditory space perception.

The association between musical consonance and pleasantness, and between musical dissonance and unpleasantness ("consonance effect") is well established [71]. Furthermore, a number of studies suggest the main involvement of the left hemisphere in the perception of dissonance and that of the right hemisphere in the perception of consonance. The consonance effect was studied in a callosotomized patient carrying a total resection and in a control group [48] (pC; Table 2). In binaural presentations, the patient did not attribute different pleasantness judgments to consonant and dissonant chords, at variance with the control group who showed the consonance effect. However, in dichotic presentations (e.g. a chord in one ear and white noise in the other ear), a trend towards the consonance effect was found in the patient too, but only when chords were presented in his right ear (left hemisphere), whereas the control group confirmed the known hemispheric asymmetry in labeling the pleasantness of consonant and dissonant chords. These results suggest that the right-hemispheric superiority in appreciating consonance might hide the inability of the right hemisphere to classify dissonant chords as unpleasant in the split-brain, whereas the left hemisphere seems capable of different labeling of the pleasantness of consonant and dissonant chords, even though it is more sensitive to dissonance.

8 Interhemispheric Coordination of Motor Responses (s8, s17, s18)

A relevant effect that has been studied for almost a century is the redundant-target effect. When multiple copies of the same stimulus are presented to subjects, in choice, go/no-go and even a simple reaction time task, reaction times tend to be faster than RTs to a single copy of the stimulus [9, 72]. This effect has been explained in two ways: according to the statistical facilitation (probabilistic) model, trials with two stimuli are facilitated because a response can be initiated as soon as either stimulus is detected; at variance, according to the coactivation model, facilitation occurs because the activation produced by the presentation of two stimuli gets multiplied in some way. Recently, several studies have assumed that coactivation is equivalent to neural summation [9].

Four subjects from the Ancona group, two with partial anterior resection (pI and pQ), one with partial posterior (pG) and one with complete section of the corpus callosum (pB; Table 2), were tested on simple RT to visual stimuli presented either singly in one or other visual field, or simultaneously in both visual fields [31]. One subject, with a posterior callosal section, showed evidence of "enhanced" redundancy gain with bilateral stimuli, i.e., an effect beyond that attributable to probability summation, associated with a prolonged interhemispheric transfer. One of the two subjects with anterior section, similar to controls, showed little evidence of enhanced redundancy gain, and no evidence of prolonged interhemispheric transfer. The other did show some enhanced redundancy gain at the fast end of the RT distribution. These and other results suggest that the posterior corpus callosum provides the principal route of interhemispheric transfer of the information required for simple visuomotor responses, and it is also responsible for the much reduced redundancy gain in normal subjects relative to that in split-brained subjects. The subject with complete callosal resection was unusual in that he responded only very rarely to stimuli in the left visual field (LVF), yet he showed markedly reduced RTs to bilateral relative to right visual field (RVF) stimuli. This finding shows that the prolonged crossed–uncrossed difference and enhanced redundancy gain, which are characteristic of split-brained subjects, are observed primarily after section of the posterior corpus callosum, perhaps including the splenium and posterior body as parallel channels. Subjects with anterior section do not show these effects. On the other hand, subjects with anterior callosal lesions do appear to show dual attention to stimuli in the left and right visual fields, which is also characteristic of subjects with complete callosal section [28]. This implies that enhanced redundancy gain is not due to dual attention. This is further confirmed by the fact that redundancy gain violating at least one version of the race model (see above for the explanation, pp. 22–23) occurred in patient pB, who only rarely responded at all to LVF stimuli, and this has been reported also in normal subjects when one of the stimuli is below threshold [73]. Overall, the results suggest that callosal section slows responses to unilateral stimuli, and in the case of patient pB virtually eliminated responses to LVF stimuli altogether. Nevertheless, there may also be some callosal inhibition, particularly in the hand-area of the cortex, associated with bilateral stimuli in the intact brain, since normal subjects typically respond more slowly to bilateral stimuli than predicted by the race model (see discussion in [31]).

Two other studies investigated, in nine split-brain patients (seven from Ancona, four of them with anterior section sparing the splenium, pL, pP, pU and pV; three with total resection, pA, pB and pE) and ten neurologically intact individuals, the redundant target effect [39] and interhemispheric transfer, through measures of manual asynchrony and bimanually recorded crossed–uncrossed difference [40]. In the first study [39], all subjects completed an RTE protocol in which targets were presented on the midline or in an inter- or intrahemispheric manner. Stimuli of different nature (luminance, equiluminant color, and global motion) were used separately in three experiments in order to investigate the contribution of subcortical versus cortical pathways. Despite the preservation of the splenium (the portion of the corpus callosum assumed to transfer visual information), partial split-brain individuals showed an enhanced RTE pattern as compared to neurologically intact individuals. Total split-brain individuals showed a tendency toward larger RTEs with the luminance stimuli than with the color and motion stimuli, whereas this was not the case for partial split-brain individuals, suggesting a contribution of the posterior portion of the corpus callosum in the RTE. It is therefore likely that both sensory and motor processes contribute to the enhanced RTE in split-brain individuals.

The second study carried out in the same patients [40] investigated the CUD, which relies on the difference between crossed and uncrossed responses, whereas the asynchrony measure relies on the reaction time difference between the two responding hands. Manipulations of sensory and attention factors were assessed for both measures. A normal CUD (3.8 ms) along with an exacerbated and more variable asynchrony was found in partial split-brain individuals (40.8 ms) compared to normal individuals (CUD: 0.4 ms, asynchrony: 13.8 ms). In turn, the CUD of total split-brain individuals (20.4 ms) was larger than that of partial split-brain and normal individuals. Also, the asynchrony of total split-brain individuals (57.6 ms) was larger and more variable than that of normal individuals, and more variable than that of partial split-brain individuals. These results were interpreted as behavioral evidence of independent mechanisms underlying the CUD and bimanual synchronization, as well as evidence of the joint involvement of both the anterior and the posterior portions of the corpus callosum in bimanual coordination.

9 Allocation of Attention and Learning-Memory Functions (Studies s3, s4, s13)

Subjects lacking the corpus callosum usually show impairments on tasks requiring bilateral interdependent motor control. However, few studies have assessed the ability of these subjects to learn a skill that requires the simultaneous contribution of each hemisphere in its acquisition.

Since the corpus callosum links cortical areas of both hemispheres, including the dorsolateral frontal areas, the study of interhemispheric integration and transfer of a visuomotor procedural skill in patients, whose interhemispheric communication is interrupted, provides an interesting model for the investigation of the cortical systems involved in this kind of procedural memory. Furthermore, considering the importance of the corpus callosum in bimanual coordination, the specific role of this structure in motor learning also needs to be defined. A corollary objective of this study was to assess if procedural and declarative memory processes could be dissociated in the intra- and interhemispheric conditions of the visuomotor task.

Three adult patients from Ancona, pB, pC and pP, participated in a study investigating whether callosotomized subjects could learn a visuomotor skill that involved a motor control from either both or a single hemisphere [27] (Table 2). The performance of the experimental subjects was compared with that of 11 matched control subjects, on a modified version of a serial reaction time task developed and applied previously [27]. This skill acquisition task involved bimanual or unimanual key-pressing responses to a sequence of ten visual stimuli that was repeated 160 times. A declarative memory task was then performed to assess explicit knowledge of the sequence. None of the experimental subjects learned the task in the bimanual condition. Patients with frontal epileptic foci also failed to learn the task in the unimanual condition when they were using the hand contralateral to the damaged hemisphere [27]. All other subjects, including the callosotomized patients with temporal foci, learned the visuomotor skill as well as their controls in the unimanual condition. In spite of the absence of transfer and interhemispheric integration of procedural learning, some of the callosotomized patients were able to learn the sequence explicitly. In summary, the results of this study suggest that the frontal lobes are important for unilateral procedural learning and that the anterior part of the corpus callosum, which connects these lobes, is crucial for integration and transfer of a procedural visuomotor skill. They also confirm the dissociation described by Squire between the declarative and procedural memory systems and extend this dissociation to processes involving simultaneous bihemispheric cooperation [74, 75].

Another "higher function" (other than memory) in which the corpus callosum might be involved is the attentive capacity. Four patients, three with a total callosotomy and one with an anterior callosal section participated in a study investigating the role of the corpus callosum in the interhemispheric integration of the visuospatial attention system [28]. Subjects produced simple reaction times to visual targets shown to the left or right visual hemifield. Preceding the target by an interval of 500 ms, arrow cues predicting the target location were shown left and right of the point of ocular fixation. For a majority of total and anterior callosotomy patients, results with valid focused cues (both arrows pointing to the target location) and with divided-attention cues (arrows pointing away from fixation) did not differ and both conditions produced shorter RTs than with neutral cues (equal signs). In contrast, neurologically intact subjects showed equal RTs with divided-attention and neutral cues, whereas valid focused cues produced reduced RTs relative to neutral cues. These results indicate that most split-brain subjects, in contrast to normal subjects, are capable of directing their attention to left and right visual field locations simultaneously and independently, and therefore that each cerebral hemisphere controls its own visuospatial attention mechanism. This division of visuospatial attention mechanisms in split-brains contrasts with evidence from normal subjects, who are incapable of dividing their attention across two distinct spatial locations at the same time [28]. These results therefore assign a crucial role to the corpus callosum in the functional integration of the brain areas of each hemisphere that are involved in the orientation of visuospatial attention.

The patient pB, with complete callosotomy, is one of the more studied patients of the Ancona group. He shows unusual neglect of stimuli in the LVF, manifested in simple RT to stimuli flashed in the LVF and in judging whether pairs of filled circles in the LVF are of the same or different color. It may reflect strong left-hemispheric control and consequent attention restricted to the right side of space. It is not evident in simple RT when there are continuous markers in the visual fields to indicate the locations of the stimuli, as suggested by a previous study on the same patient [33]. In this condition, his RTs are actually faster to LVF than to RVF stimuli, suggesting a switch to right-hemispheric control that eliminates the hemineglect [35]. Neglect is also not evident when pB responds by pointing to or touching the locations of the stimuli, perhaps because these responses are controlled by the dorsal rather than the ventral visual system. Despite his atypical manifestations of hemineglect, pB showed evidence of functional disconnection typical of split-brained subjects, including prolonged crossed–uncrossed differences in simple reaction time, inability to match colors between visual fields, and enhanced redundancy gain in simple RT to bilateral stimuli even when the stimulus in the LVF was neglected.

10 Hemispheric Collaboration in Higher Brain Functions (Studies s20, s21)

This line of research groups apparently different studies, all however dealing with mental rotation of letters, or "abstract" mental rotation, i.e., imaging one's self in the anothers' shoes. These functions have been shown to be ascribed mainly to the right hemisphere, and therefore a correct performance is supposed to require hemispheric cooperation.

The first study described here had the primary aim of testing whether mirror-image discrimination and identification of letters could also be dissociated in the split brain. Two complete callosotomized patients (pB and pC) were tested, along with a group of neurologically normal participants, on discrimination of normal and backward letters presented upright briefly in one or the other visual hemifield. In one task they were required to determine whether the letters were normal or backward, and in the other to discriminate the letters themselves. The tasks were kept simple, since split-brained patients typically have difficulty with right-hemisphere reading and naming. Responses were manual, since split-brained patients typically have little if any production of speech from the right hemisphere. Further, the stimuli were restricted to the letters F and R in one task, and b and d in another, so that all decisions were binary [42]. Where discrimination of the letters F and R by name either showed a left-hemisphere advantage or no hemispheric effect, discrimination of whether the same letters were normal or backward showed a right-hemisphere advantage. These results suggest that discrimination of mirror-image letters depends on matching to an exemplar, for which the right-hemisphere is dominant, while letter naming depends on abstract category recognition. One callosotomized patient, pB, showed systematic left–right reversal of the letters in the left visual field, classifying the normal letters as reversed and reversed ones as normal, and persisted with this reversal when the letters were shown in free vision. This suggests that reversed exemplars of the letters may be laid down in the right cerebral hemisphere. There was no such reversal in the other patient (pC). The results from both the split-brained, compared with those from the normal subjects, confirm previous evidence that the right hemisphere is dominant for mirror-image discrimination even of verbal material. In the case of pB, there was a clear dissociation between identification and mirror-image discrimination, since identification was superior in the left hemisphere. The left hemisphere may well have been dominant for identification in pC and the normal controls as well, but masked by ceiling effects. These results suggest different mechanisms for the two tasks, with identification dependent on category assignment, and mirror-image discrimination dependent on matching to exemplars. The fact that pB systematically reversed the normal-backward judgments further suggests that the two processes are independent, and that exemplars were mirror reversed in his right hemisphere.

Recent functional neuroimaging studies indicate that processes for ascribing beliefs and intentions to other people are lateralized to the right temporal parietal junction (TPJ; see literature in [43]), in that TPJ activity in the right hemisphere, but not the left, is correlated with moral judgments of accidental harm. These findings suggest that patients with disconnected hemispheres would provide abnormal moral judgments on accidental harm and failed attempts to harm, since normal judgments in these cases require information about beliefs and intentions from the right brain to reach the judgmental processes in the left brain. The present study examines this hypothesis by comparing the performance of 22 normal subjects to four Ancona patients, one with the corpus callosum completely severed (pB), one with the central portion resected (pT), and two with only anterior portions severed [43] (pP and pS; Tables 1 and 2; Fig. 1). If normal moral judgments require transfer of information regarding an agent's beliefs from the right TPJ, full split-brain patients should be abnormal in their judgments. Partial split-brain patients with an intact splenium and isthmus, however, might show normal moral reasoning because the fibers connecting the right TPJ with the left hemisphere are intact. Patients and controls made moral judgments about scenarios, in each of them the agent's action either caused harm or not, and the agent believed that the action would either cause harm or cause no harm. The crucial scenarios involved accidental harm (where the agent falsely believed that harm would not occur, but the outcome was harmful) and failed attempts (where the agent falsely believed that harm would occur but the outcome was not harmful). After each scenario was read, subjects were asked to judge the agent's action by vocally responding "permissible" or "forbidden." For the patients, this testing did not require any lateralized procedures as only the left hemisphere was assumed to be responding verbally. Previous studies show that normal subjects typically base their moral judgments on agents' beliefs even when these are inconsistent with the actions' outcomes (see literature in [43]). For example, if Grace believed the powder was sugar but it was really poison, normal subjects judge Grace's action to be morally permissible because of her neutral belief. In contrast, when Grace falsely believed the powder was poison, normal subjects judge Grace's action to be morally impermissible because of her negative belief even though no harm ensues. Given prior evidence that processing of agents' beliefs is supported by regions in the right hemisphere, whereas the left hemisphere is responsible for the verbalization of moral judgments, we hypothesized that agents' beliefs would have less impact on moral judgments when the hemispheres are disconnected in only full callosotomy patients. We found that full and partial callosotomy patients based their moral judgments primarily on actions' outcomes, disregarding agents' beliefs. The present study demonstrates that full and partial callosotomy patients fail to rely on agents' beliefs when judging the moral permissibility of those agents' actions. This finding confirms the hypotheses that specialized belief-ascription mechanisms are lateralized to the right hemisphere and that disconnection from those mechanisms affects normal moral judgments. Moreover, the neural mechanism by which interhemispheric communication occurs between key left and right hemisphere processes seems complex. Since the partial anterior callostomy patients also showed the effect, it would appear the right TPJ calls upon right frontal processes before communicating information to the left speaking hemisphere.

Another similar function could be considered the strategy adopted in imitating intransitive gestures performed by a model in a third person perspective. Preliminary observations on callosotomy subjects seem to point to a difference from control subjects, thus suggesting that some high order abstract functions require hemispheric cooperation ([76] preliminary report).

11 Emotional Hemispheric Specialization (Studies s1, s24, s25, s27)

Hemispheric lateralization in emotion processing remains a controversial issue in the field of cognitive neuroscience despite the number of studies that have delved into the issue for decades. Remarkably, opposite patterns of hemispheric superiority have been suggested, although a number of studies have failed in finding cerebral asymmetries. The two currently leading hypotheses are the "right hemisphere hypothesis" (RHH) and the "valence hypothesis" (VH) (see literature in [49]). According to the first, the right hemisphere is superior to the left in the analysis of all emotions; according to the second, the right hemisphere is specialized in negative emotion processing and the left in positive emotion processing.

The first study on the emotional hemispheric specialization on an Ancona split-brain patient (pA) was performed by Làdavas and coworkers [26]. Two main aspects of emotional behavior were addressed: the first concerned the role of conscious identification of the stimulus in producing emotional activity, and the second searched for a possible hemispheric specialization for an emotional response. These two issues were addressed by analyzing the cognitive and physiological emotional activity of a split-brain patient, who received subliminal presentations of emotional scenes to either left (LVF) or right visual field (RVF), and above threshold presentation of emotional and nonemotional stimuli. The results showed that the patient was able to discriminate emotional from neutral stimuli in both LVF and RVF presentations, leading to the conclusion that the emotional processing took place in both disconnected hemispheres, but only the right hemisphere produced an autonomic response to emotional stimuli [26].

A more recent study, aimed at better disentangling the two accounts for right hemisphere and/or valence hypotheses (see above), was carried out by presenting healthy participants and an anterior callosotomized patient (pP) with "hybrid faces," stimuli created by superimposing the low spatial frequencies of an emotional face to the high spatial frequencies of the same face in a neutral expression [49]. All participants were asked to judge the friendliness level of stimuli, which is an indirect measure of the processing of emotional information, despite this remaining "invisible." In a first experiment, hybrid faces were shown in a divided visual field paradigm, by using different tachistoscopic presentation times; in a second experiment, hybrid chimeric faces were presented in canonical view and upside-down. Results from the first experiment confirmed previous results from healthy participants: stimuli presented in the LVF, and therefore to the right hemisphere, were judged as less friendly than those presented in the RVF (left hemisphere), thus supporting the valence hypothesis (see above) also in the partial split-brain patient; on the other hand, results of the second experiments were consistent with the right hemisphere hypothesis. Actually, only for the upside-down stimuli, the patient's judgments were better when he used the left hand (activating the right hemisphere) rather than when he used the right; this observation seems to suggest a higher right-hemispheric ability in face processing or a more positive right hemispheric appreciation for emotional face in general. Importantly, patient pP seemed to attribute appropriate evaluations to the upright hybrid faces only when he was required to respond with the right (dominant) hand. In this condition, the sole in which his friendliness ratings were congruent with the emotional content of stimuli and with the responses of healthy participants, pP's results added support for the right hemisphere hypothesis. Specifically, the friendliness judgments of chimeric faces were mainly modulated by the emotional content of the left hemifaces (right hemisphere), showing a dominant role of the right hemisphere in detecting both positive and negative emotional content of stimuli. This study confirms that the low spatial frequencies of emotional faces influence the social judgments of observers. In the light of the results obtained in the callosotomized patient, the two main hypotheses on hemispheric asymmetry in emotion processing could be considered as complementary rather than competing, at least in the case of implicit emotion processing. In particular, the specific left/right hemisphere superiority in positive/negative emotion analysis could disappear in case of simultaneous processing of two subliminally emotional (hemi)faces. In other words, it can be hypothesized that the right hemisphere is clearly dominant when more than one subliminal emotional "unit" have to be processed at the same time, thus suggesting the coexistence of both types of cerebral organization for processing of subliminal emotion, i.e., the emotional content of facial expression due to the low spatial frequencies. Therefore, the right hemisphere and the valence accounts are not mutually exclusive, at least in the case of subliminal emotion processing [49].

Another issue analyzed in the Ancona patients was the relation between hemispheric specialization and processing of explicit and implicit facial expressions of emotion, assuming a right-hemispheric superiority in detecting both positive and negative emotions [46]. Two split-brain patients (one with a complete, pC, and one with a partial anterior resection, pP) and a group of healthy participants were tested using tachistoscopic stimuli consisting of hybrid or implicit/explicit expressive faces, presented both centrally and bilaterally (in this case, the facial expressions presented in the left and right hemifields could be either identical or different). The stimuli were created using photographs of two male and two female faces, in expressive and neutral poses, selected in order to obtain both an original set of faces ("explicit faces") and a modified set of faces "implicit or hybrid faces" [46]. From the explicit faces, "hybrid faces" were obtained following this procedure described in [46]: first, happy and angry faces were low-pass filtered and neutral faces were high-pass filtered. Subsequently, for each individual, both the happy and the angry low-pass filtered faces were superimposed to the neutral high-pass filtered face of that individual, resulting in a "hybrid happy" and a "hybrid angry" face respectively. All participants were asked to evaluate the friendliness level of faces. Half of the photographic stimuli were "hybrid faces," i.e., an amalgamation of filtered images which contained emotional information only in the low range of spatial frequency, blended to a neutral expression of the same individual in the rest of the spatial frequencies. The other half of the images contained unfiltered faces. With the hybrid faces the patients and a matched control group were more influenced in their social judgments by the emotional expression of the face shown in the LVF, confirming the right hemisphere superiority. When the expressions were shown explicitly, the control group and the partially callosotomized patient based their judgment on the face shown in the LVF (to the right hemisphere); the complete split-brain patient based his ratings mainly on the face presented in the right visual field, i.e., to the left hemisphere. We conclude that the processing of implicit emotions does not require the integrity of callosal fibers and can take place within subcortical routes lateralized in the right hemisphere. The results obtained seem to suggest that both the hemispheres can process subliminal expressions, but the social judgments they attribute to these stimuli is different, revealing a right-hemispheric superiority in emotion processing. In addition, the data confirm a difficulty of split-brain patients in integrating visual information from the two visual hemifields [46].

Finally, in another set of experiments, hemispheric asymmetries in audiovisual integration and the way in which lateralized visual and acoustic emotional stimuli are processed by the two hemispheres were investigated [47]. The performance of a partially callosotomized patient, a total split-brain patient and a control group was analyzed during the evaluation of the emotional valence of chimeric faces and dichotic syllables (an emotional syllable in one ear and white noise in the other ear), presented unimodally (only faces or only syllables) or bimodally (faces and syllables presented simultaneously). Stimuli could convey happy and sad expressions and participants were asked to evaluate the emotional content of each presentation, using a five-point scale (from very sad to very happy). In the unimodal presentations, the partially callosotomized patient, pP, judged the emotional valence of the stimuli depending upon those processed by the right hemisphere; the total split-brain patient, pB, showed the opposite lateralization; in the same conditions, the control group did not show asymmetries. Moreover, in bimodal presentations, results provided support for the valence hypothesis (i.e., left asymmetry for positive emotions and vice versa) in both the control group and the partially callosotomized patient; the total split-brain patient showed a tendency to evaluate the emotional content of the right hemiface even when asked to focus on the acoustic modality. We conclude that partial and total hemispheric disconnections reveal opposite patterns of hemispheric asymmetry in auditory, visual, and audiovisual emotion processing.

The performances of the two patients were very different from one another. In general, the total split-brain patient (pB) showed a rightward bias; moreover, his evaluations in bimodal presentations were mainly based on visual rather than acoustic stimuli, even when he was explicitly required to judge the auditory input. On the other hand, the results of the partially callosotomized patient (pP) largely resembled those of the control group and he did not show difficulties in paying attention to either visual or acoustic stimuli, when required. Differently from the control group, however, pP showed a trend for a leftward bias in the emotional evaluation of both acoustic and visual stimuli presented in isolation. This pattern of results confirms a right-hemispheric superiority in emotion detection, already known in the visual domain, and adds new evidence in the auditory domain.

To sum up, the results suggest an articulate balance between the two cerebral hemispheres in auditory, visual, and audiovisual processing. The results of pP in the two unimodal sessions seem to support the right hemisphere hypothesis; however, a more complete examination of the results shows that this is the sole evidence in line with the right hemisphere hypothesis, whereas the other findings substantially support the valence hypothesis, especially when the cognitive load becomes heavy. In fact, opposite hemispheric specializations for positive and negative emotion processing are found in bimodal presentations in the control group (connected hemispheres), in bimodal presentations in pP (anteriorly disconnected hemispheres), and in acoustic presentation in pB (totally disconnected hemisphere). As regards bimodal presentations, a strong visual capture in the case of totally disconnected hemispheres was observed. Importantly, when visual stimuli were presented (alone or together with acoustic stimuli), only the total split-brain patient showed a predominant role of the left hemisphere, probably due to an attention bias, rather than asymmetric emotional processing, which was not present in auditory processing. The debate on the validity of the right hemisphere/valence hypothesis remains controversial, but the described study added evidence in support of both theories depending on the task, and suggests that the valence hypothesis is more able to account for results as the cognitive load becomes heavier [47].

12 Concluding Remarks

Our almost 20-years-long research on patients with various degrees of split-brain have led to some conclusive results in some fields, mainly relating to simple sensory or motor functions brain. It has been clarified which is the callosal route by which different kinds of sensory and motor information can be shared between the two hemispheres.

Other issues, on the other hand, require further experiments. This is mainly the case for higher cognitive functions, such as emotional processing, memory, and learning, attention control and mental rotation ability, for instance.

Although observation from split brain patients could provide much important information, it should also be taken into account that these patients usually develop compensatory strategies to solve everyday life situations, strategies which could lead to a misinterpretation of split-brain performance results.

Nevertheless, this research field provides a great opportunity to study lateralized and diffuse brain functions, which can interact positively with other research methods in Neuroscience.

Acknowledgements

The authors wish to thank Professors Tullio Manzoni and Ugo Salvolini for establishing the collaboration, and for providing helpful criticism and support during the research; Drs. Angelo Quattrini, Maria Del Pesce, and Aldo Paggi for encouraging callosotomized patients to participate in the studies; Dr. Giulia Mascioli for her great fMRI processing work and neuropsychological testing; Ms. Gabriella Venanzi for scheduling patient examinations; the technical staff of the Istituto di Radiologia for their invaluable assistance during the scan acquisition; the patients and their families, all the volunteers who participated in the research.

References

1. Amaral DG (2000) The anatomical organization of the central nervous system. In: Kandel ER, Schwartz JH, Jessell TM (eds) Principles of neural science, 4th edn. McGraw-Hill, New York, pp 317–336
2. Rogers LJ, Vallortigara G, Andrew RJ (2013) Evolution. In: Rogers LJ, Vallortigara G, Andrew RJ (eds) Divided brains: the biology and behavior of brain asymmetries (Chapter 3). Cambridge University Press, New York, pp 62–97
3. Kaas JH, Jain N, Qi H-X (2002) The organization of somatosensory system in primates. In: Nelson R (ed) The somatosensory system: deciphering the brain's own body image. CRC, Boca Raton, FL, pp 1–25
4. Gazzaniga MS (2000) Cerebral specialization and interhemispheric communication: does the corpus callosum enable the human condition? Brain 123:1293–1326
5. Funnell MG, Corballis PM, Gazzaniga MS (2000) Cortical and subcortical interhemispheric interactions following partial and complete callosotomy. Arch Neurol 57:185–189
6. Fabri M, Del Pesce M, Paggi A, Polonara G, Bartolini M, Salvolini U, Manzoni T (2005) Contribution of posterior corpus callosum to the interhemispheric transfer of tactile information. Cogn Brain Res 24:73–80
7. Fabri M, Polonara G, Mascioli G, Salvolini U, Manzoni T (2011) Topographical organization of human corpus callosum: an fMRI mapping study. Brain Res 1370:99–111
8. Sperry RW (1974) Lateral specialization in the surgically separated hemispheres. In: Schmitt F, Worden F (eds) Neurosciences third study program. MIT Press, Cambridge, MA, pp 1–12
9. Zaidel E, Iacoboni M, Berman SM, Zaidel DW, Bogen JE (2011) Callosal syndromes. In: Heilman KM, Valenstein E (eds) Clinical neuropsychology. Oxford University Press, Oxford, pp 349–416
10. Fabri M, Pierpaoli C, Barbaresi P, Polonara G (2014) Functional topography of the corpus callosum investigated by DTI and fMRI. World J Radiol 6:895–906
11. Gazzaniga MS (2005) Forty-five years of split-brain research and still going strong. Nat Rev Neurosci 6:653–659
12. Fabri M, Polonara G, Quattrini A, Salvolini U, Del Pesce M, Manzoni T (1999) Role of the corpus callosum in the somatosensory activation of the ipsilateral cerebral cortex: an fMRI study of callosotomized patients. Eur J Neurosci 11:3983–3994
13. Fabri M, Polonara G, Del Pesce M, Quattrini A, Salvolini U, Manzoni T (2001) Posterior corpus callosum and interhemispheric transfer of somatosensory information: an fMRI and neuropsychological study of a partially callosotomized patient. J Cogn Neurosci 13:1071–1079
14. Westerhausen R, Hugdahl K (2008) The corpus callosum in dichotic listening studies of hemispheric asymmetry: a review of clinical and experimental evidence. Neurosci Biobehav Rev 32:1044–1054
15. Marzi CA (1999) The Poffenberger paradigm: a first, simple, behavioural tool to study interhemispheric transmission in humans. Brain Res Bull 50:421–422
16. Zaidel E, Iacoboni M (2003) Introduction: Poffenberger's simple reaction time paradigm for measuring interhemispheric transfer time. In: Zaidel E, Iacoboni M (eds) The parallel brain. The cognitive neuroscience of the corpus callosum. MIT Press, Cambridge, MA, pp 1–7

17. Marzi CA, Bisiacchi P, Nicoletti R (1991) Is interhemispheric transfer of visuomotor information asymmetric? Evidence from a meta-analysis. Neuropsychologia 29:1163–1177
18. Lassonde MC, Sauerwein HC, Lepore F (2003) Agenesis of the corpus callosum. In: Zaidel E, Iacoboni M (eds) The parallel brain. The cognitive neuroscience of the corpus callosum. The MIT Press, Cambridge, MA, pp 357–369
19. Omura K, Tsukamoto T, Kotani Y, Ohgami Y, Minami M, Inoue Y (2004) Different mechanisms involved in interhemispheric transfer of visuomotor information. NeuroReport 15:2707–2711
20. Tettamanti M, Paulesu E, Scifo P, Maravita A, Fazio F, Perani D, Marzi CA (2002) Interhemispheric transmission of visuomotor information in humans: fMRI evidence. J Neurophysiol 88:1051–1058
21. Weber B, Treyer V, Oberholzer N, Jaermann T, Boesiger P, Brugger P, Regard M, Buck A, Savazzi S, Marzi CA (2005) Attention and interhemispheric transfer: a behavioural and fMRI study. J Cogn Neurosci 17:113–123
22. Papo I, Quattrini A, Ortenzi A, Paggi A, Rychlicki F, Provinciali L, Del Pesce M, Cesarano C, Fioravanti P (1997) Predictive factors of callosotomy in drug-resistant epileptic patients with a long follow-up. J Neurosurg Sci 41:31–36
23. Quattrini A, Papo I, Paggi A, Ortensi A, Rychlicki F, Fronzoni M, Recchioni MA, Marchioro R, Pauri GL, Bonaparte A, Mancini S (1989) Anterior callosotomy in drug-resistant epilepsy. Adv Epileptol 17:4245
24. Quattrini A, Papo I, Cesarano R, Fioravanti P, Paggi A, Ortensi A, Foschi N, Rychlicky F, Del Pesce M, Pistoli E, Marinelli M (1997) EEG patterns after callosotomy. J Neurosurg Sci 41:85–92
25. Oldfield RC (1971) The assessment and analysis of handedness: the Edinburgh inventory. Neuropsychologia 9:97–113
26. Làdavas E, Cimatti D, Del Pesce M, Tuozzi G (1993) Emotional evaluation with and without conscious stimulus identification: evidence from a split-brain patient. Cognit Emot 7:95–114
27. De Guise E, Del Pesce M, Foschi N, Quattrini A, Papo I, Lassonde M (1999) Callosal and cortical contribution to procedural learning. Brain 122:1049–1062
28. Arguin M, Lassonde M, Quattrini A, Del Pesce M, Foschi N, Papo I (2000) Divided visuospatial attention systems with total and anterior callosotomy. Neuropsychologia 38:283–291
29. Aglioti SM, Tassinari G, Fabri M, Del Pesce M, Quattrini A, Manzoni T, Berlucchi G (2001) Taste laterality in the split brain. Eur J Neurosci 13:195–200
30. Fabri M, Polonara G, Quattrini A, Salvolini U (2002) Mechanical noxious stimuli cause bilateral activation of parietal operculum in callosotomized subjects. Cereb Cortex 1:446–451
31. Corballis MC, Corballis PM, Fabri M (2004) Redundancy gain in simple reaction time following partial and complete callosotomy. Neuropsychologia 42:71–81
32. Hausmann M, Corballis MC, Fabri M (2003) Line bisection in the split brain. Neuropsychology 17:602–609
33. Savazzi S, Marzi CA (2004) The superior colliculus subserves interhemispheric neural summation in both normals and patients with a total section or agenesis of the corpus callosum. Neuropsychologia 42:1608–1618
34. Corballis MC, Barnett KJ, Fabri M, Paggi A, Corballis PM (2004) Hemispheric integration and differences in perception of a line-motion illusion in the divided brain. Neuropsychologia 42:1852–1857
35. Corballis MC, Corballis PM, Fabri M, Paggi A, Manzoni T (2005) Now you see it, now you don't: variable hemineglect in a commissurotomized man. Brain Res Cogn Brain Res 25:521–530
36. Hausmann M, Corballis MC, Fabri M, Paggi A, Lewald J (2005) Sound lateralization in subjects with callosotomy, callosal agenesis, or hemispherectomy. Brain Res Cogn Brain Res 25:537–546
37. Fabri M, Polonara G, Mascioli G, Paggi A, Salvolini U, Manzoni T (2006) Contribution of the corpus callosum to bilateral representation of the trunk midline in the human brain: a fMRI study of callosotomized patients. Eur J Neurosci 23:3139–3148
38. Savazzi S, Fabri M, Rubboli G, Paggi A, Tassinari CA, Marzi CA (2007) Interhemispheric transfer following callosotomy in humans: role of the superior colliculus. Neuropsychologia 45:2417–2427
39. Ouimet C, Jolicoeur P, Miller J, Ptito A, Paggi A, Foschi N, Ortenzi A, Lassonde M (2009) Sensory and motor involvement in the enhanced redundant target effect: a study comparing anterior- and totally split-brain individuals. Neuropsychologia 47:684–692
40. Ouimet C, Jolicoeur P, Lassonde M, Ptito A, Paggi A, Foschi N, Ortenzi A, Miller J (2010) Bimanual crossed-uncrossed difference and asynchrony of normal, anterior- and totally-split-brain individuals. Neuropsychologia 48:3802–3814

41. Pizzini FB, Polonara G, Mascioli G, Beltramello A, Foroni R, Paggi A, Salvolini U, Tassinari G, Fabri M (2010) Diffusion tensor tracking of callosal fibers several years after callosotomy. Brain Res 1312:10–71
42. Corballis MC, Birse K, Paggi A, Manzoni T, Pierpaoli C, Fabri M (2010) Mirror-image discrimination and reversal in the disconnected hemispheres. Neuropsychologia 48:1664–1669
43. Miller MB, Sinnott-Armstrong W, Young L, King D, Paggi A, Fabri M, Polonara G, Gazzaniga MS (2010) Abnormal moral reasoning in complete and partial callosotomy patients. Neuropsychologia 48:2215–2220
44. Fabri M, Polonara G (2013) Functional topography of human corpus callosum: an fMRI mapping study. Neural Plast 2013:251308. doi:10.1155/2013/251308, 15p
45. Polonara G, Mascioli G, Foschi N, Salvolini U, Pierpaoli C, Manzoni T, Fabri M, Barbaresi P (2014) Further evidence for the topography and connectivity of the corpus callosum: an FMRI study of patients with partial callosal resection. J Neuroimaging 25:465–473
46. Prete G, D'Ascenzo S, Laeng B, Fabri M, Foschi N, Tommasi L (2015) Conscious and unconscious processing of facial expressions: evidence from two split-brain patients. J Neuropsychol 9:45–63
47. Prete G, Marzoli D, Brancucci A, Fabri M, Foschi N, Tommasi L (2014) The processing of chimeric and dichotic emotional stimuli by connected and disconnected cerebral hemispheres. Behav Brain Res 271:354–364
48. Prete G, Fabri M, Foschi N, Brancucci A, Tommasi L (2015) The "consonance effect" and the hemispheres: a study on a split-brain patient. Laterality 20:257–269
49. Prete G, Laeng B, Fabri M, Foschi N, Tommasi L (2015) Right hemisphere or valence hypothesis, or both? The processing of hybrid faces in the intact and callosotomized brain. Neuropsychologia 68:94–106
50. Manzoni T, Barbaresi P, Conti F, Fabri M (1989) The callosal connections of the primary somatosensory cortex and the neural bases of midline fusion. Exp Brain Res 76:251–266
51. Manzoni T (1997) The callosal connections of the hierarchically organized somatosensory areas of primates. J Neurosurg Sci 41:1–22
52. Polonara G, Fabri M, Manzoni T, Salvolini U (1999) Localization of the first (SI) and second (SII) somatic sensory areas in human cerebral cortex with fMRI. Am J NeuroRadiol 20:199–205
53. Garcha HS, Ettlinger G (1980) Tactile discrimination learning in the monkey: the effects of unilateral or bilateral removals of the second somatosensory cortex (area SII). Cortex 16:397–412
54. Hari R, Forss N (1999) Magnetoencephalography in the study of human somatosensory cortical processing. Philos Trans R Soc Lond B 354:1145–1154
55. Fabri M, Polonara G, Salvolini U, Manzoni T (2005) Bilateral cortical representation of the trunk midline in human first somatic sensory area. Hum Brain Map 25:287–296
56. Frot M, Mauguière F (1999) Operculo-insular responses to nociceptive skin stimulation in humans. A review of the literature. Neurophysiol Clin 29:401–410
57. Ploner M, Schmitz F, Freund HJ, Schnitzler A (2000) Differential organization of touch and pain in human primary somatosensory cortex. J Neurophysiol 83:1770–1776
58. Disbrow EA, Hinkley LBN, Roberts TPL (2003) Ipsilateral representation of oral structures in human anterior parietal somatosensory cortex and integration of inputs across the midline. J Comp Neurol 467:487–495
59. Polonara G, Mascioli G, Salvolini U, Fabri M, Manzoni T (2009) Cortical representation of cutaneous receptors in primary somatic sensory cortex of man: a functional imaging study. In: Columbus F (ed) Somatosensory cortex: roles, interventions and traumas. Nova Science Publishers Inc, New York, pp 51–77
60. Lent R, Schmidt SL (1993) The ontogenesis of the forebrain commissures and the determination of brain asymmetries. Prog Neurobiol 40:249–276
61. Mihrshahi R (2006) The corpus callosum as an evolutionary innovation. J Exp Zool 306B: 8–17
62. Kaas JH (2004) Evolution of somatosensory and motor cortex in primates. Anat Rec A 281:1148–1156
63. Fabri M, Manzoni T (2004) GAD immunoreactivity in callosal projecting neurons of cat and rat somatic sensory areas. Neuroscience 123:557–566
64. Mascioli G, Berlucchi G, Pierpaoli C, Salvolini U, Barbaresi P, Fabri M, Polonara G (2015) Functional MRI cortical activations from unilateral tactile-taste stimulations of the tongue. Physiol Behav 151:221–229
65. Aglioti S, Tassinari G, Corballis M, Berlucchi G (2000) Incomplete gustatory lateralization as shown by analysis of taste discrimination after callosotomy. J Cogn Neurosci 1:238–245

66. Polonara G, Mascioli G, Paggi A, Tassinari G, Berlucchi G, Salvolini U, Manzoni T, Fabri M (2006) The cortical representation of taste in the human brain: an fMRI study on callosotomized patients. In: 12th annual meeting of human brain mapping organization, Firenze, June 11–15
67. Geffen G, Nilsson J, Quinn K (1985) The effect of lesions of the corpus callosum on finger localization. Neuropsychologia 23:497–514
68. Witelson S (1989) Hand and sex differences in the isthmus and genu of the human corpus callosum: a postmortem morphological study. Brain 112:799–835
69. Hofer S, Frahm J (2006) Topography of the human corpus callosum revisited. Comprehensive fiber tractography using diffusion tensor magnetic resonance imaging. NeuroImage 32:989–994
70. Chao Y-P, Cho K-H, Yeh C-H, Chou K-H, Chen J-H, Lin C-P (2009) Probabilistic topography of human corpus callosum using cytoarchitectural parcellation and high angular resolution diffusion imaging tractography. Hum Brain Mapp 30:3172–3187
71. Gosselin N, Samson S, Adolphs R, Noulhiane M, Roy M, Hasboun D, Baulac M, Peretz I (2006) Emotional responses to unpleasant music correlates with damage to the parahippocampal cortex. Brain 129:2585–2592
72. Iacoboni M, Zaidel E (2003) Interhemispheric visuo-motor integration in humans: the effect of redundant targets. Eur J Neurosci 17:1981–1986
73. Savazzi S, Marzi CA (2002) Speeding up reaction time with invisible stimuli. Curr Biol 12:403–407
74. Squire LR (1986) Mechanisms of memory. Science 232:1612–1919
75. Squire LR (1992) Declarative and nondeclarative memory: multiple brain systems supporting learning and memory. J Cogn Neurosci 4:232–243
76. Pierpaoli C, Ferrante L, Berlucchi G, Ortenzi A, Manzoni T, Fabri M (2011) Imitation strategies in callosotomized patients. IBRO Firenze, July 14–18
77. Fabri M, Polonara G (2008) Role of the corpus callosum in the interhemispheric tranfer of somatosensory information: an fMRI study. In: LN Bakker (ed) Brain Mapping Research Developments. Nova Science Publishers, Inc. NY, pp 77–100

Chapter 3

Eye and Ear Preferences

Lesley J. Rogers

Abstract

This chapter covers methods of measuring preferences to use one eye or ear to attend to a stimulus, which reflects lateralized processing of sensory information. It begins with monocular occlusion as a way of measuring differences in strength or nature of response elicited by particular visual stimuli. Depending on the type of stimulus presented a preference for responding using the left or right eye can be found (e.g. chicks show a right-eye preference when searching for food grains and a left-eye preference when attacking a conspecific or responding to a predator). Especially in species with their eyes positioned on the sides of their head, this reflects differences in processing by the left and right sides of the brain. In these species it is also possible to test responses to stimuli presented in the left versus right monocular visual fields without having to apply eye patches. A method of determining the extents of the monocular and binocular visual fields is explained. Then a modification of the monocular testing method involving rotation of the stimulus around the animal being tested is discussed: as shown in frogs and toads, response to prey moved in this manner differs between clockwise and anticlockwise rotation. Eye preferences can also be determined using binocular presentation of stimuli that cause the test animal to turn its head to permit monocular fixation of the stimulus before a specific response is made (e.g. before attacking a conspecific, as scored in chicks and horses). Angles adopted by fish when viewing their image in a mirror have been used to measure lateralization of attending to a conspecific. Another approach is simultaneous introduction of identical stimuli into the monocular field of each eye and assessment of side biases in responding (as in toads striking at insect prey). Visual pathways are discussed briefly to help explain how eye preferences reveal brain lateralization. Next, several methods of measuring lateralization of processing and responding to auditory stimuli are covered and finally some points are made about future directions of research along these lines. The suitability of these methods for testing different species is considered in all sections of the chapter.

Key words Eye patches, Ear occlusion, Attack responses, Visual learning, Escape responses, Response to food, Head turning, Pecking bias, Social responses, Mirror tests, Visual pathways

1 Introduction

Lateralization of processing sensory information in the visual and auditory modalities has been demonstrated in a number of vertebrate species (see Ref. 1 for summary). Such laterality can be revealed by comparing the response(s) elicited by presenting a stimulus to the left eye (or ear) with those elicited by presenting the same stimulus to the right eye (or ear). A significant difference

Lesley J. Rogers and Giorgio Vallortigara (eds.), *Lateralized Brain Functions: Methods in Human and Non-Human Species*, Neuromethods, vol. 122, DOI 10.1007/978-1-4939-6725-4_3, © Springer Science+Business Media LLC 2017

between these two measures reveals asymmetry. Alternatively, asymmetry can be revealed by measuring which eye or ear an animal uses to process a particular stimulus. In the visual modality this is achieved most easily in species with their eyes positioned laterally on the sides of their head [2] because they may turn their head to view the stimulus with the left or right eye. These asymmetries in sensory processing are not necessarily associated with motor asymmetries. In fact, laterality of sensory processing has been found in species that have no limb preferences. For example, feral horses show no limb preferences although their responses to fear-inducing stimuli are stronger on the left side and they preferentially use the left eye before attacking [3, 4].

2 Measuring Visual Lateralization

2.1 Monocular Presentation Achieved by Eye Occlusion

Probably the easiest way of showing visual lateralization is by placing a patch over one or the other eye and testing performance on a range of tasks. This method is successful in species with their eyes positioned on the sides of their head since inputs from each eye are processed mainly, although not exclusively (discussed later), by the opposite side of the brain in both the midbrain and forebrain regions. Thus, differences between the left and right sides of the brain in attention and neural processing are manifested as differences between left and right eye performance. Many studies have used this method to reveal lateralization in domestic chicks and pigeons (summarized in Refs. 1, 2, 5).

Laterally positioned eyes are the case for a wide range of vertebrate species, including fish, amphibians, reptiles, most birds, and most mammals. The first experiments showing visual lateralization were performed on chicks tested monocularly after sealing the lids of one or the other eye [6]. These experiments and their results have been summarized in Chap. 8. Suffice it to say here that the chicks were first injected with cycloheximide into the left or right hemisphere (on day 2 post-hatching) and they were tested monocularly on a battery of tests applied between days 7 and 10. One of their eyelids was sealed 1.5 h before testing. This interval allowed the chick to adapt to using one eye before testing commenced. The chick's attention to the sealed eye was also distracted by making sure that the chick was hungry (by food deprivation of a few hours) before it was tested on a food-searching task. The chick searched for grains of mash scattered on a background of pebbles, the latter adhered to the floor. This test is known as the pebble-floor test, or sometimes as the grain-pebble test (Fig. 1).

Subsequent monocular tests of chicks have used eye patches instead of sealing eyelids. In species that have dry skin, fur, or feathers, it is possible to secure an adhesive patch over an eye. Eye patching has not been performed on fish or amphibians

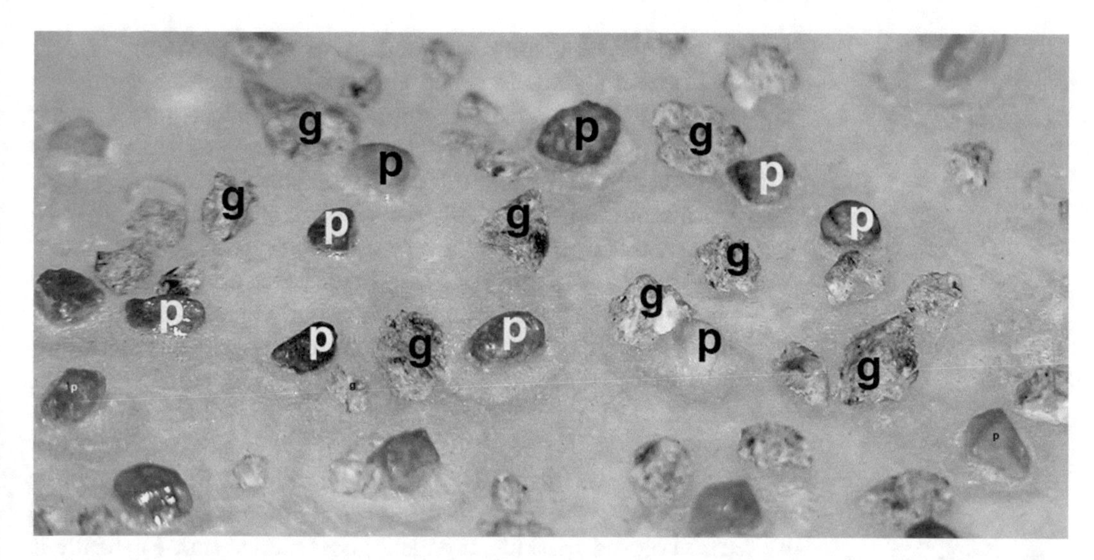

Fig. 1 The pebble-floor used to test young chicks, usually from 8 to 12 days post-hatching. The pecks at pebbles, some labeled "p" in the photograph, and at grains of chicken mash, some labeled "g," are recorded. The pebbles are stuck to the floor with a transparent adherent. Hence the floor can be cleaned between tests and washed if necessary. It is also important that the chick cannot pick up pebbles and ingest them. Learning is indicated by a shift from random choice to pecking mostly at grain, avoiding pebbles. Note that only a small and enlarged section of the floor is shown here

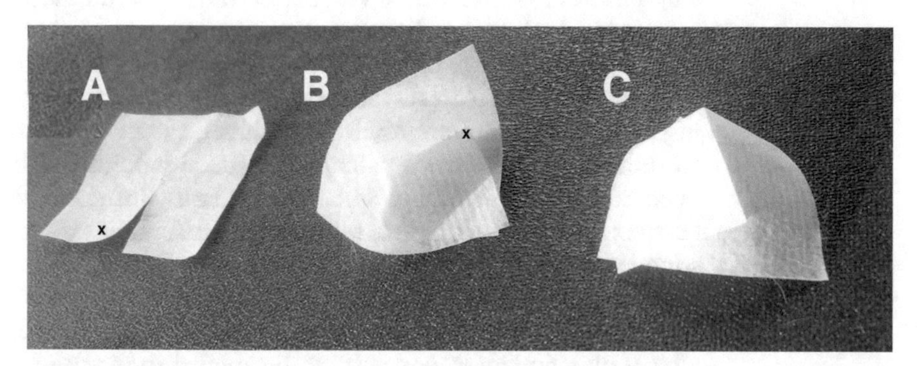

Fig. 2 Making an eye-patch for a young chick. In this example, masking tape has been used. A stronger masking tape of the same color and opacity is usually used. (**a**) First make a cut into the center of a small piece of tape. (**b**) Take the corner marked x and fold it across the other flap to form a cone. (**c**) The eye-patch turned over. The right-hand side of (**c**) is positioned next to the bird's beak

because there is no easy way of securing the patch to moist skin. However, it would be possible to use more solidly attached opaque spectacles, referred to as eye cups, as used by Güntürkün and colleagues to test pigeons monocularly (see Fig. 2 in Ref. 5).

Before testing animals with eye patches applied it is important to determine how long the species takes before it adapts to wearing the patch and choosing an interval between applying the patch and testing that minimizes distraction during testing. Some chicks, for example, may initially scratch vigorously at the patch, occasionally removing it before testing commences, or they may crouch down and close the unpatched eye. In almost all chicks this response to the eye patch wears off within minutes. In the pebble-floor task the chick can be encouraged to peck by gently tapping on the floor using the tip of a finger and out of sight of the chick. Three or four sharp and rapid taps are usually sufficient to stimulate the chick to peck but, if this does not work, repeat the tapping at short intervals. Pecking is also stimulated by having the cage in which the chick is being tested located next to another cage in which a hungry chick is pecking steadily and twittering as it does so (twittering is the usual vocalization accompanying food intake). The chick being used to stimulate pecking should not be visible to the chick being tested.

It is important to use an eye patch that is effective in occluding vision and causes the least possible distraction to the chick. White or cream colored masking tape is a good choice but, especially if the patch needs to stay on for longer, Leukplast tape with a zinc oxide adhesive can be used. The tape is cut midway along one side and to the middle of the tape and then folded into a cone shape, as illustrated in Fig. 2. When adhering the cone to the feathers around the bird's eye, it is important to position the base of the cone opposite the side with the cut so that it is next to the bird's beak. This ensures that the frontal vision is occluded and the bird's nares are not blocked. It also allows the firmest and most comfortable fit.

Note that the eye patch allows light to reach the eye but prevents any vision of patterns, shapes, or texture. However, by allowing nonpatterned light to reach the occluded eye it is less disturbing to the chick than a black patch would be, although this has not been tested systematically. Since the retina is not silent when it receives no light input (i.e., retinal cells send impulses when not stimulated by light), a black eye patch could cause more neural interference than does an opaque white patch.

In some cases and on some tasks, the eye patch can be removed after testing and another patch can be applied to the other eye for subsequent testing. Such a procedure is useful to reveal hemispheric differences only if the processing or memory of the task is not transferred readily from one hemisphere to the other. If memory of the task is transferred between hemispheres/eyes, each animal can be tested only once, and this means that the asymmetry is shown only by comparing the performance of one group tested with the left eye occluded and the other with the right eye occluded. Of course, testing first one eye and then the other of the same animal can be used to ascertain interhemispheric transfer of information and left to right versus right to left transfer may differ, as found in marsh tits [7].

As far as learning performance on the pebble-floor task goes, laterality is revealed by comparing learning by a group of chicks using the left eye with that of a group using the right eye [8, 9].

This gives a measure of group or population directionality but it does not give a measure of the strength or direction of laterality within an individual animal. Testing the same chick with one eye occluded and then with the other eye occluded is not a successful way of revealing lateralization of this particular task since memory of the task is transferred readily between hemispheres (and eyes). Nevertheless, interhemispheric transfer may differ between species and between tasks. For example, Güntürkün and colleagues [10] determined strength of lateralization in individual pigeons by applying an eyecup first to one eye and testing each pigeon on a grain-grit discrimination, followed by repeating the same test after moving the cap to the other eye. In pigeons, this method revealed degree of lateralization at the individual level, which was found to correlate with successful performance of the task.

Monocular testing of first one eye and then the other has been successful in revealing lateralization for control of copulation in chicks [11]. Testosterone-treated chicks show the expected elevation of copulation scores when they are tested with a patch over the right eye but not when the patch is on the left eye. Following removal of the patch from one eye and placement of it on the other eye of the same chick, the mirror image result is found (Fig. 3). Sequential testing of the left and right eye, or vice versa, successfully reveals lateralization possibly because copulation is not a learnt behavior involving access by the untrained as well as the trained hemisphere/eye. Activation of copulation by the right hemisphere and suppression of it by the left hemisphere is a lateralized interplay between the hemispheres. The same hemispheric asymmetry of control applies to attack behavior, as originally revealed by unilateral injection of cycloheximide [12, 13]. Monocular testing can also be used to reveal lateralization of aggression in chicks (left eye/right hemisphere activation) but attack cannot be scored as readily in monocular tests as can copulation. The reason for this seems to be a greater reliance of attack on use of binocular vision.

The data presented in Fig. 3 show a clear, and significant, decrease in copulation scores when, after testing on day 14, the eye patch is switched from the right to the left eye (blue bars in the figure). The group tested first using the right eye and then using the left eye (red bars) shows a significant elevation of copulation after the patch is switched from the left to the right eye (see Ref. 11 for statistics) but the shift in scores is not as dramatic as that of the blue group. This could have occurred as a consequence of the 9 days of repeated monocular testing prior to the eye-patch switch to the other eye. Daily testing of the blue group (eliciting elevated copulation via the left eye) may have led to unusual suppression of response by the right eye, and thus a clear and marked drop in copulation scores occurs when the eye patched is switched. In the red group (using the right eye for those first 9 days of testing) no such clear change in scores occurred on switching the eye patch: in

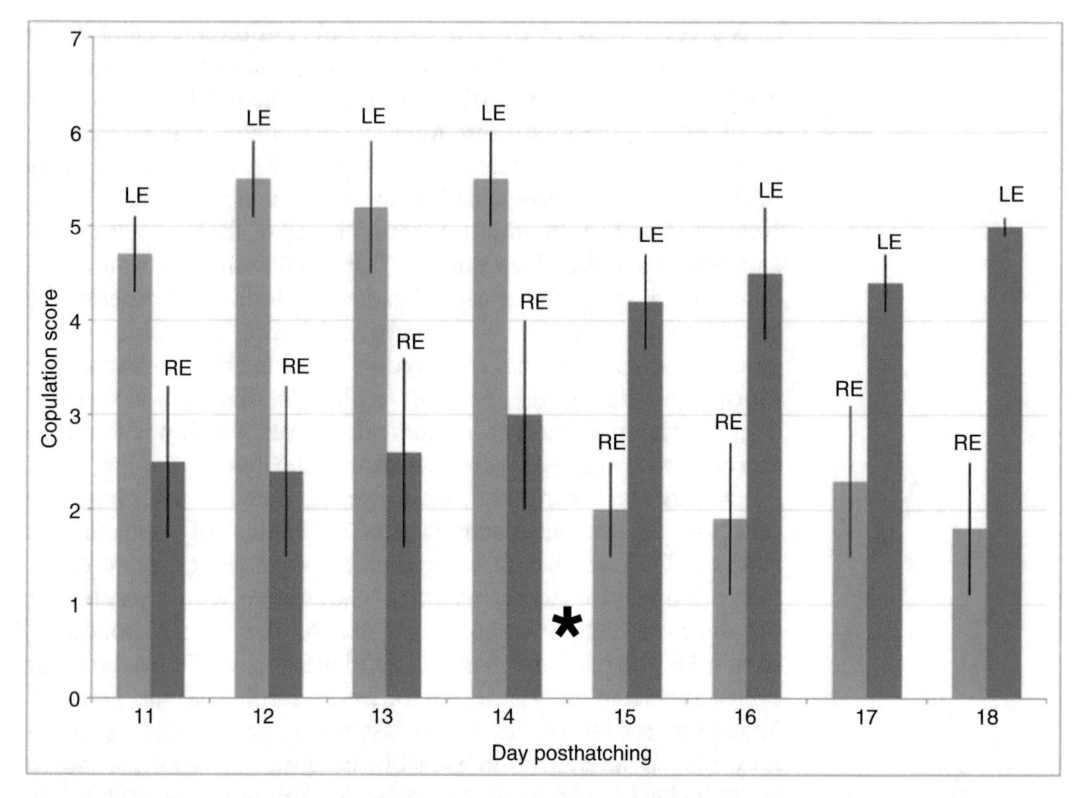

Fig. 3 Scores of copulation performed by young chicks between days 11 and 18 post-hatching are presented: the data have been taken from Fig. 1 in Ref . 11. There were 11 or 12 chicks per group and means and standard error bars have been plotted. All of the chicks were treated with testosterone oenanthate on day 2 post-hatching and their copulation responses were scored from day 6 to day 18. Treatment with testosterone elevates copulation scores in chicks tested binocularly. Here the data for two groups is shown: one group (*blue bars*) was tested until (and including) day 14 using the left eye only, LE (a patch applied to the right eye) and then the patch was switched to the left eye and the right eye was in use, RE. The other group (*red bars*) underwent the reverse regime, first using the right eye and then using the left eye. The data from day 11 on is presented (i.e., not days 6–10) during which time the scores in the *blue group*, using the LE, are elevated, whereas those of the *red group*, using the RE, remain at a mean of about 2.0. The *asterisk* marks the age at which the eye patches were switched from one eye to the other. Note that the expected elevation of copulation scores occurs only when the LE is used. See text for more discussion of these results

this case, repeated testing using the right eye seems to have led to some degree of disinhibition (or activation) of the copulation response as testing continued. Indeed, evidence of gradual disinhibition was seen also, in the same experiment, in untreated chicks (see Ref. 11). While interesting in their own right, these results should alert experimenters to potential effects on laterality of prolonged wearing of eye patches.

Monocular eye patching has been used to investigate lateralization in rats, a species that also has laterally placed eyes. Cowell et al. [14] occluded one eye of rats using small pieces of waterproof adhesive tape to hold the eye closed and then tested spatial behavior in a swim maze. Two days before this they had anesthetized each rat and shaved the hair around one of its eyes so that the tape would adhere to the skin. They tested the rats immediately after the eye had been occluded and found that rats using their left eye (right eye occluded) could find the escape platform as well as controls, but those using their right eye were unable to do so. These researchers added a useful control by putting a patch on the control subjects but not in a place where it occluded an eye. This goes some of the way to control for any disturbance caused by handling the animal or by the patch itself but, of course, it does not control for any possible disturbance caused by temporary loss of vision in one eye. Nevertheless, it is always important to include a control group tested binocularly to compare with monocular performance and to do so using animals that have received the same handling and manipulation as the monocular subjects.

Applying a patch to one eye has also been used to test lateralization in lizards [15]. Bonati et al. [15] tested lizards (*Podarcis muralis*) after attaching opaque cloth eye patches using a commercially available glue. They allowed the eye-patched lizards 24 h to become accustomed to the patches before testing them in a maze. The finding was a left eye advantage in navigating the maze, as shown by moving faster and more efficiently.

Covering one eye with a patch of cloth was also used by Wiltschko et al. [16] to test eye preference in the response of European robins to visual cues and the magnetic compass. This method revealed laterality of migration: use of the right eye (and left hemisphere) allowed correct orientation, whereas no directional preference was shown when the birds had to use their left eye.

2.2 Monocular Testing Achieved by Angle of Presentation

Monocular presentation can be achieved quite easily in species with small binocular and large monocular visual fields: that is, with their eyes positioned laterally on their head. The aim is to compare the response of the animal following introduction of a stimulus into the left or right monocular visual field. This method can be used to test species that would be difficult or impossible to test monocularly using an eye patch (e.g., to test amphibians, see [17]). Since no eye patching is required, this method is also useful for testing animals that might be very disturbed by being handled and/or by being forced to wear an eye patch. For example, this method has been used to reveal visual lateralization in horses [3] and the marsupial dunnart, *Sminthopsis macroura* [18].

The first step to take before testing is to determine the sizes of the visual fields of the selected species. For large animals, such as the horse, this is relatively easy to do in the horizontal plane at eye level. While the horse is standing quietly with its head at about the level of its withers and facing forward without head turning, the experimenter walks around the horse at a distance just greater than the rostro-caudal length of the horse. Starting with, say, the left eye, walk around and note the angle at which the left eye can no longer be seen when the experimenter is on the horse's right side and then move back and around the horse's left side until the left eye can no longer be seen. This can be repeated with the right eye, although both eyes should give the mirror-image result. Knowing these angles allows determination of the size of the monocular fields and of the binocular field. The test stimuli can then, with certainty, be introduced into the left or right monocular visual fields. For example, in the study by Austin and Rogers [3], the experimenter advanced, starting at a distance of 5 m from the horse, towards the horse's head and in a direction just more than at right angles to its flank and, while approaching, opened an umbrella. The same horse was never given more than one test per day and usually there were several days between tests. Flight responses were found to be greater when the approach was on the horse's left side, provided that the left side was tested first. If the first approach was from the right side, the heightened flight response on the left side did not occur, presumably because the horse had habituated to the stimulus.

The extent of the visual fields in small animals can be determined in a similar way. In toads and frogs, for example, Lippolis et al. [17] wanted to introduce a stimulus, a model snake, at the animal's head height. Therefore, first the visual field in a horizontal plane at about head height was determined. This can be done by holding the toad's head on a flat sheet of plastic, the head of the toad poking through a hole and the body and legs positioned below the sheet and resting on a secure surface. The experimenter, with eyes level with the plastic sheet, moves his or her head around the animal marking where a given eye can no longer be seen. As for the horse, the size of the visual fields can be calculated before the stimulus is introduced (Fig. 4). In testing the toads, the stimulus (a model snake's head mounted on a mechanically driven rod) was advanced towards the toad at an angle between 90° and 130° from the toad's midline [17]. In other words, the stimulus was presented in the monocular visual field on the toad's left or right side, and also in the binocular field. The toad was placed in the center of a featureless arena (e.g., a white or gray cylindrical-shaped arena of some 45 cm diameter). Using this method, it was found that escape and defensive responses to the model snake were stronger when the snake was introduced into the left monocular field than when it was introduced into the right monocular field.

Of course, visual fields are three dimensional and vary considerably from species to species [19, 20]. The full, three-dimensional fields can be determined approximately by placing, for example, the toad in a clear plastic hemisphere and marking many points where one eye can no longer be seen. This can be done with the half sphere above the animal and then below it to allow determination of the whole field. The difficulty of this method is holding the animal still while the measurements are made. It is important that

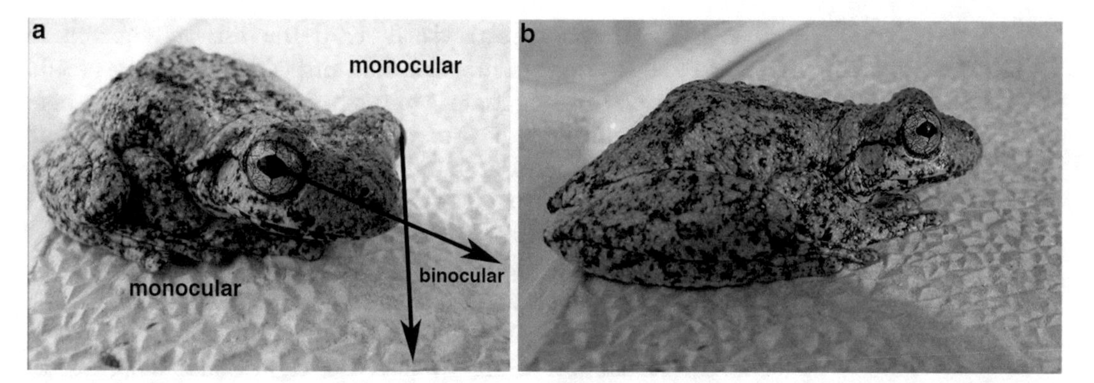

Fig. 4 Binocular versus monocular fields of vision. (**a**) indicates the areas of binocular vision versus monocular vision in a frog (*Litoria peronii*). (**b**) is the same frog viewed from the right, monocular field, showing the lateral positioning of the eyes such that only the right eye can see the observer from this position. Note that lateralization has not yet been investigated in this particular species

the animal is conscious because the eyes must be open and not at all retracted, as they are particularly in anesthetized or dead amphibians. A stereotaxic holder could be used to maintain the head steady but, if used, the visual field would have to be determined quickly to avoid stressing the animal too much. Holding the animal gently with the experimenter's hand may be a better method. The estimation of the visual field using this method is not particularly accurate but it serves as a valuable guide for making sure stimuli are introduced into the monocular field. By choosing to introduce the stimulus well within the monocular visual field and not at or near the border between the monocular and binocular fields, this method can be quite an accurate test of behavioral lateralization.

Another way of ensuring presentation of the stimulus in the lateral visual field is to carry out the test while the animal is partially constrained, for example, in a tunnel. More active animals, such as the marsupial dunnart, can be tested in this way [18]. Using a food reward the small marsupial was trained to enter a short tunnel and thereby gain entrance into a second compartment of the home-cage. As its head emerged from the tunnel the stimulus, again the model snake, was presented to the animal's left or right side, or presented frontally. Left side presentation led to higher levels of retreat than did right side.

The method of monocular presentation can be extended to test lateralization of unrestrained animals in the natural environment. Ventolini et al. [21] reported lateralized responding to visual stimuli in black-winged stilts observed in their natural habitat. They scored pecking to feed (predatory response) and sexual behavior. This species preys on fish and searches for them by moving the head to scan the water using the left or right monocular visual field, that being 170° [22]. Although the predatory peck is made using the binocular visual field, the fish is detected in the bird's left or right monocular field and which field it was could be determined from videotapes of the behavior. Hence, it was possible to determine lateralization of initial detection of the fish. The researchers [21] found a significant bias to the right side (i.e., right visual hemifield) for pecks that led to success in capturing the fish. By contrast, the males showed a bias to respond sexually to females soliciting them on their left side. From video recordings, the researchers scored response of males to females adopting the soliciting posture (head lowered and beak pointing forward) when the female was in the male's left or right visual hemifield. In this context, males showed more sexual responses to females on their left side.

Similar methods have also been used to score lateralized responses of horses tested in their natural environment, and finding a left side bias for agonistic interactions [4, 23].

2.3 Rotating the Stimulus Around the Animal

Another method developed to assess responses initialed in the left or right monocular visual field was a modification of the apparatus used by Ewert [24] to test visual responses of toads. The toad is placed in a glass cylinder through which it can see stimuli rotated clockwise or anticlockwise. Wachowitz and Ewert [25] used this apparatus to test the responses of toads to "worm" versus "anti-worm" stimuli: viz., a small rectangle moved in the direction of its longer axis, or the same rectangle moved in the direction of its shorter axis. The toads struck at the former, in attempts to feed, and adopted threat postures to the latter. Robins and Rogers [26] used a similar apparatus to present *Bufo marinus* toads with either a grasshopper, a model insect resembling a fly or a "worm" stimulus (a black rectangle moving in the direction of its longer axis). The apparatus they used was a transparent plastic cylinder of 15 cm diameter, standing on a circular stationary disc (Fig. 5). The toad was placed into the cylinder. Another disc that could be rotated freely was located underneath the stationary disc. The latter was rotated by a motor in either the clockwise or anticlockwise direction at 1.7 revolutions/min. The stimulus was attached to the outer rim of this disc (the insect stimuli with its head pointing forwards in the direction of movement) and visible to the toad as if it were moving across the surface of the upper disc. The entire set up was located inside a larger, white cylinder (40 cm diameter) and a video camera was used to record from overhead.

The responses scored as predatory acts were turns (rightwards or leftwards) made by the toad as it brings the stimulus into its binocular field. Although the toad catches sight of the stimulus when it is in the lateral field, striking at it requires it to be within in the binocular field. Turns that are predatory can be distinguished from other turns because when engaging in predation the toad lowers its head to the level of the prey. Most of such turns are

Fig. 5 The apparatus used to test toads. A cane toad is placed inside a cylinder made of glass or high-quality clear plastic. The upper disc on which the toad sits is stationary and the lower disc is rotated by a motor. In the drawing the model insect is attached to the lower disc on a wire bent so that the insect moves across the surface of the upper disc. Other stimuli presented by Robins and Rogers [26] are shown, grasshopper on the left and "worm" on the right

accompanied by striking at the prey with the tongue or lunging at the prey. Both of these responses were scored and lumped together as a single "predatory response," used also by other researchers [27, 28]. Laterality was revealed by comparing the number of strikes made by the toad at the stimulus when it was rotated clockwise or anticlockwise. Moving the stimulus in a clockwise direction elicited more predatory strikes than did moving it in the anticlockwise direction. In fact, the toads ignored the stimulus almost completely when it was rotated anticlockwise. This laterality was found for the complex stimuli, the grasshopper and the model fly, but not for the simple "worm" stimulus. The toads performed more predatory behavior in response to the "worm" than to the other two stimuli, and did so regardless of whether it was rotated clockwise or anticlockwise. Since choice of stimulus is important in revealing lateralization, testing with several stimuli is recommended.

2.4 Binocular Presentation and Head Turning

Another way of measuring lateralization of visual responses is to present the stimulus in the binocular field (frontally) and measure head turning to view it monocularly. If responses differ when the head is turned to the animal's right side (left eye use) compared to when it is turned to the left side, laterality is revealed. This method can be used only when the animal views monocularly before it responds to the stimulus. One way of achieving this is to restrain the animal being tested. Young chicks can be retrained quite easily by getting them used to sitting with their head through a small hole in a vertical screen and then recording their head movements, using a video camera placed overhead, while they look at stimuli placed a short distance in front of them [29].

A modification of this method is to place an array of stimuli in front of the bird standing with its head through a hole in a screen, and record whether the chick responds more often to those on the left or right side. This has been used by Diekamp et al. [30] to score pecking bias: the chick with its head through a hole was presented with an array of grains and, although it pecked at grain on both sides, it pecked at more on its left side than on its right side (see Chap. 19, Fig. 5).

Along similar lines, a left versus right bias has been found in adult nutcrackers and young chicks when required to locate an object amongst a row of similar objects placed in front of the animal being tested [31, 32]. The birds were trained first with the line of small pots placed in a direction sagittal to the bird's body and they received a food reward when they pecked, for example, at pot number 4. Then the line of pots was reoriented by 90° so that it projected to the left or right side of the bird. Now the bird could peck at four positions from the left end of the row or four positions from the right end. Four from the left end was preferred. Hence, there is a lateral bias and a form of spatial "hemineglect," in this test a neglect of stimuli on the right side.

Eye preference when viewing various stimuli has been tested in wild dolphins using an innovative procedure [33]. Objects were suspended from the end of a telescopic bar projecting out from the bow of a boat and a video camera was attached to the distal end of the bar and facing downwards. The direction of the boat was maintained in parallel with the direction of the swimming dolphin. The videotapes were later analyzed frame-by-frame to determine use of the dolphin's left or right eye when viewing the stimulus. It would seem possible to modify this method to present auditory stimuli and also to record the dolphin's vocalizations at the same time as testing lateralization, thus gaining more information about dolphin's interpretation of the test stimulus.

2.5 Eye Used Before Performing a Specific Behavior

Another way of determining eye preference is to record the eye used just before performing a particular response. This is particularly useful in social interactions. For example, social pecking responses of young chicks directed towards other members of their species were videotaped from overhead and scored by playing back the sequences in slow motion and frame-by-frame [34]. First the video was stopped when one chick pecked another, usually aimed at the eye or near the eye of the recipient. From that frame the video recording was moved backwards frame-by-frame and the eye used immediately before the peck was lodged could be determined. Using this method, Vallortigara et al. [34] found a preference for chicks to view their social companion/opponent with the left eye before lodging a peck.

A similar method was used by Deckel et al. [35] in testing eye preferences of *Anolis* lizards before they display aggressively to an opponent. A left eye preference was found. Eye preference to look at an opponent before attacking has also been measured in horses although for this species video recording was made at the normal lateral angle. Even using videos recorded from this angle it was possible to determine the eye used to look at the opponent before attacking and in this species also there was a significant preference to use the left eye [4]. Given the laterally positioned eyes of the horse, and the small binocular field [3], it is not difficult to score eye use from video recordings or even by direct observation.

A left side preference has also been found in agonistic interactions between gelada baboons [36] and in toads [37]. Eye preference can be accurately deduced from side biases although the size of the monocular field should be taken into consideration.

2.6 Presentation of Stimuli Simultaneously in Both Lateral Visual Fields

Another way of measuring lateralization of response to visual stimuli is to present the same visual stimulus on each side of the animal in the lateral monocular visual fields. On each presentation the animal's response to the stimulus on the left or right side is recorded. Hence, after a series of presentations a laterality index can be calculated for each animal. If habituation occurs after the first presentation, it is possible to determine only a bias in the population (i.e., not the direction or strength of laterality of each individual) and this requires testing a large sample size. Although this might be a valuable measure, it would be preferable to use stimuli to which there is no such rapid habituation.

This procedure has been used to test toads [38] by presenting worms attached to the end of the arms of a Y-shaped frame made of high-tensile copper wire. The experimenter held the Y-frame by its base positioned behind the animal so that the arms of the Y-frame projected into each of the lateral, monocular visual fields. On to the tips of the arms, model insects had been adhered securely. The arms of the Y-shaped frame were 12 cm in length and a wire brace between the arms, positioned close to the handle, ensured that the stimuli moved (vibrated) in synchrony. Vibration is desirable because it attracts the toad's attention by mimicking the movement of insects. However, this could be mechanized to ensure exactly the same presentation on each trial. The stimuli were presented to the toad in the horizontal plane at the level of the eyes while the toad was attending to an image of a "worm" moving up and down on a computer screen. A trial commenced as soon as the toad had approached the screen and was fixating the moving image, a behavior that the test animals, *Bufo marinus*, perform readily. Extra care was taken to position the model insects equidistant from the eyes. The Y-fork was rotated through 180° between each trial to control for any possible asymmetry in its construction. Repeated trials were given and the stimulus was changed once during the testing sequence [38]: if desirable, the number of different stimuli presented could be increased. The finding of Robins and Rogers [38] was a significant preference for toads to strike at novel models resembling prey to the left side. This was interpreted as a right hemispheric specialization to respond to novel stimuli, as found also in chicks [39, 40] and lizards [41].

A similar presentation of the same stimulus on both sides of the animal has been used for measuring lateralization in gull chicks [42, 43]. These researchers measured begging responses of newly hatched gull chicks to two model gull heads presented on the walls of a box on the left and right sides of the chick. These model heads were oriented with the beaks pointing towards the chick and they were oscillated in a plane parallel to the cage walls and simultaneously and identically by means of a single lever. For each trial, pecks by the chick at the model on the left or right were scored for 1 min. Each chick received two or three such trials. Only in the second trial did males show a significant bias to respond preferentially to the stimulus on their left side.

Siniscalchi et al. [44] tested lateralization of visual responding in dogs. They presented identical visual stimuli on the left and right sides of the dog as it fed from a bowl placed midway between stimuli projected onto screens at each end of an elongated box and at 200 cm to either side of the food bowl. The dog stood in an entry box with its head projecting into the test box. The inside walls of the test box were black. The stimulus to which the dog turned its head was recorded. Significant bias to turn to the stimulus on the left side was found when images of a cat or of a snake were presented.

Fish also have been presented with stimuli on their left and right sides as they emerge from a start-tank. Using this method to test zebra fish fry, Andrew et al. [45] showed that the fish were more likely to approach an image resembling a fish looking at them on their right side and avoid it on their left side. They also showed stimuli that resembled a potential region into which the fry could escape and found lateralization of response to this stimulus: the fry approached the stimulus when it was presented on their left side.

Another method using the presentation of different visual stimuli to the left and right visual fields involves measuring the recognition of chimeras of faces. This has been a method of testing face recognition in humans: recognition of images of faces made from two halves of different faces (i.e., chimera) is faster for the half-face in the left visual field. This result is consistent with the right hemisphere's specialization for face recognition. Similar tests have been used to test face recognition in primates and sheep. For example, Peirce et al. [46] trained sheep in a Y-maze and presented at the decision point two faces of familiar sheep or two faces of unfamiliar sheep. Once a sheep had been trained, it was tested with half faces or chimera. Using this method the researchers showed that, as in humans, a left visual hemifield advantage in face recognition.

Another rather similar method of testing is to present stimuli with different features (or components) visible to the left versus right visual field. This method was developed by Siniscalchi et al. [47] to test dogs' perception of lateralized tail wagging by presenting life-sized images or silhouettes of dogs projected on to a large screen. The image appeared to face the dog being tested and the tail could be seen wagging on the left or right side. In other words, the stimulus was asymmetrical in a way that might activate the left or right hemisphere differentially. Heart rate of the dog being tested was recorded. Observing the image with the tail wagging on the left side elevated heart rate and anxious behavior significantly more than did images with right-side tail wagging. In fact, the latter lowered these two measures compared to the effect of viewing a control image with a stationary tail. The results suggest that left side wagging seen by the left eye and right hemisphere elicits an aroused response, whereas right side wagging, which activated the right eye and left hemisphere, has a calming effect.

2.7 Mirror Tests of Response to the Animal's Own Image

Laterality of social behavior in fish has been demonstrated using an apparatus with mirrors as walls and measuring the side on which the fish prefers to be near its own reflection. Originally used by Sovrano et al. [48] the fish tank had two parallel mirror walls. Various modifications were made to the apparatus, one of which was a quasi-circular swim-way [49]. Several species of fish were tested, each displaying a preference to view its image, presumably seen as a conspecific, with the left eye, and therefore showing a preference to process visual information about a "social" companion using the right hemisphere.

The mirror test has also been used to score lateralization of aggressive behavior in fish [50] and in this case a right eye preference was found in the three species tested. Only one mirror was used and the test subject attacked its own image. From video recordings the researchers scored the eye used to fixate the image just before lodging an attack. A right eye preference was also found in mirror attacking by Siamese fighting fish [51].

It seems that for fish at least, and likely a number other species, response to the animal's own image is a useful way of testing lateralization of social behavior although it is always preferable to incorporate tests of interaction with a real conspecific to be sure of the function of the social interaction scored.

3 Visual Pathways

So far I have discussed laterality in terms of the eye showing an advantage in controlling a particular behavior or in responding to a particular stimulus. In birds with small binocular and large monocular fields (i.e., most avian species, but not owls) it is assumed that greater responsiveness to a stimulus viewed by one eye compared to the other means that the information is processed by the hemisphere contralateral to the preferred eye and that this hemisphere, therefore, controls the response. The same is true of studies of visual laterality in other species with large monocular visual fields (horses, rodents, marsupial dunnarts).

This interpretation of laterality revealed by eye preference is, largely speaking, correct since most of the visual input from one eye is processed by the contralateral thalamus and forebrain hemisphere and most species lack a large commissure that connects the left and right hemispheres, as does the corpus callosum in humans. Nevertheless, it must be remembered that, in animals with large monocular visual fields, some visual inputs are processed by the ipsilateral side of the brain.

Let us consider the visual pathways of chicks as an example. Inputs from one eye go to the contralateral optic tectum and the contralateral side of the thalamus (Fig. 6) and from these two relay stations they are projected to the forebrain in two visual pathways, the tectofugal and thalamofugal pathways respectively. As far as the tectofugal pathway is concerned, inputs from an eye are processed in the contralateral hemisphere since each optic tectum projects to the visual region of the forebrain on its same side (i.e. without crossing the midline). The thalamofugal pathway does not fit this pattern: although most of the projections from the thalamic region of the midbrain project from the thalamus to the ipsilateral forebrain (i.e., contralateral to the seeing eye), a substantial minority of projections goes to the contralateral side of the forebrain. To put it simply, the traditional assumption that inputs from each eye are processed by the opposite hemisphere and thus reveal the functions of that side of the brain alone is correct as far as the tectofugal system is concerned but not entirely correct for visual inputs processed by the thalamofugal system.

In birds the tectofugal visual system has been considered as the primary pathway, whereas the thalamofugal system is secondary, but that subdivision might merely reflect what visual processes have so far been studied in terms of visual perception. Studies of laterality in birds have been interpreted from this perspective and that has provided a useful foundation on which to work. Nevertheless, we should keep in mind that left eye to right hemisphere and right eye to left hemisphere is a simplification that may be important to recognize in future research on visual laterality.

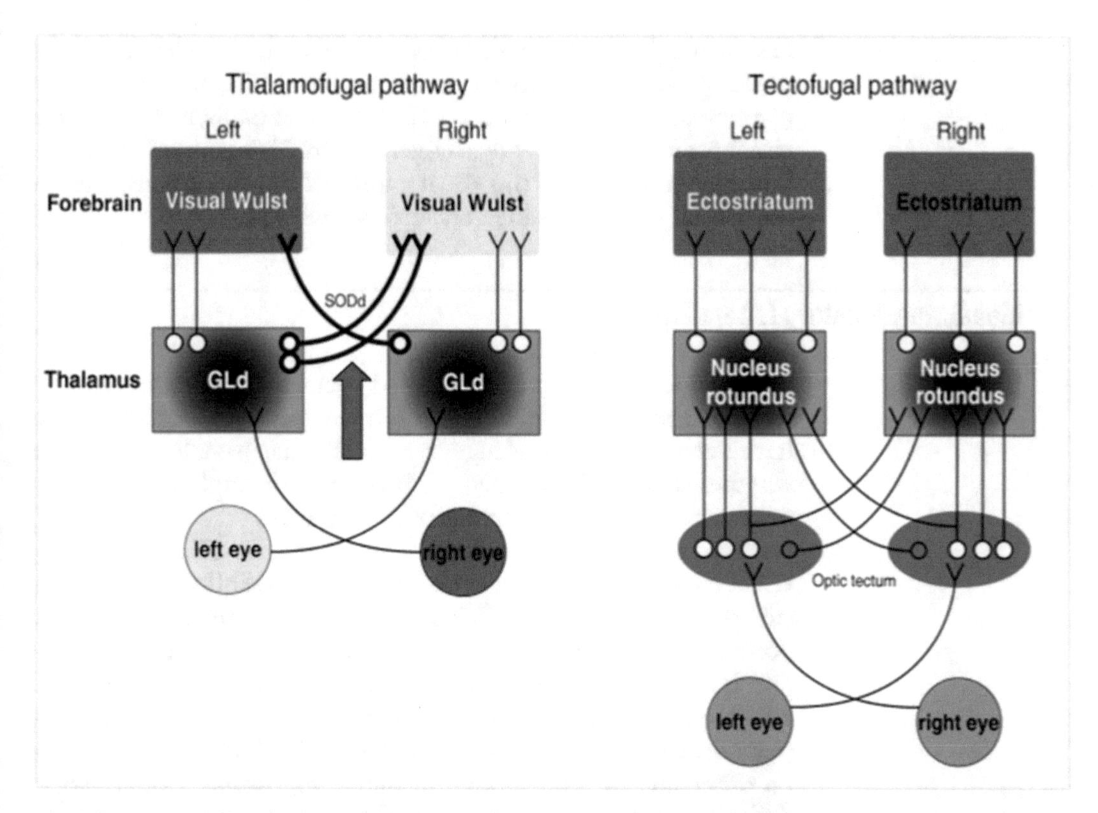

Fig. 6 Representation of the visual pathways in chicks. The thalamofugal pathway projects via the thalamus to the visual Wulst, also called the hyperpallium, the forebrain. The region of the thalamus receiving the inputs from the eye is the nucleus genticulatis lateralis pars dorsalis, GLd. There are more projections from the left GLd to the right hyperpallium than from the right GLd to the left hyperpallium. The arrow indicates the site of the asymmetry. SODd refers to the dorsal supra-optic decussation. The tectofugal visual connections are illustrated on the right. In pigeons these are asymmetrical: in this species there are more projections from the right optic tectum to the left nucleus rotundus than vice versa

Furthermore, in chicks, asymmetry is present in the visual projections from thalamus to forebrain. There are more contralateral projections from the left side of the thalamus, which receives inputs from the right eye, to the right side of the forebrain than from the right side of the thalamus (inputs from the left eye) to the left side of the forebrain [52–55]. We know this asymmetry correlates with the superiority of the right eye in the chick's performance of visual discrimination learning (grain versus pebbles) since chicks incubated in darkness in the last few days before hatching lack both asymmetry in these projections of the thalamofugal visual system [55, 56] and show no difference between performance of tasks with the left and right eye [1, 9, 57].

Moreover, species may differ in the precise organization of their visual pathways. Studies of the visual anatomy of pigeons have shown that, although they follow the general pattern, as described above, in this species it is the tectofugal visual pathway that has asymmetry in the projections to the rotundal nuclei, and these nuclei project to the forebrain hemispheres [58, 59]. Hence, in pigeons behavioral lateralization is dependent on asymmetry of the visual projections in the tectofugal system. This species difference in the organization of the visual pathways almost certainly underlies species differences in lateralized behavior [1, 60].

4 Measuring Auditory Lateralization

Fewer experiments have investigated lateralized responses to auditory than to visual stimuli in nonhuman species. Moreover, there have been, to my knowledge, no studies that have looked to see whether structural asymmetry is present in the auditory pathways or auditory processing regions of the forebrain.

Probably the main reason for fewer studies of auditory than visual lateralization is greater difficulty in presenting an auditory stimulus to just one ear. It is much more difficult to completely occlude an ear than it is an eye, and consequently lateral presentation is less simply interpreted. In humans and trained primates it is possible to use earphones to present sounds to one or both ears (e.g., dichotic listening experiments [61, 62]) but using earphones on other species is far more difficult or impossible. Despite these difficulties, some testing of auditory lateralization has been achieved without resorting to placing lesions in the regions of the brain used in auditory processing. The brain-lesion method of revealing lateralization of auditory processing was used particularly in the 1970s and 1980s for testing primates (summarized in Chap. 6 of Ref. 63), and also other species, but this approach is ethically far less defensible than the methods discussed below. In primates, there have also been imaging studies of cortical activity in response to sounds, such as that of Poremba and Mishkin [64], but these require restraint of the test animal, which limits interpretation in a functional sense.

4.1 Occlusion of an Ear

Tests of lateralized auditory responses in mice were conducted by Ehret [65], who measured the responses of adult females to ultrasonic vocalizations made by their pups after they had been isolated, and hence were distressed. Hearing in one or the other ear was markedly reduced, if not entirely prevented, by putting a plug in the ear. The mice were tested in a runway with speakers at either end, one of which was used to play recordings of the distress calls of the pups and the other to play neutral sounds. The choice of which speaker was approached was recorded. Maternal mice were first tested with no ears plugged, then with one ear plugged and again with the other ear plugged. With the right ear plugged the mice were unable to discriminate between the two sounds, whereas with the left ear plugged, or when tested without ear-plugs, they could discriminate between the sounds and approached the speaker emitting the distress calls. This asymmetry has been interpreted as a right ear and left hemisphere specialization for processing species-specific communication, and indeed that is the parsimonious explanation of the results: since auditory inputs are transmitted faster to the side of the brain contralateral the ear receiving the sound, it is probable that they are processed first by that side. It is, however, desirable to know more about the organization of the auditory pathways in the species being tested.

4.2 Head Turning to Attend to a Stimulus

Presenting sounds from speakers placed on the left and right sides of an animal is a method of testing lateralized attention to auditory stimuli in unrestrained and untreated animals. In order to do this, the test animal's head must be positioned midway between the two speakers. One way of achieving this is to place a bowl with the animal's preferred food midway between the speakers and to ensure the correct orientation of the animal, for example, by placing small transparent barriers on either side of the food bowl. The sounds are played simultaneously from the speakers while the animal has its head in position at the bowl. This method was used by Siniscalchi et al. [66] to test the response of dogs to a range of auditory stimuli. The dog being tested was not leashed and feeding was a free choice. Once the dog had taken up the correct position and was feeding, the same stimulus was played simultaneously through both speakers. Ratcliffe and Reby [67] used a similar method to test dogs but, instead of positioning the dog using a food bowl, they held it on a lead. In order to prevent (or minimize) side bias due to cues other than the auditory stimuli, the person holding the lead stood directly behind the dog.

In this method of testing, it is essential to balance the speaker outputs before commencing the experiment and to control for any slight biases by alternating the sounds played between the speakers or switching the speakers used in random sequence across multiple presentations. It is also advisable from time to time to reverse the sides on which the speakers themselves are placed in order to control for any possible biases in sound output. Siniscalchi et al. [66] played the sounds using a digital portable player and two speakers connected to an amplifier. The speakers were placed 2.5 m from the dog on each side.

Habituation of the animal to the sounds must be taken into account and it may be necessary to leave quite long intervals between tests (see Ref. 66). Exactly what the interval should be depends on the species/individuals being tested and the nature of the stimulus. Siniscalchi et al. [66] presented to the dogs various species-specific vocalizations and the sounds of a thunderstorm. Each sound was presented for 3 s. at a volume of 60–80 dB, repeated at 1-min intervals provided that the dog remained at the food bowl but frequently the dog moved away and testing could not resume until feeding recommenced. Testing time on a given day did not exceed 1 h and a break of a week occurred between testing sessions. From video recordings made using a camera placed 6 m from the bowl and facing the dog, the dog's head orienting response to the left or right speaker was scored, provided it occurred within 5 s of the playback. This allowed calculation of a Laterality Index of response to each sound, calculated using the formula (L − R/L + R), where L and R signify the number head orienting responses to the left and right speakers, respectively. The results showed significant preference to turn to the speaker on the right side, indicating use of the left hemisphere, in response to hearing species-specific vocalizations. This contrasted with a significant preference to turn to the speaker on the left side, and thus use the right hemisphere, in response to hearing the sounds of the thunderstorm.

Other behavioral scores were measured also, including time to resume feeding, vocalizations, panting, salivating, running away and hiding, defecating, urinating, and freezing. Using these scores, reactivity could be calculated and associated with the Laterality Index. Significant correlations showing increasing reactivity with increasing use of the left ear/right hemisphere were found for each stimulus. This finding is consistent with the known role of the right hemisphere in processing inputs in fear-inducing or high-arousal contexts.

The value of this test for auditory lateralization is the controlled input to each ear. It would be possible to modify this method to present, simultaneously, different auditory stimuli on the right or left sides of the animal and thus apply tests similar to the dichotic listening tasks used to test humans.

4.3 Orientation to a Single Auditory Stimulus

Orientation towards a speaker playing particular vocalizations has been measured in chicks as a way of determining ear bias and thereby deducing which hemisphere attends to the auditory stimulus [68]. Young chicks were tested in a circular arena (228 cm diameter with walls 20 cm high). Every 20 cm of the walls had 2 cm openings that were made inconspicuous to the chick by covering them with cloth and calls were played via a speaker behind one of these openings. The whole apparatus was illuminated dimly with 25 W red light and the chick's behavior was recorded using an infrared-sensitive video camera positioned above the arena. Head position relative to a line from the sound source to the chick's head was measured for the 15 s before the chick moved off towards the speaker. The measurement taken was duration in positions 10–20° from the beak, excluding positions less than 10° to the left or right of the beak, under the assumption that these positions would involve both ears to the same extent. These durations could be used to determine lateralization.

While this method of testing has some advantages for testing small animals in a setting in which they can move freely, it is less adaptable to testing larger animals than is the method of two-speaker playback described above.

5 Future Directions of this Research

Use of the methods described in this chapter is likely to continue in order to investigate laterality in more species and more functions within a particular species. In addition to these studies, it is now time to embark on research examining the integration of different sensory modalities and the potential role that lateralization may play in this. Monocular and monaural tests that have revealed laterality when applied separately could now be applied concurrently in different combinations of eye and ear use. Nostril occlusion (see Chap. 4) could also be integrated with monocular tests, and so on. The possibilities are many and various. In my opinion, they hold considerable promise for enhancing knowledge of sensory processing in different species, including invertebrates [69].

Since arousal state can influence strength, and even direction, of lateralization, of some specific functions, it is important to assess more than one behavior in any investigation of lateralization. In the dogs tested for auditory lateralization by Siniscalchi et al. [66] a number of different behavioral responses were measured and this permitted calculation of a Reactivity Index, which was interesting because this index correlated with the Laterality Index: the more reactive the dog the stronger was the bias to turn the head to the left and so attend to the heard vocalization with the right hemisphere. Hence, it is recommended that more than one type of behavior is monitored when assessing lateralization in any particular task [44] and to remember that any lateralization measure is not necessarily a fixed aspect of an individual or a species.

References

1. Rogers LJ, Vallortigara G, Andrew RJ (2013) Divided brains: the biology and behavior of brain asymmetries. Cambridge University Press, Cambridge
2. Rogers LJ, Vallortigara G (2015) When and why did brains break symmetry? Symmetry 7:2181–2194. doi:10.3390/sym7042181
3. Austin NP, Rogers LJ (2007) Asymmetry of flight and escape turning responses in horses. Laterality 12:464–474
4. Austin NA, Rogers LJ (2012) Limb preferences and lateralization of aggression, reactivity and vigilance in feral horses (*Equus caballus*). Anim Behav 83:239–247
5. Ocklenburg S, Güntürkün O (2012) Hemispheric asymmetries: the comparative view. Front Psychol 3:1–9. doi:10.3389/fpsyg.2012.00005, Article 5
6. Rogers LJ, Anson JM (1979) Lateralisation of function in the chicken fore-brain. Pharm Biochem Behav 10:679–686
7. Clayton NS (1993) Lateralization and unilateral transfer of spatial memory in marsh tits. J Comp Physiol A 171:799–806

8. Mench J, Andrew RJ (1986) Lateralisation of a food search task in the domestic chick. Behav Neural Biol 46:107–114
9. Rogers LJ (1997) Early experiential effects on laterality: research on chicks has relevance to other species. Laterality 2:199–219
10. Güntürkün O, Diekamp B, Manns M, Nottelmann F, Prior H, Schwarz A, Skiba M (2000) Asymmety pays: visual lateralization improves discrimination success in pigeons. Curr Biol 10:1079–1081
11. Rogers LJ, Zappia JV, Bullock SP (1985) Testosterone and eye-brain asymmetry for copulation in chickens. Experientia 41:1447–1449
12. Rogers LJ (1982) Light experience and asymmetry of brain function in chickens. Nature 297:223–225
13. Zappia JV, Rogers LJ (1983) Light experience during development affects asymmetry of forebrain function in chickens. Dev Brain Res 11:93–106
14. Cowell PE, Waters NS, Denenberg VH (1997) Effects of early environment on the development of functional laterality in Morris maze performance. Laterality 2:221–232
15. Bonati B, Csermely D, Sovrano VA (2013) Advantages in exploring a new environment with the left eye in lizards. Behav Proc 97:80–83
16. Wiltschko W, Traudt J, Güntükün O, Prior H, Wiltschko R (2002) Lateralization of magnetic compass orientation in a migratory bird. Nature 419:467–470
17. Lippolis G, Bisazza A, Rogers LJ, Vallortigara G (2002) Lateralization of predator avoidance responses in three species of toads. Laterality 7:163–183
18. Lippolis G, Westman W, McAllan BM, Rogers LJ (2005) Lateralization of escape responses in the striped-faced dunnart, *Sminthopsis macroura* (Dasyuridae: Marsupalia). Laterality 10:457–470
19. Martin GR (2007) Visual fields and their functions in birds. J Ornithol 148:S547–S562
20. Martin GR (2009) What is binocular vision for? A birds' eye view. J Vis 9(11):14, 1–19
21. Ventolini N, Ferrero EA, Sponza S, della Chiesa A, Zucca P, Vallortigara G (2005) Laterality in the wild: preferential hemifield use during predatory and sexual behavior in the black-winged stilt. Anim Behav 69:1077–1084
22. Martin GR, Katzir G (1994) Visual fields and eye movements in herons (Ardeidae). Brain Behav Evol 44:74–85
23. Austin NA, Rogers LJ (2014) Lateralization of agonistic and vigilance responses in Przewalski horses (*Equus przewalskii*). Appl Anim Behav Sci 151:43–50
24. Ewert J-P (1970) Neural mechanisms of prey-catching and avoidance behavior in the toad (*Bufo bufo* L.). Brain Behav Evol 3:36–56
25. Wachowitz S, Ewert J-P (1996) A key by which the toad's visual system gets access to the domain of prey. Physiol Behav 60:877–887
26. Robins A, Rogers LJ (2004) Lateralised prey catching responses in the toad (*Bufo marinus*): analysis of complex visual stimuli. Anim Behav 68:567–575
27. Burghagen H, Ewert J-P (1982) Question of "head preference" in response to worm-like dummies during prey-capture of toads, *Bufo bufo*. Behav Processes 7:295–306
28. Ewert J-P, Dinges AW, Finkenstädt T (1994) Species-universal stimulus responses, modified through conditioning, reappear after telencephalic lesions in toads. Naturwissenschaten 81:317–320
29. McKenzie R, Andrew RJ, Jones RB (1998) Lateralisation in chicks and hens: new evidence for control of response by the right eye system. Neuropsychologia 36:51–58
30. Diekamp B, Regolin L, Güntürkün O, Vallortigara G (2005) A left-sided visuospatial bias in birds. Curr Biol 15(10):R372–R373
31. Rugani R, Kelly DM, Szelest I, Regolin L, Vallortigara G (2010) Is it only humans that count from left to right? Biol Lett 6:290–292
32. Rugani R, Vallortigara G, Vallini B, Regolin L (2011) Asymmetrical number-space mapping in the avian brain. Neurobiol Learn Mem 95:231–238
33. Siniscalchi M, Dimatteo S, Pepe AM, Sasso R, Quaranta A (2012) Visual lateralization in wild striped dolphins (*Stenella coeruleoalba*) in response to stimuli with different degrees of familiarity. PLoS One 7(1):e30001. doi:10.1371/journal.pone.0030001
34. Vallortigara G, Cozzutti C, Tommasi L, Rogers LJ (2001) How birds use their eyes: opposite left-right specialisation for the lateral and frontal visual hemifield in the domestic chick. Curr Biol 11:29–33
35. Deckel AW (1995) Laterality of aggressive responses in *Anolis*. J Exp Zool 272:194–200
36. Casperd JM, Dunbar RIM (1996) Asymmetries in the visual processing of emotional cues during agonistic interactions in gelada baboons. Behav Proc 37:57–65
37. Robins A, Lippolis G, Bisazza A, Vallortigara G, Rogers LJ (1998) Lateralization of agonis-

tic responses and hind-limb use in toads. Anim Behav 56:875–881
38. Robins A, Rogers LJ (2006) Complementary and lateralized forms of processing in *Bufo marinus* for novel and familiar prey. Neurobiol Learn Mem 86:214–227
39. Andrew RJ (1991) The nature of behavioural lateralization in the chick. In: Andrew RJ (ed) Neural and behavioural plasticity: the use of the domestic chick as a model. Oxford University Press, Oxford
40. Vallortigara G, Andrew RJ (1991) Lateralization of response by chicks to a change in a model partner. Anim Behav 41:187–194
41. Robins A, Chen P, Beazley LD, Dunlop SA (2005) Lateralized predatory responses in the ornate dragon lizard (*Ctenophorus ornatus*). Neuroreport 16:849–852
42. Romano M, Parolini M, Caprioli M, Spiezio C, Rubolini D, Saino N (2015) Individual and population-level sex-dependent lateralization in yellow-legged gull (*Larus michahellis*) chicks. Behav Processes 115:109–116
43. Possenti CD, Romano A, Caprioli M, Rubolini D, Spiezo C, Saino N, Parolini M (2016) Yolk testosterone affects growth and promotes individual-level consistency in behavioral lateralization of yellow-legged gull chicks. Horm Behav 80:58–67
44. Siniscalchi M, Sasso R, Pepe AM, Vallortigara G, Quaranta A (2010) Dogs turn left to emotional stimuli. Behav Brain Res 208:516–521
45. Andrew RJ, Osorio D, Budaev S (2009) Light during embryonic development modulates patterns of lateralization strongly and similarly in both zebrafish and chick. Philos Trans R Soc Lond B 362:983–989
46. Peirce JW, Leigh AE, Kendrick KM (2000) Configurational coding, familiarity and the right hemisphere advantage for face recognition in sheep. Neuropsychologia 38:475–483
47. Siniscalchi M, Lusito R, Vallortigara G, Quaranta A (2013) Seeing left- or right-asymmetric tail wagging produces different emotional responses in dogs. Curr Biol 23:2279–2282
48. Sovrano VA, Rainoldi C, Bisazza A, Vallortigara G (1999) Roots of brain specializations: preferential left-eye use during mirror-image inspection in six species of teleost fish. Behav Brain Res 106:175–180
49. Sovrano VA, Bisazza A, Vallortigara G (2001) Lateralization of response to social stimuli in fishes: a comparison between different methods and species. Physiol Behav 74:237–244
50. Bisazza A, de Santi A (2003) Lateralization of aggression in fish. Behav Brain Res 141:131–136
51. Forsatkar MN, Dadda M, Nematollahi MA (2015) Lateralization of aggression during reproduction in male Siamese fighting fish. Ethology 121:1039–1047
52. Rogers LJ, Sink HS (1988) Transient asymmetry in the projections of the rostral thalamus to the visual hyperstriatum of the chicken, and reversal of its direction by light exposure. Exp Brain Res 70:378–384
53. Adret P, Rogers LJ (1989) Sex difference in the visual projections of young chicks: a quantitative study of the thalamofugal pathway. Brain Res 478:59–73
54. Rajendra S, Rogers LJ (1993) Asymmetry is present in the thalamofugal projections of female chicks. Exp Brain Res 92:542–544
55. Rogers LJ, Deng C (1999) Light experience and lateralization of the two visual pathways in the chick. Behav Brain Res 98:277–287
56. Rogers LJ, Bolden SW (1991) Light-dependent development and asymmetry of visual projections. Neurosci Lett 121:63–67
57. Deng C, Rogers LJ (1997) Differential contributions of the two visual pathways to functional lateralization in chicks. Behav Brain Res 87:173–182
58. Güntürkün O (1993) The ontogeny of visual lateralization in pigeons. German J Psychol 17:276–287
59. Güntürkün O (2002) Ontogeny of visual asymmetry in pigeons. In: Rogers LJ, Andrew RJ (eds) Comparative vertebrate lateralization. Cambridge University Press, Cambridge, pp 247–273
60. Deng C, Rogers LJ (2002) Factors affecting the development of lateralization in chicks. In: Rogers LJ, Andrew RJ (eds) Comparative vertebrate lateralization. Cambridge University Press, NY, pp 206–246
61. Hugdahl K (2003) Dichotic listening in the study of auditory laterality. In: Hugdahl K, Davidson RJ (eds) The asymmetrical brain. The MIT Press, Cambridge, pp 441–475
62. O'Leary DS (2003) Effects of attention on hemispheric asymmetry. In: Hugdahl K, Davidson RJ (eds) The asymmetrical brain. The MIT Press, Cambridge, pp 476–508
63. Bradshaw JL, Rogers LJ (1993) The evolution of lateral asymmetries, language, tool use, and intellect. Academic Press, San Diego
64. Poremba A, Mishkin M (2007) Exploring the extent and function of higher-order auditory cortex in rhesus monkeys. Hear Res 229:14–23

65. Ehret G (1987) Left hemisphere advantage in the mouse brain for recognizing ultrasonic communication calls. Nature 325:249–251
66. Siniscalchi M, Quaranta A, Rogers LJ (2008) Hemispheric specialization in dogs for processing different acoustic stimuli. PLoS One 3:e3349
67. Ratcliffe VF, Reby D (2014) Orienting asymmetries in dogs' responses to different communicatory components of human speech. Curr Biol 24:1–5
68. Miklósi A, Andrew RJ, Dharmaretnam M (1996) Auditory lateralization: shifts in ear use during attachment in domestic chicks. Laterality 1(3):215–224
69. Letzkus P, Boeddeker N, Wood JT, Zhang SW, Srinivasan MV (2006) Lateralization of visual learning in the honeybee. Biol Lett 4:16–18

Chapter 4

Olfactory Lateralization

Marcello Siniscalchi

Abstract

This chapter focuses on the methods by which behavior can be used to study olfactory lateralization in different animal models. Starting from the description of different testing paradigms, each paragraph reports and discusses the evolution and role of lateralized brain functions related to olfaction. The main advantages and disadvantages of behavioral techniques used to study olfactory lateralization are also discussed, as well as any behavioral and physiological validation. Furthermore, suggestions for improvements to the current methods are given in the last section, with a focus on dog species.

Key words Olfactory lateralization, Behavioral methods, Monorhinal stimulation, Sniffing behavior, Odor categorization

1 Introduction

Olfaction, as in the other sensory domains, allows external stimuli to be analyzed mainly by one brain hemisphere because only a few contralateral connections between both sides of the olfactory neural pathways via the anterior commissure have been reported [1]. The main feature that distinguishes this sense from the others (e.g. vision and audition) is that olfactory neurons project their input from each nostril to the olfactory cortex on the ipsilateral side (i.e. olfactory input to the brain is uncrossed).

Behavioral methods represent valuable and promising tools to study lateralization of olfaction since these techniques are quite easy to use and, at the same time, are minimally invasive or completely noninvasive in different animals models. In addition, in different studies, animals can be tested in completely unrestrained conditions ensuring that behavioral asymmetries (e.g. nostril preference), which reflect asymmetries of brain functions, are observed in conditions resembling as much as possible the natural ones. This means they have direct implications for both animal cognition and welfare.

Lesley J. Rogers and Giorgio Vallortigara (eds.), *Lateralized Brain Functions: Methods in Human and Non-Human Species*, Neuromethods, vol. 122, DOI 10.1007/978-1-4939-6725-4_4, © Springer Science+Business Media LLC 2017

2 Monorhinal Stimulation and Methods of Using Visual Analogue Scales to Study Olfactory Lateralization in Humans

The first behavioral studies investigating olfactory lateralization were performed in humans to explore if hedonic olfactory perception (i.e. odor pleasantness or unpleasantness) depended on whether the olfactory stimulus was presented to the left or to the right nostril [2]. Results revealed that, when subjects used the right nostril (right hemisphere analysis) the hedonic evaluation of odors (both pleasantness and unpleasantness) was higher than when they used the left nostril (left hemisphere analysis). Very similar results were subsequently reported by Herz et al. [3] and then by Dijksterhuis et al. [4]. Interestingly, in the work of Herz et al. [3], a left nostril superiority in odor naming, a more linguistic than olfactory sensory task, was observed supporting the left hemispheric specialization for language processing [5]. Although olfactory detection does not appear to be lateralized, perceptual processing of odors seems to be mainly under the control of the right hemisphere since a right nostril advantage has also been reported in a discrimination task [6]. A more recent study found that this superiority of the right side of the brain in odor discrimination was valid only for unfamiliar odors, whereas the discrimination of familiar odorants was symmetrical with respect to left-right nostril use [7]. In such studies, odorants to be discriminated or rated in terms of familiarity, etc., were usually presented in glass bottles with cotton wands in controlled, suprathreshold concentrations. Time intervals between items in pairs of odors to test (about 20 s) and inter-trial intervals (usually 1 min) were present in order to avoid the possibility that one odor was adversely modified by sensory adaptation to the previous one [3, 7]. During presentation, odor samples were placed under the subject's nostril, alternating the side and odor presentation order was balanced between nostrils. Monorhinal stimulation was performed by asking the subject to hold one nostril closed with her/his finger and to sniff only through the other open nostril. Sniffing behavior was also controlled: usually only two sniffs per item or 2 s of sniffing per item were allowed [3, 7]. Odor scoring (familiarity, hedonic rates, intensity, etc.) was then calculated using visual analogue scales (bipolar VAS) [8].

The main advantages of behavioral techniques used to study olfactory lateralization in humans came from their design simplicity, ease of testing, and economy of time. In addition, the testing apparatuses are definitely not expensive, especially compared to other experimental techniques (e.g. genetic and functional neuroimaging studies). Although a bipolar visual analogue scales score is a valid and commonly used tool for evaluating olfactory perceptual estimates [9], the measure scored using this method comes from the subject's own perspective and, as a consequence, it is highly subjective [10]. Another limitation of this method is that VAS score can be used to evaluate perceptual response only in adult humans.

Approach-withdrawal behavior in response to odors differing in affective valence have been analyzed in order to examine the relationship between emotional development and laterality in human infants [11]. Positive (vanilla and almond) and negative (rotten eggs and sour milk) smells were presented to each infant while lying supine at the mother's bedside. Head orientation (towards and away) in response to odors was video recorded. Odorants were placed on cotton balls located in sterile surgical syringes with thin tubing attached in place of the needle. The tubing ended with a thin piece of plastic that was used by the experimenter to direct the odor nebulizer under either the left or the right nostril of the infant (see Fig. 1a). An unscented air stream was used as the control and, as a common paradigm in these types of experiments, odorant presentation was counterbalanced. Subjects were tested when the infant was asleep and just prior the presentation of the odors the baby's head and body were gently shifted to a middle position by the experimenter (starting position). During each trial, left-right head turns (with respect to the starting position) were coded from video recordings by independent raters. Results revealed that presentation of positive odors to the left nostril (left hemisphere analysis) was more likely to elicit approach

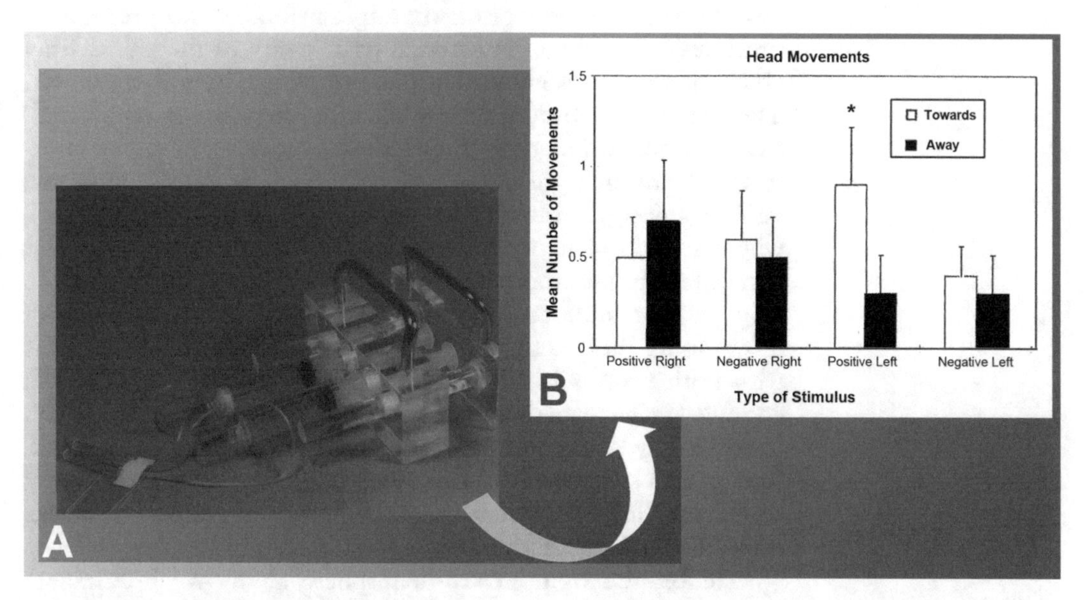

Fig. 1 (**a**) Experimental apparatus used to present odors to human infants. (**b**) Approach and withdrawal behavioral responses of infants after nebulizations of positive and negative odors into the right and left nostril (mean number of towards and away head movements for each type of stimulus with S.E.M. are showed). (Reproduced from Olko and Turkewitz [11] with permission from Taylor & Francis Ltd., www.tandfonline.com)

behavior towards the odor than was presentation of the same odor to the right nostril (see Fig. 1b), supporting the Davidson theory of the left hemisphere specialization for approach behavior [12]. Such results showed that lateralization of emotional processing is present even during early stages of human life suggesting, in an evolutionary perspective, that approach behavior may develop earlier than withdrawal behavior.

3 Asymmetric Nostril Use During Navigation: Behavioral Methods Used to Study Lateralized Olfactory Pathways in Birds and Fishes

The earliest work on olfactory lateralization in birds investigated behavioral response to an olfactory change in an imprinting object [13]. These experiments revealed an advantage of the right nostril in discriminating between familiar and unfamiliar olfactory stimuli. Chicks were reared from the first day of life in cages with a red plastic cylinder with four symmetrical holes through which scent could diffuse from a cotton wool ball inside the cylinder (imprinting object). In one experimental condition, the cylinder used as imprinting object was odorless and on the day of the test chicks were placed in the center of a runway at each end of which was located a cylinder visually similar to the rearing one but with an odor of n-amyl-acetate or orange oil (unfamiliar odors) or one that to which no odor had been added (familiar odor) (see Fig. 2a). About 30 min before test either the right or the left nostril of some chicks was temporarily blocked using a malleable wax preparation previously shaped to the external nasal cavity of the chick. Brief ether anesthesia was induced just before the application of the wax. The wax was removed a little while after the test was completed. Results revealed that only those subjects receiving olfactory information from the right nostril (right hemisphere in use) chose the familiar object (i.e. the object resembling in odor the rearing one), whereas when olfactory input was sniffed by the left nostril (left hemisphere in use) there was no effect of odor on the choice made (Fig. 2b). An additional experiment showed that the behavior of chicks using their right nostril was very similar to that of subjects with both nostrils in use (unplugged nostrils) (see Fig. 2b). In another similar experiment, half of the chicks were reared with unscented cylinders and half with a cylinder smelling of clove oil. Also in this experiment chicks using their right nostril chose the object smelling like that which they had been reared but chicks using their left nostril did not.

Hence, the right nostril-hemisphere ability to discriminate between familiar and unfamiliar odors was demonstrated also when the imprinting object was scented as shown by the fact that chicks chose to approach the cylinder with the same odor as the

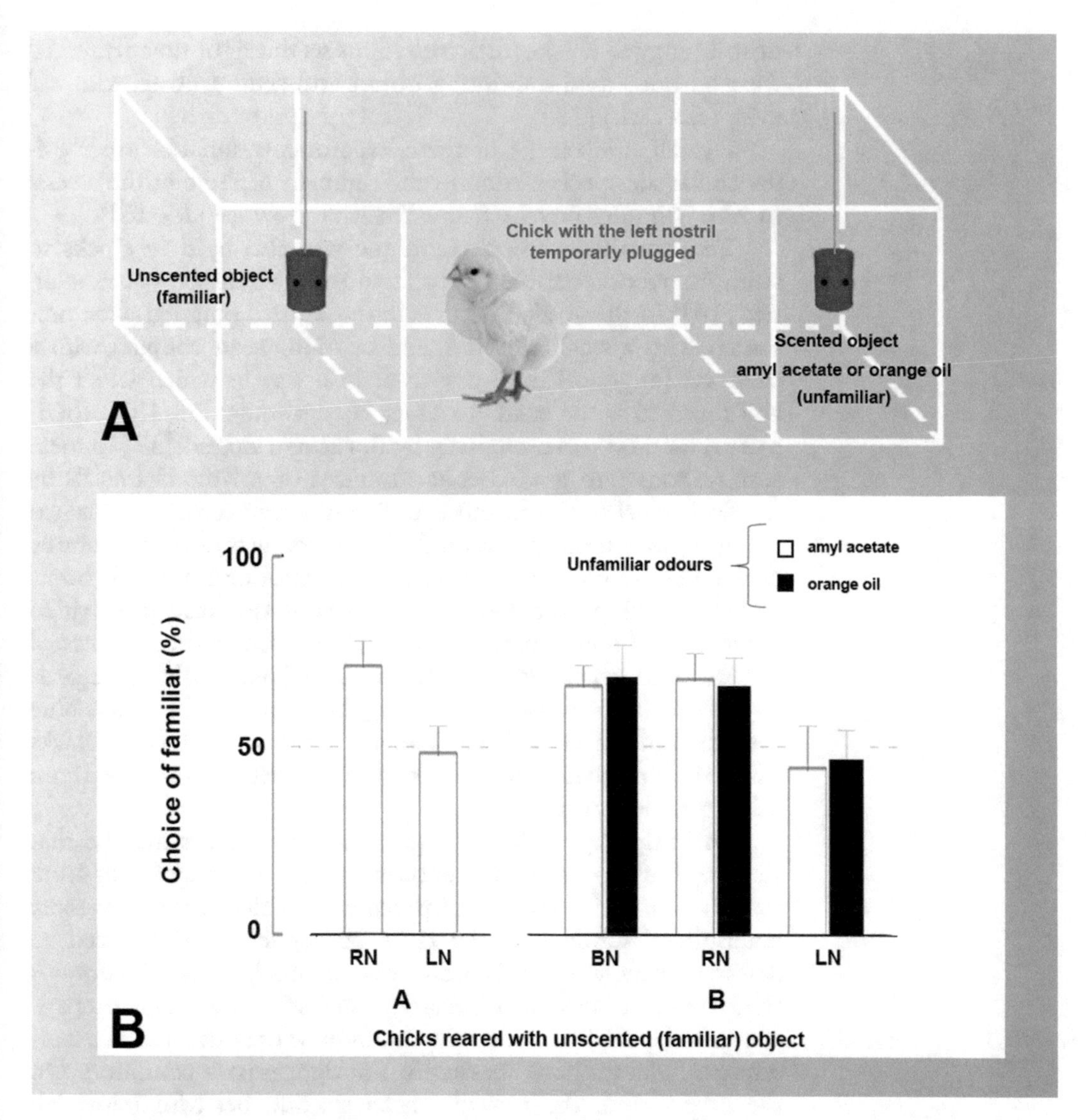

Fig. 2 (**a**) Schematic representation of the testing apparatus used to study olfactory lateralization in chicks (explanation in the text). (**b**) Choice of the familiar stimulus (i.e. the unscented object). Chicks used right (RN), left (LN), or both (BN) nostrils. In one experiment (**a**) amyl acetate was the scent (unfamiliar olfactory stimulus) used at test, whilst in a second experiment (**b**) the scent was amyl acetate for half the subjects and orange oil for the other half. (Mean percentage of total time spent near both objects with S.E.M. are shown.). *Panel b* is reproduced from Vallortigara and Andrew [13], with permission from Elsevier

one on which they had imprinted in the home cage (i.e. clove oil) in preference to the cylinder without this odor [13]. Chicks thus appear to learn odor properties of their imprinting object only when olfactory input is driven to their right hemisphere through the right nostril.

A bias toward the right nostril was observed also in detecting novel odors as shown by increased head shaking behaviors in left nostril plugged chicks (olfactory input to the right nostril) reared with un unscented imprinting object and then exposed to novel odor (clove oil) [13].

Overall, the results of these experiments support other evidence for the specialization of the right hemisphere in the assessment of stimulus novelty (for extensive review see [14, 15]).

The nostril occlusion technique was also used in chicks to study interaction between the olfactory and visual sensory modalities [16]. In this test subjects with one nostril plugged were presented with a small, colored (red or blue) bead coupled with a clove oil (eugenol) odor dispenser in a way in which when the chick pecked at the bead, it was exposed to the odor. This experimental method was used firstly by Burne and Rogers [17] to measure responses to graded concentrations of volatile chemicals by domestic chicks. Also in this experiment a thin coating of malleable wax was used (unscented depilatory wax) to temporarily occlude either the left or the right nostril. Odor detection (bouts of head shaking) and number of pecks at the bead were video recorded and then scored. Results showed that in the presence of the red bead, chicks attended to the odor (head shaking) regardless of which nostril they were using, but when the bead was blue a lateralized effect in head shaking response was seen. Chicks using their right nostril performed more head shakes than those using their left nostril.

An intriguing explanation given by the authors may be that visual activation of a hemisphere could enhance perception of other sensory modalities and neural processing carried out by the same hemisphere. Given that the right hemisphere is specialized to detect novelty, visual activation by an unusual, attractive stimulus (blue beads are unusually effective visual stimuli even in comparison to red beads) may enhance attention to the odor and as a consequence elevate head shaking in the right-nostril condition. On the other hand, the fact that head shaking behavior (clove oil detection) in chicks using their left nostril remains at a low level could be explained by the hypothesis that, in the left hemisphere, there is enhanced activation of the perceptual process that are used first to categorize a stimulus (vision in this case) followed by little or no attention to additional cues, such as the odor.

More recent studies have used unilateral olfactory input to study navigational process in homing pigeons [18, 19]. Adult naïve homing pigeons, which were continually allowed to carry out daily flights from their loft, were divided into three experimental groups (unplugged control pigeons with no plugs in their nostrils, and pigeons with either the left or right nostril plugged) and then released from one of three locations, placed north, east, and south with respect to home. The unilaterally plugged pigeons which successfully returned home from the first release were tested again from a new and more distant site after switching the side of the plugged nostril, together with the unplugged control group. A small amount of paste, which after insertion into the nostril becomes solid, was inserted some hours before releasing the bird for testing and removed after the pigeon had completed the test. Pilot tests showed that by at most 11 days after the time of nostril occlusion all pigeons were able to remove the plug by scratching the nostril with their claw. Such an event represents a clear advantage in terms of birds' welfare since it guarantees that the pigeons, which eventually did not return to their home, could unplug their nostril. The day of the test, pigeons were released singly, alternating the three experimental groups; all the experimental releases were performed in good weather conditions (sunny day with no wind or only light wind) and bird flight was observed using binoculars. An instructed experimenter blind to the testing paradigm recorded the birds' vanishing time (i.e. the time of flying until the bird disappeared from the observer's sight) and the azimuth of the vanishing bearing using a compass. Homing time was recorded by another observer who waited for the pigeon's arrival at the home loft.

Results revealed impairment in the initial orientation in subjects with their right nostril plugged and poorer homing performances in both groups smelling with one nostril compared to the unplugged control group. The latter suggested a fundamental bilateral involvement of the olfactory system when pigeons were released from unfamiliar sites where pieces of olfactory information are crucial cues for homing.

Further study analyzed flight tracks of unilateral plugged homing pigeons using GPS data loggers [19]. Differently from the previous experiment, pigeons had a single previous homing experience as they had been released once from 40 km north or south from home 1 year before testing (see [19], for specific quantitative GPS data analyses). The researchers observed that pigeons using their left nostril were more likely to be distracted by the environment cues (i.e. the flight path was more tortuous and interrupted) compared to those receiving olfactory inputs only from their right nostril and to controls with both nostrils in use. Since the occlusion of the right nostril interfered with the homing behavior of both naïve and experienced pigeons throughout the homing flight, researchers argued that the right nostril was dominant for processing olfactory information in this species.

Interestingly, this pattern of lateralization in favor of the right nostril was in contrast with a previous experiment, in which naïve pigeons released after unilateral lesioning of the piriform cortex showed a strong functional asymmetry with a left hemispheric superiority for processing olfactory cues useful for navigation [20]. The pivotal role of the left hemisphere in the pigeons' homing spatial behavior is supported also by the left hemispheric advantage in the development of the olfactory map under conditions in which map learning during navigation depended on the hippocampal formation [21]. Given that both the major ipsilateral and smaller contralateral projection from the olfactory bulb to higher brain areas are symmetrical [22], the observed reversed pattern of lateralization at periphery level (right nostril dominance) with respect to central level processing (left piriform cortex advantage) could not be explained by neural asymmetric organization and it is a topic that deserves further investigations [23].

Unilateral nostril occlusion has been used to investigate relations between navigation and lateralized olfactory process also in fish (European silver eel—*Anguilla anguilla*) during spawning migration [24]. After tagging, either the left or the right nostril of the eel was temporarily blocked by injection of silicone (perception of olfactory information was temporarily deprived by preventing water exchange into the occluded nasal cavity). All manipulation of the fish occurred during anesthesia with benzocaine. Data showed that, when the eels had their left nostril occluded, navigation behavior was disturbed (e.g. migration was slower and more errors in the route choice were manifested, see Fig. 3); on the other hand, no differences in speed of migration and orientation capacity were reported between eels with their right nostril blocked and unmanipulated controls (i.e. with both nostrils in use).

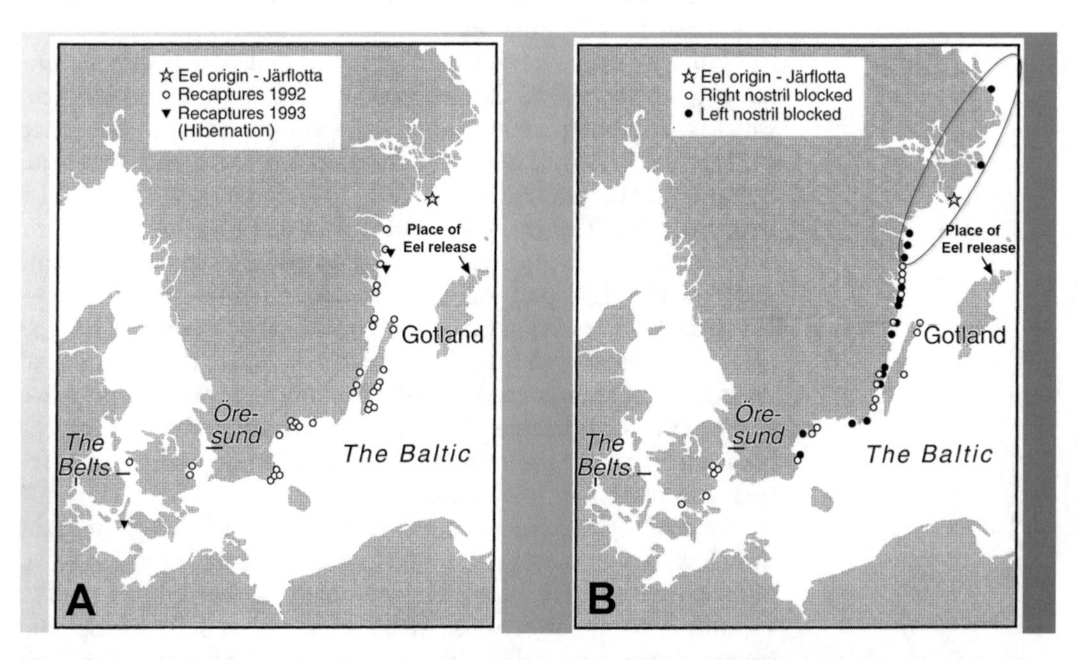

Fig. 3 Origin and recaptures of control group (**a**) and partly anosmic (**b**) (i.e. with either the left or the right nostril occluded). Silver eels released near Gotland. Note that the eels with left nostril blocked (right nostril in use) showed a higher frequency of recaptures north of the release locality (marked with *red circle*) than for natural eels and the group with the right nostril blocked. Reproduced from Westin [24] with permission from Elsevier

Contrary to what has been reported in birds, the left and not the right nostril seems to play a pivotal role in spatial navigation in fish and a fascinating hypothesis given by Rogers and colleagues [14] is that this manifested left nostril advantage could be due to the long evolutionary role of the left hemisphere in sustained attention and orientation during migration.

4 Right and Left Nostril Use in Mammals to Sniff at Arousing and Familiar Odors

Behavioral methods to study olfactory lateralization in mammals were firstly used in horses by McGreevy and Rogers [25]. These researchers investigated nostril preference (i.e. the bias to use one nostril rather than the other) to sniff stallion feces in a large population of thoroughbred horses. Approximately 30 g of thawed stallion feces were presented at noise height of unrestrained horses by an observer.

Feces were held in an open bag and the observer was instructed to stay in a midline with respect to the horse's nose and to use both hands together to hold the bag during odor presentation (in order to avoid any possible left-right handling bias). The first nostril used to sniff and the total number of inhalations through the left and the right nostril were observed until subjects ceased sniffing. Results revealed a population bias to use the right nostril first during sniffing at the stallion feces and this bias was evident especially among younger horses. As also in mammals olfactory input ascends ispilaterally to the brain [26], right nostril use represents a behavioral indicator of the prevalent activation of the horse's right hemisphere. Since the observed right nostril bias wanes with both stimulus presentation (the odor was presented on 2 consecutive days) and with horse age, authors argued that results are consistent with the right hemisphere investigation of odor novelty.

A strong correlation between the first nostril used and the total number of inhalations was found, suggesting that first nostril used could represent a valid indicator of olfactory sensory lateralization. No data on the nostril used last were available in this study.

In horses, additional behavioral investigation of olfactory perception asymmetries was performed by presenting 12 jumper horses with odors that differed in terms of emotional valence [27]. Olfactory stimuli were presented on cotton disks commonly used for make-up cleaning and comprised: food, repellent gel for mosquitoes, adrenaline, urine of an unfamiliar oestrous mare and the cotton disk without any particular odor (see Fig. 4).

Odorants were presented to the horses by a familiar observer (the horse groomer), who was instructed not to wear particular odors (e.g. deodorants and perfumes) for 2 days before the experiment and until after the end. The observer stood in the midline with respect to the horse head during presentation of the

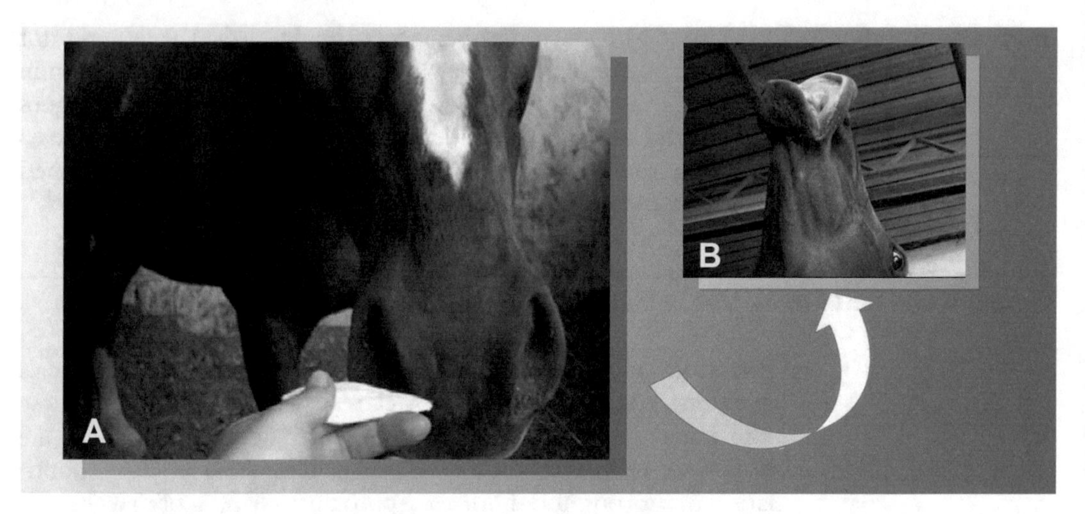

Fig. 4 Right nostril use (**a**) and flehmen behavior (**b**) during sniffing at oestrous mare urine odor

odors and did so holding in one hand the cotton disk and in the other a digital video camera by which nostril use was recorded (see Fig. 4, cotton disk and video camera positions were alternated during the experiment). A second fixed video camera at a distance of about 8 m was used to record the horse's behavioral reactivity (e.g. tail raised; head-tossing/shaking; nose tilting; abnormal oral behavior; visibility of eye-white; ears fixed backwards; head up; stamping with forelegs; stamping with back-legs; defecation; urination; yawning; stereotypy; pawing; kicking; forceful expulsion of air through the nostrils; flehmen; attempts to eat the cotton and licking). Olfactory stimuli were presented in random order during weekly sessions of about 10 min until four presentations × each odor × each subject were achieved. Two minutes was the maximum time allowed to sniff odor samples with a 1-min inter-trial interval.

Physiological reactivity to odors was also measured using a wireless device to monitor horse cardiac activity during sniffing. A significant bias in the use of the right nostril was shown during sniffing of adrenaline odor (see Fig. 5a). This pattern was consistent with the right hemisphere specialization in the analysis of arousal stimuli (see [14, 15] for an extensive review). Behavioral and physiological measures supported this hypothesis since horses displayed more reactive behaviors and higher cardiac activity during inhalations of adrenaline (see Fig. 5b).

The higher cardiac activity during presentation of adrenaline odor supported the general hypothesis that the activation of the sympathetic division of the autonomic nervous system during stressful situations associated with the expression of intense emotions (e.g. escape behavior and fear) is under the prevalent control by the right hemisphere via the hypothalamic-pituitary-adrenal-axis [28, 29].

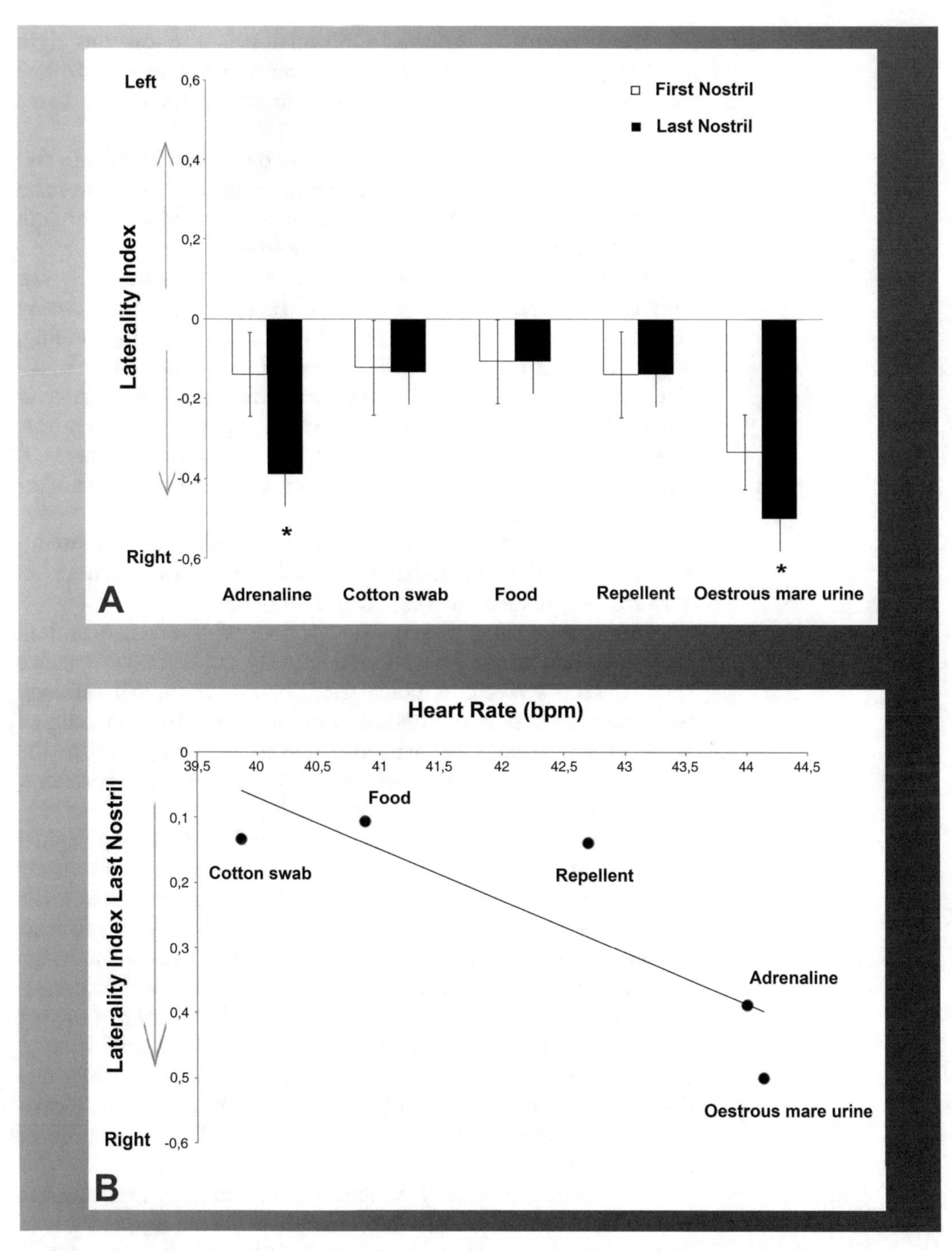

Fig. 5 (**a**) Laterality index of the first and last nostril used during inspection of different odorants (group means averaged across all four trials with S.E.M. are shown). *Asterisks* indicate significant biases (*$P < 0.05$, two-tailed one-sample *t* tests). (**b**) Correlation discussed between Laterality Index of last nostril use and the Heart Rate. Data presented are means calculated for each horse over experimental sessions. Reproduced from Siniscalchi et al. [17] with permission from Taylor & Francis Ltd., www.tandfonline.com

Furthermore in horses, additional evidence on the right hemisphere activation during processing of novel/arousal stimuli has come from functional studies of motor [30, 31] and visual laterality [32, 33].

A clear right nostril bias paralleled by higher heart rate frequency was also observed during sniffing urine of an oestrous mare (see Fig. 5a and b), supporting the main role of the right hemisphere in the control of sexual behavior [14].

Right nostril use during sniffing of adrenaline and urine was very clear in terms of the measure of the last nostril used during inhalation but less so for the nostril used initially during sniffing, suggesting that the last nostril used could represent a more sensitive indicator of nostril preference during inspecting of a particular stimulus since it occurred when olfactory scanning between nostrils, which drives olfactory analysis between brain hemispheres, is over (and as a consequence the olfactory stimulus has been categorized by the horse).

Behavioral methods have been used to study olfactory lateralization also in the species that are considered to have higher olfactory abilities, namely the dog (*Canis familiaris*).

The test involved presentation of dogs with olfactory stimuli that differed in terms of degree of familiarity and emotional valence [34]. Cotton swabs commonly used for canine vaginal cytology were impregnated with different odors and then fixed on a digital video camera located on a tripod. The tripod was placed in the center of a large room into which, on the day of the experiment, the dogs were taken separately on a leash (see Fig. 6).

Non-aversive olfactory stimuli were food, lemon, vaginal secretion of an oestrous bitch, and the cotton swab without any particular odor; arousal olfactory stimuli were adrenaline and the sweat of the veterinarian (the last was particularly arousing to dogs because of its association with routinely clinical stressful activities such as, for example, vaccine administration). To collect veterinarian sweat, the cotton swab was placed under the armpits of the vet for 10 min and after removal it was frozen at −80 °C until testing. The veterinarian was instructed not to use particular odorants deodorant/antiperspirant and to take only a shower on the morning for 2 days before the experiment until after the collection of the swab was over.

The oestrous olfactory stimulus was collected by inserting the cotton swab into the vagina of a bitch that showed typical signs of oestrous (both clinical and behavioral). Odor of lemon was simply collected from natural lemon juice.

As soon as each dog arrived at the testing area, the leash was removed and subjects were free to sniff odors in unrestrained conditions (see note section for possible faults during testing dogs under this experimental conditions). Sniffing behavior was recorded through the video camera and two plastic panels were placed

Fig. 6 Schematic representation of the testing apparatus used to study olfactory lateralization in dogs. Reproduced from Siniscalchi et al. [34], with permission from Elsevier

symmetrically on both sides of the tripod in order to guarantee a centered position of the dog's head with respect to the cotton swab and the video recording area (see Fig. 6).

Odors were presented in a random order during 1-h sessions at weekly intervals until seven presentations × each odor × each subject were achieved. Video recordings were analyzed using a frame-by-frame technique in order to measure both the first and the last nostril used to sniff, and the total time spent sniffing odor samples with each nostril. In line with results previously reported in horses, during sniffing at arousing odors, dogs showed a right nostril bias (right hemisphere activation) and it was consistent for all of the seven trials (see Fig. 7).

The observed right nostril bias during sniffing at the odor of a veterinary surgeon and adrenaline odor was paralleled by similar data on other sensory modalities (e.g. vision: [35, 36]; auditory: [37]), which indicates a general specialization of neural structures of the right side of the canine brain to respond to threatening and alarming stimuli.

Furthermore, when repeated presentations occurred, a shift from right to left nostril use was observed during sniffing at non-aversive stimuli (food, vaginal secretion, lemon, and cotton swab).

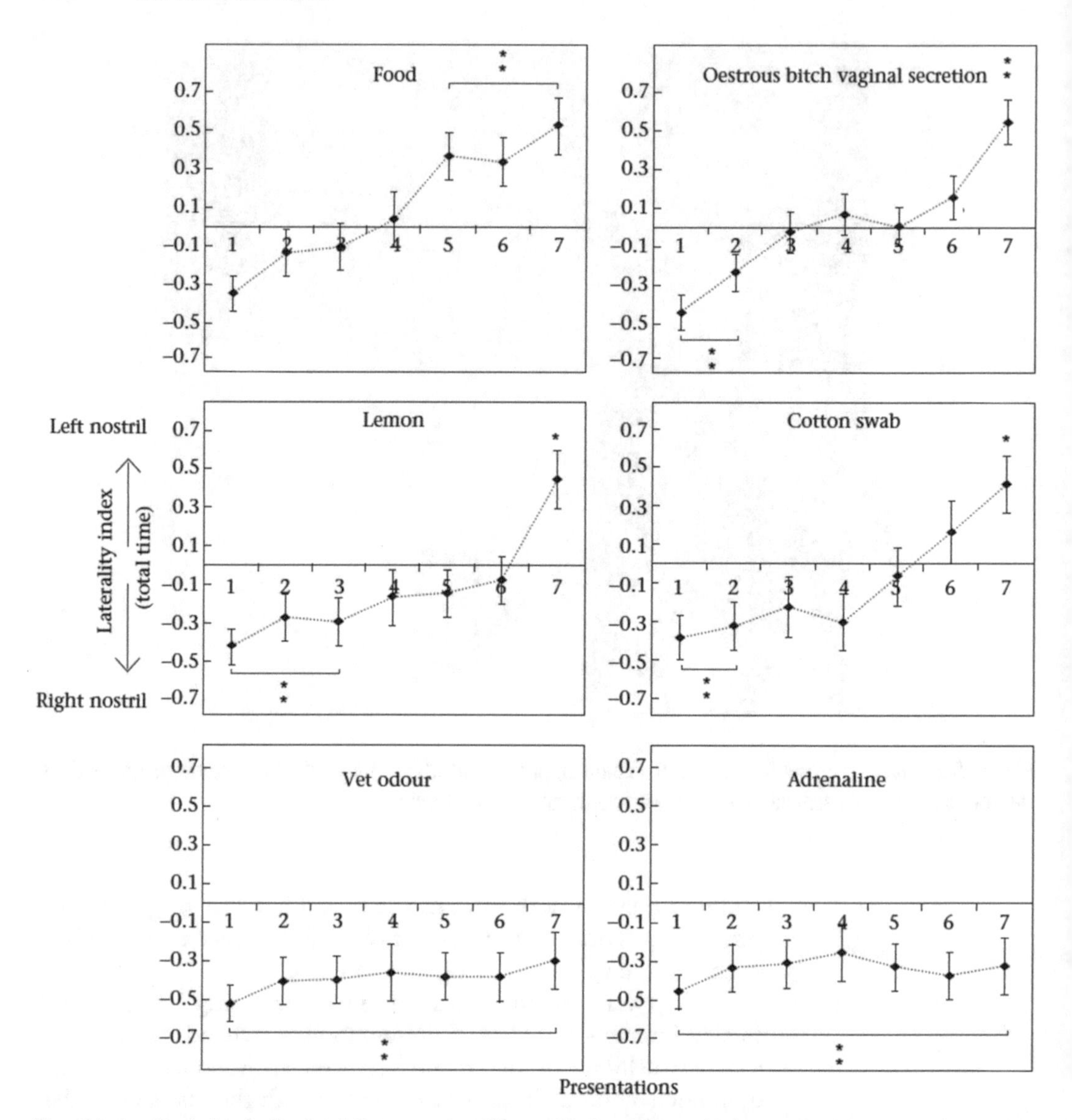

Fig. 7 Laterality index in the total time spent sniffing with the right and left nostril with repeated stimulus presentations (group means with S.E.M. are shown). *Asterisks* indicate significant biases (*$P < 0.05$; **$P < 0.01$). Reproduced from Siniscalchi et al. [34], with permission from Elsevier

This pattern has been shown in different animal models (e.g. fish: [38]; birds: [39]; insects: [40]), indicating that the right hemisphere has the ability to detect novel stimulus features (initial scanning) followed by use of the left hemisphere, which is specialized for routine control of behavior when the stimulus has become familiar (categorization).

Possible advantages of studying olfactory lateralization by measuring free sniffing behavior come from the fact that the absence of any particular devices used to occlude either the left or the right nostril reduces, or even completely eliminates, stress during the experiment (or also during the pre-experimental phase).

By contrast, given that in unrestrained conditions both nostrils are in use, some precautions should be taken by the observer in order to measure unilateral nostril use more accurately: a high-speed digital camera should be placed relatively close to the nostrils and a frame-by-frame technique is then necessary to code nostril use accurately (e.g. number of flares × nostril or nostril shift during odor inspection) especially in species such as dogs, in which nostril use is very fast.

Behavioral methods are currently in use in our laboratory in order to test the possible lateralized effects on dogs' behavioral and physiological responses to odors collected from both humans (armpit sweat) and other dogs (saliva, interdigital, and anal sac secretions) during different emotional events (e.g. dogs: an "isolation" situation in which a dog was isolated from its owner in an unfamiliar environment or a "disturbance" situation caused by an approach of a stranger to the car where the dog was resting; humans: odors samples of "fear" and "joy", for example, collected using emotional eliciting movies). We expected asymmetries in nostril use (i.e. a prevalent use either of the left or the right nostril) paralleled by changes in subjects' behavior and cardiac activity depending on the different emotional valence of odors. The results could extend our knowledge of olfactory processing of odors emotional cues, supporting the fundamental role of the olfaction in social communication.

Furthermore in future studies, the use of behavioral methods to investigate the link between olfaction, cognition, and emotions could be easily applied to a variety of species in order to establish the importance of olfactory cues in the conspecific and interspecific relational systems.

5 Notes

Although studying olfactory lateralization by measuring free sniffing behavior in unrestrained conditions represented a clear advantage since the subjects' behavior resembles as much as possible the natural ones, some aspects should be taken into account in order to avoid possible faults during testing dogs.

Firstly, subjects' tendency to sniff olfactory stimuli during testing could depend on the nature of the odor. Direct experimental evidence from our laboratory suggests that dogs are more responsive to high-intensity odors (e.g. food and oestrous vaginal secretions) and, as a consequence, when subjects are tested with these odors, they approach the testing apparatus (i.e. the cotton swab placed on a tripod) fast and they start to sniff odor sample promptly.

When less intense olfactory stimuli are used (e.g. odors collected from armpit sweat of humans during watching emotional eliciting movies, unpublished data), the tendency of dogs to approach the testing apparatus (i.e. the tripod) is weaker. In some cases, dogs completely ignored the apparatus or/and stayed freezed or curled up near the owner till the end of the test. In order to overcome this possible problem, it would be suggestable to test dogs by presenting the olfactory stimuli using an instructed experimenter. The experimenter should be instructed to stay in a midline with respect to the dog's nose and to use both hands together (i.e. without a left or right bias) to hold a digital video camera with an installed cotton swab impregnated with different odors at the dog's head height. This procedure is extremely useful in order to focus dogs' attention to the swab. The experimenter should be also instructed to abstain from heavily flavored foods (e.g. spices, asparagus, onions, garlic) and not to use deodorants, antiperspirants, and scented products which could interfere with dogs' odor perception, from 2 days before the start of the experiment until after its end. In addition, it would be suggestable before the beginning of the session, to allow dogs to explore the room and become familiar with the experimenter. If dogs are family pets living in households, their owner should bring them to the experimenter on the leash since the presence of the owner during the test reduces dogs' stress and enhances the tendency of dogs to sniff the swab (i.e. dogs' exploratory behavior is enhanced). The owner should be asked not to influence dogs' behavior (e.g. either to indicate the swab or to force sniffing behavior) and to stay alternatively to the right and to the left side of the dog to avoid the influence of his/her position on dog's performance [41]. A second aspect which should be taken into account when testing dogs is that, differently from other species, they get habituated quite quickly to odor stimuli and this should be considered if we are planning to use a repeated measure experimental design. In order to overcome possible testing faults due to fast stimulus habituation (i.e. low % of response), it should be advisable to present odor samples at weekly intervals and each single odor should be presented to the same dog no more than three times for weak odors and seven times for high-intensity odors.

References

1. Shipley MT, Ennis M (1996) Functional organization of olfactory system. J Neurobiol 30:123–176
2. Ehrlichman H (1986) Hemispheric asymmetry and positive-negative affect. In: Ottoson D (ed) Duality and unity of the brain. Kluwer, Dordrecht, The Netherlands, pp 194–206
3. Herz RS, McCall G, Cahill L (1999) Hemispheric lateralization in the processing of odor pleasantness versus odor names. Chem Senses 24:691–695
4. Dijksterhuis GB, Møller P, Bredie WL, Rasmussen G, Martens M (2002) Gender and handedness effects on hedonicity of laterally presented odours. Brain Cogn 50:272–281
5. Knecht S, Deppe M, Dräger B et al (2000) Language lateralization in healthy right-handers. Brain 123:74–81

6. Zatorre RJ, Jones-Gotman M (1991) Human olfactory discrimination after unilateral frontal or temporal lobectomy. Brain 114:71–84
7. Savic I, Berglund H (2000) Right-nostril dominance in discrimination of unfamiliar, but not familiar, odours. Chem Senses 25:517–523
8. Murphy C, Cain WS, Gilmore MM, Skinner RB (1991) Sensory and semantic factors in recognition memory for odors and graphic stimuli: elderly versus young persons. Am J Psychol 104(2):161–192
9. Hashimoto Y, Fukazawa K, Fujii M et al (2004) Usefulness of the odor stick identification test for Japanese patients with olfactory dysfunction. Chem Senses 29:565–571
10. Wewers ME, Lowe NK (1990) A critical review of visual analogue scales in the measurement of clinical phenomena. Res Nurs Health 13:227–236
11. Olko C, Turkewitz G (2001) Cerebral asymmetry of emotion and its relationship to olfaction in infancy. Laterality 1:29–37
12. Davidson RJ (2004) Well-being and affective style: neural substrates and biobehavioural correlates. Philos Trans R Soc Lond B Biol Sci 359(1449):1395–1411
13. Vallortigara G, Andrew RJ (1994) Olfactory lateralization in the chick. Neuropsychologia 32:417–423
14. Rogers LJ, Vallortigara G, Andrew RJ (2013) Divided brains. The biology and behaviour of brain asymmetries. Cambridge University Press, Cambridge
15. Rogers LJ, Andrew RJ (2002) Comparative vertebrate lateralization. Cambridge University Press, New York
16. Rogers LJ, Andrew RJ, Burne TH (1998) Light exposure of the embryo and development of behavioural lateralisation in chicks: I. Olfactory responses. Behav Brain Res 97(1–2):195–200
17. Burne TH, Rogers LJ (1996) Responses to odorants by the domestic chick. Physiol Behav 60(6):1441–1447
18. Gagliardo A, Pecchia T, Savini M, Odetti F, Ioalè P, Vallortigara G (2007) Olfactory lateralization in homing pigeons: initial orientation of birds receiving a unilateral olfactory input. Eur J Neurosci 25(5):1511–1516
19. Gagliardo A, Filannino C, Ioalè P, Pecchia T, Wikelski M, Vallortigara G (2011) Olfactory lateralization in homing pigeons: a GPS study on birds released with unilateral olfactory inputs. J Exp Biol 214(Pt 4):593–598
20. Gagliardo A, Odetti F, Ioalè P, Pecchia T, Vallortigara G (2005) Functional asymmetry of left and right avian piriform cortex in homing pigeons' navigation. Eur J Neurosci 22: 189–194
21. Gagliardo A, Ioalè P, Odetti F, Bingman VP, Siegel JJ, Vallortigara G (2001) Hippocampus and homing in pigeons: left and right hemispheric differences in navigational map learning. Eur J Neurosci 13:1617–1624
22. Patzke N, Manns M, Güntürkün O (2011) Telencephalic organisation of the olfactory system in homing pigeons (Columba livia). J Neurosci 194:53–56
23. Pecchia T, Gagliardo A, Filannino C, Ioalè P, Vallortigara G, Csermely D, Regolin L (eds) (2013) Navigating trough an asymmetrical brain: lateralization and homing in pigeon. In: Behavioral lateralization in vertebrates. Springer-Verlag, Berlin, Heidelberg
24. Westin L (1998) The spawning migration of European silver eel (Anguilla anguilla L.) with particular reference to stocked eel in the Baltic. Fish Res 38:257–270
25. McGreevy PD, Rogers LJ (2005) Motor and sensory laterality in thoroughbred horses. Appl Anim Behav Sci 92:337–352
26. Royet JP, Plailly J (2004) Lateralization of olfactory processes. Chem Sens 29:731–745
27. Siniscalchi M, Padalino B, Aubé L, Quaranta A (2015) Right-nostril use during sniffing at arousing stimuli produces higher cardiac activity in jumper horses. Laterality 20(4):483–500
28. Henry JP (1993) Psychological and physiological responses to stress: the right hemisphere and the hypothalamo-pituitary-adrenal axis, an inquiry into problems of human bonding. Integr Physiol Behav Sci 28:369–387
29. Sullivan RM, Gratton A (1999) Lateralized effects of medial prefrontal cortex lesions on neuroendocrine and autonomic stress responses in rats. J Neurosci 19:2834–2840
30. Austin NP, Rogers LJ (2012) Limb preferences and lateralization of aggression, reactivity and vigilance in feral horses, Equus caballus. Anim Behav 83:239–247
31. Siniscalchi M, Padalino B, Lusito R, Quaranta A (2014) Is the left forelimb preference indicative of a stressful situation in horses? Behav Proc 107:61–67
32. Austin NP, Rogers LJ (2007) Asymmetry of flight and escape turning responses in horses. Laterality 12:464–474
33. Sankey C, Henry S, Clouard C, Richard-Yris M-A, Hausberger M (2011) Asymmetry of behavioral responses to a human approach in young naive vs. trained horses. Physiol Behav 104:464–468

34. Siniscalchi M, Sasso R, Pepe AM, Dimatteo S, Vallortigara G, Quaranta A (2011) Sniffing with right nostril: lateralization of response to odour stimuli by dogs. Anim Behav 82:399–404
35. Siniscalchi M, Sasso R, Pepe AM, Vallortigara G, Quaranta A (2010) Dogs turn left to emotional stimuli. Behav Brain Res 208:516–521
36. Siniscalchi M, Lusito R, Vallortigara G, Quaranta A (2013) Seeing left- or right-asymmetric tail wagging produces different emotional responses in dogs. Curr Biol 23:2279–2282
37. Siniscalchi M, Quaranta A, Rogers LJ (2008) Hemispheric specialization in dogs for processing different acoustic stimuli. PLoS One 3(10):e3349
38. Sovrano VA (2004) Visual lateralization in response to familiar and unfamiliar stimuli in fish. Behav Brain Res 152:385–391
39. Vallortigara G, Regolin L, Pagni P (1999) Detour behaviour, imprinting, and visual lateralization in the domestic chick. Cogn Brain Res 7:307–320
40. Rogers LJ, Vallortigara G (2008) From antenna to antenna: lateral shift of olfactory memory in honeybees. PLoS One 3:e2340
41. Siniscalchi M, Bertino D, Quaranta A (2013) Laterality and performance of agility trained dogs. Laterality 19(2):219–234

Chapter 5

Hand, Limb, and Other Motor Preferences

Gillian S. Forrester

Abstract

Historically, many cognitive competencies have been considered human unique. An anthropomorphic mindset has thwarted a better understanding of inherited behaviors that have emerged as the result of millions of years of shared evolution with other vertebrates. The behavioral sciences stand at the forefront of a paradigm shift to view human cognition within the framework of an evolutionary trajectory. We are beginning to develop and apply measures of increasing validity to cross-species investigations of motor behavior. Recent findings have revealed that lateralized motor behaviors, once thought unique to humans, are present in other animal species, suggesting inheritance from common ancestors. Population-level motor biases may reflect an early evolutionary division of primary survival functions for the left and right hemispheres of the brain. Moreover, these primary functions may still influence modern human behavior and provide a necessary platform for the development of higher cognitive functions. In this way, the evolution and development of cognition are inextricably linked. This chapter considers hand, limb, and other motor preferences, and focuses on common methods used to investigate lateralized motor behavior across species within an evolutionary framework.

Key words Lateralized motor behavior, Evolution, Development, Humans, Animals

1 Introduction

For all animals, motor behaviors are more than muscle responses: they involve sequences of goal-oriented behavioral units [1] to perform biologically adaptive behaviors. In humans, primary sensory and motor functions represent the building blocks of mental processing [2] and are underpinned by specialized neural mechanisms [3]. Motor actions for specific behaviors tend to be dominant on one side of the body, resulting from cerebral lateralization of brain function. Cerebral lateralization refers to differences in contributions of the left and right hemispheres of the brain for controlling sensory, motor, and higher cognitive processes (for a review, see [4]). The nerve fibers of the motor cortices are contralaterally innervated. Therefore, dominant hemisphere processes can manifest as contralateral motor behaviors [5] and act as informative behavioral markers of brain organization and function (e.g. [6]).

Lesley J. Rogers and Giorgio Vallortigara (eds.), *Lateralized Brain Functions: Methods in Human and Non-Human Species*, Neuromethods, vol. 122, DOI 10.1007/978-1-4939-6725-4_5, © Springer Science+Business Media LLC 2017

Historically, cerebral lateralization had been considered a unique characteristic of the human brain. However, examples of cerebral lateralization of function tied to contralateral motor biases have recently been identified in vertebrates [7, 8] and invertebrates [9, 10]. A range of species has demonstrated left hemisphere and right-biased motor actions for well-learned sequences of motor actions. For example, right motor biases have been revealed in: fish and toads for prey capture [11], in birds for foraging and manipulating food items [11] and in birds [12], monkeys (e.g. [13]) and apes (e.g., [14]) for object manipulation. Conversely, many species have demonstrated right hemisphere and left-biased motor actions for novel and urgent stimuli. For example, birds [15–17], lizards (in the laboratory: [18]; in the wild: [19]) and toads [20] exhibited left eye preferences for predator monitoring, precipitating a rightward bias in predator escape behaviors.

Patterns of motor dominance in animals are consistent with the hypothesis that throughout the evolution of the vertebrate brain, the right hemisphere became dominant for urgent responses to the environment (e.g. predators), while the left hemisphere emerged as dominant for routine and structured sequences of actions (e.g. feeding) (but see marsupials for evidence of reversed laterality, [21]) [22, 23]. Asymmetric behavioral activity might have an adaptive value, facilitating simple reflexive behavioral responses to increase the survival of individuals [24]. Additionally, the dissociation of specialized processing of left and right hemispheres may also increase neural efficiency by: allowing different functions to operate in parallel, decreasing the duplication of functioning across hemispheres and eliminating the initiation of simultaneous and potentially incompatible responses [25, 26]. One theory is that the two hemispheres became dominant for primary survival strategies (feeding and monitoring for predators) early in vertebrate evolution. In its simplest form, the vertebrate brain functions as an "eat and not be eaten" parallel processor.

To date, empirical evidence suggests varying degrees of lateralized motor behaviors across species. Also, population-level motor asymmetries of the forelimb appear sensitive to a range of evolutionary and environmental factors (for a review see [27]). Nonhuman animals demonstrate less skewed population-level hand dominance compared with humans. Additionally, nonhuman animal populations possess a greater number of nonlateralized individuals (without a preference) compared to humans (e.g. [28]). For some motor asymmetries (e.g. handedness), it has been speculated that human and nonhuman motor asymmetries may reflect stages along a phylogenetic trajectory of the vertebrate brain [29] and correlates with cognitive function (e.g. [30]). To address evolutionary hypotheses, investigations require standard and systematic experimental parameters that allow for valid comparisons across species [31].

Systematic, cross-species measures of lateralized motor behavior in humans and nonhuman animals have been lacking. This may, in part, be due to an historic propensity to investigate human behavior in isolation from other animals. Today, we recognize that during human development, higher cognitive abilities scaffold, build and bootstrap upon a foundation of evolutionarily early perceptual and motor capabilities (e.g. [32]). Thus, it is likely that modern human behavior builds upon evolutionarily early vertebrate brain organization and function. For example, it has been proposed that a left hemisphere dominance for routine sequences of motor actions was extended to support language functions (e.g. [33]), while a right hemisphere dominance for predator avoidance behavior was extended to support human social-emotional processes (e.g. [22]) (for a review, see [34]).

Placing human cognition within an evolutionary framework requires that researchers "re-visit" human-centric experimental designs and interpretations of results. The presence of lateralized motor function in humans and other animals provides a unique strategy to investigate the evolution and development of cognition within and between species under a common framework of a shared evolutionary history. To date, many forms of lateralized motor behavior go largely unrecorded across species (e.g. social positioning). The following sections will concentrate on lateralized motor actions in humans and nonhuman animals with a particular focus on humans and great apes.

2 Lateralized Manual Behavior in Humans

2.1 Human Population-Level Right-Handedness

Population-level right-handedness is the most extreme example of cerebral lateralization in humans. Approximately 90% of the human population demonstrates a bias to use the right hand more than the left hand (e.g. [35]). Human handedness data are often collected using surveys and self-report (e.g. Edinburgh Handedness Inventory, [36]) making the method difficult to adapt to nonhuman populations. Within the right-handed human population, approximately 95% of individuals have dominance for language processing in the left hemisphere of the brain [37]. The high correlation between left hemisphere language regions and right-handedness in humans has led scientists to argue that the two are causally linked [38–41] e.g. right-handedness is the manifestation of left hemisphere dominant brain regions for language production (Broca's region: [42]) (e.g. [43]) (Fig. 1).

Evidence from multiple disciplines now propose that language abilities result from a direct adaptation of a more primitive apparatus for planning goal-directed motor action sequences. Hierarchical structure, a distinctive component of language (e.g. [44]), has been identified in nonlinguistic domains (e.g.

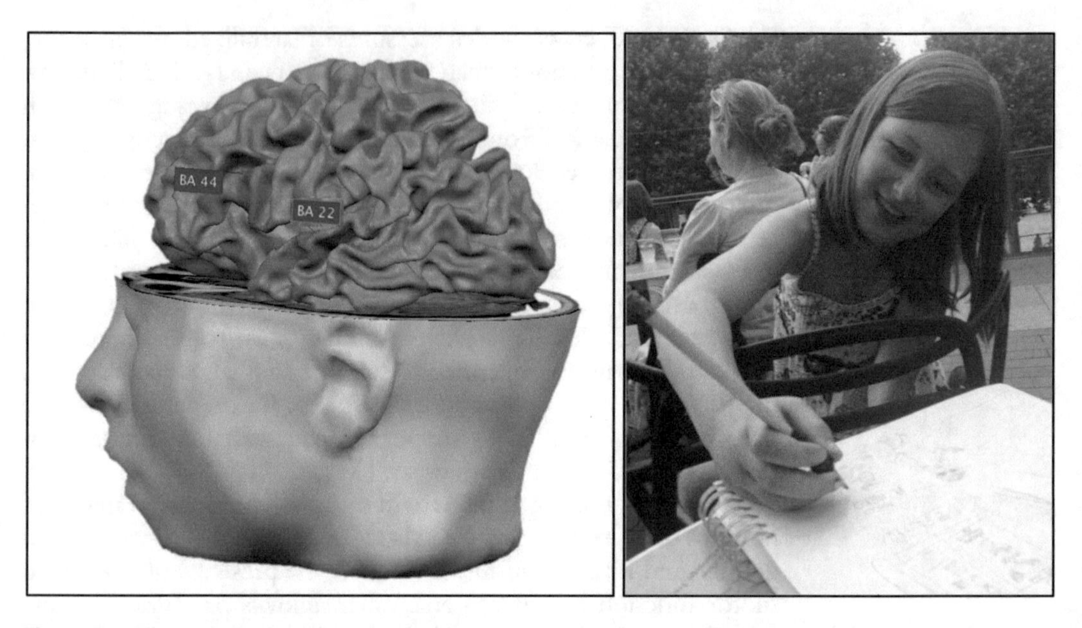

Fig. 1 This figure indicates approximate locations of Broca's area (Brodmann 44) and Wernicke's area (Broadmann 22), both dominant in the left hemisphere, implicated in language processing and significantly correlated with right-handedness

goal-oriented motor sequences) (e.g. [45]). Gestural origins of language proponents [46] argue that a left hemisphere dominance to guide the right hand in fine articulation of motor action sequences primed the brain for language syntax [33]. Moreover, tool use theory advocates argue that tool use was a critical precursor to language, through the emergence of a motor syntax [47]. An action-based proto-syntax afforded by tool use would have been useful in evolutionary history to scaffold a gestural communication system [48] prior to the emergence of speech. It is hypothesized that right-handed tool use provided an evolutionary bridge between left-hemisphere-dominant action sequences and language processes [49].

Theoretical proposals of left hemisphere dominance to guide right-handed, goal-directed actions is also supported by neurobiological evidence. Studies have revealed an overlap in left hemisphere neural regions that support language production and goal-oriented manual action sequences. Broca's area may act as a "supramodal hierarchical processor" [50], performing the computational processing required to produce the structured action sequences that supports both tool use and speech production [51, 52]. Higuchi and colleagues reported an activity in the neural region associated with Broca's area for tool use in great apes and for language in humans, suggesting that Broca's area did not emerge as a language-specific human-unique brain region [53]. Based on evidence across fields, it is reasonable to consider that structured action sequences, dominant within the left hemisphere underpins both tool use and language [54], driving population-level right-handedness in humans. To learn more about how this motor bias emerged in humans within an evolutionary framework, it is necessary to consider lateralized limb behavior in nonhuman animals.

2.2 Limb Preferences in Nonhuman Animals

A growing body of evidence has reported right limb preferences in nonhuman animals for motor action sequences across cognitive domains. For instance, Leatherback sea turtles (*Dermochelys coriacea*) preferentially uses the right flipper to shift sand to conceal their eggs [55]; European toads (*Bufo bufo*) preferentially use the right forelimb to remove foreign objects adhered to their bodies [56], chicks (*Gallus gallus*) use the right foot preferentially to step on a platform [57] and the Australian parrot (*Platycerus elegans*) holds food in the right foot. The findings provide additional evidence that right-biased feeding behaviors (for a review see [8]) are supported by a left hemisphere dominance for behaviors comprised of sequences of routine motor actions. However, motor asymmetry patterns across species are far from straightforward. Left forelimb biases have also been revealed in different species for a variety of activities. For example, eight of nine Australian parrots species investigated, demonstrated a significant left foot preference for food holding [58]; chimpanzees and gorillas have demonstrated a left hand preference for face touching [59] and bipedal marsupials have shown left forelimb dominance for unimanual tasks [21]. While cerebral lateralization can manifest as contralateral motor dominance, it is not the only contributing factor. Posture [60–62], bipedalism [63, 64] and task attributes (e.g. [65]) also influence motor behavior. To place humans within the phylogenetic trajectory of cerebral lateralization it is necessary to investigate the behaviors of ancient man.

2.3 Great Apes as a Proxy for Ancient Human Behavior

Although we cannot directly investigate the cognitive capabilities of our extinct hominid ancestors, extant great apes offer a valuable functional model to study the evolution of both handedness and human cognition. Great apes have close phylogenetic proximity to humans and possess anatomical human-like features, including the morphology and the articulation of the hands [66]. Additionally, like humans, great apes have the ability and motivation to locomote bipedally [67]; produce and use tools both in captivity (e.g. [68]) and in the wild (e.g. [69, 70]); culturally transmit information across and down generations creating group-specific cultures (e.g. [71]) and use intentional communication signals (e.g. [72–74]). The neural organization of great ape brains also share common structural and processing capabilities with the human brain. Neuroimaging studies have indicated that all great ape species display a homologous human Broca's region [75, 76] that is asymmetrically larger in the left hemisphere. The fact that apes possess a brain region known for language function in humans, but are not language users, further supports the hypothesis that Broca's region is neither language-specific nor unique to humans. Apes have now been extensively investigated for lateralized manual behavior.

3 Lateralized Manual Behavior in Great Apes: Methods and Findings

Historically, investigations of captive and wild nonhuman primates revealed no clear evidence of species-level manual lateralization (e.g., [77–81]), perpetuating the idea that population-level right-handedness is a human-unique characteristic. However, experimental parameters (e.g., terminology, behavioral tasks, rearing histories, analysis procedures) have varied across laboratories, potentially contributing to inconsistent cross-species findings (e.g., [29, 31, 82–85]). We have yet to develop and introduce standard practices for sampling and evaluating motor behavior, hindering direct comparisons across laboratories and species. In the future, the adoption of a standard and systematic approach will facilitate cross-species comparisons of motor behavior. The following sections will introduce common methods for investigating lateralized manual behavior.

3.1 Terminology

Although investigations of human handedness often treat manual behavior with dichotomous distinctions (e.g. left, right), hand laterality is not a dichotomous variable. In fact, handedness exists on a gradient and can be influenced by a number of identified factors including: task type, task complexity, posture, hand use (unimanual versus bimanual) setting (captive versus wild) and rearing history (encultured versus non-encultured) (for reviews, see [29, 85]). Therefore, both the labeling and the measures of hand behaviors become important considerations for the validity of comparisons within and between individuals, populations, and species.

The current literature reflects a myriad of labels for discussing and evaluating manual laterality in human and nonhuman primates. It is possible that these inconsistencies have led to unintended and confounded comparisons of different phenomena. To increase the validity of comparisons, Marchant and McGrew suggested a basic category framework that defines manual laterality at both the individual and group levels [29]. They consider two main variables, each with two levels (Table 1).

This framework generates four distinct types of manual laterality. "Handedness" is reserved for congruence across subjects across tasks (e.g. multiple tasks performed by multiple individuals). Handedness is distinct from "hand preference," which refers to

Table 1
Marchant and McGrew [29] provided a framework for defining manual motor biases

	Subject(s)		
		Within	**Across**
Task(s)	Within	Hand preference	Task specialization
	Across	Manual specialization	Handedness

Adapted from Marchant and McGrew [29]

within subject and task (e.g. a single task performed by a single subject). "Manual specialization" is considered appropriate for cases of within subject and across task (multiple tasks performed by a single individual), while "task specialization" refers to across subjects and within task (e.g. a single task performed by multiple individuals). The framework demonstrates the multidimensionality of lateralized manual behavior and provides a consistent approach for valid comparisons across tasks and individuals.

3.2 Common Methods for Data Sampling

The most frequently applied data sampling methods are focal follows (e.g. [86]) within and across individuals, although some studies employ temporal interval sampling (e.g. [87]). Many researchers also agree that valid datasets, whether coded in vivo or off-line from video, should represent the normal repertoire of behaviors demonstrated by the species in a setting during daily life (as opposed to a selective set of idiosyncratic behaviors). Additionally, sample sizes, both for subjects and data sets should be of sufficient statistical power (e.g. [29]).

3.2.1 One- and Two-Handed Actions

Unimanual actions are those that require only a single hand to perform a task. Much of the manual laterality literature on nonhuman primates that has revealed a lack of hand dominance has concentrated on the unimanual action of "simple reaching," feeding behavior (e.g. [88]). Simple reaching behaviors are argued to be vulnerable to extraneous factors (e.g. position, posture) because the activity can be so simple that it can be performed efficiently with either hand. Therefore, unimanual actions, like simple reaching during feeding, are not considered to be a sensitive measure of hand dominance (for a review, see [89]).

Coordinated bimanual manipulation appears to be a more sensitive measure of hand dominance (e.g. [89, 90]), resulting in fewer ambi-preferent individuals. These actions require the participation of both hands such that one hand holds an object (non-dominant hand) while the opposite hand performs manipulations of the object [91] (Fig. 2).

Fig. 2 This figure demonstrates the process of a bimanual coordinated action. Photo by Davila-Ross, adapted from Forrester et al. [92]

Bimanual actions tend to minimize postural factors, as the individual must adopt a bipedal or seated posture in order to maintain the freedom of both hands [93].

3.2.2 Bouts and Events

The sampling of "bouts" versus "events" remains a point of contention across laboratories sampling animal motor action. Marchant and McGrew have argued that data points should be statistically independent so as to avoid pseudo-replication leading to false positive results from skewed datasets [29]. Independence of data points refers to the probability of using one hand for a response based on the occurrence of the same hand being used on the previous response, in a sequence or bout of actions (for a discussion, see [94]). For example, in the case where an ape repetitively slaps the ground with the same hand only the first hand action in the sequence would be considered as an independent choice. The remaining actions in the sequence would be argued to be dependent on the first action. The inclusion of only the first action in a sequence of actions involving the same hand is referred to as sampling "bouts." Whereas including all actions in the sequence in a dataset is referred to as sampling "events." It is vigorously argued that empirical studies demonstrate that hand use is consistent whether recorded as events (non-independent data points) or bouts (independent data points) [89]. Additionally only collecting bouts can result in a loss of statistical power and a greater risk of skewing datasets towards the experimental hypothesis. One solution to this debate is for researchers to collect both measurements for statistical evaluation [95] for further consideration of the issue.

3.3 Common Methods for Data Analyses

Once terminology and data sampling is considered, it is the evaluation of the datasets that reveals whether or not individuals or populations express significantly lateralized manual behavior.

It is common practice for investigations of great ape manual behavior to tally the frequency counts of bouts and/or events and evaluate the totals of individuals and groups using descriptive measures and statistical tests. Handedness Index (HI) scores are commonly employed to descriptively assess an individual's strength of lateral hand bias. HI scores are calculated for each subject to establish the degree of hand asymmetry, using the formula [HI = (R − L)/(R + L)], with R and L being the frequency counts for right and left hand dominance for frequency counts of bouts or events. HI values vary on a continuum between −1.0 and +1.0, where the sign indicates the direction of hand preference. Positive values reflect a right hand preference while negative values reflect a left hand preference. When R = L, the HI is taken to be zero. To statistically assess individual patterns of hand dominance, it is common to calculate: *z*-scores and the binomial approximations of the *z*-scores. The direction of hand preference for each subject is also calculated using *z*-scores. Individuals are considered left handed when $z \geq -1.96$, right handed when $z \geq 1.96$ and ambiguously handed when $-1.96 < z < 1.96$. The binomial approximation of the *z*-score reveals the significance level of the *z*-score. When tests are two-tailed, then no hand preference (ambi-laterality) acts as the null hypothesis (i.e., $P = q = 0.50$).

To reveal descriptive group results, Mean Handedness Index scores (MHI) can be calculated. Additionally, Absolute Handedness Index (ABHI) scores can be calculated to indicate the strength of handedness, not considering the direction of laterality. One-sample *t*-tests are often used to statistically evaluate group-level handedness using HI scores (e.g. [92]), although two-sample *t*-tests have also been employed to test for significant differences in the frequencies of two samples (e.g. [65, 96, 97]). Additionally, some studies have employed proportions as an alternative to frequencies. Proportions allow for individual scores to contribute equal weighting to a dataset, such that results are not skewed by those with higher frequencies of bouts/events (e.g. [30]). Proportions are calculated by dividing in turn, the total number of left and right motor actions by the total of all left and right actions. Table 2 provides a summary of common data sampling and evaluation techniques.

3.3.1 Limitations of Statistical Calculations

The statistical evaluation of lateralized motor behavior has some limitations. Great ape investigations of manual laterality support the concept that handedness is not a dichotic variable (left or right), but rather exists on a scale of strength that may represent domain specific hemispheric contributions. While descriptive evaluations represent such a scale (e.g. HI scores), the statistical *z*-score test treats lateralized manual behavior in a dichotomous manner. This makes the use of *z*-scores an extremely conservative approach to estimating lateralized manual behavior [89] and can mask latent

Table 2
The table below offers a summary of methods commonly used to sample and evaluate lateralized manual behavior

Types of handedness	Data sampling	Data compiling	Individual data evaluation	Group data evaluation
Unimanual hand actions	*Events*	*Frequencies* of left/right actions (derived from bouts or events)	*Handedness index (HI)* HI = L + R/L − R (derived from frequencies) Descriptive measure to assess the strength and direction of laterality	*Mean Hand Index (MHI)* (derived from frequencies) Descriptive measure to assess strength and direction of laterality
Bimanual hand actions	*Bouts*	*Proportions* of left/right actions (derived from bouts or events)	*Absolute Hand Index (ABHI)* (derived from HI scores) Descriptive measure to assess the strength of laterality	*Mean Absolute Hand Index (MABHI)* (derived from HI scores) Descriptive measure to assess the strength of laterality
			z-Scores (derived from frequencies) A statistical measure of an individual's score in relationship to the mean of the group	*One-Sample T-Test* or nonparametric equivalent (derived from HI scores) A statistical test to assess if a sample comes from a particular population
			Binomial Sign Test (derived from frequencies) A statistical test to assess the probability of deviation from the expected distribution of observations in two categories	*Paired-sample T-Test* or nonparametric equivalent (derived from frequencies or proportions) A statistical test to compare two population means

laterality patterns. Moreover, human handedness has not been evaluated based on a similarly strict criterion. Thus, it is possible that human populations possess a larger proportion of ambi-preferent individuals than has been reported in the literature [89]. Finally, it is important to note that *z*-scores are influenced by sample size. Smaller samples that are proportionately equal to larger samples may fail to reveal lateral motor patterns.

3.3.2 Descriptive Scales

Researchers are finding different ways to reveal patterns of laterality using descriptive measures (e.g. HI scores). Yet, there is reluctance from the scientific community to consider nonsignificant results. For example, Hopkins used *z*-scores to classify individuals as strongly left-handed (SL), mildly left-handed (ML), ambiguously handed (AH), mildly right-handed (MR), and strongly right-handed (SR). Subjects with *z*-scores ≤ -1.96 were classified as SL, −1.95 to −1.0 (ML), −0.99 to 0.99 (AH), 1.0 to 1.95 (ML), and ≥1.96 (SR) [89]. Strongly lateralized individuals (left or right) expressed statistically significant lateralized behavior. Mildly handed (left or right) categories were established to represent one standard deviation from the mean in a normally distributed population. However, these categories do not represent statistically significant lateralized manual behavior. Descriptive gradients have also been considered to reveal group patterns of lateral bias. For example, Forrester and colleagues introduced binned ABHI scores ($-1 < L < -.33 < A < 0.33 < R < +1$). For this type of evalution, R and L equal the laterality index score of bouts, to consider group-level laterality [92]. Under this framework, binomial tests were applied to distinguish lateralized from ambi-preferent individuals within a group.

3.4 Factors that Influence the Laterality of Manual Behavior

A multitude of factors are now known to influence manual motor biases in primates. Three main components to be discussed in the following sections are: setting, ecological validity, and task.

3.4.1 Setting

Behavior is multidimensional and is produced within space, time, and context (e.g. [98]). The more variables that are matched in comparative studies of lateralized motor behavior, the greater the ecological validity. Marchant and McGrew have argued that the environmental setting within which behavior occurs requires careful consideration [29]. Moreover, they have argued that the setting cannot simply be defined as "wild" or "captive." Rather, setting should be described on a continuum between the two extremes. Marchant and McGrew suggest a scale of settings with seven gradations ranging from "nature" (no human influence) to "laboratory" [29]. The ability to compare like tasks across different species and settings is important to our understanding of species-specific capabilities in addition to how environmental pressures may have contributed to the shaping of behavior.

3.4.2 Ecological Validity

Nonhuman primate studies have rarely considered lateral motor dominance in natural behaviors. This is a serious omission because natural behaviors are likely to be the conditions under which hemispheric specialization for motor dominance evolved. Therefore, the investigation of lateral biases of natural behaviors provides a necessary element of ecological validity. Natural food preparation and feeding sequences of apes provide an excellent opportunity to investigate cerebral lateralization and motor biases. To date, reported findings have been inconsistent. However, the vast array of observed behaviors, coding criteria, and assessment parameters make direct comparisons difficult across species and laboratories. Some studies have found hand biases only at the individual level in wild chimpanzees for food consumption (e.g. [99]) and anvil use [100], while others report no lateral bias for nontool-using feeding behaviors (e.g. [79, 101]). Studies focusing on subsistence tool-use (the use of a tool to obtain food) or object manipulation in wild apes have also failed to reveal evidence of population-level lateral biases [102–104].

Only a few investigations of natural behavior have evaluated bimanually coordinated sequences of actions. Of these studies, one investigation demonstrated a right-handed population-level preference for nettle processing in mountain gorillas (*Gorilla beringei beringei*) [105]. Two further studies of captive gorillas (*Gorilla gorilla gorilla*) also noted a significant population-level right hand preference for bimanual foraging behaviors ([91], but see [88]) and in bimanually coordinated honey-dipping and nettle processing (e.g., [30]). A recent study by Forrester et al. found that at the individual-level, few subjects demonstrated statistically significant manual biases (left or right) for strychnos fruit eating [92]. However, at the population-level, bout frequencies and proportions both revealed a statistically significant right hand bias for coordinated bimanual actions during fruit consumption. The finding was in contrast to previous studies that did not demonstrate population-level right-handedness for bimanual behaviors in wild chimpanzees (e.g., [104, 106]). However, there were significant differences in the investigation parameters. The earlier studies did not isolate a specific behavior to evaluate. Instead, a combination of the 15 most frequent behaviors were pooled and assessed for hand dominance. It is possible that the pooling of different manual tasks obscured patterns that may have been revealed for specific tasks. The findings from Forrester and colleagues [92] were consistent with previous investigations that have strictly focused on bimanual coordinated feeding behavior in wild gorillas [105] and in captive gorillas [91].

3.4.3 Task

There is no standard task used for comparisons of lateralized manual behavior across individuals, populations, and species. However, investigations employing subsistence tool use contribute to a growing body of literature that has demonstrated population-level right-handedness in great apes (e.g. [107]). The nature of the task is paramount because it has been shown to influence hand choice even for unimanual actions [96–98]. For example, pooled simple reaching behaviors of gorillas, chimpanzees, and children directed towards objects or the self and social partners, revealed inconsistent results across species. However, when hand actions were analyzed within tasks, all three species demonstrated right hand biases for manual actions to objects and ambi-laterality to social stimuli. The authors argued that reaching to objects elicited the dominant hand required for object manipulation.

Tool Use

Over the past decade, a subsistence tool use task was developed and employed across an extensive number of captive great ape populations. The tube task necessitates coordinated bimanual actions to obtain a food reward from inside a plastic tube. Subjects grip the tube with one hand and use a finger to extract a high quality food reward from inside the tube. Robust and significant population-level right-handedness has been consistently reported across ape groups (e.g. [89, 91, 106, 108, 109]). Some have suggested that population-level handedness is restricted to captive apes because they are modeling the right-handedness of the humans to which they are exposed (see [82, 84]). Interestingly, population-level right-handedness patterns have now been revealed in apes across a variety of rearing histories, suggesting that, for this specific task, exposure to humans does not significantly impact population-level right-handedness in great apes [110, 111].

Gesture

It is widely accepted that humans possess a population-level right hand bias for communicative gesturing as opposed to non-communicative gestures (e.g. self touching) (e.g. [112]). Humans are also prone to right-hand-biased manual movements produced simultaneously when talking [113, 114]. Right-biased intentional communication gestures (e.g. pointing) during vocalization are evident early in infant development [115]. However, only recently have lateral biases been reported for communicative behaviors in nonhuman primates. Captive baboons (*Papio anubis*) demonstrated a population-level right hand bias for both coordinated bimanual actions and communicative gestures [116]. Additionally, captive chimpanzees preferentially gestured for food with the right hand [117], and also produced a right-handed bias for species-typical gestures when directed to both humans and conspecifics [118]. Studies of gestural communication in captive chimpanzees have reported both individual and population-level right-handedness (e.g. [108, 117]). Moreover, the first report of lateralized manual behavior for communicative gestures in wild chimpanzees found a right hand preference for intentional gestures, suggesting that exposure to humans cannot fully account for the right-biased lateralization of gestural behavior in great apes [119].

Findings from investigations of tool use and gesture provide evidence of population-level right-handedness in both human and great ape species. The findings suggest that these behaviors may be underpinned by shared neural processing properties of the left hemisphere. Moreover, the findings reflect a property of an early evolutionary division of labor between the hemispheres (e.g. [8]) inherited from a last common ancestor of humans and apes. To date, investigations indicate that the strength of great ape population-level right-handedness does not equal those reported for humans (e.g., 90% right-handed at the population-level, e.g., [35]). However, it is difficult to know if disparate testing methods across species have confounded comparisons of data sets. One hypothesis is that the strength of great ape hand dominance represents an intermediate stage along the phylogenetic trajectory of human manual lateralization. Great ape handedness may represent hand strength inherited by a last common ancestor before sophisticated tool use and modern language skills may have exaggerated the extreme manual laterality found in our own evolutionary lineage [29].

3.5 Incorporating Cross-Species Methodologies

Matching experimental parameters and measures across laboratories and species is not a trivial task. Hopkins performed the monumental task of re-assessing data sets from great ape laterality studies throughout the literature under the same evaluation criteria [89]. This investigation revealed five significant findings. First and foremost, it was reported that the collective group (based on HI scores) exhibited a significant right hand bias. Second, species differences were found in the distribution of handedness. Specifically, bonobos (*Pan paniscus*) and chimpanzees demonstrated population-level right-handedness, but gorillas and orangutans did not. Third, handedness was sensitive to the task type. Fourth, captive apes were more right-handed than wild apes, even though both settings for these species revealed population-level right-handedness. Finally, although not discussed within this chapter, it was noted that both the laterality and strength of hand preference for the genus *Pan*, was heritable. A further study with a substantially increased sample size of great apes (n = 774; 536 chimpanzees, 76 gorillas, 118 bonobos, 47 orangutans) has also investigated lateralized manual behavior employing the tube task [76]. Chimpanzees, bonobos, and gorillas all expressed population-level right-handedness, whereas orangutans demonstrated population-level left-handedness. The authors suggested that lateralized manual behavior is influenced by ecological adaptations associated with posture and locomotion and argue for hemispheric specialization in primates to be addressed within an evolutionary framework.

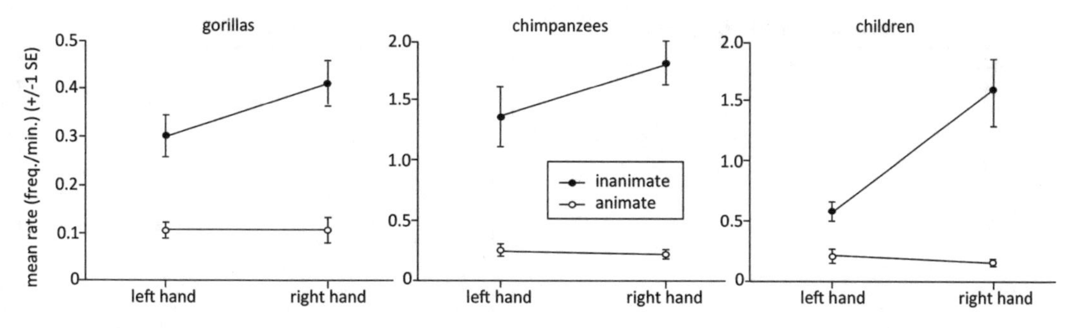

Fig. 3 This figure illustrates the similarity in pattern of hand dominance across human and great ape populations (figure adapted from Forrester et al. [65])

A triad of studies published in 2011, 2012 and 2013 sought to address comparative validity across human and nonhuman primates. Three investigations compared gorilla, chimpanzee, and child unimanual actions to objects and social partners under matched experimental conditions. Unimanual bouts were defined as those that incorporate an independent choice of hand for an action towards a target, and any instance in which one hand is occupied was not included for statistical analysis. Findings from these experiments revealed that gorillas [96], chimpanzees [97] and children [65] all presented right hand biases for manual actions towards objects, but not for manual actions for social stimuli (e.g. social partners or the self) (Fig. 3).

The results suggest that hand dominance is not human-unique, but likely to be inherited from an ancestor common to humans and apes. The authors suggested that right-handed object manipulation is not the direct manifestation of language processing, but for rudimentary processing of structured sequences of motor actions. The findings suggest that human population-level right-handedness is not a direct byproduct of language, but rather it is the manifestation of a left hemisphere dominance for controlling foundational motor sequencing. The results highlight the importance of cross-species investigations that place human behavior within an evolutionary framework.

3.5.1 Developing New Measures

With the knowledge that lateralized manual behavior can be influenced by a number of contextual factors, new measures are required to reveal the sensitivities of these motor biases. Tabiowo and Forrester noted that gorillas used coordinated bimanual actions to obtain the edible portion of food items (e.g. removal of shells or skins) [30]. Prior to the manipulation, gorillas often performed a "hand transfer." This behavior occurred when an item grasped by the dominant hand required manipulation. Gorillas were likely to transfer the item to the nondominant hand to be used as a support while the dominant hand performed the more

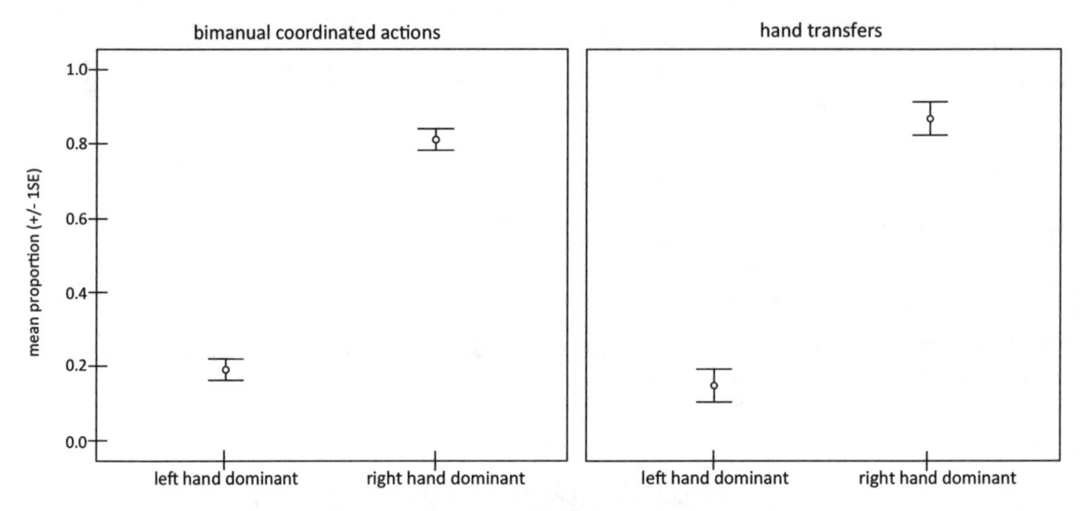

Fig. 4 This figure illustrates mean proportions of left and right hand dominance during bouts of bimanual coordinated hand actions and hand transfers (figure adapted from Tabiowo and Forrester [30])

dexterous motor activity. A significant bias was found for transfers from the right hand to the left hand compared to transfers from the left hand to the right hand. Moreover, hand transfers were shown to be a comparatively sensitive a measure of lateralized manual behavior as coordinated bimanual hand dominance for object manipulation (Fig. 4).

The authors suggested that it was more efficient for gorillas to take the time to transfer the object to the nondominant hand, freeing the dominant hand to perform the manipulation, than to attempt the manipulation with the nondominant hand. The investigation indicated that the development of additional measures is an untapped methodological resource that has the potential to unveil previously uncovered behavioral phenomena.

The studies discussed in the previous sections highlight the multidimensionality of lateralized manual behavior. Investigations of hand biases in great ape populations have indicated that lateralized manual biases fall on a gradient of strength and direction and is likely to be influenced by both phylogenetic and ontogenetic properties. Systematic experimental methods are required to tease out patterns of motor biases to gain a better understanding of the evolutionary origins of neural function and organization.

4 Lateralized Manual Behavior and Cognitive Development

Fine motor capabilities are considered to be a key precursor to the development of language function. While evolutionary psychology provides evidence for a common cognitive system underpinning goal-oriented object manipulation and language (e.g.

[53]), developmental psychology suggests that structured motor actions that underlie object manipulation and gesture set the stage for the acquisition of symbolic systems required for typical language acquisition [120]. It has not been suggested that there is a direct correlation between phylogeny and ontogeny (e.g. [121]). However, if higher cognitive functions build, at least in part, upon primary motor function, then we must acknowledge that primary motor processes will continue to play an important role in the development of cognitive abilities in humans. For example, typical language development requires the mastering of fine motor coordination prior to language production. Any anomalies present in early motor processes, due to genetic or experiential factors, will cascade to influence emerging higher cognitive functions [32].

4.1 Handedness in Children

Handedness is generally established by the time typically developing children start school (e.g. [122]). Handedness is often assessed through subjective self-reporting and surveys (e.g. Edinburgh Handedness Inventory) [36]. Human handedness is typically categorized as right, left, or mixed, although it is acknowledged that manual lateralization exists along a gradient that ranges from strongly left-handed to strongly right-handed (e.g. [123]). Handedness is considered to be the hand that is preferred for a specific task, regardless of performance; however, in some studies, the term can also reflect the hand that expresses greater efficiency for speed and accuracy (e.g. [124]). As discussed at the beginning of this chapter, associations have been drawn between hemispheric asymmetries associated with language and hand biases in children [125]. Specifically, the development of hand dominance in children has been linked with the successful hemispheric specialization for language [6]. Conversely, reduced right-handedness has been linked with increased incidence of clinical pathology (e.g. depression: [126]; schizophrenia: [127]) and reduced immune response [128]. Interestingly, animals with left forelimb biases also appear to possess less effective immune responses compared with animals demonstrating right forelimb biases (e.g. dogs: [129]; mice: [130]), suggesting that lateralized motor behaviors may act as an indirect marker for the efficiency of a range of foundational developmental processes.

4.2 Mixed-handedness

While some studies of human hand dominance argue that left-handedness can be an indicator of decreased cerebral lateralization (e.g. [131, 132]), others suggest that strong hand dominance (left or right) correlates with typical language acquisition [133]. The presence of hand dominance in children for manipulative tasks, for example in drawing, has been associated with typical neurodevelopment. Conversely, inconsistent hand dominance or "ambi-preference" has been associated with significantly lower developmental assessment scores in children [134, 135]. A growing body of evidence indicates that reduced cortical lateralization is associated with impaired cognitive function and can manifest as mixed-handedness (e.g. [136–138]).

Mixed-handedness makes up approximately 3–4% of the general population (e.g. [139]) but appears to rise to between 17% and 47% in populations of children diagnosed with autistic spectrum condition (ASC) (for a review see [140]). Additionally, children with ASC tend to present impairments in receptive and expressive language [141]. Language deficits are often the most obvious symptom of the disorder, but they typically lead to a relatively late diagnosis [142]. However, a lack of manual bias in these children suggests that ASC is likely to have an early developmental onset characterized by disrupted lateralization of brain function for language processes [e.g. 143] or the computational processes that underpin language function [144]. For example, some have suggested that ASC may result from aberrant neural pruning during development, disrupting the lateralization of early sensory and motor processes [145]. Infants with a familial risk for developing ASC have demonstrated significantly lower motor scores as early as 7 months of age [146] than children without a familial risk for developing ASC. Therefore, disrupted lateralization may be present and visible in these children as decreased manual dominance long before any language impairment becomes evident [147]. Behavioral investigations of motor action biases in child populations provide one avenue to investigate how motor processes support higher cognitive function in typically developing children and how early motor anomalies cascade to influence higher cognitive function in children with cognitive dysfunction.

4.3 Incorporating Cross-Species Methodologies

A recent case-by-case investigation of child manual behavior compared task-specific hand actions in children with and without autism [144]. The experimental parameters were adapted from earlier cross-species investigations of gorillas [96], chimpanzees [97] and children [65]. Video recordings were taken of each child as they participated in a clinical battery of tests used to diagnose ASC (Autistic Diagnostic Observational Schedule, ADOS) [148]. Lateralized manual behavior was coded offline and evaluated for task type. Neurotypical (NT) children demonstrated a statistically significant right hand bias for actions towards objects and a left hand dominance for actions directed to the self (e.g. self directed behaviors, SDB). Although a right hand bias for actions directed towards objects is consistent with previous studies of NT children [65], a left hand dominance for SDB illustrated a previously undetected pattern of motor bias.

The authors suggest that SDB reflect increased contribution of the right hemisphere for arousing situations (see Sect. 3). On the other hand, children with ASC presented ambi-preference for manual actions directed towards objects and the self. The findings

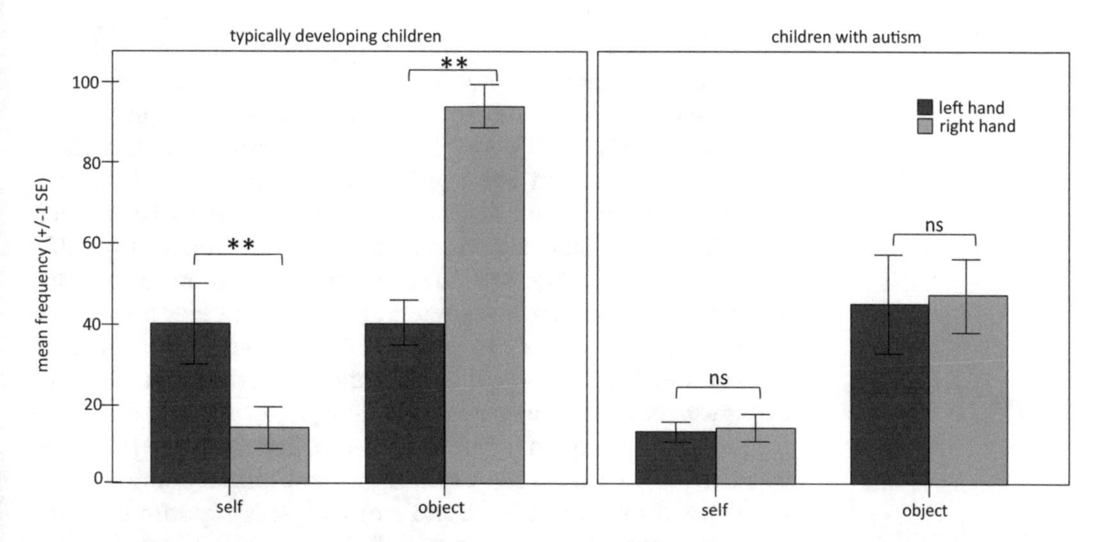

Fig. 5 This figure highlights the significant patterns of right hand dominance for hand actions towards objects and left hand dominance for hand actions towards the self. Conversely, children with autism did not demonstrate hand dominance for actions towards objects or the self (figure adapted from Forrester et al. [144])

are consistent with the hypothesis that children with ASC have disrupted cerebral lateralization compared with NT children [144]. Group patterns were mirrored by the individual participant patterns (Fig. 5).

The results of the investigations within this section indicate that motor actions can serve as a valuable indirect marker of cerebral lateralization. Lateralized motor behavior could allow for the longitudinal evaluation of cognitive developmental, providing a valuable indicator of when behavior diverges form typicality. Motor deficits occur across a range of psychopathologies [149]. Thus, a firm understanding of motor biases during early cognitive development may offer a valuable method for targeting and treating children at risk for the development of a range of cognitive deficits, and offer a link between behavior, neuropathology, and prognosis.

5 Right Hemisphere Specialization and Contralateral Motor Biases

5.1 Lateralized Manual Behavior in Humans and Nonhuman Animals

Most investigations of cerebral lateralization focus on assoications between left hemisphere dominance for right-sided motor actions. However, humans and nonhuman animal investigations of high social arousal, have demonstrated increased right hemisphere involvement and right-biased motor actions. For example, right-handed adults responded more quickly to unexpected stimuli with their left hand compared with their right hand (e.g. [150]). Additionally, NT right-handed children have demonstrated a significant left hand bias for SDB during a battery of challenging cognitive tasks [144]. In great apes, left-biased manual actions have also been associated with SDB. Specifically, rehabilitated orangutans exhibited a significant group-level lateralized preference for left-handed scratching and for the fine manipulation of parts of the face [151]. Gorillas and chimpanzees have demonstrated an increase in left hand activity during naturalistic unimanual SDB and hand actions directed towards social partners compared with hand actions directed towards objects [96, 97]. Additionally, although self-directed scratching showed no hand preference in chimpanzees, there was a significant bias for scratching on the left side of the body [152]. The findings from both human and nonhuman primate studies are consistent with the evolutionary theory that the right hemisphere emerged dominant for novel and urgent stimuli (for a review, see [23]). Taken together, the findings support a right hemisphere dominant role in the processing of emotive and arousal-increasing stimuli generally associated with negative stimuli. However, left-biased motor behaviors extend beyond manual actions in social contexts. The following sections consider lateralized motor behavior during social detection and social positioning.

5.1.1 Lateralized Social Detection

The ability to detect novel or urgent stimuli (e.g. predators) is paramount to the survival of an organism. Approaching and avoiding behaviors are well documented across animal species [153, 154]. Left eye dominance for predator avoidance and species recognition has been reported in fish [155], toads [156], lizards [157], pigeons [158], chicks [159], beluga whales [160] and in the striped-faced dunnart Sminthopsis macroura (*Dasyuridae*: *Marsupalia*) [161]. Studies of nonhuman primates that consider spontaneous species-specific encounters have also reported a left visual preference (right hemisphere dominant) during aggressive encounters in gelada baboons [162] and in a zoo-housed group of mangabeys [163] during spontaneous approach behaviors. Additional evidence from great apes report that high-ranking chimpanzees were approached significantly more frequently from their left visual hemifield, suggesting a right hemisphere facilitation for rapid identification of facial expressions and predicting behaviors [164]. These findings may reflect a common vertebrate characteristic of right hemisphere dominance for processing social stimuli to facilitate escape behavior and increased survival rates (e.g. [23, 165]).

A dominant role for the right hemisphere in face perception could have emerged from early evolutionary abilities to monitor conspecifics (see for a review [22]). A left gaze bias for face perception (e.g. looking time of centrally presented faces) has been reported in: sheep [166], dogs and rhesus monkeys [167] and chimpanzees [168]. Additionally, the left side of the face in nonhuman primates has been reported to display emotive expression earlier and more intensely than the right side of the face (chimpanzees:

[163]; macaques: [169]; marmosets: [170]; baboons [171]). These findings suggest dominant right hemisphere control for both the perception and production of emotion in these animals.

5.1.2 Lateralized Social Positioning

A left visual preference for detecting and monitoring conspecific behavior has ramifications for social positioning during natural animal behavior. Left lateralized motor behavior has been detected in the physical positioning strategies of animals within their social groups. A recent study of natural behavior in great apes revealed that individuals navigate around conspecifics with a bias for keeping social partners to their left side. Quaresmini and colleagues found group-level biases in both gorillas and chimpanzees for keeping conspecifics proximally situated to the left side of the observed individual compared with the right side [172] (Fig. 6).

The findings suggest that the right hemisphere may provide an advantage for monitoring the threat levels of conspecifics. Additionally, it has been proposed that social positioning can facilitate bonding with conspecifics, based on identity and facial expression (for a review, see [165]). In support of this theory, the majority of great ape investigations considering maternal cradling have revealed a preference for left-side cradling. For example, population-level left-sided cradling biases have been reported in chimpanzees (e.g. [173]) and gorillas (e.g. [174]), and this behavior appears to be largely unaffected by setting. Cradling bias is typically measured by considering the ventral position of the infant with a focus on the position of the infant's head in relation to the midline of the mother's chest. When the infant's head is orientated to the left of the mother's midline (regardless of the torso and limbs), the bout is

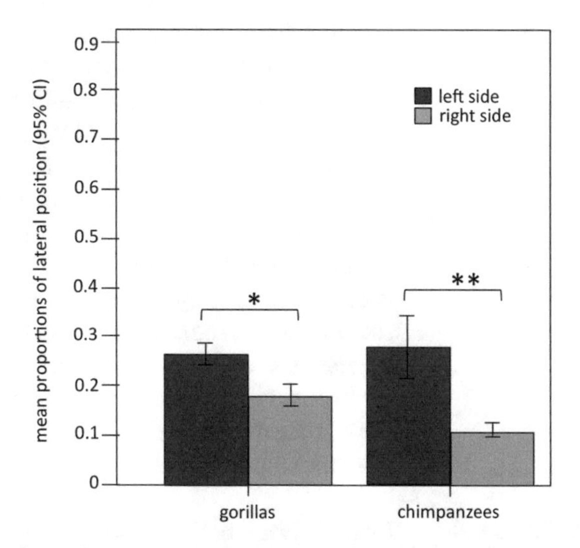

Fig. 6 This figure illustrates biases in the social positioning of gorillas and chimpanzees (figure adapted from Quaresmini et al. [172])

recorded as left-sided (e.g. [175]). The social positioning is likely to create an advantage of the right eye/left hemisphere for both infant and mother in support of rapid identification of facial identity and expression.

Based on our shared evolutionary history with great apes, it is critical that we consider human behavior as a continuum to natural animal behavior. Placing human behavior within an evolutionary framework will allow us to gain valuable insight into how modern human social-emotional processing emerged from foundational right hemisphere dominant neural processing. Some researchers have suggested that the right hemisphere possesses a sensitive attentional system that responds selectively for novel and dangerous stimuli in the environment (for a review, see [150]). Others have suggested that urgent responses to environmental threats may have contributed to the emergence of negative emotions [8, 34].

5.2 Lateralized Social Behavior in Humans

5.2.1 Lateralized Social Detection

Humans also exhibit left lateralized motor biases for processing social stimuli. For example, left gaze biases have been reported for face perception in humans (e.g. [176, 177]). Neuroimaging evidence has reported greater activation in the right hemisphere than the left hemisphere (in the region of the superior temporal sulcus) associated with the processing of approaching strangers with directed gaze compared with averted gaze [178]. Additionally, nonverbal, emotional vocalizations (e.g. cries and shouts) elicited greater right-hemisphere activation compared with the left hemisphere (for a review, see [179]). Moreover, individuals known to be right hand dominant demonstrate faster responses to unexpected stimuli with their left hand compared with their right hand (e.g. [150]).

Human findings for processing social stimuli are relatively consistent with nonhuman animal studies, however, they tend to be discussed in the light of two prevailing theories of cerebral lateralization unique to human emotion. The right hemisphere hypothesis (e.g. [180, 181]) proposes that the right hemisphere is solely responsible for the processing of emotion. Alternatively, the valence hypothesis (e.g. [182]) purports that both the right and the left hemispheres are involved in affect processing, such that the left hemisphere is dominant for positive affect and the right hemisphere is dominant for negative affect. Although there is debate over which theory more accurately reflects human social processing, the right hemisphere and the valence theories share a common theme with evolutionary theories of vertebrate brain function. Specifically, they both support the notion that the right hemisphere is dominant for primary responses to novel and threatening stimuli.

5.2.2 Lateralized Social Positioning in Humans

Little is known about how a right hemisphere bias for processing social stimuli influences natural human behavior. There is also a paucity of data relating to how humans navigate their social space.

Existing evidence comes mainly from cradling behaviors of parents with newborn children. Like great apes, human mothers and fathers prefer to position their offspring on left side of their bodies [183, 184]. The physical positioning is thought to facilitate the processing of social-emotional stimuli (e.g. gaze, facial expression) by establishing a direct route to the right hemisphere (for a review, see [185]). A recent longitudinal investigation found that a left side bias for infant cradling decreased below significance by 12 weeks of age. The authors argue that a left side cradling bias facilitates maternal monitoring of the infant state (see for a review, see [186]). Left-sided maternal cradling may even facilitate typical social development in children during the early weeks of development. One study revealed that babies who were held with a left arm preference developed a typical left visual field (right hemisphere) bias for faces on chimeric face tests, whereas babies reared with a right-arm cradling bias did not develop a visual field bias [187]. The findings suggest that experiential factors have the capacity to influence the development of cerebral lateralization of cognitive function. While disruption to left cerebral lateralization for fine motor dominance is associated with weaker language acquisition in young children (e.g. [138]), the ramification of disruption of right cerebral lateralization assoicated with social-emotional processing has yet to be explored within the scope of cognitive development.

Like great apes, human physical positioning within a social group appears to be influenced by cerebral lateralization for the processing of social stimuli. A recent study has provided the first evidence of lateralized social navigation in the natural behaviors of children. Extending upon animal studies of social positioning (e.g. [172]), this study observed the spontaneous navigational routes of children around adults, peers, and objects during play [188]. A focal individual was monitored for their directional path (left or right) around a stationary individual when there was equal opportunity to pass on either side (Fig. 7, left panel). In the control condition, the focal individual was monitored for their directional path around an inanimate object (e.g. a bin). Analyses revealed that children expressed a significant bias for choosing a right versus left navigational path around another human. Moreover, children expressed no lateral bias for navigating around an object (see Fig. 7, right panel).

A right navigational bias for locomoting around other people suggests that the navigating child presents the stationary individual with the left side of their body. A bias to keep conspecifics on the left side provides the navigator with an advantage for viewing the stationary individual with the left visual field. The left visual field would provide the most efficient route to the right hemisphere for processing identity, intention, and angry or fearful facial expressions in order to expedite escape behaviors. The pattern of findings

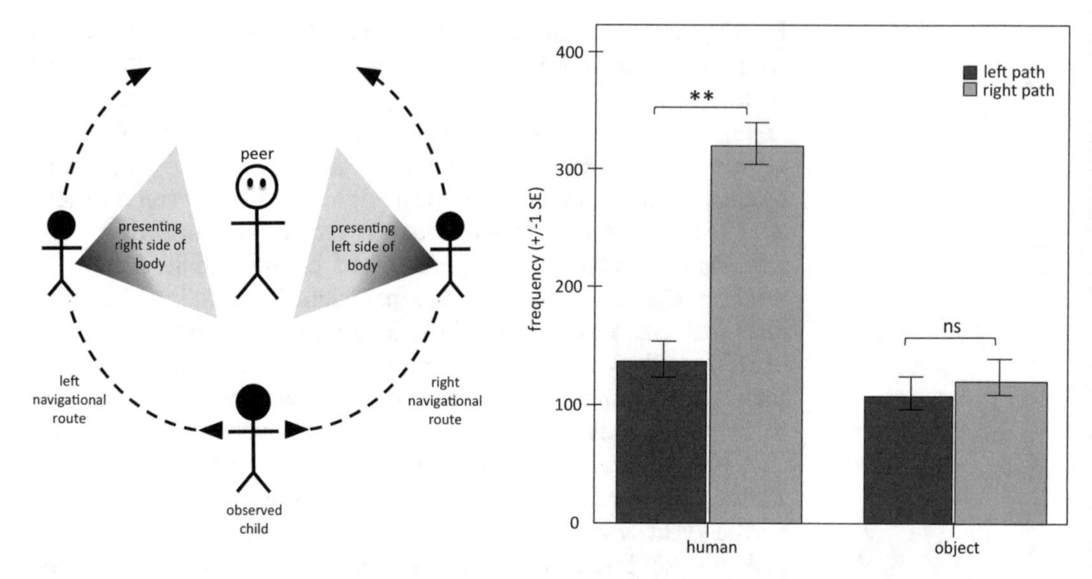

Fig. 7 This figure demonstrates the navigational path options around a stationary individual (*left panel*) and a significant bias for choosing a rightward compared to a leftward path (*right panel*) (figures adapted from Forrester et al. [189])

are consistent with animal studies that have demonstrated a left eye/right hemisphere preference bias for to monitoring familiar versus unfamiliar conspecifics in non-primate animals (e.g. chicks: [189]; fish: [190]). The results are also consistent with investigations of social positioning in chimpanzees and gorillas that indicated significant biases to expose the left side of the body to conspecifics [172].

Taken together, investigations of human motor actions to social stimuli suggest that the environment elicits predictable left-biased motor behavior. Future investigations may attempt to identify individuals who lack right hemisphere dominance for social processing in order to understand how motor behaviors cascade to influence cognitive development. Tracking the lateral biases for behavioral responses to the social environment may also shed light on how and which environmental pressures helped to shape modern human social behavior. When considering population-level motor biases, the environment may have been a critical pressure in aligning population behavior for social actions requiring cooperation (e.g. [191]).

6 Conclusions

The human and nonhuman animal findings discussed within this chapter suggest that context-dependent behaviors of vertebrates are dictated by evolutionarily old cerebral lateralization of brain function. An early delineation of hemisphere dominance may have served for carrying out basic survival strategies in parallel. Although the impetus to place human behavior and cognitive ability within an evolutionary framework may appear obvious in hindsight, we are only just beginning to practice well-controlled cross-species comparisons of behavior.

However, findings from investigations of cerebral lateralization and contralateral motor action must be treated with some caution. Studies indicate that cerebrally dominant functions are not solely processed by one hemisphere. For example, studies of social stimuli processing in monkeys (e.g. [192]), dogs [167], and sheep [193] also indicate small contributions from the left hemisphere. Thus, there may be limitations to the extent with which we can attribute a single function to a single hemisphere. Nevertheless, as we have seen throughout this chapter, contralateral motor dominance can act as a valuable and informative behavioral marker of brain organization (e.g. [22]) affording insight into how primary motor systems evolve and develop to support higher cognitive function.

Studies have indicated that a variety of factors can influence the laterality of motor behavior. Future investigations may extend a focus on nature (e.g. genetic) and nurture (e.g. social learning) contributions to lateralized motor behavior. Additionally, future investigations may consider why motor biases arise at the population-level rather than being equally distributed across individuals within a population. Authors from the present investigations have hypothesized that the social environment might have played a key role in aligning individual-level asymmetries, to coordinate the behavior among individuals for predator defense and for cooperation. Effectively, population-level motor asymmetries may have evolved to facilitate stable strategies for coordinated behavior [8, 191, 194].

In order to fully address evolutionary and developmental research questions regarding lateralized motor behavior we require the development of standard and systematic cross-species methodologies. Further research is necessary to understand the factors that have influenced the evolution, and are likely to also influence the development of species-specific lateralized motor behavior.

References

1. Sommerville JA, Decety J (2006) Weaving the fabric of social interaction: articulating developmental psychology and cognitive neuroscience in the domain of motor cognition. Psychon Bull Rev 13(2):179–200
2. Hommel B, Müsseler J, Aschersleben G, Prinz W (2001) The Theory of Event Coding (TEC): a framework for perception and action planning. Beh Br Sci 24(5):849–878
3. Rizzolatti G, Craighero L (2004) The mirror-neuron system. Annu Rev Neurosci 27: 169–192. doi:10.1146/annurev.neuro.27.070203.144230
4. Bradshaw JL, Rogers LJ (1993) The evolution of lateral asymmetries, language, tool-use and intellect. Academic, San Diego
5. Hellige JB (1993) Unity of thought and action: varieties of interaction between the

left and right hemispheres. Curr Dir Psychol Sci 2:21–25. doi:10.1111/1467-8721.ep10770559
6. Toga AW, Thompson PM (2003) Mapping brain asymmetry. Nat Rev Neurosci 4:37–48
7. Rogers LJ, Andrew RJ (2002) Comparative vertebrate lateralization. Cambridge University Press, Cambridge
8. Vallortigara G, Rogers LJ (2005) Survival with an asymmetrical brain: advantages and disadvantages of cerebral lateralization. Behav Brain Sci 28:575–589
9. Anfora G, Rigosi E, Frasnelli E, Ruga V, Trona F, Vallortigara G (2011) Lateralization in the invertebrate brain: left-right asymmetry of olfaction in bumble bee, *Bombus terrestris.* PLoS One 6(4):18903. doi:10.1371/journal.pone.0018903
10. Frasnelli E, Vallortigara G, Rogers LJ (2012) Left-right asymmetries of behaviour and nervous system in invertebrates. Neurosci Biobehav Rev 36:1273–1291
11. Alonso Y (1998) Lateralization of visual guided behaviour during feeding in zebra finches (*Taeniopygia guttata*). Behav Process 43:257–263
12. Rutldige R, Hunt G (2003) Lateralized tool use in wild New Caledonian crows. Anim Behav 67:327–332
13. Westergaard GC, Suomi SJ (1996) Hand preference for a bimanual task in tufted capuchins (*Cebus paella*) and rhesus macaques (*Macaca mulatta*). J Comp Psychol 110(4):406–411
14. Hopkins WD (2007) Evolution of hemispheric specialization in primates. Academic, Oxford
15. Franklin WE III, Lima SL (2001) Laterality in avian vigilance: do sparrows have a favourite eye? Anim Behav 62:879–885. doi:10.1006/anbe.2001.1826
16. Koboroff A, Kaplan G, Rogers LJ (2008) Hemispheric specialization in Australian magpies (*Gymnorhina tibicen*) shown as eye preferences during response to a predator. Brain Res Bull 76:304–306. doi:10.1016/j.brainresbull.2008.02.015
17. Rogers LJ (2000) Evolution of hemispheric specialization: advantages and disadvantages. Brain Lang 73:236–253. doi:10.1006/brln.2000.2305
18. Bonati B, Csermely D, Sovrano VA (2013) Looking at a predator with the left or right eye: asymmetry of response in lizards. Laterality 18:329–339. doi:10.1080/1357650X.2012.673623
19. Martín J, López P, Bonati B, Csermely D (2010) Lateralization when monitoring predators in the wild: a left eye control in the common wall lizard (Podarcis muralis). Ethology 116:1226–1233. doi:10.1111/j.1439-0310.2010.01836.x
20. Lippolis G, Bisazza A, Rogers LJ, Vallortigara G (2002) Lateralisation of predator avoidance responses in three species of toads. Laterality 7:163–183. doi:10.1080/13576500143000221
21. Giljov A, Karenina K, Ingram J, Malashichev Y (2015) Parallel emergence of true handedness in the evolution of marsupials and placentals. Curr Biol 25:1878–1884
22. MacNeilage PF, Rogers LJ, Vallortigara G (2009) Origins of the left and right brain. Sci Am 301:60–67
23. Rogers LJ, Vallortigara G, Andrew RJ (2013) Divided brains. The biology and behaviour of brain asymmetries. Cambridge University Press, New York. doi:10.1017/CBO9780511793899
24. Rutherford HJV, Lindell AK (2011) Author reply: More than evaluation: lateralization of the neural substrates supporting approach and avoidance motivational systems. Emot Rev 3:347–348. doi:10.1177/1754073911402404
25. Vallortigara G (2000) Comparative neuropsychology of the dual brain: a stroll through animals' left and right perceptual worlds. Brain Lang 73:189–219
26. Rogers LJ (2002) Lateralization in vertebrates: its early evolution, general pattern and development. Adv Stud Behav 31:107–162
27. Versace E, Vallortigara G (2015) Forelimb preferences in human beings and other species: multiple models for testing hypotheses on lateralization. Front Psychol 6:233. doi:10.3389/fpsyg.2015.00233
28. Ströckens F, Güntürkün G, Ocklenburga S (2013) Limb preferences in non-human vertebrates. Laterality 18(5):536–575
29. Marchant LF, McGrew WC (2013) Handedness is more than laterality: lessons from chimpanzees. Ann N Y Acad Sci 1288:1–8
30. Tabiowo T, Forrester GS (2013) Structured bimanual actions and hand transfers reveal population-level right-handedness in captive gorillas. Anim Behav 86:1049–1057
31. Cashmore L, Uomini N, Chapelain A (2008) The evolution of handedness in humans and great apes: a review and current issues. J Anthropol Sci 86:7–35
32. D'Souza D, Karmiloff-Smith A (2011) When modularization fails to occur: a developmental perspective. Cogn Neuropsychol 28(3–4):276–287
33. Greenfield PM (1991) Language, tools, and brain: the ontogeny and phylogeny of hierar-

chically organized sequential behavior. Behav Brain Sci 14:531–550
34. Vallortigara G, Chiandetti C, Sovrano VA (2011) Brain asymmetry (animal). Cogn Sci 2:146–157. doi:10.1002/wcs.100
35. McManus IC (2002) Right hand, left hand: the origins of asymmetry in brains, bodies, atoms, and cultures. Weidenfeld and Nicolson, London
36. Oldfield RC (1971) The assessment and analysis of handedness: the Edinburgh inventory. Neuropsychologia 9:97–113
37. Foundas AL, Leonard CM, Heilman KM (1995) Morphologic cerebral asymmetries and handedness: the pars triangularis and planum temporal. Arch Neurol 52(5):501–508
38. Crow T (2004) Directional asymmetry is the key to the origin of modern Homo sapiens (the Broca-Annett axiom): a reply to Rogers' review of The Speciation of Modern Homo Sapiens. Laterality 9:233–242
39. Ettlinger GF (1988) Hand preference, ability and hemispheric specialization. How far are these factors related in the monkey? Cortex 24:389–398
40. Warren JM (1980) Handedness and laterality in humans and other animals. Physiol Psychol 8:351–359
41. Williams NA, Close JP, Giouzeli M, Crow TJ (2006) Accelerated evolution of Protocadherin 11X/Y: a candidate gene-pair for cerebral asymmetry and language. Am J Med Genet B Neuropsychiatr Genet 141B:623–633
42. Broca P (1865) Sur le siège de la faculté du langage articulé. Bull Soc Anthropol Paris 6:377–393. doi:10.3406/bmsap.1865.9495
43. Annett M (2002) Handedness and brain asymmetry the right shift theory. Psychology Press, Sussex
44. Hauser MD, Chomsky N, Fitch WT (2002) The faculty of language: what is it, who has it, and how did it evolve? Science 298:1569–1579. doi:10.1126/science.298.5598.1569
45. Lashley KS (1951) The problem of serial order in behavior. In: Jeffress LA (ed) Cerebral mechanisms in behavior: the Hixon Symposium. Wiley, New York, pp 112–136
46. Armstrong DF, Stokoe WC, Wilcox SE (1995) Gesture and the nature of language. Cambridge University Press, Cambridge, MA
47. Bradshaw JL, Nettleton NC (1982) Language lateralization to the dominant hemisphere: tool use, gesture and language in hominid evolution. Curr Psychol 2:171–192. doi:10.1007/BF02684498
48. Corballis MC (2002) From hand to mouth: the origins of language. Princeton University Press, Princeton
49. Hamzei F, Rijntjes M, Dettmers C, Glauche V, Weiller C, Buchel C (2003) The human action recognition system and its relationship to Broca's area: an fMRI study. Neuroimage 19(3):637–644
50. Tettamanti M, Weniger D (2006) Broca's area: a supramodal hierarchical processor? Cortex 42:491–494. doi:10.1016/S0010-9452(08)70384-8
51. Pulvermüller F, Fadiga L (2010) Active perception: sensorimotor circuits as a cortical basis for language. Nat Rev Neurosci 11:351–360. doi:10.1038/nrn2811
52. Petersson KM, Folia V, Hagoort P (2012) What artificial grammar learning reveals about the neurobiology of syntax. Brain Lang 120:83–95. doi:10.1016/j.bandl.2010.08.003
53. Higuchi S, Chaminadeb T, Imamizua H, Kawatoa M (2009) Shared neural correlates for language and tool use in Broca's area. NeuroReport 20:1376–1381. doi:10.1097/WNR.0b013e3283315570
54. Nishitani N, Schurmann M, Amunts K, Hari R (2005) Broca's region: from action to language. Physiology 20(1):60–69. doi:10.1152/physiol.00043.2004
55. Sieg AE, Zandonà E, Izzo VM, Paladino FV, Spotila JR (2010) Population level "flipperedness" in the eastern Pacific leatherback turtle. Behav Brain Res 206:135–138
56. Bisazza A, Cantalupo C, Robins A, Rogers LJ, Vallortigara G (1996) Right-pawedness in toads. Nature 379:408. doi:10.1038/379408a0
57. Casey MB, Martino C (2000) Asymmetrical hatching behaviours influence the development of postnatal laterality in domestic chicks (*Gallus gallus*). Dev Psychol 34:1–12
58. Rogers LJ (1980) Lateralisation in the avian brain. Bird Behav 2:1–12. doi:10.3727/015613880791573835
59. Dimond S, Harries R (1984) Face touching in monkeys, apes and man: evolutionary origins and cerebral asymmetry. Neuropsychologia 22:227–233. doi:10.1016/0028-3932(84)90065-4
60. MacNeilage PF, Studdert-Kennedy MG, Lindblom B (1987) Primate handedness reconsidered. Behav Brain Sci 10:247–303. doi:10.1017/S0140525X00047695
61. MacNeilage PF (2007) The evolution of hemispheric specialization in primates. Spec Top Primatol 5:58–91
62. Tommasi L, Vallortigara G (1999) Footedness in binocular and monocular chicks. Laterality 4:89–95
63. Westergaard GC, Kuhn HE, Suomi SJ (1998) Bipedal posture and hand preference in humans

and other primates. J Comp Psychol 112: 55–64. doi:10.1037/0735-7036.112.1.55
64. Corbetta D (2003) Right-handedness may have come first: evidence from studies in human infants and nonhuman primates. Behav Brain Sci 26:217–218. doi:10.1017/S0140525X03320060
65. Forrester GS, Quaresmini C, Leavens DA, Mareschal D, Thomas MSC (2013) Human handedness: an inherited evolutionary trait. Behav Brain Res 237:200–206. doi:10.1016/j.bbr.2012.09.037
66. Byrne RW, Corp N, Byrne JME (2001) Manual dexterity in the gorilla: bimanual and digit role differentiation in a natural task. Anim Cogn 4:347–361. doi:10.1007/s100710100083
67. Videan EN, McGrew WC (2002) Bipedality in chimpanzee (Pan troglodytes) and bonobo (Pan paniscus): testing hypotheses on the evolution of bipedalism. Am J Phys Anthropol 118:184–190. doi:10.1002/ajpa.10058
68. Roffman I, Savage-Rumbaugh S, Rubert-Pugh E, Stadler A, Ronen A, Nevo E (2015) Preparation and use of varied natural tools for extractive foraging by bonobos (Pan Paniscus). Am J Phys Anthropol 158(1):78–91
69. Boesch C, Boesch-Achermann H (2000) The chimpanzees of the Taï Forest: behavioural ecology and evolution. Oxford University Press, Oxford, p 192
70. Breuer T, Ndoundou-Hockemba M, Fishlock V (2005) First observation of tool use in wild gorillas. PLoS Biol 3:380. doi:10.1371/journal.pbio.0030380
71. Hobaiter C, Poisot T, Zuberbuhler K, Hoppitt W, Gruber T (2014) Social network analysis shows direct evidence for social transmission of tool use in wild chimpanzees. PLoS Biol 12(9):e1001960. doi:10.1371/journal.pbio.1001960
72. Bard K (1992) Intentional behavior and intentional communication in young free-ranging Orangutans. Child Dev 62:1186–1197
73. Leavens DA, Hopkins WD (1998) Intentional communication by chimpanzees: a cross-sectional study of the use of referential gestures. Dev Psychol 34:813–822
74. Hobaiter C, Byrne RW (2011) The gestural repertoire of the wild chimpanzee. Anim Cogn 53:285–295. doi:10.1007/s10071-011-0409-2
75. Cantalupo C, Pilcher DL, Hopkins WD (2003) Are planum temporale and sylvian fissure asymmetries directly correlated? A MRI study in great apes. Neuropsychologia 41:1975–1981. doi:10.1016/S0028-3932(02)00288-9
76. Hopkins WD, Russell JL, Cantalupo C (2007) Neuroanatomical correlates of handedness for tool use in chimpanzees (Pan troglodytes) implication for theories on the evolution of language. Psychol Sci 18:971–977. doi:10.1111/j.1467-9280.2007.02011.x
77. Finch G (1941) Chimpanzee handedness. Science 94:117–118
78. Marchant LF, Steklis HD (1986) Hand preference in a captive island group of chimpanzees (*Pan troglodytes*). Am J Primatol 10:301–313
79. Parnell RJ (2001) Hand preference for food processing in wild western lowland gorillas (Gorilla gorilla gorilla). J Comp Psychol 115(4):365–375
80. Fletcher AW, Weghorst JA (2005) Laterality of hand function in naturalistically housed chimpanzees (Pan troglodytes). Laterality 10:219–242
81. Marchant LF, McGrew WC (2007) Ant fishing by wild chimpanzees is not lateralised. Primates 48:22–26
82. McGrew WC, Marchant LF (1997) On the other hand: current issues in and meta-analysis of the behavioral laterality of hand function in nonhuman primates. Yearb Phys Anthropol 40:201–232
83. Palmer AR (2002) Chimpanzee right-handedness reconsidered: evaluating the evidence with funnel plots. Am J Phys Anthropol 118:191–199
84. Palmer AR (2003) Reply to Hopkins and Cantalupo: chimpanzee right-handedness reconsidered. Sampling issues and data presentation. Am J Phys Anthropol 121: 382–384
85. Hopkins WD (2013) Comparing human and nonhuman primate handedness: challenges and a modest proposal for consensus. Dev Psychol 55(6):621–636
86. Altmann J (1974) Observational study of behaviour: sampling methods. Behaviour 49:227–267
87. Sovrano VA (2003) Visual lateralization in response to familiar and unfamiliar stimuli in fish. Behav Brain Res 152:385–391
88. Lambert M (2012) Hand preference for bimanual and unimanual feeding in captive gorillas: extension in a second colony of apes. Am J Phys Anthropol 148:641–647
89. Hopkins WD (2006) Comparative and familial analysis of handedness in great apes. Psychol Bull 132:538–559. doi:10.1037/0033-2909.132.4.538
90. Vauclair J, Meguerditchian A (2007) Perceptual and motor lateralization in two species of baboons. In: Hopkins WD (ed) Evolution of hemispheric specialization in primates. Academic, Oxford, pp 177–198

91. Meguerditchian A, Calcutt SE, Lonsdorf EV, Ross SR, Hopkins WD (2010) Brief communication: captive gorillas are right-handed for bimanual feeding. Am J Phys Anthropol 141:638–645
92. Forrester GS, Rawlings B, Davila-Ross M (2016) An analysis of bimanual actions in natural feeding of semi-wild chimpanzees. Am J Phys Anthropol 159(1):85–92
93. Roney LS, King JE (1993) Postural effects on manual reaching laterality in squirrel monkeys (Saimiri sciureus) and cotton-top tamarins (Saguinus oedipus). J Comp Psychol 107:380–385
94. Hopkins WD (1999) On the other hand: statistical issues in the assessment and interpretation of hand preference data in nonhuman primates. Int J Primatol 20:851–866
95. Hopkins WD (2013) Independence of data points in the measurement of hand preferences in primates: statistical problem or urban myth? Am J Phys Anthropol 151:151–157
96. Forrester GS, Leavens DA, Quaresmini C, Vallortigara G (2011) Target animacy influences gorilla handedness. Anim Cogn 14:903–907. doi:10.1007/s10071-011-0413-6
97. Forrester GS, Quaresmini C, Leavens DS, Spiezio C, Vallortigara G (2012) Target animacy influences chimpanzee handedness. Anim Cogn 15(6):1121–1127
98. Forrester GS (2008) A multidimensional approach to investigations of behaviour: revealing structure in animal communication signals. Anim Behav 76:1749–1760
99. Sugiyama Y, Fushimi T, Sakuro O, Matsuzawa T (1993) Hand preference and tool use in wild chimpanzees. Primates 34:151–159
100. McGrew WC, Marchant LF, Wrangham RW, Kleinm H (1999) Manual laterality in anvil use: wild chimpanzees cracking Strychnos fruits. Laterality 4:79–87
101. Marchant LF, McGrew WC (1996) Laterality of limb function in wild chimpanzees of Gombe National Park: comprehensive study of spontaneous activities. J Hum Evol 30:427–443
102. Boesch C (1991) Handedness in wild chimpanzees. Int J Primatol 6:541–558
103. Matsuzawa T, Yamakoshi G (1996) Comparison of chimpanzee material culture between Bossou and Nimba, West Africa. In: Russon AE, Bard KA, Parker ST (eds) Reaching into thought: the minds of the great apes. Cambridge University Press, Cambridge, pp 211–232
104. McGrew WC, Marchant LF (1996) On which side of the apes? Ethological studies of laterality of hand use. In: McGrew WC, Marchant LF, Nishida T (eds) Great ape societies. Cambridge University Press, Cambridge, pp 255–272
105. Byrne RW, Byrne JM (1991) Hand preferences in the skilled gathering tasks of mountain gorillas (*Gorilla gorilla berengei*). Cortex 27:521–536
106. McGrew WC, Marchant LF (2001) Ethological study of manual laterality in the chimpanzees of the Mahale mountains, Tanzania. Behaviour 138:329–358
107. Hopkins WD, Phillips KA, Bania A, Calcutt SE, Gardner M, Russell J, Schaeffer J, Lonsdorf EV, Ross SR, Schapiro SJ (2011) Hand preferences for coordinated bimanual actions in 777 great apes: implications for the evolution of handedness in Hominins. J Hum Evol 60:605–611
108. Hopkins WD, Cantalupo C, Freeman H, Russell J, Kachin M, Nelson E (2005) Chimpanzees are right-handed when recording bouts of hand use. Laterality 10:121–130
109. Meguerditchian A, Gardner MJ, Schapiro SJ, Hopkins WD (2012) The sound of one-hand clapping: handedness and perisylvian neural correlates of a communicative gesture in chimpanzees. Proc R Soc B 79:1959–1966
110. Llorente M, Mosquera M, Fabré M (2009) Manual laterality for simple reaching and bimanual coordinated task in naturalistic housed chimpanzees (Pan troglodytes). Int J Primatol 30:183–197
111. Llorente M, Riba D, Palou L, Carrasco L, Mosquera M, Colell M, Feliu O (2011) Population-level right-handedness for a coordinated bimanual task in naturalistic housed chimpanzees: replication and extension in 114 animals from Zambia and Spain. Am J Primatol 73:281–290
112. Dalby TJ, Gibson D, Grossi V, Schneider RD (1980) Lateralized hand gesture during speech. J Motor Behav 12:292–297. doi:10.1080/00222895.1980.10735228
113. Kimura D (1973) Manual activity during speaking: I. Right-handers. Neuropsychologia 11:45–50
114. Kimura D (1973) Manual activity during speaking: II. Left-handers. Neuropsychologia 11:51–55
115. Blake J, O'Rourke P, Borzellino G (1994) Form and function in the development of pointing and reaching gestures. Infant Behav Dev 17:195–203
116. Meguerditchian A, Molesti S, Vauclair J (2011) Right-handedness predominance in 162 baboons (*Papio anubis*) for gestural communication: consistency across time and groups. Behav Neurosci 25(4):653–660

117. Hopkins WD, Leavens DA (1998) Hand use and gestural communication in chimpanzees (*Pan troglodytes*). J Comp Psychol 112(1): 95–99
118. Meguerditchian A, Vauclair J, Hopkins WD (2010) Captive chimpanzees use their right hand to communicate with each other: implications for the origin of the cerebral substrate for language. Cortex 46:40–48
119. Hobaiter C, Byrne RW (2013) Laterality in the gestural communication of wild chimpanzees. Ann N Y Acad Sci 1288:9–1. doi:10.1111/nyas.12041
120. Carlson NR, Buskist W, Enzle ME, Heth CD (2005) Psychology: the science of behaviour, 3rd edn. Pearson Education, Canada, p 384
121. Gilbert SF (2006) "Ernst Haeckel and the biogenetic law". Developmental biology, 8th edn. Sinauer Associates, Sunderland, MA
122. Gudmundsson E (1993) Lateral preference of preschool of primary school children. Percept Mot Skills 77:819–828
123. Beaton AA (2003) The determinants of handedness. In: Hugdahl K, Davidson RJ (eds) Brain asymmetry, 2nd edn. MIT Press, Cambridge, MA, pp 105–158
124. Healy JM, Liederman J, Geschwind N (1986) Handedness is not a unidimensional trait. Cortex 22(1):33–53
125. Hervé PY, Crivello F, Perchey G, Mazoyer B, Tzourio-Mazoyer (2006) Handedness and cerebral anatomical asymmetries in young adult males. Neuroimage 29:1066–1079
126. Denny K (2009) Handedness and depression: evidence from a large population survey. Laterality 14:246–255. doi:10.1080/13576500802362869
127. Dragovic M, Hammond G (2005) Handedness in schizophrenia: a quantitative review of evidence. Acta Psychiatr Scand 111:410–419. doi:10.1111/j.1600-0447.2005.00519.x
128. Stoyanov Z, Decheva L, Pashalieva I, Nikolova P (2011) Brain asymmetry, immunity, handedness. Cent Eur J Med 7:1–8. doi:10.2478/s11536-011-0121-2
129. Quaranta A, Siniscalchi M, Frate A, Vallortigara G (2004) Paw preference in dogs: relations between lateralised behaviour and immunity. Behav Brain Res 153:521–525. doi:10.1016/j.bbr.2004.01.009
130. Neveu PJ, Barnéoud P, Vitiello S, Betancur C, Le Moal M (1988) Brain modulation of the immune system: association between lymphocyte responsiveness and paw preference in mice. Brain Res 457:392–394. doi:10.1016/0006-8993(88)90714-7
131. Soper HV, Satz P, Orsini DJ, Henry RR, Zvi JC, Schulman M (1986) Handedness patterns in autism suggest subtypes. J Autism Dev Disord 16(2):155–167
132. Dane S, Balci N (2007) Handedness, eyedness and nasal cycle in children with autism. Int J Dev Neurosci 25:223–226
133. Leask SJ, Crow TJ (2001) Word acquisition reflects lateralization of hand skill. Trends Cogn Sci 5(12):513–516
134. Kastner-Koller U, Deimann P, Bruckner J (2007) Assessing handedness in preschoolers: construction and initial validation of a hand preference test for 4–6-year-olds. Psychol Sci 49(3):239–254
135. Gérard-Desplanches A, Deruelle C, Stefanini S, Ayoun C, Volterra V, Vicari S et al (2007) Laterality in persons with intellectual disability: II. Hand, foot, ear, and eye laterality in persons with trisomy 21 and williams-beuren syndrome. Dev Psychobiol 48:82–91. doi:10.1002/dev.20163
136. Crow TJ, Crow LR, Done DJ, Leask S (1998) Relative hand skill predicts academic ability: global deficits at the point of hemispheric indecision. Neuropsychology 36(12):1275–1282
137. Yeo RA, Gangestad SW, Thoma RJ (2007) Developmental instability and individual variation in brain development: implications for the origin of neurodevelopmental disorders. Curr Dir Psychol Sci 16(5):245–249
138. Rodriguez A, Kaakinen M, Moilanen I, Taanila A, McGough JJ, Loo S et al (2010) Mixed-handedness is linked to mental health problems in children and adolescents. Pediatrics 125(2):340–348
139. Satz P, Soper HV, Orsini DL, Henry RR, Zvi JC (1985) Handedness subtypes in autism. Psychiatr Ann 15:447–450
140. Lindell AK, Hudry K (2013) Atypicalities in cortical structure, handedness, and functional lateralization for language in autism spectrum disorders. Neuropsychol Rev 23:257–270
141. Tager-Flusberg H, Caronna E (2007) Language disorders: autism and other pervasive developmental disorders. Pediatr Clin North Am 54(3):469–481
142. Wetherby A, Woods J, Allen L, Cleary J, Dickson H, Lord C (2004) Early indicators of autism spectrum disorders in the second year of life. J Autism Dev Disord 34(5):473–493
143. Knaus TA, Silver AM, Kennedy M, Lindgren KA, Dominick KC, Siegel J et al (2010) Language laterality in autism spectrum disorder and typical controls: a functional volumetric, and diffusion tensor MRI study. Brain Lang 112(2):113–120
144. Forrester GS, Pegler R, Thomas MSC, Mareschal D (2014) Handedness as a marker of cerebral lateralization in children with and

without autism. Behav Brain Res 15:4–21. doi:10.1016/j.bbr.2014.03.040
145. Thomas MSC, Knowland VCP, Karmiloff-Smith A (2011) Mechanisms of developmental regression in autism and the broader phenotype: a neural network modeling approach. Psychol Rev 118(4):637–654
146. Leonard HC, Elsabbagh M, Hill EL, the BASIS team (2014) Early and persistent motor difficulties in infants at-risk of developing autism spectrum disorder: a prospective study. Eur J Dev Psychol 11(1):18–35
147. Elsabbagh M, Johnson MH (2010) Getting answers from babies about autism. Trends Cogn Sci 14:81–87
148. Lord C, Risi S, Lambrecht L, Cook EH Jr, Leventhal BL, DiLavore PC et al (2000) The autism diagnostic observation schedule-generic: a standard measure of social and communication deficits associated with the spectrum of autism. J Autism Dev Disord 30(3):205–223
149. Scharoun SM, Bryden PJ (2014) Hand preference, performance abilities and hand selection in children. Front Psychol 5:82
150. Fox MD, Corbetta M, Snyder AZ, Vincent JL, Raichle ME (2006) Spontaneous neuronal activity distinguishes human dorsal and ventral attentional systems. Proc Natl Acad Sci U S A 103(35):10046–10051. doi:10.1073/pnas.0606682103
151. Rogers LJ, Kaplan G (1996) Hand preferences and other lateral biases in rehabilitated orangutans, *Pongo pygmaeus pygmaeus*. Anim Behav 51:13–25. doi:10.1006/anbe.1996.0002
152. Hopkins WD, Russell JL, Freeman H, Reynolds EAM, Griffis C, Leavens DA (2006) Lateralized scratching in chimpanzees (*Pan troglodytes*): evidence of a functional asymmetry during arousal. Emotion 6(4):553–559
153. Quaranta A, Siniscalchi M, Vallortigara G (2007) Asymmetric tail-wagging responses by dogs to different emotive stimuli. Curr Biol 17:199–201
154. Siniscalchi M, Lusito R, Vallortigara G, Quaranta A (2013) Seeing left- or right-asymmetric tail wagging produces different emotional responses in dogs. Curr Biol 23:2279–2282
155. De Santi A, Sovrano A, Bisazza G, Vallortigara G (2001) Mosquitofish display differential left- and right-eye use during mirror-image scrutiny and predator- inspection responses. Anim Behav 61:305–310
156. Robins A, Lippolis G, Bisazza A, Vallortigara G, Rogers LJ (1998) Lateralized agonistic responses and hindlimb use in toads. Anim Behav 56:875–881
157. Hews DK, Worthington RA (2001) Fighting from the right side of the brain: left visual field preference during aggression in free-ranging male lizards (*Urosaurus ornatus*). Brain Behav Evol 58:356–361
158. Nagy M, Àkos Z, Biro D, Vicsek T (2010) Hierarchical group dynamics in pigeon flocks. Nature 464:890–893. doi:10.1038/nature08891
159. Vallortigara G, Andrew RJ (1991) Lateralization of response to change in a model partner by chicks. Anim Behav 41:187–194
160. Karenina K, Giljov A, Baranov V, Osipova L, Krasnova V, Malashichev Y (2010) Visual laterality of calf-mother interactions in wild whales. PLoS One 5(11):e13787. doi:10.1371/journal.pone.0013787
161. Lippolis G, Westman W, McAllan BM, Rogers LJ (2005) Lateralization of escape responses in the striped-faced dunnart, Sminthopsis macroura (*Dasyuridae*: *Marsupalia*). Laterality 10:457–470
162. Casperd JM, Dunbar RIM (1996) Asymmetries in the visual processing of emotional cues during agonistic interactions by gelada baboons. Behav Processes 37:57–65. doi:10.1016/0376-6357(95)00075-5
163. Baraud I, Buytet B, Bec P, Blois-Heulin C (2009) Social laterality and 'transversality' in two species of mangabeys: influence of rank and implication for hemispheric specialization. Behav Brain Res 198:449–458
164. Fernández-Carriba S, Loeches A, Morcillo A, Hopkins WD (2002) Functional asymmetry of emotions in primates: new findings in chimpanzees. Brain Res Bull 57:561–564. doi:10.1016/S0361-9230(01)00685-2
165. Rosa Salva O, Regolin L, Mascalzoni E, Vallortigara G (2012) Cerebral and behavioural asymmetry in animal social recognition. Comp Cogn Behav Rev 7:110–138. doi:10.3819/ccbr.2012.70006
166. Peirce JW, Leigh AE, Kendrick KM (2000) Configurational coding, familiarity and the right hemisphere advantage for face recognition in sheep. Neuropsychologia 38:475–483. doi:10.1016/S0028-3932(99)00088-3
167. Guo K, Meints K, Hall C, Hall S, Mills D (2009) Left gaze bias in humans, rhesus monkeys and domestic dogs. Anim Cogn 12:409–418. doi:10.1007/s10071-008-0199-3
168. Morris RD, Hopkins WD (1993) Perception of human chimeric faces by chimpanzees: evidence for a right hemisphere advantage. Brain Cogn 21:111–122. doi:10.1006/brcg.1993.1008
169. Hauser MD (1993) Right hemisphere dominance for the production of facial expression in monkeys. Science 261:475–477. doi:10.1126/science.8332914

170. Hook-Costigan MA, Rogers LJ (1998) Lateralized use of the mouth in production of vocalizations by marmosets. Neuropsychologia 36:1265–1273. doi:10.1016/S0028-3932(98)00037-2
171. Wallez C, Vauclair J (2011) Right hemisphere dominance for emotion processing in baboons. Brain Cogn 75:164–169. doi:10.1016/j.bandc.2010.11
172. Quaresmini C, Forrester GS, Speizio C, Vallortigara G (2014) Social environment elicits lateralized behaviours in gorillas and chimpanzees. J Comp Psychol 128:276–284
173. Nishida T (1993) Left nipple suckling preference in wild chimpanzees. Ethol Sociobiol 14:45–52
174. Manning JT, Heaton R, Chamberlain AT (1994) Left-side cradling: similarities and differences between apes and humans. J Hum Evol 26:77–83
175. Manning JT, Chamberlain AT (1990) Left-side cradling preference in great apes. Anim Behav 39:1224–1226
176. Burt DM, Perret DI (1997) Perceptual asymmetries in judgments of facial attractiveness, age, gender, speech and expression. Neuropsychologia 35:685–693. doi:10.1016/S0028-3932(96)00111-X
177. Kanwisher N, Tong F, Nakayama K (1998) The effect of face inversion on the human fusiform face area. Cognition 68:B1–B11. doi:10.1016/S0010-0277(98)00035-3
178. Pelphrey KA, Viola RJ, McCarthy G (2004) When strangers pass: processing of mutual and averted social gaze in the superior temporal sulcus. Psychol Sci 15(9):598–603
179. Scott SK, Sauter D, McGettigan C (2009) Brain mechanisms for processing perceived emotional vocalizations in humans. In: Brudzynski SM (ed) Handbook of mammalian vocalization: an integrative neuroscience approach. Academic, London, pp 187–198
180. Campbell R (1982) Asymmetries in moving faces. Br J Psychol 73:95–103. doi:10.1111/j.2044-8295.1982.tb01794.x
181. Borod JC, Obler KL, Erhan HM, Grunwald IS, Cicero BA, Welkowitz J, Santschi C, Agosti RM, Whalen JR (1998) Right hemisphere emotional perception: evidence across multiple channels. Neuropsychology 12:446–458. doi:10.1037/0894-4105.12.3.446
182. Davidson RJ (1995) Cerebral asymmetry, emotion, and affective style. In: Davidson RJ, Hugdahl K (eds) Brain asymmetry. MIT Press, Cambridge, MA, pp 361–389
183. Scola C (2009) The importance of mother–infant relationship for holding-side preferences. Enfance 61:433–457
184. Scola C, Vauclair J (2010) Infant's holding side biases by fathers in maternity hospitals. J Reprod Infant Psychol 28:3–10
185. Scola C, Vauclair J (2010) Is infant holding-side bias related to motor asymmetries in mother and child? Dev Psychobiol 52: 475–486
186. Todd BK, Banerjee R (2015) Lateralization of infant holding by mothers: a longitudinal evaluation of variations over the first 12 weeks. Laterality 21(1):12–33
187. Vervloed MPJ, Hendriks AW, van den Eijnde E (2011) The effects of mothers' past infant-holding preferences on their adult children's face processing lateralisation. Brain Cogn 75(3):248–254. doi:10.1016/j.bandc.2011.01.002
188. Forrester GS, Crawley M, Palmer C (2014) Social environment elicits lateralized navigational paths in two populations of typically developing children. Brain Cogn 91: 21–17
189. Vallortigara G, Cozzuti C, Tommasi L, Rogers LJ (2001) How birds use their eyes: opposite left-right specialization for the lateral and frontal visual hemifield in the domestic chick. Curr Biol 11:29–33. doi:10.1016/S0960-9822(00)00027-0
190. Brown C, Western J, Braithwaite VA (2007) The influence of early experience on, and inheritance of, cerebral lateralization. Anim Behav 74:231–238. doi:10.1016/j.anbehav.2006.08.014
191. Ghirlanda S, Frasnelli E, Vallortigara G (2009) Intraspecific competition and coordination in the evolution of lateralization. Philos Trans R Soc B Biol Sci 364:861–866. doi:10.1098/rstb.2008.0227
192. Pinsk MA, DeSimone K, Moore T, Gross CG, Kastner S (2005) Representations of faces and body parts in macaque temporal cortex: a functional MRI study. Proc Natl Acad Sci U S A 102:6996–7001. doi:10.1073/pnas.0502605102
193. Peirce JW, Kendrick KM (2002) Functional asymmetry in sheep temporal cortex. Neuroreport 13(18):2395–2399
194. Ghirlanda S, Vallortigara G (2004) The evolution of brain lateralization: a game theoretical analysis of population structure. Proc R Soc B Biol Sci 271:853–857. doi:10.1098/rspb.2003.2669

Chapter 6

Lateralization in Invertebrates

Elisa Frasnelli

Abstract

This chapter discusses the methods of studying behavioral lateralization in invertebrate animals. Although to date not a great deal is known about lateralized behavior and cognitive function in invertebrates, a number of studies have provided evidence of lateralization in a range of invertebrate species. Behavioral asymmetries have been shown in phyla such as Arthropoda (Insecta, Arachida, and Malacostraca), Mollusca (Gastropoda and Cephalopoda) and Nematoda, and in a variety of behaviors. Here I report the findings of research conducted on lateralization in invertebrates with a specific focus on the methodology adopted. Behavioral asymmetries in the invertebrate line have been investigated by observing biases in different types of behavior that can be classified in six main groups corresponding to the six sections of the chapter (summarized in Table 1). These six sections analyze the methods used to investigate lateral biases in (1) catching prey and foraging behavior; (2) escape response; (3) interactions with conspecifics (aggressive and sexual behavior); (4) spontaneous motor behavior (preferential choice in a T-maze); (5) sensory modalities (olfaction, vision, and hearing); and (6) recall of memory associated with conditioning in one of these sensory modalities. For each method the advantages and disadvantages of using it are examined and the main findings are reported and discussed.

Key words Behavioral lateralization, Invertebrates, Catching prey, Foraging behavior, Escape response, Aggressive behavior, Sexual behavior, Motor bias, Sensory modalities, Memory recall

1 Introduction

Functional specialization of the right and left sides of the brain is a characteristic feature of the vertebrate brain. In fact, brain and behavioral lateralization have traditionally been considered uniquely human characteristics, subserving handedness, and linguistic functions [1]. Over the past four decades, research in comparative psychology, neuroscience, and developmental biology has demonstrated that lateralization is a widespread phenomenon among vertebrates [2]. It is not only restricted to limb preference (see for a recent review [3]), but it can also manifest itself as functional side biases in other motor behavior (e.g. a turning preference of an individual persisting under different experimental conditions, see [4–6]) and in sensory perception

Lesley J. Rogers and Giorgio Vallortigara (eds.), *Lateralized Brain Functions: Methods in Human and Non-Human Species*, Neuromethods, vol. 122, DOI 10.1007/978-1-4939-6725-4_6, © Springer Science+Business Media LLC 2017

(e.g. a side preference when detecting or learning about visual stimuli, see [7–11]). Having a lateralized brain confers several advantages by (1) avoiding competition between the hemispheres in their control of behavior, (2) using neuronal tissue sparsely by avoiding the duplication of functions and (3) enabling separate and parallel processing to increase processing power and speed [12]. All of this is thought to enhance cognition so that humans and vertebrate animals can display flexible behavior and solve complex problems.

Table 1
Summary table of the types of behavior investigated and the relative methods applied to study behavioral asymmetries in different invertebrate species

Behavior	Method	Species
Prey catching/ Fora ging	Observation of the direction adopted on natural and laboratory foraging trails	Ants *Lasius* spp.
	Recording of the antennae used in antennal contacts during food exchange	Ants *F. aquilonia*
	Observations of leg lost in the field and leg use during prey handling in the laboratory	Spiders (Araneae) *S. globula*
	Observations of the direction in circling behavior while foraging on vertical inflorescences in the field	Bumblebees *Bombus* spp.
	Laboratory recordings of eye use during prey attack (plastic crab or real crab) and of the eye and arm used in food retrieve and exploration in the tank	Octopus *O. vulgaris*
Escape response	Laboratory tests of the direction of escape response to a vibratory stimulus	Atyid shrimp *N. denticulata*
	Laboratory tests of the direction of escape response induced by tapping the rostrum with a piece of foam on a stick	Crayfish *P. clarkia*
	Observation of the direction of escape response when individuals are alarmed around natural nest and in the laboratory	Ants *L. niger*
Aggressive behavior	Field experiments on left- and right-clawed males to test their success in defending the burrow, fighting, and mating	Fiddler crab *U. v. vomeris*
	Laboratory recordings of social and aggressive interaction in dyads of individuals belonging to the same or different colony and presenting only the right, only the left, or both antennae	Honeybee *A. mellifera*
	Laboratory recordings of social and aggressive interaction in dyads of individuals belonging to the same or different sex and presenting only the right, only the left, or both antennae	Mason bee *O. bicornis*

(continued)

Table 1 (continued)

Behavior	Method	Species
	Laboratory recordings of the body side used in aggressive interactions in dyads of individuals belonging to the same or different sex	Fruit fly *C. capitata* Olive fruit fly *B. oleae*
	Laboratory recordings of the body side used in aggressive male–male interactions	Blowfly *C. vomitoria*
Sexual behavior	Laboratory recordings of the side males approach females during courtship and mating	Olive fruit fly *B. oleae*
	Observation of the body side exposed to partners in the laboratory	Ants *L. niger*
	Laboratory recordings of the body side used by females in aggressive response against undesired males	Tiger mosquito *A. albopictus*
	Laboratory observation of the direction in circling precopulatory behavior and correlation with the body chirality and its heritability	Snail *L. stagnalis*
	Laboratory observation of lateral bias in mating and correlation with the body side of reproductive organs and brain correlates	Snail *H. aspersa*
	Observation of penis use in individuals collected in the field and in generations raised in the laboratory	Earwig *L. riparia*
	Laboratory observation of asymmetry in wing use during calling songs to attract females	Field cricket *C. campestris*
Spontaneous motor bias	Laboratory observation of preferential turning in one of the two arms of a T-maze	Giant water bug *B. flumineum Say*
	Laboratory observation of preferential turning in one of the two arms of a T-maze in animals of different ages and by providing or not visual cues	Cuttlefish *S. officinalis*
	Laboratory observation of preferential turning in one of the two arms of a T-maze	Octopus *O. vulgaris*
	Laboratory observation of lateral biases in exploring an unknown nest	Ants *T. albipennis*
	Laboratory test for preferential choice in a Y-maze with odors at the end of each arm in animals presenting both antennae or with one antenna amputated	Cockroach *P. americana*
	Laboratory tracking of the right and left turns in individual Y-maze; Laboratory tracking of the circling direction in a open arena	Fruit fly *D. melanogaster*

(continued)

Table 1 (continued)

Behavior	Method	Species
	Laboratory recordings of the forelimb used to reach across a gap	Locust *S. gregaria*
Sensory modalities	Classical conditioning experiments (PER paradigm) to study the ability to learn the association between an odor (or a color) and a food reward in animals with the right or left antenna (or eye) coated	Honeybee *A. mellifera*
	Laboratory experiments to determine how landmarks are used in navigation	Ants *T. albipennis*
	Laboratory tests in an olfactory visual flight simulator on tethered animals with one antenna occluded	Fruit fly *D. melanogaster*
	Laboratory recordings of chemotaxis in larvae genetically manipulated to have a single functional olfactory sensory neuron on the right or left side	Fruit fly *D. melanogaster*
	Laboratory tests of the ability to orient to conspecific call in females with reduced sensitivity of either the right or the left ear	Bushcricket *R. verticalis*
Learning and memory	PER paradigm to study the recall of olfactory memory at different times in animals with one antenna coated with a silicon compound	Honeybee *A. mellifera*
	PER paradigm to study the recall of olfactory memory at different times in intact animals by lateral odor presentation	Different species of bees
	Conditioning experiment (odors associate with or without an electric shock) and recall of memory tests by presenting both odors in a T-maze in animals presenting asymmetrical and symmetrical brain structures	Fruit Fly *Drosophila*
	Conditioning and test for odor-aversion memory in animals with unilateral ablation of the procerebrum	Slug Limax

More recently, the assumption that lateralization is an exclusive feature of large-sized brains in the vertebrate line has been challenged. Behavioral side biases and underlying anatomical asymmetries have been found in invertebrates (summarized by [13]) suggesting that lateralization should be conceptualized as an essential process in brains of different sizes and complexity. Investigating behavioral and brain asymmetries in invertebrates is extremely important for a number of reasons. Firstly, invertebrates are relatively simple organisms, which are easily available and can be studied in a variety of controlled behavioral and neurophysiological experiments. Secondly, they are very good models to investigate the genetic bases, developmental mechanisms, and environmental conditions that cause and shape the expression of lateralized behavior. Thirdly, some invertebrate species (e.g. some insect species and the nematode *C. elegans*) provide many advantages for uncovering principles of brain organization and evolution since much is known about the anatomical and functional organization of their nervous systems, as well as how they generate and control various behaviors. Finally and very importantly, comparative studies are of vital importance in order to provide essential evidence of common or/and different lateralization patterns between vertebrates and invertebrates and to shed light on the evolution of brain and behavioral asymmetries (see [14]).

2 Catching Prey and Foraging Behavior

Studies conducted on several vertebrate species have shown that the left hemisphere is specialized in controlling behavioral routines and establishing behavioral patterns (summarized in [15, 16]). Catching prey (or foraging) is a typical routine behavior. Nearly every class of vertebrate animal tested so far (from mammals to birds, from amphibians to fishes) displays a right-side—hence left-brain—bias for controlling behavior patterns under ordinary circumstances such as feeding. Recently, biases in prey catching and foraging behavior have been shown in invertebrates by simply observing their behavior in natural conditions or in the laboratory.

Heuts and Brunt [17] recorded the direction of 13 temperate and tropical ant species *Lasius* spp. on natural and laboratory foraging trails. They showed that 12 ant species kept mainly to the right side of their foraging "streets", whereas there was only one species that kept to the left. Further observations were made for several *Lasius niger* colonies on tree trails. Results showed that 49 *Lasius niger* colonies kept to the right versus 26 to the left. Moreover, when running slowly around the natural nest at more than 10 cm from the closest nest-mate, lone foraging *L. niger* ants turn to the right significantly more often to the left (the ratio was 14 to 2).

The black-meadow ant *Formica pratensis* has been used as model organism to identify traffic rules on ant trails in the field [18]. To organize the traffic between nest and food source, *F. pratensis* establishes permanent trunk trails, which are maintained by the ants. By following 74 ants (37 inbound and outbound each) on a 50-cm long part of the trail including the focus section and 35 cm of the trail outwards, Hönicke et al. [18] examined if the ants show an individual-level lateralization whilst leaving or avoiding an encounter. Whenever a focal ant was confronted with another ant, it could either make contact or evade to the right or to the left. The evading direction was identified from the point of view of the focal ant. From the number of left and right decisions a laterality index L was computed for every individual ant defined as the number of left turns over the number of total left and right turns. Hence, L ranged from L = 1 (the ant always turned to the left) to L = 0 (the ant turned only right). In- and out-bound ants showed different side preferences at the individual level when facing another ant. Inbound ants had a laterality index of Lin = 0.57, whereas outbound ants showed no significant lateralization Lout = 0.47. Thus, whilst walking towards the nest, foragers more often moved to the left side than to the right, whereas outbound ants (with no food) did not show individual lateralization. It is interesting to notice that *F. pratensis* does not use trail pheromones on trunk trails. This means that inbound ants preferred to use significantly more frequently the left side of the trail not as a consequence of following trail pheromones but as a motor bias. The fact that outbound ants did not display the same lateralization might suggest that conditional lateralization depends on the status of the ant (with or without food).

Red wood ants *Formica aquilonia* exchange food through trophallaxis. In these "feeding" contacts, a "donor" (D) ant exchanges food mouth-to-mouth with a "receiver" (R) ant. The food exchange is accompanied by antennal contacts between D and R. Frasnelli and colleagues [19] examined possible asymmetry in antennal use during ant "feeding" contacts. The study was run in the laboratory on a colony of red wood ants *Formica aquilonia* Yarrow of about 2000 individuals. The foraging ants were individually marked with colored paint. The researchers applied the "binary tree" paradigm that consists of a maze "binary tree" where each "leaf" of the "tree" ends with an empty trough, except one with syrup (Fig. 1a).

A paradigm was designed to study the process of information transfer in ants. The basic idea was that ants were forced to communicate information about the location of a food source within the maze and experimenters could measure and control the quantity of this information [20]. The use of the binary tree in the study by Frasnelli et al. [19] allowed observation of a strong division of labor, in red wood ants, between scouting individuals and members of their constant teams ("foragers"), and to distinguish "feeding" contacts from other types of antennal contacts (i.e. contacts to activate foragers and contacts to transfer concrete information). Antennal contacts between the D ant returning from the binary tree and the R ant waiting outside the nest were recorded. Two experimenters scored all video recording using VLC Media player 2.0.0 (VideoLAN Project, France) at 0.12× speed. The number of contacts made by the right and the left antenna of the R ant on the D ant's head and the number of contacts made by the right and the left antenna of the D ant on the R ant's head were counted (see Fig. 1b).

Fig. 1 (**a**) A laboratory arena divided into two parts, one containing a laboratory nest, and another with a binary tree with three forks. (**b**) Pair feeding contacts between the "receiver" (R) ant and the "donor" (D) ant. The D is left on top. Photograph by N. Bikbaev. From Frasnelli et al. [19]

Video analyses showed that on average R ants used their right antenna significantly more often than their left antenna in begging for food during "feeding" contacts with D ants.

Observations in the natural environment are a good method to assess asymmetrical use of legs in simple behaviors. Heuts and Lambrechts [21] studied leg loss in 18 families of spiders (Arachnida, Araneae) in their natural habitat without touching or catching them in order to exclude the possibility that the missing legs were the result of human interference. The study was conducted over all season in 2 years, i.e. 1995 and 1996. In each case the position of the missing legs and on the front-rear and on the left-right axis was noted. Only specimens with a single missing leg were considered (a small minority of individuals that missed more than one leg was discarded). The most extensive data came from three araneid spider species, i.e. *Zygiella x-notata* (Clerck) ($N=29$), *Araneus diadematus* (Clerck) ($N=32$), and *Lariniodes sclopetarius* ($N=34$). The data of these three species investigated in the wild did not show a significant laterality of leg loss as a group. However, in one of them, *Z. x-notata*, the laterality was significant and consisted of higher left than right leg loss (21 versus 8).

Ades and Ramires [22] collected female *Scytodes globula* (Arachnida, Araneae, Scytodidae) for 2 years in the woods of the

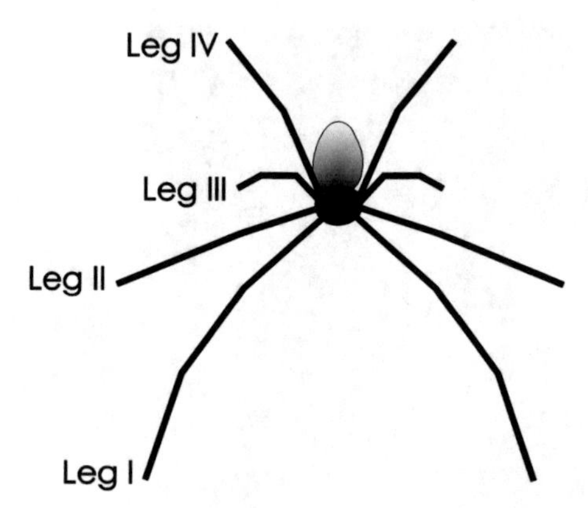

Fig. 2 Schematic representation of the numbers of the spider legs in the rostrocaudal direction

western part of Sao Paulo in Brazil. This species catches its prey by rapidly ejecting a mixture of glue and venom from the chelicerae. On direct contact during the handling of the prey, *S. globula* uses legs I and/or II more often than the other legs (numbers of the legs in the rostrocaudal direction—see Fig. 2).

Observations made in the field and in the laboratory show that *S. globula* can prey on *Loxosceles* spiders with which they share their habitat. During predatory interactions, *Loxosceles* spiders can also bite *S. globula* causing autotomy (i.e. self-amputation) of one or more legs or even death [23]. In a sample of 162 spiders, the researchers looked at the frequency of spiders with missing legs, a measure that can reveal the way these spiders use their legs during prey handling. Ades and Ramires [22] observed that of 162 field-collected spiders, 36 had one leg missing; among these 36 the frequency of spiders missing leg I or leg II was significantly higher than that of spiders missing leg III or leg IV. Interestingly they found a significant difference in the number of the spiders missing the left compared to those missing the right leg I (15 missing the left I leg and four missing the right I leg). Although it is unlikely that the loss of legs is a consequence of intraspecific agonistic episodes, given the tolerance *S. globula* displays towards conspecifics, the higher proportion of missing left legs in *S. globula* collected in the field it is not sufficient evidence of the spiders' preferential use of left legs in predatory contexts. This is one of those cases where field observations lead to interesting evidence and studies in the laboratory are needed to provide further confirmation of the observations made in the field. In addition to the field observations, Ades and Ramires [22] conducted an experiment in the laboratory where they observed directly the leg use of *S. globula* spiders when they interacted with their prey. They looked at touches during predatory encounters of individual intact adult females *S. globula* collected in the field with prey from three species of *Loxosceles* spiders (*L. intermedia*, *L. gaucho*, and *L. laeta*). In experimental sessions, each *S. globula* spider was placed in a plastic box that had been occupied for the 12 preceding hours by one *Loxosceles intermedia*, *L. gaucho*, or *L. laeta* and was positioned approximately 10 cm from *Loxosceles*. A camera situated above the box recorded the 30-min session. From the videotape records of the encounters in which predation of *Loxosceles* occurred, the researchers recorded the number of times *S. globula* touched its prey or a region very close to it (touching movements), by raising and lowering leg I or leg II. Recording of leg touches was started just after ejection of the adhesive substance and stopped when *S. globula* moved away from its prey. The results revealed that touching movements with the left anterior legs were significantly more frequent than with the right anterior legs. Each *S. globula*, however, was tested only once, although repeated testing of the same individuals is required in order to define consistency of use of a preferred leg. Using a similar methodology it would be also interesting to look at the way *S. globula* spiders use their legs in handling non-spider prey as well as in other behavioral contexts to assess the generality or specificity of leg preference.

By observing four different species of bumblebees *Bombus* spp. while foraging in a field of *O. viciifolia*, Kells and Goulson [24] reported that most bees display a consistent bias in circling while visiting florets arranged in circles around a vertical inflorescence. The individuals of each species (*B. lapidaries* (*N*= 17), *B. terrestris* (*N*= 11), *B. lucorum* (*N*= 11)), and *B. pascuorum* (*N*= 13) were selected at random and observed as they foraged on ten successive inflorescences. The direction in which they rotated around each inflorescence was recorded. A different bee species was chosen each time to minimize the likelihood of observing the same bee twice, but the bee population was so large that this is likely to have occurred very rarely if at all. The observer also moved around the field when making observations, so risk of pseudo-replication was minimal. Where a bee probed fewer than three florets on an inflorescence, or was forced to alter direction due to obstacle (leaves, stem of another flower, interference from another pollinator), that individual was excluded from the data set. For the duration of the data collection period, weather conditions were approximately uniform. In three (*B. lapidarius*, *B. lucorum*, and *B. pascuorum*) out of four species examined the majority of bumblebees circled in the same direction: two species circled anticlockwise and one clockwise, whereas no directional asymmetry was revealed in *B. terrestris*. The fact that the three species displayed different direction of bias suggests that the asymmetry is not function of the structure of the florets; otherwise they would have all circled anticlockwise or clockwise. Bumblebees observe and copy the behavior of others with regard to floral choices [25] and they can learn to make nectar-robbing holes in flowers as a result of encountering them [26].

Recently, Goulson and colleagues [27] investigated side bias in nectar-robbing bumblebees (*Bombus wurflenii* and *Bombus lucorum*) feeding on *Rhinanthus minor*, a flower that can be robbed from either the right-hand side or the left-hand side, and they looked at a possible effect of social learning on handedness. To determine the patterns of robbing a total of 28 patches of *R. minor* spread across an alpine landscape at altitudes ranging from 470 to 1850 m in Switzerland were analyzed in June of 2009, 2010, and 2011. Distances between patches ranged from 330 m to 15.2 km (mean, 5.6 km). Since not all patches were present in every year, data were collected from 20, 16, and 22 patches in 2009, 2010, and 2011, respectively. In each patch, 20 racemes were chosen at random and each floret was scored according to whether it had been robbed on the left-hand side or the right-hand side, or on both sides. Scoring began with the highest and hence youngest open floret, and worked downwards to the oldest floret, so that the score is correlated with the age of the floret. The frequency of robbing on the left versus the right of the floret varied greatly between patches and years. Of 51 patch × year combinations for which sufficient robbing was present to test for a bias, 38 exhibited a significant bias to either left or right (14 biased to the left and 24 biased to the right). There was no relationship between bias and altitude in any year and bias within patches was not correlated across any pair of years. Interestingly, there was no spatial autocorrelation in robbing strategy, i.e. the pattern of robbing on the left or the right did not tend to be more similar in patches located nearer to one another.

Bee behavior was recorded in a subset of 13 patches in 2009 and 2011 (determined by suitability of weather for bee activity when visiting patches). Bees were followed until they had visited up to 20 florets or were lost from sight. Wherever possible, the bee was captured to confirm identity and to minimize the frequency with which the same bee was observed foraging more than once. The side of the floret that the bee first approached was recorded for primary robbing (involving the active use of mandibles to bite through one side of the calyx and corolla to access nectar) and secondary robbing (whereby an existing robbing hole is used to access nectar). Observations were made on the foraging behavior of 168 individual bumblebees, visiting a total of 906 florets of *R. minor*. The intensity of side bias increased through the season and was strongest in the most heavily robbed patches. Bees within a particular patch tended to approach florets on the side which was most likely to have a hole, and this correlation was stronger for secondary robbers compared to primary robbers. When acting as primary robbers, 91.7% of *B. wurflenii* approached florets on the same side as the majority of existing holes (77 out of 84). For *B. lucorum*, all 20 observations of primary robbing involved an approach on the side of the majority of existing holes. Bees within patches seemed to learn robbing strategies (including handedness) from one another, either by direct observation or from experience with the location of holes, leading to rapid frequency-dependent selection for a common strategy, i.e. adopting the same handedness within particular flower patches.

Octopuses *Octopus vulgaris* (Cephalopoda, Octopoda, Octopodidae) have their eyes placed laterally on the head [28], so they show a preference for monocular eye use in many situations. Octopuses generally use only one eye at a time and turn sideways to look at prey, although the visual fields of the two eyes overlap slightly in front of and behind the animal [29]. Muntz [30] observed that in visual attacks during hunting, the octopus uses only one eye to view the target while approaching it. Some laboratory studies [31–33] have investigated whether octopuses show a preference to use the right or the left eye in prey detection and catching. Byrne et al. [31, 32] measured preferential eye use in *O. vulgaris* by recording the time animals spent watching a stimulus presented to them outside their tanks while they were holding on to the front glass of the tank. The stimulus used was a life-sized plastic model of a crab (resembling the crabs usually fed to the octopuses) mounted on a transparent stick (Fig. 3b). This was presented at five equidistant positions along the length of the tank and was held at each position for 3 min according to a random sequence, so that the time at each of the five positions was

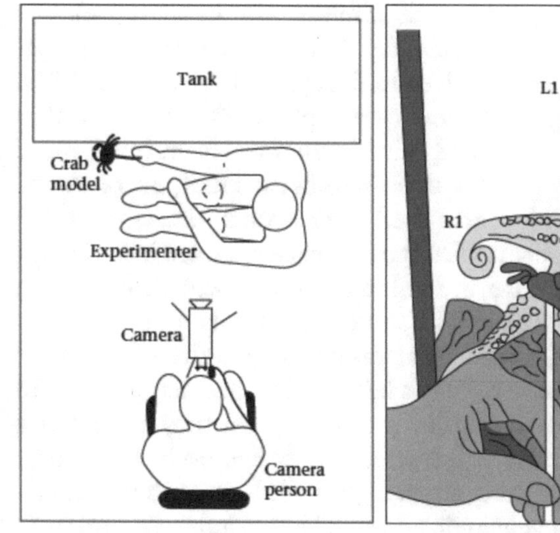

a.

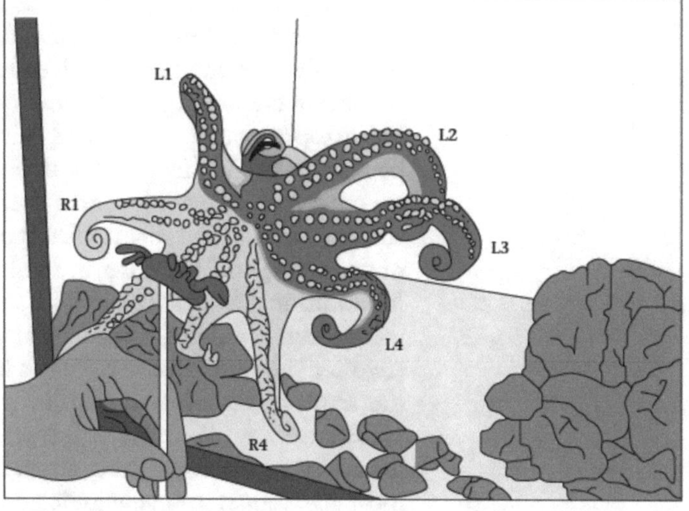

b.

Fig. 3 (**a**) Schematic representation of the set-up used by Byrne and colleagues to test eye-preference in *O. vulgaris*. The octopus was filmed with both the experimenter and the camera person visible to the octopus. The experimenter moved a plastic crab along the front of the tank to attract the subject's attention. (**b**) An example of an octopus holding on to the front glass of its tank actively watching the stimulus outside in monocular eye use position (left eye). From Byrne et al. [32]

the same in order to avoid directional bias (Fig. 3a). Octopuses were tested in individual tanks containing a sandy bottom, small rocks, and large rocks, which provided building material for dens and shelters.

When presented with the stimulus, each octopus moved from the den position toward the tank's frontal glass and the eye used to approach the stimulus was scored. Since Byrne and colleagues did not report in either of their studies what was the position of the den inside the tank, it is difficult to determine whether or not the eye used to approach the stimulus was to any extent influenced by the starting position of the animal in the den (see Fig. 3b where the big rock constituting the den is placed on the left side of the tank—octopus' point of view—and the octopus displays a monocular left attack). Results indicated an individual lateral preferences for either the left or right eye: five for the left and three for the right in a sample of eight animals [31]; 13 for the left and ten for the right in a sample of 25 animals [32], but at the population level no systematic bias towards left or right was observed. The method adopted by Byrne and colleagues could be improved by minimizing any potential environmental asymmetry, such as the position of the den or other objects in the tank, that may induce or influence the bias in the animal behavioral reaction to the stimulus. Moreover, the position of the animal prior the presentation of the stimulus should be recorded in order to determine a possible correlation with the eye use during the approach toward the stimulus (e.g. if the animal is looking outside the den in a monocular right position it is likely that it will approach the stimulus with its right eye).

Byrne et al. [34] also tested the limb use of eight octopuses, seven of which had been involved in the eye preference study [32]. A T-maze containing a food reward was placed in the tank so that the octopus could explore and retrieve the food from the T-maze by inserting only one arm at a time. Within each trial, the researchers scored (1) the contact arm used to first touch the T-maze, (2) the arm used in the first choice (i.e. to make a choice into the left or the right) and (3) the arm used in the second choice (to search for additional food). The results showed than only four octopuses out of eight showed a significant preference on the first contact (two for the right and two for the left). These four octopuses were the same animals tested for eye preference and they exhibited a preference for the arms on the same side of the body as the preferred eye: two right-eyed animals preferred the anterior rightmost arm (conventionally indicated as R1); two left-eyed preferred L2 arm. An explanation for this could be that the food item was visible to the octopuses through a transparent T-maze and thus the arm choice could have been influenced by the eye preference.

To explore how strongly eye and arm choice influence each other, Byrne et al. [35] positioned plastic bottles at three different depths in the tank. One bottle had a small rock inside and was resting on the bottom of the tank and extended 9.5 cm up, one was tied to a small rock with a nylon string and floated in mid-water 12–17 cm above the bottom, and the third one floated on the water surface. One object at a time was put into the tank at the beginning of each 1 h trial and taken out at the end every day for 8 days in a row. After each of these sessions every octopus had 1 week of break until a new session with the next different object began. The presentation order of objects to the octopuses was randomized. The octopuses were observed during approach, contact initiation and exploration of objects. None of the seven subjects showed a left/right bias for approaching the objects, irrespective of the position of the objects in the water column. A strong association was found between the direction of approach and the first arm used to touch the objects. In 99% of all cases, if the object was in front of or above the octopus, the octopus used an anterior arm and, if the object was behind or below the octopus, a posterior arm was used. If the object was to the right (or left) of the octopus, a right (or left) arm was used.

The results of this study indicate that octopuses most commonly use an arm to initiate contact with an object that is in a direct line between the eye used and the object. Anatomically this is a logical solution, because it would be more complicated for an octopus to use an arm that is on the other side of its body to grab an object during monocular visual exploration. A limit of the methodology may depend on the simplicity of the task since, in vertebrates at least, simple tasks such as basic unimanual reaching induce weaker lateralized responses than do high-level tasks such as bimanual coordination [36]. Thus it would be interesting to investigate whether the same eye and arm coordination shown in octopuses [35] occurred in a more demanding task.

A more sophisticated task employing a visually cued three-choice maze was recently employed by Gutnick and colleagues [37] to test learning that required single-arm control in *O. vulgaris*. The maze shape, consisting of a narrow central tube opening into three choice compartments, was designed in order to test the natural movement that octopus arms often perform when exploring and hunting in small crevices and under rocks. In the task octopuses had to reach a single arm through the tube, out of the water (thus preventing chemical cueing), and into the water of the goal compartment to reach the food reward. A small piece of food, visually marked by a black disk in the goal compartment, was moved between choice compartments in a random sequence. Eight animals were tested for ten trials a day in which they had to make a choice within 3 min and were not allowed a second choice. Off-line analyses of the videotapes indicated that seven out of the eight octopuses tested reached the criterion for learning of five correct trials in a row within 61–211 trials. Completing the operant task required the animals to associate a visual cue with their own voluntary motor actions. In the first 20 trials of the experiment, animals performed at chance level, whereas in the last 20 trials up to reaching the criterion they performed significantly above chance level. When solving the task, octopuses used different sitting positions in the tank, which offered them different views of the maze and its compartments. A strong correlation between not seeing the target and failing to complete the task was observed. Interestingly, in the last third of the experiment, successful trials were significantly longer than unsuccessful trials suggesting that animals needed more time to correctly position themselves for a clear view of the maze and to visually control the search movements that were more prevalent when they had learned. The study by Gutnick et al. [37] clearly shows that octopuses are able to learn to complete an operant task requiring them to control the movement of a single arm, and that in doing so they can determine the position of their arm and learn to visually guide it to a location. The task used by Gutnick et al. [37] will be extremely suitable to measure laterality in the eye and arm use in order provide further evidence to the results of eye and arm coordination shown by Byrne and colleagues [35].

Previous studies investigated eye preference of octopuses in prey detection when a plastic silhouette of a crab was used (Fig. 3a—[31, 32]). Recently, Frasnelli et al. [33] looked at the eye used preferentially by *O. vulgaris* while attacking a real crab. Fifteen individuals were tested in the morning while they were resting in their den, a shelter constituted of three bricks positioned on the side of the tank opposite the front glass. Each octopus had its own tank where a constant flow of seawater was provided. Seven individuals had their den positioned centrally, whereas the remaining eight had their den position in the right corner of the same side of the tank (Fig. 4). This was done in order to assess whether the position of the den could influence the bias in the eye used to look at and subsequently attack the crab. Each octopus was tested in its own tank when it was in the den. A live crab attached to a nylon wire was placed into the tank centrally on the side of the front glass, and thus opposite to the den, once a day for a total of 14 trials (Fig. 4b). The eye used by the octopus to look outside of its den before the crab was placed into the tank, and the eye used to attack the crab was recorded every time (Fig. 4d).

The experiments are still running. So far data revealed that most individuals do not show any consistent preference through the trials, but they seem to use their right or left eye a similar number of times or to have a weak preference for one eye. Only four individuals out of 15 showed a strong significant preference to use one eye consistently: two for the right and two for the left eye. Compared to the previous studies run by Byrne et al. [31, 32], in which the animals were observed for hours while looking at the plastic crab outside the tank, here the number of observations is

Fig. 4 (**a**) Photograph of an octopus staying in its den centrally placed in the tank in the laboratory of the Stazione Zoologica Anton Dorhn (Naples, Italy). (**b**) An octopus attacking a live crab (monocular left). (**c**) An octopus looking outside the tank (monocular left). (**d**) Video sequence of an octopus molecular attack (monocular right) to a live crab placed in the tank as in the experiment by Frasnelli et al. [33]. In this case the den is positioned in the right corner of the tank. Photograph courtesy of Dr. Graziano Fiorito and Dr. Giovanna Ponte

limited to one trial a day for a total of 13–14 trials. This may explain the lack of individual bias observed in most individuals. Although the study is not completed, the methodological approach used by Frasnelli et al. [33] is interesting because it uses live crabs (which are common prey for octopuses) placed in the same tank with the octopus, and thus it recreates a natural context where the animal can display its predatory behavior. This allows assessment of real predatory behavior and not exploration or curiosity as in the studies by Byrne et al. [31, 32].

3 Escape Responses

Several studies have investigated lateral biases in escape responses of a variety of vertebrate species (summarized in [2]). As said above, the left hemisphere is specialized for controlling routine behavior, whereas the right hemisphere is involved in responding to unexpected and novel stimuli, such as predators, and in controlling escape reactions [2]. This functional specialization of the right hemisphere (left eye in vertebrates with their eyes positioned on each side of their head) is reflected as a lateral bias in the escape response after the predator has been detected (with the left eye).

Moreover, most of the times this laterality occurs at the population level, i.e. most individuals have the same direction of bias. Several species of fish, for example, show a consistent escape response on a particular side, usually the left side [38–42].

Atyid shrimps *Neocaridina denticulata* (Malacostraca, Decapoda, Atyidae) show lateral asymmetry in escape responses when presented with a vibratory stimulus [43]. Given that the animals appear to perceive the direction of a stimulus, the use of directional stimuli, such as a visual stimulus, is usually avoided in tests of lateral biases in escape behavior. In fact, it may be difficult to present such stimuli rigorously to the free-moving animal without introducing a possible bias. As a consequence, vibratory stimuli are commonly used in the laboratory to evoke escape responses in fishes (e.g., [38]). Using a large petri dish "arena," Tekeuchi et al. [43] recorded with a high-speed video camera located above the arena the directions in which *N. denticulata* individuals attempted to escape when exposed to a vibratory stimulus (Fig. 5). Each shrimp was tested only once. At the beginning of each test the shrimp was settled at the center of the arena and enclosed within

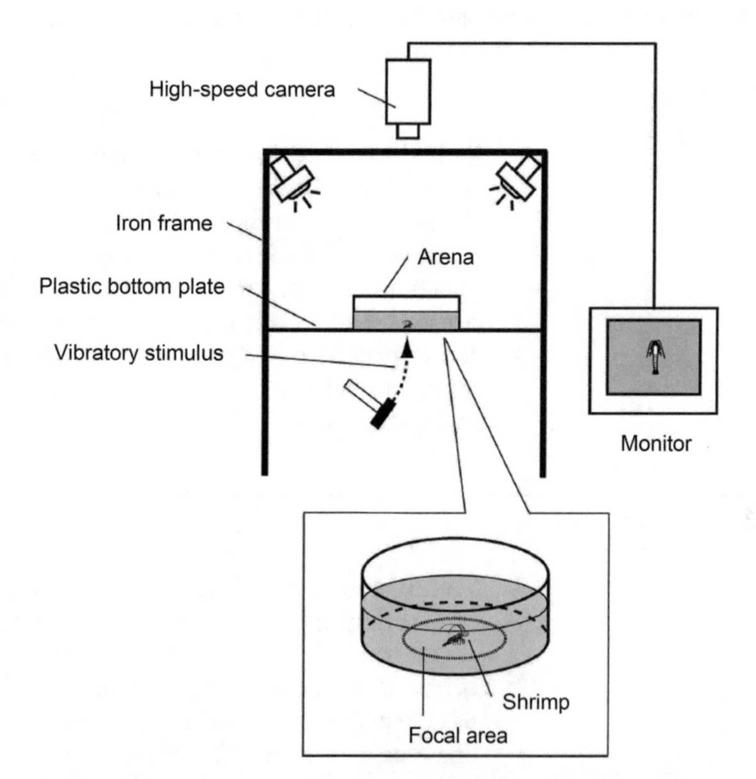

Fig. 5 Schematic representation of the apparatus used by Takeuchi et al. [43] to observe escape responses in atyid shrimps. The shrimp was stimulated with vibrations by striking the bottom plate of the arena with a hammer from below. From Takeuchi et al. [43]

an opaque plastic cylinder for 7 min so that it could be familiarized with the environment. After this pre-test phase, the opaque cylinder was removed and the plastic bottom plate that supported the arena was struck from below with a hammer to evoke an escape response (see Fig. 5). All hammer strikes landed just below the arena center, which was not visible to the shrimp. In most cases, the shrimp made a swift jump backwards, either to the left or the right within the focal area.

Takeuchi et al. [43] also measured the curvature of the abdomen of the same *N. denticulata* individuals used for the behavioral experiment and for another species (*Limnocaridina latipes* Calman), revealing asymmetry in the curvature. The frequency distributions of the angle of the abdominal curvature in both species were bimodal, suggesting that the two populations are composed of both left- and right-type individuals. In *N. denticulata*, the escape direction of each individual after the presentation of the vibratory stimulus was correlated with the abdominal curvature: left- (right-) type shrimps jumped backwards to the left (right) significantly more often than expected by chance. However, in the escape experiment conducted by these authors [43], each shrimp was tested only once. Replication of at least ten individual trials is necessary in order to determine whether the lateralized direction of an escape jump shows a consistent side bias within an individual.

A similar experimental approach was adopted by Tobo et al. [44] to investigate lateralization escape behavior and morphological asymmetry in the red swamp crayfish *Procambarus clarkia* (Malacostraca, Decapoda, Cambaridae). The study was conducted in the laboratory using an arena equipped with one "starting booth" and seven "safety pockets" placed behind the starting booth in an arc format and designated L3, L2, L1, 0, R1, R2, and R3 from left to right (for details see [44]). During each trial a crayfish was placed in the starting booth and left to familiarize for 10 min before the stimulation. In this case the escape response was induced by the experimenter, who stood alternately on the left or right side of the arena, and tapped the crayfish by the rostrum using a piece of black urethane foam bound to a thin wooden stick. As soon as stimulated, usually the crayfish made a sequence of jumps backward. The direction of the first jump and location of the safety pocket (L3 to R3) were recorded for each trial. Ten individuals were tested for a total of 20 trials each. Results revealed that each individual jumped toward one side far more times than toward the other side, revealing lateralized escape behavior at the individual, but not at the population level. The position of the experimenter did not affect the direction of either the first jump or the choice of the safety pocket.

In addition to the behavioral experiment, the morphological asymmetry was determined in 411 crayfish based on the distance between the left (or right) orbital cavity to the central point of the line that divides the carapace into the cephalic and thoracic regions. Similarly to shrimps [43], the population of crayfish seems to be composed of two types of individuals: one type has the right side of the carapace larger than the left side, and the other type has the left side was larger than the right side. Tobo et al. [44] also measured the morphological asymmetry of the same individuals tested for behavioral escape response and found the two measurements correlated significantly. In individuals with the left side of the carapace longer than the right side, escape behavior entails a rightward jump, whereas in individuals with the right side longer, escape entails a leftward jump.

L. niger and four other ant species display a specific behavioral lateralization when running on unknown ground in an "alarm" situation. As Heuts and Brunt [17] observed in eight *L. niger* colonies disturbed (alarmed) near their natural nest, all colonies turned mainly to the left. Specimens from five species of the two main ant subfamilies showed a left-versus right-turn increase when put onto an unfamiliar surface (and thus alarmed) in the laboratory (24 left versus 4 right turns). The leftward turning following alarm situations and the rightward bias in foraging observed in several ant species [17] suggest that invertebrates may possess the same behavioral bias as vertebrates but in the opposite direction of specialization since invertebrates are characterized by ipsi-lateral projections between the body and the brain, whereas vertebrates are characterized by cross-lateral projections. This means that the left-hemifield (left eye) use in escape response is reflected by a left-nervous system specialization in invertebrates and by a right-hemisphere specialization in vertebrates. In the same way, the primary use of the right eye in foraging corresponds to a right-side specialization in invertebrates and to a left-hemisphere specialization in vertebrates. However, lateralization in escape responses may be due to a morphological asymmetry in the body rather than in the brain, as described above for the atyid shrimp [43] and red swamp crayfish [44].

4 Interactions with Conspecifics

A simple method to reveal the presence of behavioral asymmetries in social behavior consists in observing social responses (e.g. aggressive or sexual behaviors) that are directed toward conspecifics (for a general review see also [45]). This can be assessed by recording the frequency of acts of aggressive or sexual behavior when a conspecific appears on the left versus the right side of the subject (i.e. is detected by the left or the right eye). Another way of investigate lateralization in social contexts is to look at which part of the body (left versus right) is used or exposed in interspecific interactions of aggression or mating. In vertebrates a left eye bias (right hemisphere control) during aggressive or courtship behavior has been demonstrated in lizards [46], toads (e.g. *Bufo bufo* [10]), in avian species (e.g. [47]), in horses (e.g. [48]) and nonhuman primates (e.g. [49]; see also [16]).

4.1 Aggressive Behavior

Fiddler crabs (genus *Uca*) males have a single enlarged claw on either the left or right. This major claw is used as a weapon and, in most species, is also waved at females during courtship. In 90 species, left- and right-clawed males are equally common and their fighting and mating behavior is indistinguishable, implying either no selection or negative frequency-dependent selection on laterality (e.g. [50]). However, in two species (*Uca vocans* and *Uca tetragonon*) 91–99 % of males are right-clawed [51]. Backwell et al. [52] looked at behavioral differences between left- and right-clawed males in the fiddler crab *Uca vocans vomeris*. This species is reported to have 96–99 % right-clawed males [51]. Given their rarity, left-clawed males should be relatively more experienced at fighting opposite-clawed males. On the other hand, this rarity suggests that there may be frequency-independent costs associated with being left-clawed. Each *U. v. vomeris* crab owns a burrow that provides a refuge during high tide. At low tide, they emerge to feed and males visit females and mate at the burrow entrance. When a male courts a female at the burrow entrance of the female, it either approaches a nearby female and later returns to its burrow ("sallying"), or permanently leaves its burrow to seek out females ("aggressive wanderer"; [53]). Aggressive wanderers spend most of the low-tide period moving through the population, fighting resident males and searching for females, leading them to a higher mating success [53], but also to more contacts with males. In their study, Backwell and colleagues [52] tested whether a left-clawed male is (1) better at defending its burrow (burrow tenancy); (2) more likely to win fights when wandering (success as a wanderer); (3) a better fighter or has lower success at gaining mates (success as a resident) and (4) more successful in fighting (fighting outcome). To test this, left-clawed and right-clawed crabs were collected in the field in Darwin, Australia. Crabs were selected by size in order to match males by size, which affects fighting behavior. To test whether left-clawed males are better at defending their burrow (1), the researchers [52] located 26 pairs of size-matched males of opposite clawedness that held burrows less than 2 m from each other but were not immediate neighbors so they did not interact. The male who first left its burrow was noted, assuming that longer residency indicates greater fighting ability. Results showed that a left-clawed male retains a burrow for a period that is significantly shorter than a size-matched right-clawed male. To test whether left-clawed males are more likely to win fights when wandering (2), 19 size-matched pairs of left- and right-clawed males were captured, measured, and released onto the mudflat. Each male was monitored until it acquired a burrow, by documenting how it obtained a new burrow (fought a male, evicted a female, or occupied an empty burrow), how long it took to obtain a burrow, and how many times it (1) fought with resident males, (2) had a non-contact interaction with another male, (3) interacted with females, and (4) entered burrows. Results indicated no difference in the tactics used by left- and right-clawed individuals to obtain a new burrow. However, right-clawed males were significantly more likely than left-clawed males to initiate fights with resident males. To test whether left-clawed males are better fighters, or have lower success at gaining mates (3), another 35 pairs of left- and right-clawed males were located using the same criteria as for the burrow tenancy study. Male behavior was recorded and the proportion of males that (1) engaged in mating activities (courted/mated), (2) fought and (3) retreated into their burrow was compared. There was no difference either in the proportion of left- and right-clawed resident males that attempted to mate or that fought, or in the rate at which left- and right-clawed males attempted to mate. In the same way as in (2), however, right-clawed residents engaged in significantly more fights than left-clawed residents. This is in line with the previous observation that a significantly greater proportion of left-clawed males retreated into their burrow to observe and avoid fights. However, if one considers the fighting outcome when fights occur (4), no significant difference in the proportion won by left- and right-clawed residents is observed. Overall left-clawed males appeared to be less likely to fight. An explanation for that may be that it is easier to assess an opponent's strength when it has the same direction of clawedness [54]. As such, left-clawed males might be generally inclined to avoid fights as they are more likely to escalate an encounter with a larger opponent which they are unlikely to defeat.

Rogers et al. [55] investigated how honeybees from the same or different hives interact when they are placed in dyads. The following dyads were tested: two bees with the left antennae removed (using their right antennae, RA), two bees with the right antennae removed (using their left antennae, LA) and two intact bees (using both antennae, BA). For each antennal condition, dyads of bees from the same hive and dyads of bees from different hives were tested (six different types of dyads). The researchers used a simple set-up consisting of two up-turned petri dishes, each with a small opening in the side that could be closed by turning the lid. The two dishes were placed alongside each other and secured so that the openings in the lids were next to each other. The apparatus was placed inside a white, featureless, circular arena. At the beginning of testing a bee was placed in each of the dishes with the opening holes closed, where it remained for 5 min (pre-test period). When the lids were turned so that the openings were juxtaposed and the bees could gain access to either dish (Fig. 6), the 5 min test commenced.

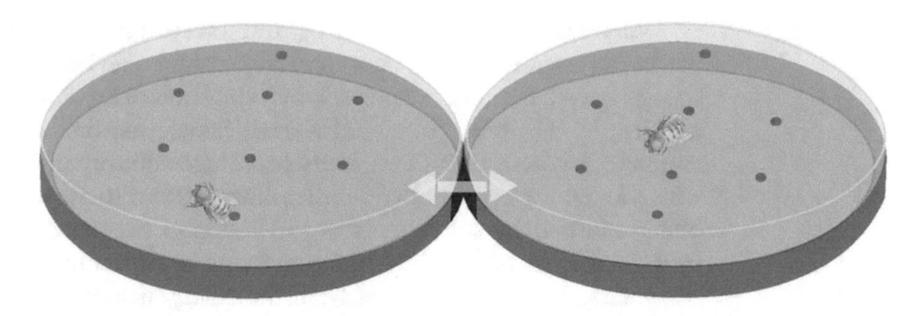

Fig. 6 Testing apparatus used by Rogers et al. [55] to test social and aggressive interactions between honeybees of the same and different colonies and with only the right antenna (RA), the left antenna (LA), or both antennae (BA). The apparatus consisted of two up-turned petri dishes, each with a small opening in the side (colored *black* in the figure) that allowed the dishes to be separated from each other (pre-test), or interconnected by aligning the holes (test). The *small circles* represent the air holes. The *arrow* indicates the potential of bees to move from one dish to another through the openings on the side of each dish. From Rogers et al. [55]

A webcam placed over the dishes recorded the pre-test and test periods. By playing back the recordings of the tests the following behavior was scored: the latency from the start of testing to the time when the bees first contacted each other; the number of proboscis extension responses (PER—extension of the proboscis as in feeding and trophallaxis); the number of C-responses (i.e. arching the abdomen into a C-shape so that the sting and the mandibles are pointed to the other bee); and the number of times the mandibles were opened (mandibulation).

In bees from the same colony, dyads of bees using their right antennae (RA) took significantly less time to make the first contact with each other than did dyads using their LA. Moreover, RA dyads scored more PER (social) in mutual interaction than did LA dyads, but at the same amount as BA dyads (intact bees), suggesting that the right antenna controls this social behavior in intact bees. The fewer C-responses (i.e. aggressive interactions) in RA dyads compared to LA dyads indicate that the right antenna elevates explorative social interactions and suppresses aggressive social interactions between members of the same colony. In bees from different colonies, although there was no lateralized difference in latency to contact, the latency in the RA dyads was significantly longer compared to latency of RA dyads of the same colony. Unexpectedly, LA dyads of bees from different colonies performed fewer C-responses than RA dyads. This suggests that the right antenna controls social behavior appropriate to context: LA dyads do not adjust their agonistic behavior (C-responses) according to the social context, whereas RA dyads do.

Very recently a similar experiment was carried out in the same laboratory to investigate how lateralization may affect interactions in a solitary species of bee, the mason bee *Osmia bicornis* [56].

O. bicornis is considered solitary because every female is fertile and makes her own nest [57], contrary to eusocial honeybees in which species only the queen is fertile and all the females are its offspring and live in the same colony. However, female mason bees often compete for nest sites. This leads to antagonistic interactions in which females engage to defend their burrows and to prevent them being usurped by other females [58]. Mason bee males emerge from their nests before the females and tend to cluster in at the nest site and on flowers while they await the emergence of the females. During this time and during mating itself at least some degree of social behavior occurs, but in general males do not engage in direct social competition [59]. Rogers et al. [56] tested whether mason bees would exhibit lateralization in performing any type of social behavior when tested in a similar apparatus to that used previously with honeybees (Fig. 6—[55]) and with one or the other antenna removed. Mason bees were tested in dyads of the following combinations: two males, or two females, with both antennae intact (B + B), both using the left antenna only (L + L), both using the right antenna only (R + R) or one using the left antenna and the other using the right antenna (L + R). Another three types of dyad were a male plus a female in the following combinations: both antennae intact (B + B), both using the left antenna (L + L), both using the right antenna (R + R). There were at total of 11 different types of dyad and between 10 and 12 dyads of each type. Similar to the previous experiment [55] but with some differences given that mason bees behave differently compared to honeybees, the researchers [56] scored the latency to first physical contact, the number of physical contacts not involving aggression, and the number of highly aggressive interactions (in which one bee rammed the other causing the latter to overturn followed by buzzing and moving in a chaotic manner).

Results revealed that female–female pairs contacted after shorter latency than did male–male pairs and female–male pairs, but there was no significant effect of the antennae in use. Similarly, the number of contacts without aggression was significantly higher in dyads of females than in dyads of males, but no significant effect of dyad type was found. The number of aggressive interactions was lateralized: in females, there were much higher scores of aggression in L + L dyads than in R + R dyads. The same result was found in male–male dyads, but the difference was less strong than in females. Overall results suggest that the left antenna leads to aggressive behavior in mason bees.

A leftward bias in aggression has been recently reported in two species of tephritid flies, *Ceratitis capitata* [60] and *Bactrocera oleae* [61]. Benelli and colleagues [60] investigated lateralization of aggressive displays (boxing with forelegs and wing strikes) in the Mediterranean fruit fly *Ceratitis capitata*, a ubiquitous fruit pest in subtropical and tropical regions worldwide. *C. capitata* is a good model to investigate lateralization of agonistic displays since, in this fly, aggressive interactions have been shown both in males competing for territories [62] and in females maintaining preferred oviposition sites [63]. Fighting in both sexes of the Mediterranean fruit fly is characterized by escalating levels of aggression going from avoidance, through wing waving, chasing, head rocking, mouthparts extending, to wing striking, diving, and boxing (see [60] for more details). Two of the main aggressive displays, wing strikes (i.e. the attacker brings forward a wing and strikes the opponent) and boxing (i.e. the attacker raises a foreleg, hitting the opponent on the head and/or thorax) can be performed with either left or right body parts. The researchers [60] examined aggressive interactions in dyads composed of two females or two males in a Plexiglas testing arena (diameter 150 mm, length 200 mm) for 30 min. Only contests characterized by high aggression levels (i.e. wing strikes or boxing) were considered for data analysis. In both male–male and female–female dyads, the wing used for wing strikes or the foreleg used for boxing against the opponent and the outcome of the fight (i.e. which male/female left the twig at the end of the contest) were scored. For both sexes, wing strike and boxing data came from different individuals. For each observation period, only the first contest was recorded since, after a boxing/wing strike event, the contest usually terminated with the immediate abandonment of the territory by one of the two flies. To evaluate the constancy of use of the preferred body parts, each pair of flies was tested a second time under the same conditions. Results showed left-biased population-level lateralization of aggressive displays in both male–male and female–female pairs. When the aggressive interactions were analyzed in relation to subsequent fighting success, aggressive behaviors performed with left body parts were reported to lead to greater fighting success than those performed with right body parts. This left-biased preferential use of body parts for both wing strikes and boxing suggests that the left foreleg/wing may be quicker in exploring/striking than the right counterpart. To test this hypothesis, Benelli et al. [60] calculated the mean velocity of aggressive displays wing strikes and boxing performances in both sexes, using high-speed video recordings. For both sexes, aggressive displays that led to success were faster than unsuccessful ones. However, left wing/legs were not faster than right ones while performing aggressive acts.

Using the same experimental approach, Benelli et al. [61] looked at lateralization of aggression in another tephritid species, the olive fruit fly *Bactrocera oleae*. As for *C. capitata* a left-biased population-level lateralization of aggressive displays was observed in both female–female and male–male dyads and aggressive behaviors performed with left body parts were shown to lead to greater fighting success than those performed with right body parts.

Interestingly, the left bias displayed by the two species of tephritid flies in aggressive interactions is not a common characteristic of other species of the Diptera order. A study performed by the same group of researchers [64] on male blowflies *Calliphora vomitoria* (Diptera: Calliphoridae) showed that in this species boxing behavior is lateralized at the population-level but in the other direction compared to tephrid flies. In aggressive displays *C. vomitoria* males displayed a consistent preferential use of right legs, whereas both males and females of *C. capitata* [60] and *B. oleae* [61] showed a preferential use of left legs. Analogous to observations of tephrid species [60, 61], in *C. vomitoria* the use of right legs in boxing acts leads to higher fighting success over males using left legs.

4.2 Sexual Behavior

In the same study of olive fruit fly males discussed above [61], the researchers also looked at possible side biases during courtship and mating behavior, by placing a *B. oleae* male to the same cylindrical arena used to test aggressive responses. The arena contained a twig of olive *Olea europea* L. with about ten leaves and four ripe fruits. Once the male established a territory and was stationed on an olive leaf for more than 3 min, a female was released at the opposite end of the arena, and both flies were observed for 60 min. When the male noticed the female, he moved toward her, started courtship wing vibration, and then attempted copulation. For each male courting a female, the side of the female approached by the male (left, right, front, or back), the duration of the male courtship (wing vibration plus directional walking toward the female), the mating success (if successful intromission of the aedeagus and copulation), and the duration of copulation (from the intromission of the aedeagus to genital disentanglement after copulation) were scored. Results revealed that males court females more frequently from the left than from the right, front, or back. Interestingly, the duration of courtship and copulation were similar in all the male directional approaches, suggesting that male mating success was independent the side of approaching the female.

A left-side bias in courtship was also observed in ants [17]. In *Lasius niger* colonies a significant majority of couples in the laboratory had the left side of their bodies exposed to their partners when resting.

A right-biased lateralization of kicking behavior was observed in the Asian tiger mosquito *Aedes albopictus* [65] in a context of female–male interaction. In this species mosquito females display aggressive responses against undesired males, performing rejection kicks with the hind legs. When dyads composed of a female and a male were placed in the testing arena, females used preferentially their right legs to kick undesired males. However, when they used left legs, the mean number of kicks per rejection event was not different from that performed with right legs, and both left and right kicking behavior led to successful displacement of undesired partners.

Lateralization in mating behavior has been investigated in the pond snail *Lymnaea stagnalis* (Gastropoda, Lymnaeidae, Lymnaeinae), a self-fertilizing hermaphrodite that can take the male role or the female role in any single mating [66, 67]. The coiling direction (or chirality) of the shell of the snail can be sinistral or dextral. This primary asymmetry of *L. stagnalis* is determined by the maternal genotype at a single nuclear locus where the dextral allele (D) is dominant to the sinistral allele (S) (for details see [66]). In dextral *L. stagnalis*, the individual playing the male role first climbs onto the shell of the one playing the female role, and circles over the shell in an anticlockwise direction until the female gonophore is reached [68]. Davison and colleagues [67] investigated this chiral precopulatory mating behavior in more detail by looking at the possible correlation with both body chirality and nervous system asymmetry. Individual snails were tested for behavioral lateralization by placing virgin pairs together in small watch glasses. Their movements were observed until mating began and the direction of the circling was recorded. Each snail was used only once. Results showed that all dextral "male" snails circled in a counter-clockwise manner, no matter whether they were paired with another dextral or a sinistral snail. Similarly, all the sinistral snails, both those paired with dextral and those paired with sinistral, circled in a clockwise manner.

Chirality in mating behavior is matched by an asymmetry in the brain. *L. stagnalis* has a ring of nine ganglia that form a central nervous system around the esophagus, with two more distant buccal ganglia on the buccal mass. In all dextral individuals, the right parietal ganglion is fused with the visceral ganglion and the left visceral ganglion is unpaired. By contrast, in all sinistral individuals, the reverse is observed; the left parietal ganglion is formed by fusion with a visceral ganglion. The central nervous system in sinistral pond snails, therefore, has an asymmetry that is the reverse of that of dextral snails. As the coil of the shell is determined by the maternal chirality genotype and the asymmetry of the behavior is in accordance with this, it is likely that the same genetic locus, or a closely linked gene, determines the behavior. The findings of Davison et al. [67] suggest that the lateralized behavior of the snails is established early in development and is a direct consequence of the asymmetry of the body.

The snail *Helix aspersa* (Gastropoda, Helicidae) also performs lateralized sexual behavior linked to an asymmetry in the nervous system [69]. *H. aspersa* is a simultaneous reciprocal hermaphrodite with most of the reproductive organs located on the right side of the animal. The mesocerebrum of *H. aspersa*, a region of the brain that controls sexual behavior, has 23% more neurons in the right than in the left lobe, and these neurons are 24% larger [69]. The excitatory synaptic inputs derive predominately from neurons on the right side, and both the axons of right- and left-side mesocerebral neurons travel mostly in right-side connecting nerves. This asymmetry in the mesocerebrum and in the position of the reproductive organs causes an asymmetry in the sexual behavior that is executed almost entirely on the animal's right side [69].

A right bias in mating behavior has been also shown in the earwig *Labidura riparia* (Insecta, Dermaptera) [70]. The males of this species have two penises. In the laboratory, Kamimura [70] looked at the penis use in mating behavior by releasing adult pairs of field-collected and laboratory-reared earwig into a mating arena, and by recording the male mating posture (direction of abdominal twisting). One minute after the initiation of copulation, the mating pairs were instantaneously fixed and the samples were later dissected to determine which penis had been used for insemination. Results showed that nearly 90% of males hold their intromittent organs in the "right-ready" state when not mating as well as when mating. Looking at the phylogenetic relationships in this species, Kamimura [70] suggested that male earwigs might have evolved from a primitive state in which they held both penises in the "not-ready" orientation when not mating, to a stage in which they always held one penis (either the right or left at random) in the "ready" orientation. Males that still possessed two morphologically indistinguishable penises, but which preferentially held the right in the "ready" orientation represent a second evolutionary step. Finally, the less-preferred (left) penis disappeared, leaving only traces of a closed and nonfunctional ejaculatory duct. This is an example of a phenotype-precedes-genotype mode of evolution where a behavioral asymmetry might have facilitated the evolution of a complete morphological asymmetry.

Male field crickets *Gryllus campestris* (Insecta, Orthoptera, Gryllidae) perform calling songs to attract females [71]. A high intensity sound production can only be produced if the wings are oriented in a specific manner during singing: the right forewing needs to lie over the left wing [72]. If this wing orientation is inverted (i.e. the left wing lies over the right) during singing, almost no sound is produced although the wings are morphologically identical [73]. Moreover, a special wing-spreading behavior is used by the insect to restore the normal right over left orientation and thus to ensure an effective acoustical communication. Elliott and Koch [72] showed that a set of two hair fields on each wing have a stabilizing effect on wing orientation. Elliot and Koch [72] developed a set-up for recording the wing movements during stridulation, in which the cricket sited in a high frequency magnetic field, and carried three small coils. The coils mounted on the wing gave the position of the wings relative to the body of the male. The difference between the signals from the two wing coils gave the position of the wings relative to each other. The sound produced by the cricket was picked up by a condenser microphone, the signal was rectified, and the sound envelope displayed. After removal of the inner hair plates, that is the lower on the right forewing and the upper on the left forewing, Elliott and Koch [72] measured a 13% increase in the wing opening and the jitter in the opening position increased by 100%. Since the wings are then fully separated in only 2% of the calling song syllables observed, the chance of accidental wing inversion rose dramatically as result of removal of the inner hair plates. The results of the study run by Elliott and Koch [72] showed that sensory feedback from these hair fields stabilizes the wing movement amplitude, preventing wing inversion during calling, which ensures the high intensity sound production necessary to attract females.

5 Spontaneous Motor Biases

In behavioral science, a T-maze (or the variant Y-maze) is a simple maze used in animal spatial cognition experiments [74]. As the name suggests, it is shaped like the letter T (or Y), providing the subject with a straightforward choice. T-mazes are used to study cognitive abilities in animals (i.e. memory and spatial learning in rodents [75]), but they can also offer a simply way of investigating spontaneous motor biases. In these types of experiments, however, it is very important to control for external environmental cues, by rotating for instance the apparatus by 180°, in order to ensure that the motor bias is not due or influenced to an asymmetrical distribution of light or other external factors.

Using a T-maze apparatus a population-level preferential bias to turn left was observed in giant water bugs, *Belostoma flumineum Say* (Insecta, Hemiptera, Belostomatidae [6]). Giant water bugs are large aquatic insects, predators of other aquatic invertebrates and small fishes. The initial idea of the study performed by Kight et al. [6] was to train bugs, previously collected as adults from ponds in Sussex County (New Jersey, USA) to swim left or right in a T-maze using different kinds of ecologically relevant reinforcers. Surprisingly, the researchers noticed a significant preference of the water bugs to turn left and they decide to investigate this naïve bias further. The behavior of ten males and ten females randomly selected was observed by releasing each water bug at the entrance of a translucent plastic cylindrical T-maze immersed in a white plastic container filled up with 25 °C tap water (for more details see [6]). After swimming forward 20 cm, the water bug reached the intersection of the maze, where it swam either left or right for a distance of approximately 10 cm until it reached the terminus of the chosen arm. At the end of each maze arm, a 2 cm depression containing aquarium gravel that the water bug invariably clung to on reaching the end of the maze. Ten seconds later, the subject was removed from the maze and immediately replaced at its entrance. This process was repeated 20 times per individual, and between each experiment the apparatus was cleaned and water changed. To control for environmental cues that might bias the turning direction of water bugs in the maze, a second separate experiment on an independent group of 20 water bugs was run. The sole difference between the two experiments was that after the first group of 20 water bugs was tested, the T-maze apparatus was rotated 180° in the laboratory. Each experiment therefore served as a control for the other. This is a necessary operation in studies that investigate lateral biases, because rotating the apparatus presumably allows reversal of the polarity of all directional environmental cues such as lighting or electromagnetic fields. In the second experiment the same left turn tendency was observed, confirming the presence of a population-level leftwards swimming behavior in water bugs.

A similar unexpected phenomenon was discovered in cuttlefish *Sepia officinalis* while being trained to enter a dark, sandy compartment at the end of one arm of a T-maze [76]. T-maze learning was assessed in a cross-shaped maze (Fig. 7). The apparatus was constructed entirely from white PVC (200 cm long × 110 cm wide × 30 cm high) and was illuminated by a 300-W halogen lamp located 1 m above the center of the maze. Two arms were used as start boxes (S1 and S2; top and bottom arms of the cross-shaped maze in Fig. 7) while the other two arms were used as goal arms (left and right arms of the maze in Fig. 7). At the end of each goal arm were two goal compartments, one on the left and one on the right. The goal compartments were dark and covered with an opaque sliding PVC top. The bottom of each goal compartment was entirely covered with sand. To form a T-maze, one start box and the immediately adjacent set of goal compartments were excluded by closing the three opaque sliding doors connecting them to the maze alley. In this T-maze configuration, goal com-

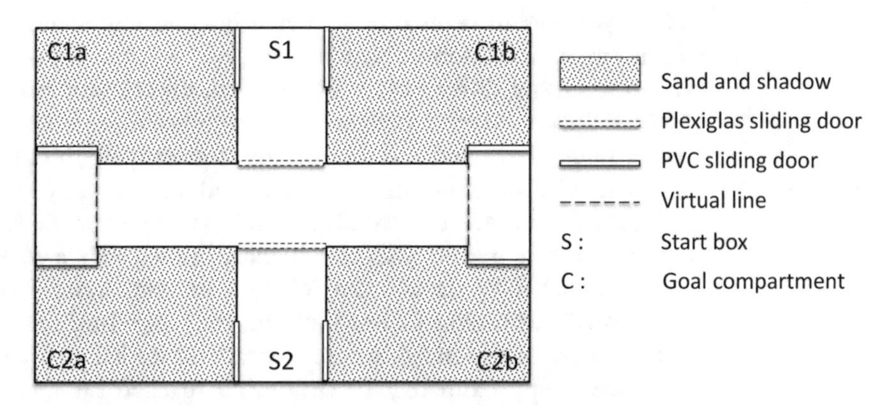

Fig. 7 Schematic representation of the cross-maze apparatus used by Alves and colleagues [76]. Modified from Alves et al. [76]

partments C1a and C1b were used when the start box was S1, and C2a and C2b when the start box was S2 (Fig. 7). From the starting point (e.g. S1), the cuttlefish could not see if the entrance of the corresponding goal compartments were open or closed (e.g. C1a and C2a if the start box was S1) and had to turn twice in the same direction to access one of the two goals (e.g. right-right if the goal was C1a and the starting box S1). Sliding doors between goal compartments and start boxes permitted experimenters to move cuttlefish from a goal compartment to its corresponding start box with the minimum amount of handling. Seawater in the T-maze was 30 cm deep with water flow provided between trials to reduce water heating.

The alternate use of the two starting boxes (S1 and S2) allowed control for external cues by avoiding rotation of the entire apparatus. Out of 15 cuttlefish (six adults and nine sub-adults) tested, eight were tested from start box S1, whereas the other seven were tested from start box S2. Each cuttlefish was placed individually in the start box for 15 s before the clear sliding door to the maze alley was removed. The cuttlefish was allowed to move freely out of the start box and into either arm of the maze. As soon as the cuttlefish chose an arm of the maze (movement of any part of the animal beyond the virtual line at the far end of an arm, Fig. 7), it was gently lifted out of water with a net and placed back into the start box. The arm-choice for each trial was recorded. If a cuttlefish did not turn into one of the two arms within 10 min, the cuttlefish was removed and placed back into the start box. This procedure was repeated until 20 choices were made. Side-turning preference was determined using a criterion of 15 or more choices of the same arm (significant side-turning preference at 5% level of significance with a binomial test). In this way, cuttlefish were categorized into "left preference," "right preference" and "no preference" (the remaining cuttlefish). Three of the 15 cuttlefish (20%) showed a significant left preference and eight (53%) showed a significant right preference, whereas the remaining four (27%) cuttlefish did not show any preference. Of the 11 cuttlefish showing a significant preference, all preferred the side of their initial choice.

To find out whether or not visual perception plays a role in determining the direction of turning and whether this lateral bias is age-dependent, Jozet-Alves and colleagues [77] tested cuttlefish in a T-maze during post-embryonic development (3, 7, 15, 30, and 45 days) in two different configurations of the same apparatus. The T-maze used in this study (Fig. 8—[77]) was different from the T-maze used in the previous studies (Fig. 7—[76]), since it had a single start box and the right and left arms did not provide access to any goal. In this case, to reduce uncontrolled environmental cues as a source of bias, the apparatus was regularly rotated 90° so that there were four possible orientations

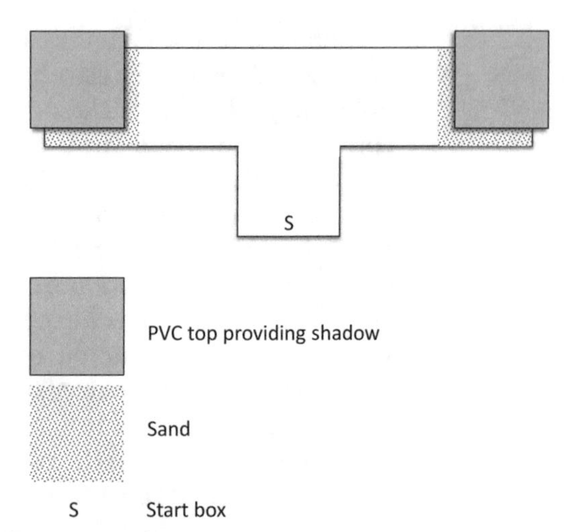

Fig. 8 Schematic representation of the T-maze apparatus used by Jozet-Alves and colleagues [77]. Modified from Jozet-Alves et al. [77]

of the apparatus in the laboratory. Six animals were tested for each age group. For each orientation, either one or two cuttlefish were tested in each group.

In one of the two configurations of the apparatus, shelters (consisting of sand and shade created by an opaque sliding PVC top) were provided at the end of the right and left arms of the T-maze (Fig. 8). Since these shelters are highly attractive to cuttlefish when placed in a T-shaped apparatus [76], this study aimed to see whether or not their presence has an influence on the preferred turning direction. Each cuttlefish was tested using a procedure similar to the previous studies [76] but only until ten choices were made (instead of 20). Results showed that cuttlefish developed a left-turning bias from 3 to 45 days post-hatching (no bias at 3 or 7 days, bias at 15, 30, and 45 days) but only when shelters were provided in the apparatus [77]. The left-turning preference observed only in the presence of shelters may be due to lateralization of the visual system and of the corresponding brain side. If that is the case, when the cuttlefish left the start box, they would have seen the two shelters simultaneously, but the visual information received via the left eye would have been dominant over the input received via the right eye, determining the subsequent motor direction (i.e. turning to the shelter situated on the left). The different results of the two experiments (with or without the shelters) suggest that motivational differences can alter the turning preference. However, it is difficult to assess what is the motivation of the cuttlefish leaving the start box of the maze and thus to infer something about the hemispheric specialization for the turning behavior. In fact, when cuttlefish leave the start box of the maze, they may be motivated by exploratory behavior, they may be looking for prey, or they may just be fearful and looking for a shelter to escape from the light.

The last may be the most likely explanation since cuttlefish usually avoid open and lit areas when they cannot bury themselves in the substrate [76].

Cerebral correlates of this visual lateralization have been found by looking at anatomical (volumes of vertical lobe—VL, peduncle lobe—PL, inferior buccal, and optical lobe—OL) and neurochemical (concentration of serotonin, dopamine, and noradrenaline in OL) brain asymmetries, and at their correlation with turning behavior in cuttlefish at 3 and 30 days post-hatching [78]. Brain and behavior asymmetries were present only at 30 days post hatching: a population level bias towards a larger right PL and higher monoamine concentration in the left OL was observed [78]. Interestingly, there was a correlation with the behavioral results in the T-maze: the larger the right OL and the right part of the VL, the stronger the bias to turn leftwards. Since cuttlefish display a left turning preference and have ipsilateral projections between the eyes and the optic lobes [79], the left optic lobe is expected to be larger than the right (and not the other way around). However, possessing a larger brain structure on one side does not necessarily mean that the preferential eye use is in the same direction. Indeed, it is possible that for other visual stimulation (e.g. prey or conspecific), cuttlefish with a left-turning preference in the T-maze would favor information collected by the right eye (and maybe even turn rightward). As mentioned before, it is difficult to know the real motivation of cuttlefish leaving the start box of the maze. Thus, it is plausible that different motivations (escape versus prey catching/exploration) may favor the input from one eye (and side of the brain) or the other (as happens in vertebrates with laterally placed eyes), resulting in a turning in one or the other direction.

Cuttlefish tested for turning bias in previous experiments by Jozet-Alves and colleagues [76–78] were either trawled or came from eggs laid by several females originally fished by traps in the vicinity of Luc-sur-Mer (France). The eggs were maintained in strainers floating in tanks supplied with running oxygenated seawater and no lateralization in turning was observed in cuttlefish until they reach 15 days. Jozet-Alves and Hébert [80] investigated whether the perception of chemical signals of a predator by cuttlefish embryos could modulate visual asymmetry in newly hatched cuttlefish (at 3 days of life). Three groups of embryos were exposed to a different type of olfactory stimulation prior to hatching. The first group was incubated with a predator fish odor: the strainer containing eggs was floating in the tank (150 × 80 × 35 cm) of a European seabass (*Dicentrarchus labrax*; predator group). A perforated gate located next to the outflow allowed the strainer to be isolated from direct contact with the fish. The second group was incubated with a nonpredator odor: the strainer was floating in a tank (80 × 60 × 40 cm) containing three sea urchins (*Paracentrotus lividus*; nonpredator group). The third group was incubated in a strainer floating in a tank (80 × 60 × 40 cm) with no other animal (control group). Hatchlings were collected at 9.00 from eggs that had hatched during the night (Day 0) and cuttlefish that hatched from eggs exposed for less than 10 days to predator or nonpredator odor were excluded. Cuttlefish were tested 3 days after hatching in the same T-maze apparatus used in previous studies (Fig. 8—[77, 78]) and with the same procedure until ten choices were made. Results revealed that cuttlefish exposed to predator odor prior to hatching show a left-turning bias in the T-maze, whereas embryos exposed to nonpredator odor and embryos incubated with no odor did not show any bias. Interestingly, when tested with predator odor in the T-maze apparatus all cuttlefish display a left-turning preference. The findings by Jozet-Alves and Hébert [80] suggest that cuttlefish possess an ability to innately recognize predator odor and that a predator chemical signal induces a leftward bias in newly hatched cuttlefish. Cuttlefish from embryos not exposed to predator odor would develop this preferential leftward turning bias anyway after they reach the age of 15 days. Taking all together the results of the studies conducted by Jozet-Alves and her collaborators suggest that the spontaneous motor bias of cuttlefish to turn leftward in a T-maze may be driven by a motivation to escape, revealing a dominance of the left side of the brain to control this type of behavior (in line with findings in other invertebrates species—see Sect. 3).

Recently Frasnelli et al. [33] employed a paradigm similar to that used by Alves and colleagues with cuttlefish, to investigate spontaneous lateral bias in *Octopus vulgaris*. Fifteen octopuses collected in the Gulf of Naples were tested until they made 20 choices in a T-maze apparatus (200 cm long × 80 cm wide × 40 cm high) immersed in a tank supplied with constant seawater flow. The apparatus presented only one start box that could be opened through an opaque sliding door and two exits were provided on one side on each lateral arm, so that the animal had to turn in the same direction twice in order to exit the maze (Fig. 9). Since the T-maze presented only one start box, the entire apparatus was rotated 180° half way through the experiment so that each animal was tested for ten trials with the apparatus in one position and for the other ten trials with the apparatus rotated. Seven individuals started the experiment with the apparatus in the original position and eight individuals started the experiment with the apparatus rotated 180°. This differs from the protocol of Jozet-Alves' group where half of the individuals were tested with the T-maze in one position and the other half with the maze rotated 180°. Moreover, in contrast to the studies performed on cuttlefish, in which the motivation of the individuals leaving the start box of the maze was to escape and find a shelter, here possible biases in exploring a new environment were investigated. Since the aim of the study was to test for spontaneous exploratory behavior in the T-maze, and not

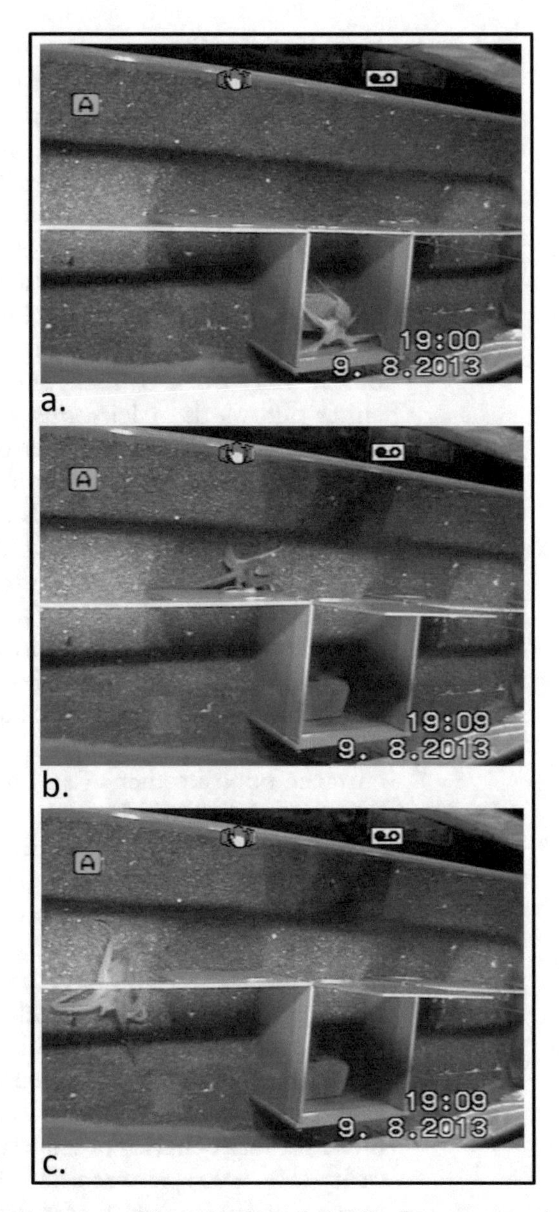

Fig. 9 Video sequence of the octopus exploring the T-maze apparatus as in the experiment by Frasnelli et al. [33]. (**a**) An octopus on top of the den in the start box. (**b**) The octopus makes a first turn to the left. (**c**) The octopus exits the T-maze by turning left a second time. Photograph courtesy of Dr. Graziano Fiorito and Dr. Giovanna Ponte

for escape response, the environment was created so that the octopus could feel as comfortable as possible. Sand was placed all over the T-maze surface, not only at the goals, and a den composed by three bricks as the one used in the tank (Fig. 4a) was placed in the start box (Fig. 9a). The octopus was given 5 min in the start box in order to acclimatize, after which the sliding door was opened and the octopus had 10 min to complete the trial, i.e. to exit the maze on one side by turning twice in the same direction. Figure 9 shows an example of leftward sequential turns. To avoid habituation and stress to the animal each individual received a maximum of three trails per day.

Results revealed that 10 out of 15 individuals turned rightwards in the first trial, independently of the position of the apparatus, revealing that no environmental cues affected the choice of one T-maze arm over the other one. However, when tested over the 20 trials none of them showed a significant bias in turning either rightwards or leftwards. Specifically, two individuals chose the right arm the same number times as the left arm (i.e. 10), five individuals chose the left arm slightly more often than the right arm (11 vs. 9), two individuals chose the right arm slightly more often than the left arm (9 vs. 11), while the remaining six individuals chose the right arm more often than the left one (3 individuals 12 vs. 8; 3 individuals 13 vs. 7).

A consistent leftward bias in exploring unknown nest sites was recently observed in *Temnothorax albipennis* ants [81]. In their study, Hunt et al. [81] placed ant colonies in a large square Petri dish (230 × 230 × 19 mm) and positioned the colony's nest entrance opposite that of an unknown nest (Fig. 10a). Since *T. albipennis* inhabits dark rock crevices in the wild, the unknown nest was darker than the starting nest in order to make it more attractive to scouting ants. Exploration of the unknown nest was induced by destroying the starting nest in the first experiment, and by removing a temporary cardboard cover from the starting nest to increase its light level in the second experiment. After ants had explored and exited the unknown nest, they were removed to prevent them from participating in a second trial. The apparatus was partially replaced or washed with water after each trial to prevent accumulation of pheromones. In the first experiment, the initial turning behavior of scouting ants entering an unknown nest cavity (Fig. 10b) was recorded. Between 5 and 15 scouting ants were recorded from each of eight colonies, for a total of 89 observations. In the second experiment, the consistency of the lateral bias was investigated using nest cavities with four branches and two decision points (Fig. 10c). Ten colonies of 64–166 workers (average 100) were used. Given the different colony activity levels between 8 and 19 observations per colony were made for a total of 113 observations. The entry direction was annotated for 80 trials from seven colonies.

Results of the first experiment indicate that ants were significantly more likely to turn left than right. The second experiment showed that ants tend to persist in incidental thigmotactic behavior, i.e. wall-following (favoring a repeated choice of left–left or right–right) when they make the first choice at the first branching

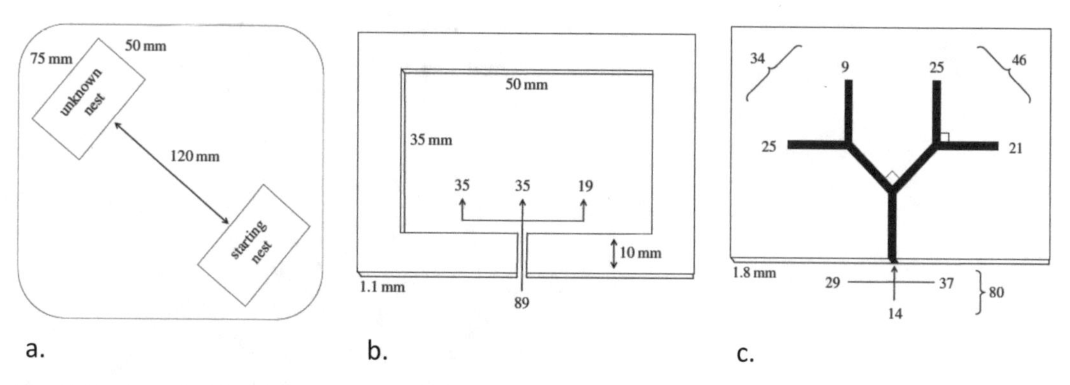

Fig. 10 The layout of the study performed by Hunt and colleagues [81]. (**a**) The experimental arena layout. (**b**) Ants entering an unfamiliar nest cavity prefer to turn left. (**c**) In a branching cavity there is a left choice bias interacting with a tendency to wall-follow (entry direction numbers left/right or unaligned and choices shown). From Hunt et al. [81]

nest cavity (Fig. 10c). However, where thigmotaxis is absent or otherwise diminished in importance before the second choice, because the ants were observed to (necessarily) detach from the wall at the first choice (left entry, right first choice; right entry, left first choice; unaligned entry), a significant leftward turning bias at the population-level becomes evident. Although the researchers did not control for the effect of environmental factors by rotating the apparatus half way through the experiment, the following considerations should be made: (1) for lighting, the ants were exploring a darkened nest, not navigating outside, hence the lighting should be constant, and (2) ants were in a small dish, and a 180° rotation would have resulted in a small 12 cm displacement, which should be within the same local electromagnetic field. This is probably sufficient to avoid problems of control due to not rotating the dish. However, in future it would be recommended and in general a good practise to rotate the apparatus half way through the experiment.

A Y-maze has been used to determine whether the common American cockroach, *Periplaneta americana* (Insecta, Blattaria, Blattidae) exhibits a bias to turn left or right in reaching an odor plume [82]. Cockroaches were tested in the dark to run through a Y-tube and make a choice of which direction to take. To prevent any chemical influence from previously tested *P. americana* a Y-tube with a replaceable bottom was made. The Y-tube channels had a square cross-section of 100 × 100 mm made of Plexiglas 5 mm thick. The bottomless Y-tube was placed on a flat surface covered with butchers' paper. The paper was changed at the end of each test. The interior walls of the channels were coated with Vaseline to prevent insects from walking on the walls. The Y-tube and its dimensions are shown in Fig. 11.

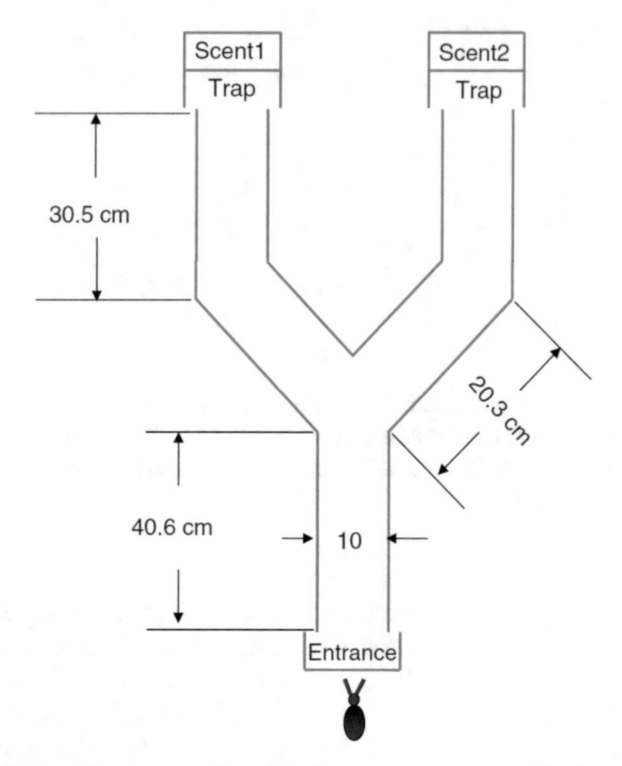

Fig. 11 Schematic representation of the Y-maze apparatus used by Cooper et al. [82] with entrance, exit traps, and scent sources. From Cooper et al. [82]

Two chemicals with distinct odors, both attractive to cockroaches, were used in the experiments: undiluted vanilla extract and 99.5 % pure ethanol. A cotton ball was immersed to saturation in each of the chemicals to be tested and placed at the end of each of the Y-tube's arms. All tests were performed with both scents placed at one of the ends of the Y-tube to stimulate the cockroaches to reach the end of the tubes. The location of each scent was randomized in a way that each scent was placed on both tubes the same number of trails. A fan from a used computer that pulls air was fitted on the entrance cage that extracted air from the Y-tube, forcing air over the chemicals and into the Y-tube at a rate of 5CFM. Special care was taken to make the holes on the cages of the same size to provide a uniform airflow through both channels. To prevent habituation, each insect was allowed to complete the Y-tube only once. Prior to every third test the Y-tube was rotated 180° in order to control for any variables from the room such as disparities in air flow, discrepancies in lighting, and other environmental effects. Thirty-eight adult cockroaches were tested only once in each of the five conditions: both antennae intact, half of the left antenna cut, the whole left antenna cut, half of the right antenna cut, and the whole right antenna cut. Antennal amputation was performed on CO_2 anesthetized individuals to reduce trauma and facilitate handling. Results showed that scent location had no effect on path choice. Injury of one antenna affected the choice of direction, but not in any consistent way: while 93% of the cockroaches with no left antenna went right, only 70% of the cockroaches with no right antenna went left. Similar results were obtained when either antenna was cut in half. Individuals with both antennae intact chose the right path in 57% of the trials. Interestingly, when the entire right antenna was cut, 70% of the individuals tended to follow the more volatile scent, i.e. ethanol. This result is difficult to interpret since both odors used have a positive valence for cockroaches. Moreover, each individual was only tested once and the bias showed by intact individuals is not very strong. Although the scent location was shown to have no effect on path choice [82], for future research using this paradigm it would be interesting to use odors belonging to very different families since they are likely to activate different classes of behavior. More importantly, replication of at least ten trials per individual is necessary in order to determine whether the bias showed is consistent within each individual and persist within the population.

A sophisticated method to monitor the behavior of fruit flies in a Y-maze has been recently developed by Buchanan et al. [83]. Hundreds of 4- to 8-day-old individual flies from seven different genetic lines were placed into individual Y-mazes or arenas and allowed to walk freely for 2 h. Mazes were illuminated from below with white LEDs and imaged with digital cameras. The X-Y positions of the flies' centroids were automatically tracked using background subtraction and recorded with custom-written software. By looking at the frequency of right and left turns in individuals that made more than 50 turns (data from flies making fewer than 50 turns were discarded) Buchanan and colleagues showed that during exploratory walking, individual flies display significant bias in their left versus right locomotor choices, with some flies being strongly left or right biased. This idiosyncrasy was present in all genotypes examined, including wild-derived populations and inbred isogenic laboratory strains. The researchers [83] investigated two additional lateralized behaviors: the direction of spontaneous exploration in circular arenas and the folding arrangement of the wings at rest. Results indicated that individual flies have a preference in the direction in which they circle, with some individuals often showing strong tendencies to circle clockwise or counter-clockwise, but no directional bias within the population was revealed. In the same way, individual flies exhibit preferences in which wing is placed on top of the other at rest, some fold left on top of right, others right on top of left. For the same individuals, turn bias scores in the Y-maze correlate positively with a clockwise circling bias in the arena, but circling bias was completely uncorrelated to wing-folding bias, suggesting that the turning biases in the Y-maze likely reflect an apparatus independent locomotor handedness phenomenon. Buchanan et al. [83] also looked at possible morphological asymmetries in leg segment and body lengths in flies tested previously in the Y-maze. They tested 28 metrics of leg length asymmetry and found that just one correlated weakly with turning bias. Finally, using transgenics and mutants, Buchanan et al. [83] showed that the magnitude of locomotor side bias is under the control of the central complex, a brain region implicated in motor planning and execution. When these neurons are silenced, exploratory laterality was shown to increase, with more extreme leftwards and rightwards bias.

Bell and Niven [84] reported individual level asymmetries in the forelimb used to reach across a gap in the desert locust *Schistocerca gregaria*. Individual locusts ($N = 29$) of the same age were tested in a rectangular white Perspex arena where two horizontal elevated platforms were placed opposite one another separated by 25 mm to create a gap. Each locust was tested 20 times, but not in consecutive order to ensure an appreciable time interval between replicates and minimize possible effects of task familiarity on the locusts' behavior. Moreover, to exclude any possible asymmetrical bias in the arena, in half the trials the locusts crossed from left to right, whereas in the other half they crossed from right to left. From each trial the forelimb first placed into the gap as well as the subsequent reach across the gap was recorded. A second cohort of locusts ($N = 29$) was tested to assess the forelimb used to initiate walking on the platforms. Finally, a transparent glass surface between the platforms bridging the gap was inserted to allow locusts to cross to the other platform by walking whilst maintaining the visual impression of the gap. A third cohort of locusts ($N = 30$) was tested with this set-up. The forelimb first placed onto the glass slide and the forelimb used to step from the glass onto the opposite platform were scored.

The results of the first cohort suggested that during gap-crossing many individuals have a strong bias to use one forelimb first, but the strength and direction of this bias differed significantly among individuals, with no consistent bias among the population. However, the same cohort of locusts did not display a bias in the forelimb placed into the gap immediately prior to reaching, and the forelimb placed into the gap did not significantly influence the forelimb subsequently used to reach across the gap. In the second cohort of locusts, individuals did not show any bias in forelimb use to initiate walking on the platform. Finally, the third cohort of locusts tested with the glass slide bridging the gap did not show any preference for the forelimb placed first onto the glass, but they did show an individual preference for the forelimb used to step from the glass slide onto the opposite platform. The findings by Bell and Niven [84] indicate that lateralization in the use of forelimbs in *S. gregaria* is context-dependent: movements are unbiased when the locusts step onto the bridge but are biased when stepping off it, suggesting that visual perception of the gap influences the switch.

6 Sensory Asymmetries

Hemispheric specialization can be studied by investigating the asymmetric use of paired sensory organs such as eyes, nostrils (or antennae in animals such as insects) and ears.

Honeybees have provided evidence of lateralization in olfaction and vision. Letzkus et al. [85] first showed that honeybees *Apis mellifera* display laterality in learning to associate an odor with a sugar reward. The researchers used the proboscis extension reflex (PER) paradigm [86], a classical conditioning paradigm in which tethered bees are conditioned to extend their proboscis when they perceive a particular odor associated with a food reward. Bees are able to learn the association between an odor and a drop of sugar solution after a few trials (Fig. 12a). Subsequently, when presented with the odor only, they extend the proboscis in anticipation to the sugar reward (Fig. 12b).

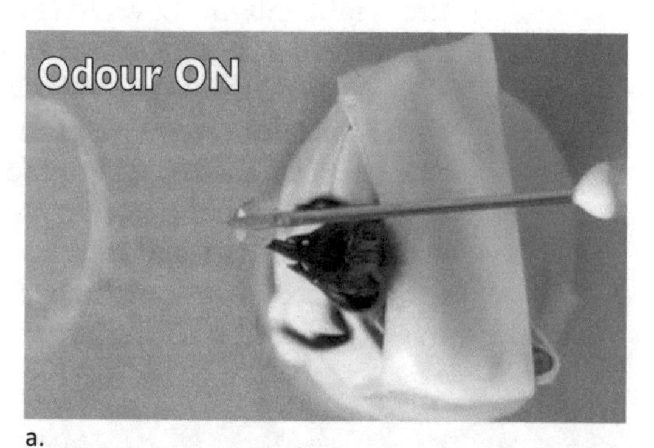

a.

b.

Fig. 12 Proboscis Extension Reflex (PER) paradigm. (**a**) During the training phase the bee is conditioned to associate an odor delivered by a olfactory stimulus controller with a sugar droplet presented with a syringe. (**b**) During the test phase the bee is presented with the odor and extends its proboscis (PER) in anticipation to the sugar reward. Photograph courtesy of Dr. Elisa Rigosi

Letzkus et al. [85] tested bees in two versions of the PER paradigm. In both versions the conditioned stimulus (odor) and the unconditioned stimulus (sugar or salt water solution) were administered together as a scented drop of sugar solution emerging from a syringe needle. In the first version, honeybees were conditioned to extend their proboscis to a scented drop of 1 M sugar solution (positive stimulus) but not to an unscented drop of saturated salt solution (negative stimulus). In the second version they were conditioned to extend their proboscis to one odor (dissolved in a 1 M sugar solution—positive stimulus) but not to another odor (dissolved in a saturated salt solution—negative stimulus). The scents used were lemon and vanilla in the form of commercially available flavoring agents for food and thus not toxic for the bees. In the first version either vanilla or lemon (10 μl of the essence dissolved in 3 ml of the sugar solution) was used as positive stimulus on two separated cohorts of bees, while in the second version lemon was used as the positive stimulus and vanilla as the negative stimulus. Each version of the learning task was carried out on three groups of bees comprised of at least 70 two-week-old bees. The bees in one group had their left antenna covered with a silicone compound (Exaflex, GC America Inc.), which prevents detection of odor, those in the second group had their right antenna covered (Fig. 13), and those in the third group constituted a control in which both antennae were left uncovered. Bees received three training trials each separated by 6 min. During each trail, the positive stimulus was presented for 5 s to the bee by touching her antennae with the drop of scented sugar solution, so that the bee could perceive both the odor and the sugar solution with her olfactory and taste receptors on the antennae and extend the proboscis to receive the sugar solution. The same procedure was performed with the negative stimulus. On the morning after

Fig. 13 A honeybee with her right antenna coated with a silicone compound to prevent odor detection. From Letzkus et al. [85]

the training, each bee was tested once by presenting each stimulus for 5 s by holding the drop at a distance of 1–2 cm in front of the bee, without touching the antennae.

The percentage of bees that produced a fully correct response (i.e. extension of proboscis to the positive stimulus only), a partially correct response (i.e. extension of proboscis to both stimuli or no extension of proboscis to either stimulus), and a fully incorrect response (i.e. extension of proboscis to the negative stimulus only) was scored. The probability of a correct response was calculated as the ratio of the number of fully correct responses to the sum of all responses. Results revealed that the bees with the right antenna covered performed worse (lower probability of correct response) than the bees with their left antenna covered and bees with both antenna uncovered. In fact, the bees trained with only their right antenna in use performed just as well as the untreated controls, suggesting that the right antenna is necessary to learn the association. However, it is hard to determine whether it is really learning or it is recall of memory on the day following training since the three groups of bees were trained and tested in the same condition (i.e. with one or both antennae in use). Thus, it may be that the right antenna is not essential in learning the association between the odor and the reward (or punishment), but it is essential in recalling the memory the day after the training. This point is discussed further in the next section.

Letzkus et al. [85] also looked at the number of olfactory sensilla placodea on ten right antennae and ten left antennae (seven of these left-right pairs originated from the same individuals) using scanning electron micrographs. They found that mean number of olfactory receptors per segment was 401 in the right antenna, as opposed to 365.5 in the left antenna and suggested that this may be the reason why learning/recalling is better with the right than the left antenna. A study by Frasnelli et al. [87] on 14 antennal pairs from the same animals looked at all the types of olfactory and non-olfactory sensilla and confirmed that on average right antennae present more olfactory sensilla that left antennae.

Letzkus et al. [88] also used a PER paradigm to compare visual learning of honeybees with only the left or right eye in use. Bees were tested with their antennae removed since conditioning to visual stimuli is easier to obtain in bees without antennae [89]. The bees were tested with both eyes covered (BEC), both eyes exposed (BEE), their right eye exposed (REE), or their left eye exposed (LEE). The conditioned stimulus (CS) was a large yellow rectangle presented on a computer-controlled display. The bees were conditioned to extend their proboscis in anticipation of a drop of sugar solution (unconditioned stimulus, US) when presented with the CS. Each experiment consisted of two 10-training sessions: the first one was conducted the morning after the eyes had been covered with a light-occluding silicon material and the second one the following morning. The two training sessions were subdivided into four 5-trial blocks. In each trial the CS was presented for 15 s and after 7 s of stimulus presentation, the bee's mouthparts were touched with a drop of 1 M sugar solution from a size 23 syringe to motivate a proboscis extension, and a drop sugar solution was given. A correct response was scored for proboscis extension to the CS alone in the first 7 s of a trial. After 7 s the mouthparts were touched and a response/no response to the US was scored. The proportion of correct responses was calculated as the ratio of responses to CS alone to the total number of responses (to CS alone and to CS + US). As expected, control BEC bees showed no learning throughout the entire training. The BEE bees' performance rose steadily, with a mean response rate of 39 % in the last five trials. REE bees also showed an increase in learning performance, but the response rate was slightly (but not significantly) lower than that of BEE bees throughout the training. Only in the last five trials REE bees showed a performance level of 33 %. LEE bees reached a mean learning performance of only 13 %, which was significantly lower than that of BEE and REE groups. LEE bees were statistically different from BEE and REE bees in trials 6–10 and from BEE bees in trials 11–15. Thus, at least in the last part of the test, bees seem to primarily use their right eye for learning and/or detecting objects (for associating a visual stimulus with a food reward, in this case), and therefore they are better at responding to a visual object when using their right eye than when using their left eye. As for the previous study [85], since the second ten-training session was performed the day after the first ten-training session, it is unclear whether the lateralization was associated primarily with learning or with memory recall.

Recent research has found that workers of the house-hunting ant *Temnothorax albipennis* seem to rely more on their right eye to recognize landmarks for navigation [90]. Basari et al. [90] tracked ants in a large arena during tandem running in which one ant leads another ant to an important resource such as food or a new nest [91]. They compared their performance in the presence of different types of landmarks to determine how they learn landmarks and use this information in their search strategy. Surprisingly Basari et al. [90] found that ants exhibit behavioral lateralization in which they possibly use their right eye more than their left eye to recognize landmarks for navigation. *T. albipennis* ants showed a significant deviation to the right during the search phase when a large landmark located behind the nest was moved to the right, suggesting a bias towards their right visual field. In addition, when the large landmark was moved to the left side and ants return to the old nest, their path deviation to the left was significant. This is what would be expected if the ants use vision of their right eye to navigate because their right eye would see the large landmark placed on the left side of the arena as they move back towards the old nest. As a result, mean bearings were biased to the left. Hence, the results indicate that ants use their right eye (and right brain side) to detect the large landmark behind the new nest, which could help them to remember the paths they had explored. However, more studies are needed in order to assess more about lateralization in spatial behavior in insects. In vertebrates the right hemisphere (left eye) attends to global, geometric spatial cues. However, when, for example, binocularly trained birds are tested in a spatial task with only one eye in use, left-eye birds perform the same as binocular birds, relying on large-scale spatial information (enclosure), whereas right-eye birds rely on local information (landmarks) [92].

Evidence of olfactory lateralization has been reported in tethered fruit flies *Drosophila melanogaster* tested in an olfactory visual flight simulator [93]. Flies readily steer directly toward a laterally positioned odor plume and do so by orienting saccades directly up an odor gradient. Duistermars et al. [93] investigated whether this ability is abolished when odor detection is precluded in one antenna and thus eliminating bilateral spatial comparison. To remove olfactory input to the antennae, a small drop of UV dental glue (Cas-Ker) was spread over the third antennal segment and cured after annealing with a 10 s burst of UV light (ELC-410, Electro-Lite). Cold-anesthetized flies were tethered to a tungsten pin in the apparatus, so that they could rotate freely in the yaw axis (for details see [93]). Then they were presented with a wide-field rotating high-contrast checkerboard pattern and a continuous water vapor stimulus, which was interrupted with either a 5 s pulse of apple cider vinegar, or a 0.1 s pulse, to provide the mechanosensory equivalent of switching to odor while minimizing the impact of the odor stimulus. Odor presentation resulted in a significant bias in flight oriented toward the intact antenna, but intriguingly occluding the left antenna had a stronger effect than occluding the right antenna. Sensory signals from the left antenna were sufficient to elicit a significant odor-mediated decrease in saccade frequency and were able to generate a higher proportion of left turns in response to odor than input to the right antenna, revealing a consistent asymmetry in antenna-mediated flight control. This suggests that asymmetrical olfactory and mechanosensory signals in *Drosophila* may facilitate stable odor tracking in complex multisensory environments, and underlies the importance of a lateralized nervous system in cognitive demanding tasks.

Chemotaxis behavior in *Drosophila melanogaster* larvae genetically manipulated to have a single functional olfactory sensory neuron (OSN) on either the left or right side of the head has been studied [94]. Larval behavior was tested in a new behavioral assay developed in the laboratory, where the quantity and distribution of the olfactory stimulus could be controlled and precisely quantified in space. The single and multiple-odor-source devices consisted of three 96-well plate lids stacked on top of each other. Either one or six of the 96 condensation rings imprinted in the top lid were filled with an odor dissolved in paraffin oil (10-ml droplets). The odors used were anisole (Chemical Abstracts Service (CAS): 100-66-3), (E)-2-hexenal (6728-26-3), isoamyl acetate (CAS: 123-92-2) and isopropyl acetate (CAS: 108-21-4). Single larvae were positioned in the middle lid of a multiple-odor-source device a minute after the odor was loaded in the top lid. The locomotor activity of each larva was tracked with a CCD camera for 3 min. Results indicated that a single functional olfactory neuron provides sufficient information to permit larval chemotaxis and that left-right comparison is not essential, although it enhances the accuracy of larval chemotaxis. More interestingly here, animals with a right-functional OSN displayed significantly better chemotaxis than the corresponding left-functional animals, revealing that the right bias in orienting chemotaxis toward an odor source is an inherent feature of the larval olfactory system.

Asymmetrical hearing has been investigated in insects. Bailey and Yang [95] determined the degree of the fluctuating left-right natural asymmetry in the hearing system of the Australian bush-cricket *Requena verticalis* (Insecta, Orthoptera) by measuring the size of the auditory spiracle of females ($N=29$). Right-left differences ranged from 0% to 8%, which translates to maximum area differences of 0.03 mm^2 and approximate 0.8 dB differences in threshold, but no directional bias emerged from the data. To investigate the effect of this morphological asymmetry on behavior, Bailey and Yang [95] tested the ability to orient to conspecific calls in an open arena in females with reduced sensitivity of one ear by packing cotton wool into either the right or left auditory spiracle. Packing cotton wool into one of the spiracles created an asymmetry far exceeding naturally occurring asymmetry. Free-moving females were allowed to orient towards a speaker emitting male calls at near-natural call intensities. Results indicated no variation in angle or vector between experimental and control insects, nor any difference in acuity between call intensities, demonstrating that the natural bilateral asymmetry of the auditory spiracle is so small that it has a negligible effect on a female's ability to find a sound source. The method could potentially be modified and extended to study the effects of this morphological asymmetry in the hearing system of *Requena verticalis* on its behavior. Instead of manipulate the auditory sensitivity of one ear, for example, the animal could be presented with the same (or different) male call played simultaneously on its right and left side. Such an experiment would put the two ears (and halves of the brain) in competition in the control of response, revealing a potential dominant role of the bigger auditory spiracle in controlling a specific behavior.

The water bug *Corixa punctata* (Insecta, Hemiptera, Corixidae) is a water-dwelling insect which uses trapped air in its physical gill to convert water-borne sounds into airborne sounds that it can hear. The *C. punctata* presents two physiologically asymmetrical ears that cause bilaterally asymmetric rocking movements of the clubbed process, a structure of the mesothoracic tympanal organ [96]. Prager and Larsen [96] examined this asymmetry with laser vibrometry at physiological sound pressure levels. Results showed that in the right tympanic organ the vibrations were larger at 1.73 kHz than at 2.35 kHz, whereas in the left organ the vibrations were larger at 2.35 kHz than at 1.73 kHz. The higher velocity of the clubbed process of the left tympanum at 2.35 kHz is due to a greater sensitivity of the left receptor cell A1 at this frequency. This ensures high sensitivity over the entire range of stridulation-sound carrier frequencies [97]. Although the biological significance of this physiological asymmetry has been explained [97], its effects on behavior are still unknown. It is expected that the right ear and the left ear respond to different sound frequencies. As a consequence of this hearing specialization some behavior may also display laterality. This could be assessed by presenting sounds of different frequencies to the animal and looking at its directional behavior, and/or by manipulating the hearing sensitivity of one ear.

7 Learning and Recall of Memory

In the previous section, the use of the PER paradigm to condition bees and investigate lateralization was discussed. Since in the studies by Letzkus et al. [85, 88] the bees were both trained and tested with only one antenna (or one eye) in use, it is difficult to assess whether the performance in the test was a measurement of how well they learned the association or how good they were in recalling the memory of the association. To overcome this problem, Rogers and Vallortigara [98] trained the bees with both antennae in use and tested them with only one antenna in use at various times after the training. After PER training with both antennae in use, using lemon in sucrose solution as the positive stimulus and vanilla in saturated salt solution as the negative stimulus, bees were tested for recall 1–2 and 23–24 h later and with the left or right antenna coated with the silicone compound. At 1–2 h after training, bees with the right antenna in use showed excellent (short-term memory) recall, but bees with the left antenna in use showed poor or no recall. By contrast, 23–24 h after training (long-term memory) recall was good in bees with the left antenna in use but not in bees with the right antenna in use.

Rogers and Vallortigara [98] also checked whether the laterality was manifested as side biases to odors presented to the left or right side of the bee without any covering of the antennae, and hence in a more natural condition than in the paradigm requiring

Fig. 14 Procedure used by Rogers and Vallortigara [98] to test honeybees with lateral presentation of odors without coating the antennae with the silicone compound. The odor is dissolved in the droplet held to the right or left side of the bee. From Rogers and Vallortigara [98]

an antenna to be coated with silicone compound. Bees were trained using both antennae and the same odors as in the previous experiment. Recall was tested at several intervals after training (1, 3, 6, or 23 h) without coating of one antenna but by presenting the positive and negative stimulus to the right or left side of the bee (Fig. 14).

Results revealed significantly more correct responses (PER to the positive stimulus but not to the negative stimulus) to odors presented to the right side than to the left side of the bee at 1 h after training. No significant left/right difference was observed at 3 h after training. At both 6 and 23 h after training, correct responses were higher when the odors were presented to the left side than to the right side. This study suggests that the retrieval of olfactory learning is a time-dependent process involving asymmetrical use of neuronal circuits.

This evidence for a preferred side during olfactory retrieval (right side for shorter-term memory and left side for long-term memory recall) has been recently confirmed [99]. Using several odor combinations, recall of olfactory memory was tested on different groups of honeybees at 1 and 6 h after training. In line with the finding of Rogers and Vallortigara [98], after training with lemon(+)/vanilla(−) or cineol(+)/eugenol(−) recall at 1 h was better when the odor was presented to the right side of the bee than to the left side, and recall at 6 h was better when the odor was presented to the left than the right side.

Overall, the results of the studies conducted on lateralization in conditioned honeybees [85, 87, 98, 99] indicate that the right antenna is involved in learning the odor-reward association and in the recall of short-term olfactory memory. Recall of long-term memory via the left antenna occurs only if both antennae are involved in learning. If the bee is forced to use only its right antenna, it can recall long-term memory using this antenna, but if the bee is forced to use only its left antenna, it cannot learn or recall at all. Hence, it seems that both antennae are used in learning but in a way that enables either short- or long-term memory traces to be established separately and then accessed in recall tests. There is an apparent shift in memory from the right to the left side, with the short-term memory being recalled through the right-side circuits and the long-term memory through the left-side circuits. This could be explained by encoding of the memory in different time frames (short- and long-term) on each side of the brain rather than it being transferred from one side to the other [99]. This lateralization would allow the right antenna to learn about new odors without interference from odor memories in long-term stores. In fact, since bees visit different flowers at different times of the day, as nectar becomes available, the formation of different odor associations during the course of the day would be required, and this is a process that might be aided if recall of older odor memories is avoided on the side of the brain undergoing new learning. A shift of recall access from one to the other brain side has been noted previously in birds [100–102], suggesting that lateralized events in memory formation may be common in invertebrates and vertebrates.

A strong odor dependence of lateralization of short-term olfactory memory has been reported in honeybees by Rigosi and colleagues [103]. A series of behavioral experiments evaluated response asymmetry of odor recall at 1 h after PER conditioning using three different odors (1-octanol, 2-octanone, and (−)-linalool). The training and the following test were performed on three groups of bees (the bees in the first group had both antennae in use, the bees in the second and third groups had respectively the right and the left antenna coated with silicon compound). Recall of short-term memory at 1 h after training demonstrated odor dependence of the lateralization. All bees trained with 1-octanol and 2-octanone performed well in the recall test regardless of the antenna in use. In contrast, bees trained with (−)-linalool showed a significant effect of the antenna in use: bees trained (and tested) with their right antenna in use performed as well as the bees with both antennae in use, and significantly better than bees with only their left antenna in use.

The behavioral experiments performed by Rigosi et al. [103] added a new aspect to the previous results [85, 87, 98, 104], showing that, in the recall test at 1 h after conditioning, different types of plant odor volatiles manifest asymmetries or not, probably depending on the biological relevance of the plant compound. The (−)-linalool is one of the most common derivates of floral scents playing a crucial role as cue for pollinators [105]. The 1-octanol and 2-octanone are unspecific and ubiquitous volatiles released from the green organs of the plants and thus of minor importance in pollinator plant interaction. It has been demonstrated that honeybees are able to learn complex odor mixtures by using a subset of key odors, such as (−)-linalool [106] and that, after conditioning bees to a mixture of odors, (−)-linalool elicits higher levels of responding than do other components of the mixture presented singly [107]. Since bees are selective in their responses to odors, the different biological relevance of the odor compounds used by Rigosi et al. [103] might be a reason for the observed difference in lateralization. Another explanation may be the fact that the bees tested in the study by Rigosi et al. [103] had unknown previous experience. Thus, it is plausible that they had been already exposed to 1-octanol and 2-octanone (these being ubiquitous volatiles released from the plants), and that these two odors were already present in the bees' long-term memory, allowing the bees with the left antenna in use to perform as well as the other two groups in the recall of memory.

Olfactory lateralization has been investigated in bumblebees *Bombus terrestris* [108]. Bumblebees were trained to associate (−)-linalool with a reward (10 ml of (−)-linalool dissolved in 3 ml of the 1 M sucrose solution) using the PER paradigm and recall of memory was tested 1 h after. As for honeybees [85, 99, 104], bumblebees with the left antenna coated performed as well as those with both antennae in use, whereas bumblebees with the right antenna coated performed significantly less well.

The PER paradigm has been also adapted to study whether primitive social bees, stingless bees (Insecta, Hymenoptera, Apidae, Apinae, Meliponini) present the same laterality as honeybees in the recall of olfactory memories [109]. Three species of Australian native, stingless bees (*Trigona carbonaria*, *Trigona hockingsi*, and *Austroplebeia australis*) were trained to discriminate the same two odors used for honeybees [98, 99], i.e. lemon(+)/vanilla(−), by presenting the odors on a cotton bud held on the left or right side of the bee. Recall of the olfactory memory at 1 h after training was better when the odor was presented to the right than to the left side of the bees. In contrast, recall at 5 h after training was better when the odor was presented to the left than to the right side of the bees. Hence, stingless bees (Meliponini) show the same laterality as honeybees (Apini), which may suggest that olfactory lateralization is likely to have evolved prior to the evolutionary divergence of these species.

In the fruitfly *Drosophila melanogaster* a structure located near the fan-shaped body connects the right and the left hemispheres [110]. This structure is an asymmetrical round body (called AB) and is not characteristic of all flies since some flies have symmetry in this region. In a sample of 2550 wild-type flies, 92.4% of individuals were found to have the AB in the right side of the brain [111]. Wild-type flies were trained to associate an odor with an electric shock: a single training cycle was used for short-term memory testing, and five individual training sessions (15-min rest intervals) for long-term memory testing. Flies were conditioned by exposure to an odor paired with electric shocks and subsequent exposure to a second odor without shock. To exclude any odor effect, a reciprocal conditioning experiment was carried out with different flies, in which the second odor was associated with the electric shock. During the memory test, flies were exposed simultaneously to both odors, each provided in one of the two arms of a T-maze. After 2 min of test, flies were trapped in either arm of the T-maze. Flies that made correct and incorrect choices during the test were processed separately and memory performance index was calculated for wild-type flies with asymmetric and symmetric brains, before knowing whether they had a symmetrical or an asymmetrical brain. This had the advantage that all flies were subjected to equal treatment, conditioning, and testing, and that the experiment was performed blind with respect to brain anatomy. After the memory test, *Drosophila* brains were dissected, fixed and the asymmetry or symmetry of their brains was determined using confocal analysis. Among the 1248 flies tested for memory recall at 3 h, analyses revealed that 80 flies had a symmetrical brain, whereas among the 1302 flies tested for 4-days memory 115 flies presented a symmetric brain. Pascual et al. [111] observed no evidence of 4-day long-term memory in wild-type flies with a symmetrical brain structure, although their short-term memory was intact. Long-term memory was formed only by flies with the asymmetrical structure, suggesting that the brain asymmetry is not necessary for the *Drosophila* to establish short-term memory but it is important in the formation or retrieval of long-term memory.

Studies on olfactory memory have been conducted on the terrestrial slug *Limax* (Gastropoda, Limacidae, Limacinae [112]). This species is capable of acquiring odor-aversion memories; i.e. when presented with the odor of food in combination with an aversive odor, it avoids that food. Previous studies using bilateral ablation of the procerebrum (PC), a secondary olfactory center of terrestrial molluscs, demonstrated that PC is necessary for this type of learning [113] and that PC is the site of memory storage. Friedrich and Teyke [114] demonstrated that the same side of the tentacles (the main olfactory pathway comprises the olfactory epithelia located on the two posterior tentacles) must be functional during both acquisition and retrieval of appetitive olfactory memory. In fact, conditioned behavior is not observed when the tentacles on one side are inactivated during conditioning and those on the other side are inactivated during recall of the memory, indicating that the memory is stored only on the side of the brain receiving olfactory input from the tentacles that were functional during conditioning.

Matsuo and colleagues [112] investigated further this unilateral memory storage in *Limax* through PC unilateral ablation experiments. Slugs were conditioned to associate carrot juice with 1% quinidine sulfate solution. Only the slugs that reached the carrot juice within 3 min during the conditioning were considered. Different groups of slugs underwent surgery before or after conditioning and either the right or left PC was surgically destroyed. In post-conditioning PC ablation, either the right or the left PC was ablated 24 h after the conditioning, and memory retention was tested 7 days following the surgery. In pre-conditioning PC ablation, either the right or the left PC was ablated 7 days prior to the conditioning, and memory retention was tested 24 h after conditioning. In the memory retention test, if the slug did not reach the carrot juice within 3 min, the experimenter concluded that the odor-aversion memory was had been formed and recalled. Results showed that, when the PC is surgically ablated only unilaterally before or after conditioning, approximately half of the slugs are unable to form an odor-aversion association, whereas the other half retain unimpaired memory performance. This indicates that only the PC on one side, either the left or right PC, is used for the storage of odor-aversion memory in *Limax* and which one is used in any individual is random.

Matsuo and colleagues [112] also tested whether the unilaterally stored memory is transferred to the other side at 7 days from the conditioning. If this is the case, unilaterally PC-ablated slugs should show memory performance comparable to that of the sham-operated slugs in the memory retention test. This was not the case, revealing that the unilateral memory storage in *Limax* continues for at least a week without lateral transfer.

8 Conclusions and Future Directions

In this chapter the different methods employed to investigate behavioral asymmetries in invertebrates have been reported and discussed. Different types of behavior have been analyzed and biases have been shown in foraging behavior, escape response, aggressive or sexual interactions with conspecific, spontaneous motor biases, sensory modalities, and recall of memory. However, a problem has emerged in that often it is difficult to classify the behavior under a specific class. A couple of examples follow. *S. globula* spiders use their left legs more than their right legs when they interact with their prey. However, even if this is prey handling, these interactions are aggressive and can lead to serious damage of the predator. Thus, it is problematic to infer whether the preferential use of the left leg reflects a specialization of the left side of the brain (remembering that invertebrates have ipsilateral projections between the brain and the body) in prey catching or in aggressive behavior. T-maze apparatus provides another example. T-mazes have been used to investigate spontaneous motor biases, but the true motivation of the animal remains unknown. Thus, it becomes hard to infer something more specific about the lateral specialization underlying the bias in turning toward one or the other side. This problem may be overcome by giving the animal food before the test to exclude the possibility of predatory behavior (but this may also lead to inactivity) or by presenting a more comfortable environment if the aim is to study spontaneous exploratory behavior.

A general rule before carrying out a study on lateralization should be to have a clear question about the behavior to study, in order to develop a specific experimental protocol and eliminate other possible factors that can make the results difficult to interpret. Typically, when studying animal behavior, field studies may be better than laboratory studies since they permit the experimenter to investigate behavior in natural conditions. Unfortunately, not always it is possible to study animal behavior in nature. Moreover, laboratory studies are needed to control for experience and allow manipulating a number of variables (e.g. number of repetitions per subject) in order to assess something more precise about a specific behavior. More detailed comments about each section follow:

Prey catching/Foraging: This is a natural behavior but it may be difficult to observe in nature. It can be investigated in the laboratory by observing the animal's strategy used to detect, catch, and handle a prey. In the laboratory hunger can be easily controlled (by feeding or not feeding the animals prior the test) and preferential side biases in prey catching/foraging can be examined using simple experimental protocols (e.g. by presenting the prey on one side or the other). Single observations are not enough and repetitions are essential in order to make the finding convincing. It is important not to stress the animal, as well as to avoid habituation effects. To do so, trials should be spread over some time.

Escape responses: The same considerations made for foraging behavior are valid. Often this behavior is more difficult to study in the laboratory given the limitations of the laboratory spaces or the lack of availability of the predator. To overcome this problem, silhouettes of the predator or vibratory stimuli can be employed. Several trials, well distributed in time, are necessary to confirm the existence of a bias.

Interactions with conspecifics (aggressive and sexual behavior): This behavior has to be forced in the laboratory (e.g. Petri dishes for bees—see Sect. 4). It would be better to study it in a more natural context when it is possible (e.g. putting IR cameras in hives for bees).

Spontaneous motor biases: As discussed before the use of a T-maze apparatus may have several limitations. Firstly, it is difficult to control for environmental cues. The apparatus needs to be rotated not only between different individuals but also between different trials of the same individual. Secondly, the results are often difficult to interpret because the relevant motivation of the animal (escape, feeding, or exploration) is unknown. These types of experiments need several replicates separated by an appreciable time interval to minimize possible effects of task familiarity on the animal's behavior.

Sensory modalities: This area of investigation has the advantage that sensory organs are often paired and anatomically identical. Studying them allows separate understanding of the specialization of each of the paired sensory organs (and in some cases of the corresponding brain side) for a specific task. It is quite easy to restrict the contribution of the external input to one or the other of the paired sensory organs. This can be achieved by lateral presentation of stimuli or by coating one of the two sensory organs, but the latter is invasive and may affect the animal's performance. In both cases, conditioning of tethered animals is often necessary. In order to studying the asymmetry in more natural conditions, it would be better to allow the animal to move freely when possible. To avoid tethered animals, conditioning may be done in a T-maze.

Memory: This is one of the most fascinating topics where lateralized circuits seem to be involved. More studies are needed to shed light on the functioning of the asymmetric dynamics that characterize memory formation and consolidation. Invertebrates are very good models for this aim. A difficulty may rely in the separation between learning and memory recall that is not always easy.

To conclude, all the evidence about behavioral lateralization in invertebrates suggest that lateralization provides substantial advantages, since it has persisted, or evolved many times, in both vertebrate and invertebrate animals. The complex issue is whether homologous genes in invertebrates and vertebrates determine lateralization or whether there has been analogous evolution of lateralized function in vertebrates and invertebrates. For this reason, studying lateralization in invertebrates is becoming essential to expand our current knowledge about the evolution of behavioral and brain asymmetries.

References

1. McManus IC (1999) Handedness, cerebral lateralisation and the evolution of language. In: Corballis MC, Lea SEG (eds) The descent of mind: psychological perspective on hominid evolution. Oxford University Press, Oxford
2. Rogers LJ, Vallortigara G, Andrew R (2013) Divided brains: the biology and behaviour of brain asymmetries. Cambridge University Press, Cambridge
3. Versace E, Vallortigara G (2015) Forelimb preferences in human beings and other species: multiple models for testing hypotheses on lateralization. Front Psychol 6:233
4. Dadda M, Koolhaas WH, Domenici P (2010) Behavioural asymmetry affects escape performance in a teleost fish. Biol Lett 6:414–417
5. Rogers LJ (2002) Lateralized brain function in anurans: comparison to lateralization in other vertebrates. Laterality 7:219–239

6. Kight SL, Steelman L, Coffey G, Lucente J, Castillo M (2008) Evidence of population level in giant water bugs, *Belostoma flumineum Say* (Heteroptera: Belostomatidae): T-maze turning is left biased. Behav Proc 79:66–69
7. Lippolis G, Joss J, Rogers LJ (2009) Australian lungfish (*Neoceratodus forsteri*): a missing link in the evolution of complementary side biases for predator avoidance and prey capture. Brain Behav Evol 73:295–303
8. Tommasi L, Andrew RJ, Vallortigara G (2000) Eye use is determined by the nature of task in the domestic chick (*Gallus gallus*). Behav Brain Res 112:119–126
9. Rogers LJ, Kaplan G (2006) An eye for a predator: lateralization in birds, with particular reference to the Australian magpie. In: Malashichev Y, Deckel W (eds) Behavioral and morphological asymmetries in vertebrates. Landes Bioscience, TX, pp 47–57
10. Vallortigara G, Rogers LJ, Bisazza A, Lippolis G, Robins A (1998) Complementary right and left hemifield use for predatory and agonistic behaviour in toads. NeuroReport 9:3341–3344
11. Robins R, Rogers LJ (2006) Complementary and lateralized forms of processing in *Bufo marinus* for novel and familiar prey. Neurobiol Learn Mem 86:214–227
12. Vallortigara G, Rogers LJ (2005) Survival with an asymmetrical brain: advantages and disadvantages of cerebral lateralization. Behav Brain Sci 28:575–633
13. Frasnelli E, Vallortigara G, Rogers LJ (2012) Left-right asymmetries of behavioural and nervous system in invertebrates. Neurosci Biobehav Rev 36:1273–1291
14. Frasnelli E (2013) Brain and behavioral lateralization in invertebrates. Front Psychol 4(939):1–10
15. Rogers LJ (2014) Asymmetry of brain and behavior in animals: its development, function, and human relevance. Genesis 52(6): 555–571
16. Vallortigara G, Versace E (2015) Laterality at the neural, cognitive and behavioural levels. In: Snowdon C, Burghardt G, Pepperberg I, Call J, Zentall T (eds) APA handbook of comparative psychology. American Psychological Association Press, Washington, DC
17. Heuts BA, Brunt T (2005) Behavioural left-right asymmetry extends to arthropods. Behav Brain Sci 28:601–602
18. Hönicke C, Bliss P, Moritz RF (2015) Effect of density on traffic and velocity on trunk trails of *Formica pratensis.* Sci Nat 102(3–4):17
19. Frasnelli E, Iakovlev I, Reznikova Z (2012) Asymmetry in antennal contacts during trophallaxis in ants. Behav Brain Res 32:7–12
20. Reznikova Z (2007) Animal intelligence: from individual to social cognition. Cambridge University Press, Cambridge
21. Heuts BA, Lambrechts DYM (1999) Positional biases in leg loss of spiders and harvestmen (*Arachnida*). Entomol Ber (Amst) 59:13–20
22. Ades C, Ramires EN (2002) Asymmetry of leg use during prey handling in the spider *Scytodes globula* (Scytodidae). J Insect Behav 15:563–570
23. Ramires EN (1999) Uma abordagem comparativa ao comportamento defensivo, agonístico e locomotor de três espécies de aranhas do gênero *Loxosceles (Sicariidae).* Unpublished doctoral dissertation, Institute of Psychology, University of São Paulo, Brazil
24. Kells AR, Goulson D (2001) Evidence for handedness in bumblebees. J Insect Behav 14:47–55
25. Kawaguchi LG, Ohashi K, Toquenaga Y (2007) Contrasting responses of bumble bees to feeding conspecifics on their familiar and unfamiliar flowers. Proc R Soc B 274: 2661–2667
26. Leadbeater E, Chittka L (2008) Social transmission of nectar-robbing behaviour in bumble-bees. Proc R Soc B 275:1669–1674
27. Goulson D, Park KJ, Tinsley MC, Bussière LF, Vallejo-Marin M (2013) Social learning drives handedness in nectar robbing bumblebees. Behav Ecol Sociobiol 67: 1141–1150
28. Wells MJ (1978) Octopus: physiology and behaviour of an advanced invertebrate. Chapman & Hall, London
29. Wells MJ (1962) Brain and behaviour in cephalopods. Heinemann, London
30. Muntz WRA (1963) Interocular transfer and the function of the optic lobes in octopus. Q J Exp Psychol 15:116–124
31. Byrne RA, Kuba M, Griebel U (2002) Lateral asymmetry of eye use in *Octopus vulgaris.* Anim Behav 64:461–468
32. Byrne RA, Kuba MJ, Meisel DV (2004) Lateralized eye use in *Octopus vulgaris* shows antisymmetrical distribution. Anim Behav 68:1107–1114
33. Frasnelli E, Ponte G, Fiorito G, Vallortigara G (2014) Investigating lateralization in octopuses: first evidence of asymmetry in the optic lobes. In: Fourth workshop on cognition and evolution, Rovereto, Italy

34. Byrne RA, Kuba MJ, Meisel DV, Griebel U, Mather JA (2006) Does *Octopus vulgaris* have preferred arms? J Comp Psychol 3:198–204
35. Byrne RA, Kuba MJ, Meisel DV, Griebel U, Mather JA (2006) Octopus arm choice is strongly influenced by eye use. Behav Brain Res 172:195–201
36. Fagot J, Vauclair J (1991) Manual laterality in non human primates: a distinction between handedness and manual specialization. Psychol Bull 109:76–89
37. Gutnick T, Byrne RA, Hochner B, Kuba M (2011) *Octopus vulgaris* uses visual information to determine the location of its arm. Curr Biol 21:460–462
38. Heuts BA (1999) Lateralization of trunk muscle volume, and lateralization of swimming turns of fish responding to external stimuli. Behav Processes 47:113–124
39. Bisazza A, Rogers LJ, Vallortigara G (1998) The origins of cerebral asymmetry: a review of evidence of behavioural and brain lateralization in fishes, amphibians, and reptiles. Neurosci Biobehav Rev 22:411–426
40. Bisazza A, De Santi A, Vallortigara G (1999) Laterally and cooperation: mosquitofish move closer to a predator when the companion is on the left side. Anim Behav 57:1145–1149
41. Vallortigara G, Bisazza A (2002) How ancient is brain lateralization? In: Andrew RJ, Rogers LJ (eds) Comparative vertebrate lateralization. Cambridge University Press, Cambridge, pp 9–69
42. Vallortigara G, Rogers LJ, Bisazza A (1999) Possible evolutionary origins of cognitive brain lateralization. Brain Res Rev 30:164–175
43. Takeuchi Y, Tobo S, Hori M (2008) Morphological asymmetry of the abdomen and behavioral laterality in atyid shrimps. Zool Sci 25:355–363
44. Tobo S, Takeuchi Y, Hori M (2011) Morphological asymmetry and behavioural laterality in the crayfish, *Procambarus clarkia*. Ecol Res 27(1):53–59
45. Rosa Salva O, Regolin L, Mascalzoni E, Vallortigara G (2012) Cerebral and behavioural asymmetry in animal social recognition. Comp Cogn Behav Rev 7:110–138
46. Hews DK, Castellano M, Hara E (2004) Aggression in females is also lateralized: left-eye bias during aggressive courtship rejection in lizards. Anim Behav 68:1201–1207
47. Ventolini N, Ferrero EA, Sponza S et al (2005) Laterality in the wild: preferential hemifield use during predatory and sexual behaviour in the black-winged stilt. Anim Behav 69:1077–1084
48. Austin NA, Rogers LJ (2012) Limb preference and lateralization of aggression, reactivity and vigilance in feral horses (*Equus caballus*). Anim Behav 83:239–247
49. Casperd JM, Dunbar RIM (1996) Asymmetries in the visual processing of emotional cues during agonistic interactions in gelada baboons. Behav Processes 37:57–65
50. Pratt AE, McLain DK, Lathrop GR (2003) The assessment game in sand fiddler crab contests for breeding burrows. Anim Behav 65:945–955
51. Jones DS, George RW (1982) Handedness in fiddler crabs as an aid in taxonomic grouping of the genus *Uca* (Decapoda, Ocypodidae). Crustaceana 43:100–102
52. Backwell PRY, Matsumasa M, Double M, Roberts A, Murai M, Keogh JS, Jennions MD (2007) What are the consequences of being left-clawed in a predominantly right-clawed fiddler crab? Proc R Soc B 274:2723–2729
53. Salmon M (1984) The courtship, aggressive and mating system of a "primitive" fiddler crab (*Uca vocans*). Trans Zool Soc Lond 37:1–50
54. Hyatt GW, Salmon M (1978) Combat in fiddler crabs *Uca pugilator* and *Uca pugnax*-quantitative analysis. Behaviour 65:182–211
55. Rogers LJ, Rigosi E, Frasnelli E, Vallortigara G (2013) A right antenna for social behaviour in honeybees. Sci Rep 3:2045
56. Rogers LJ, Frasnelli E, Versace E, Vallortigara G (2016) Lateralized social behaviour in a "solitary" red mason bee, *Osmia bicornis*. Sci Rep 6:29411. doi:10.1038/srep29411
57. Nepi M, Cresti L, Maccagnani B, Ladurner E, Pacini E (2005) From the anther to the proctodeum: Pear (*Pyrus communis*) pollen digestion in *Osmia cornuta* larvae. J Insect Physiol 51:749–757
58. Tepedino VJ, Torchio PF (1994) Founding and ussuroing: equally efficient paths to nesting success in *Osmia lignaria propinqua* (Hymenoptera: Megachilidae). Ann Entomol Soc Am 87:946–953
59. Seidelmann K (1999) The race for females: the mating system of the red mason bee, Osmia rufa (L.) (Hymenoptera: Megachilidae). J Insect Behav 12:13–25
60. Benelli G, Donati E, Romano D, Stefanini C, Messing RH, Canale A (2015) Lateralization of aggressive displays in a tephritid fly. Sci Nat Naturwiss 102(1–2):1251
61. Benelli G, Romano D, Messing RH, Canale A (2015) Population-level lateralized aggressive and courtship displays make better fighters not lovers: evidence from a fly. Behav Processes 115:163–168

62. Shelly TE (2000) Aggression between wild and laboratory-reared sterile males of the Mediterranean fruit fly in a natural habitat (Diptera: Tephritidae). Fla Entomol 83:105–108
63. Papadopoulos NT, Carey JR, Liedo P, Muller G, Senturk D (2009) Virgin females compete for mates in the male lekking species *Ceratitis capitata*. Physiol Entomol 34:238–245
64. Romano D, Canale A, Benelli G (2015) Do right-biased boxers do it better? Population-level asymmetry of aggressive displays enhances fighting success in blowflies. Behav Processes 113C:159–162
65. Benelli G, Romano D, Messing RH, Canale A (2015) First report of behavioural lateralisation in mosquitoes: right-biased kicking behaviour against males in females of the Asian tiger mosquito, *Aedes albopictus*. Parasitol Res 114:1613–1617
66. Asami T, Gitternberger E, Falkner G (2008) Whole-body enantiomorphy and maternal inheritance of chiral reversal in the pond snail *Lymnaea stagnalis*. J Heredity 99(5):552–557
67. Davison A, Frend HT, Moray C, Wheatley H, Searle LJ, Eichhorn MP (2009) Mating behaviour in pond snails *Lymnaea stagnalis* is a maternally inherited lateralized trait. Biol Lett 5:20–22
68. Van Duivenboden YA, Ter Maat A (1988) Mating behaviour of *Lymnaea stagnalis*. Malacologia 28:53–64
69. Chase R (1986) Brain cells that command sexual behavior in the snail *Helix aspersa*. J Neurobiol 17(6):669–679
70. Kamimura Y (2006) Right-handed penises of the earwig *Labidura riparia* (Insecta, Dermaptera, Labiduridae): evolutionary relationships between structural and behavioral asymmetries. J Morphol 267:1381–1389
71. Regen J (1913) Ueber die Anlockung des Weibchens von *Gryllus campestris* L. durch telephonisch uebertragene Stridulationslaute des Maennchens. Pfliigers Arch 155: 193–200
72. Elliott CJH, Koch UT (1983) Sensory feedback stabilizing reliable stridulation in the field cricket *Glyllus campestris* L. Anim Behav 31:887–901
73. Stärk AA (1958) Untersuchungen am Lautorgan einiger Grillen- und Laubheuschrecken-Arten, zugleich ein Beitrag zum Rechts-Links-Problem. Zool Jahrb Anat 77:9–50
74. Olton DS (1979) Mazes, maps, and memory. Am Psychol 34:583–596
75. O'Keefe J, Dostrovsky J (1971) The hippocampus as a spatial map. Preliminary evidence from unit activity in the freely-moving rat. Brain Res 34(1):171–175
76. Alves C, Chichery R, Boal JG, Dickel L (2007) Orientation in the cuttlefish *Sepia officinalis*: response versus place learning. Anim Cogn 10:29–36
77. Jozet-Alves C, Viblanc VA, Romagny S, Dacher M, Healy SD, Dickel L (2012) Visual lateralization is task- and age-dependent in cuttlefish (*Sepia officinalis*). Anim Behav 83:1313–1318
78. Jozet-Alves C, Romagny S, Bellanger C, Dickel L (2012) Cerebral correlates of visual lateralization in *Sepia*. Behav Brain Res 234:20–25
79. Nixon M, Young JZ (2003) The brains and lives of cephalopods. Oxford University Press, Oxford
80. Jozet-Alves C, Hébert M (2013) Embryonic exposure to predator odour modulates visual lateralization in cuttlefish. Proc R Soc B 280(1752):2012–2575
81. Hunt ER, O'Shea-Wheller TA, Albery GF, Bridger TH, Gumn M, Franks NR (2014) Ants show a leftward turning bias when exploring unknown nest sites. Biol Lett 10(12):20140945
82. Cooper R, Nudo N, Gonzales JM, Vinson SB, Liang H (2010) Side-dominance of *Periplaneta americana* persists through antenna amputation. J Insect Behav 24:175–185
83. Buchanan SM, Kain JS, de Bivort BL (2015) Neuronal control of locomotor handedness in *Drosophila*. Proc Natl Acad Sci U S A 112(21):6700–6705
84. Bell ATA, Niven JE (2014) Individual-level, context-dependent handedness in the desert locust. Curr Biol 24:R382–R383
85. Letzkus P, Ribi WA, Wood JT, Zhu H, Zhang SW, Srinivasan MV (2006) Lateralization of olfaction in the honeybee *Apis mellifera*. Curr Biol 16:1471–1476
86. Bitterman ME, Menzel R, Fietz A, Schafer S (1983) Classical conditioning of proboscis extension in honeybees (*Apis mellifera*). J Comp Psychol 97:107–119
87. Frasnelli E, Anfora G, Trona F, Tessarolo F, Vallortigara G (2010) Morpho-functional asymmetry of the olfactory receptors of the honeybee (*Apis mellifera*). Behav Brain Res 209:221–225
88. Letzkus P, Boeddeker N, Wood JT, Zhang SW, Srinivasan MV (2007) Lateralization of visual learning in the honeybee. Biol Lett 4:16–18
89. Hori S, Takeuchi H, Arikawa K, Kinoshita M, Ichikawa N, Sasaki M, Kubo T (2006)

Associative visual learning, color discrimination, and chromatic adaptation in the harnessed honeybee *Apis mellifera* L. J Comp Physiol A Neuroethol Sens Neural Behav Physiol 192:691–700
90. Basari N, Bruendl AC, Hemingway CE, Roberts NW, Sendova-Franks AB, Franks NR (2014) Landmarks and ant search strategies after interrupted tandem runs. J Exp Biol 217:944–954
91. Möglich M (1978) Social organization of nest emigration in Leptothorax (Hym., Form.). Insectes Soc 25:205–225
92. Tommasi L, Vallortigara G (2001) Encoding of geometric and landmark information in the left and right hemispheres of the avian brain. Behav Neurosci 115:602–613
93. Duistermars BJ, Chow DM, Frye MA (2009) Flies require bilateral sensory input to track odour gradients in flight. Curr Biol 19:1301–1307
94. Louis M, Huber T, Benton R, Sakmar TP, Vosshall LB (2007) Bilateral olfactory sensory input enhances chemotaxis behavior. Nat Neurosci 11:87–199
95. Bailey WJ, Yang S (2002) Hearing asymmetry and auditory acuity in the Australian bush-cricket *Requena verticalis* (Listroscelidinae; Tettigoniidae; Orthoptera). J Exp Biol 205:2935–2942
96. Prager J, Larsen ON (1981) Asymmetrical hearing in the water bug *Corixa punctata* observed with laser vibrometry. Naturwissenschaften 68:579–580
97. Prager J, Streng R (1982) The resonance properties of the physical gill of *Corixa punctata* and their significance in sound reception. J Comp Physiol A 148:323–335
98. Rogers LJ, Vallortigara G (2008) From antenna to antenna: lateral shift of olfactory memory in honeybees. PLoS One 3:e2340
99. Frasnelli E, Vallortigara G, Rogers LJ (2010) Response competition associated with right-left antennal asymmetries of new and old olfactory memory traces in honeybees. Behav Brain Res 209:36–41
100. Cipolla-Neto J, Horn G, McCabe BJ (1982) Hemispheric asymmetry and imprinting: the effect of sequential lesions of the hyperstriatum ventrale. Exp Brain Res 48:22–27
101. Clayton NS, Krebs JR (1994) Lateralization and unilateral transfer of spatial memory in marsh tits: are two eyes better than one? J Comp Physiol A 174:769–773
102. Andrew RJ (1999) The differential roles of right and left sides of the brain in memory formation. Behav Brain Res 98:289–295
103. Rigosi E, Frasnelli E, Vinegoni C, Antolini R, Anfora G, Vallortigara G, Haase A (2011) Searching for anatomical correlates of olfactory lateralization in the honeybee antennal lobes: a morphological and behavioural study. Behav Brain Res 221(1):290–294
104. Anfora G, Frasnelli E, Maccagnani B, Rogers LJ, Vallortigara G (2010) Behavioural and electrophysiological lateralization in a social (*Apis mellifera*) but not in a non-social (*Osmia cornuta*) species of bee. Behav Brain Res 206:236–239
105. Knudsen JT, Tollsten L, Bergström LG (1993) Floral scents-a checklist of volatile compounds isolated by head-space techniques. Phytochemistry 33(2):253–280
106. Reinhardt J, Sinclair M, Srinivasan MV, Claudianos C (2010) Honeybees learn odour mixtures via a selection of key odorants. PLoS One 5(2):e9110
107. Laloi D, Bailez O, Blight M, Roger B, Pham-Delegue M-H, Wadhams LJ (2000) Recognition of complex odors by restrained and free-flying honeybees, *Apis mellifera*. J Chem Ecol 26:2307–2319
108. Anfora G, Rigosi E, Frasnelli E, Ruga E, Trona F, Vallortigara G (2011) Lateralization in the invertebrate brain: left-right asymmetry of olfaction in bumble bee, *Bombus terrestris*. PLoS One 6(4):e18903
109. Frasnelli E, Vallortigara G, Rogers LJ (2011) Right-left antennal asymmetry of odour memory recall in three species of Australian stingless bees. Behav Brain Res 224(1): 121–127
110. Heisenberg M (1994) Central brain function in insects: genetic studies on the mushroom bodies and central complex in *Drosophila*. Neural basis of behavioural adaptations. Fortschr Zool 39:61–79
111. Pascual A, Huang K-L, Nevue J, Préat T (2004) Brain asymmetry and long-term memory. Nature 427:605–606
112. Matsuo R, Kawaguchi E, Yamagishi M, Amano T, Ito E (2010) Unilateral memory storage in the procerebrum of the terrestrial slug *Limax*. Neurobiol Learn Mem 93:337–342
113. Kasai Y, Watanabe S, Kirino Y, Matsuo R (2006) The procerebrum is necessary for odor-aversion learning in the terrestrial slug *Limax valentianus*. Learn Mem 13(4): 482–488
114. Friedrich A, Teyke T (1998) Identification of stimuli and input pathways mediating food-attraction conditioning in the snail, *Helix*. J Comp Physiol A 183:247–254

Part II

Neurobiological Methods

Chapter 7

Unilateral Lesions

Martina Manns

Abstract

Lesion studies represent a core approach to investigate the functional role of specific brain regions. Especially asymmetry research profits from this method since the impact of left- and right-hemispheric lesions can be directly compared to unravel the functional significance of each brain side. A large number of permanent or reversible lesion methods are available that differ in applicability and reliability depending on size and developmental stage of the neuronal target structure. This chapter presents a brief overview of the different methods and provides a short and general methodological guide. It also discusses principle advantages, drawbacks, and problems of lesion studies that should be considered when selecting the appropriate method or interpreting results.

Key words Ablation, Aspiration, Cooling, Electrolytic lesion, Excitotoxin, Glutamate, Ibotenic acid, Kainic acid, Neurotoxin, Radiofrequency lesion, Stereotaxy, Tetrodoxoxin, Thermocoagulation, Transection

1 Brain Lesions as the Cornerstone of Asymmetry Research

1.1 Asymmetry Research Is Historically Based on Lesion Patients

It was the groundbreaking claim "Nous parlons avec l'hemisphere gauche" that demarcated the dawn of asymmetry research in 1861. The French surgeon Paul Broca summarizes with this sentence his neuroanatomical observations of aphasic lesion patients during meetings of the French Anthropological Society and later of the Anatomical Society of Paris [1–5]. He demonstrated that only destruction of the left posterior inferior frontal gyrus leads to loss of speech. This result was the first generally accepted support of the "localization theory" although the French neurologist Marc Dax reported equivalent though not well-documented observations already 30 years earlier [6]. The localization theory assumes that the human brain is organized in functionally segregated building-blocks associated with specific cortical areas. This idea was intensively discussed during the nineteenth century when many scientists favored the concept of a brain that only acts as an inextricable unity. But Broca's carefully described case studies

Lesley J. Rogers and Giorgio Vallortigara (eds.), *Lateralized Brain Functions: Methods in Human and Non-Human Species*, Neuromethods, vol. 122, DOI 10.1007/978-1-4939-6725-4_7, © Springer Science+Business Media LLC 2017

demonstrated that cognitive functions depend on the integrity of identifiable brain structures whose impact may differ between left and right brain side. Thus, studies of patients verified brain lesions as a critical method to explore structure–function interrelation in general and in detecting left-right differences in particular—an approach that has inspired a long tradition of animal research.

Analyzing the behavioral outcome of accidental lesions in neurological patients was the only data source for investigating lateralized brain functions for about 100 years. In 1949, the Japanese neurologist Wada introduced a technique that transiently silences the left or right hemisphere as part of diagnostic assessment before surgery for epilepsy and to eventually minimize severe cognitive side effects of neurological treatment [7, 8]. Injection of the barbiturate sodium amobarbital (or amytal) into the left or right internal carotid artery anesthetizes the left or right brain side, respectively. Since the non-injected hemisphere remains active, the awake patient can be engaged in different cognitive tests that may directly demonstrate outage of functions depending on the anesthetized brain hemisphere. This method allowed exploration of hemisphere-specific processing before functional brain imaging methods had been established as modern noninvasive techniques. Studies using the WADA test support the left-hemispheric dominance for speech processing within the human population (e.g. [9–12]) and demonstrate asymmetry of memory (e.g. [11]) or self-identification [13]. In principle, the WADA test is a blueprint for later approaches in animal research inducing temporary brain lesions via selective pharmacological injections on the left or right side of the brain.

Around the same time as Wada published his method, a very special patient group provided exciting new insight into the lateralized organization of the human brain. A group of neurosurgeons and neuroscientists around Roger Sperry at the Caltech in California dared to transect the major forebrain commissure in the human brain—the corpus callosum—to control seizures of patients with intractable epilepsy [14–16]. They subjected their patients to sophisticated neuropsychological tests and could show for the first time how the hemispheres act independently from each other. Split-brain research during the last 50 years has given rise to deep insights into the differential functioning of the left and right hemispheres and how the brain halves interact ([17, 18]; see Chap. 2 for studies on split-brain). This research exemplifies that cutting intra- and interhemispheric fiber tracts provide important clues about how far neuronal structures process information independently, how they exchange information and how they communicate.

The short historic overview illustrates that lesion studies in human patients constitute a core approach of asymmetry research: Carefully analyzing the outcome of accidental or intended brain damages provides critical insight into the lateralized brain organization. It was again a lesion approach with which the scientific community was persuaded of neuronal asymmetries in the animal kingdom, burying the assumption that cerebral lateralization is a unique human trait [19]. The pioneering work of Nottebohm [20] demonstrated a left bias for song production in chaffinches by means of nerve transections. Follow-up lesion studies in different species and with different techniques have unraveled the complex lateralized organization of the avian song system [21]. It is presumably more than incidental that this system was so convincing since it is considered as a worthwhile model to explore the functional organization and ontogeny of language processing—one of the most strongly lateralized functions of the human brain.

1.2 Lesion Approaches Exemplified by the Avian Song System

Birdsong is produced by the coordinated action of vocal, supravocal, and respiratory muscles. The vocal organ of birds is the bipartite syrinx, which is positioned at the junction of the two bronchial passages at the bottom of the trachea. The two halves of the syrinx are innervated by ipsilateral tracheosyringealis branches of the hypoglossal XIIth brain nerve, which is part of a unilaterally organized neuronal network- the song system. Accordingly, the two halves of the syrinx can act as independently controlled sound sources. Different species produce species-typical sounds using the two syrinx halves in patterns of unilateral, bilateral, alternating, or sequential phonation to achieve their differing temporal and spectral song characteristics [22, 23]. Song control arises from the forebrain starting with a projection from the main vocal center (HVC) to the robust nucleus of the arcopallium (RA). HVC is a sensorimotor integration area that is innervated by auditory centers, including field L, which in turn receives input from the nucleus ovoidalis of the auditory thalamus. RA forms part of the descending motor pathway and sends output projections to the nucleus hypoglossus and to brainstem nuclei responsible for vocal-respiratory control. The lateralized functional organization of this pathway has been identified by different lesion experiments. These studies have investigated how the left and right hemispheres contribute to vocal perception, central as well as peripheral song production and learning by disrupting auditory input, motor output, or by destroying central processing areas. Lesion effects are quantified by changes in the complexity of a bird's song after lesioning left- and/or right-hemispheric structures and in comparison to sham operations. An avian song is composed of a limited number of sound elements that are joined together as syllables. Repetition of the same syllable constitutes a "phrase" [24]. The complexity of a bird's song repertoire is measured in terms of numbers of different syllables included in the song. Changes in complexity are' hence, quantified by the number of syllables lost or retained [25].

Asymmetries in peripheral song production have been demonstrated by cutting the left or right tracheosyringealis nerves.

Unilateral denervation inactivates the ipsilateral syrinx half and hence, can demonstrate its specific contribution to the species-typic song in most passerine birds. Corvid species (e.g. [26]) and non-oscine birds' like mallards [27] or budgerigars [28] display a bilateral innervation of the syringeal muscles. In the chaffinch (*Fringilla coelebs*; [20]) as well as in the Waterslager canary (*Serinus canaries*), the white-crowned (*Zonotrichia leucophrys*; [25, 29]) and white-throated sparrows (*Zonotrichia albicollis*; [30]) denervation of the left syrinx leads to loss of most syllables in their song, while blocking the right syrinx has only minor effects. This pattern is interpreted as a left hypoglossal dominance for song production. In zebra finches on the other hand, it is the right syrinx that dominates song production since lesion effects are larger after right hypoglossal transections. (*Taeniopygia guttata*; [31–33]). Song production in the domestic canary (*Serinus canaria domestica*) involves both syringeal halves. Unilateral denervation of the left or right syrinx demonstrates that some syllables are produced entirely by the left or by the right side of the syrinx whereas others are produced by sequential contributions from each side [34].

Studies lesioning central forebrain structures indicate that asymmetries of peripheral song production are likely caused by lateralized central processing. Unilateral lesions lead to impairments that are very similar to peripherally induced ones. In the Waterslager canary, destroying the left HVC induces significant deterioration of the song while right hemisphere lesions cause only minor effects. Similar impairments are observed when lesions are placed within the left RA and left field L [24]. In the domestic canary, HVC lesions demonstrate that both left- and right-hemispheric pathways mediate particular temporal and spectral features of the song [35]. The dominance of the right hemisphere in zebra finches is supported by stronger song degradation after lesioning HVC within the right hemisphere [31].

There are only a few lesion-based data on the lateralization of song perception and learning. Lateralization of vocal perception can be detected by disrupting sensory input from the left or right side. Cynx et al. [36] lesioned the nucleus of the auditory thalamus what interrupted auditory input either to the right or left hemisphere of the song control system of zebra finches. After left-thalamic lesions, the birds displayed stronger impairments in discriminating own from conspecific song than after right-sided lesions but they were less impaired in detecting subtle changes in the structure of the heard song [36]. There is moreover a left-hemispheric dominance of the higher order auditory cortex for processing song memories in juvenile zebra finches [37] while the left HVC displays enhanced activation when exposed to tutor or unfamiliar song [38]. In the Bengalese finch (*Lonchura striata domestica*), there is evidence that the left HVC is not only dominant for motor control but also for the perceptual discrimination of song. Birds with a left HVC lesion required significantly more time learning to discriminate between Bengalese finch song and a zebra finch song than did birds with a right HVC lesion or intact control birds [39].

1.3 Intra- and Interhemispheric Processing

Lesion studies in the avian song system exemplify two different experimental approaches namely damage of neurons within a target structure or the disruption of axons within specific fiber tracts. Destruction of a cell population leads to the functional outage of the area within which the neurons are localized and therefore provides evidence for the functional impact of this area. A comparison of left- and right-hemispheric lesion effects can therefore indicate functional asymmetries that are presumably based on left-right differences in the structural organization of the target area. Cutting nerve fibers on the other hand interrupts communication between neuronal structures and thus, can point to the relevance of distinct intra- and interhemispheric pathways. Dynamic network activity comes more and more into the focus of asymmetry research [40–42]. Transecting fiber tracts disrupt these processes and point to aspects of lateralization that are not necessarily based on hardwired left-right differences but on synaptic activity-regulated processes [43]. Such a complex lateralized neuronal network develops in a tight interplay between geno- and enviro-typic factors and lesion studies have also contributed to unravelling the underlying neuronal mechanisms [42, 43].

Three different approaches provide hints of lateralized processing: (a) transection of efferent nerves prevents control of motor output, (b) transection of afferent nerves deprives brain areas from sensory input, (c) transection of intra- and inter-hemispheric fiber tracts interrupts communication between brain structures.

Cutting efferent nerves restricts the impact of neuronal cell populations on descending motor control and therefore their behavioral relevance. Differential behavioral impairments after left- or right-side transections provide evidence for the lateralized action of the neurons, as exemplified by hypoglossal nerve transections in song birds [20, 21, 44]. The left-hemispheric dominance of visuomotor control in pigeons has been shown in a study by Güntürkün and Hoferichter [45]. In birds, one major fiber bundle, the tractus occipitomesencephalius (TOM), connects premotoric, arcopallial forebrain areas with the midbrain. While lesions of the right TOM have no effect on the number of pecking responses in a visual discrimination task, left-sided lesions profoundly impair pecking performance [45]. The general left-hemispheric dominance in top-down control on visual processing and response selection is further supported by studies using the sodium channel blocker tetrodoxin (TTX) to temporarily silence forebrain areas [46, 47]. In a similar way, unilateral but not bilateral damage of a visual forebrain area, the visual Wulst that is connected to the visual midbrain via the tractus septomesencephalicus (TSM), impairs discrimination performance of pigeons [48]. At the physiological level, the left Wulst exerts more enhanced forebrain control of visual processing than the right Wulst and dominates choice behavior in case of conflict situations as indicated by reversible inhibition of the Wulst ([46, 47, 49]; see also Chap. 8 for electrophysiology). This lateralized top-down influence is not based on hard-wired left-right differences in the number of descending fibers. The Wulst presumably exerts its lateralized forebrain control via the mesencephalic midbrain commissures that are known to be asymmetrically organized [46, 50]. The functional dominance of the midbrain commissure was demonstrated by a lesion study showing that transecting the mediating intertectal commissure results in a reversal of behavioral asymmetries [51].

Cutting afferent fibers limits input to one brain area and allows analysis of left-right differences in the sensitivity to such deprivation. This provides evidence for the functional role of isolated structures on the left and right sides of the brain. In most vertebrate species, the optic nerves cross completely so that occlusion of one eye restricts visual input primarily to the contralateral brain side. This contralateral organization allows separate testing of the hemispheres just by occluding one eye [52]. In the mammalian visual system, however, only a part of retinal fibers cross within the optic chiasm so that each hemisphere receives input from the left as well as from the right visual field. Consequently, monocular testing is not sufficient to investigate hemispheric-specific performance. Roger Sperry and his colleagues therefore introduced a lesion approach that restricts visual information to one hemisphere by transecting the optic chiasm. As a consequence, visual information present in the left visual field enters the right hemisphere, while information entering the right visual field is projected to the left side of the brain [53]. Nevertheless, information is quickly exchanged in the mammalian brain via the corpus callosum. Therefore, cutting the optic chiasm must be combined with a transection of this fiber bundle to enable the investigation of isolated hemispheres [53, 54]. This approach has demonstrated for the first time that a functional corpus callosum is necessary for interhemispheric information transfer [55–57] and uncovers hemispheric specializations in humans [17] monkeys [58] and even mice [59].

The differential transection of callosal and subcallosal commissures has demonstrated the role of different systems for interhemispheric information exchange and communication (e.g. [60–62]). Moreover, cutting commissural fiber systems allows us to gain insight into the dynamic interactions that are critical for the establishment and maintenance of cerebral asymmetries [63, 64].

Disruption of information transfer by cutting or silencing neuronal input during ontogeny provides important insight into the plasticity of mechanisms generating cerebral asymmetries. In birds for instance, critical aspects of asymmetry formation in the visual system depend on asymmetric photic stimulation during ontogeny ([52, 65]; see Chap. 19). In pigeons, temporary disruption of retinal activity by single TTX injections into the left or right eye leads to the functional dominance of the nondeprived eye. The resulting adult lateralization pattern points to the critical role of interhemispheric processes for the establishment of cerebral asymmetries [66]. Ascending crossing systems mediate lateralized input as indicated by cutting the relevant commissures. In chickens, transection of the supraoptic decussation through which visual fibers re-cross the midline, affects the development of functional asymmetries [67]. Biased sensory input in turn contributes to an asymmetrical functional organization of commissural systems in birds [68–70] and mammals [71].

Combined lesions of more than one target can also indicate interactions between neuronal structures. This is exemplified by the Sprague effect demonstrating that damage of multiple brain areas does not necessarily result in greater behavioral deficits than damage within only one structure [49]. Unilateral removal of the occipito-temporal cortex leads to hemianopia within the contralateral visual field of cats—a phenomenon that is known as cortical blindness. However, additional damage of the superior colliculus contralateral to the cortical ablation or transection of the midbrain commissures restores visual orientation to the visual field that had appeared to be permanently blind. These observations demonstrate how projections arising from the forebrain interact with interhemispheric systems to modulate visual attention and orientation behavior [72, 73].

This short overview illustrates how worthwhile lesion studies are for our understanding of the functional organization of a lateralized brain. In the following, major lesion approaches useful to investigate lateralized functions are discussed.

2 Lesion Techniques

In principal, a brain lesion is the disruption of neuronal tissue caused by disease or surgical intervention. In human subjects, brain damage is mostly the result of strokes or trauma although neurosurgeons sometimes conduct planned lesions for medical treatment (e.g. removal of tumors or epileptic foci). But in general, only animal research allows carefully planned lesion experiments to interrelate behavioral deficits with a subsequent histological analysis of brain damage directly after the end of behavioral experiments.

When, after damage of a particular area, an individual does not show a behavior anymore, this target area must be critically involved in the generation of this behavior. It is therefore the primary lesion approach to systematically damage neuronal structure and to test whether and how this changes behavior. A systematic comparison of effects after disrupting different parts of a neuronal network provides important insight into the functional organization generating a specific behavior (e.g. [74, 75]). Asymmetry research has especially benefited from this approach. Comparing the outcome of left- and right-hemispheric lesions sheds light on left- and right-hemispheric contributions on cognitive task performances. Hemispheric-specific deficits point to the functional dominance of the lesioned brain side.

There are a number of different procedures to generate lesions, which differ in their applicability depending on general brain size or dimensions and localization of the target structure. Moreover, methods differ in their selectivity. Nonselective lesion methods include surgical and electrolytic techniques, which are simple and generally cheap. They completely destroy the tissue and do not differentiate between neurons, their axons, or glia. Accordingly, these methods can be accompanied by unintended side effects that may impede the interpretation of functional effects. To overcome this problem, more selective pharmacological methods can be used that cannot damage large brain areas but affect only (specific) neuronal populations. Other pharmacological or genetic interventions allow the transient and reversible inactivation of neuronal structures. This approach overcomes critical problems arising by permanent brain damage.

The following section provides an overview of commonly used methods, as well as their advantages and problems together with examples from asymmetry research (for an overview see Table 1).

2.1 Unspecific Methods

2.1.1 Ablation

Removal or ablation with a surgical knife is the simplest method when a neuronal structure is easily accessible or when large brain areas such as cortical fields or complete hemispheres should be dissected. This approach was the first used in early studies attempting to challenge the localization theory (e.g. [87, 88]). In neurosurgery, extirpation of larger brain areas is still conducted to treat abnormal behavior or to rescue integrity of healthy tissue in epileptic patients. The patients in turn provide important insight into the functional role of ablated areas. For example, extirpation of human motor cortex in order to control abnormal involuntary movements inspired ideas concerning the role of the motor cortex in movement control [89]. Ablation during ontogeny allows conclusions about the genetically determined predisposition, and also plasticity

Table 1
Summary of lesion techniques useful for asymmetry research

Method	Principle	Parameters defining lesion size	Advantages	Drawbacks	Exemplary studies from asymmetry research
Ablation, Transection	• Mechanic dissection of a complete brain area • Cutting of fiber tracts or nerves using a surgical knife	• Visual control	• Simple and fast	• Not neuron-specific • Unspecific side effects due to damage of en-passent fibers or bleeding • Impaired antero- and retrograde trophic support affect afferent and efferent neuronal populations	[20, 25, 32, 33, 51, 54, 57, 59, 61, 62, 67, 72, 76]
Aspiration	• Suction of neuronal tissue through a pipette eventually attached to a vacuum pump	• Visual control	• Simple and fast	• Only for superficial brain tissue • Not neuron-specific, • Unspecific side effects e.g. due to damage of en-passent fibers	[77–80]
Electrolytic Lesion	• Electric current passing between two electrodes leads to physical damage of neuronal tissue that acts as an electrolyte	• Material and length of uninsulated tip of the lesion electrode • Intensity, duration, direction of applied current	• Simple and fast • Suitable for deep brain structures • Focused and reliable lesions	• Not applicable in large areas Not neuron-specific • Induction of only small lesions • Damage of en-passent fibers • Hyperactivation/seizure induction results in secondary cell death in distant cell populations	[24, 31, 36, 39, 81]
Radiofrequency Lesion	• Thermocoagulation • Radiofrequency current passing through an electrode converts electrical energy into heat within the surrounding tissue	• Size, material of the electrode • Duration of electric current • Peak tissue temperature	• Simple and fast • Suitable for deep brain structures • Highly focused and reliable lesions • Only marginal seizure induction	• Not applicable in large areas • Not neuron-specific • Induction of only small lesions	[45, 82, 83]

(continued)

Table 1 (continued)

Method	Principle	Parameters defining lesion size	Advantages	Drawbacks	Exemplary studies from asymmetry research
Neurotoxic Lesion	• Injection of a neurotoxic substance • Excitotoxins (e.g. kainin/ibotenic acid) bind to glutamatergic receptors leading to hyperexcitation and finally cell death	• Concentration, volume of the injected neurotoxin • Pressure and velocity of the injection • Tissue spread • Sensitivity of the affected cell population	• Focused and reliable lesions • Neuron-specific sparring en-passent fibers	• Not applicable in large areas • Stimulation of neurons distant from target • Differential degree of seizure induction • Differential degree of secondary neuronal damage	[35, 84, 85]
Cooling	• Inhibition of neuronal activity by lowering tissue temperature by means of a cryoelement • Cold is generated electrically or by circulating coolant	• Shape of the cooling device • Temperature of the cooled tissue	• Quickly reversible • No obvious side effects • No obvious permanent tissue damage • Reduced plastic reorganization within the affected brain area	• Difficult application in small brain or deep brain areas • Movement restriction (when using a cryprobe)	[86]
Pharmacological inhibition	• Application of a drug inhibiting neuronal activity (e.g. TTX) through a permanently implanted cannula	• Concentration, volume of the injected drug • Pressure and velocity of the injection • Sensitivity of the affected cell population	• Quickly reversible • Highly neuron-specific • Highly localized application • Limited side effects • Reduced plastic reorganization within the affected brain area	• Not applicable in large areas • Cannula-induced additional tissue damage • Appearance of permanent long-term effects • Reduced spread over time	[46, 47, 66]

of brain areas. A differential malleability of the left- and right hemisphere may indicate inborn asymmetries and contributes to the general nature–nurture debate in lateralization research (e.g. [90–93]). Neurosurgeons sometimes remove or disconnect an entire cerebral hemisphere when young children suffer from intractable epilepsy [94]. General speech and language functions are often rescued even after removal of the left hemisphere but specific cognitive tests can demonstrate a differential degree of compensatory plasticity of the two brain halves. Liégeois et al. analyzed verbal abilities of children hemispherectomized due to pathology underlying an epileptic disorder arising in the pre- or postnatal period [95]. Accordingly, dysfunctionality of the left or right (later disconnected) hemisphere originated before or after birth. This study detected domain-specific impairments that are dependent on the onset of pathology. Although both hemispheres have the potential for developing an adequate level of receptive vocabulary, right-hemispheric damage results in stronger impairments before birth while left-hemispheric loss is more deleterious after birth. Concerning grammatical and semantic aspects of language processing, postnatal hemispheric loss leads to better language scores compared to prenatal one. Prenatal brain damage results in substantial language deficits in either hemisphere while postnatal right-hemispheric loss allows functioning of the left hemisphere within the normal range. Despite a genetically-predetermined dominance of the left hemisphere for language processing, fully-developed language competence also depends on the right hemisphere, depending on the developmental time course of pathology.

2.1.2 Aspiration

This method means suction of neuronal tissue through a glass or metal pipette attached to a vacuum pump. It is applicable for the removal of tissue on the surface of the brain and hence, can be conducted under visual control.

In asymmetry research, this method has been used in studies ablating the hippocampus in pigeons and chicks.

In birds, the hippocampal formation (HF) is located superficially and hence, easily accessible. Like in mammals, it is critical for learning map-like representations of the environmental space what may guide navigation. Thereby, the left and right HF differ in using environmental landmarks or stimulus features as well as their spatial relationships for encoding the environmental space and localization of a goal in space [43, 77, 96–99]. In chicks, encoding of geometric features of an enclosure depends on an intact right hippocampus [77]. In pigeons, a study using electrolytic lesions indicates that only the left HF is sensitive to landmarks that are located within the boundaries of an experimental environment. The right HF is indifferent to such landmarks but sensitive to global environmental features of the experimental space [81, 96].

Hippocampal aspiration before or after navigational map learning demonstrates differences in the participation of the left and right HF to spatial memory formation. When the right HF is ablated before learning, pigeons are as good as intact controls at orienting homeward from distant, unfamiliar locations whereas left HF-ablated pigeons are significantly impaired [78]. But neither left- nor right-hemispheric HF lesions lead to navigational map retention deficits when ablations are conducted only after learning [79]. Thus, ablation studies indicate that navigational map learning but not retention depends on an intact left HF. The differential contribution of the left and right HF for spatial orientation has also been shown in a sun compass-based spatial learning task. Control, left-, and right-HF lesioned pigeons are trained in an outdoor arena to locate a food reward using their sun compass in the presence or absence of alternative feature cues. Impairments of pigeons in using the sun compass occur only when the left but not the right HF is aspirated indicating that an intact left HF is sufficient to support sun compass-based learning. In contrast both hemispheres are able to use feature cues for goal localization [80].

2.2 Stereotactic Methods

Methods described above can be conducted easily under visual control but they are not very precise since they nonspecifically damage the complete neuronal tissue. Especially in cases, when small and/or deeply located brain areas should be lesioned, more specific methods are necessary. These methods generally insert an electrode or needle into the neuronal structure under stereotactic control. Via these devices, electric or chemical lesion-inducing agents can be applied that allow relatively precise control of lesion position and size. Stereotactic surgery requires a stereotactic apparatus with which the head of an individual can be fastened in a fixed position relative to prominent standard features of the skull. From these “landmarks”, the approximate location of hidden brain structures can be determined by means of a stereotactic atlas. A stereotaxic atlas provides a species-typical map of brain regions in three-dimensional stereotactic coordinates (see also Chap. 9). If no stereotactic coordinates are available or when individual variability in the shape and length of specific neuronal areas is large, structural brain imaging like magnetic resonance tomography can be used in advance to determine correct localization of lesions.

2.2.1 Transections

A surgical knife can also be used to transect isolated fiber tracts. This lesion interrupts communication between intra- or interhemispheric brain structures and therefore allows exploration of how a neuronal network generates a specific function. The vertebrate brain possesses a limited number of longitudinal intrahemispheric and commissural interhemispheric fiber systems [64, 100].

Cutting these tracts separately or in combination allows investigating their role in lateralized processing.

The accessibility of intra- and inter-hemispheric fiber tracts, and hence difficulty of the surgery can differ profoundly. In the case of superficially located systems like the corpus callosum, the two brain hemispheres simply have to be pulled apart to see the commissure that can be easily distinguished from other neuronal tissue. Then, insertion of a microknife or simply a resized razor blade cuts fibers under visual control. When fibers located more deeply in the brain, and hence covered by other neuronal tissue, have to be transected, the use of a stereotactic apparatus is necessary to guide the cut. In this case, the knife should be fixed to a mircomanipulator allowing exact cuts according to stereotactic coordinates. Mechanical bending of the blade may be useful to achieve complete transections (e.g. [51]). In case of commissurotomy, transections should be conducted along the midline of the brain to avoid damage of neighboring neuronal structures. This is, however, complicated by the presence of blood sinuses that are often located on the brain surface. Since extensive bleeding may lead to traumatic brains responses and additional necrosis of neuronal tissue, damage of the sinus should be avoided. Therefore, the dura has to be incised lateral to the sinus that, in turn, must be carefully retracted before a knife can be inserted. Moreover, injury of the ventricles can increase intracranial pressure (Fig. 1), which in turn leads to tissue distortion and necrosis. In case of peripheral nerves, it is possible that the fibers regenerate. It is therefore not sufficient to simply cut the nerve; larger parts of the nerve have to be excised (see e.g. hypoglossal nerve transections in song birds [25, 33, 34]).

Although split-brain research provides groundbreaking insight for an understanding of cerebral asymmetries [17, 53, 54], only a small number of studies have investigated the role of neuronal fiber tracts in cerebral lateralization using a lesion approach. This might be due to the difficult surgery often accompanied by unintended side effects caused by disruption of close-by neuronal tissue. In mammals, separate or combined transection of cortical and subcortical commissures unravels their differential role for interhemispheric information exchange [60, 61]. Cutting the major midbrain commissures in chickens [76] and pigeons [51] has demonstrated the role of mesencephalic interhemispheric projection in dynamic controlling lateralized visuomotor processing.

2.2.2 Electrolytic Lesions

Neuronal tissue can be destroyed by passing direct current through two electrodes. Since electrolytic effects are confined to the area around the electrode tip, lesions never become very large [74]. Therefore, this method is very effective in damaging deep brain nuclei but not applicable when large brain areas should be lesioned.

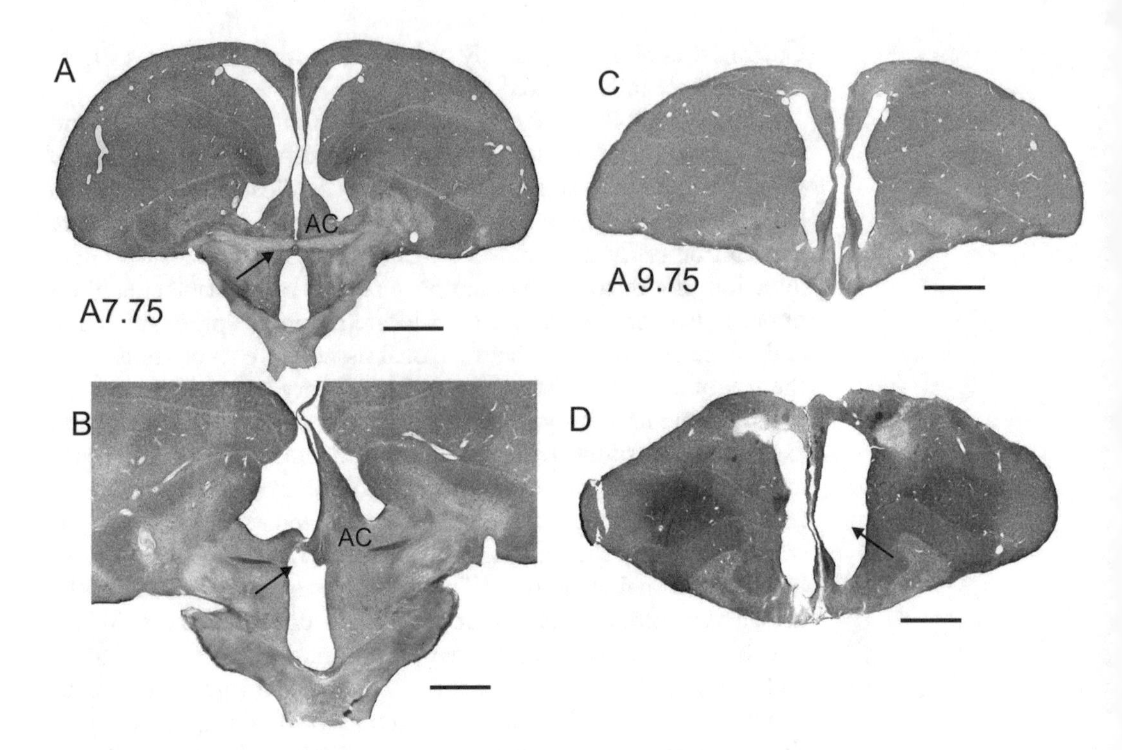

Fig. 1 Localization and transection of the anterior commissure (AC) in pigeons [101]. (**a**) AC is deeply localized within the forebrain (*arrow*), but can be nevertheless transected (*arrows* in **b**). Presumably due to inevitable violation of the ventricular epithelium, the right ventricle is enlarged (*arrow* in **d**) compared to normal conditions (**c**). Scale bars: 1 mm in **a**, **b**, **d**, 50 μm in **c**. Image courtesy: Sara Letzner

Since electrolysis damages all neuronal tissue, en-passant fibers (those traversing the lesion site) are also disrupted.

For electrolytic lesions an electrode is inserted to the envisaged position within the brain. A second electrode can be an ear clip or anal plug. Apart from suitable electrodes, only a current generator is necessary whereby it must be set to unipolar current mode during surgery. When current is turned on, the brain acts as an electrolyte. Neuronal damage is caused by the diffusion of metallic ions from the electrode tip into the surrounding tissue and by the formation of gas bubbles on the surface of the electrode tip leading to mechanic deformation of the tissue. Lesion size and reliability depends on the material of the electrode and length of its uninsulated tip, on intensity and duration of the applied current and on direction of the current, meaning that the electrode is used as the anode or cathode [102]. Careful checking of relevant parameters allows the induction of reliably focused lesions. Electric current, however, produces widespread excitation of nerve cells and therefore, (hyper)activation of neurons efferent to the lesion site (e.g. [103]) and induction of seizures (e.g. [104]) that lead to unintended and eventually detrimental side effects. They are less pronounced in a related method using high-frequency currents (see below) and hence, an electrolytic approach is seldom the method of choice. In asymmetry research, it was used in song birds [24, 31, 36], and pigeons [81]. This method is routinely used to confirm the position of electrodes after electrophysiological recordings (e.g. [49]).

2.2.3 Radio frequency Lesion

Another method that uses electric current for neuronal damage is thermocoagulation. Death of neurons and nerve fibers can be induced by the heat generated by electrical high-frequency current. Comparable to the electrolytic method, this approach is useful for small and/or deeply embedded brain structures and generally in cases, when a lesion of predetermined size and location is desired. This method is also useful when focused lesions within a larger brain structure are planned to investigate functional subdivisions.

For the induction of radiofrequency lesions, an insulated electrode (e.g. simple stainless steel insect pins; e.g. [105, 106]) with a small uninsulated tip is placed stereotactically in the brain while a ground electrode can be implanted elsewhere. A lesion generator generates the radiofrequency current. When current passes through the lesioning electrode, electrical energy at the tip is converted into heat within the surrounding tissue. The size of the lesion is dependent on peak tissue temperature, size of the lesion electrode, and duration of the lesion. Temperatures above 43 °C destroy neuronal cells while temperatures above 100 °C boil and char the tissue. Thus, for a given electrode configuration the peak tissue temperature is the parameter, which has the largest impact on the size of the lesion [107]. Evaluating optimal parameters in preceding experiments allows generation of highly reproducible lesions. Lesion size can be accurately controlled allowing disruption of small nerves without damage of other nearby motor and sensory nerves. It is therefore often used in neurological clinics for nerve lesioning to treat chronic pain (e.g. [108]). Compared to electrolytic lesions, undesirable side effects are less frequent (e.g. [104]) and hence, this method is preferred when possible.

In asymmetry research, this technique has been applied, for example, in chickens to explore the role of the hyperpallium [82] or arcopallium [83] in one-trial-passive avoidance learning. It also has been used for the thermocoagulation of fiber tracts [45].

2.2.4 Neurotoxic Lesions

In this approach, neuronal tissue is destroyed by the infusion of a neurotoxic pharmacological substance through a stereotactically positioned cannula. Since neurotoxins generally interfere with chemical mechanisms regulating neuronal excitation, they destroy cell bodies sparring en-passant fibers. But they only indirectly lead to cell death; lesion effects do not appear instantaneously. Most substances belong to the so-called "excitotoxic" amino-acids. Other types of neurotoxins induce degeneration of fibers with particular chemical properties. They do not completely destroy the neuronal tissue so that the morphological integrity of the target structure is maintained. This allows determining the size of the affected cell population relatively precisely [74].

Excitotoxic Lesions

Injection of excitotoxins utilizes the effect that nerve cells degenerate when they are overstimulated. Overstimulation can be induced by excessive amounts of glutamate, the major excitatory neurotransmitter within the central nervous system of vertebrates. A number of structurally related substances, such as kainic or ibotenic acid, are even more excitatory and therefore more toxic than glutamate itself. The excitotoxic action is dependent on binding to specific receptors. Since neurotransmitter receptors are confined to synapses located on cell bodies or dendrites, nearby axons are typically not affected. Nevertheless, receptors can be expressed on pre- as well as postsynaptic elements of a synaptic contact (e.g. [109]). Thus, pre- as well as postsynaptic effects can be responsible for the physiological outcome of excitotoxins and hence, retro- as well as anterograde effects must be considered for the interpretation of observed behavioral effects.

The stimulating effects of excitotoxins range from enhanced synaptic transmission up to a detrimental overstimulation that culminates into cell death. Excitatory effects depend on the general neuronal excitability of the affected cell population, which in turn is determined by the number and composition of excitatory receptors at the synapse and by inhibitory interactions with GABAergic interneurons [109–111]. Both are regulated by developmental and dynamic processes. This means that the severity of detrimental effects depends on maturity and composition of the neuronal networks.

Glutamate

Glutamate activates different classes of receptors comprising ionotrophic AMPA (α-amino-3-hydroxy-5-methyl-4-isoxazole-propionic acid)-, Kainate-, and NMDA (*N*-methyl-d-aspartate)—as well as metabotrophic receptors [112]. Binding at glutamatergic receptors induces depolarization of the postsynaptic cell due to changes of ionic current across the cell membrane. Excessive glutamate release permanently depolarizes central neurons, increases the frequency of neuronal discharge, and augments synaptic activity that finally leads to disrupted ionic regulatory mechanisms [113, 114]. NMDA-receptors play a central role in glutamate excitotoxicity since excessive NMDA-receptor mediated calcium influx activates calcium-dependent enzymes, which in turn initiate programmed cell death pathways. Spillover of glutamate can be exacerbated by a reversal of astrocytic glutamate uptake [115]. This is dramatically exemplified by the deleterious effects of acute central nervous system insults like ischemia or traumatic brain injury and of long lasting and repetitive seizures in epileptic individuals [116].

In asymmetry research, application of glutamate has been used to unravel the impact of the left and right forebrain in controlling visuomotor behavior in chickens. Single and unilateral glutamate injections demonstrate that only disruption of the left but not right visual forebrain elevates attack and copulation behavior in young chickens [84]. The left hemisphere is also more strongly impaired than the right one in discriminating pebbles from grains. The observed differences in susceptibility to glutamate-induced neuronal disruption indicate a dominant left-hemispheric inhibitory control over these behaviors [84, 85, 117].

Kainic Acid

Kainic acid is a direct agonist of the glutamatergic kainate receptors and a powerful neurotoxin. Its excitatory potency can be hundred times higher than glutamate itself [118]. In low concentrations, excitatory effects are reversible but higher doses lead to an irreversible depolarization of neurons and hence, to cell death that is not confined to the target area [119]. There are several reports detecting neuronal degeneration distant from the injection side in regions interconnected with the lesioned area. This secondary damage is mainly caused by the high potential of kainic acid to induce post-operative seizures [110, 119]. Due to these problematic side effects, kainic acid it is not very suitable for lesion studies investigating the functional relevance of brain areas but it has been established as a neuropathological lesion model for studies investigating excitotoxic processes in neurologic diseases (e.g. [116]).

Ibotenic Acid

One of the widely used neurotoxic substances is ibotenic acid (α-amino-3-hydroxy-5-isoxazoleacetic acid), which acts as a non-selective glutamate receptor agonist with a preference for NMDA as well as metabotropic glutamate receptors. It was originally extracted from *Amanita* mushrooms, which are notorious for their psychotropic effects. Together with the $GABA_A$ receptor agonist muscimol, ibotenic acid represents the principal psychoactive component of this fungi class (e.g. [120]).

The neurotoxic potency of ibotenic acid appears to be lower by at least 1 order of magnitude than that of kainic acid, and considerably higher doses must be used in order to obtain lesions comparable in size. Ibotenic acid exerts its toxic effect only over a restricted area and is therefore especially suitable for animals with small brains. The spatial specificity of ibotenic acid action allows highly selective damage of even small components of neuronal structures (e.g. substructures of the hippocampal formation; [110, 121]). Although multiple injections are possible, in large animals with big brains, eventually not all neurons within a large structure can be removed by ibotenic acid injections [110, 121]. In general, ibotenic acid is preferable to kainic acid because of its low disposition of seizure induction and absence of secondary lesions outside the locus of application [110, 119, 121]. Due to this selectivity, behavioral effects can be more subtle than those induced by other lesion techniques.

Ibotenic acid has been used to induce lesions in the songbird brain to investigate the asymmetrical role of the HVC [35]. Injections into the rat forebrain uncovered the dominance of the right medial prefrontal cortex in activating neuroendocrine and autonomic stress responses activation [122].

Neurochemical Lesions

Specific amino acid analogues lead to quick degeneration of fibers when injected into the brain ventricles or directly into neuronal tissue. The best-known example is 6-hydroxydopamine (6-OHDA) that accumulates in neurons that have a membrane uptake mechanism for catecholaminergic neurotransmitters. As a "false transmitter", 6-OHDA binds with high affinity to dopamine and norepinephrine (or: noradrenalin) transporter. Accumulation of 6-OHDA provokes destruction of catecholaminergic uptake storage mechanisms leading to quick depletion of catecholamines and complete breakdown of synaptic transmission (e.g. [123]). As a consequence, axon and fiber terminals start to degenerate within 1–4 h after 6-OH-DA administration. In a similar way, 5,6- and 5,7-dihydroxytryptamine leads to selective degeneration of serotonin-containing axons and terminals [124].

In case of 6-OH-DA, the neurotoxin has to reach a critical cellular threshold concentration of about 50–100 mM to induce degeneration [125]. Thereby low doses only affect axon terminals, moderate concentrations lead to axon degeneration while high doses finally lead to death of the denervated cells (e.g. [126]). 6-OHDA oxidizes rapidly within the cell and thereby activates cytotoxic metabolites, such as free radicals and hydrogen peroxide, which in turn damage the neuron by affecting proteins, membrane lipids, and DNA. The temporal dynamic of transmitter depletion may differ between noradrenergic and dopaminergic cells. While noradrenalin is gradually reduced over time after intraventricular application of 6-OHDA, dopamine level increases within the first 24 h, followed by a delayed reduction [123, 127]. In order to focus neurotoxicity of 6-OHDA on dopaminergic or noradrenergic cells, it is possible to apply additional pharmacological substances. Combining 6-OHDA injection with selective catecholaminergic reuptake inhibitors (e.g. desipramine for noradrenaline, or benztropine for dopamine) largely restricts damage to dopaminergic or noradrenergic fibers, respectively [123, 128]. However, selectivity and degree of lesion effects should be always controlled e.g. by immunohistochemistry [125].

6-OHDA-induced neurodegeneration has been successfully used to unravel neuroanatomy, neurochemistry, and neurophysiology of nigrostriatal pathways as well as their role in motor control

[123, 129]. Accordingly, 6-OHDA is widely used to investigate the pathogenesis and progression of Parkinson's disease [128] and has been proposed as a putative neurotoxic factor in the pathogenesis of Parkinson [127]. Unilateral application of 6-OHDA affects primarily ipsilateral nigrostriatal projections and hence, induces ipsilateral motor impairments that may differ between the left and right side of the brain. Actually, 6-OHDA injections indicate that asymmetrical motor behavior in rodents [130, 131] might be related to neurochemical left-right differences within the dopaminergic nigrostriatal system [132, 133]. This approach might also be useful for investigating how far imbalances between left- and right-hemispheric processing account for lateralized cognitive impairments in Parkinson disease [134].

2.2.5 Temporary Lesions

All techniques described above induce permanent destruction of neuronal structures. This approach can go along with side effects that may limit the functional interpretation of observed lesion effects. Some of the problems can be avoided by reversible procedures that induce only temporary lesions [74, 135]. These techniques do not completely destroy cells but inhibit neuronal activity or transmission only over a restricted time period. This reduces plastic adaptation processes of the nervous system that may confound observed lesion effects. Reversible inactivation allows comparing the impact of specific brain areas switched on and off during different experimental steps or between different tasks. Accordingly, this approach is also used during electrophysiological recordings when the activity pattern of neuronal cell populations should be analyzed depending on the activity of afferent input (e.g. [49] and Chap. 8).

In the following, two major methods—cryogenic and pharmacological inhibition of neuronal activity—are described.

Cooling

Lowering the temperature of a brain region by means of a cryo-element silences neuronal activity since neuronal transmission is blocked when nerve cells are cooled. When the cryo-element is turned off or removed, the nerve cells warm up and function normally again. This means that the induced lesion is reversible within minutes, does not compromise en passant fibers and generally allows high control over onset, duration, and recovery [74]. This method is particularly applicable in superficially located, large brain regions (e.g. [135, 136]) but usage in deeper brain areas is also possible. Since cryo-elements can be implanted at several loci in parallel, this method allows investigating of the functional impact of different areas within one animal (e.g. [136]). Different kinds of cryo-elements are available, which are differentially suitable depending on the shape and position of the area that needs to be inactivated. A *croyprobe* or (Peltier) chip is a thermoelectric device that can be positioned above the target area. Cold is generated electrically utilizing the so-called Peltier effect. To this end, the device must be connected to a power supply and requires fixed-head restraint [74]. Due to the generally inflexible shape it is only suitable for cooling superficial structures like cortex or midbrain. Other devices—cryoloops and cryotips—do not utilize electricity. Cooling of neuronal tissue is mediated by circulating a coolant through the cryo-elements (e.g. chilled methanol, helium, or liquid nitrogen). A *cryoloop* consists of hypodermic stainless steel tubing that can be adapted to the shape of the regions in focus (e.g. cortical sulci and gyri). Moreover, it does not require head restraint, and so allows a high degree of freedom to conduct behavioral tasks [74]. *Cryotips* are concentric tubes that can be implanted into deeper brain areas. They consist of a shaft and a tip. A coolant is passed down inside the shaft to confine cooling to the tip of the shaft whereby the shaft itself must be insulated or even heated to prevent cooling of overlying tissue, which it is passing by [74].

The temperature of cooled tissue is critical for the magnitude of the cooling effect. For example, in cats, cooling of cryoloops to 1–3 °C is sufficient to deactivate the full thickness of the cortex [137]. In order to control temperature within the cooled tissue, microthermocouples can be additionally implanted. Cooling leads to a selective and reproducible inactivation so that neuronal activity switched on and off can be quickly compared. Implanted cryoelements can be maintained over long time periods (in rats, cats, and monkeys for more than 2 years) without any signs of local or distant degeneration. Accordingly, the observed effects are stable even over long time periods and might not be compromised by neuronal compensation mechanisms [138, 139]. This is a strong advantage compared to other lesioning methods. However, the cooling approach is not appropriate for all applications. It is not possible to avoid damage of overlying structures when a cryotip has to be implanted into a deep brain area. As mentioned above, the shaft must be insulated and this increases the size of the element and, in turn, increases damage of the overlying tissue. In cases in which critical neuronal structures, such as major fiber bundles, might be affected by these lesions, approaches requiring only very small cannulae would be preferred [74].

To best of my knowledge, this method has not been used in asymmetry research yet but it might be useful to induce repeated alternate inhibition of brain regions over short time periods. This approach might be especially interesting for studies investigating the functional significance of commissural systems (e.g. [86]).

Pharmacological Interruption of Neuronal Activity

There are several pharmacological substances available that inhibit neuronal activity or transmission. Stereological application by means of small diameter cannulas allows silencing of areas deep within the brain, thereby minimizing damage of overlying tissue and confining application to small areas. Cannulas are permanently implanted to enable repetitive injections of the pharmacological substance or its control. Application of the vehicle (e.g. saline) controls for traumatic effects simply induced by the injection itself. Since chemical inactivation can be reversed quickly, repeated testing of the animals is possible with only short time intervals between single sessions.

Neuronal inhibition can be elicited via different pharmacological mechanisms. On the one hand, application of the inhibitory neurotransmitter GABA or its agonists like Muscimol leads to reduced network activity and, hence, represents a kind of reversible lesion. Since this approach does not completely prevent neuronal activity, it is more useful in studies requiring modulation of neuronal activity. Pharmacological substances that are commonly used to induce a complete transient lesion prevent the generation of action potentials by inhibiting voltage-gated sodium- or potassium- channels. Local anesthetics such as lidocaine and procaine belong to this class of nerve blocking agents. But the most widely used agent is tetrodotoxin (TTX). TTX is a highly potent, naturally occurring toxin that can be visualized by means of immunohistochemistry in histological sections [140]. Its name is derived from Tetraodontidae, the family of puffer fish from which it was first isolated, but it is also found in a variety of marine and even in some terrestrial species [141]. TTX selectively blocks voltage-gated sodium channels; therefore it completely prevents depolarization and propagation of action potentials in nerve cells without affecting their resting-potentials [142]. The TTX-molecule contains a guanidinium-group, which binds to the external orifice of sodium-channels. Binding occludes the pore completely and hence, prevents sodium currents [143]. The duration of inactivation and the time required to recover is quite variable and may take several hours. Immunohistochemical detection of TTX is possible even up to 48 h after injections [140]. Therefore, in experiments with several cycles of turning on and off, sufficient time to recover must be included. Moreover, some kind of scar tissue around the implant can prevent diffusion of the agent, reducing the extent of inhibition (Fig. 2). Accordingly, it is useful to reconstruct postmortem the extent of TTX diffusion in histological sections [140]. Additionally, it must be considered that repeated TTX-injections into a particular brain region can produce permanent damage of neuronal populations what means that the action of TTX is not completely reversible over time [74].

In the pigeon's visual system, alternating TTX-injections into the visual Wulst demonstrate the functional relevance of a left-hemispheric top-down control on visual processing [49]. Silencing the left but not the right Wulst impairs transfer of unilaterally learnt color discriminations [46] and level the left-hemispheric dominance of response selection in conflict situations [47].

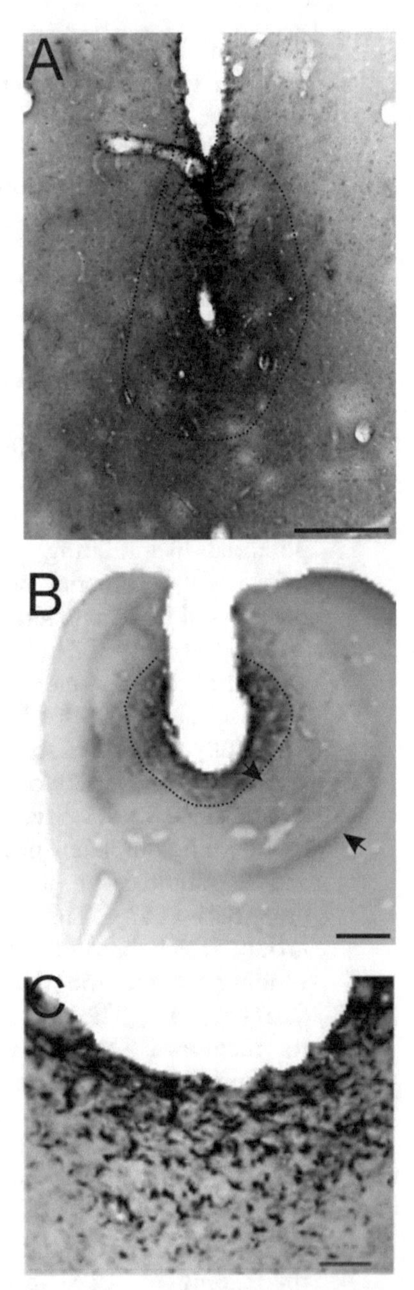

Fig. 2 Immunohistochemical detection of TTX. (**a**) Two weeks after implantation of the canula maximal spread of TTX 1 h after injection can be observed (*dotted line*); (**b**) several months later, spread of TTX (*dotted line*) is reduced due to the development of scar tissue that presumably limits TTX diffusion (area between *arrows*). (**c**) Labeling pattern indicates that TTX is incorporated into cells. Scale bars: 50 μm in **a**, 1 mm in **b**, 20 μm in **c**. Image Courtesy: Nadja Freund, Martina Manns

Optogenetics

A modern approach to transiently silencing nerve cells is the usage of optogenetics. This neurogenetic method uses photic stimulation to control neurons, which have been genetically sensitized to light [144]. This method allows precise measuring of the effects of activating or inactivating neuronal cell populations in real-time. It has been recently introduced as a tool in asymmetry research. Shipton et al. investigated lateralized processing within the hippocampus of mice. They found that silencing the CA3 area of the left but not right hippocampus impaired associative spatial long-term memory, whereas the equivalent manipulation in the right hippocampus did not [145]. This method is the topic of Chap. 11. It is mentioned here as a new way to induce spatially, functionally, and temporarily transient blockade of neuronal processing leading to fine-tuned temporary lesions overcoming many of the drawbacks of other lesioning approaches.

3 Problems in Planning and Interpreting Lesion Experiments

Although planned damage of specific brain structures is a straightforward method to explore the functional role of these areas, there a several problems limiting the interpretation of observed behavioral effects. As indicated above, lesions techniques differ in applicability depending on (a) the localization and size of the lesioned structure, (b) the experimental paradigm that will be used (e.g. single versus repeated, free moving versus stationary, acute versus delayed testing), or (c) the age of the experimental animals. Moreover, it is possible that it is not only the physical tissue damage itself that is responsible for an observable lesion effect; the neuronal mechanism through which a technique generates a lesion has also some impact. Methods, for example, differ in the amount to which they induce secondary lesions in areas distant from the target structure, which in turn might account for inconsistencies in the literature regarding behavioral effects (e.g. [103]). Therefore, the functional outcome of lesioning one area may differ depending on the technique applied. All this means that any given method has both advantages and drawbacks, and these must be carefully considered when choosing the best method to approach a specific research question.

In the following, major aspects that should be considered when planning and interpreting lesion experiments are summarized. These points may help to select the most appropriate lesioning technique and to plan necessary control experiments. Some of the problems also tackle tract tracing experiments as discussed in Chap. 9.

1. *Lesion position*: An appropriate interpretation of lesion-induced behavioral impairments is primarily dependent on the exact

localization of the lesion. Therefore, a postmortem histological analysis is always necessary to verify the lesion position. Even with stereotactic surgery, lesions cannot be always applied to the planned position especially when the structure is small and/or deeply located within the brain. This is simply caused by the individual variability of the brain morphology so that stereotactic coordinates do not perfectly fit in every individual case. Moreover, lesioning devices can be deflected when traversing ventricles or meningeal tissue. The deeper a target area is located within the brain, the greater the angular deviation from target position is. Apart from problems with the localization itself, secondary lesions must be checked. First of all, additional mechanical damage is caused by the inserted lesion devices, such as electrodes or cannulas. Accordingly, the diameter of these devices should be as small as possible. In general, sham lesions should be conducted that replicate all steps of the surgery except the one that finally causes neuronal cell damage. Furthermore, secondary lesions of cell populations connected with the target area must be considered, as well as damage caused by bleeding during surgery or disruption of the ventricles (Fig. 1). In cases, when secondary effects cannot be excluded, control experiments might be necessary to differentiate which of the affected neuronal structures actually caused the observed effects.

Typically, histological investigations are only conducted postmortem and hence, after long lasting behavioral testing. In cases when no or unexpected functional effects have been observed, this might be problematic since training and testing of animals with inadequate lesioning may waste money and time of the experimenters. For studies requiring time consuming behavioral tasks or when the lesion position is difficult to replicate, it might be preferable to choose a temporary lesion approach. This allows direct comparison of the functional effects of one area switched on and switched off. Differences in performance provide evidence whether or not the inhibited area has some impact on the investigated function (independently of its exact position).

2. *Lesion size*: In order to explore the role of one neuronal structure, a lesion must be large enough to impair or even completely disrupt functioning of the affected structure. In most cases, the target area must, therefore, be totally damaged. But this, on the other hand, enhances the risk to damage neighboring structures that may entail their own and additional effects. Thus, it might be difficult to ascribe a behavioral function solely to the target region. In experiments in which the risk of damaging neuronal tissue outside the target area is high it is useful to estimate in pilot studies the minimal lesion size that

induces outage of the target structure. This simply reduces the risk of generating unintended lesions that complicate interpretation of the data.

Variation in lesion size can be used to relate lesion size to behavioral effects. This supports conclusions about the functional role of the investigated system. Güntürkün and Böhringer demonstrated, for instance, that the asymmetry of visually controlled pecking responses in pigeons is reversed to a degree proportional to the extent of midbrain commissurotomy [51]. An analysis of relations between lesion size and behavioral outcome is especially necessary in studies comparing the impact of left- and right-hemispheric structures. It must be determined how far asymmetrical lesion effects are not simply due to a bias in the amount of tissue damage.

3. *Damage of en-passant fibers*: Neuronal structures often contain crossing fibers that neither originate nor terminate in that region. If a considerable number of these fibers is lesioned, observed behavioral impairments can be caused by disturbances of the areas connected by these fibers. Thus, one must differentiate between behavioral effects that are caused by destruction of the target area or by lesions of traversing fibers. This problem can be minimized by the use of selective neurotoxic substances that only affect cell bodies and spare traversing fibers.
4. *Secondary effects*: As indicated above, the interpretation of brain lesion studies is complicated by the fact that outage of the target area is not necessarily the only source of behavioral impairments. The involvement of structures distant from the lesion site may account for additional behavioral deficits. When the experimenter does not control for secondary effects, the functional role of the target might be overemphasized.

Apart from disrupting en-passant fibers and hence affecting neuronal circuitries that are independent from the target structure, lesioning a neuronal structure can cause damage within areas directly connected to the target. This in turn means that the severity of behavioral deficits differs depending on the tendency of different techniques to induce secondary damage. This is exemplified by the differential over-excitatory effects of excitotoxins like kainic and ibotenic acid that result in a differential degree of cell loss distant from the target. Kainic acid that is well known for its strong seizure-inducing activity generates more severe behavioral deficiencies than ibotenic acid [110, 121]. Similarly, application of electric current does not only destroy the target area but also induces abnormal excitation and hence, may damage neurons in structures efferent to the lesion site. As a consequence behavioral deficits can be more severe than those induced by aspiration—a method that tends to produce larger primary but virtually no secondary lesions [103].

A related problem arises on account of the complex networks that mediate a specific neuronal function. Focal damage may lead to a disorganization of the complete network and, hence, behavioral deficits must be interpreted as an impairment of network activity. Ascribing deficits only to the lesion target may overemphasize the role of this area. On the other hand, modified network activity tends to compensate for loss of one component what in turn may disguise the functional role of a lesioned area. In general, direction of deleterious effects must be considered (Fig. 3). Neuronal cell death can

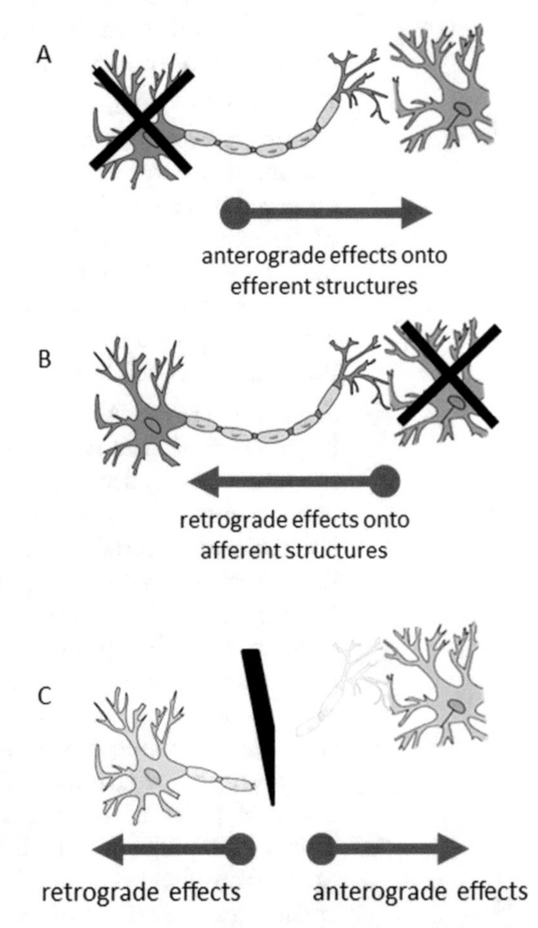

Fig. 3 The target structure is not the only area where lesion effects can be expected—Damage of neuronal cells can lead to effects within the efferent target area due to reduced anterograde trophic support and reduced synaptic activation (**a**), but can also lead to retrograde effects within afferent cell population due to disrupted retrograde trophic support (**b**). Transection of neuronal fibers leads to the degeneration of the distant part of an axon inducing anterograde effects within the efferent target but eventually also to cell death of neurons that are deprived from retrograde support following axonal injury (**c**; e.g. [146, 147])

generate anterograde effects within the efferent cell population due to reduced synaptic activity and trophic support. In a similar way, retrograde effects within the innervating cell population can be expected due to reduced retrograde trophic support. Cutting neuronal fibers initiates retro- as well as anterograde effects [146, 147]. This means that the consequences of the outage of one neuronal structure are only understandable as a modulation of afferent and efferent connections.

5. *Plasticity and regeneration*: Even when removal or disruption of a neuronal structure leads to significant behavioral deficits, it is possible that the observed effects diminish or even disappear over time. First of all, it is possible that a directly observed deficit is caused by a general trauma after surgery. This problem can be controlled for by the use of a "sham" control group. This group receives the identical surgical manipulation as the experimental group except the final lesion step [74]. In cases, when a neuronal area is silenced by (repeated) injections of a pharmacological substance, an appropriate control consists of only applying the vehicle. When the behavioral outcome does not differ between the experimental and the control group, the initial deficit can be ascribed to the surgical manipulation and not to neuronal inactivation. In this case, a sufficiently long time interval is necessary prior to postoperative testing thus allowing the brain to recover from traumatic responses. However, a long delay increases the risk of missing the appearance of critical deficits since the function may have recovered.

   Long lasting recovery is promoted by spontaneous or experience-dependent adaptation mechanisms allowing adjacent areas to progressively take over the function previously mediated by the damaged neurons. This plastic reorganization is age-dependent and largely subtends clinical recovery of motor, language, or sensorimotor integration after stroke. It includes both local as well as distributed changes of processing and leads to better compensation of impaired function the younger an individual is (e.g. [148–151]). Accordingly, chronic lesion effects may differ profoundly from acute ones and may mask the normal impact of the lesioned brain area. Therefore, the critical role of one area for a specific function might be detected only in the short term, and acute and chronic effects must be carefully compared. Concerning research on lateralization, it is possible that an impaired function is adopted by the originally subdominant hemisphere because it is not inhibited anymore. For example, injection of glutamate into the left Wulst of young chickens increases aggressive as well as sexual behavior indicating that a left-hemispheric inhibitory control function [84]. This change may have important consequences for the interpretation of lateralized effects. The use of a reversible inactivation approach reduces the problem of long-term adaptations after permanent lesions.

6. *Reversibility of effects*: Due to the high potential of neuronal systems to reorganize, it must be considered to what extent a permanent lesion approach is suitable for a specific research question. Permanent outage of function may induce compensatory adaptations within a neuronal network. Therefore, long lasting experiments requiring extensive training or repeated testing over several experimental steps should be generally conducted with a reversible lesion approach. When a neuronal system is only temporarily inactivated, synaptic transmission recovers and this should reduce any plastic reorganization that complicates data interpretation [74]. Moreover, this approach directly allows comparison of individual performances with one brain region turned on and off. This design reduces the number of experimental animals since each animal serves as its own control and hence, increases reliability of the data. This approach eventually supersedes control or sham lesioned groups. In case of asymmetry research, direct comparison of the left- and right-hemispheric performances within one animal is possible. Nevertheless, long-term effects cannot be excluded (e.g., see permanent TTX effects). Inhibitory effects can decrease due to reduced diffusion of the pharmacological substances (Fig. 2). Accordingly, the degree of behavioral deficits may decrease over time and, in turn, this diminishes mean effects after repeated testing. It is obvious that such an effect is especially problematic when individual left- and right-hemispheric performances are compared.

7. *Limited functional interpretation*: Lesion studies can demonstrate the general impact of a particular structure in the generation of a specific behavior by correlating behavioral changes with the amount of lesion, and in this way may indicate a differential contribution of the left or right brain side. Fiber transection studies can suggest intra- and inter-hemispheric processes as critical for the (lateralized) generation of a cognitive function. But they cannot elucidate the underlying neuronal mechanisms (neither at the anatomical nor the physiological level). Thus, they only provide a starting point to understand the functional organization of a specific cognitive process [75].

4 Short Experimental Guide

Although the experimental details differ depending on the lesion approach, there are steps that always have to be considered when planning and conducting lesion experiments. These steps are summarized below accompanied by practical tips and hints to typical pitfalls (see also Table 1).

4.1 Planning the Experiment

1. *Behavioral paradigm*: A lesion approach can only demonstrate left-right differences in the behavioral significance of a neuronal structure when the animals are tested in appropriate behavioral tasks (e.g. [152]). Therefore, you have to consider carefully the experimental paradigms and you should develop clear-cut predictions about the behavioral outcome [75]. Depending on the selected paradigm, you have to plan the optimal time for lesion surgery. A task that does not require specific training or learning can be conducted before and after surgery so that each animal can serve as its own control. In this case, only the number of left- and right-hemispheric lesions must be balanced. The experimental design is more complex in experiments requiring several training steps. You eventually need different experimental groups that are lesioned only after achieving successive experimental steps. In a learning task for instance, it might be necessary to lesion the target structure before or after learning to distinguish its role in acquisition or memory.
2. *Lesion technique*: Depending on the position and size of the target area, the appropriate lesion approach must be considered. Some techniques like ablation or aspiration are only suitable for superficial and/or large brain regions. The use of more selective methods, on the other hand, can be limited by the maximal size of an induced lesion (e.g. since electrolytic effects are confined to the area around the electrode tip, they do not become larger than about 5 μm^3; [74]). Moreover, the age of the experimental animals must be considered. Changing neuronal sensitivity for pharmacological substances or changing capacities for regeneration and plastic reorganization may lead to differences in lesion size and behavioral effects. Finally, the behavioral paradigm determines whether a permanent or reversible approach is necessary.
3. *Localization of the lesion*: If necessary, determine the stereotactic coordinates of the intended lesion target. Localization as well as size of the target can be calculated with the help of a stereotactic atlas. Both parameters limit suitable lesion techniques (see above). When you have selected a technique, conduct a pilot study to define parameters that guarantee reliable

lesions with a minimum of undesired side effects. Since side effects typically increase with the severity of interventions, minimal parameters leading to a lesion should be determined. In case of electrolytic or radiofrequency lesions, electric parameters (strength and length of current flow through a given electrode type) can be regulated to confine neuronal damage and to reduce nonspecific activation of distant neuronal populations. In case of pharmacological approaches, degree and spread of neuronal damage depends on the concentration, volume, and diffusion of the injected drug. Estimate the lowest concentration and volume of the injected drug to limit nonspecific effects. While suitable concentrations of a pharmacological substance do not vary between species, the injection volume should be carefully determined. The area of spread after a single injection is affected by several variables including extracellular space, diffusion characteristics of the injected substance, and the presence of fiber tracts that prevent spread of the substance. Moreover, the extent of cell loss within the target area decreases with decreasing concentrations. Thus, it can be necessary to increase number of injection sites within one area to completely disrupt one structure [110]. Physical characteristics of the injection device, such as diameter of the cannula and injection pressure influence the degree of mechanic tissue damage. Speed of injection must also be considered. Injections should be made at slow rates to prevent pressure-induced trauma within the tissue. In sum, it can be better to perform several injections with small volumes compared to a single injection of a large volume. When an antibody against the substance is available, immunohistochemical staining allows direct estimation of diffusion over time and therefore adjustment of the amount of injected drug for the kind of tissue, injection-procedure, and cannula [140].

4. *Control groups*: Consider adequate controls depending on the behavioral paradigm and lesion method. In case of a permanent lesion approach, an independent control group that receives sham lesions is necessary to control for nonspecific effects simply induced by the surgery itself. When a neurotoxic substance is applied in the experimental group, inject the vehicle in control animals (= the solution in which the toxin is dissolved (e.g. saline)). These injections control for traumatic effects caused by the injection itself. In case of using a reversible lesion approach, each animal can serve as its own control by subsequent and balanced injections of the active (neurotoxic) and inactive (control) solution. Balance the order of active and control session on the left and right side for each animal.

4.2 Surgery

Irrespective of the actual lesion technique, surgery follows similar steps that are described below.

1. *Preparation and anesthesia of the animal*: Food and water deprive the animal for about 6–24 h to avoid emesis. When surgery is conducted during a training phase, weight of the animals should not be too low. Depending on the species, age and/or condition of the animals or on length of the surgery different anesthetics are preferable. Often a combination of anesthetics, muscle-relaxing drugs, and analgesics are used. For longer surgeries, a heating pad is useful to maintain body temperature to prevent hypothermia.
2. *Preparation of the lesion device/cannula*: Depending on the technique, preparation of devices might be necessary. Electrodes have to be isolated and/or sterilized. Glass capillaries have to be pulled and cut to desired length. In case of pharmacological injections, the cannula must be filled with the drug.

   All drugs are more or less harmful, handle them carefully and only after reading all relevant safety instructions. TTX for instance is a very potent neurotoxin; it is over a thousand times more toxic to humans than is cyanide. Also consider that TTX has no known antidote [141]. Take care that the drug is appropriately dissolved (no sediment in the tubes within which the drug is stored). Accordingly, consider appropriate dilution. Take care that the drug remains stable (e.g. by preparing fresh solution or by adequate freezing). Pharmacological substances can convert to inactive derivatives. TTX for example converts into nontoxic tetrodonic acid [142]. TTX is a base that does not dissolve well in lipids, organic solvents, or water. Toxicity is lost in alkaline solutions and is less stable in strong acids. Therefore, it should be dissolved in weak acids or buffered to pH 4–5 (e.g. citrate buffer) so that it does not precipitate and remains stable if kept refrigerated. Dispose of the drug and contaminated instruments appropriately according to safety guidelines. It might be useful to inactivate the drug. TTX, for instance, can be rapidly destroyed by boiling at pH2.

   After filling, fix the cannula to the stereotactic apparatus.
3. *Placement of the animal into the stereotactic apparatus*: Place the animal in the stereotactic apparatus and secure it (typically by means of maxillary and ear bars) but only when it is deeply anesthetized. Depth of anesthesia can be controlled by checking reflexes (e.g. withdrawal reflexes of extremities). Check the depth of anesthesia regularly during surgery.
4. *Opening the skull*: crop hair or feathers when necessary, cut the skin and retract it to expose the skull. Check stereotactic coordinates and drill holes into the cranial bone at the stereotactically estimated positions. Size of the holes must be adapted to

the size of the lesion device and extent of the lesions. In principal, the opening should be as small as possible. Avoid damage of the dura and/or blood vessels since blood or liquor impairs ability to see the neuronal tissue. Application of gelatin sponges is helpful in drying the area of surgery.

5. *Insertion of the lesion device*: Open carefully the dura mater (in most cases puncturing with a surgical needle is sufficient). Insert the lesioning device into the brain according to estimated stereotactic coordinates by means of a micromanipulator and lower it at the predetermined depth (an electrode in case of an electric approach, a cryotip in case of a cooling approach, a guiding cannula or syringe in case of injecting a neurotoxic drug). Guiding cannulas must be implanted when several injections are planned at corresponding positions on the left and right sides of the brain. Secure them to the skull with dental cement; additional stainless steel screws support and hold the cannula. After the cement has dried, cap the cannulas with dummy stylets.
6. Lesion *induction*: When using a knife method, cut the neuronal tissue. Prepare hemostatic agents to stop bleeding when necessary with pressure application. Strong bleeding may require the use of electrocauterization. When using a current to induce the lesion, turn on the lesion generator according to the estimated parameters and retract the electrode. In case of applying a neurotoxic drug, the injections should be made at a slow rate by hand or with the help of a micromanipulator. Following infusion, the needle must be left in place to avoid spread of the injected drug up the needle tract.
7. *End of surgery*: suture the skin around the lesion or implant, and then protect the wound margin with antibiotic powder or salve. Remove the animal from the stereotactic apparatus and allow it to recover from anesthesia while warming under a warming lamp or on a heating pad. Place the animal back to the home cage only when it is able to move. Allow the animal to recover for some days. Do not wait too long when you are interested in acute lesion effects. But when using a neurotoxic substance, consider that cell death is at the earliest induced 24 h after injection [74].
8. *Behavioral testing*: After sufficient time to recover, start food deprivation when necessary. Conduct all steps of the planned behavioral task. In case of a reversible lesion approach, each animal is tested several times. To this end, familiarize the animals with the lesion procedure. In case of using a cooling approach, place the animal into the behavioral set up, connect the cooling device, wait until temperature in the target region

has decreased sufficiently and immediately start the behavioral experiment.

In the case of pharmacological injections, administrate the drug by means of a Hamilton syringe attached to a microliter infusion pump that allows controlled injection with predefined constant infusion rate. To this end, connect an injection needle of adequate length to the needle of the Hamilton syringe (connected via a short length of polyethylene tubing). Fill the whole system with distilled water and fix the Hamilton syringe on the microinjection pump. Carry out filling of the injection needle. Start with a small air bubble that helps to verify drug application (when the bubble moves, the system is working), then draw up the drug. Place and restrain the animal in front of the infusion pump. Remove the dummy stylet and insert the Hamilton needle. Inject the intended amount of the drug via the microinjection pump with the previously determined injection rate [140, 153]. Wait for about 60 s before withdrawing the injection needle from the cannula, and cap the cannula with the dummy stylet. Wait about 5–10 min before putting the animal in the behavioral setup and start the behavioral experiment. Wait sufficient time before starting the next experimental session to ensure that the action of the injected drug has finished. TTX-effects for example remain about 24 h [154].

4.3 Post Hoc Analysis of Brain Damage

After finishing the behavioral tests, verify localization and extent of damaged brain tissue.

1. *Brain preparation*: In order to prepare histological brain sections, the brains must be adequately processed. This may include fixation of the brains (mostly by perfusion) and preparation of brain slices (see also Chap. 9).
2. *Histological staining of brain slices*: several staining methods are available that provide adequate analysis of lesion size. It might be useful to combine soma (Nissl stain) and fiber or myelin staining to differentiate between perikaryal and fiber degeneration. Additionally, immunohistochemical stainings can be conducted to investigate loss of specific cell populations (e.g. tyrosinehydroxylase antibody staining allows detection of loss of catecholaminerigc cells). In the case of temporary lesions, neurons should not be damaged so that you can only indirectly evaluate the extent of neuronal inhibition. In case of TTX, the availability of antibodies allows detection of TTX by means of immunohistochemistry [140]. This technique can be used to visualize the maximal diffusion of TTX within a neuronal structure, the time up to which TTX is detectable and, hence, presumably active, and the diffusion at the end of an experiment.

3. *Histological examination*: Analyze carefully brain sections to estimate size and localization of the lesions. In case of implanted cannula, visible needle tracks allow estimation of (symmetric) injections on the left and right sides of the brain. Exact reconstruction of the lesions is necessary for a comparison of left- and right-sided brain damage and the calculation of relations between lesion size and behavioral effects. This may require the use of a neuroimaging system. Analysis should not be confined to the target area. The possibility of distant secondary effects requires the careful inspection of all areas connected to the target structure. Necrotic (hence, dying) cells and degenerating fibers outside the primary lesion site indicate additional lesion areas.

References

1. Broca P (1861) Nouvelle observation d'aphémie produite par une lésion de la troisième circonvolution frontale. Bull Soc Anat 6:398–407
2. Broca P (1861) Perte de la parole: ramollissement chronique et destruction partielle du lobe anterieur gauche du cerveau. Bull Soc Anthropol 2:235–238
3. Broca P (1861) Remarques sur le siège de la faculté du langage articulé, suivies d'une observation d'aphémie (perte de la parole). Bull Soc Anat 6:330–357
4. Broca P (1865) Sur le siège de la Faculté du langage articulé. Bull Soc Anthropol 6:377–393
5. Dronkers NF, Plaisant O, Iba-Zizen MT, Cabanis EA (2007) Paul Broca's historic cases: high resolution MR imaging of the brains of Leborgne and Lelong. Brain 130:1432–1441
6. Dax M (1865) Lésions de la moitié gauche de l'encéphale coïncident avec l'oubli des signes de la pensée (lu à Montpellier en 1836). Bull Hebd Med Chir 2:259–262
7. Wada J (1949) A new method for determination of the side of cerebral speech dominance. A preliminary report on the intra-carotid injection of sodium amytal in man. Igaku Seibutsugaku 4:221–222
8. Wada JA, Rasmussen T (1960) Intracarotid injection of sodium amytal for the lateralization of cerebral speech dominance. J Neurosurg 17:266–282
9. Loring DW, Meador KJ, Lee GP, Murro AM, Smith JR, Flanigin HF, Gallagher BB, King DW (1990) Cerebral language lateralization: evidence from intracarotid amobarbital testing. Neuropsychologia 28:831–838
10. Oxbury SM, Oxbury JM (1984) Intracarotid amytal test in the assessment of language dominance. Adv Neurol 42:115–123
11. Powell GE, Polkey CE, Canavan AG (1987) Lateralisation of memory functions in epileptic patients by use of the sodium amytal (Wada) technique. J Neurol Neurosurg Psychiatry 50:665–672
12. Zatorre RJ (1989) Perceptual asymmetry on the dichotic fused words test and cerebral speech lateralization determined by the carotid sodium amytal test. Neuropsychologia 27:1207–1219
13. Keenan JP, Nelson A, O'Connor M, Pascual-Leone A (2001) Self-recognition and the right hemisphere. Nature 409:305
14. Gazzaniga MS, Bogen JE, Sperry RW (1962) Some functional effects of sectioning the cerebral commissures in man. Proc Natl Acad Sci U S A 48:1765–1769
15. Bogen JE, Fisher ED, Vogel PJ (1965) Cerebral commissurotomy: a second case report. JAMA 194:1328–1329
16. Bogen JE, Gazzaniga MS (1965) Cerebral commissurotomy in man: minor hemisphere dominance for certain visuospatial functions. J Neurosurg 23:394–399
17. Gazzaniga MS (2000) Cerebral specialization and interhemispheric communication: does the corpus callosum enable the human condition? Brain 123:1293–1326
18. Gazzaniga MS (2005) Forty-five years of split-brain research and still going strong. Nat Rev Neurosci 6:653–659
19. Walker SF (1980) Lateralization of functions in the vertebrate brain: a review. Br J Psychol 71:329–367

20. Nottebohm F (1971) Neural lateralization of vocal control in a passerine bird: I. Song. J Exp Zool 177:229–261
21. George I (2010) Hemispheric asymmetry of song birds. In: Hugdahl K, Westerhausen R (eds) The two halves of the brain—information processing in the cerebral hemispheres. MIT Press, Cambridge
22. Suthers RA (1997) Peripheral control and lateralization of birdsong. J Neurobiol 33:632–652
23. Suthers RA, Zollinger SA (2004) Producing song: the vocal apparatus. Ann N Y Acad Sci 1016:109–129
24. Nottebohm F, Stokes TM, Leonard CM (1976) Central control of song in the canary, Serinus canarius. J Comp Neurol 165:457–486
25. Nottebohm F, Nottebohm M (1976) Left hypoglossal dominance in the control of canary and white-crowned sparrow song. J Comp Physiol 108:171–192
26. Tsukahara N, Kamata N, Nagasawa M, Sugita S (2009) Bilateral innervation of syringeal muscles by the hypoglossal nucleus in the jungle crow (Corvus macrorhynchos). J Anat 215:141–149. doi:10.1111/j.1469-7580.2009.01094.x
27. Lockner FR, Youngren OM (1976) Functional syringeal anatomy of the mallard: I. In situelectromyograms during ESB elicited calling. Auk 93:324–342
28. Wild JM (1997) Neural pathways for the control birdsong production. J Neurobiol 33:653–670
29. Hartley RS, Suthers RA (1990) Lateralization of syringeal function during song production in the canary. J Neurobiol 21:1236–1248
30. Lemon RE (1973) Nervous control of the syrinx in white-throated sparrows (Zonotrichia albicollis). J Zool 171:131–140
31. Williams H, Crane LA, Hale TK, Esposito MA, Nottebohm F (1992) Right-side dominance for song control in the zebra finch. J Neurobiol 23:1006–1020
32. Floody OR, Arnold AP (1997) Song lateralization in the zebra finch. Horm Behav 31:25–34
33. Liao C, Li D (2012) Effect of vocal nerve section on song and ZENK protein expression in area X in adult male zebra finches. Neural Plast. doi:10.1155/2012/902510
34. Suthers RA, Vallet E, Tanvez A, Kreutzer M (2004) Bilateral song production in domestic canaries. J Neurobiol 60:381–393
35. Halle F, Gahr M, Kreutzer M (2003) Effects of unilateral lesions of HVC on song patterns of male domesticated canaries. J Neurobiol 56:303–314
36. Cynx J, Williams H, Nottebohm F (1992) Hemispheric differences in avian song discrimination. Proc Natl Acad Sci U S A 89:1372–1375
37. Chirathivat N, Raja SC, Gobes SM (2015) Hemispheric dominance underlying the neural substrate for learned vocalizations develops with experience. Sci Rep 5:11359. doi:10.1038/srep11359
38. Moorman S, Gobes SM, Kuijpers M (2012) Human-like brain hemispheric dominance in birdsong learning. Proc Natl Acad Sci U S A 109:12782–12787
39. Okanoya K, Ikebuchi M, Uno H, Watanabe S (2001) Left-side dominance for song discrimination in Bengalese finches (Lonchura striata var. domestica). Anim Cogn 4:241–245. doi:10.1007/s10071-001-0120-9
40. Ocklenburg S, Güntürkün O (2012) Hemispheric asymmetries: the comparative view. Front Psychol 3:5
41. Hervé PY, Zago L, Petit L, Mazoyer B, Tzourio-Mazoyer N (2013) Revisiting human hemispheric specialization with neuroimaging. Trends Cogn Sci 17:69–80. doi:10.1016/j.tics.2012.12.004
42. Ocklenburg S, Hugdahl K, Westerhausen R (2013) Structural white matter asymmetries in relation to functional asymmetries during speech perception and production. Neuroimage 83:1088–1097
43. Manns M, Ströckens F (2014) Functional and structural comparison of visual lateralization in birds—similar but still different. Front Psychol 5:206
44. Nottebohm F (2005) The neural basis of birdsong. PLoS Biol 3:e164. doi:10.1371/journal.pbio.0030164
45. Güntürkün O, Hoferichter HH (1985) Neglect after section of a left telencephalotectal tract in pigeons. Behav Brain Res 18:1–9
46. Valencia-Alfonso CE, Verhaal J, Güntürkün O (2009) Ascending and descending mechanisms of visual lateralization in pigeons. Philos Trans R Soc Lond B Biol Sci 364:955–963
47. Freund N, Valencia-Alfonso CE, Kirsch J, Brodmann K, Manns M, Güntürkün O (2015) Asymmetric top-down modulation of ascending visual pathways in pigeons. Neuropsychologia 83:37–47. doi:10.1016/j.neuropsychologia.2015.08.014
48. Nau F, Delius JD (1981) Discrepant effects of unilateral and bilateral forebrain lesions on the visual performance of pigeons. Behav Brain Res 2:119–124
49. Folta K, Diekamp B, Güntürkün O (2004) Asymmetrical modes of visual bottom-up and top-down integration in the thalamic nucleus rotundus of pigeons. J Neurosci 24:9475–9485
50. Keysers C, Diekamp B, Güntürkün B (2000) Evidence for physiological asymmetries in the intertectal connections of the pigeon

(Columba livia) and their potential role in brain lateralisation. Brain Res 852:406–413
51. Güntürkün O, Böhringer PG (1987) Lateralization reversal after intertectal commissurotomy in the pigeon. Brain Res 408:1–5
52. Manns M, Güntürkün O (2009) Dual coding of visual asymmetries in the pigeon brain: the interaction of bottom-up and top-down systems. Exp Brain Res 199:323–332
53. Sperry RW (1961) Cerebral organization and behavior: the split brain behaves in many respects like two separate brains, providing new research possibilities. Science 133:1749–1757
54. Crowne DP, Novotny MF, Steele Russell I (1991) Completing the split in the split-brain rat: transection of the optic chiasm. Behav Brain Res 43:185–190
55. Myers RE (1956) Function of the corpus callosum in interocular transfer. Brain 79:358–363
56. Myers RE, Sperry RW (1958) Interhemispheric communication through the corpus callosum: mnemonic carry-over between the hemispheres. Arch Neurol Psychiatry 80:298–303
57. Schrier AM, Sperry RW (1959) Visuomotor integration in split-brain cats. Science 129:1275–1276
58. Hamilton CR, Vermeire BA (1988) Complementary hemispheric specialization in monkeys. Science 242:1691–1694
59. Shinohara Y, Hosoya A, Yamasaki N, Ahmed H, Hattori S, Eguchi M, Yamaguchi S, Miyakawa T, Hirase H, Shigemoto R (2012) Right-hemispheric dominance of spatial memory in split-brain mice. Hippocampus 22:117–121. doi:10.1002/hipo.20886
60. Berlucchi G, Buchtel HA, Lepore F (1978) Successful interocular transfer of visual pattern discriminations in split-chiasm cats with section of the intertectal and posterior commissures. Physiol Behav 20:331–338
61. Hottman TJ, Sheridan CL, Levinson DM (1981) Interocular transfer in albino rats as a function of forebrain or forebrain plus midbrain commissurotomy. Physiol Behav 27:279–285
62. Glickstein M (2009) Paradoxical interhemispheric transfer after section of the cerebral commissures. Exp Brain Res 192:425–429. doi:10.1007/s00221-008-1524-4
63. Bloom JS, Hynd GW (2005) The role of the corpus callosum in interhemispheric transfer of information: excitation or inhibition? Neuropsychol Rev 15:59–71
64. Van der Knaap LJ, Van der Ham IJ (2011) How does the corpus callosum mediate interhemispheric transfer? A review. Behav Brain Res 223:211–221
65. Rogers LJ (2014) Asymmetry of brain and behavior in animals: its development, function, and human relevance. Genesis 52:555–571. doi:10.1002/dvg.22741
66. Prior H, Diekamp B, Güntürkün O, Manns M (2004) Post-hatch activity-dependent modulation of visual asymmetry formation in pigeons. Neuroreport 15:1311–1314
67. Rogers LJ, Robinson T, Ehrlich D (1986) Role of the supraoptic decussation in the development of asymmetry of brain function in the chicken. Brain Res 393:33–39
68. Chiandetti C, Regolin L, Rogers LJ, Vallortigara G (2005) Effects of light stimulation of embryos on the use of position-specific and object-specific cues in binocular and monocular domestic chicks (Gallus gallus). Behav Brain Res 163:10–17
69. Manns M, Römling J (2012) The impact of asymmetrical light input on cerebral hemispheric specialization and interhemispheric cooperation. Nat Commun 3:696. doi:10.1038/ncomms1699
70. Letzner S, Patzke N, Verhaal J, Manns M (2014) Shaping a lateralized brain: asymmetrical light experience modulates access to visual interhemispheric information in pigeons. Sci Rep 4:4253. doi:10.1038/srep04253
71. Cynader M, Leporé F, Guillemot JP (1981) Inter-hemispheric competition during postnatal development. Nature 290:139–140
72. Sprague JM (1966) Interaction of cortex and superior colliculus in mediation of visually guided behavior in the cat. Science 153:1544–1547
73. Wallace SF, Rosenquist AC, Sprague JM (1989) Recovery from cortical blindness mediated by destruction of nontectotectal fibers in the commissure of the superior colliculus in the cat. J Comp Neurol 284:429–450
74. Lomber SG (1999) The advantages and limitations of permanent or reversible deactivation techniques in the assessment of neural function. J Neurosci Methods 86:109–117
75. Wurtz RH (2015) Using perturbations to identify the brain circuits underlying active vision. Philos Trans R Soc Lond B Biol Sci 370:1677. doi:10.1098/rstb.2014.0205
76. Parsons CH, Rogers LJ (1993) Role of the tectal and posterior commissures in lateralization of the avian brain. Behav Brain Res 54:153–164
77. Tommasi L, Gagliardo A, Andrew RJ, Vallortigara G (2003) Separate processing mechanisms for encoding of geometric and landmark information in the avian hippocampus. Eur J Neurosci 17:1695–1702

78. Gagliardo A, Ioalè P, Odetti F, Bingman VP, Siegel JJ, Vallortigara G (2001) Hippocampus and homing in pigeons: left and right hemispheric differences in navigational map learning. Eur J Neurosci 13:1617–1624
79. Gagliardo A, Ioalè P, Odetti F, Kahn MC, Bingman VP (2004) Hippocampal lesions do not disrupt navigational map retention in homing pigeons under conditions when map acquisition is hippocampal dependent. Behav Brain Res 153:35–42
80. Gagliardo A, Vallortigara G, Nardi D, Bingman VP (2005) A lateralized avian hippocampus: preferential role of the left hippocampal formation in homing pigeon sun compass-based spatial learning. Eur J Neurosci 22:2549–2559
81. Kahn MC, Bingman VP (2004) Lateralization of spatial learning in the avian hippocampal formation. Behav Neurosci 118:333–344
82. Davies DC, Taylor DA, Johnson MH (1988) The effects of hyperstriatal lesions on one-trial passive-avoidance learning in the chick. J Neurosci 8:4662–4666
83. Lowndes M, Davies DC (1994) The effects of archistriatal lesions on one-trial passive avoidance learning in the chick. Eur J Neurosci 6:525–530
84. Howard KJ, Rogers LJ, Boura AL (1980) Functional lateralization of the chicken forebrain revealed by use of intracranial glutamate. Brain Res 188:369–382
85. Deng C, Rogers LJ (2002) Prehatching visual experience and lateralization in the visual Wulst of the chick. Behav Brain Res 134:375–385
86. Orton LD, Poon PW, Rees A (2012) Deactivation of the inferior colliculus by cooling demonstrates intercollicular modulation of neuronal activity. Front Neural Circuits 6:100. doi:10.3389/fncir.2012.00100
87. Flourens MJP (1842) Recherches expérimentales sur les propriétés et les functions du système nerveux dans les animaux vertébrés. JB Ballière, Paris
88. Lashley KS (1931) Mass action in cerebral function. Science 73:245–254
89. Vilensky JA, Gilman S (2003) Using extirpations to understand the human motor cortex: Horsley, Foerster, and Bucy. Arch Neurol 60:446–451
90. Manns M (2006) The epigenetic control of asymmetry formation—lessons from the avian visual system. In: Malashichev Y, Deckel W (eds) Behavioral and morphological asymmetries in vertebrates. Landes Bioscience, Georgetown
91. Bishop DVM (2013) Cerebral asymmetry and language development: cause, correlate, or consequence? Science 340:1230531. doi:10.1126/science.1230531
92. Fagard J (2013) The nature and nurture of human infant hand preference. Ann N Y Acad Sci 1288:114–123
93. Francks C (2015) Exploring human brain lateralization with molecular genetics and genomics. Ann N Y Acad Sci 1359:1–13. doi:10.1111/nyas.12770
94. Griessenauer CJ, Salam S, Hendrix P, Patel DM, Tubbs RS, Blount JP, Winkler PA (2015) Hemispherectomy for treatment of refractory epilepsy in the pediatric age group: a systematic review. J Neurosurg Pediatr 15:34–44. doi:10.3171/2014.10.PEDS14155
95. Liégeois F, Cross JH, Polkey C, Harkness W, Vargha-Khadem F (2008) Language after hemispherectomy in childhood: contributions from memory and intelligence. Neuropsychologia 46:3101–3107. doi:10.1016/j.neuropsychologia.2008.07.001
96. Bingman VP, Hough GE 2nd, Kahn MC, Siegel JJ (2003) The homing pigeon hippocampus and space: in search of adaptive specialization. Brain Behav Evol 62:117–127
97. Bingman VP, Gagliardo A (2006) Of birds and men: convergent evolution in hippocampal lateralization and spatial cognition. Cortex 42:99–100
98. Bingman VP, Siegel JJ, Gagliardo A, Erichsen JT (2006) Representing the richness of avian spatial cognition: properties of a lateralized homing pigeon hippocampus. Rev Neurosci 17:17–28
99. Tommasi L, Chiandetti C, Pecchia T, Sovrano VA, Vallortigara G (2012) From natural geometry to spatial cognition. Neurosci Biobehav Rev 36:799–824. doi:10.1016/j.neubiorev.2011.12.007
100. Suárez R, Gobius I, Richards LJ (2014) Evolution and development of interhemispheric connections in the vertebrate forebrain. Front Hum Neurosci 8:497. doi:10.3389/fnhum.2014.00497
101. Letzner S, Simon A, Güntürkün O (2016) Connectivity and neurochemistry of the commissura anterior of the pigeon (Columba livia). J Comp Neurol 524:343–361
102. King BM (1991) Comparison of electrolytic and radio-frequency lesion methods. In: Conn P (ed) Lesions and transplantation. Academic, London, San Diego
103. Glenn MJ, Lehmann H, Mumby DG, Woodside B (2005) Differential fos expression following aspiration, electrolytic, or exci-

totoxic lesions of the perirhinal cortex in rats. Behav Neurosci 119:806–813
104. Silver JM, Bonhaus DW, McNamara JO (1986) Method of lesioning brainstem determines seizure probability. J Neurosci Methods 17:303–310
105. Aronow S (1960) The use of radio-frequency power in making lesions in the brain. J Neurosurg 17:431–438
106. Olton DS (1975) Technique for producing directionally specific brain lesions with radio frequency current. Physiol Behav 14:369–372
107. Strickland BA, Jimenez-Shahed J, Jankovic J, Viswanathan A (2013) Radiofrequency lesioning through deep brain stimulation electrodes: a pilot study of lesion geometry and temperature characteristics. J Clin Neurosci 20:1709–1712
108. Soloman M, Mekhail MN, Mekhail N (2010) Radiofrequency treatment in chronic pain. Expert Rev Neurother 10:469–474
109. Pinheiro PS, Mulle C (2008) Presynaptic glutamate receptors: physiological functions and mechanisms of action. Nat Rev Neurosci 9:423–436. doi:10.1038/nrn2379
110. Jarrard LE (2002) Use of excitotoxins to lesion the hippocampus: update. Hippocampus 12:405–414
111. Kantamneni S (2015) Cross-talk and regulation between glutamate and GABAB receptors. Front Cell Neurosci 9:135. doi:10.3389/fncel.2015.00135
112. Greenamyre JT, Porter RH (1994) Anatomy and physiology of glutamate in the CNS. Neurology 44:S7–S13
113. Rothman SM (1985) The neurotoxicity of excitatory amino acids is produced by passive chloride influx. J Neurosci 5:1483–1489
114. Rothman SM (1992) Excitotoxins: possible mechanisms of action. Ann N Y Acad Sci 648:132–139
115. Fujikawa DG (2015) The role of excitotoxic programmed necrosis in acute brain injury. Comput Struct Biotechnol J 13:212–221. doi:10.1016/j.csbj.2015.03.004
116. Lau A, Tymianski M (2010) Glutamate receptors, neurotoxicity and neurodegeneration. Pflugers Arch 460:525–542. doi:10.1007/s00424-010-0809-1
117. Deng C, Rogers LJ (1997) Differential contributions of the two visual pathways to functional lateralization in chicks. Behav Brain Res 87:173–182
118. Contractor A, Mulle C, Swanson GT (2011) Kainate receptors coming of age: milestones of two decades of research. Trends Neurosci 34:154–163. doi:10.1016/j.tins.2010.12.002
119. Contestabile A, Migani P, Poli A, Villani L (1984) Recent advances in the use of selective neuron-destroying agents for neurobiological research. Experientia 40:524–534
120. Stebelska K (2013) Fungal hallucinogens psilocin, ibotenic acid, and muscimol: analytical methods and biologic activities. Ther Drug Monit 35:420–442. doi:10.1097/FTD.0b013e31828741a5
121. Jarrard LE (1989) On the use of ibotenic acid to lesion selectively different components of the hippocampal formation. J Neurosci Methods 29:251–259
122. Sullivan RM, Gratton A (1999) Lateralized effects of medial prefrontal cortex lesions on neuroendocrine and autonomic stress responses in rats. J Neurosci 19:2834–2840
123. Schwarting RK, Huston JP (1996) The unilateral 6-hydroxydopamine lesion model in behavioral brain research. Analysis of functional deficits, recovery and treatments. Prog Neurobiol 50:275–331
124. Nobin A, Björklund A (1978) Degenerative effects of various neurotoxic indoleamines on central monoamine neurons. Ann N Y Acad Sci 305:305–327
125. Jonsson G (1980) Chemical neurotoxins as denervation tools in neurobiology. Annu Rev Neurosci 3:169–187
126. Segura Aguilar J, Kostrzewa RM (2004) Neurotoxins and neurotoxic species implicated in neurodegeneration. Neurotox Res 6:615–630
127. Soto-Otero R, Méndez-Alvarez E, Hermida-Ameijeiras A, Muñoz-Patiño AM, Labandeira-Garcia JL (2000) Autoxidation and neurotoxicity of 6-hydroxydopamine in the presence of some antioxidants: potential implication in relation to the pathogenesis of Parkinson's disease. J Neurochem 74:1605–1612
128. Breese CR, Breese GR (1998) The use of neurotoxins to lesion catecholamine-containing neurons to model clinical disorders: approach for defining adaptive neural mechanisms and role of neurotrophic factors in brain. In: Kostrzewa RM (ed) Highly selective neurotoxins: basic and clinical applications. Humana Press, Totowa
129. Schwarting RKW, Huston JP (1996) Unilateral 6-hydroxydopamine lesions of meso-striatal dopamine neurons and their physiological sequelae. Prog Neurobiol 49:215–226
130. Glick SD, Zimmerberg B, Jerussi TP (1977) Adaptive significance of laterality in the rodent. Ann N Y Acad Sci 299:180–185
131. Miller R, Miller R, Beninger RJ (1991) On the interpretation of asymmetries of posture and locomotion produced with dopamine agonists in animals with unilateral deple-

tion of striatal dopamine. Prog Neurobiol 36:229–256
132. Xu ZC, Ling G, Sahr RN, Neal-Beliveau BS (2005) Asymmetrical changes of dopamine receptors in the striatum after unilateral dopamine depletion. Brain Res 1038:163–170
133. Ren Y, Li X, Xu ZC (1997) Asymmetrical protection of neostriatal neurons against transient forebrain ischemia by unilateral dopamine depletion. Exp Neurol 146:250–257
134. Cronin-Golomb A (2010) Parkinson's disease as a disconnection syndrome. Neuropsychol Rev 20:191–208. doi:10.1007/s11065-010-9128-8
135. Payne BR, Lomber SG, Villa AE, Bullier J (1996) Reversible deactivation of cerebral network components. Trends Neurosci 19:535–542
136. Lomber SG, Payne BR (2004) Cerebral areas mediating visual redirection of gaze: cooling deactivation of 15 loci in the cat. J Comp Neurol 474:190–208
137. Lomber SG, Payne BR (2000) Translaminar differentiation of visually guided behaviors revealed by restricted cerebral cooling deactivation. Cereb Cortex 10:1066–1077
138. Lomber SG, Payne BR, Horel JA (1999) The cryoloop: an adaptable reversible cooling deactivation method for behavioral and electrophysiological assessment of neural function. J Neurosci Methods 86:179–194
139. Yang XF, Kennedy BR, Lomber SG, Schmidt RE, Rothman SM (2006) Cooling produces minimal neuropathology in neocortex and hippocampus. Neurobiol Dis 23:637–643
140. Freund N, Manns M, Rose J (2010) A method for the evaluation of intracranial tetrodotoxin injections. J Neurosci Methods 186:25–28. doi:10.1016/j.jneumeth.2009.10.019
141. Bane V, Lehane M, Dikshit M, O'Riordan A, Furey A (2014) Tetrodotoxin: chemistry, toxicity, source, distribution and detection. Toxins 6:693–755
142. Evans MH (1972) Tetrodotoxin, saxitoxin, and related substances: their applications in neurobiology. Int Rev Neurobiol 15:83–166
143. Fozzard HA, Lipkind GM (2010) The tetrodotoxin binding site is within the outer vestibule of the sodium channel. Mar Drugs 8:219–234. doi:10.3390/md8020219
144. Boyden ES (2015) Optogenetics and the future of neuroscience. Nat Neurosci 18:1200–1201. doi:10.1038/nn.4094
145. Shipton OA, El-Gaby M, Apergis-Schoute J, Deisseroth K, Bannerman DM, Paulsen O, Kohl MM (2014) Left-right dissociation of hippocampal memory processes in mice. Proc Natl Acad Sci U S A 111:15238–15243. doi:10.1073/pnas.1405648111
146. Korsching S (1993) The neurotrophic factor concept: a reexamination. J Neurosci 13:2739–2748
147. Oppenheim RW (1989) The neurotrophic theory and naturally occurring motoneuron death. Trends Neurosci 12:252–255
148. Allred RP, Kim SY, Jones TA (2014) Use it and/or lose it-experience effects on brain remodeling across time after stroke. Front Hum Neurosci 8:379. doi:10.3389/fnhum.2014.00379
149. Xerri C, Zennou-Azogui Y, Sadlaoud K, Sauvajon D (2014) Interplay between intra- and interhemispheric remodeling of neural networks as a substrate of functional recovery after stroke: adaptive versus maladaptive reorganization. Neuroscience 283:178–201. doi:10.1016/j.neuroscience.2014.06.066
150. Witte OW, Buchkremer-Ratzmann I, Schiene K, Neumann-Haefelin T, Hagemann G, Kraemer M, Zilles K, Freund HJ (1997) Lesion-induced network plasticity in remote brain areas. Trends Neurosci 20:348–349
151. Witte OW (1998) Lesion-induced plasticity as a potential mechanism for recovery and rehabilitative training. Curr Opin Neurol 11:655–662
152. Glickstein M, Berlucchi G (2008) Classical disconnection studies of the corpus callosum. Cortex 44:914–927
153. Ambrogi Lorenzini CG, Baldi E, Bucherelli C, Sacchetti B, Tassoni G (1997) Analysis of mnemonic processing by means of totally reversible neural inactivation. Brain Res Protoc 1:391–398
154. Zhuravin IA, Bures J (1991) Extent of the tetrodotoxin induced blockade examined by pupillary paralysis elicited by intracerebral injection of the drug. Exp Brain Res 83:687–690

Chapter 8

Pharmacological Agents and Electrophysiological Techniques

Brian McCabe

Abstract

Pharmacological and electrophysiological study of hemispheric asymmetry has been particularly fruitful in systems in which anatomical, physiological, and behavioral asymmetries are readily demonstrated. In the domestic chick, several behavioral phenomena meet these conditions, including the learning processes of filial imprinting and passive avoidance learning. The chick possesses special advantages for studies of this kind, such as a wide range of readily measured behaviors, a highly permeable blood–brain barrier and amenability to strict control of the developmental environment. Moreover, modification of the visual environment during embryonic development can produce hemispheric asymmetry in visual pathways. A restricted region of the chick forebrain, the intermediate and medial mesopallium (IMM), has been shown to be a memory region, and shows marked functional hemispheric asymmetry during learning and memory processing.

Whenever hemispheric asymmetry in the control of behavior is being sought, significant statistical interaction of experimental treatment with side is generally required for a secure inference of lateralization. When drugs are directly applied to brain tissue, topographical spread may be measured directly, or inferred from diffusion models and drug application at control sites; the effects of systemic injections may be estimated using pharmacokinetic modeling and local or systemic application of agonists and antagonists.

A whole hemisphere may be inactivated reversibly by intra-carotid injection of anesthetic. Many commissural projection neurons are glutamatergic, allowing inter-hemispheric communication to be modified by intra-cerebral application of appropriate drugs. Much behavior in the chick has been shown pharmacologically (e.g. by activating or blocking glutamate receptors) to be under asymmetrical control, including visual discrimination, auditory habituation, copulation, and attack behavior, as well as behavior resulting from learning and memory associated with imprinting and passive avoidance learning. Learning- and memory-related hemispheric symmetries have been found in neuronal activity and neurotransmitter systems in the IMM. Neural systems participating in birdsong have long been known to exhibit hemispheric asymmetry and this is reflected in neuronal activity and sensitivity to electrical stimulation. Hemispheric asymmetry in olfactory pathways has been found in mammals and birds, and has been investigated pharmacologically and electrophysiologically.

The study of hemispheric asymmetry is likely to be enhanced by specific pharmacological targeting of proteins and nucleic acids and increasingly specific drug design, complemented by electrophysiological analysis.

Key words Androgens, Birdsong, Cycloheximide, Glutamate receptors, Domestic chick, Imprinting, Learning, Memory, Passive avoidance learning

Lesley J. Rogers and Giorgio Vallortigara (eds.), *Lateralized Brain Functions: Methods in Human and Non-Human Species*, Neuromethods, vol. 122, DOI 10.1007/978-1-4939-6725-4_8, © Springer Science+Business Media LLC 2017

1 Introduction

"One might suppose that Dr. Jekyll had discovered a drug that blocked neurotransmission between the two cerebral hemispheres." Professor Sir Gabriel Horn, in a lecture to medical students.

Robert Louis Stephenson evidently insisted that he was not influenced by medical research when writing "The Strange Case of Dr. Jekyll and Mr. Hyde," published in 1886. Despite this avowal, it has been argued [1] that the idea of the "split personality," in conjunction with contemporary neurological research (cf. [2]) may have at least subconsciously influenced Stevenson in his tale of a drug that reversibly transformed the benign Dr. Jekyll into the psychopathic Edward Hyde. Given that the story was published before the discovery of the first hormone and at a time when the very idea of the synapse was in gestation, it would be far-fetched to propose too much of a scientific basis for Stevenson's novella. However, once hemispheric asymmetry in the control of behavior was established, pharmacology has made an important contribution.

This review surveys the contributions made to our understanding of hemispheric asymmetry using pharmacological and electrophysiological methods. It has been necessary to describe some of the processes that are represented asymmetrically but these descriptions have been kept brief to minimize duplication of recent comprehensive accounts (e.g. [3, 4]). Much of the research in this field has been performed on birds and this research has accordingly been emphasized to illustrate the methods employed.

2 General Methodological Considerations

It is appropriate first, as background, to describe some types of behavior that have been studied, and the neurobiological techniques employed, before discussing their application in the study of asymmetry.

2.1 Filial Imprinting

Precocial animals such as the domestic chick are attracted to conspicuous stimuli during a sensitive period over the first few days after birth or hatching. The young animal learns the characteristics of a stimulus, which as a result becomes more attractive. The learning process is known as filial imprinting; for convenience it will just be referred to here as "imprinting." Social behavior directed towards the stimulus reflects the learning that has occurred and the strength of imprinting may be determined in a preference test, measuring approach to the stimulus, time spent near it, or some other behavior reflecting a preference, relative to the same measure directed towards a dissimilar stimulus that is novel (for reviews see [5–7]).

In the course of extensive studies of the neural basis of imprinting in the chick, a number of hemispheric asymmetries have been discovered. A restricted part of the chick forebrain is critical for visual imprinting, and a large body of evidence has led to the conclusion that it is a site at which information about the imprinting stimulus is stored (for reviews see [8–10]). This region is the intermediate and medial mesopallium (IMM), formerly known as the intermediate and medial hyperstriatum ventrale (IMHV; see [11]). The left and right IMM have different roles in imprinting ([10], cf. [12]). It is known from lesion studies that imprinting can occur when only either the left or the right side of the IMM is present [13]. In addition, a preference acquired through imprinting can be supported by either side of the IMM after exposure to an imprinting stimulus—referred to subsequently as "training" [14]. After training, the right IMM (but not the left) is necessary to establish a supplementary memory region (S′) outside the IMM. Region S′ and the IMM have different, albeit overlapping, roles in memory. Whereas the IMM is necessary for acquisition and retention for at least 6 h after training, S′ is also necessary for flexible use of the learned information, such as classification together of different perceptual representations of the same object; this function of S′ has been called "mediational" by Honey et al. [15, 16].

Imprinting training gives rise to an increase in the mean area of postsynaptic densities of axospinous (presumed excitatory) synapses. This change is unilateral, occurring in the left IMM [17, 18]. Also restricted to the left IMM, and consistent with the morphological change in synapses, is an increase in the numerical density of the *N*-methyl-D-aspartate (NMDA) subtype of glutamate receptor [19]. There is evidence for other lateralised, learning-related changes in levels of functionally important proteins in the IMM [20–27]; where asymmetries have been found, they are biased towards the left side but the lateralization is often not absolute. During a study of learning-related changes in proteins critical for oxidative phosphorylation (subunits I and II of cytochrome c oxidase), a remarkable correlation between the amounts of these proteins was found, superimposed upon and independent of the effects of learning. The correlation was strongly lateralised and restricted to the left IMM, suggestive of a specialization of this side of the IMM in preparation for efficient learning during the sensitive period for imprinting [25].

2.2 Passive Avoidance Learning (PAL) in the Domestic Chick

This type of learning has been used extensively to study neural changes associated with memory formation and many such changes have been shown to be lateralised. Domestic chicks tend to avoid an aversive-tasting object after a single trial. This behavior may be exploited by the use of an experimental bead coated with a bitter-tasting substance such as methyl anthranilate [28] or quinine [29]. Training and testing procedures vary between laboratories but in essence a bead coated with aversant is presented individually to chicks, which are strongly disposed to peck the bead ("training"). Subsequently, the chicks' avoidance of an identical test bead without aversant is compared with their avoidance of a bead having a different appearance [28, 30], or with avoidance of the test bead by control chicks trained with a neutral water-coated bead [31]. The difference in avoidance based on these comparisons is taken as a measure of memory for the association between the visual and aversive gustatory stimuli.

Three regions of the chick brain were found by Kossut and Rose [32] to be sites of metabolic change following PAL. These workers trained chicks on a passive avoidance task and measured incorporation of the neural activity marker 2-deoxyglucose 30 min later. The regions showing changes were the IMM, the medial striatum, and the lateral striatum; the last two regions were previous known as the lobus parolfactorius (LPO) and palaeostriatum augmentatum (PA) respectively (cf. [11]). Lateralization of neural changes in the IMM following PAL was found by Stewart et al. [33] in a quantitative electron microscope study of synaptic structure. Many biochemical, morphological, or electrophysiological changes have subsequently been found after PAL in one or more of the three regions implicated by Kossut and Rose, giving rise to a scheme for molecular mechanisms of memory formation (for review see [34]). Many of the changes in the IMM following passive avoidance training are restricted to the left side. In complementary ablation studies, pre-training lesions in the left IMM, but not the right, were found to impair PAL [35, 36]. However, even bilateral IMM lesions did not affect retention of the avoidance response when placed >1 h after training [36]; retention at these later times was impaired by bilateral, but not unilateral, lesions of the medial striatum. Changes were found in the right IMM several hours after training, namely increases in spontaneous neuronal bursting activity [37] and in calcium-dependent potassium-stimulated release (i.e. presumed releasable synaptic pool) of L-glutamate and γ-aminobutyric acid (GABA) [38]. The right IMM is also the site of decreased net glycoprotein synthesis in synaptic plasma membranes and postsynaptic densities 6 h after training, and in postsynaptic densities 24 h after training [39]. The 6 h time point is significant because it is a time at which synaptic re-modeling is proposed to occur, apparently with differential involvement of the left and right hemispheres.

2.3 Detection of Laterality

Statistical considerations are relevant when studying laterality. Matters are straightforward when trying to find out whether there is a structural or physiological difference between corresponding points on the two sides of the brain. However, when investigating functional laterality it is more complicated when experimental and control treatments need to be applied on each side of the brain.

One would then seek a statistically significant interaction between experimental effect and side. This can prove problematic when, as in many biological situations, experimental error is substantial; moreover a statistical interaction carries greater error variance than the corresponding main effects. One may then be left with the unsatisfactory situation in which an effect is significant on one side and not the other, but with no significant interaction between side and experimental effect. One may, however, live with this ambiguity until further evidence resolves the matter one way or another.

2.4 Drug Administration and Specificity

It is sometimes convenient to inject drug or vehicle into a particular target region under stereotaxic control and subsequently confirm the position of the injection site either histologically or by micro X-ray computed tomography [40]. Behavior and physiological processes may then be studied with unilateral or bilateral drug injection.

It is important to estimate, as precisely as possible, concentration of drug at various distances from the site of administration, and its temporal profile. Although the theory of diffusion is well established for homogeneous media [41], estimates of the extent to which a drug moves away from the site of application must take account of the complex nature of the tissue through which the drug passes. The use of an electrophysiological bioassay (neuronal firing rate) to estimate the effective diffusion coefficients of the neurotransmitters L-glutamate and GABA in brain tissue put these quantities at approximately 40% of their values in water [42]. Autoradiographic measurements (see e.g. [43]) indicate that spread can be sufficiently limited for useful discrimination between injection sites in adjacent neural structures. Numerous reports are now available in which the distribution of locally applied substances have been measured; see Syková and Nicholson [44] for a comprehensive discussion of the relevant considerations. As well as diffusion in a multicompartment space, aqueous and lipid solubility, loss, clearance, uptake, and binding are likely to influence the extent of spread of pharmacological agents. Despite these potential complexities, it is noteworthy that, for many compounds, two important parameters are fairly constant across different types of neural tissue. These are the fractional volume of extracellular space α and tortuosity λ, where

$$\lambda = \sqrt{D / D^{*}}$$

(D is the free diffusion coefficient and D^* the effective diffusion coefficient).

The values of α and λ are generally in the region of 0.2 and 1.6 respectively [44]. The considerations summarized in such studies, as well as informing pharmacological studies targeted at receptors, may also be applied to more recently developed techniques involving DNA transfection (e.g. [45]), whereby the promoter sequence for a gene (e.g. the neuronal activity marker *c-fos*) is cloned into the cDNA for a bioluminescent agent such as firefly luciferase. This construct is then inserted into a target brain region by electroporation and expression of the gene measured in vivo as a bioluminescent signal.

Anatomical specificity of drug action may be estimated by local administration when it is possible to obtain a dissociation between drug action and site of application: that is, an effect of the drug (but not the vehicle) at one injection site and an effect of neither agent at another, control site. If the experimental question involves two different actions and two specific sites of action, a double dissociation between injection site and drug effect may be attempted (cf. e.g. [46]).

Systemic administration of drugs can be helpful in analyzing lateralization of neuronal processes and behavior when the processes are directly observed or the behavior is known to be under lateralised control. Although, when administering agents systemically, one might not have all the pharmacokinetic information one would wish for, some such information (cf. [47]) is valuable. Drug specificity may be evaluated by the use of agonists or antagonists, bearing in mind their ability to penetrate the blood–brain barrier. It is noteworthy that the blood–brain barrier in the domestic chick is permeable to many substances that do not cross the barrier in adult vertebrates [48], a feature that has been exploited in some of the neuropharmacological studies discussed below.

2.5 Receptor Ligand Binding

The binding of labeled ligands to receptors may be measured autoradiographically or in tissue suspensions; each technique has contributed to our understanding of hemispheric asymmetry and both are well understood methodologically (cf. [49]).

2.6 Sensory Modulation of Hemispheric Asymmetry

A number of behaviorally expressed hemispheric asymmetries in birds depend on whether there is illumination of the egg, or different levels of illumination of the two eyes, at sensitive periods during development. Such effects have been studied in detail in the domestic chick (see e.g. [4]) and pigeon [50, 51] see Chap. 3 and 19. The underlying mechanisms appear to differ between the two species [52]. Illumination of the domestic chicken egg during the period encompassing days 18–21 of incubation stimulates the right eye more than the left in almost all embryos because the position of the head causes the left eye to be shielded by the embryo's body [53, 54]. The asymmetric illumination of the eyes produces anatomical asymmetry in the thalamofugal visual pathway up to and beyond day 12 post-hatch: the *right* eye (preferentially illuminated by light on the egg) projects to the left dorsal lateral geniculate nucleus of the thalamus, which in turn projects to both sides of the forebrain. The *left* eye (shielded from illumination) also projects to its contralateral (i.e. right) lateral geniculate but the great majority of downstream fibers from the lateral geniculate remain ipsilateral [55, 56]. Reversal of the natural pattern of egg illumination, i.e. selective illumination of the left eye instead of the right on days 20 and 21 of incubation, reverses the asymmetry [56]. So far, neuroanatomical consequences of light stimulation during incubation of the chicken egg have been described in the crossed projections of the dorsal lateral geniculate nucleus, but this is not to say that other use-dependent structural asymmetries do not occur. In the pigeon, functional and anatomical asymmetries caused by unilateral visual stimulation in the late pre-hatch and early post-hatch phase have been ascribed to centrifugal activity emanating from the telencephalon [52].

3 Pharmacological and Electrophysiological Studies of Hemispheric Asymmetry

3.1 Transient Unilateral Inactivation of a Whole Hemisphere

The technique of temporary selective inactivation of one cerebral hemisphere was developed by Wada in the 1940s (cf. [57]), for investigation of epileptic foci and to estimate the separate contributions of the left and right cerebral hemispheres to cognitive function. Typically, a catheter is introduced via a femoral artery into the corresponding left or right internal carotid artery and an anesthetic drug such as amobarbital (e.g. [58]) or propofol [59], injected so as to temporarily inactivate the side of the brain supplied by the catheterized artery. Psychological and physiological tests are then conducted to determine the functional losses resulting from hemispheric inactivation.

3.2 Neuro transmission in Commissural Neurons

Axons in the corpus callosum are thought to be largely excitatory [60] but direct inhibitory commissural connections employing GABA also exist [61]. Kumar and Hugenard [62] stimulated the corpus callosum electrically and found pharmacological evidence that such stimulation monosynaptically activates at least two subtypes of L-glutamate receptor on cortical neurons, confirming observations by Vogt and Gorman [63]. Both α-amino-3-hydroxy-5-methyl-4-isoxazolepropionate (AMPA) and *N*-methyl-D-aspartate (NMDA) subtypes of glutamate receptor were implicated. The excitatory responses to callosal stimulation had several components, some associated with inhibitory GABAergic and excitatory glutamatergic inputs from ipsilateral local circuit neurons.

3.3 Intracerebral Drug Administration

Much research, particularly on the domestic chick, has investigated functional hemispheric asymmetry by direct injection of drugs separately into each hemisphere in the unanesthetised animal. This procedure is simplified in the chick because the dorsal surface of the skull in the first few days post-hatch is incompletely ossified and can readily be penetrated by a needle [64].

3.3.1 Visual Discrimination, Auditory Habituation, Copulation, and Attack Behavior

Unilateral injection of cycloheximide into the chick forebrain has revealed hemispheric asymmetries in the expression of learning. Rogers and Anson [65] subjected chicks with one eye occluded to a visual discrimination task, in which the chicks learned to discriminate food grains from inedible, immovable pebbles [66]: see Fig. 6 in Chap. 3. The optic nerve in the post-hatch chick brain is virtually completely crossed [67, 68], so that visual input from the right eye projects almost entirely to the left side of the brain and vice versa (see Fig. 1). Interaction between the hemispheres in processing the visual information may then occur via several commissural pathways [11]. When cycloheximide had been injected into the *left*

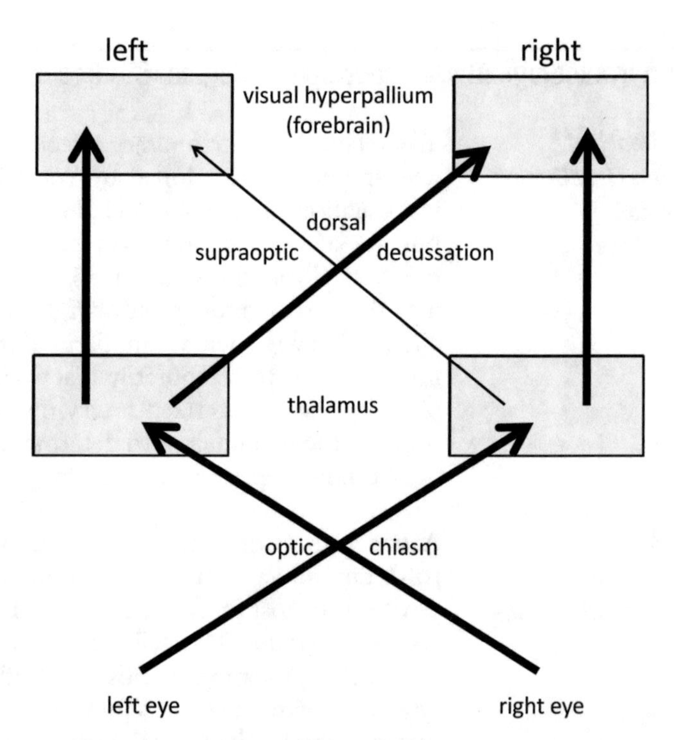

Fig. 1 Thalamofugal visual pathways in the domestic chick up to day 12 after hatching (modified from [69]). Visual input from the retinae is completely crossed at the optic chiasm and projects to the contralateral thalamus. There are then both ipsi- and contralateral projections from the thalamus to the visual regions of the forebrain, in the hyperpallium. In chicks hatched from eggs that received light exposure between days 19 and 21 of incubation, the crossed projection from left thalamus to right forebrain contains more fibres than the other crossed pathway, as indicated here. This is the result of stronger illumination of the right eye than the left eye because the latter is shielded by the chick's body in the egg. Experimental illumination of the left eye rather than the right eye between days 19 and 21 of incubation reverses the anatomical asymmetry in the crossed thalamic projection [55, 56]. This asymmetry is transient, disappearing by day 21 post-hatch [56]

hemisphere on day 2 post-hatch, later performance of the visual discrimination (on day 9) was impaired when either eye was used. In contrast, chicks that had received cycloheximide in the *right* hemisphere were impaired only when the left eye was used, suggesting a special, lateralised function in the left hemisphere necessary for visual discrimination via either eye [65]. It was found in the same study that injection into the left hemisphere, but not the right, attenuated auditory habituation. Performance of other tasks investigated in this study (visual habituation and visual detection) did not depend significantly on the hemisphere injected.

Although cycloheximide is a protein synthesis inhibitor, it is not necessarily exclusively through this action that this substance affects behavior: Hambley and Rogers [70] found that intracerebral injection of cycloheximide increased levels of several neuroactive amino acids and provided evidence that these amino acids, rather than modification of protein synthesis per se, were responsible for the behavioral effects. Work on mammals indicates that cycloheximide can also substantially alter monoamine release [71]. Moreover anisomycin, another protein synthesis inhibitor, may influence learning by altering levels of monoamines [72] and may even block propagated electrical activity [73]. Whatever its mechanism of action, locally applied cycloheximide is likely to modify neuronal activity at the site of injection and nearby.

On the hypothesis that cycloheximide acts through elevation of extracellular excitatory amino acid concentrations, Howard et al. [74] investigated the effect of unilateral injection of the neurotransmitter L-glutamate (164 mM in 5 μl) on chicks' performance of the visual discrimination and auditory habituation tasks. Attack and copulatory behavior were also studied. All of these behaviors were impaired by L-glutamate, but only when it was injected into the *left* hemisphere. The effect of the amino acid, unlike that of cycloheximide [65], did not depend on which eye was occluded during the visual discrimination test; nevertheless, the results of Howard et al. [74] confirm the importance of the left hemisphere in the control of certain types of behavior.

Focal injection of glutamate in smaller quantities into the visual wulst (the avian homologue of the mammalian visual cortex) has revealed interhemispheric differences in the control of visually guided behavior, differences arising from asymmetrical visual stimulation in ovo. Unilateral illumination of either the left or the right eye in ovo resulted in asymmetric sensitivity to glutamate (cf. Fig. 1): injection into the visual wulst contralateral to the illuminated eye significantly reduced the accuracy with which chicks visually discriminated food from nonfood items. This asymmetry was not found when the eggs were incubated in darkness, a condition producing no anatomical asymmetry in the visual pathway [75]. Injection of glutamate into the left visual wulst (but not the right) when the right eye is illuminated during incubation (the normal condition) also enhanced attack and copulation behavior. The effect was localized: unilateral glutamate injection into the entopallium (receiving input from the tectofugal visual pathway), was without effect on pecking accuracy and copulation [76]; attack behavior, however, was increased by injection of glutamate into the left entopallium (i.e. the side opposite the illuminated eye).

3.3.2 Filial Imprinting

Intracerebral injection of L-glutamate was used to investigate memory arising from other types of learning and has revealed both laterality and time-dependent changes following training. When injected unilaterally through the skull of the chick at a concentration of 100 mM in 5 μl, L-glutamate had an asymmetric effect on retention of a preference acquired through filial imprinting. Injection into the right hemisphere 1, 3, or 6 h after imprinting training blocked retention 8 h after training; no such effect was found in chicks receiving injections in the left hemisphere at those times [77]; see Table 1. Because it was effective when administered after training, the amino acid affected either consolidation or retrieval processes. The amnesia was temporary: it was not observed 48 h after training. The vulnerability of the right hemisphere to injection was itself transient: injection into this hemisphere was ineffective on retention if delayed until 9 h after training [77].

Johnston and Rogers [78] found that selective left-eye illumination in ovo (i.e. reversing the normal effect of light on the egg) also reversed hemispheric vulnerability to L-glutamate: retention after imprinting was now disrupted by injection into the left hemisphere. The usual vulnerability of the right hemisphere to

Table 1
Summary of time-courses of pharmacological changes associated with imprinting discussed in text

Action/measurement	Time after start of training	Result
Ser831 Phosphorylation of AMPA receptors in IMM [26]	2 h	Learning-related increases in left and right IMM
Unilateral forebrain injection of L-glutamate [77, 78]	3–9 h	*Dark-reared or right eye illuminated on day 19 of incubation*: temporary impairment of retention after injection into *right* hemisphere. *Left eye illuminated on day 19 of incubation*: temporary impairment of retention after injection into *left* hemisphere
NMDA-receptor binding in IMM [19, 79]	10–11 h	Learning-related increase in left IMM
Synaptic release of GABA and taurine from IMM [80, 81]	4.5–11 h	Learning-related increase in left IMM

glutamate [77] thus depends on visual experience before hatching. It is not only vulnerability of memory to L-glutamate that is determined by light exposure in ovo: hemispheric asymmetry of radioligand binding to glutamate and GABA neurotransmitter receptors in the chick forebrain can be reversed by monocular exposure of the left eye to light on day 19 of incubation [82].

3.4 Radioligand Binding Studies

Quantitative electron microscopy showed that imprinting gave rise to an increase in the mean profile length of postsynaptic densities of axospinous synapses in the left IMM, suggesting a change in synaptic connectivity [17, 18]. Since the morphology of these synapses suggested that they were excitatory, McCabe and Horn [19] targeted a class of excitatory neurotransmitter receptors by studying NMDA-sensitive binding of ^{3}H-L-glutamate to the NMDA glutamate receptor subtype in membranes prepared from the IMM. A saturating concentration of radioligand was used, so that any experimentally-induced change in binding could be attributed to a change in the numerical density of receptors; since only one concentration of radioligand was used, the experiments could not measure receptor affinity. There was a correlation between NMDA receptor number and strength of learning measured behaviorally. This change, like the result of the previous morphological studies, was restricted to the left side of the IMM (Table 1). The magnitude of the increase in binding was, to a good approximation, that predicted from the previously observed increase in postsynaptic density area on the assumption that NMDA receptor number per unit area of plasma membrane remained constant. The eggs from which the chicks in these experiments were hatched were incubated in darkness, so the hemispheric asymmetry observed could not be attributed to asymmetric illumination of the eyes during incubation as could the results of intracerebral L-glutamate administration [77, 78, 82].

The question then arose as to whether NMDA receptors in the IMM are necessary for imprinting, as might be expected from the dependence on NMDA receptors of the type of synaptic plasticity known as long-term potentiation (LTP) [83]. It has frequently been proposed that LTP and certain types of learning share the same neural mechanism (see e.g. [84]) and LTP has been demonstrated in the IMM of the chick [85]. The increase in receptor number was delayed, occurring 7–8 h after training and therefore did not occur at the same time as the learning. If the NMDA receptors are sub-serving synaptic plasticity, the learning-related effect following imprinting may reflect synaptic modification through an LTP-like mechanism, although this is not the only possible explanation [19]. Notwithstanding, there were grounds justifying proposal of the hypothesis that learning the features of the imprinting stimulus depends on NMDA receptors in the IMM. This hypothesis was tested by infusing the specific competitive NMDA receptor antagonist (2R)-amino-5-phosphopentanoate (D-AP5) into the IMM. Since the right IMM can support acquisition in the absence of the left IMM [13] and since training increases NMDA receptor number in the left IMM only [19], chicks were anesthetised and the right IMM ablated before infusing D-AP5 into the left IMM of awake chicks after recovery. Chicks were then exposed to a visual imprinting stimulus. Infusion was performed in the unanesthetised, restrained chick, through an indwelling guide cannula. See Davey et al. [86] for a description of the infusion procedure. Infusion of 0.7 nmol D-AP5 into the IMM prevented imprinting without detectable effects on other aspects of behavior, and infusion of the same amount into the hyperpallium apicale (part of the homologue of the mammalian visual cortex) was without effect on learning [87]. The drug was probably acting specifically on NMDA receptors, although it is not possible to estimate the concentration of drug precisely in an in vivo study. The authors' calculations indicated that the concentration was unlikely to exceed 50 μM, at which specific blockade of NMDA receptors has been reported in mammalian brain tissue. It is noteworthy that infusion of 0.2 nmol D-AP5 into the IMM was without significant effect on imprinting whereas 0.7 nmol gave significant impairment [87]. The concentration in brain tissue may have been over-estimated in view of findings by Morris et al. [88] indicating that the concentration of free D-AP5 in brain tissue can be lowered by a calcium-dependent sequestration process.

Johnston et al. [89] found a unilateral increase in glutamate-sensitive glutamate binding in the left IMM of domestic chicks that had been imprinted, relative to dark-reared, handled controls. This change, ascribed to NMDA receptors by Johnston et al. [79], in a binding study using the specific NMDA receptor-specific ligand MK-801, was found at approximately the same time after training as the correlation between NMDA receptor number and strength of imprinting reported by McCabe and Horn [19]; see Table 1. Phosphorylation at serine-831 of the GluA1 subunit of another major glutamate receptor subtype, the AMPA receptor, increased in the IMM in a learning-related manner earlier than the change in NMDA receptor binding, namely approximately 1 h after the end of training (2 h after the start of training) [26]. Phosphorylation of serine-831 of the GluA1 subunit causes an increase in channel conductance of the AMPA receptor and promotes its insertion into the postsynaptic membrane; these events have been identified as contributing to LTP [90, 91]. A clear learning-related increase in phosphorylation was found in the left IMM and a similar but non-significant trend in the right IMM [26]—one of the many examples of a nonsignificant bias towards the left side of this structure in memory-related changes.

3.5 Neurotransmitter Release Following Imprinting

The capacity of synapses to release neurotransmitter may be studied by determining the calcium-dependent releasable neurotransmitter pool; that is, by superfusing brain tissue ex vivo with artificial cerebrospinal fluid and subjecting neurons in the tissue to strong depolarization with e.g. 45 mM KCl in the presence or absence of calcium. The difference between release of neurotransmitter with calcium, and release of neurotransmitter without calcium, is then taken as the releasable pool. Imprinting gives rise to a significant correlation between preference score (a behavioral measure of the strength of imprinting) and release of GABA and taurine from tissue taken from the left IMM; no such correlation was found in the right IMM [80]. The correlation in the left IMM was subsequently found to persist for at least 10 h after training (Table 1) but was no longer present at 24 h after training [81]. There was thus evidence of enhanced efficacy of presynaptic origin in inhibitory synapses for several hours after the end of training with an imprinting stimulus.

3.6 Electro physiological Studies of Imprinting

The IMM, as well as being a site of synaptic modification and other biochemical changes as a result of imprinting, contains neurons that respond to visual and auditory stimuli. If chicks undergo visual imprinting, the proportion of multiple unit recording sites in the IMM responding to the visual imprinting stimulus increases approximately threefold when recorded a day after the start of training [92, 93]. This type of increase occurs on both sides of the IMM, but subtle differences between the sides were described by Nicol et al. [93] when comparing neuronal responses from the right IMM with responses in the left IMM recorded in the previous study [92]; see Table 2. The proportion of recording sites in trained birds responding to the training stimulus did not differ significantly between hemispheres. Nor, in untrained birds, was there a significant difference between the two sides with respect to the proportion of recording sites responding to either of the two visual imprinting stimuli used. However, in trained birds, significantly fewer sites in the right IMM were responsive to an alternative stimulus (i.e. not used for training and therefore novel) as compared to the left IMM. That is, the disparity between responsiveness to the training stimulus and to the alternative stimulus was greater in the right IMM than in the left IMM of trained chicks. There was also a number of differences between responsiveness of the left and right IMM irrespective of training experience.

Since either the left or the right IMM alone is capable of supporting retention of a preference acquired through imprinting [14], it is to be expected that neuronal responsiveness in both sides of this region is tuned to the training stimulus. Moreover, learning-related increases in expression of the neuronal activity marker Fos have been found in both the left and right IMM after imprinting [97]. It is also not surprising that there are subtle differences between the hemispheres at a time after training when the

Table 2
Electrophysiological measures according to side in IMM of untrained chicks and at stated times after the start of imprinting training

	Untrained (dark-reared)	3–4 h after start of training	9–10 h after start of training	~24 h after start of training
Background firing rate	L>R [94]	L~=R [94]	L>R [94]	L, indep. of tr. stimulus R, dep. on tr. stimulus [92, 93]
% sites responsive to model fowl (novel stimulus)	L>R [92, 93]			L>R [92, 93]
% responses excitatory	R>L [92, 93]			R>L [92, 93]
Responsiveness to maternal call	R>L [92, 93]			R>L [92, 93]
% sites responsive to novel visual stimulus				R>L [92, 93]
Range of stimuli to which neurons are responsive				R>L [95]
Number of neurons responding exclusively to visual imprinting stimulus				L>R [95]
Differentiation between novel and familiar chicks				R>L [96]

supplementary memory region S′ is being established under the control of the right IMM [14] and widespread left-bias of learning-related synaptic changes [10]. How these inter-hemispheric differences might indicate the different roles of the right and left IMM after training is a matter of speculation. Nicol et al. [93] point out that the change in synaptic efficacy implied by the unilateral increase in mean length of postsynaptic density profiles in the left IMM following training [17, 18] could reflect memory in the left IMM; a similar process in the right IMM would be masked if, in addition, a process active only on the right side caused some postsynaptic density profiles to shrink, giving no overall change in mean size. Such a combination of events is consistent with the reduction of responsiveness to an unfamiliar stimulus in the right IMM reported by Nicol et al. [93].

More recent electrophysiological analyses of imprinting from single unit recordings have revealed further differences between the left and right sides of the IMM resulting from imprinting training. Neurons in the right IMM are significantly more likely to respond across a range of stimuli, both familiar and novel, than neurons in the left IMM. Neurons responding exclusively to the familiar visual imprinting stimulus are more common in the left IMM [95].

Responses of neurons in the IMM to conspecifics were recorded in the left and right IMM in chicks that had first been imprinted to an artificial stimulus and then socially reared in a group of six chicks for 9 h. Neuronal responsiveness to familiar and unfamiliar conspecifics was then recorded from the left and right IMM in freely moving individuals. Responsiveness to unfamiliar conspecifics was significantly greater than responsiveness to familiar conspecifics and this disparity was significantly greater in the right than in the left IMM. There was thus a hemispheric asymmetry in the neuronal expression of a recognition process [96]; see Table 2 for a summary.

It is known from behavioral experiments [98–100] that the chick's left eye system (projecting to the right hemisphere) shows better discrimination between familiar and novel individuals than the complementary right eye system. The laterality in electrophysiological responsiveness to individual animals [96] is consistent with these results and implicates the IMM in the underlying process of recognition.

3.7 Passive Avoidance Learning (PAL)

3.7.1 Nitric Oxide and Laterality

Interest in neural plasticity has stimulated research in the chick brain on nitric oxide (NO), a compound proposed as a retrograde messenger contributing to use-dependent plasticity at glutamatergic and GABAergic synapses (see e.g. [101]). Experiments have been conducted to block NO synthesis in the chick forebrain by focal injection of inhibitors. Focal injection of NO synthase inhibitors into the chick forebrain both before and after training impairs PAL and/or its associated memory (for review see [34, 102, 103]). Unilateral injection of NO synthase inhibitors revealed hemispheric asymmetry and effects that depended on the inhibitor employed. Diphenyleneiodonium chloride (assumed selective for endothelial NO synthase) impaired retention of a passive avoidance task when injected into the left forebrain hemisphere at the time of training, and into the right hemisphere 10–20 min after training. In contrast, *N*-propyl-L-arginine (assumed selective for neuronal NO synthase) was only effective when injected into the left hemisphere around the time of training [104, 105]. There is thus evidence of hemispheric asymmetry in the roles of the two types of NO synthase. As with all experiments entailing application of drugs to the nervous system, the interpretation of any effect relies on assumptions of pharmacological specificity (which receptors are affected?) and anatomical specificity (where is the drug acting?). There is some localization of effect within hemispheres, since Nω-nitro-L-arginine methyl ester (L-NAME, an inhibitor of both endothelial and neuronal NO synthase) mimicked the effects of the more specific inhibitors when injected into the region of the left or right intermediate and medial hyperpallium, but had no significant effect when applied at the same times, relative to training, to the region of the medial striatum. Pharmacological specificity is apparent from the results of Rickard and Gibbs [104]. Both types of specificity are important: anatomical because of the differing roles of regions within a single hemisphere and pharmacological, because blockade of epithelial NO synthase would be expected to influence blood flow. Neuronal activity might thus be affected through modification of oxygen supply rather than via a direct action by NO on synaptic efficacy.

3.7.2 NMDA Receptors

Reflecting a change found in binding to NMDA glutamate receptors in the IMM after PAL [106], Steele and Stewart [107] found that 7-chlorokynurenate, a noncompetitive inhibitor of the neurotransmitter L-glutamate at its NMDA site, impaired PAL when injected 30 min before training and when retention of the avoidance response was tested 0.5, 1, and 3 h after training. The drug was effective when injected into the left forebrain hemisphere but not the right; it was not effective when injected 5 min after training. In contrast, both unilateral and bilateral injection of the noncompetitive NMDA receptor blocker MK-801 into the forebrain 5 min after training, although impairing retention, showed no significant laterality in its effect [108]. An asymmetry was, however, revealed after PAL, in which NMDA-sensitive glutamate binding was measured: this binding, although not that of radiolabelled MK-801, was significantly increased in the left IMM and not the right. A further region implicated in PAL, the medial striatum (area LPO), showed a significant decrease in binding on its right side [106].

Binding of L-glutamate to its NMDA receptor is enhanced by the binding of glycine to a regulatory site on the receptor [109]. Steele et al. [110] trained chicks on a passive avoidance task using a 10% aqueous solution of methyl anthranilate, which is a weak aversant and causes correspondingly weak passive avoidance learning. Injection of D-cycloserine, a partial agonist of glycine [111] into the left hemisphere 5 min after passive avoidance training on 10% methyl anthranilate, significantly enhanced the level of the learned passive avoidance response 1 and 6 h after training; injection into the right IMM did not do so [110].

To obtain evidence as to whether the changes in receptor binding were specifically related to learning, chicks were trained on the passive avoidance task and subjected to sub-convulsive trans-cranial electric shock 5 min after training. The effect of this procedure was to divide the chicks into two groups: chicks avoiding the bead (inferred recall of the training procedure) and those that pecked the bead (inferred amnesia). Binding in the left IMM and left medial striatum 30 min after training was significantly greater in the "recall" group than in the "amnesia" group, relating the changes in binding to memory formation [112].

3.7.3 AMPA Receptors

The effect of passive avoidance training on the AMPA subtype of glutamate receptor in chick brain was investigated in a binding study by Steele and Stewart [113], who measured binding to AMPA receptors 6.5 h after training. No hemispheric asymmetry in the effect of training was found in the IMM but there was a unilateral reduction in binding in the right lateral striatum, attributable to reductions both in affinity (K_D) and maximum binding capacity (B_{max}).

Incorporation of radioactive fucose into glycoproteins was shown by Bourne et al. [29] to be significantly greater in the left forebrain than in the right in untrained chicks and chicks when quinine, rather than the more strongly tasting methyl anthranilate, was used as an aversant. No such asymmetry was found in chicks trained with methyl anthranilate, so it is possible that net glycoprotein synthesis was selectively increased in the left IMM following training. Alternatively, the effect of training could have been to abolish an asymmetry that was present in the untrained and undertrained chicks. The latter possibility implies interaction between the hemispheres to maintain similar levels of net synthesis after learning.

3.8 Bird Song

There is abundant evidence of lateralization of vocalization in birds and this phenomenon has been studied in particular detail in songbirds [114, 115]: see also Chap. 7. Auditory perception may also be lateralized in birds and this discussion will be limited to the use of electrophysiological techniques in the study of the underlying neural mechanisms of perception and motor output.

Field L, the primary auditory projection area in the avian forebrain, was reported by George et al. [116] to show hemispheric asymmetry with respect to neuronal responsiveness to species- or individual-characteristic sounds when studying neuronal activity in male starlings. The right hemisphere contained more sound-sensitive neurons than the left and seemed to be more responsive to species-characteristic sounds. In contrast, neurons on the left side of Field L responded more to the unfamiliar vocalizations of individuals. Hemispheric specialization was also found in the higher vocal center (HVC) of male starlings where, in a state-dependent way, neurons in opposite hemispheres were differentially sensitive to particular species-specific sounds. There were, however, only four birds in this study, which is likely to be insufficient to make reliable generalizations [117]. Recording from neurons in Field L of female zebra finches also showed hemispheric asymmetry, manifest as a lower response to artificial sounds, as opposed to conspecific birdsong, in the left hemisphere, giving rise to a greater selectivity for song in the left hemisphere [118]. A further region in the songbird forebrain, the nidopallium caudale pars medialis (NCM), a candidate region for memory of learned song [119], also shows hemispheric asymmetry in adult males and female zebra finches that were tutored with conspecific song when young. In these animals, implanted chronically to yield multiple unit recordings in the adult, neuronal responsiveness to species-typical vocalizations was significantly greater in the right NCM than in the left. Untutored males showed a similar asymmetry. Asymmetry in the type of call responded to, and in the rate of adaptation to auditory stimuli depended on several features—sex, experience when young, and whether birds were devocalised by surgical bypassing of the syrinx [120].

Hemispheric asymmetries can be highly dynamic. Electrical stimulation of the HVC of adult male zebra finches can impair song production, unsurprisingly in view of its important role in song production (see e.g. [121] for review). Wang et al. [122] found that electrical stimulation of one side of the zebra finch HVC could severely disrupt song production but only at certain times: when stimulation of the right HVC was most effective, stimulation of the left HVC was much less effective and vice versa. Vulnerability to stimulation changed sides with an average period of several tens of milliseconds. The results imply that laterality of song control is highly dynamic. This is remarkable in view of the high level of synchronization between neurons in the left and right HVC during song production [123]. Such rapid interhemispheric switching must be the result of re-routing of impulse traffic by excitatory and inhibitory interneurons but it is unclear where in the song system this control is exerted.

3.9 Hemispheric Asymmetry in Olfactory Processing

Mammalian olfactory learning typically involves both left and right olfactory pathways. Both left and right pyriform cortices (PCX) receive a projection from the ipsilateral olfactory bulb but can also communicate with each other [124]. Learning of an olfactory discrimination by a rat is accompanied by hemispheric asymmetries in local field potentials in the PCX, in particular with respect to power of activity in the beta frequency band (16–31 Hz, thought to reflect recurrent activity between olfactory bulb and cortex and between cortex and hippocampus). Inter-hemispheric coherence was also measured. As learning of the discrimination proceeded, asymmetry was found to emerge in the beta band, with more power in the left PCX than on the right side. The difference was, however, temporary and had disappeared when learning was complete. At the same time, inter-hemispheric coherence progressively increased as learning progressed. As in avian song learning therefore, hemispheric asymmetry in the olfactory system can be transient. It may, moreover, signal processes underlying learning.

Olfactory lateralization has been found in the domestic chick, in terms of improved discrimination between familiar and unfamiliar odors [125], and increased behavioral sensitivity to a strong concentration of eugenol [126], when the right rather than the left nostril was used. The navigation of homing pigeons was also better when olfactory input was restricted to the right rather than the left nostril [127, 128]. Both left and right olfactory bulbs project to the ipsilateral and contralateral pyriform cortex; a lesion study [129] has indicated a stronger navigational role for the left pyriform cortex, but homing was impaired by unilateral ablation of either side. Coordination between hemispheres therefore seems to be necessary for normal function [see also Chap. 4].

3.10 Lateralization and Systemic Pharmacological Action

Laterality of brain and behavior may result from humoral influences [see Chap. 20]. It is well established that sex hormones can influence lateralised neural processes and hence behavior. Much of this work has been conducted on the domestic chick, with its advantages for behavioral neuropharmacology that are apparent from the previous discussion.

Asymmetry in the use of the left and right eye by domestic chicks was described by Andrew et al. [130] (reviewed in [131–134]). Both differential illumination of the eyes during incubation [135] and steroid hormones can affect brain asymmetry. Testosterone treatment on day 2 post-hatch was found to induce an asymmetry in the medial habenular nuclei of female chicks (right nucleus larger than the left), creating a pattern seen in untreated males [136]. However, testosterone (12.5 mg) injected into the egg on day 16 of incubation abolished the asymmetry of the crossed thalamofugal visual pathway that occurs when the right eye is illuminated during days 18–21 of incubation as a result of exposing the egg to light [137]. Oestradiol treatment of the embryo had a similar effect [138], raising the possibility that the lesser degree of asymmetry found in female chicks is due to their higher plasma levels of this hormone. The asymmetry was also disrupted by injection of corticosterone into the embryo on day 18 of incubation [139], suggesting that stress, through release of corticosteroids, may influence the development of sensory and other neural pathways. Such treatment was also found to have behavioral effects after hatching: the dose of corticosterone effective on asymmetry of the visual pathway decreased the readiness with which the image of a predator (a hawk) was detected by chicks tested on day 8 post-hatch [140].

4 Conclusion

Pharmacology and electrophysiology have distinct, complementary advantages for the investigation of neural function, and accordingly have been productive in the study of hemispheric asymmetry. Birds have been useful experimental subjects in such studies because of their readily demonstrated hemispheric asymmetries and anatomical separation of sensory pathways on the left and right sides. The domestic chick possesses the additional advantages of well-characterized behavior and neural circuits, amenability to unilateral modification of visual input during embryonic development and unusual permeability of the blood–brain barrier to many drugs.

The quality of the information obtained about hemispheric asymmetry in a pharmacological investigation is influenced by a number of factors, such as dosage, pharmacological specificity, localization of drug action, choice of control sites for local drug administration, pharmacokinetic principles, and statistical evidence for disparate effects of drugs on the two sides of the brain.

When probed pharmacologically and electrophysiologically, a number of neural changes in the forebrain following learning have been described in the domestic chick, with distinct time courses. Many such changes are lateralised. The IMM region (a memory store) shows time-dependent lateralization.

Lateralization has been described in the song system of birds, and in olfactory processing in birds and mammals. Some of the effects observed appear to be dynamic, involving time-dependent interaction between left and right sides of the brain.

Pharmacological and electrophysiological methods continue to be refined, offering more opportunities for the study of hemispheric asymmetry. New techniques used in combination are likely to possess particular advantages as, for example, in the use of behavioral experimentation, targeted optogenetic stimulation and pharmacological activation of discrete neuronal assemblies via specific cell-surface receptors [141].

Acknowledgements

Supported by the Biotechnology and Biological Sciences Research Council, the Isaac Newton Trust, and the Balfour Trust, Department of Zoology, University of Cambridge.

References

1. Stiles A (2006) Robert Louis Stevenson's "Jekyll and Hyde" and the double brain. Stud Engl Lit 1500–1900(46):879–900. doi:10.1353/sel.2006.0043
2. Fechner GT (1860) Elemente der Psychophysik. Breitkopf und Hartel, Leipzig
3. Rogers LJ, Vallortigara G, Andrew RJ (2013) Divided brains: the biology and behaviour of brain asymmetries. Cambridge University Press, Cambridge
4. Rogers LJ (2014) Asymmetry of brain and behavior in animals: its development, function, and human relevance. Genesis 52:555–571. doi:10.1002/dvg.22741
5. Bateson PPG (1966) The characteristics and context of imprinting. Biol Rev 41:177–220
6. Sluckin W (1972) Imprinting and early learning, 2nd edn. Methuen, London
7. Bolhuis JJ (1991) Mechanisms of avian imprinting: a review. Biol Rev 66:303–345
8. Horn G (1985) Memory, imprinting, and the brain. Oxford University Press, Oxford, http://www.oxfordscholarship.com/view/10.1093/acprof:oso/9780198521563.001.0001/acprof-9780198521563. Accessed 31 July 2014
9. Horn G (2004) Pathways of the past: the imprint of memory. Nat Rev Neurosci 5:108–120
10. McCabe BJ (2013) Imprinting. Wiley Interdiscip Rev Cogn Sci 4:375–390. doi:10.1002/wcs.1231

11. Reiner A, Perkel DJ, Bruce LL, Butler AB, Csillag A, Kuenzel W et al (2004) Revised nomenclature for avian telencephalon and some related brainstem nuclei. J Comp Neurol 473:377–414
12. McCabe BJ (1991) Hemispheric asymmetry of learning-induced changes. In: Andrew RJ (ed) Neural and behavioural plasticity: the use of the domestic chick as a model, 1st edn. Oxford University Press, Oxford, pp 262–276
13. Horn PG, McCabe BJ, Cipolla-Neto J (1983) Imprinting in the domestic chick: the role of each side of the hyperstriatum ventrale in acquisition and retention. Exp Brain Res 53:91–98. doi:10.1007/BF00239401
14. Cipolla-Neto J, Horn G, McCabe BJ (1982) Hemispheric asymmetry and imprinting: the effect of sequential lesions to the hyperstriatum ventrale. Exp Brain Res 48:22–27
15. Honey RC, Bateson P, Horn G (1994) The role of stimulus comparison in perceptual-learning—an investigation with the domestic chick. Q J Exp Psychol B Comp Physiol Psychol 47:83–103
16. Honey RC, Horn G, Bateson P, Walpole M (1995) Functionally distinct memories for imprinting stimuli—behavioral and neural dissociations. Behav Neurosci 109:689–698
17. Bradley P, Horn G, Bateson P (1981) Imprinting: an electron microscopic study of chick hyperstriatum ventrale. Exp Brain Res 41:115–120
18. Horn G, Bradley P, McCabe BJ (1985) Changes in the structure of synapses associated with learning. J Neurosci 5:3161–3168
19. McCabe BJ, Horn G (1988) Learning and memory: regional changes in *N*-methyl-d-aspartate receptors in the chick brain after imprinting. Proc Natl Acad Sci U S A 85:2849–2853
20. Solomonia RO, McCabe BJ, Jackson AP, Horn G (1997) Clathrin proteins and recognition memory. Neuroscience 80:59–67
21. Solomonia RO, McCabe BJ, Horn G (1998) Neural cell adhesion molecules, learning and memory. Behav Neurosci 112:646–655
22. Solomonia RO, Morgan K, Kotorashvili A, McCabe BJ, Jackson AP, Horn G (2003) Analysis of differential gene expression supports a role for amyloid precursor protein and a protein kinase C substrate (MARCKS) in long-term memory. Eur J Neurosci 17:1073–1081
23. Solomonia RO, Kotorashvili A, Kiguradze T, McCabe BJ, Horn G (2005) Ca2+/calmodulin protein kinase II and memory: learning-related changes in a localized region of the domestic chick brain. J Physiol (Lond) 569:643–653
24. Solomonia RO, Apkhazava D, Nozadze M, Jackson AP, McCabe BJ, Horn G (2008) Different forms of MARCKS protein are involved in memory formation in the learning process of imprinting. Exp Brain Res 188: 323–330
25. Solomonia RO, Kunelauri N, Mikautadze E, Apkhazava D, McCabe BJ, Horn G (2011) Mitochondrial proteins, learning and memory: biochemical specialization of a memory system. Neuroscience 194:112–123
26. Solomonia RO, Meparishvili M, Mikautadze E, Kunelauri N, Apkhazava D, McCabe BJ (2013) AMPA receptor phosphorylation and recognition memory: learning-related, time-dependent changes in the chick brain following filial imprinting. Exp Brain Res 226:297–308. doi:10.1007/s00221-013-3435-2
27. Meparishvili M, Nozadze M, Margvelani G et al (2015) A proteomic study of memory after imprinting in the domestic chick. Front Behav Neurosci 9:319. doi:10.3389/fnbeh.2015.00319
28. Cherkin A (1969) Kinetics of memory consolidation: role of amnesic treatment parameters. Proc Natl Acad Sci U S A 63:1094–1101. doi:10.1073/pnas.63.4.1094
29. Bourne RC, Davies DC, Stewart MG, Csillag A, Cooper M (1991) Cerebral glycoprotein synthesis and long-term memory formation in the chick (Gallus domesticus) following passive avoidance training depends on the nature of the aversive stimulus. Eur J Neurosci 3:243–248. doi:10.1111/j.1460-9568.1991.tb00086.x
30. Gibbs ME, Ng KT (1977) Psychobiology of memory: towards a model of memory formation. Biobehav Rev 1:113–136. doi:10.1016/0147-7552(77)90017-1
31. Rose SPR, Gibbs ME, Hambley J (1980) Transient increase in forebrain muscarinic cholinergic receptor binding following passive avoidance learning in the young chick. Neuroscience 5:169–172. doi:10.1016/0306-4522(80)90083-4
32. Kossut M, Rose SPR (1984) Differential 2-deoxyglucose uptake into chick brain structures during passive avoidance training. Neuroscience 12:971–977. doi:10.1016/0306-4522(84)90184-2
33. Stewart MG, Rose SPR, King TS, Gabbott PLA, Bourne R (1984) Hemispheric asymmetry of synapses in chick medial hyperstriatum ventrale following passive avoidance training: a stereological investigation. Dev

Brain Res 12:261–269. doi:10.1016/0165-3806(84)90048-8
34. Rose SPR (2000) God's organism? The chick as a model system for memory studies. Learn Mem 7:1–17
35. Patterson T, Alvarado M, Warner I, Bennett E, Rosenzweig M (1986) Memory stages and brain asymmetry in chick learning. Behav Neurosci 100:856–865
36. Patterson T, Gilbert D, Rose S (1990) Pre-training and post-training lesions of the intermediate medial hyperstriatum-ventrale and passive-avoidance learning in the chick. Exp Brain Res 80:189–195
37. Gigg J, Patterson TA, Rose SPR (1993) Training-induced increases in neuronal activity recorded from the forebrain of the day-old chick are time dependent. Neuroscience 56:771–776. doi:10.1016/0306-4522(93)90373-N
38. Daisley JN, Rose SPR (2002) Amino acid release from the intermediate medial hyperstriatum ventrale (IMHV) of day-old chicks following a one-trial passive avoidance task. Neurobiol Learn Mem 77:185–201
39. Bullock S, Rose SPR, Zamani R (1992) Characterisation and regional localisation of pre- and postsynaptic glycoproteins of the chick forebrain showing changed fucose incorporation following passive avoidance training. J Neurochem 58:2145–2154. doi:10.1111/j.1471-4159.1992.tb10957.x
40. de Crespigny A, Bou-Reslan H, Nishimura MC, Phillips H, Carano RAD, D'Arceuil HE (2008) 3D micro-CT imaging of the postmortem brain. J Neurosci Methods 171:207–213. doi:10.1016/j.jneumeth.2008.03.006
41. Crank J (1979) The mathematics of diffusion, 2nd edn. Oxford University Press, Oxford
42. Herz A, Zieglgänsberger W, Färber G (1969) Microelectrophoretic studies concerning the spread of glutamic acid and GABA in brain tissue. Exp Brain Res 9:221–235. doi:10.1007/BF00234456
43. Cooke SF, Attwell PJE, Yeo CH (2004) Temporal properties of cerebellar-dependent memory consolidation. J Neurosci 24:2934–2941. doi:10.1523/JNEUROSCI.5505-03.2004
44. Syková E, Nicholson C (2008) Diffusion in brain extracellular space. Physiol Rev 88:1277–1340. doi:10.1152/physrev.00027.2007
45. Yamaguchi S, Iikubo E, Hirose N, Kitajima T, Katagiri S, Kawamori A et al (2010) Bioluminescence imaging of c-fos gene expression accompanying filial imprinting in the newly hatched chick brain. Neurosci Res 67:192–195. doi:10.1016/j.neures.2010.02.007
46. Attwell PJE, Cooke SF, Yeo CH (2002) Cerebellar function in consolidation of a motor memory. Neuron 34:1011–1020. doi:10.1016/S0896-6273(02)00719-5
47. Ratain MJ, Plunkett WK (2003) Principles of pharmacokinetics. http://www.ncbi.nlm.nih.gov/books/NBK12815/. Accessed 11 Oct 2015
48. Saunders NR, Dreifuss J-J, Dziegielewska KM, Johansson PA, Habgood MD, Møllgård K et al (2014) The rights and wrongs of blood-brain barrier permeability studies: a walk through 100 years of history. Front Neurosci 8:404. doi:10.3389/fnins.2014.00404
49. Yamamura H, Enna S, Kuhar M (1985) Neurotransmitter receptor binding, 2nd edn. Yamamura HI; Dep Pharmacol, Coll Med, Univ Ariz Health Sci Cent, Tucson, AZ, USA
50. Manns M, Gunturkun O (1999) "Natural" and artificial monocular deprivation effects on thalamic soma sizes in pigeons. Neuroreport 10:3223–3228. doi:10.1097/00001756-199910190-00018
51. Manns M, Gunturkun O (1999) Monocular deprivation alters the direction of functional and morphological asymmetries in the pigeon's (Columba livia) visual system. Behav Neurosci 113:1257–1266. doi:10.1037/0735-7044.113.6.1257
52. Manns M, Stroeckens F (2014) Functional and structural comparison of visual lateralization in birds—similar but still different. Front Psychol 5:206. doi:10.3389/fpsyg.2014.00206
53. Hamburger V, Oppenheim R (1967) Prehatching motility and hatching behavior in the chick. J Exp Zool 166:171–203. doi:10.1002/jez.1401660203
54. Kovach JK (1968) Spatial orientation of the chick embryo during the last five days of incubation. J Comp Physiol Psychol 66:283–288. doi:10.1037/h0026374
55. Boxer MI, Stanford D (1985) Projections to the posterior visual hyperstriatal region in the chick: an HRP study. Exp Brain Res 57:494–498. doi:10.1007/BF00237836
56. Rogers LJ, Sink HS (1988) Transient asymmetry in the projections of the rostral thalamus to the visual hyperstriatum of the chicken, and reversal of its direction by light exposure. Exp Brain Res 70:378–384. doi:10.1007/BF00248362
57. Wada JA (1997) Youthful season revisited. Brain Cogn 33:7–10. doi:10.1006/brcg.1997.0879
58. Trenerry MR, Loring DW (1995) Intracarotid amobarbital procedure. The Wada test. Neuroimaging Clin N Am 5:721–728

59. Takayama M, Miyamoto S, Ikeda A, Mikuni N, Takahashi JB, Usui K et al (2004) Intracarotid propofol test for speech and memory dominance in man. Neurology 63:510–515. doi:10.1212/01.WNL.0000133199.65776.18
60. Bloom JS, Hynd GW (2005) The role of the corpus callosum in interhemispheric transfer of information: excitation or inhibition? Neuropsychol Rev 15:59–71. doi:10.1007/s11065-005-6252-y
61. Caputi A, Melzer S, Michael M, Monyer H (2013) The long and short of GABAergic neurons. Curr Opin Neurobiol 23:179–186. doi:10.1016/j.conb.2013.01.021
62. Kumar SS, Huguenard JR (2003) Pathway-specific differences in subunit composition of synaptic NMDA receptors on pyramidal neurons in neocortex. J Neurosci 23:10074–10083
63. Vogt B, Gorman A (1982) Responses of cortical neurons to stimulation of corpus callosum in vitro. J Neurophysiol 48:1257–1273
64. Andrew RJ (1991) Drug administration. In: Andrew RJ (ed) Neural and behavioural plasticity: the use of the domestic chick as a model, 1st edn. Oxford University Press, Oxford, pp 34–35
65. Rogers LJ, Anson JM (1979) Lateralisation of function in the chicken fore-brain. Pharmacol Biochem Behav 10:679–686. doi:10.1016/0091-3057(79)90320-4
66. Rogers LJ, Drennen HD, Mark RF (1974) Inhibition of memory formation in the imprinting period: irreversible action of cycloheximide in young chickens. Brain Res 79:213–233. doi:10.1016/0006-8993(74)90412-0
67. Drenhaus U, Rager G (1992) Organization of the optic chiasm in the hatched chick. Anat Rec 234:605–617. doi:10.1002/ar.1092340416
68. Jeffery G (2001) Architecture of the optic chiasm and the mechanisms that sculpt its development. Physiol Rev 81:1393–1414
69. Rogers L, Deng C (1999) Light experience and lateralization of the two visual pathways in the chick. Behav Brain Res 98:277–287. doi:10.1016/S0166-4328(98)00094-1
70. Hambley JW, Rogers LJ (1979) Retarded learning induced by intra-cerebral administration of amino-acids in the neonatal chick. Neuroscience 4:677–684. doi:10.1016/0306-4522(79)90144-1
71. Freedman LS, Judge ME, Quartermain D (1982) Effects of cycloheximide, a protein synthesis inhibitor, on mouse brain catecholamine biochemistry. Pharmacol Biochem Behav 17:187–191
72. Canal CE, Chang Q, Gold PE (2007) Amnesia produced by altered release of neurotransmitters after intraamygdala injections of a protein synthesis inhibitor. Proc Natl Acad Sci U S A 104:12500–12505. doi:10.1073/pnas.0705195104
73. Sharma AV, Nargang FE, Dickson CT (2012) Neurosilence: Profound suppression of neural activity following intracerebral administration of the protein synthesis inhibitor anisomycin. J Neurosci 32:2377–2387. doi:10.1523/JNEUROSCI.3543-11.2012
74. Howard KJ, Rogers LJ, Boura ALA (1980) Functional lateralization of the chicken forebrain revealed by use of intracranial glutamate. Brain Res 188:369–382. doi:10.1016/0006-8993(80)90038-4
75. Deng C, Rogers LJ (2002) Prehatching visual experience and lateralization in the visual Wulst of the chick. Behav Brain Res 134:375–385. doi:10.1016/S0166-4328(02)00050-5
76. Deng C, Rogers LJ (1997) Differential contributions of the two visual pathways to functional lateralization in chicks. Behav Brain Res 87:173–182. doi:10.1016/S0166-4328(97)02276-6
77. Johnston ANB, Rogers LJ (1998) Right hemisphere involvement in imprinting memory revealed by glutamate treatment. Pharmacol Biochem Behav 60:863–871
78. Johnston ANB, Rogers LJ (1999) Light exposure of chick embryo influences lateralized recall of imprinting memory. Behav Neurosci 113:1267–1273. doi:10.1037//0735-7044.113.6.1267
79. Johnston AN, Rogers LJ, Dodd PR (1995) [H-3] Mk-801 binding asymmetry in the IMHV region of dark-reared chicks is reversed by imprinting. Brain Res Bull 37:5–8. doi:10.1016/0361-9230(94)00249-5
80. McCabe BJ, Kendrick KM, Horn G (2001) Gamma-aminobutyric acid, taurine and learning: release of amino acids from slices of chick brain following filial imprinting. Neuroscience 105:317–324
81. Meredith RM, McCabe BJ, Kendrick KM, Horn G (2004) Amino acid neurotransmitter release and learning: a study of visual imprinting. Neuroscience 126:249–256
82. Johnston AN, Bourne RC, Stewart MG, Rogers LJ, Rose SP (1997) Exposure to light prior to hatching induces asymmetry of receptor binding in specific regions of the chick forebrain. Brain Res Dev Brain Res 103:83–90

83. Bliss TV, Collingridge GL (1993) A synaptic model of memory: long-term potentiation in the hippocampus. Nature 361:31–39
84. Morris RGM (2003) Long-term potentiation and memory. Philos Trans R Soc Lond B Biol Sci 358:643–647
85. Bradley PM, Burns BD, King TM, Webb AC (1993) NMDA-receptors and potentiation in an area of avian brain essential for learning. Neuroreport 5:313–316
86. Davey JE, McCabe BJ, Horn G (1991) Technique for cannulation to allow for subsequent microinjection into the IMHV of the awake chick. In: Andrew RJ (ed) Neural and behavioural plasticity: the use of the domestic chick as a model. Oxford University Press, Oxford, pp 43–44
87. McCabe BJ, Davey JE, Horn G (1992) Impairment of learning by localized injection of an *N*-methyl-d-apartate receptor antagonist into the hyperstriatum ventrale of the domestic chick. Behav Neurosci 106:947–953. doi:10.1037//0735-7044.106.6.947
88. Morris R, Davis S, Butcher S (1990) Hippocampal synaptic plasticity and NMDA receptors—a role in information-storage. Philos Trans R Soc Lond Biol Sci 329:187–204. doi:10.1098/rstb.1990.0164
89. Johnston AN, Rogers LJ, Johnston GAR (1993) Glutamate and imprinting memory—the role of glutamate receptors in the encoding of imprinting memory. Behav Brain Res 54:137–143
90. Barria A, Muller D, Derkach V, Griffith LC, Soderling TR (1997) Regulatory phosphorylation of AMPA-type glutamate receptors by CaMKII during long-term potentiation. Science 276:2042–2045
91. Lee HK, Barbarosie M, Kameyama K, Bear MF, Huganir RL (2000) Regulation of distinct AMPA receptor phosphorylation sites during bidirectional synaptic plasticity. Nature 405:955–959
92. Brown M, Horn G (1994) Learning-related alterations in the visual responsiveness of neurons in a memory system of the chick brain. Eur J Neurosci 6:1479–1490. doi:10.1111/j.1460-9568.1994.tb01009.x
93. Nicol AU, Brown MW, Horn G (1995) Neurophysiological investigations of a recognition memory system for imprinting in the domestic chick. Eur J Neurosci 7:766–776. doi:10.1111/j.1460-9568.1995.tb00680.x
94. Davey J, Horn G (1991) The development of hemispheric asymmetries in neuronal-activity in the domestic chick after visual experience. Behav Brain Res 45:81–86. doi:10.1016/S0166-4328(05)80183-4
95. Nicol AU, Horn G, Brown MW (2017) Polymodal responsiveness and functional lateralisation of neuronal encoding in a visual recognition memory system. Submitt. Publ
96. Town SM (2011) Preliminary evidence of a neurophysiological basis for individual discrimination in filial imprinting. Behav Brain Res 225:651–654. doi:10.1016/j.bbr.2011.08.018
97. McCabe BJ, Horn G (1994) Learning-related changes in Fos-like immunoreactivity in the chick forebrain after imprinting. Proc Natl Acad Sci U S A 91:11417–11421
98. Vallortigara G, Andrew RJ (1991) Lateralization of response by chicks to change in a model partner. Anim Behav 41:187–194. doi:10.1016/S0003-3472(05)80470-1
99. Vallortigara G (1992) Right hemisphere advantage for social recognition in the chick. Neuropsychologia 30:761–768. doi:10.1016/0028-3932(92)90080-6
100. Vallortigara G, Andrew RJ (1994) Differential involvement of right and left hemisphere in individual recognition in the domestic chick. Behav Processes 33:41–58
101. Hardingham N, Dachtler J, Fox K (2013) The role of nitric oxide in pre-synaptic plasticity and homeostasis. Front Cell Neurosci 7:190. doi:10.3389/fncel.2013.00190
102. Holscher C, Rose SPR (1992) An inhibitor of nitric-oxide synthesis prevents memory formation in the chick. Neurosci Lett 145:165–167
103. Holscher C, Rose S (1993) Inhibiting synthesis of the putative retrograde messenger nitric-oxide results in amnesia in a passive-avoidance task in the chick. Brain Res 619:189–194. doi:10.1016/0006-8993(93)91611-U
104. Rickard NS, Gibbs ME (2003) Hemispheric dissociation of the involvement of NOS isoforms in memory for discriminated avoidance in the chick. Learn Mem 10:314–318. doi:10.1101/lm.59503
105. Rickard NS, Gibbs ME (2003) Effects of nitric oxide inhibition on avoidance learning in the chick are lateralized and localized. Neurobiol Learn Mem 79:252–256. doi:10.1016/S1074-7427(03)00004-2
106. Stewart MG, Bourne RC, Steele RJ (1992) Quantitative autoradiographic demonstration of changes in binding to NMDA-sensitive [3H]glutamate and [3H]MK801, but not [3H]AMPA receptors in chick forebrain 30 min after passive avoidance training. Eur J Neurosci 4:936–943. doi:10.1111/j.1460-9568.1992.tb00120.x
107. Steele RJ, Stewart MG (1993) 7-Chlorokynurenate, an antagonist of the glycine

binding site on the NMDA receptor, inhibits memory formation in day-old chicks (Gallus domesticus). Behav Neural Biol 60:89–92. doi:10.1016/0163-1047(93)90145-8

108. Burchuladze R, Rose SPR (1992) Memory formation in day-old chicks requires NMDA but not non-NMDA glutamate receptors. Eur J Neurosci 4:533–538. doi:10.1111/j.1460-9568.1992.tb00903.x
109. Thomson A, Walker V, Flynn D (1989) Glycine enhances NMDA-receptor mediated synaptic potentials in neocortical slices. Nature 338:422–424. doi:10.1038/338422a0
110. Steele RJ, Dermon CR, Stewart MG (1996) d-Cycloserine causes transient enhancement of memory for a weak aversive stimulus in day-old chicks (Gallus domesticus). Neurobiol Learn Mem 66:236–240. doi:10.1006/nlme.1996.0064
111. Hood W, Compton R, Monahan J (1989) d-Cycloserine—a ligand for the *N*-methyl-d-aspartate coupled glycine receptor has partial agonist characteristics. Neurosci Lett 98:91–95. doi:10.1016/0304-3940(89)90379-0
112. Steele RJ, Stewart MG, Rose SPR (1995) Increases in NMDA receptor-binding are specifically related to memory formation for a passive-avoidance task in the chick—a quantitative autoradiographic study. Brain Res 674:352–356. doi:10.1016/0006-8993(95)00014-H
113. Steele RJ, Stewart MG (1995) Involvement of AMPA receptors in maintenance of memory for a passive avoidance task in day-old domestic chicks (Gallus domesticus). Eur J Neurosci 7:1297–1304. doi:10.1111/j.1460-9568.1995.tb01120.x
114. Ocklenburg S, Stroeckens F, Guentuerkuen O (2013) Lateralisation of conspecific vocalisation in non-human vertebrates. Laterality 18:1–31. doi:10.1080/1357650X.2011.626561
115. Moorman S, Nicol AU (2015) Memory-related brain lateralisation in birds and humans. Neurosci Biobehav Rev 50:86–102. doi:10.1016/j.neubiorev.2014.07.006
116. George I, Vernier B, Richard JP, Hausberger M, Cousillas H (2004) Hemispheric specialization in the primary auditory area of awake and anesthetized starlings (Sturnus vulgaris). Behav Neurosci 118:597–610. doi:10.1037/0735-7044.118.3.597
117. George I, Cousillas H, Richard JP, Hausberger M (2005) State-dependent hemispheric specialization in the songbird brain. J Comp Neurol 488:48–60. doi:10.1002/cne.20584
118. Hauber ME, Cassey P, Woolley SMN, Theunissen FE (2007) Neurophysiological response selectivity for conspecific songs over synthetic sounds in the auditory forebrain of non-singing female songbirds. J Comp Physiol Neuroethol Sens Neural Behav Physiol 193:765–774. doi:10.1007/s00359-007-0231-0
119. Bolhuis JJ, Zijlstra GGO, denBoerVisser AM, VanderZee EA (2000) Localized neuronal activation in the zebra finch brain is related to the strength of song learning. Proc Natl Acad Sci U S A 97:2282–2285
120. Phan ML, Vicario DS (2010) Hemispheric differences in processing of vocalizations depend on early experience. Proc Natl Acad Sci U S A 107:2301–2306. doi:10.1073/pnas.0900091107
121. Bolhuis JJ, Moorman S (2015) Birdsong memory and the brain: in search of the template. Neurosci Biobehav Rev 50:41–55. doi:10.1016/j.neubiorev.2014.11.019
122. Wang CZH, Herbst JA, Keller GB, Hahnloser RHR (2008) Rapid interhemispheric switching during vocal production in a songbird. PLoS Biol 6:2154–2162. doi:10.1371/journal.pbio.0060250
123. Schmidt MF (2003) Pattern of interhemispheric synchronization in HVc during singing correlates with key transitions in the song pattern. J Neurophysiol 90:3931–3949. doi:10.1152/jn.00003.2003
124. Cleland TA, Linster C (2003) Central olfactory structures. In: Doty RL (ed) Handbook of olfaction and gustation, 2nd edn. Marcel Dekker, New York, pp 165–180
125. Vallortigara G, Andrew R (1994) Olfactory lateralization in the chick. Neuropsychologia 32:417–423. doi:10.1016/0028-3932(94)90087-6
126. Rogers LJ, Andrew RJ, Burne THJ (1998) Light exposure of the embryo and development of behavioural lateralization in chicks: I. Olfactory responses. Behav Brain Res 97:195–200
127. Gagliardo A, Pecchia T, Savini M, Odetti F, Ioalè P, Vallortigara G (2007) Olfactory lateralization in homing pigeons: initial orientation of birds receiving a unilateral olfactory input. Eur J Neurosci 25:1511–1516
128. Gagliardo A, Filannino C, Ioale P, Pecchia T, Wikelski M, Vallortigara G (2011) Olfactory lateralization in homing pigeons: a GPS study on birds released with unilateral olfactory inputs. J Exp Biol 214:593–598. doi:10.1242/jeb.049510
129. Gagliardo A, Odetti F, Ioale P, Pecchia T, Vallortigara G (2005) Functional asymmetry of left and right avian piriform cortex in homing pigeons' navigation. Eur J Neurosci

22:189–194. doi:10.1111/j.1460-9568.2005.04204.x

130. Andrew RJ, Mench J, Rainey C (1982) Left-right asymmetry of response to visual stimuli in the domestic chick. In: Ingle J, Goodale M, Mansfield RJ (eds) Analysis of visual behavior. MIT Press, Cambridge, MA, pp 225–236
131. Vallortigara G (2000) Comparative neuropsychology: a stroll through animals' left and right perceptual worlds. Brain Cogn 43:15–16
132. Rogers LJ (2002) Lateralization in vertebrates: its early evolution, general pattern, and development. Adv Study Behav 31(31): 107–161. doi:10.1016/S0065-3454(02)80007-9
133. Rogers LJ, Andrew RJ (2002) Comparative vertebrate lateralization. Cambridge University Press, Cambridge
134. Concha ML, Bianco IH, Wilson SW (2012) Encoding asymmetry within neural circuits. Nat Rev Neurosci 13:832–843
135. Rogers LJ (1982) Light experience and asymmetry of brain function in chickens. Nature 297:223–225
136. Gurusinghe C, Zappia J, Ehrlich D (1986) The influence of testosterone on the sex-dependent structural asymmetry. J Comp Neurol 253:153–162. doi:10.1002/cne.902530203
137. Schwarz IM, Rogers LJ (1992) Testosterone: a role in the development of brain asymmetry in the chick. Neurosci Lett 146:167–170. doi:10.1016/0304-3940(92)90069-J
138. Rogers L, Rajendra S (1993) Modulation of the development of light-initiated asymmetry in chick thalamofugal visual projections by estradiol. Exp Brain Res 93:89–94. doi:10.1007/BF00227783
139. Freire R, van Dort S, Rogers LJ (2006) Pre- and post-hatching effects of corticosterone treatment on behavior of the domestic chick. Horm Behav 49:157–165. doi:10.1016/j.yhbeh.2005.05.015
140. Rogers LJ, Deng C (2005) Corticosterone treatment of the chick embryo affects light-stimulated development of the thalamofugal visual pathway. Behav Brain Res 159:63–71. doi:10.1016/j.bbr.2004.10.003
141. Madroñal N, Delgado-García JM, Fernández-Guizán A, Chatterjee J, Köhn M, Mattucci C et al (2016) Rapid erasure of hippocampal memory following inhibition of dentate gyrus granule cells. Nat Commun 7:10923. doi:10.1038/ncomms10923

Chapter 9

Tract Tracing and Histological Techniques

Felix Ströckens and Onur Güntürkün

Abstract

Since the 1970s, a multitude of studies has proven that brain asymmetries are not unique to humans, but a common feature in vertebrate and even in invertebrate species. While the majority of these studies have focused mainly on the behavioral aspect of these asymmetries, an increasing number of studies have also investigated the underlying brain structure of lateralized behavior. In this chapter, we will concentrate on summarizing studies that have applied histological methods to unravel the cellular basis of neuronal lateralization. In recent years, two methods have been of particular importance to this field. The first method, neuronal tract tracing, can be used to analyze possible asymmetries in the connectivity pattern of a given area. The second method, immunohistochemistry, has been developed to localize specific neurochemically specified components within tissue slices and has been used to identify asymmetries in number, position, or morphology of specific neuron types, or in the distribution of certain receptors. By providing a detailed description of the theoretical background, applications, required materials, as well as a troubleshooting guide for the most common problems for both methods, we would like to encourage scientists working in lateralization research to perform more studies on the anatomical basis of brain asymmetries.

Key words Neuronal tracing, Immunohistochemistry, Histology, Lateralization

Abbreviations

ABC	Avidin-biotin complex
CtB	Cholera toxin B
DAB	3,3′-Diaminobenzidine
DMSO	Dimethyl sulfoxide
FG	FluoroGold
GABA	Gamma-aminobutyric acid
HIER	Heat-induced epitope retrieval
HRP	Horseradish peroxidase
IEG	Immediate early gene
IHC	Immunohistochemistry
PB	Phosphate buffer
PBS	Phosphate buffered saline

Lesley J. Rogers and Giorgio Vallortigara (eds.), *Lateralized Brain Functions: Methods in Human and Non-Human Species*, Neuromethods, vol. 122, DOI 10.1007/978-1-4939-6725-4_9, © Springer Science+Business Media LLC 2017

PFA	Paraformaldehyde
PHA-L	Phaseolus vulgaris Leucoagglutinin
PIER	Protease-induced epitope retrieval
RITC	Rhodamine B isocyanate
WGA	Wheat Germ Agglutinin

1 General Introduction

It is widely accepted that neuronal lateralization is a common phenomenon in vertebrate and invertebrate clades [1]. There is a vast body of evidence showing lateralization of traits like hand/foot use [2, 3], eye and ear preferences [4, 5], turning direction [6, 7], and even on some more unusual traits like trunk usage in elephants [8], or tail wagging in dogs [9]. All of these asymmetries have been investigated in either purely observational studies or behavioral experiments under laboratory conditions, often using quite sophisticated experimental setups (compare Chaps. 3 and 4). Various experimental paradigms have also been devised to record and measure lateralized behavior in humans (compare Chap. 1 and 2). These experiments serve to identify and test novel lateralized behavior, thereby providing insight into the phylo- and ontogenetic scenario of functional asymmetries. It is clear, however, that behavioral studies do not identify the neuronal underpinnings of these lateralizations. To understand the neuronal correlates of behavioral asymmetries, we have to perform anatomical studies at the level of the macroscopic brain and at the microscopic cellular level. In some cases, studies on brain anatomy can even reveal left-right differences that are not directly observable on the behavioral level.

After the asymmetry of language functions was discovered by Pierre Paul Broca in 1865 [10], a large number of studies tried to reveal the neural foundations of left-right differences. Eberstaller [11] and Cunningham [12, 13] analyzed in detail the cortical surface anatomy of humans and other primates and discovered the asymmetry of the Sylvian fissure. Subsequent macro- and microscopic studies revealed much more intricate left-right differences in the language areas of the human cortex [14]. Much later the Golgi method was employed for asymmetry research. This technique visualizes all the dendrites and axons of a small fraction of neurons in the brain. Using this method, Scheibel et al. and Seldon revealed morphological cellular asymmetries of the anterior and posterior language areas, respectively [15–18].

These studies were primarily conducted on human brains, and occasionally on nonhuman primate brains. This was mainly due to the reason that until 1970s, it was generally believed that brain lateralization was a trait unique to humans, and therefore histological analysis was restricted to the limited anatomical knowledge gathered from the autopsied brains of naturally deceased humans.

Although gross anatomical asymmetries were known to be present in the diencephalon of some vertebrate species already in the beginning of the twentieth century, the majority of scientists ignored these findings during this time (for review see [19]). As a consequence, anatomical asymmetry research simply missed major technical breakthroughs of neuroanatomical analysis that can only be used on animal brains that have been properly fixed. These novel techniques were, for example, fluorescence microscopy (introduced in 1904 by August Köhler and Moritz von Rohr [20]), electron microscopy (1933 by Ernst Ruska [21]), phase contrast microscopy (1935 by Frits Zernike [22]), and confocal microscopy (1955 by Marvin Minsky [23]), which were further refined and are still frequently used in modern histological research. Also, labeling techniques took a large leap forward in allowing the marking of specific neuron types, proteins, and nucleotides within the brain. One of these methods, immunohistochemistry, will be explained in more detail later in this chapter. Despite these advances, the majority of lateralization research in humans has focused on less invasive techniques such as behavioral paradigms (see Chap. 1), EEG (Chap. 13) and more recently MRI, fMRI, and DTI (see Chap. 14). The only invasive technique used on humans was the famous sodium amytal studies by Juhn Atsushi Wada [24] that was employed to reveal language asymmetry in epileptic patients before brain surgery.

However, in 1971, experiments by Fernando Nottebohm in songbirds [25] and, less than a decade later, by Lesley Rogers in chicken [26] revealed that neuronal lateralization can be found in nonhuman species as well. In songbirds, Nottebohm found a lateralization in the production of vocalization in the peripheral nervous system. Shortly after, Rogers and colleague could for the first time show lateralization on brain level in a bird species, revealing an asymmetry in the visual system of young chickens [25–27]. These findings led to a surge in the studies focusing on lateralization in nonhuman species, proving the existence of lateralization in traits like limb preferences and conspecific vocalization in many vertebrate species (for review see [2, 28]).

Due to the improved availability of fresh brain tissue and the possibility of performing invasive experiments on the brains of these species, histological techniques also began to be applied more frequently starting in the 1980s. The follow-up studies by Rogers and colleagues were particularly noteworthy, in that they were able to perform extended tracing studies in the visual system of chickens and show a lateralization in the projection pattern of this system ([29, 30], see Sect. 2.1 for further details). In addition to tracing, immunohistochemistry was used to identify a variety of brain asymmetries in different species, ranging from left-right differences in cell size in pigeons [31] and lateralized neuronal activity in zebra finches [32] to receptor density asymmetries in rats

[33]. Other studies employed histological techniques like in situ hybridization to show, for example, an asymmetrical distribution of neurotransmitters in the habenula complex of zebrafish [34]. Also, classic histological staining techniques like cresyl violet, silver nitrate, or luxol blue were and are still being used in lateralization research and have provided many important insights into asymmetric organization of the vertebrate brain [35–37].

In this chapter we would like to describe in detail two major techniques that have been very successfully used in the microstructural analysis of asymmetries in animal models. The first method is tract tracing, which can be used to unravel fiber connections between two areas, or in case of trans-synaptic tracing, more than two areas. The second method we are going to describe is immunohistochemistry. With this technique, the distribution of proteins or peptides within brain tissue can be visualized. Both techniques have been used extensively to study the anatomical foundations of behavioral asymmetries, and we hope this chapter will enable the reader to perform additional studies in different animal models based on our descriptions.

2 Tract Tracing

2.1 Introduction

When trying to explain what causes a behavioral asymmetry, such as the dominance of one eye for a given task, one of the most plausible explanations would be a difference in innervation density. In this case, it is conceivable that either the dominant eye sends more projections to the brain, or that the areas processing information from the dominant eye are more strongly interconnected, leading to a more efficient information analysis than on the contralateral side. The tool to answer such questions is neuronal tract tracing. In this method a substance, the tracer, is injected into a target area to identify its connectivity. In general, tracers can be roughly divided in two groups: anterograde and retrograde tracers. Anterograde tracers, such as Phaseolus vulgaris Leucoagglutinin (PHA-L, for an overview of commonly used tracers see Table 1), are taken up by neurons or their axons and are transported towards axon terminals. These tracers will accumulate over time in these terminals and can later be visualized by either immunohistochemical techniques or by a fluorescence tag attached to the tracer. By mapping the marked terminals, the efferent connections of neurons in the injected area can be revealed. The second group is constituted by retrograde tracers (e.g. Cholera toxin B). Here the tracer is transported towards the soma of the neuron by being taken up by axons innervating the area, labeling the soma of cells projecting to the injected area, and effectively marking all afferents of the target region. Few tracers have anterograde as well as retrograde properties and therefore mark both efferent and afferent connections, for example Wheat-germ agglutinin (WGA).

Table 1
Selection of tracers commonly used for neuroanatomical studies

Tracer	Transport direction	Commonly used concentration	Citation
Cholera toxin B (CtB)	Retrograde	1 % in distilled water	[38, 39]
Dextran amines (biotinylated or fluorophor coupled)	Anterograde (+[a]) Retrograde	10 % in distilled water (+2–3 % DMSO)	[39–41]
DiI	Anterograde Retrograde	100 % (solid crystals) or in ethanol (saturated solution)[b]	[42]
Fast Blue (FB)	Retrograde	1–2 % in distilled water (+1 % DMSO)	[40, 43]
FluoroGold (FG)	Retrograde	2 % in distilled water	[38, 40]
Horseradish Peroxidase (HRP)	Anterograde Retrograde	20–50 % in Tris buffer (+2 % DMSO)	[44, 45]
Phaseolus vulgaris Leucoagglutinin (PHA-L)	Anterograde	2.5 % in PBS or TBS	[46, 47]
Rhodamine B isocyanate (RITC)	Anterograde Retrograde	1–2 % in distilled water (+1 % DMSO)	[43, 48]
Wheat Germ Agglutinin (WGA)	Anterograde Retrograde	2 % in saline	[49]

The direction of transport in relation to the injection site, as well as the most commonly used concentration of the tracers as described in the literature is given

[a]Dextran amines have better anterograde than retrograde tracing properties

[b]DiI is normally not injected as a fluid, but is placed in its crystalline form directly into the desired target area. It is often used in in vitro studies, in which brain tissue can be accessed more easily

In addition to transport direction, the mechanism by which a tracer is transported within an axon is of great importance, since it defines the speed of the tracing as well as properties of the labeling. After uptake of the tracer into an axon, which can happen either actively by endocytosis or passively by diffusion, there are principally two ways that a tracer can be transported. During active transport, the tracer accumulates in vesicles that are transported along the cytoskeleton towards either the soma or the axon terminal. Lipophilic tracers, like WGA, can bind to membrane molecules, which are reinserted into the cell membrane. This results in labeling of the cell membrane, including processes like dendrites [44]. Tracers like CtB stay in the vesicle and therefore produce granular-like labeling in the soma. Active transport tends to be fast, ranging from 100–400 mm/day. In contrast, passive transport is rather slow. It relies on lateral diffusion within the plane of the axonal membrane and tends to transport only a few millimeters per day. This mechanism is used by tracers such as DiI and DiA.

The advantage here is that in contrast to active transport, passive transport is not ATP dependent, and can also take place in dead tissue. In addition, the labeling is membrane bound due to the membrane binding properties of the tracer. For the interested reader, detailed reviews of tracer uptake and transport mechanism have been published by Köbbert et al. [44] and Oztas [50].

In addition to these classic tracers, a new generation of viral tracers has been developed, which have the distinct advantage of crossing synapses and therefore being able to mark circuits of neurons [51–53]. However, we are not going to cover the specifications of viral tracings in this book chapter, because this would fill a whole chapter on its own. For interested readers, we can recommend a very recent article by Nassi et al. [54], in which viral tracing is reviewed in detail.

First, let us return to the eye dominance example in the beginning of this section. Some bird species like chickens, pigeons, quails, and parrots do indeed show a dominance of one eye for specific tasks [4, 55–57]. Especially in chickens and pigeons, a large number of studies have been performed to investigate the anatomical basis of this lateralization. Both chickens and pigeons take an asymmetrical position inside the egg during embryonic development, causing a dominance of the right eye/left hemisphere in post-hatch chickens and adult pigeons for various discrimination tasks [58, 59]. Several tracing studies have shown that these behavioral lateralizations are accompanied by an asymmetry in the ascending visual projections within the brain. In chickens, the thalamofugal visual system, which is comparable to the geniculo-cortical visual system in mammals [60], is lateralized [30, 48]. Rogers and colleagues injected the retrograde tracers Fluoro-Gold (FG) and True Blue into the Hyperpallium of juvenile chickens and counted labeled cells within the thalamic nucleus geniculatus lateralis pars dorsalis (Gld). They found that projections from left Gld to the right Hyperpallium were in greater number than from right Gld to the left Hyperpallium [30]. The second avian visual system, the tectofugal system (comparable to the extrageniculo-cortical system in mammals), however, showed no lateralization [48]. The corresponding experiments in pigeons were performed by Ströckens et al. [38]. In contrast to chickens, injections of the tracers CtB and FG into the Hyperpallium of pigeons did, however, not reveal a lateralization within in the thalamofugal pathway ([38], see Fig. 1). Instead, tracer injections of retrograde tracer into the nucleus rotundus proved an asymmetry in the projections from the optic tectum to this area. These tectofugal asymmetries comprise more fibers running from the right tectum opticum to the left nucleus rotundus than vice versa [61]. Together, these neuroanatomical studies revealed that in chickens the thalamofugal system is lateralized, while in pigeons the tectofugal system shows a lateralization. Thus, although the behavioral

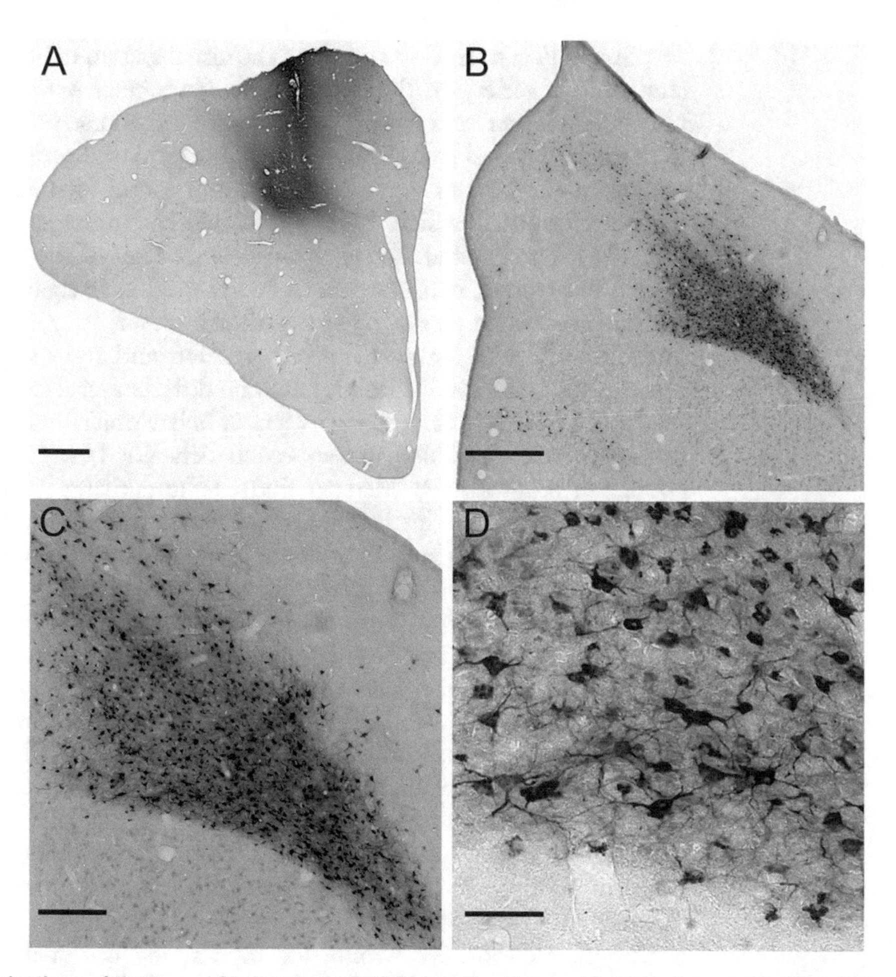

Fig. 1 Injections of the tracer Cholera toxin B (CtB) into the Hyperpallium of pigeons (**a**). After 48 h, animals were perfused and brain slices were immunohistochemically labeled against CtB. Labeling revealed numerous labeled cells in the diencephalic nucleus geniculatus lateralis pars dorsalis (Gld) on the ipsi- (**b**) and contralateral side. (**c**) depicts a higher magnification of the labeling shown in (**b**). Note the clear border between Gld containing labeled neurons and the nucleus rotundus directly below the Gld containing no labeled cells at all. Although both the Gld and the nucleus rotundus are visual relay areas projecting to the telencephalon, only the Gld has projections to the Hyperpallium. (**d**) shows the typical labeling pattern after accumulation of CtB in traced neurons. CtB is taken up by endocytosis and vesicles containing CtB are transported retrogradely to the neurons soma. Depending on the amount of CtB-filled vesicles in the soma, the labeling appears either granular (few vesicles) or can fill the whole soma (soma filled with vesicles). Since the vesicles cannot enter the nucleus, no nuclear labeling will occur. Scale bars indicate 1000 μm (**a**), 500 μm (**b**), 200 μm (**c**), and 50 μm (**d**)

asymmetries between chickens and pigeons appear comparable at first glance, the underlying structural asymmetries differ tremendously—possibly even indicating different processing strategies for similar tasks in between these species [62]. This underlines how important studies of the neuronal architecture are for the understanding of cerebral asymmetries.

In addition to tracing studies in the visual system of birds many studies in a variety of different species have been performed to uncover whether and which asymmetries in connectivity could underlie behavioral lateralizations. Galuske et al. did post mortem tracing in the human temporal cortex and found asymmetries in neuron clustering, which might be related to language asymmetries [63]. Patzke et al. analyzed the connectivity of the olfactory system in pigeons, which seems to be involved in navigation tasks with a dominance of the right nostril. However, they could not find differences in connectivity between left and right side, suggesting that this lateralization is based on different structural asymmetries [39]. Zebra fish show a variety of behavioral lateralizations, in particular for eye use and approach behavior [64–66], and a couple of studies have tried to find the underlying anatomical mechanisms causing this lateralization. It was discovered that the habenular complex seems to play a major role in these behavioral lateralizations, and it was furthermore shown that this complex shows several structural asymmetries (for review see [67]). As an example, Hendricks and Jesuthasan performed tracing studies on the connectivity of the habenula to the forebrain in larval zebra fish. They found that subsets of forebrain neurons terminate only in the right habenula, irrespective of the hemisphere these neurons are located in, and postulated that this could have an impact on the lateralized behavior in adult animals [42].

These examples make it obvious that neuronal tracing is a powerful method in lateralization research, enabling scientists to uncover the foundations of behavioral lateralizations. In the remaining sections, we would like to describe in detail what is required to run such a study, and how to perform neuronal tracings. We will also provide some tips on how to deal with problems during application of the method, which are not typically mentioned in published manuscripts. These hints are based on years of experience with neuronal tracing and hundreds of tracing experiments. Although these studies were mainly performed in pigeons, the method can be easily altered for use in other species, and we would like to encourage readers to apply these techniques in other model species to broaden our knowledge on brain asymmetries.

2.2 Materials

For a precise injection of the tracer into a specific brain area, a brain atlas is needed—or at least the coordinates of the target area of injection. Furthermore, a stereotactic apparatus is required in almost all cases. In a large-brained species with a broader target area, making injections manually only using visible landmarks might be possible, but we do not recommend this. For most model species, like mice, rats, or monkeys, stereotactical devices are commercially available. These devices allow injections with a deviation from the desired coordinates of only a few micrometers in all three dimensions and are therefore even suited for injections into small

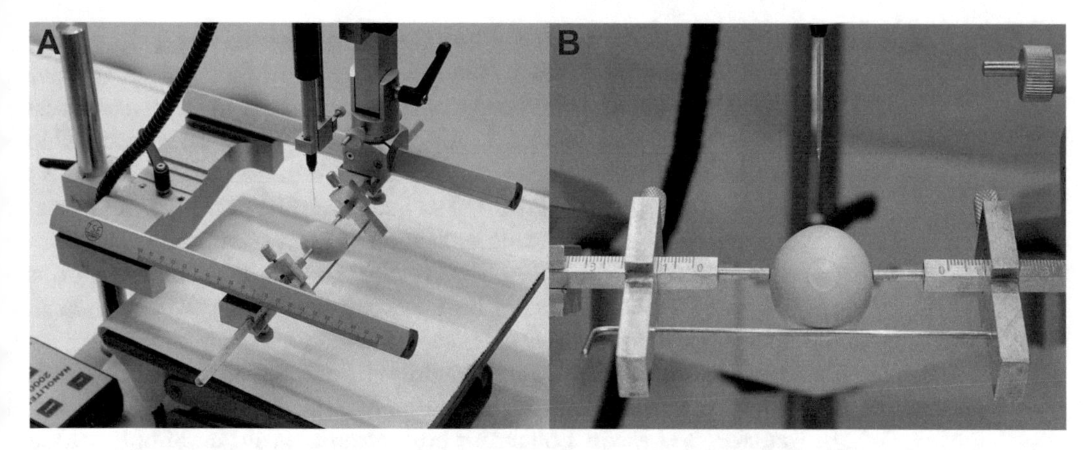

Fig. 2 (**a**) Stereotactic device used for the injection of tracers into the brain of pigeons in our lab. A plastic spacer was used to simulate the positioning of the pigeons head in the device. Since stereotactic devices for nonmodel species like pigeons are in most cases not commercially available, we purchased a device built for rats and modified it to fit our needs (parts made out of brass). By exchanging the standard ear bars to the modified version which fits the pigeon ears and by adding a beak holder, the device is now perfectly suited to perform very precise tracer injections in pigeons. (**b**) shows a closer view of the custom made adjustments. The brass ear bars touching the plastic spacer are well visible. We attached an adaptor made out of brass in a 45° angle to the individual ear bars. This adaptor also attaches a stainless steel rod which accommodates the beak between the upper and lower parts of the beak. The 45° angle strategically positions the injections according to the plane used in the standard pigeon brain atlas [68]

nuclei. In our lab, we use a frame designed for rats, and then adapted the size of the ear bars and replaced the head holder with a pigeon-specific beak holder (Fig. 2).

For the injection itself, we recommend using a nanoliter injector in conjunction with glass capillaries for small-brained species like pigeons, chickens, rats, and mice. For species with bigger brains, Hamilton syringes can be used, because the injection volume is likely to be higher. However, it is not advisable to do the injections manually. As described in the method section below, the tracer should be injected stepwise in small volumes, which is very difficult to achieve for nanoliter volumes without technical help. There are dozens of different pumps or injectors for exactly this purpose on the market, which should be used to guarantee optimal results. When using a nanoliter injector, the application of tracer is done with glass capillaries. These glass capillaries can either be bought or self-made from glass templates using a micropipette puller. The advantage of self-made capillaries is that one can adjust the diameter of the capillary according to tracer properties and desired injection volumes. For tracers like CtB or dextran amines, inner capillary diameters of 10–15 μm are sufficient, while tracers with a higher viscosity like Fluoro-Gold require diameters up to 20 μm. In general, one should perform a test run with the injector before the surgery, to check if the diameter chosen allows exact application without leaking or clotting of the capillary.

The concentration and handling of the tracer is highly dependent on the tracer properties. Giving an overview of all available tracers would go beyond the scope of this chapter. In most cases tracers are injected in a concentration between 1% and 10% in distilled water, with some tracers requiring the addition of a solvent like Dimethyl sulfoxide (DMSO) to facilitate diffusion into the tissue. An overview of applicable concentrations of commonly used tracers is given in Table 1. Furthermore, comprehensive descriptions of many available tracers have been published by Köbbert et al. [44] and Schofield [69].

The amount of tracer is mainly defined by the area one intends to trace. Injections into small areas like the mouse's lateral geniculate or the pigeon's olfactory bulb require a volume of only 100 nl [39, 70]. The visual Wulst of chickens or pigeons requires an amount of between 100 and 200 nl [38, 48], while injections covering whole cortical areas, even in small animals like mice, require 300 nl [70].

For surgery, anesthetize the animal and monitor the animal's vital signs during surgery. Because surgery needs can vary drastically between species, we will only briefly outline the general surgical equipment we use in pigeons. For anesthesia, we use a 7:3 mixture of ketamine/xylazine in combination with isoflurane gas for anesthesia induction (see below). In addition to an electric medical bone drill, we use a standard set of surgical tools including scalpel, forceps, scissors, needles, etc. It is advisable to use a surgery microscope with at least 4x magnification during the whole surgery procedure in order to have a clear view of the entry point of the capillary and to also check the tracer volume within the capillary. The animal's body temperature typically drops during lengthy surgeries, so the use of a heat blanket placed under the animal is advisable. After the surgery, the application of painkillers (in pigeons we use Caprofen) and, if necessary, antibiotics are highly recommended.

For collection of the brain, we strongly recommend a transcardial perfusion to avoid protein degeneration. For this, a perfusion pump suited to the animal of choice is helpful. However, an IV-set or a big syringe with an appropriately sized needle can substitute for a perfusion pump. For perfusion, a 0.9% Sodium chloride (NaCl) solution in water and a 4% paraformaldehyde solution (PFA) in phosphate buffer (PB, 0.12 M, pH 7.4) are required. Furthermore, a post fixation solution (4% PFA + 30% sucrose) and a cryoprotection solution (PBS + 30% sucrose) are needed. For the perfusion itself, a surgical tool set, including appropriate tools for opening of the breast cavity and the skull, are necessary. To obtain brain slices suited for labeling of the tracer, we suggest using a microtome or cryostat with the ability to cut slices with a thickness of 30–40 μm. The chemical TissueTek, or a comparable product, is used to freeze the brain on to the microtome. A set of brushes

(#2–#8 round, soft hair) is needed to collect and transport slices. To prevent fungal growth, slices should be stored in PBS + 0.1% Sodium azide. The equipment needed to visualize the tracer depends greatly on the tracer type. For tracers already containing a fluorescent tag, a fluorescence microscope with a suitable filter set is sufficient. However, if the tracer has to be detected by immunohistochemical means, additional processing and equipment are required. A detailed description of immunohistochemistry, including a paragraph on the needed materials, is provided in the second part of this chapter.

2.3 Method

When planning to perform a tracer experiment, the first step should always be a precise determination of the target area. In the case of a small target with a volume of 1 mm^3, a single injection site is generally sufficient. The injection site should be centered in the middle of the targeted area to avoid tracer spreading to neighboring areas. However, if the area is larger than 1 mm^3, one should split injections over several injection sites with a distance of 1 mm between each of the three dimensional directions. It is advisable to keep a safe distance from neighboring areas to ensure specific tracing. Aside from the actual injection sides, the entry points of the capillary into the brain should also be chosen with care. In general, the distance the capillary has to traverse should always be planned to be as short as possible. The reason for this is that leaking of the tracer from the capillary can never be prevented completely, resulting in unintended labeling of projections in bypassed areas. By reducing the distance the capillary has to go, false positive tracing results are reduced. However, in some cases, the most direct route from the brain surface to the target is blocked by structures that should not be damaged (like eyes and ears) and thus forces one to look for a longer, indirect injection path to the target area. In this case, the chosen injection path should traverse areas with a different projection pattern than the target area. Also avoid hitting ventricles, because tracer leakage into the ventricle can lead to uncontrolled spread of the tracer to all areas next to the ventricular space. Also, avoid hitting big blood vessels in order to prevent bleeding, which can cause a multitude of issues (e.g. brain damage, uncontrolled distribution of tracer, problems during perfusion, nonspecific staining during immunohistochemistry). If injections at multiple locations are planned, try to reduce the number of reentries of the capillary into the brain to an absolute minimum, since every reentry bears the risk of additional tracer leakage and permanent damage to the brain. It is also worth noting that it is important to check whether the used stereotactic apparatus is actually able to reach the designated entry points since very laterally set areas often cannot be reached by many commercially available setups.

Although there are many methods of injecting tracer in the brain (Iontophoretic injections, insertion of dye crystals), we will focus here on pressure injections using a nanoliter injector (for information on other injection techniques see [44, 50]). First, connect a glass capillary (either self-made or bought) with an inner diameter that fits the injector. The diameter is determined by the tracer that will be used, and should be around 10–15 μm for tracers with low viscosity and around 20 μm for tracers with high viscosity. Fill the capillary with a fluid that does not mix with the tracer. If the tracer is dissolved in an aqueous solution, a hydrophobic fluid should be used. We typically use a commercially available mineral oil. Avoid air bubbles at all costs, because air is compressible and can have an impact on the volume to be injected. Release a small amount of oil from the capillary and refill the empty volume with tracer. The amount of tracer should be a little higher than needed, in order to avoid injections of oil into the brain. When working with viral tracers, additional preparation steps are required as outlined by Beier and Cepko [70].

Anesthetize the animal, fixate the animal's head in the stereotactic system, remove the skin covering the skull over the designated injection site and use a bone drill to drill a hole until the dura mater is reached. The hole should be large enough to still see the tip of the capillary when it touches the brain. This is necessary to calculate the depth of penetration to hit the target area. If necessary, carefully incise the dura mater to allow access to the brain. In small animals, the dura mater is so thin that the capillary can traverse it easily such that no incision is needed. Take care not to damage any major blood vessels, because bleeding can block the view on the injection site. If required, blood vessels can be carefully pulled away by using a blunt hook. Then, the capillary should be moved slowly into the brain until the desired depth is reached. Before starting with the injection, one should wait for at least 10 min in order to relieve the sheering of brain tissue due to the entry of the capillary. Never inject the entire volume of tracer at once, because the mechanical pressure caused by large volumes of liquid can damage tissue. Inject 10 nl at a time, pausing for 5 min between each injection. After injections are done, wait for another 10 min and then slowly remove the capillary from the brain. If needed, make further injections following the same injection procedure. Close the wound and administer an analgesic and/or antibiotic if needed. The amount of time required between surgery and sacrifice of the animal depends on the speed of the tracer transportation (see above) and the distance of the injected area to its projection targets. A tracer such as CtB, with a transport speed of 200 mm/day can basically reach all areas in a small brain (such as pigeon or mouse) within 2 days. When calculating survival time, remember that axons usually do not take the most direct route, effectively increasing the distance traveled by the tracer. Survival time also has an effect on the labeling pattern of most tracers. While short survival times lead to a rather weak labeling within the projection targets and more labeling of fibers running to this target, long survival times will lead to strong labeling of the projection target and fewer labeled projection fibers (this is only the case for tracers being actively transported and not incorporated into the membrane). One should not extend the survival time for much longer than the tracer needs to reach its target, because some tracers will degrade over time.

For removal of the brain, deeply anesthetize the animal and perform a transcardial perfusion using saline followed by 4% PFA. Again, this procedure is highly dependent on the species and should be performed according to species-specific protocols. After successful perfusion, place the brain in a 30% sucrose solution for at least 24 h for cryoprotection. Then, use a microtome or cryostat to make 30–40 μm wide brain slices. If the tracer has a fluorescent tag, slices can be directly embedded in a fluorescence-compatible embedding medium, cover slipped, and analyzed using a fluorescence microscope. Tracer with no such labeling requires additional techniques such as immunohistochemistry for visualization, which is described below. Due to inter-individual differences between animals and slight differences in equipment between laboratories, there is, of course, no guarantee that every tracing study performed as described above will work out perfectly.

2.4 Troubleshooting

In most publications, only successful applications of a given technique are reported, neglecting potentially dozens of trials in which the method did not work perfectly. This is also true of tracing experiments, where a multitude of problems can occur that reduce the quality of the tracing or entirely prevent successful tracing. The following sections are aimed at identifying such problems and to help in overcoming them in an efficient manner.

2.4.1 Tracer Leakage

One significant problem is the risk of tracer leakage from the capillary during lowering of the capillary into the brain or the spread of the tracer into areas neighboring the injection site. In both cases, the tracer will also be taken up by neurons and the neuropil outside of the injection site. This can result in false-positive result patterns, which could potentially skew the interpretation of data.

Leaking of tracer from the capillary can have several causes, which in most cases can be prevented quite easily. The most common reason is that the inner diameter of the capillary is too large. In particular, when tracers in an aqueous solution with a very low viscosity, are being used, capillary action and hydrogen bonds might not be strong enough to resist gravity in capillaries with a large diameter. This will cause the tracer to slowly leak from the capillary in an uncontrolled way. Before starting the experiment, check the tip of the filled capillary with a microscope. If drops are forming, switch to a different capillary with a smaller diameter. In addition, it is advisable to use the injector's fill function to create a negative pressure inside the capillary. In doing so, one has to be careful to avoid suctioning in of any air. It is also advisable to reduce handling of the capillary to an absolute minimum, as even a small amount of jarring can result in tracer loss.

As described above, a more indirect way to control for tracer leakage is intensive planning of the injection pattern. Try to reduce the number of entries into the brain, take the most direct path to the injection site, and avoid paths through areas that have similar projection patterns. In addition, a combination of retrograde and anterograde tracing can be used to double check the specificity of labeled connections.

2.4.2 Tracer Spread not Sufficient

Sometimes, a tracer spread within the tissue is insufficient for proper labeling. In such a case, only a few projections will be labeled, and the portion of labeled projections might not be representative of the projection pattern for the whole area. The most obvious reason for this could be an injection volume that is too small, and this can be taken care of quite easily. Most of the time, however, this will not be the case if the injection pattern was planned adequately beforehand. A common reason for this problem is that the tip of the capillary becomes blocked and the tracer cannot exit the capillary. This can happen either because of clotting of the tracer—especially with high-viscosity tracers, or because of remnants of brain tissue, blood, dura, or cerebrospinal fluid during multiple injections. To prevent this from happening, it is advisable to observe the capillary during injections with a surgery microscope. If the meniscus of the tracer does not move while injecting, it is likely that the capillary is blocked. In some cases, it is possible to remove the block by very carefully wiping the capillary tip with soft tissue and some water without breaking the tip. If this does not work, it is also possible to carefully cut off just a few micrometers of the tip using microscissors. However, this bears the risk of tracer leakage (see above) and pieces of broken glass or sharp edges damaging the brain. Therefore, always check the tip after cutting with a surgical microscope for leakage and broken glass particles.

In addition, the characteristics of the injected tissue can have an impact on tracer distribution. In our experience, high cell densities, bypassing fiber tracts, and glial-rich lamina all hinder tracer diffusion. Furthermore, glial scars occurring after brain tissue damage [71], for example after long-term electrophysiological recording, can prevent tracer diffusion quite significantly. Adding a low percentage of the solvent DMSO to the tracer solution can reduce this problem, as DMSO enhances permeability of membranes [72]. As mentioned in Table 1, adding DMSO has been applied successfully in a variety of studies.

2.4.3 Problems with Visibility of Tracers

Even after successful tracer injection it is possible that traced cells or fibers are not visible as expected. This can have a couple of causes. It is possible that the survival time of the animal was not long enough, and the tracer simply did not reach its target destination. This can be checked quite easily, because the tracer should be visible in fibers somewhere between the injection site and target area. In general, this problem can be avoided if the transportation speed of the tracer and the distance between injection site and projection areas are known, as this allows calculation of the survival time needed. It is advisable to add some additional time to the estimated survival time to ensure that most of the tracer reaches its target.

If tracers with a fluorescent tag are used, it is possible that the fluorescence has partly faded, either through exposure to light or by using chemicals that destroy the tag or quench the fluorophore [73]. In addition to being careful to prevent fading, additional immunohistochemical labeling of the tracer can bypass this problem. For most tracers, antibodies detecting the tracer are available commercially and can be used to make the tracer visible (see Sect. 3 for further information).

2.4.4 Issues When Using Two Tracers

Some experiments require the use of two tracers at the same time. This can, for example, be the case when trying to figure out if two nuclei have the same projection area or receive input from the same area [74]. Furthermore, in lateralization research one frequently tackled question is whether projection strength of a given area differs between hemispheres. In this case, simultaneous injections of two different tracers, one into the left hemisphere and one into the right, allows intra-individual comparisons and reduces the number of animals needed [38, 48]. Usage of two tracers at the same time though does create some challenges. Because tracer properties such as uptake speed and transportation rate vary greatly (see Sect. 2.2 and [44]), the number of labeled cells/axon terminals between two tracer injections can differ strongly, even if the same amount of tracer was injected in the exact location on left and right side [38, 43]. This can be prevented altogether by simply using the same tracer in both locations, each with a different fluorescent tag. Although this should produce comparable numbers of labeled cells on the left and right sides, it has the disadvantage of fading fluorescence, which cannot be coped with by later immunohistochemistry (the antibody will label both tracers irrespective of injection site). A different approach to minimize this problem is the introduction of correction factors. Repeated injections of a tracer into different animals under identical conditions will lead to a stable average number of labeled cells/axons in a given area, such that the acquired mean number of labeled cells/axons between two tracers can then be used to calculate a correction factor. As an example: An injection of tracer A results in a mean of 1.2 times more labeled cells than tracer B. By multiplying the cells labeled by tracer B with 1.2, a direct comparison of the number of cells labeled by tracer A and B is possible. However, such correction factors should be acquired very carefully using a high number of injections to validate them [38, 43].

3 Immunohistochemistry

3.1 Introduction

As described in the beginning of this chapter, even basic histological techniques were and still are powerful tools in lateralization research. Amongst others, commonly used stains like cresyl violet for general neuron labeling (by labeling of the rough endoplasmic reticulum), or silver nitrate applications and luxol fast blue staining for visualization of fibers (for review see [73]) can be useful for unraveling structural lateralizations in less studied species as well [35]. However, these techniques suffer from the problem of producing a rather nonspecific staining. Cresyl violet, for example, labels neurons irrespective of morphology, localization, and function. Furthermore, it also labels non-neuronal cells [75]. A solution to this problem was first developed in 1941 by Coons and colleagues [76]. Based on John Marrack's work on antigen-antibody interactions [77], they coupled the fluorescent compound anthracence isothiocyanate to an antibody against *Streptococcus pneumoniae* and were able to successfully label pneumococcus strains with it as a result [76]. Further experiments showed that these antibodies were also able to label tissue infected with this bacterium [78]. This was the first time a very specific cell type could be reliably labeled within tissue, with precise enough labeling to even identify individual pneumococcus strains. Subsequent studies used antibodies against a variety of different antigens, and the introduction of enzyme-coupled antibodies has since allowed cheaper and more stable labeling (for review see [79, 80]). These tissue stains, based on antibody-antigen interactions, are referred to in general as immunohistochemistry, or if applied in cell cultures, immunocytochemistry. In the final sections of this chapter, we would like to focus on this method to explain its theoretical background, its relevance for lateralization research and to give instructions on how to perform immunohistochemistry-based experiments.

3.1.1 Theoretical Background of Immunohistochemistry

In principle, immunohistochemistry (IHC) is based on the highly specific binding of an antibody to a single antigen in order to label this binding [81]. In most cases, an antigen is a protein or peptide that can be recognized by the antigen-binding fragment of an antibody [80]. In other words, the antigen represents the target of a given antibody. The antigen can be a receptor for a neurotransmitter, the neurotransmitter itself, components of the cytoskeleton, or even substances artificially injected into the brain [31, 38, 41, 82, 83]. By choosing an antigen which is specific for a given cell population, IHC allows labeling of only this cell population (e.g. all GABAergic neurons could be labeled by using an antibody against the neurotransmitter gamma-aminobutyric acid (GABA) [31]). Normally, the antibody does not bind to the whole antigen, but to a small sequence of 5–6 amino acid residues, called epitope [84]. Because antigens are typically much larger than this, they can in theory be targeted by many different antibodies, each specific for a single epitope on the antigen. Currently, antibodies against almost all antigens can be manufactured and are commercially available for any known antigen amino acid sequence. In general there a two major types of antibodies: polyclonal and monoclonal antibodies. Polyclonal antibodies are produced by injections of a given antigen into a host animal (mostly mice, rabbits, or goats). The immune system of the host animal will start to produce antibodies against the foreign agent, and these antibodies will accumulate in the blood serum. The serum can be collected, purified by various methods, and then be used for IHC. Under optimal conditions the serum will contain a mixture of antibodies directed against different epitopes of the antigen with which the host was injected. These polyclonal antibodies are inexpensive to produce and have a high sensitivity for the antigen, because they can recognize more than one epitope. However, their specificity is not as narrow as that of monoclonal antibodies. The reason for this is that not all epitopes are antigen specific (e.g. some membrane binding domains can be found in a couple of different proteins) and insufficient purification in production can lead to remnants of host antibodies in the serum, which are not directed against the injected antigen, but might label other antigens instead [85, 86].

Monoclonal antibodies are also produced by injecting a host animal with an antigen. But instead of collecting serum, antibody-producing lymphocytes are collected and fused with myeloma cells. The hybrid cells produced as a result are immortal, and can be cultivated in cell cultures to produce antibodies against a specific epitope. By selecting cell cultures that produce antibodies against the injected antigen, extremely pure antibody solutions can be harvested. Because the production of monoclonal antibodies is much more complex, and the acquired number of antibodies is rather low, monoclonal antibodies are typically much more expensive. However, monoclonal antibodies are much more specific for their antigen, because they bind to only one epitope. This comes at the cost of lower sensitivity, because obstruction of this single epitope, for example during fixation, can prevent any antibody binding at all [85, 86].

Although specific binding of the antibody to the antigen is a crucial part for successful IHC, the second part of the method—the labeling of the antigen-antibody binding—is no less important. As described above, the first applications of IHC used fluorophores directly coupled to the antibody [78]. This simple method is still

Fig. 3 Neurons in the pigeons substantia nigra pars reticulata, containing both GABA and parvalbumin (*yellow labeling*). Neurons were stained by IHC using polyclonal antibodies against GABA and parvalbumin simultaneously. Labeling was achieved by the indirect labeling method using secondary antibodies coupled to a fluorescent tag. For GABA, a secondary antibody coupled to the fluorophore Alexa488 (*green*) was used, while the antibody against parvalbumin was labeled using a secondary antibody coupled to the fluorophore Alexa594 (*red*). Courtesy of picture by Martin Stacho

applied in many modern IHC experiments [87], although the variety and quality of the fluorophores has increased drastically since the 1940s. Antibodies with different colored fluorescent tags are available for thousands of different antigens, allowing labeling and visualization of multiple antigens at the same time (see Fig. 3).

The direct coupling of fluorescent tags to antibodies, however, does present several disadvantages. As described in Sect. 2.4.3, fluorophores can fade over time and are therefore not suited for permanent labeling or long term analysis. A second problem is low signal intensity. Especially for monoclonal antibodies, the number of antibodies bound to an antigen is very low, because only one antibody can bind per epitope and monoclonal antibodies only bind to one specific epitope. To solve such problems of direct antibody labeling, more indirect labeling methods have been developed [88]. The indirect labeling simply adds a further layer of antibodies to the method. These secondary antibodies are directed at the primary antibody (in contrast to primary antibodies, which are directed at epitopes of the antigen). The secondary antibodies are usually polyclonal antibodies aimed at detecting a variety of epitopes specific to the host species of the primary antibody. As several secondary antibodies can bind to the primary antibody, signal intensity can be increased drastically. Presently, the majority of IHC experiments use indirect labeling to visualize antigens (e.g. [32, 38, 89, 90], compare Fig. 4).

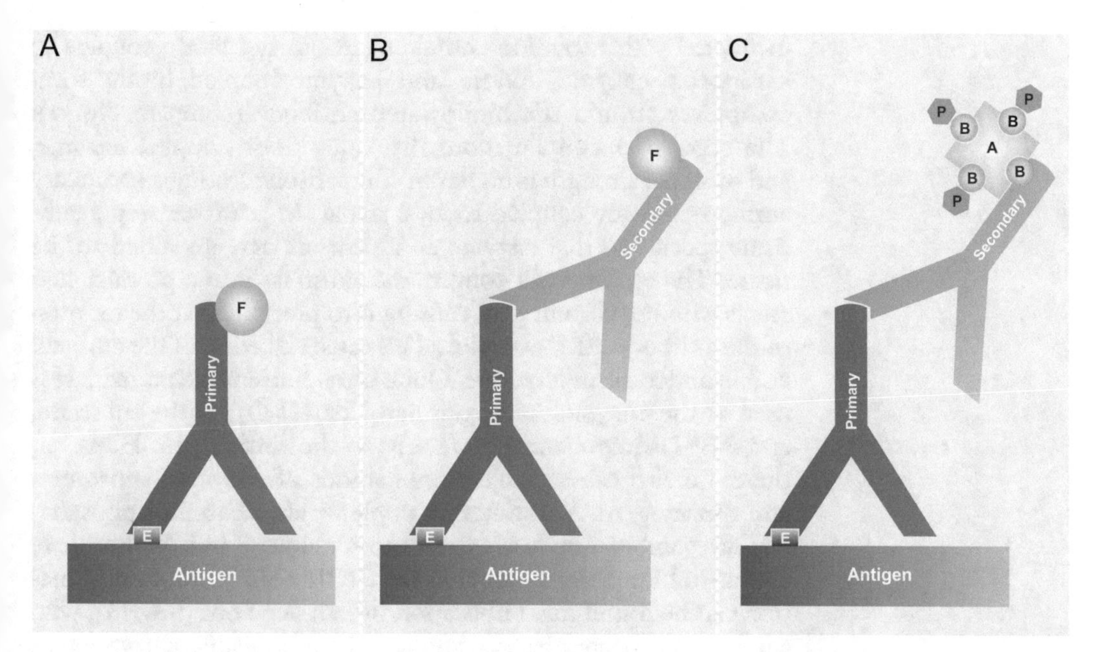

Fig. 4 Schematic, simplified depiction of three IHC labeling methods. (**a**) depicts the direct labeling method using a fluorescent labeling. A fluorophore is directly coupled to the primary antibody to visualize the position of the antibody bound to an epitope of the antigen. In (**b**) and (**c**), two indirect labeling methods are shown. The indirect method relies on a secondary antibody carrying a label which is able to detect the primary antibody. In (**b**), the secondary antibody is directly coupled to a fluorophore. In (**c**) the ABC method is used to label the antibody complex. Here, the antibody is coupled to a biotin molecule. By adding avidin and peroxidase-coupled biotin to the antibodies, an avidin-biotin-peroxidase complex will form around the secondary antibody. In presence of hydrogen peroxide, the peroxidases will trigger precipitation of specific chromogens (e.g. DAB) at the location of the antibodies, thus labeling the antigen. Note that for the sake of clarity, the image was simplified. In reality, each primary antibody can be bound by multiple secondary antibodies. Furthermore, the ABC complex consists of many biotin-coupled avidin molecules. Modified from Ramos-Vara and Miller [84]. *A* avidin, *B* biotin, *E* epitope, *F* fluorophore, *P* peroxidase

To avoid signal loss due to fading of the fluorophore, several antibody visualization methods have been developed that do not rely on fluorescence. Reviewing all of these methods here would be beyond the scope of this chapter, so we will only focus on a few of them. A detailed overview can be found in Ramos-Vara and Miller [84]. In general, these methods all rely on coupling an enzyme or enzyme complex to the secondary antibody, which catalyzes the reaction between a substrate and a chromogen to produce a distinct color, which precipitates at the location of the antigen-antibody complex. One of the most commonly used procedures is the avidin-biotin-complex method (ABC method [91]). Avidin is a glycoprotein with four binding sites for the molecule biotin, also known as vitamin B_7. Biotin itself can be bound via a different binding site to antibodies and reporter enzymes. For IHC, a secondary antibody coupled to biotin (a biotinylated antibody) is incubated with a solution containing avidin and biotin coupled to a reporter enzyme. Avidin and enzyme-coupled biotin form complexes around the biotinylated antibody (compare Fig. 4). The complexes contain more than one biotin-coupled enzyme, and so signal strength is higher in comparison to using a secondary antibody directly coupled to an enzyme. In a further step a substrate specific to that enzyme and a chromogen are added to the tissue. The enzyme will convert the substrate into a product that reacts with the chromogen, causing it to precipitate at the location of the antibody-ABC complex. This causes labeling of the antigen, visible under the microscope. Quite often horseradish peroxidase is used as the enzyme, hydrogen peroxide (H_2O_2) as the substrate, and 3,3′-Diaminobenzidine (DAB) as the chromogen. However, there are also other viable combinations of enzymes, substrates, and chromogens. A common example would be alkaline phosphatase in combination with 5-bromo-4-chloro-3-indolylphosphate and nitro blue tetrazolium chloride (BCIP/NBT). More information can be found in a publication by van der Loos [92, 93], who summarized properties of the majority of available chromogens. For more detailed information on the theoretical background of IHC, we can highly recommend the recent review articles by deMatos et al. [80] and Ramos-Vara and Miller [84].

3.1.2 Immunohistochemistry in Lateralization Research

Let us use the example of the GABAergic neuron labeling experiment given above to clarify the IHC procedure further: Manns and Güntürkün investigated whether asymmetrical light stimulation of a pigeon embryo before hatching causes not only projection asymmetries, but also asymmetries in GABAergic neuron sizes in the visual system ([31], refer to the first half of this chapter). Left-right differences in soma size were used as a proxy for lateralized ramifications of GABAergic neurons within the local neuropil. To identify GABAergic neurons, they applied a polyclonal anti-GABA antibody, raised in a rabbit, to pigeon brain slices. For labeling, they employed the indirect method by using a secondary biotinylated anti-rabbit antibody, able to detect the anti-GABA antibody hosted in rabbits. To visualize the antibody complex, the ABC method was used. After sequential incubation of the tissue with the antibodies, a mixture of avidin and biotin-coupled peroxidase was added to the slices. The substrate H_2O_2 and the chromogen DAB were chosen to work with. Hydrogen peroxide, H_2O_2, was not added directly, however, but supplied via an additional reaction using β-D-glucose and glucose oxidase. This can be done to achieve a slower, more controlled staining reaction. Furthermore, the heavy metals cobalt and nickel were added to the reaction to intensify the staining. IHC caused GABAergic neurons to be labeled in black, allowing assessment of the soma size. They found that GABAergic cells displayed larger somata in the left optic tectum in comparison to the right. A simultaneous IHC to label parvalbumin-positive cells revealed

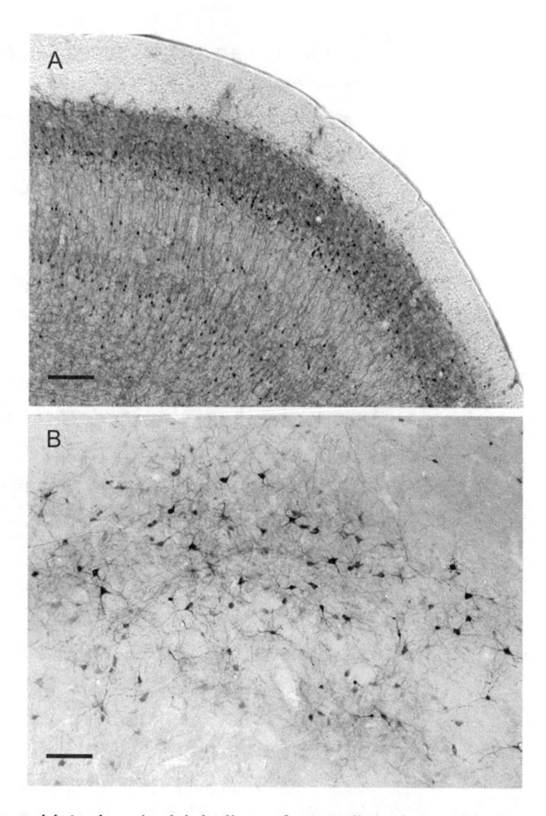

Fig. 5 Immunohistochemical labeling of parvalbumin positive neurons in the optic tectum of pigeons (**a**) and calbindin containing neurons in the pigeons pretectal area (**b**). Manns et al. [31, 82] used monoclonal antibody to detect both peptides and labeled them using the ABC method with a nickel/cobalt intensified DAB reaction. They found that the soma of parvalbumin containing neurons in the optic tectum of pigeons are bigger on the right than on the left side. Animals hatched from dark incubated eggs did not show this lateralization. Furthermore they could show that ocular injections of the neurotrophic factor BDNF can affect the cell size of parvalbumin and calbindin positive cells, suggesting a role of this factor in during the establishment of visual asymmetries. Scale bars indicate 100 μm

larger somata of these cells within the right tectum opticum compared to the left (refer to Fig. 5). In pigeons hatched from dark incubated eggs, these asymmetries were absent. They concluded that asymmetrically applied environmental factors can not only promote cell growth on the stimulated side but can also have growth promoting effects on the contralateral side. This finding is of great importance since it shows that unilateral stimulation does not only affect one hemisphere but triggers interhemispheric processes shaping the structure of both hemispheres.

The scope of application of IHC in lateralization research is so broad that we cannot give examples for the entire range. Instead, we will review a few studies to demonstrate the versatility of IHC to encourage other researchers to apply this method in their experiments. Besides the already mentioned studies, there are further studies that investigate asymmetry by means of IHC. Manns et al. used IHC to map the distribution of the brain-derived neurotrophic factor (BDNF) receptor TrkB in pigeon hatchlings directly after hatching and found that cells containing the receptor showed an enlarged soma size in the right optic tectum [92, 93]. This could indicate an effect of neurotrophic factors on the development of structural asymmetries. Furthermore, several studies that applied tracing methods in order to unravel lateralization in connectivity of the visual systems of birds used IHC to detect their tracer or enhance tracer signal [38, 61, 94]. In a more recent study, Hami et al. found a lateralization in receptor distribution within the hippocampus of male rat pups. They used antibodies to localize the insulin-like growth factor-I receptor (IGF-1) and found a significantly increased number of IGF-1 positive cells in CA1, CA3, and dentate gyrus on the left side [33].

Aside from just localizing antigens in tissue, IHC can also be used for activity imaging. George et al. analyzed activity in the optic tectum of male zebra finches when they were singing in front of female birds by using antibodies against the proteins of the immediately early genes (IEG) egr-1 (also known as ZENK) and c-fos [32]. It has been shown that such genes act as transcription factors and are up-regulated in cells in response to stimulation [95, 96]. Because IEG protein products can be visualized by IHC, they can serve as reliable activity markers. George et al. found that IEG expression was higher in the left optic tectum than in the right when the bird was singing in presence of a female. Furthermore, male birds sang more when they could use only their right eye than when they could use only their left. Thus it was concluded that there is a right eye/left hemispheric dominance for courtship singing in zebra finches [32].

The principles of IHC can also be combined with other histological techniques, further widening its scope. DeCarvalho et al. applied in situ hybridization to map the mRNA distribution of several neurotransmitters in the habenula of zebra fish. To label the hybridized RNA probes, they used antibodies directed against a digoxigenin tag coupled to the probe. This revealed a profound asymmetry in the distribution of neurotransmitters between left and right side, present in both larval and adult zebra fishes, suggesting a possible influence on lateralized behavior in zebra fishes [34].

This section has given a brief, but by no means complete, overview of the theoretical background of IHC and its applications in lateralization research. We hope that this might clarify how versatile and beneficial IHC experiments can be for the field of lateralization research. In the remaining sections of this chapter, we would like to explain in as much detail as possible how an actual IHC experiment can be performed and what is needed for such an experiment. Just as in the sections on tract tracing, we would also like to tackle some of the most typical problems faced during an IHC study, and how to prevent or solve them. In doing so, we would like to help other researchers starting new IHC studies and provide assistance for the first, often troublesome, experiments.

3.2 Materials

When planning to perform IHC experiments, tissue quality is of great importance (see Sects. 3.3 and 3.4). It is highly recommended to use tissue fixated by perfusion if possible (see Sect. 2.2). If perfusion is not possible, immersion fixation is an option. This requires 4% PFA in PB or another fixative of choice. For a list of the equipment required to prepare brain slices, please refer to Sect. 2.2 as well. IHC can either be performed with free-floating sections or slices already mounted on slides. Free-floating IHC has the advantage of incubation solutions being able to access the tissue from all sides, which can facilitate tissue penetration of the antibody and allows more thorough rinsing. Performing IHC on mounted slides reduces the chance of slices being lost or intermingled and is advantageous when working with thin or fragile slices such as retinal sections. For free-floating IHC, a compartmentalized fine-meshed sieve in a fitting glass cuvette is ideally suited for all rinsing steps. Smaller containers like the wells of cell culture plates should be used for incubation with antibodies, serum, and ABC solution, as these solutions are rather expensive. The transfer of slices should be performed with soft brushes or blunt hooks for larger sections. For IHC on mounted slices, we would recommend glass Coplin staining jars for all rinsing steps and staining reactions, while the remaining incubations are performed on the slide. For keeping solutions on the slide, it is advisable to draw a line with a hydrophobic substance around the edges of the slide (pens for this purpose are commercially available but nail polish and other sticky, hydrophobic substances work as well). Regular rinsing of the brain slices is necessary throughout the whole procedure. Many IHC studies use phosphate buffer or phosphate buffered saline at a neutral pH with a molarity of around 0.1 M [97]. However, depending on tissue and antibody, other buffers might be more suitable (e.g. TRIS-based buffers [98, 99]). For the antibody incubation, we suggest adding 0.1% Triton-X (or another detergent, e.g. Tween-20) to the buffer [97]. If it is planned to label the antibody with peroxidase-based staining reaction, we would recommend incubation in a solution of 0.3% hydrogen peroxide in distilled water before antibody application in order to saturate endogenous peroxidases [99]. If epitope retrieval is necessary (see below), a heating block is required. The remaining equipment is highly dependent on the selection of antibody and labeling method. In almost all cases we would recommend using an indirect labeling method with either a fluorophore or biotin-coupled secondary antibody. For antibody selection, the primary antibody is defined by the antigen planned to be labeled, while the secondary antibody has to detect the host species of the primary antibody.

A proper selection of the antibodies is crucial for successful IHC, especially when not working with a standard model species. For common antigens, well-established primary antibodies are commercially available, and these generally work well and produce reliable results. For these antibodies, specifications on working concentrations, optimal buffer systems and specificity of the labeling are known and can be found in the literature or even in the manual of the manufacturer. However, if it is planned to investigate less studied antigens in nonmodel species, antibodies established in model species might not work. This can be due to changes in the amino acid sequence or slightly different folding patterns of the antigen in the studied species in comparison to the model species for which the antibody was made for. If the amino acid sequence of the antigen of choice is known, several companies offer the manufacture of custom made antibodies. However, we highly recommend running a series of control experiments when using such antibodies to prove their specificity. This should include: a western blot, to check if the antibody is binding the correct antigen; blocking the antibody with the antigen before application, to test for nonspecific binding; and control staining once without the primary and once without the secondary antibody to check for specificity of the antibody and labeling chain. Furthermore, antibody concentrations delivering optimal staining patterns have to be figured out before reliable results can be gathered. If more than one antigen has to be labeled in one staining by, for example, fluorescent labeling, it has to be ensured that antibody combinations for each antigen do not interact with each other. This should also hold true for the serum, if serum blocking is performed.

In addition to the antibodies, serum from a species that is not detected by either the primary or secondary antibody in order to block nonspecific binding sides might be needed as well. Such serum blocks, usually done with serum from the species in which the secondary antibody was hosted, are applied by the majority of IHC studies to reduce nonspecific binding of the antibodies [38, 98, 99]. It should be noted however, that a study by Buchwalow et al. on human tissue showed that serum blocking might have no effect at all [89]. We have still incorporated this step into this description of IHC since using a serum block is still common practice in the majority of histology laboratories. If it is planned to choose a different serum, one should remember that as a general rule, the serum should always be from a species that can neither be detected by the primary nor the secondary antibody. In addition, if no serum is available, milk powder diluted in buffer can work as well [100].

In case enzyme-based labeling is planned, the relevant enzyme, substrate and chromogen are required. In most cases, functional combinations of these chemicals can be bought in kits, making labeling substantially simpler. For this book chapter, we will use ABC labeling with a nickel-cobalt DAB reaction as an example (for other chromogen-based labeling methods, see [84]). In addition to the primary and biotinylated secondary antibodies, avidin and peroxidase-coupled biotin are needed for this method. In our laboratory, we use a commercially available ABC kit, but single components can be purchased as well. Furthermore, DAB, ammonium nickel(II) sulfate hexahydrate ($(NH_4)_2Ni(SO_4)_2 \cdot 6H_2O$) and Cobalt(II) chloride ($CoCl_2$) are needed as chromogens. We suggest using DAB tablets, as they are safer to handle (DAB is carcinogenic, refer to material safety data sheet, CAS number 91-95-2) than powder or solutions. Instead of using pure H_2O_2 to run the staining reaction, we frequently use glucose oxidase in combination with β-D-glucose to ensure a slow, more controlled reaction. The staining reaction is run in 0.1 M sodium acetate buffer (pH 6.0). We would recommend running the staining reaction in separate containers, to prevent contamination of other labware with DAB.

Dehydration and coverslipping of slides requires ethanol, isopropanol, xylene, a xylene miscible mountant, and coverslips. The use of fluorescent labeling in IHC requires specific embedding media, which do not interfere with the fluorophore, and should not be dehydrated.

3.3 Method

At the beginning of an IHC experiment, one should first decide on using a monoclonal or a polyclonal antibody (specificity vs. sensitivity), on employing direct or indirect labeling (speed vs. signal strength), and on the usage of fluorophore or chromogen-based labeling (nonpermanent and allows multiple labels in one slice vs. permanent and allows only a single label per slice). Depending on the decisions made, the IHC procedure can differ significantly. In addition, tissue preparation has an impact on IHC results. For most antibodies, PFA perfusion-fixated tissue is the material of choice. However, some experiments do not allow a lengthy perfusion (e.g. immediate early gene studies). In these cases tissue being post-fixated by immersion fixation in PFA or another fixative will suffice, but will present a number of potential challenges (see Sect. 3.4.1). If slice thickness is not predetermined by another experiment, a slice thickness of 30–40 μm is recommended. This thickness allows easy handling as well as good penetration of the antibody into the tissue. Before starting with the actual experiment, we suggest rinsing the slices for at least 5 min in PBS to remove possible remainders of storing buffer, cutting medium etc. (or use your preferred rinsing buffer—if you use a different buffer, exchange PBS with your buffer each time we refer to PBS in the following). As a general note, rinsing should always be performed thoroughly to remove leftovers of the previous incubation steps. Furthermore, all rinsing and incubation steps should be performed on a shaker at low speed to facilitate intermixing. After the initial rinse, if peroxide dependent labeling is used, slices should be transferred to 0.3% H_2O_2 for 30 min in order to saturate endogenous peroxidases which could have remained active after fixation [99]. After incubation, slices should be rinsed three times for 10 min each in PBS. The last two steps can be skipped if the labeling does not rely on peroxidases. In the next step, slices should be incubated for 30 min in a 10% serum solution of PBS (see Sect. 3.2 for the selection of serum and [89] for criticism on serum blockade). After additional rinsing (3 × 10 min in PBS), slices are transferred to the primary antibody solution. The solution consists of PBS, 0.1% Triton-X, and the primary antibody in the desired concentration. Concentrations for different primary antibodies differ drastically. Often, values are provided by the supplier or can be found in prior publications using the antibody. If this is not the case, optimal concentrations must be acquired empirically using a dilution series with different concentrations. In our laboratory, incubation of slices with the primary antibody is done at 4 °C overnight. Other protocols use shorter incubation times (e.g. 1 h at room temperature [99]). Following primary antibody incubation, slices should be rinsed (3 × 10 min in PBS) and then transferred into the secondary antibody in PBS for 1 h at room temperature. As for the primary antibody, necessary concentrations of secondary antibodies are often provided by the manufacturer. If using secondary antibodies with a fluorescent tag, the IHC protocol is essentially completed and slices can be mounted and coverslipped with a fluorescence compatible embedding medium. However, it is important to control that light exposure of such slices is minimal to prevent bleaching of the fluorophore. If chromogen-based labeling of antibodies is used, slices have to be rinsed again (3 × 10 min in PBS) before continuing with the labeling reaction. In this chapter, we will describe only the nickel-cobalt intensified ABC labeling method [31]. For other methods, refer to Ramos-Vara and Miller [84].

After rinsing, slices are transferred to the ABC solution. The solution should be prepared at least 30 min before usage to allow aggregation of the avidin-biotin-peroxidase complex (compare Fig. 4). In most cases, concentration of the ABC solution is provided by the supplier. In our laboratory, we use a concentration of 1:100 in PBS. Incubation in ABC is normally done for 1 h at room temperature. Then, slices are rinsed two times for 10 min each in PBS and once for 10 min in a 0.1 M sodium acetate buffer. In the next step, slices are transferred into the DAB solution (20 mg DAB, 400 mg ß-D-glucose, 2.5 g ammonium nickel(II) sulfate hexahydrate, 40 mg cobalt(II) chloride, 40 mg ammonium-chloride ad 100 ml sodium acetate buffer, 0.1 M). Ammonium nickel(II) sulfate hexahydrate, cobalt(II) chloride, and ammonium chloride can be omitted if a nonintensified reaction is preferred. Likewise, ß-D-glucose can be omitted if the reaction is intended to be maintained by pure H_2O_2. Incubate the slices for 10 min in DAB solution and add 80 μl of 0.1% glucose oxidase in distilled water per 50 ml DAB solution used. If pure H_2O_2 is used for a faster but less controlled reaction, 0.03% H_2O_2 has to be added [101]. After 10 min, the DAB solution should be exchanged for a fresh one and fresh glucose oxidase needs to be added as well. This step has to be repeated until a sufficient staining level is achieved. In our laboratory, we typically need to exchange the DAB solution 3–4 times before the desired staining is visible. The progress of the reaction has to be monitored constantly. Structures containing the antigen will turn black, while the remaining structures in the slice should remain unlabeled. The reaction should be stopped by transferring the slices to a sodium acetate buffer as soon as the antigen containing structures are sufficiently stained. Rinse the slices for 10 min in sodium acetate buffer and transfer them into PBS for additional rinsing (twice for 10 min each time). After this, the slices can be mounted on slides. We would recommend allowing the slices to dry for several hours before dehydration in order to prevent them from floating off of the slide. For dehydration, incubate slides in alcohol of increasing concentration (70% ethanol, 90% ethanol, twice in 100% isopropanol) for 5 min each, followed by incubation for 5 min in 100% xylene, twice. After xylene incubation, coverslip the slides immediately with a xylene miscible mountant. Let the mountant dry for 24 h and analyze the slices under a microscope.

3.4 IHC Troubleshooting

Although IHC is a well-established and very reliable technique, it is unfortunately prone to problems when experiments are not designed carefully. Small details, which might seem irrelevant to the inexperienced experimenter, can have a strong impact on the staining result. This can range from no staining at all, to overstaining of the whole tissue and even false positive labeling. In the last few sub-sections (Sects. 3.4.1–3.4.3) of this chapter section on IHC, we will address some of the most common problems during IHC, and offer solutions for these problems. However, we can only focus on the core problems. A more detailed IHC troubleshooting guide has been published by Bussolati and Leonardo, which might be helpful if the information presented here does not assist in overcoming your IHC challenges [102].

3.4.1 Tissue Fixation and Epitope Retrieval

One of the deciding factors between failure and success of an IHC experiment is proper tissue fixation. Tissue fixation is crucial in preventing autolysis, which can lead to degradation of cellular structures, and to prevent displacement of these components. Furthermore, it hardens the tissue, which makes it less prone to mechanical damage [103]. Thus, long delays between death of the animal and fixation of the tissue increases the chance of unrecoverable epitopes and antigen loss due to autolysis. If this happens, IHC is not applicable anymore. As stated above, we therefore strongly suggest performing transcardial perfusions with a fixative whenever possible to assure fast and thorough fixation of brain tissue. If perfusion cannot be applied, tissue samples should be immersion fixated as soon as possible. In most studies, PFA or PFA containing fixatives are used, but fixation can also be achieved by using alcohols or other aldehydes like glutaraldehyde or acrolein [73, 84, 104]. Although fixation is required for successful IHC, the fixation process can impair IHC as well. PFA, as well as glutaraldehyde and acrolein, are cross-linking fixative forming methylene bridges between peptides and proteins, which immobilizes them and changes their conformation. Cross-linking inactivates enzymes (preventing autolysis), but can also mask or damage epitopes, interfering with IHC by preventing access of the primary antibody [80]. It has even been reported that cross-linking can create new epitopes that are detected by the antibody and then are falsely interpreted as the location of the antigen [105]. To circumvent these problems, several techniques have been developed to reverse the changes induced by the fixative directly before IHC. The two most relevant of these techniques are protease-induced epitope (and antigen) retrieval and heat-induced epitope retrieval (PIER and HIER [102]). In PIER, proteases like trypsin, proteinase K, or pepsin are added to the tissue before commencing IHC [84, 106]. These proteases cleave proteins and can therefore weaken the interlinking of the tissue, possibly unmasking epitopes. However, proteases might also damage the epitope, making it impossible for the primary antibody to bind. As such, the type of the protease, incubation time, and buffer system all have to be carefully matched to the antigen in use, to prevent damage to the epitopes. The HIER method relies on heating of sample tissue to unmask epitopes [107]. This can either be achieved by using a microwave or incubation in a preheated water bath. Although the exact function of HIER is still unclear, it is known to reliably unmask epitopes and is the most commonly used epitope retrieval technique [84]. However, heating for too long or at too warm temperatures can denaturate the tissue, leading to epitope loss and general tissue instability.

In case the IHC experiment does not produce sufficient labeling, we would recommend using one of these epitope retrieval techniques. This holds especially true when immersion fixated tissue was used or if the tissue rested for a long period of time in PFA. In general, HIER is easier to apply, as most of the required equipment is inexpensive and already present in most labs. In our lab, we apply HIER after H_2O_2 and before serum incubation. During HIER we incubate slices stored in 0.01 M sodium citrate buffer (pH 6.0) sealed in plastic tubes in a water bath for 30 min at 90 °C. This and very similar approaches are being used in many studies and seem to be very effective [84]. From personal experience, we cannot recommend longer incubation times, as this will significantly denaturate the tissue. If no water bath is available, a microwave can work as well. After HIER, slices should be rinsed in PBS and the normal IHC protocol should be continued. In case the slices become too fragile during HIER, reduce the incubation time and try to find a balance between staining quality and tissue integrity. A more detailed overview on HIER and PIER, including the pitfalls of these methods, can be found in Bussolati and Leonardo and Ramos-Vara and Miller [84, 102].

3.4.2 Broken Tissue and Lost Slices

A rather annoying problem during IHC is breaking of slices or the loss of whole sections during rinsing or incubation. If slices are very fragile and tend to break during IHC or while mounting, one should consider to cut slightly thicker slices or to run a more thorough fixation. The latter can be achieved by prolonging post-fixation or using a higher concentration of fixative. This, however, might lead to over-fixation, which then must be counteracted by epitope retrieval. Epitope retrieval, though, can also increase fragility of the tissue. Unfortunately there is no perfect protocol to balance accessibility of the epitope and structural stability of the tissue. In case of broken tissue we would suggest performing a series of test IHC experiments to balance fixation and epitope retrieval until a satisfactory result is achieved. Furthermore, one should always try to reduce mechanical manipulation of tissue to an absolute minimum, for example by using soft paintbrushes for section manipulation.

Loss of tissue can happen during IHC procedures performed on slides, such as sections disconnecting from the slide during rinsing or tissue incubation. In some cases this can be prevented by letting the sections thoroughly dry before starting IHC. However, if this happens regularly, using slides coated with adhesives should be considered. Such slides are commercially available, with different coatings for specific applications (e.g. coating without autofluorescence for fluorescent IHC). A less expensive solution is coating slides manually with gelatin. In this simple method, dip clean glass slides into a 37 °C warm, 2 % gelatin solution in distilled water for 30 s and dry for 24 h [108]. It should be noted that gelatin is autofluorescent which renders these slides suboptimal for fluorescent IHC.

3.4.3 Lack of Stain and Abnormal Staining Patterns

An unsuccessful IHC is often characterized by complete lack of staining, low signal strength, high background staining, or nonspecific staining. To our knowledge, the most common reasons for a complete lack of staining are either blocked epitopes due to over fixation (especially when using monoclonal antibodies, which only recognize one specific epitope) or problems with the enzyme-based labeling. Whereas the first can be counteracted by epitope retrieval (see Sect. 3.4.1), the latter is often caused by inactive enzymes. Most enzymes, like glucose oxidase or peroxidase used in the ABC labeling method, need to be stored at low temperatures at all times to prevent degradation of the enzyme, which causes it to lose its function. Frequent usage of enzyme containing vials at room temperature can cause a drop and even a complete loss of enzyme activity, preventing the labeling reaction from producing any labels. To avoid this, we suggest storing enzymes in a 50% glycerol solution in small volumes in separate vials and using a cooling rack to keep vials at low temperature.

Low signal strength is often caused by inadequate concentrations of antibody, insufficient incubation times, partial epitope loss due to weak fixation or over-fixation (see Sect. 3.4.1) or fading of the fluorescence signal. When a new IHC protocol is established, we would always suggest testing different antibody concentrations, even if a working concentration is given by the manufacturer of the antibody. Only this way an optimal signal to noise ratio can be achieved. The same holds true for incubation times. Although incubation durations given in this chapter and other protocols work well for most IHC, some IHC experiments require longer incubation times to achieve sufficient signal strength. Also, changes in incubation temperature can boost signal strength in some instances. In most cases fading of fluorophores coupled to either the primary or secondary antibody is caused by prolonged periods of light exposure. Reduce light levels as much as possible and try to analyze or digitalize sections as soon as possible. Furthermore, there are a couple of specific mounting media for fluorescent experiments commercially available that can at least reduce the rate of fading.

The most common reasons for high background staining, causing an unfavorable signal to noise ratio, are weak fixation (see Sect. 3.4.1), insufficient blocking of endogenous enzymes, autofluorescence of tissue, and extended labeling reactions. As stated above, when using an enzyme-based labeling reaction, it is pivotal to block or inactivate endogenous remainders of this enzyme within the tissue. Otherwise, all regions containing the endogenous enzyme will stain during the labeling reaction. This holds especially true for nonperfused, weakly immersion-fixated tissue, as it contains unfixed erythrocytes, rich in enzymes [102, 109]. In addition to enhancing the fixation process, longer incubations with the endogenous enzyme substrate at higher concentrations than standard can reduce the problem. It should however be noted that some substrates, such as H_2O_2, can damage the tissue itself, limiting the effective maximum duration and concentration. Autofluorescence of tissue can cause severe problems during detection of a fluorescence signal after IHC labeling, where the signal of the label is subsumed by the autofluorescence. Incubation of tissue before IHC in $CuSO_4$ or $NaBH_4$ solutions, or irradiation of tissue with light can reduce autofluorescence effectively [110–112]. Based on our experience, extended incubation in the chromogen solution after addition of the substrate during the labeling reaction can cause massive background noise as well. The challenge here is to find the optimal time when the desired signal-to-noise ratio is reached. Unfortunately there is no ideal time to stop the reaction that works for all IHC experiments, because reaction speed also relies on the density of antigen (and therefore amount of bound antibodies) in the tissue. Empirical observations are needed for every single IHC experiment in order to figure out the optimal time to stop the labeling reaction.

Nonspecific staining is caused by the primary antibody binding to epitopes similar to the antigen's epitopes. Common reasons for nonspecific binding are antibody concentrations that are too high (facilitates binding to similar epitopes) or a target antigen containing an amino acid sequence common in several other proteins. The chances for nonspecific staining are higher when using polyclonal antibodies, because polyclonal sera contain several antibodies binding to different epitopes of the antigen (see Sect. 3.1.1). Without knowledge of the expected staining pattern, nonspecific staining can easily be misinterpreted as a positive signal, jeopardizing the whole IHC experiment. Therefore, some general knowledge of the antigen, like cell types containing the antigen, localization of the antigen within the cell (e.g. nucleus, soma, dendrites, axons, membrane bound), or distribution of the antigen within the brain, is needed. As described under Sect. 3.3, control experiments like a western blot or a peptide block of the primary antibody are strongly advised during establishment of a new IHC experiment to detect nonspecific antibody binding. If available, control staining in tissue of KO animals lacking the antigen can be helpful as well [113]. If control experiments reveal that the primary antibody is detecting more targets than just the antigen, exchange of the primary antibody is necessary. If the control experiments show that the primary antibody binds specifically to the antigen, but results from IHC procedures on tissue deviate from the expected staining pattern derived from antigen properties, the following approaches might solve the problem. If the nonspecific staining is only caused by antibody concentrations that are too high, a simple series of IHCs using different dilutions of the primary antibody can help to figure out optimal antibody concentrations. Furthermore, prolonged serum block, or adding 1% blocking serum to the primary antibody incubation can reduce nonspecific binding (but see [89]). In some cases, changing the fixative, application of epitope retrieval, or switching to a different buffer system can decrease nonspecific binding as well. Note that these parameters have to be investigated empirically for each primary antibody, and that there is no perfect protocol for removing nonspecific staining from an IHC experiment. However, Fritschy, as well as Lorincz and Nusser, collected some data on nonspecific labeling during IHC that we would recommend as an additional source for troubleshooting [98, 113].

The previous few sections should have made clear that successful IHC experiments require a lot of empirical work to acquire optimal conditions for tissue fixation, antibody incubation, and labeling reaction. We nonetheless hope that we have supplied enough information to help others to set up and run their own IHC experiments and to assist during troubleshooting. The advantage of using a method invented more than 70 years ago, and constantly improved in the meanwhile, is that there are already a number of excellent review articles that can provide further help if obstacles occur during an IHC experiment (for example [80, 84, 98, 102]).

4 Summary and Outlook

Within this chapter, we tried to give an overview over the history of general histological techniques and their application in lateralization research. Within this, we focused in particular on two very prominent methods in the field: neuronal tracing and immunohistochemistry. We explained the theory and application of these methods in order to encourage future generations of lateralization researchers to apply them in their model species. We also gave a few examples on how these techniques can and have been used to unravel the foundations of behavioral asymmetries. Although these studies have provided valuable insights into the neuronal structures involved in lateralized behavior, it is still true that for most behavioral lateralizations observed, the underlying neuronal mechanisms are still unknown or have been insufficiently investigated. As such, the main intention of this chapter was not deliver a comprehensive overview of every single aspect of the given method, but to supply a detailed start-to-finish procedural overview that allows readers not experienced in these techniques to start planning their own tracing and IHC experiments. We hope that, in this way, we can encourage lateralization researchers to perform more anatomical studies, which in turn will contribute to urgently needed data on the neuronal correlates of brain asymmetries. Only research using a combination of behavioral, physiological, and anatomical experiments can truly provide the information needed to understand brain asymmetries thoroughly.

Acknowledgements

We would like to thank Brendon Billings, Alexis Garland, Rena Klose, Sara Letzner, Martina Manns, Sebastian Ocklenburg, Annika Simon, Martin Stacho, and John Tuff for fruitful discussions, helpful comments, and aid during acquisition of images.

References

1. Rogers LJ, Vallortigara G, Andrew RJ (2013) Divided brains: the biology and behaviour of brain asymmetries. Cambridge University Press, Cambridge
2. Ströckens F, Güntürkün O, Ocklenburg S (2013) Limb preferences in non-human vertebrates. Laterality 18:536–575
3. Versace E, Vallortigara G (2015) Forelimb preferences in human beings and other species: multiple models for testing hypotheses on lateralization. Front Psychol 6:233
4. Brown C, Magat M (2011) Cerebral lateralization determines hand preferences in Australian parrots. Biol Lett 7:496–498
5. Böye M, Güntürkün O, Vauclair J (2005) Right ear advantage for conspecific calls in adults and subadults, but not infants, California sea lions (Zalophus californianus): hemispheric specialization for communication? Eur J Neurosci 21:1727–1732
6. Vallortigara G, Regolin L, Pagni P (1999) Detour behaviour, imprinting and visual lateralization in the domestic chick. Brain Res Cogn Brain Res 7:307–320
7. Bisazza A, Sovrano VA, Vallortigara G (2001) Consistency among different tasks of left-right asymmetries in lines of fish originally selected for opposite direction of lateralization in a detour task. Neuropsychologia 39:1077–1085
8. Keerthipriya P, Tewari R, Vidya TN (2015) Lateralization in trunk and forefoot movements in a population of free-ranging Asian elephants (Elephas maximus). J Comp Psychol [Epub ahead of print]
9. Quaranta A, Siniscalchi M, Vallortigara G (2007) Asymmetric tail-wagging responses by dogs to different emotive stimuli. Curr Biol 17:R199–R201
10. Broca P (1866) Sur la siège de la faculté du langage articulé. Bull Soc Anthropol 6:337–399
11. Eberstaller O (1890) Das Stirnhirn. Urban und Schwarzenberg, Wien, Leipzig
12. Cunningham DJ (1892) Contribution to the surface anatomy of the cerebral hemispheres. Royal Irish Academy, Dublin
13. Cunningham DJ (1902) Right-handedness and left-brainedness. J Anthropol Inst G B Irel 32:273–296
14. von Economo C, Horn L (1930) Über Windungsrelief, Masse und Rindenarchitektonik der Supratemporalfläche, ihre indivi-duellen und ihre Seitenunterschiede. Z Ges Neurol Psychiatr 130:687–757
15. Scheibel AB, Paul LA, Fried I, Forsythe AB, Tomiyasu U, Wechsler A, Kao A, Slotnick J (1985) Dendritic organization of the anterior speech area. Exp Neurol 87:109–117
16. Seldon HL (1981) Structure of human auditory cortex: I. Cytoarchitectonics and dendritic distribution. Brain Res 229:277–294
17. Seldon HL (1981) Structure of human auditory cortex: II. Axon distributions and morphological correlates of speech perceptions. Brain Res 229:295–310
18. Seldon HL (1982) Structure of human auditory cortex: III. Statistical analysis of dendritic trees. Brain Res 249:211–221
19. Vallortigara G, Chiandetti C, Sovrano VA (2011) Brain asymmetry (animal). WIREs Cogn Sci 2:146–157
20. Köhler A, von Rohr M (1904) Eine mikrophotographische Einrichtung für ultraviolettes Licht. Z Instnun Kde 24:341–349
21. Ruska E (1933) Die elektronenoptische Abbildung elektronenbestrahlter Oberflächen. Z Phys 83:492–497
22. Zernike F (1935) Das Phasenkontrastverfahren bei der mikroskopischen Beobachtung. Z Tech Phys 16:454–457
23. Minsky M (1988) Memoir on inventing the confocal scanning microscope. Scanning 10:128–138
24. Wada J, Rasmussen T (1960) Intracarotid injection of sodium amytal for the lateralization of cerebral speech dominance. J Neurosurg 106:1117–1133
25. Nottebohm F (1971) Neural lateralization of vocal control in a passerine bird: I. Song. J Exp Zool 177:229–261
26. Rogers LJ (1978) Functional lateralization in the chicken fore-brain revealed by cycloheximidine treatment. In: XVII international congress of ornithology, pp 653–659
27. Rogers LJ, Anson JM (1979) Lateralisation of function in the chicken fore-brain. Pharmacol Biochem Behav 10:679–686
28. Ocklenburg S, Ströckens F, Güntürkün O (2013) Lateralisation of conspecific vocalisation in non-human vertebrates. Laterality 18:1–31
29. Rogers LJ, Sink HS (1988) Transient asymmetry in the projections of the rostral thalamus to the visual hyperstriatum of the chicken, and reversal of its direction by light exposure. Exp Brain Res 70:378–384
30. Rogers LJ, Bolden SW (1991) Light-dependent development and asymmetry of visual projections. Neurosci Lett 121:63–67
31. Manns M, Güntürkün O (2003) Light experience induces differential asymmetry pattern

of GABA- and parvalbumin-positive cells in the pigeon's visual midbrain. J Chem Neuroanat 25:249–259

32. George I, Hara E, Hessler NA (2006) Behavioral and neural lateralization of vision in courtship singing of the zebra finch. J Neurobiol 66:1164–1173
33. Hami J, Kheradmand H, Haghir H (2014) Gender differences and lateralization in the distribution pattern of insulin-like growth factor-1 receptor in developing rat hippocampus: an immunohistochemical study. Cell Mol Neurobiol 34:215–226
34. deCarvalho TN, Subedi A, Rock J, Harfe BD, Thisse C, Thisse B, Halpern ME, Hong E (2014) Neurotransmitter map of the asymmetric dorsal habenular nuclei of zebrafish. Genesis 52:636–655
35. Sherwood CC, Raghanti MA, Wenstrup JJ (2007) Is humanlike cytoarchitectural asymmetry present in another species with complex social vocalization? A stereologic analysis of mustached bat auditory cortex. Brain Res 1045:164–174
36. Bezchlibnyk YB, Sun X, Wang JF, MacQueen GM, McEwen BS, Young LT (2007) Neuron somal size is decreased in the lateral amygdalar nucleus of subjects with bipolar disorder. J Psychiatry Neurosci 32:203–210
37. Arpini M, Menezes IC, Dall'Oglio A, Rasia-Filho AA (2010) The density of Golgi-impregnated dendritic spines from adult rat posterodorsal medial amygdala neurons displays no evidence of hemispheric or dorsal/ventral differences. Neurosci Lett 469:209–213
38. Ströckens F, Freund N, Manns M, Ocklenburg S, Güntürkün O (2013) Visual asymmetries and the ascending thalamofugal pathway in pigeons. Brain Struct Funct 218:1197–1209
39. Patzke N, Manns M, Güntürkün O (2011) Telencephalic organization of the olfactory system in homing pigeons (Columba livia). Neuroscience 194:53–61
40. Novikova L, Novikov L, Kellerth JO (1997) Persistent neuronal labeling by retrograde fluorescent tracers: a comparison between Fast Blue, Fluoro-Gold and various dextran conjugates. J Neurosci Methods 74:9–15
41. Letzner S, Simon A, Güntürkün O (2015) Connectivity and neurochemistry of the commissura anterior of the pigeon (Columba livia). J Comp Neurol [Epub ahead of print]
42. Hendricks M, Jesuthasan S (2007) Asymmetric innervation of the habenula in zebrafish. J Comp Neurol 502:611–619
43. Güntürkün O, Melsbach G, Hörster W, Daniel S (1993) Different sets of afferents are demonstrated by the fluorescent tracers fast blue and rhodamine. J Neurosci Methods 49:103–111
44. Köbbert C, Apps R, Bechmann I, Lanciego JL, Mey J, Thanos S (2000) Current concepts in neuroanatomical tracing. Prog Neurobiol 62:327–351
45. Paton JA, Manogue KR, Nottebohm F (1981) Bilateral organization of the vocal control pathway in the budgerigar, Melopsittacus undulatus. J Neurosci 1:1279–1288
46. Wouterlood FG, Groenewegen HJ (1985) Neuroanatomical tracing by use of Phaseolus vulgaris-leucoagglutinin (PHA-L): electron microscopy of PHA-L-filled neuronal somata, dendrites, axons and axon terminals. Brain Res 326:188–191
47. Moon EA, Goodchild AK, Pilowsky PM (2002) Lateralisation of projections from the rostral ventrolateral medulla to sympathetic preganglionic neurons in the rat. Brain Res 929:181–190
48. Rogers LJ, Deng C (1999) Light experience and lateralization of the two visual pathways in the chick. Behav Brain Res 98:277–287
49. Marfurt CF (1988) Sympathetic innervation of the rat cornea as demonstrated by the retrograde and anterograde transport of horseradish peroxidase-wheat germ agglutinin. J Comp Neurol 268:147–160
50. Oztas E (2003) Neuronal tracing. Neuroanatomy 2:2–5
51. Kelly RM, Strick PL (2000) Rabies as a transneuronal tracer of circuits in the central nervous system. J Neurosci Methods 103:63–71
52. Callaway EM (2008) Transneuronal circuit tracing with neurotropic viruses. Curr Opin Neurobiol 18:617–623
53. Ugolini G (2010) Advances in viral transneuronal tracing. J Neurosci Methods 194:2–20
54. Nassi JJ, Cepko CL, Born RT, Beier KT (2015) Neuroanatomy goes viral! Front Neuroanat 9:80
55. Rogers LJ (2002) Advantages and disadvantages of lateralization. In: Rogers LJ, Andrew R (eds) Comparative vertebrate lateralization. Cambridge University Press, Cambridge, pp 126–154
56. Valenti A, Sovrano VA, Zucca P, Vallortigara G (2003) Visual lateralisation in quails (Coturnix coturnix). Laterality 8:67–78
57. Manns M, Güntürkün O (2009) Dual coding of visual asymmetries in the pigeon brain: the interaction of bottom-up and top-down systems. Exp Brain Res 199:323–332
58. Rogers LJ (1982) Light experience and asymmetry of brain function in chickens. Nature 297:223–225

59. Skiba M, Diekamp B, Güntürkün O (2002) Embryonic light stimulation induces different asymmetries in visuoperceptual and visuomotor pathways of pigeons. Behav Brain Res 134:149–156
60. Güntürkün O (2005) How asymmetry in animals starts. Eur Rev 13:105–118
61. Güntürkün O, Hellmann B, Melsbach G, Prior H (1998) Asymmetries of representation in the visual system of pigeons. NeuroReport 9:4127–4130
62. Manns M, Ströckens F (2014) Functional and structural comparison of visual lateralization in birds—similar but still different. Front Psychol 5:206
63. Galuske RA, Schlote W, Bratzke H, Singer W (2000) Interhemispheric asymmetries of the modular structure in human temporal cortex. Science 289:1946–1949
64. Sovrano VA, Andrew RJ (2006) Eye use during viewing a reflection: behavioural lateralisation in zebrafish larvae. Behav Brain Res 167:226–231
65. Facchin L, Argenton F, Bisazza A (2009) Lines of Danio rerio selected for opposite behavioural lateralization show differences in anatomical left-right asymmetries. Behav Brain Res 197:157–165
66. Dadda M, Domenichini A, Piffer L, Argenton F, Bisazza A (2010) Early differences in epithalamic left-right asymmetry influence lateralization and personality of adult zebrafish. Behav Brain Res 206:208–215
67. Roussigne M, Blader P, Wilson SW (2012) Breaking symmetry: the zebrafish as a model for understanding left-right asymmetry in the developing brain. Dev Neurobiol 72:269–281
68. Karten HJ, Hodos W (1967) A stereotaxic atlas of the brain of the pigeon (Columba livia). The Johns Hopkins University Press, Baltimore
69. Schofield BR (2008) Retrograde axonal tracing with fluorescent markers. Curr Protoc Neurosci. Chapter 1: Unit 1.17
70. Beier K, Cepko C (2012) Viral tracing of genetically defined neural circuitry. J Vis Exp. pii: 4253
71. Stichel CC, Müller HW (1998) The CNS lesion scar: new vistas on an old regeneration barrier. Cell Tissue Res 294:1–9
72. Notman R, Noro M, O'Malley B, Anwar J (2006) Molecular basis for dimethylsulfoxide (DMSO) action on lipid membranes. J Am Chem Soc 128:13982–13983
73. Mulisch M, Welsch U (eds) (2010) Romeis—Mikroskopische Technik. Springer, Heidelberg
74. Kuypers HG, Bentivoglio M, Catsman-Berrevoets CE, Bharos AT (1980) Double retrograde neuronal labeling through divergent axon collaterals, using two fluorescent tracers with the same excitation wavelength which label different features of the cell. Exp Brain Res 40:383–392
75. Pilati N, Barker M, Panteleimonitis S, Donga R, Hamann M (2008) A rapid method combining Golgi and Nissl staining to study neuronal morphology and cytoarchitecture. J Histochem Cytochem 56:539–550
76. Coons AH, Creech HJ, Jones RN (1941) Immunological properties of an antibody containing a fluorescence group. Proc Soc Exp Biol Med 47:200–202
77. Marrack J (1934) Nature of antibodies. Nature 133:292–293
78. Coons AH, Creech HJ, Jones RN, Berliner E (1942) The demonstration of pneumococcal antigen in tissues by the use of fluorescent antibody. J Immunol 45:159–170
79. Brandtzaeg P (1998) The increasing power of immunohistochemistry and immunocytochemistry. J Immunol Methods 216:49–67
80. deMatos LL, Trufelli DC, de Matos MG, da Silva Pinhal MA (2010) Immunohistochemistry as an important tool in biomarkers detection and clinical practice. Biomark Insights 5:9–20
81. Ramos-Vara JA (2005) Technical aspects of immunohistochemistry. Vet Pathol 42:405–426
82. Manns M, Freund N, Güntürkün O (2008) Development of the diencephalic relay structures of the visual thalamofugal system in pigeons. Brain Res Bull 75:424–427
83. Legent K, Tissot N, Guichet A (2015) Visualizing microtubule networks during drosophila oogenesis using fixed and live imaging. Methods Mol Biol 1328:99–112
84. Ramos-Vara JA, Miller MA (2014) When tissue antigens and antibodies get along: revisiting the technical aspects of immunohistochemistry—the red, brown, and blue technique. Vet Pathol 51:42–87
85. Ritter AM (2000) Polyclonal and monoclonal antibodies. In: George AJT, Urch CE (Eds.), Methods in molecular medicine, vol 40: Diagnostic and therapeutic antibodies. Humana Press Inc., Totowa
86. Lipman NS, Jackson LR, Trudel LJ, Weis-Garcia F (2005) Monoclonal versus polyclonal antibodies: distinguishing characteristics, applications, and information resources. ILAR J 46:258–268
87. Schellingerhout D, LeRoux LG, Hobbs BP, Bredow S (2012) Impairment of retrograde

neuronal transport in oxaliplatin-induced neuropathy demonstrated by molecular imaging. PLoS One 7:e45776

88. Coons AH, Leduc EH, Connolly JM (1955) Studies on antibody production: I. A method for the histochemical demonstration of specific antibody and its application to a study of the hyperimmune rabbit. J Exp Med 102:49–60
89. Buchwalow I, Samoilova V, Boecker W, Tiemann M (2011) Non-specific binding of antibodies in immunohistochemistry: fallacies and facts. Sci Rep 1:28
90. Ströckens F, Güntürkün O (2015) Cryptochrome 1b—a possible inductor of visual lateralization in pigeons? Eur J Neurosci [Epub ahead of print]
91. Hsu SM, Raine L, Fanger H (1981) Use of avidin-biotin peroxidase complex (ABC) in immunoperoxidase techniques: a comparison between ABC and unlabeled antibody (PaP) procedures. J Histochem Cytochem 29:577–580
92. van der Loos CM (2010) Chromogens in multiple immunohistochemical staining used for visual assessment and spectral imaging: the colorful future. J Histotechnol 33:31–40
93. Manns M, Güntürkün O, Heumann R, Blöchl A (2005) Photic inhibition of TrkB/Ras activity in the pigeon's tectum during development: impact on brain asymmetry formation. Eur J Neurosci 22:2180–2186
94. Manns M, Freund N, Patzke N, Güntürkün O (2007) Organization of telencephalotectal projections in pigeons: Impact for lateralized top-down control. Neuroscience 144:645–653
95. Sheng M, Greenberg ME (1990) The regulation and function of c-fos and other immediate early genes in the nervous system. Neuron 4:477–485
96. Long KD, Salbaum JM (1998) Evolutionary conservation of the immediate-early gene ZENK. Mol Biol Evol 15:284–292
97. Buchwalow I, Böcker W (2010) Immunohistochemistry: basics and methods. Springer Science & Business Media, Berlin
98. Fritschy JM (2008) Is my antibody-staining specific? How to deal with pitfalls of immunohistochemistry. Eur J Neurosci 28:2365–2370
99. Chen X, Cho DB, Yang PC (2010) Double staining immunohistochemistry. N Am J Med Sci 2:241–245
100. Duhamel RC, Johnson DA (1984) Use of nonfat dry milk to block nonspecific nuclear and membrane staining by avidin conjugates. J Histochem Cytochem 33:711–714
101. Fung KM, Messing A, Lee VM, Trojanowski JQ (1992) A novel modification of the avidin-biotin complex method for immunohistochemical studies of transgenic mice with murine monoclonal antibodies. J Histochem Cytochem 40:1319–1328
102. Bussolati G, Leonardo E (2008) Technical pitfalls potentially affecting diagnoses in immunohistochemistry. J Clin Pathol 61: 1184–1192
103. Thavarajah R, Mudimbaimannar VK, Elizabeth J, Rao UK, Ranganathan K (2012) Chemical and physical basics of routine formaldehyde fixation. J Oral Maxillofac Pathol 16:400–405
104. Howat WJ, Wilson BA (2014) Tissue fixation and the effect of molecular fixatives on downstream staining procedures. Methods 70:12–19
105. Josephsen K, Smith CE, Nanci A (1999) Selective but nonspecific immunolabeling of enamel protein-associated compartments by a monoclonal antibody against vimentin. J Histochem Cytochem 47:1237–1245
106. Huang SN, Minassian H, More JD (1976) Application of immunofluorescent staining on paraffin sections improved by trypsin digestion. Lab Invest 35:383–390
107. Shi S-R, Key ME, Kalra KR (1991) Antigen retrieval in formalin fixed, paraffin-embedded tissue: an enhancement method for immunohistochemical staining based on microwave heating of tissue sections. J Histochem Cytochem 39:741–744
108. Goding JW (1996) Monoclonal antibodies: principles and practice, 3rd edn. Academic, London
109. Li CY, Ziesmer SC, Lazcano-Villareal O (1987) Use of azide and hydrogen peroxide as an inhibitor for endogenous peroxidase in the immunoperoxidase method. J Histochem Cytochem 35:1457–1460
110. Clancy B, Cauller LJ (1998) Reduction of background autofluorescence in brain sections following immersion in sodium borohydride. J Neurosci Methods 83:97–102
111. Neumann M, Gabel D (2002) Simple method for reduction of autofluorescence in fluorescence microscopy. J Histochem Cytochem 50:437–439
112. Viegas MS, Martins TC, Seco F, do Carmo A (2007) An improved and cost-effective methodology for the reduction of autofluorescence in direct immunofluorescence studies on formalin-fixed paraffin-embedded tissues. Eur J Histochem 51:59–66
113. Lorincz A, Nusser Z (2008) Specificity of immunoreactions: the importance of testing specificity in each method. J Neurosci 28:9083–9086

Chapter 10

Brain Mapping Using the Immediate Early Gene *Zenk*

Tadd B. Patton, Ahmet K. Uysal, S. Leilani Kellogg, and Toru Shimizu

Abstract

Several lines of evidence show that induction of immediate early genes (IEGs) is a crucial step in the regulation of long-term plasticity-related changes at the cellular level. The term immediate early gene aptly characterizes the fact that these genes are induced rapidly, within minutes, and transiently following neuronal activation. This is due to the fact that their activation is not dependent on de novo protein synthesis as other, late response genes require. Such rapid induction of these genes offers a unique and powerful way to examine how a specific stimulus can lead to long-term cellular changes. By mapping IEG mRNA activity or the resulting protein expression, researchers have been able to investigate patterns of brain activation in response to various stimuli and in different behavioral paradigms. In addition, IEG activity can be used to examine structural and functional aspects of brain hemisphere lateralization. Several IEGs will be discussed in this chapter but our specific methods will focus on *Zenk*, a widely studied IEG in birds that has been used to identify specific brain regions critical for song learning, memory, homing behavior, and conspecific recognition. Here, we describe how standard immunohistochemical techniques and widely available imaging software can be used to visualize ZENK protein expression at the cellular level and at a macro level, such as whole sagittal or coronal brain sections. These visualization techniques provide the investigator with a powerful tool to examine asymmetries in the brain.

Key words Immediate early gene, Avian, EGR-1, Immunohistochemistry, Plasticity

1 Introduction

Immediate early gene (IEG) induction can be detected and visualized as a means to examine plasticity-related long-term changes at the cellular level. By mapping IEG activation, functionally different brain regions that might otherwise be cytoarchitectonically difficult to define can be characterized. In this chapter we: (1) provide a concise review of the discovery of IEGs and discuss some notable technical considerations when examining their expression, (2) present a detailed, yet easy to follow protocol that can be used to visualize IEG activity at the micro (cellular) level and macro (intra and inter-hemispheric) level allowing the investigator neuroanatomical asymmetries as well as functional hemisphere lateralization, and (3) discuss the strengths and weaknesses associated with this method.

Lesley J. Rogers and Giorgio Vallortigara (eds.), *Lateralized Brain Functions: Methods in Human and Non-Human Species*, Neuromethods, vol. 122, DOI 10.1007/978-1-4939-6725-4_10,

In the late 1980s, researchers began reporting on a class of transcription factor genes that could be rapidly induced by neuronal stimulation [1, 2]. Research showed that these early response genes, or IEGs, were inducible by applying various agents such as nerve growth factor or inducing seizure activity by administering convulsants or by electrical kindling. These genes include: *c-fos*, *zif/268*, *NGFI-A*, *c-jun*, and *jun-B* [1]. Several subsequent studies have revealed vital roles for IEGs in the maintenance of long-term neuronal changes [3–6] and have been used to investigate neural activity patterns in fish [7, 8], amphibians [9], birds [10], mammals [11], and invertebrates [12–17]. Furthermore, coding regions (sequences) and expression patterns of some IEG transcription factors have been examined and compared in humans, rodents, and birds and show that these factors are highly conserved across these phylogenetically diverse groups [18].

The exact function of many IEGs remains unclear; yet the most commonly studied, *c-fos* and *zif/268*, act as inducible transcription factors that subsequently influence neuronal physiology by regulating late response genes whereas other IEGs have more direct influence on cellular function [19]. Regardless of their function, mapping IEG expression is widely used as a technique for examining neuronal activation and can be used to investigate anatomical and functional asymmetry in the brain [20, 21]. IEGs are among the first class of genes to be induced following neuronal activation because unlike late response genes, they are not dependent on de novo protein synthesis [3, 4, 22–26]. In fact, most IEGs show peak induction within 30 min of a given stimulus. Figure 1 shows the time course for mRNA induction and protein product expression of an IEG (*zenk*) in a songbird brain after song presentation.

Furthermore, most IEGs are not induced by general stimulation and thus, offer a powerful way to examine, at the molecular level, how specific environmental stimuli can initiate cellular changes [27, 28]. By mapping the induced IEGs and their protein products, researchers are able to investigate patterns of brain activity in response to various stimuli and in numerous behavioral paradigms [4].

In birds, IEG expression has been used to identify which neurons are involved with specific environmental events and delineate which areas of the brain are activated in response to stimuli. For example, researchers used IEG expression to study the brain areas associated with sexual imprinting. Some songbirds learn to remember specific phenotypic features of parents when they are young and later use the memory to select a potential mate when they are adults. This behavior, termed sexual imprinting, is an early learning process by which a stable preference for a sexual partner is established.

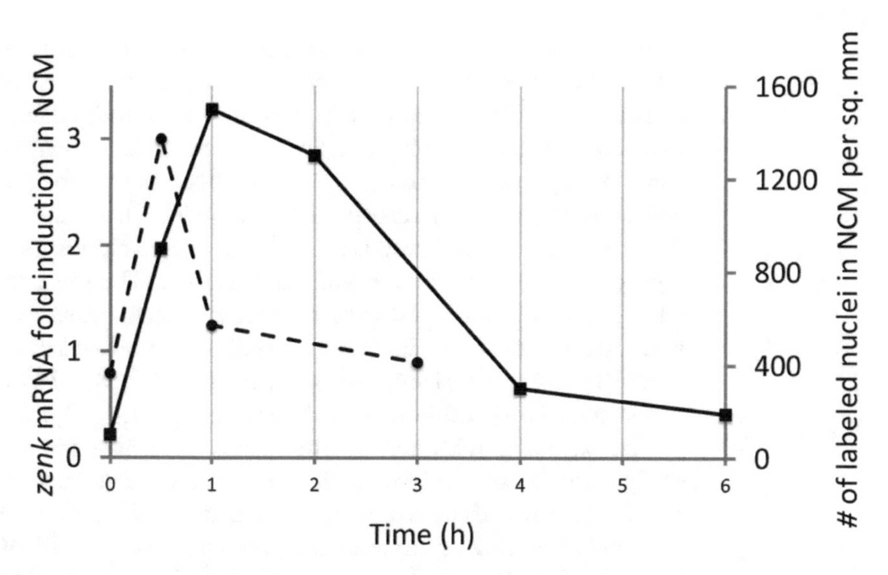

Fig. 1 ZENK expression at the protein level in the caudomedial neostriatum (NCM) after song presentation. Quantification of ZENK protein expression in NCM after song presentation. Shown are the average densities (number of nuclei per unit area 6 S.E.M.; y axis on the right) of ZENK-labeled nuclei (n 5 3 per time point). For comparison, the time course of ZENK mRNA induction by song in NCM (adapted from [49]) is represented diagrammatically by the dashed line (*y* axis on the left). Reproduced from [49] with permission from (Wiley)

In order to better understand the neural mechanisms underlying sexual imprinting, Sadananda and Bischof [29] examined the expression *c-fos* protein, as well as radioactive, modified glucose (14C-2-deoxyglucose, 2-DG), after first courtship. The 2-DG experiments revealed no difference in activation of brain areas between the first courtship condition and the group simply being chased to be aroused. However, the *c-fos* experiment showed different results. In the case of chasing, there was an intense signal in the brain area called meso nido- and hyper-pallium (formerly known as medial neo- and hyperstriatum; MNH) and a sparse signal in another area called lateral nido- and hyperpallium (formerly known as lateral neo- and hyperstriatum; LNH), whereas there was a strong signal in LNH and no signal in MNH after first courtship. The results suggest that the *c-fos* distribution may long term neuronal changes associated with sexual imprinting [29]. This is an important point. Neuronal activation does not always result in IEG expression. There are many highly active brain regions where IEGs are not expressed suggesting that they are expressed exclusively, or primarily, in neurons that maintain the feature of plasticity [29].

Among various IEGs, *zenk* has been widely used in avian brain research. Its name is an acronym of four gene names of the avian homologue: *zif268*, *EGR-1*, *NGFI-A*, and *krox24* (*zenk* will be used in this chapter when referring to the gene, ZENK when referring to the protein). *Zenk* induction activity and expression of its protein product have been extensively used to identify brain regions involved in specific aspects of behavior. For instance, when adult canaries and zebra finches were presented with audio playbacks of their conspecific or heterospecific song, the *zenk* expression in the song learning regions of the brain was much higher for a conspecific song than for a heterospecific song [30]. These findings suggest that neurons in these regions were selectively responding to biologically and socially relevant stimuli [30]. The same methods have been used to investigate the neural mechanisms associated with various behaviors, including sexual behavior in quail and starlings [31, 32], sexual imprinting in finches [29, 33, 34], species recognition [35] and homing behavior in pigeons [10, 36].

Zenk, along with a few other IEGs have also been used to study lateralization of brain processes (for a comprehensive review see [21]). Many of the studies are aimed at identifying lateral biases and corresponding asymmetrical neural circuitry [34, 37–40]. For instance, Lieshoff, Große-Ophoff, Bischof [34] studied the expression pattern of ZENK activity in brain regions associated with sexual imprinting. The results did show that, like *c-fos*, more ZENK protein was observed in the forebrain regions after first courtship. In addition, the authors reported finding a more robust expression of ZENK in the Optic Tectum (TeO) on the left side of the brain compared to the right. This difference was evident regardless of the experimental condition adding further evidence of a lateralized visual system, even at the early stages of processing [34]. ZENK-immunoreactivity has also been used to the degree to which brain regions involved with song learning and vocalization in songbirds are lateralized [39]. In this study, the authors conducted a series of experiments in which ZENK activity patterns were examined in juvenile and adult songbirds in response to tutor song playbacks. Their findings revealed left-hemispheric dominance in the areas called caudomedial nidopallium (NCM) and (HVC, a letter-based name) in zebra finches (*Taeniopygia guttata*). The authors point out the functional and anatomical similarity between these structures in the songbird and the neural basis of human spoken language [39, 40]. Specifically, the HVC and NCM in songbirds appears to be analogous to structures in the human brain involved in speech production and perception, Broca's area and Wernicke's area respectively. Further, these structures in both species are robustly left-sided dominant.

Zenk-immunoreactivity has also been used to clarify the role of olfaction in homing behavior of pigeons [10]. In a recent study, researchers examined the activity patterns of ZENK in the olfactory bulb and piriform cortex of homing pigeons. In some instances, the left or right nostril of the pigeons was occluded while it performed the homing task. The results showed that pigeons with right nostril occlusions had significantly reduced levels of ZENK-immunoreactivity in the piriform cortex, whereas no effect was observed when the left nostril was occluded. These findings indicate some degree of lateralized sensory (olfactory) processing is involved when pigeons navigate from a novel release site [10].

In our lab, we have used IEG immunohistochemical methods to study the interactions of the avian central visual pathways between the two hemispheres. In birds, retinal fibers from one eye project almost completely to the contralateral hemisphere [41, 42]. In addition, unlike placental mammals, birds do not have a corpus callosum. With exception of a few small commissures and decussations, information from one visual field is processed in the opposite hemisphere [43–45]. Because of this neural design, it is rather easy to block the visual input from one eye reaching each hemisphere. Much of our research has focused on ZENK expression in the avian forebrain and our findings have confirmed and extended our knowledge of how visual information is processed between the hemispheres [35] after they had received either a unilateral thalamic lesion or a monocular occlusion.

2 Materials

This section outlines the equipment and materials needed to prepare tissue sections for immunohistochemistry.

2.1 Tissue

We used brain tissue from an adult male pigeon for this procedure. The pigeon was anesthetized with ketamine and xylazine (cocktail) and perfused transcardially with 0.9% saline solution followed by 4% ice-cold paraformaldehyde (PFA) fixative. The tissue was sectioned using a freezing sliding microtome, but a cryostat may also be used.

2.2 Equipment and Software

Tissues on slides were digitized using a Mirax Scan 150 BF and BF/BL Digitizer (V1.12; Carl Zeiss, Toronto, ON, Canada).

NIH freeware image analysis program (ImageJ IJ 1.46r).

GNU Manipulation Program (GIMP) (V.2.6.11).

Panoramic Viewer 1.15.4 by 3DHISTECH Ltd.

2.3 Chemicals and Reagents

Hydrogen peroxide (H_2O_2)—Concentrated hydrogen peroxide is generally sold as a 30% solution.

4% paraformaldehyde (PFA) fixative.

Phosphate-buffered saline (PBS).

Triton X-100.

Primary antibody (*see* **Note 1**).

Biotinylated secondary antibody (species specific).

Avidin biotinylated horseradish peroxidase reagent kit.

Permount™ mounting medium.

3,3′ Diaminobenzidene (DAB) substrate solution (*see* **Note 2**).

3 Methods

3.1 Basic Procedure for ZENK IHC

There are several ways tissue can be processed using immunohistochemistry (IHC). The procedures detailed below utilize the indirect method of IHC in which the primary antibody signal is amplified by a labeled secondary antibody binds to it. The signal is amplified further by using avidin or streptavidin with biotinylated secondary antibodies. Figure 2 summarizes the process of indirect labeling and the end product of these steps. ZENK by itself is not easy to be visualized under microscope. With antibodies, and protein complexes, it is going to be easy to detect and count ZENK expressing neurons. These procedures also assume that the brain is fixed via transcardial perfusion and tissue will be sectioned and free-floating. These steps can be modified for pre-mounted sections.

1. Wash tissue 1 × 10 min in PBS (*see* **Note 3**).
2. Block endogenous peroxidase by incubating tissue in 0.25% hydrogen peroxide (H_2O_2) in Phosphate-buffered saline (PBS) for up to 30 min at room temperature (*see* **Note 4**).

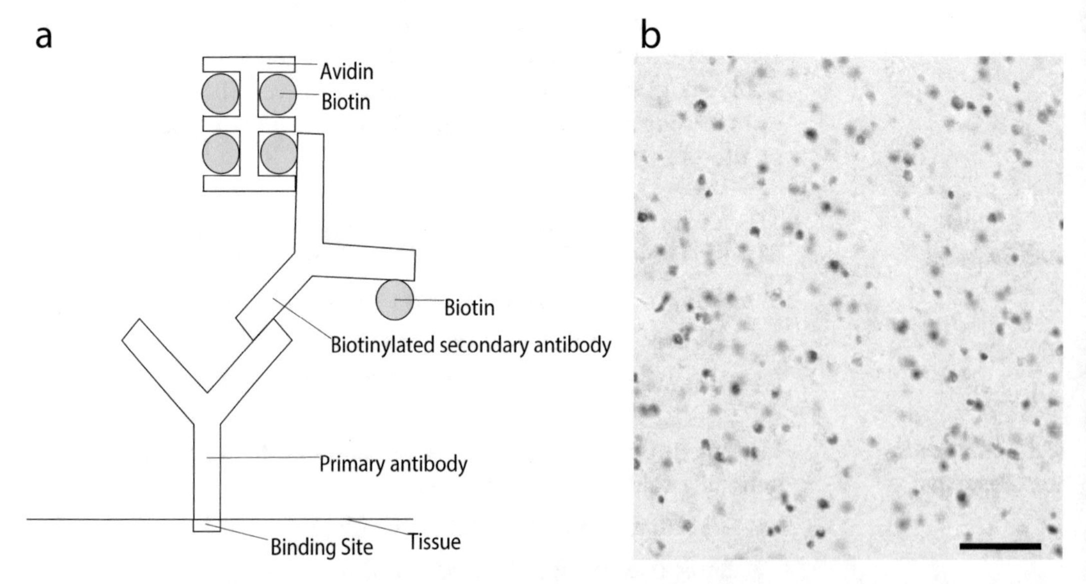

Fig. 2 Example of ZENK expression at high magnification. (**a**) A diagram of the indirect method for IHC, showing the chemicals and reagents described in this IHC protocol [50, 51]. (**b**) A sample of high magnification of ZENK from a tissue processed with the protocol explained. Picture was extracted from a digital image acquired with Mirax Scan; Carl Zeiss. Scale bar equals 50 μm

3. Wash tissue 3 × 10 min in PBS.
4. Block unspecific binding sites with 3 % normal goat serum in 0.3 % Triton/PBS for 30 min. Blocking with serum of the animal in which the secondary antibody is obtained is not recommended, as this may result in nonspecific binding and blocking of antigen sites.
5. Incubate tissue in primary antibody solution: polyclonal rabbit anti-EGR-1 + PBS + 0.3 % Triton X-100 at 4 °C for 12 h (recommended dilution 1:2000).
6. Wash tissue 3 × 10 min in PBS.
7. Incubate tissue in secondary antibody solution: (biotinylated anti-rabbit + PBS + 0.3 % Triton X-100) at room temperature for 1 h.
8. Wash tissue 3 × 10 min in PBS.
9. Incubate tissue in avidin-biotin reagent + PBS + 0.3 % Triton X-100 (NaCl) at room temperature for 1 h (*see* **Note 5**).
10. Wash tissue 3 × 10 min in PBS.
11. Visualize immunoreactively labeled neurons by using a standard DAB procedure (0.025 % 3,3′ diaminobenzidine + PBS solution + H_2O_2) (*see* **Note 6**).
12. Wash tissue 3 × 10 min in PBS.
13. Apply mounting medium and coverslip, gently pushing off any air bubbles that form and allow covered slides to dry.
14. Recommended: counterstain one series of tissue sections with cresyl violet.

3.2 Image Capturing

While ZENK protein expression is easily recognized at high magnification, it is often difficult to visualize the overall distribution pattern close up. However, low-power images can be taken of entire tissue sections which give one the ability to examine overall patterns of ZENK protein expression. For example, by studying the distribution pattern of ZENK across both hemispheres, an entire picture of plasticity-related neuronal activity in a whole brain can be observed (*see* Fig. 3a).

When ZENK proteins are stained, the chromatin remains in the nucleus of the neuron but not in the dendrites or axons. Therefore, ZENK protein expression is morphologically highly similar from cell to cell; ZENK-immunoreactive cells can be differentiated from artifact staining and counted quickly (*see* Fig. 3b).

3.3 Image Analysis

Using existing optical analysis functions in ImageJ (IJ 1.46r) we developed a new image analysis technique that uses a "Find Maxima" function to identify and label ZENK-immunoreactive cells as localized spikes of chromatic value on an image. This function identifies the largest value of image, either within a given

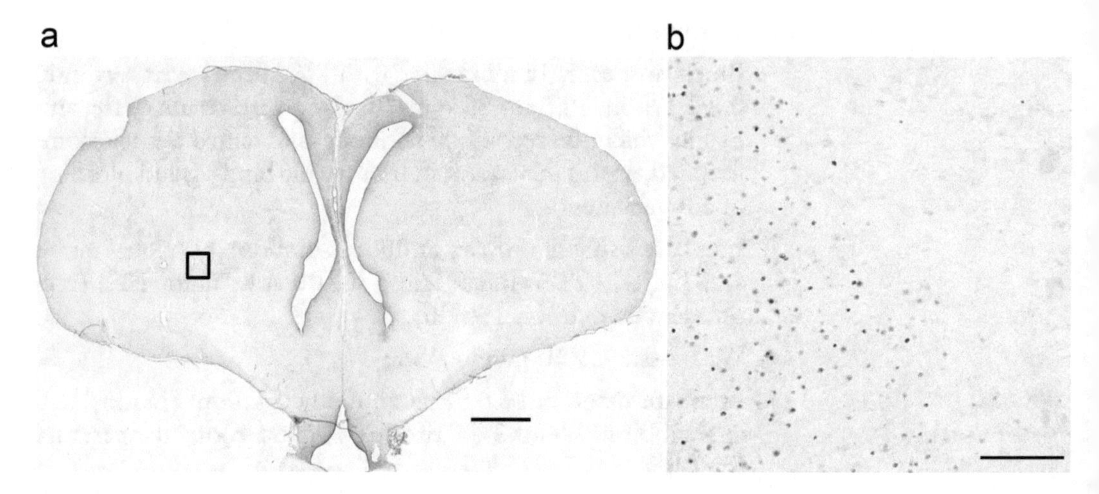

Fig. 3 Example of a digital slide from macroscanner. (**a**) Picture of a pigeon brain section at low magnification. The right eye of the bird was covered to study the effect of monocular occlusion in the left hemisphere. Section corresponds to A 7.00 according to the Karten and Hodos Atlas of the Pigeon Brain [52]. Scale bar equals 2 mm. (**b**) Region outlined on left hemisphere (*black box*) magnified 20 times. Scale bar equals 100 μm. Pictures were extracted from a digital image acquired with Mirax Scan; Carl Zeiss

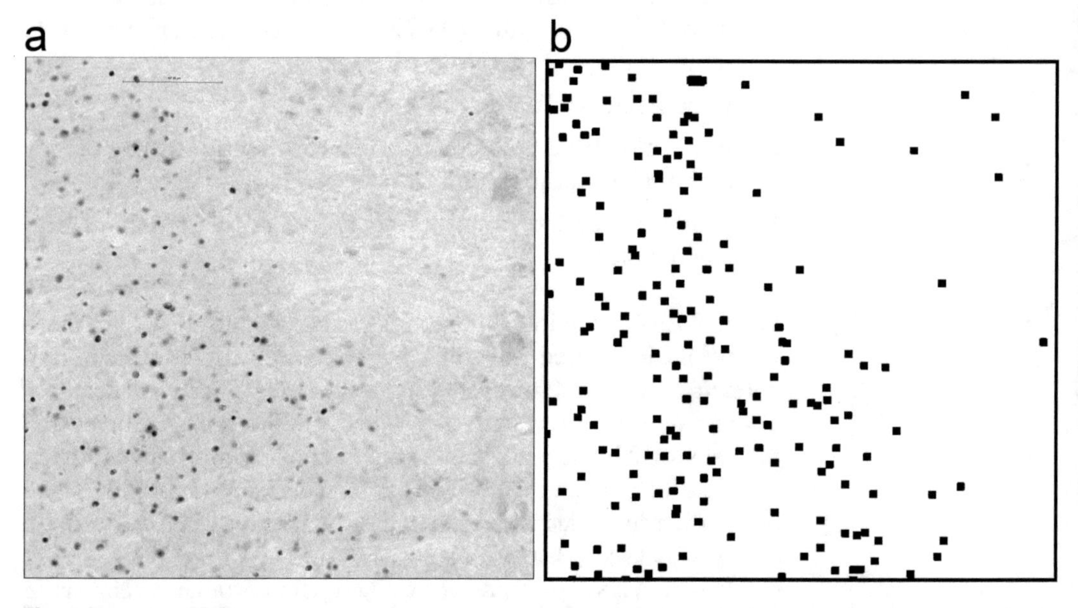

Fig. 4 Detecting ZENK protein expressing neurons with Local Maxima. (**a**) Image in Fig. 3b converted into grayscale. (**b**) Binary image shows the ZENK detection as localized spikes of chromatic value

range (relative extrema) and creates a binary image of the same size with the maxima highlighted [46]. Thus, the function works by recognizing any pixel with a high chromatic value relative to its neighboring pixels. The Find Maxima function is used here to detect signals based on spikes of chromatic values present in the tissue sample (as shown in Fig. 4).

Because signals are traditionally localized dark spots against a lighter background the Maxima function only looks at local color values, it is not subject to overall optical inconsistencies such as uneven staining, lighting, or weak signal strength. For instance, staining patterns often clump up near the edges of samples masking signals in a generally dark color that visual and even threshold-based analyses cannot differentiate. The threshold function actually views the entire region as one dark signal due to its relative darkness in comparison with the rest of the slide instead of a region of multiple signals. Once the Find Maxima function is executed, the image can be reproduced with only the maxima highlighted. Then, a function known as Dilation can be applied to this reproduced image, which simulates a form of signal extraction. Once the maxima are dilated, the signals have a finite black area. Therefore, the more signals that are present in any given space, the more area is colored black in those squares. Once the squares are averaged, darker squares can be correlated with higher signal count.

Detection and Distribution Analysis Protocol Instructions

1. Open ImageJ (IJ 1.46r).
2. Open Digital Photograph of Stained Slide.
3. If Noise Tolerance Calibration is necessary follow step 4—if not go to step 12.
4. Zoom in on picture until signals are visible.
5. Using the "Rectangle Select Tool" select a segment within the image.
6. Manually count and make note of every signal.
7. Select "Find Maxima" under Binary submenu within Process menu.
8. Select "Preview Point Selection".
9. Select "Light Background" if slide has a lighter background compared to the signals.
10. Find highest "Noise Tolerance" value until every signal manually counted is correctly labeled (no extra or missing signals)—make note of the value.
11. Close Maxima Window.
12. Deselect Selection (single-click any part of picture).
13. Select "Find Maxima" under Binary submenu within Process menu.
14. Enter value in "Noise Tolerance" based on the previous calibration step.
15. Select "Single Points" under the Output Type drop-down menu.
16. Click "OK".

17. Select "Dilate" under Binary submenu within Process menu.
18. Repeat step 17 until dots are the same size as a typical signal.
19. Save the Maxima image as either a TIFF or JPEG file.

3.4 Pseudocolor Imaging

With the GNU Manipulation Program (GIMP, V.2.8.14), single points are averaged by Gaussian blur, resulting in a gradient of gray colors from white to black. The gradient, made by Gaussian blur, is then converted into a pseudocolor image to differentiate levels of protein expression (as shown in Fig. 5). The original image and pseudocolor image can then be combined to understand the locations of protein expression.

Pseudocolor Imaging Protocol

1. Open GIMP.
2. Open Maxima image from step 19 of "Detection and Distribution Analysis Protocol".
3. Select "Gaussian" under Blur submenu within Filters menu.
4. Set and Select Blur radius depending on your purpose. *Lower numbers will result in a detailed picture, whereas higher numbers will result in less detail.
5. Click "OK".
6. Select "RGB" under Mode submenu within Image menu.
7. Select "Gradient" in Gradient menu of Toolbox.
8. Select "Gradient Map" under Map submenu within Color menu and save and close final image.
9. Open Digital Photograph of Stained Slide in step 2 of "Detection and Distribution Analysis Protocol Instructions" as a new file in GIMP, and make it grayscale from "Mode" in Image menu of Toolbox.

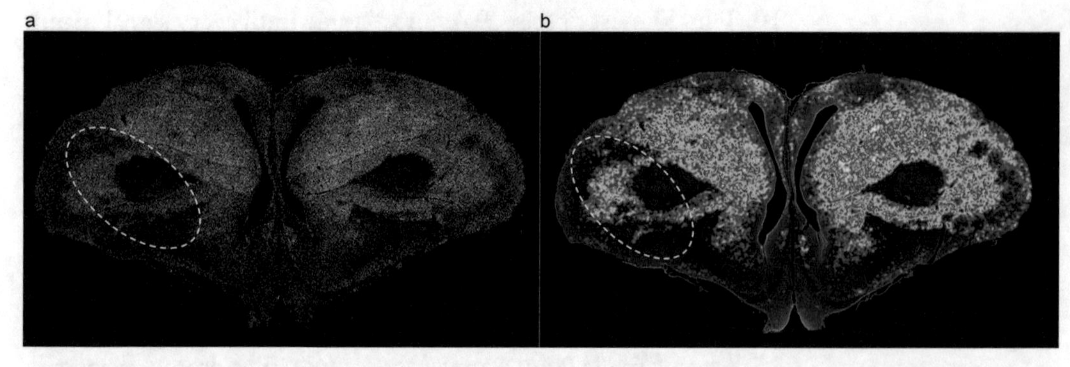

Fig. 5 Conversion of detected ZENK protein expressing cells into pseudocolor image. (**a**) Overlay of all signals detected in image in Fig. 3a. To ease the visualization, signals are converted to *green color*. (**b**) Pseudocolor image of signals after averaged by Gaussian blur function and overlaid on the original brain section

10. Invert colors of grayscale image with "Invert" in Colors menu of Toolbox.
11. Make the image mode back to "RGB" under Mode submenu within Image menu of Toolbox.
12. Open image saved in step 9 as a new layer and adjust transparency of new layer.

3.5 Implications

Our results show that this simple technique can be used to provide overall detection of certain signals that might otherwise be difficult to differentiate from tissue artifact and/or background staining. The conventional method of applying "threshold" to tissue images for analysis does not always give an accurate picture of the distribution and has problems with visually ambiguous areas, such as artifacts and edges. As shown in Fig. 6, this technique provides more accurate information of the number and density when compared to counting by hand and can be readily applied to large amounts of tissue (i.e., whole brain analysis).

This method is not intended to offer the exact number of cells. However, it will be useful to draw an overall distribution pattern and identify regions of interest for further analysis. Ability to visualize ZENK protein expression at a low power magnification can be used to examine distribution pattern of ZENK across both hemispheres as shown in Fig. 7.

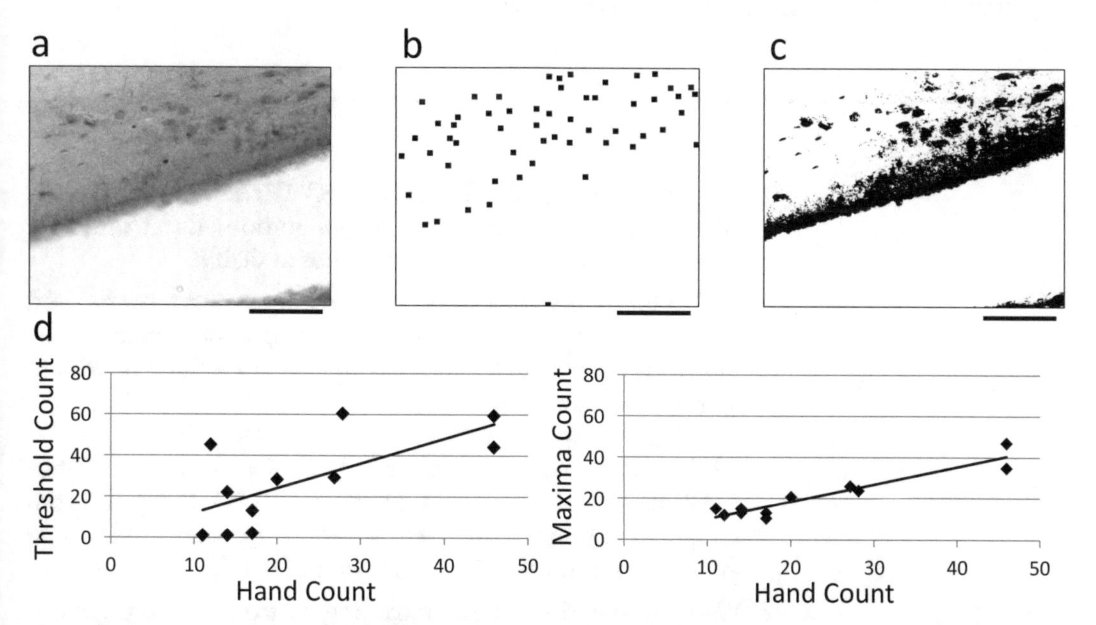

Fig. 6 Comparison of detection sensibility of Maxima and Threshold functions in heterogeneous regions. (**a**) High magnification image of the edge of a brain section that has been mounted and coverslipped. (**b**) Maxima function detects ZENK proteins while ignoring darker edges. (**c**) Threshold function sets a value for whole image resulting in failure to separate cells in dark regions. (**d**) Comparison of number of cells detected with threshold and maxima function to hand count in ten different clear and dark regions. Scale bars equal 100 μm

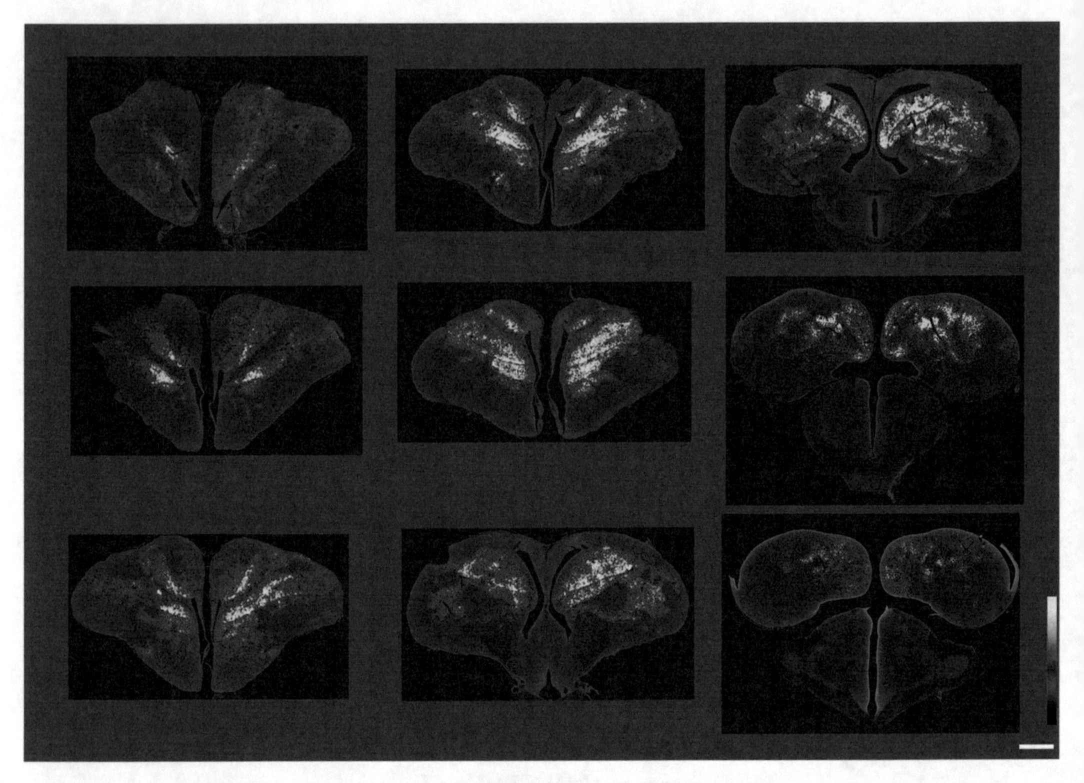

Fig. 7 Overall distribution of ZENK. Sections are from the same pigeon in Fig. 3. Methods explained in this chapter aid in the observation of lateralization in brain activity due to eye occlusion. Scale bar equals 1 mm. High expression of ZENK is shown (*yellow regions*)

4 Notes

1. Primary antibody: Egr-1 Antibody (C-19) sc-189 (Santa Cruz Biotechnology). At the time of publication, Egr-1 antibody from Santa Cruz Biotechnology is not available.
2. CAUTION: DAB is a suspected carcinogen and must be used with great care. Exposure to dry DAB may be minimized by preparing large batches of DAB solution in a fume hood and storing as frozen aliquots.

   DAB aliquot preparation: Dissolve 60 mg 3,3′ Diaminobenzidene in 200 mL PBS. Divide solution into 5–10 mL aliquots until use. Store at −20 °C. Bring to room temperature prior to use. Defrost DAB aliquot and add hydrogen peroxide to a final concentration of 0.1%.
3. Washing the tissue after retrieving it from the refrigerator (4 °C) prevents the tissue from curling onto itself when first introduced to the hydrogen peroxide solution.
4. The duration of hydrogen peroxide blocking can be decided based on the success of perfusion and/or the background staining already achieved in previous tissue staining.

5. Adding NaCl to the Triton X-100 + PBS solution tends to cause shrinkage in the tissue sections. Although this addition is a personal preference, be sure to be consistent throughout all samples within one experiment. Do NOT add NaCl to either the primary or secondary antibody incubations.
6. Nickel ammonium sulfate (NAS) + DAB + H_2O_2 results in a black signal which is useful for double-labeling or for signal enhancement. The NAS solution should be the same concentration as the DAB solution (e.g., 0.025 % for NAS and DAB).

4.1 Troubleshooting

Immunohistochemistry requires very careful attention to detail and the results can be affected by the myriad of steps from the beginning of tissue processing to the end. A complete guide to immunohistochemistry, that would provide a comprehensive section on troubleshooting, is beyond the scope of this chapter. Many articles providing strategies for improving the likelihood of IHC success are available to the reader, but we suggest the following publication as an excellent starting point [47]. This document provides a useful introduction to the processes of immunohistochemistry. Here, we cite some of the more common issues that can interfere with successful signal detection.

If you receive little or no positive staining (signal) in the tissue, check that all steps were followed and no chemicals/reagents were omitted, used in the wrong order, or incorrectly. If you are certain that all steps were followed correctly, proceed by answering the following questions:

- Are all reagents including primary antibodies, secondary antibodies, and all additional reagents (buffers, Triton, etc.) mixed properly, defective, or expired?
- Are all chemicals and solutions stored according to the manufacturer's specifications?
- Did you check the dilution of the primary antibody?
  - Try different dilutions on practice tissue.
  - Incubation settings (time and temperature).
- Did you dilute the primary antibody with the correct buffer? Changes of ion or pH of the antibody diluent might affect the antibody. Thus, the wrong buffer can diminish the sensitivity of the antibody. For ZENK, PBS (pH 7.4) appears to work well.

There are many procedures, in addition to the specific steps outlined in our IHC protocol, which will influence the intensity of signal to noise ratio. Three major procedures are:

- Perfusion of the animal and fixation of the brain.
- Cryoprotection of the brain/tissue (sucrose for frozen sectioning).
- Sectioning of the tissue.

Immediate early genes such as ZENK can be used to investigate brain asymmetries and functional lateralizations. However, it is important to run the appropriate control conditions in order to be certain that the observed differences truly represent an asymmetry and not the result of procedural artifact, such as uneven staining or section thickness. For example, the following conditions should be implemented when investigating IEG lateralization in the avian visual system:

- No Eye Occlusion (Control): ZENK is observed at basal levels in several brain regions. Thus, differences must be compared to the baseline expression observed in the absence of stimulation. With no eye occlusions, depending on existence of lateralization, two different results are possible. If there is no lateralization in the brain, then ZENK would equally be expressed in both hemispheres [35]. However, it is also possible to find difference in ZENK expression between two hemispheres, as in [34], which will suggest lateralization of functions between hemispheres. Therefore, it is also important to quantify a brain region known to show no potential bilateral differences in ZENK-immunoreactivity. For example, ZENK level does not change in hippocampal and parahippocampal areas in response to auditory stimulation [30, 48]. Another candidate brain region that shows no differences between hemispheres would be lateral striatum (LST), which is involved in motor functions [36]. These stable regions (in terms of ZENK activity) can be used as a "control" or baseline measure of ZENK expression.
- Single Eye Occlusion: Expression patterns in brain regions of interest observed after single (left or right) eye occlusion then compared to the basal expression patterns of the control group. One possible outcome includes an increase or decrease in IEG activity in the visual structures located ipsilateral to the occluded eye suggesting that visual processing is dependent on stimulus input that does not cross the midline. On the other hand, a change in IEG expression in brain regions contralateral to the occluded eye relative to the control group would indicate that visual information from the opposing eye does cross the midline (at least to some degree).
- It is also possible that IEG expression change would differ between left and right eye occlusions; therefore, a final comparison should be between opposite eye occlusion groups. If there is a difference in change in IEG expressions, it would suggest asymmetrical connections similar to differences observed after left and right nostril occlusions [43]. By measuring the differences between the control group and single eye occlusions, one can determine if there is an actual brain asymmetry regarding visual processing.

- Bilateral Eye Occlusion: Blocking visual input to both eyes may also result in asymmetrical expression of IEGs. Specifically, differential IEG expression between the two hemispheres when there is no visual stimulation would indicate that neural activity in the brain regions of interest are dependent (to a separate degree) on external visual stimulation.

References

1. Cole AJ et al (1990) Rapid rise in transcription factor mRNAs in rat-brain after electroshock-induced seizures. J Neurochem 55(6):1920–1927
2. Morgan JI et al (1987) Mapping patterns of c-fos expression in the central nervous system after seizure. Science 237(4811):192–197
3. Beckmann AM, Wilce PA (1997) Egr transcription factors in the nervous system. Neurochem Int 31(4):477–510
4. Kaczmarek L, Chaudhuri A (1997) Sensory regulation of immediate-early gene expression in mammalian visual cortex: implications for functional mapping and neural plasticity. Brain Res Rev 23(3):237–256
5. Okuno H (2011) Regulation and function of immediate-early genes in the brain: beyond neuronal activity markers. Neurosci Res 69(3):175–186
6. Perez-Cadahia B, Drobic B, Davie JR (2011) Activation and function of immediate-early genes in the nervous system. Biochem Cell Biol 89(1):61–73
7. Burmeister SS, Fernald RD (2005) Evolutionary conservation of the Egr-1 immediate-early gene response in a teleost. J Comp Neurol 481(2):220–232
8. Harvey-Girard E et al (2010) Long-term recognition memory of individual conspecifics is associated with telencephalic expression of egr-1 in the electric fish Apteronotus leptorhynchus. J Comp Neurol 518(14):2666–2692
9. Chakraborty M, Mangiamele LA, Burmeister SS (2010) Neural activity patterns in response to interspecific and intraspecific variation in mating calls in the tungara frog. PLoS One 5(9):e12898
10. Patzke N et al (2010) Navigation-induced ZENK expression in the olfactory system of pigeons (Columba livia). Eur J Neurosci 31(11):2062–2072
11. Zangenehpour S, Chaudhuri A (2002) Differential induction and decay curves of c-fos and zif268 revealed through dual activity maps. Mol Brain Res 109(1–2):221–225
12. Alaux C, Robinson GE (2007) Alarm pheromone induces immediate-early gene expression and slow behavioral response in honey bees. J Chem Ecol 33(7):1346–1350
13. Alaux C et al (2009) Honey bee aggression supports a link between gene regulation and behavioral evolution. Proc Natl Acad Sci U S A 106(36):15400–15405
14. Kiya T, Kunieda T, Kubo T (2007) Increased neural activity of a mushroom body neuron subtype in the brains of forager honeybees. PLoS One 2(4):e371
15. Ugajin A, Kunieda T, Kubo T (2013) Identification and characterization of an Egr ortholog as a neural immediate early gene in the European honeybee (Apis mellifera L.). Febs Lett 587(19):3224–3230
16. Lutz CC, Robinson GE (2013) Activity-dependent gene expression in honey bee mushroom bodies in response to orientation flight. J Exp Biol 216(11):2031–2038
17. McNeill MS, Robinson GE (2015) Voxel-based analysis of the immediate early gene, c-jun, in the honey bee brain after a sucrose stimulus. Insect Mol Biol 24(3):377–390
18. Long KD, Salbaum JM (1998) Evolutionary conservation of the immediate-early gene ZENK. Mol Biol Evol 15(3):284–292
19. Terleph TA, Tremere LA (2006) The use of immediate early genes as mapping tools for neuronal activation: concepts and methods. In: Pinaud R, Tremere LA (eds) Immediate early genes in sensory processing, cognitive performance and neurological disorders. Springer US, Boston, MA, pp 1–10
20. Mehlhorn J, Haastert B, Rehkaemper G (2010) Asymmetry of different brain structures in homing pigeons with and without navigational experience. J Exp Biol 213(13):2219–2224
21. Vallortigara G, Rogers LJ (2005) Survival with an asymmetrical brain: Advantages and disad-

vantages of cerebral lateralization. Behav Brain Sci 28(4):575

22. Goelet P et al (1986) The long and the short of long-term memory-a molecular framework. Nature 322(6078):419–422
23. Morgan JI, Curran T (1989) Stimulus-transcription coupling in neurons: role of cellular immediate-early genes. Trends Neurosci 12(11):459–462
24. Sheng M, Greenberg ME (1990) The regulation and function of c-fos and other immediate early genes in the nervous system. Neuron 4(4):477–485
25. Tischmeyer W, Grimm R (1999) Activation of immediate early genes and memory formation. Cell Mol Life Sci 55(4):564–574
26. Clayton DF (2000) The genomic action potential. Neurobiol Learn Mem 74(3):185–216
27. Guzowski JF et al (1999) Environment-specific expression of the immediate-early gene Arc in hippocampal neuronal ensembles. Nat Neurosci 2(12):1120–1124
28. Rosen JB et al (1998) Immediate-early gene expression in the amygdala following footshock stress and contextual fear conditioning. Brain Res 796(1–2):132–142
29. Sadananda M, Bischof HJ (2002) Enhanced fos expression in the zebra finch (Taeniopygia guttata) brain following first courtship. J Comp Neurol 448(2):150–164
30. Mello CV, Vicario DS, Clayton DF (1992) Song presentation induces gene expression in the songbird forebrain. Proc Natl Acad Sci U S A 89(15):6818–6822
31. Ball GF, Balthazart J (2001) Ethological concepts revisited: immediate early gene induction in response to sexual stimuli in birds. Brain Behav Evol 57(5):252–270
32. Can A, Domjan M, Delville Y (2007) Sexual experience modulates neuronal activity in male Japanese quail. Horm Behav 52(5):590–599
33. Sadananda M, Bischof HJ (2004) c-fos is induced in the hippocampus during consolidation of sexual imprinting in the zebra finch (Taeniopygia guttata). Hippocampus 14(1):19–27
34. Lieshoff C, Grosse-Ophoff J, Bischof HJ (2004) Sexual imprinting leads to lateralized and non-lateralized expression of the immediate early gene zenk in the zebra finch brain. Behav Brain Res 148(1–2):145–155
35. Patton TB, Husband SA, Shimizu T (2009) Female stimuli trigger gene expression in male pigeons. Soc Neurosci 4(1):28–39
36. Shimizu T et al (2004) What does a pigeon (Columba livia) brain look like during homing? Selective examination of ZENK expression. Behav Neurosci 118(4):845–851
37. Avey MT, Phillmore LS, MacDougall-Shackleton SA (2005) Immediate early gene expression following exposure to acoustic and visual components of courtship in zebra finches. Behav Brain Res 165(2):247–253
38. George I, Hara E, Hessler NA (2006) Behavioral and neural lateralization of vision in courtship singing of the zebra finch. J Neurobiol 66(10):1164–1173
39. Moorman S et al (2012) Human-like brain hemispheric dominance in birdsong learning. Proc Natl Acad Sci U S A 109(31):12782–12787
40. Moorman S et al (2015) Learning-related brain hemispheric dominance in sleeping songbirds. Sci Rep 5
41. Shimizu T, Karten HJ (1993) The avian visual system and the evolution of the neocortex. In: Zeigler HP, Bischof H-J (eds) Vision, brain, and behavior in birds. MIT Press, Cambridge, MA, pp 103–114
42. Shimizu T & Karten HJ (1991) Central visual pathways in reptiles and birds: evolution of the visual system. In: J Cronly-Dillon & R Gregory (Eds) Vision and visual dysfunction: Vol. 2. Evolution of the eye and visual system. McMillan Press, London pp 421–441
43. Keysers C, Diekamp B, Gunturkun O (2000) Evidence for physiological asymmetries in the intertectal connections of the pigeon (Columba livia) and their potential role in brain lateralisation. Brain Res 852(2):406–413
44. Weidner C et al (1985) An anatomical study of ipsilateral retinal projections in the quail using radioautographic, horseradish peroxidase, fluorescence and degeneration techniques. Brain Res 340(1):99–108
45. Saleh CN, Ehrlich D (1984) Composition of the supraoptic decussation of the chick (Gallus gallus). A possible factor limiting interhemispheric transfer of visual information. Cell Tissue Res 236(3):601–609
46. Rasband WS. ImageJ. 1997–2016 [cited 2016 6/12/2016]. Available from: http://imagej.nih.gov/ij/
47. Seidman J (2012) In situ hybridization and immunohistochemistry. Current protocols in molecular biology. 98:14.0:14.0.1–14.0.3.
48. Mello CV, Clayton DF (1994) Song-induced ZENK gene expression in auditory pathways of songbird brain and its relation to the song control system. J Neurosci 14(11):6652–6666
49. Mello CV, Ribeiro S (1998) ZENK protein regulation by song in the brain of songbirds. J Comp Neurol 393(4):426–438
50. Coons AH, Creech HJ, Jones RN (1941) Immunological properties of an antibody

containing a fluorescent group. Proc Soc Exp Biol Med 47(2):200–202

51. Coons AH, Kaplan MH (1950) Localization of antigen in tissue cells; improvements in a method for the detection of antigen by means of fluorescent antibody. J Exp Med 91(1): 1–13
52. Karten HJ et al (1967) A stereotaxic atlas of the brain of the pigeon (*Columba livia*). John Hopkins University Press, Baltimore, MD, pp 1–193

Chapter 11

Optogenetic Methods to Study Lateralized Synaptic Function

Mohamady El-Gaby, Michael M. Kohl, and Ole Paulsen

Abstract

Lateralization of brain function has primarily been studied at the macroscopic level. Studies have detected gross anatomical differences in neuron types, numbers, distribution, and connectivity and related these to functional divisions between the two hemispheres. Comparatively little is known about lateralization of synaptic function. A notable exception to this is the recently uncovered lateralization of receptor composition, structure, and function of hippocampal *Cornu Ammonis* (CA)3–CA1 synapses in rodents. Electrophysiological and electron microscopic studies have revealed that synapses made by the left CA3 onto CA1 neurons on either side of the hippocampus exhibit distinct receptor composition and morphology from those made by the right CA3. Here, we discuss how optogenetic activation and silencing methods have been employed to investigate the consequences of this lateralization for synaptic physiology and circuit organization during learning. The results suggest two general conclusions. First, the spatiotemporal precision of optogenetic tools enables the dissection of lateralization of circuit function in unprecedented detail. Second, seemingly subtle molecular and subcellular lateralizations can translate into prominent differences in circuit function across the hemispheres.

Key words Optogenetics, Synapses, Lateralization, Hippocampus, CA3–CA1, Long-term potentiation, Behavior, Memory, Activation, Silencing

Abbreviations

AAV	Adeno-associated virus
AMPAR	α-Amino-3-hydroxy-5-methyl-4-isoxazolepropionic acid receptor
Arch	Archaerhodopsin from *Halorubrum sodomense*
ArchT	Archaerhodopsin T from *Halorubrum* strain TP009
CaMKII	Calcium/calmodulin-dependent protein kinase II
ChAT	Choline acetyltransferase
ChR2	Channelrhodopsin 2 from *Chlamydomonas reinhardtii*
CNS	Central nervous system
EPSCs	Excitatory postsynaptic currents
EPSPs	Excitatory postsynaptic potentials
fMRI	Functional magnetic resonance imaging
GABA	γ-aminobutyric acid

Lesley J. Rogers and Giorgio Vallortigara (eds.), *Lateralized Brain Functions: Methods in Human and Non-Human Species*, Neuromethods, vol. 122, DOI 10.1007/978-1-4939-6725-4_11, © Springer Science+Business Media LLC 2017

LTD	Long-term depression
LTM	Long-term memory
LTP	Long-term potentiation
mGluR	Metabotropic glutamate receptor
mRNA	Messenger RNA
MWM	Morris water maze
NpHR	Halorhodopsin from *Natronomonas pharaonis*
NMDAR	*N*-methyl-D-aspartic acid receptor
P2X	Purinoreceptor 2X
PSD	Postsynaptic density
sCRACM	Subcellular ChR2-assisted circuit mapping
STM	Short-term memory
TRPV1	Transient receptor potential cation channel subfamily V member 1
VGAT	Vesicular GABA transporter
VHC	Ventral hippocampal commissure

1 Introduction

The classical hippocampal trisynaptic circuit ends with one of the most extensively studied synapses in the mammalian central nervous system: the CA3-to-CA1 synapse ([1]; Fig. 1). The experimental accessibility of these synapses and their potential importance in memory processing [2] have made them a popular target for

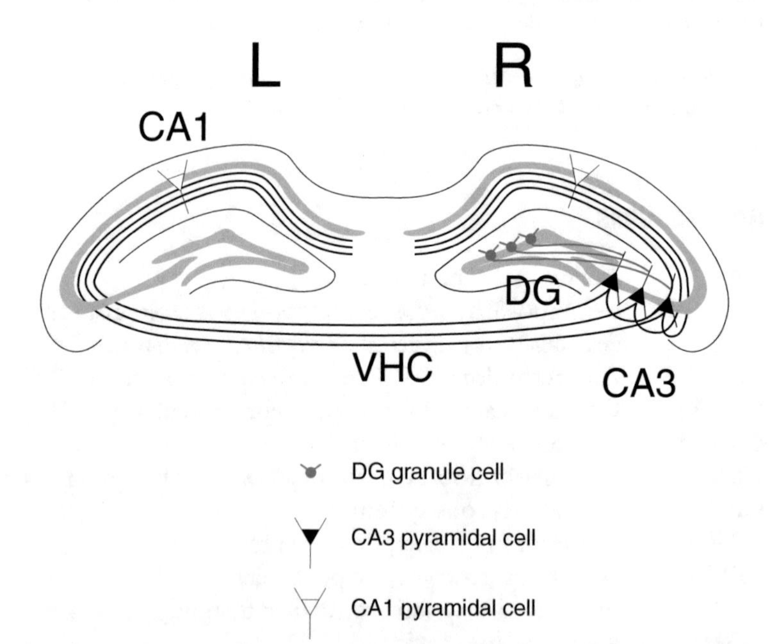

Fig. 1 Schematic of the rodent hippocampal circuit. Granule cells in the dentate gyrus (DG) project onto pyramidal neurons in the CA3. These in turn send bilateral excitatory glutamatergic projections to pyramidal neurons in the CA1

molecular, electrophysiological, and behavioral studies. In rodents, CA3 pyramidal neurons send excitatory glutamatergic projections both to the ipsilateral CA1 (Schaffer collaterals) and contralateral CA1 (commissural fibers; [1]). Intriguingly, recent data from the mouse hippocampus indicate that inputs originating in the left CA3 form synapses onto CA1 neurons that are functionally distinct from those originating in the right CA3 (henceforth referred to as left-CA3-to-CA1 and right-CA3-to-CA1 synapses respectively; [3–6]). This provides a unique opportunity to investigate the mechanisms and consequences of lateralization at the synaptic level. Before discussing this, however, we begin by briefly introducing the fundamentals of synaptic transmission at CA3–CA1 synapses and describe how synaptic efficacy can change in response to prior activity. We then discuss some of the early evidence for molecular and subcellular lateralization at CA3–CA1 synapses, before motivating the need for more refined techniques, including optogenetics, in order to investigate lateralization of synaptic function.

2 CA3–CA1 Synapses

Glutamate is released from *en passant* presynaptic boutons from CA3 pyramidal cells and activates ionotropic glutamate receptors on postsynaptic spines of CA1 pyramidal cells [1]. The cation permeability of these receptors results in depolarizing currents that can summate to excite CA1 pyramidal neurons. The majority of the resultant depolarization is mediated by the tetrameric α-amino-3-hydroxy-5-methyl-4-isoxazolepropionic acid receptors (AMPARs), while the additional contribution of *N*-methyl D-aspartate receptors (NMDARs) increases with increasing depolarization of the postsynaptic membrane ([7]; Fig. 2). Unlike AMPARs, NMDARs require coincident depolarization-induced relief of a magnesium block and glutamate binding for activation. The high calcium permeability and cytoplasmic C-terminal tail interactions of NMDARs place them at the center of several intracellular signaling networks that are critical for modulating synaptic function [7]. In particular, NMDARs are implicated in the induction of activity-dependent changes in synaptic efficacy, known as synaptic plasticity. Strong activation of NMDARs triggers signaling processes that ultimately mediate long-lasting increases in the efficacy of CA3–CA1 synapses (long-term potentiation; LTP), while weaker activation of these receptors recruits separate mechanisms that reduce synaptic efficacy (long-term depression; LTD; [8]). These processes involve the recruitment of additional AMPARs into the postsynaptic density of CA3–CA1 synapses in the case of LTP, and the removal of these receptors when LTD is induced [8]. Concomitant changes in spine morphology may occur, with potentiated spines increasing

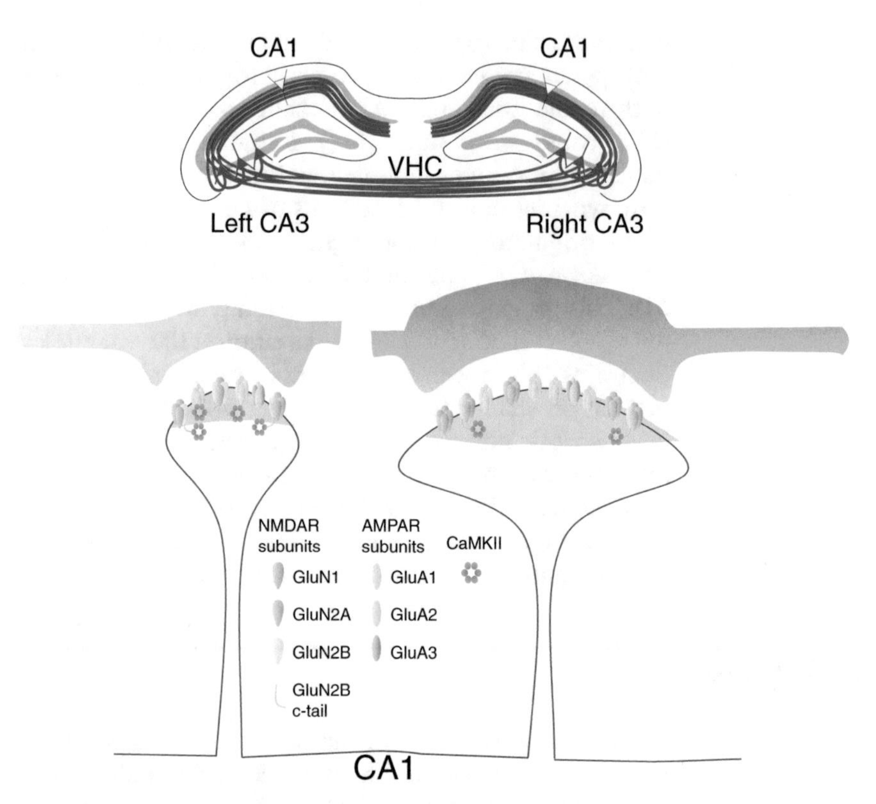

Fig. 2 Composition and lateralization of CA3–CA1 synapses. Schematic representation of glutamatergic synapses at CA1 pyramidal neurons innervated by left or right CA3 axons. Left CA3 inputs innervate dendritic spines of lower volumes than their right counterparts. Furthermore, the postsynaptic densities (PSDs) of spines receiving right CA3 inputs are larger in area than those receiving left CA3 inputs. The PSDs of spines innervated by left CA3 inputs have a higher density of GluN2B-containing NMDARs and a lower density of GluA1-containing AMPARs than those receiving right CA3 inputs. The C-terminal tails of GluN2B subunits contain high-affinity binding sites for CaMKII. These can be occupied at the basal state or can serve as docking sites for the recruitment of CaMKII in response to LTP-inducing stimuli, thereby bringing CaMKII close to its substrates and enabling LTP. The densities of GluA2 and GluA3 subunits do not differ between the two types of inputs and hence their total numbers scale with the area of the PSD

in size and assuming a more mature, mushroom-shaped structure, while depressed spines may shrink [9].

A large number of interconnected signaling molecules have been implicated in synaptic plasticity at CA3–CA1 synapses. Of particular note is the calcium/calmodulin-dependent protein kinase II (CaMKII), a multimeric enzyme that plays a central role in NMDAR-dependent LTP at CA3–CA1 synapses. Activation of NMDARs and the concomitant calcium influx result in the binding of calcium-calmodulin to CaMKII, which is critical for its activation. CaMKII in turn maintains autonomous activity beyond the initial calcium signal through autophosphorylation, and subsequently phosphorylates and directly binds to a number of targets that mediate the recruitment of AMPARs to the postsynaptic density (PSD) of CA1 spines [10]. CaMKII is brought to the PSD in part through binding to the C-terminal tail of the GluN2B subunit of NMDARs [10]. Such an interaction brings CaMKII closer to the transient high calcium microdomains near the channel opening of NMDARs and closer to its target substrates and interacting partners in the PSD. The GluN2B-CaMKII interaction is therefore predicted to be an important modulator of the ability of CA3–CA1 synapses to undergo LTP. In support of this, LTP is impaired in transgenic mice in which the GluN2B-CaMKII interaction is disrupted ([11]; reviewed in [12]). Furthermore, GluN2B levels are higher early in postnatal development, when synapses are in an immature, highly plastic state and diminish with age as synapses mature and become less plastic [13]. However, a subset of synapses may retain the immature state into adulthood. Consistent with this, a greater contribution of GluN2B subunit-containing NMDARs to the total NMDAR excitatory postsynaptic currents is seen in small plastic spines compared to large stable spines in the same adult CA1 neurons [14, 15]. Thus, a high density of GluN2B subunit-containing NMDARs may serve as a marker of highly plastic synapses.

3 Lateralization of CA3–CA1 Synapses

The first indications of a lateralization of CA3–CA1 synapses came from experiments in which the ventral hippocampal commissure (VHC), a bundle of axons connecting hippocampal subfields across the hemispheres ([1]; Fig. 2), was transected to reduce commissural CA3–CA1 connections [3, 4]. A combination of electrophysiological recordings and immunogold electron microscopy in VHC-transected mice revealed that the left CA3 innervates CA1 spines in which the GluN2B-containing NMDARs contribute more to the total NMDAR current and are of a higher density than in spines innervated by the right CA3 ([3, 4]; Fig. 2). Conversely, CA1 spines innervated by the right CA3 exhibit a higher density of GluA1 subunit-containing AMPARs compared to those innervated by the left CA3 ([4]; Fig. 2). The morphology of left and right innervated CA1 spines is also different, with the latter being on average larger and more mushroom-shaped ([4]; Fig. 2). Together, these findings suggest that left-CA3-to-CA1 synapses, with their higher GluN2B density, lower GluA1 density, and smaller spine heads, may be in a more immature, plastic state than right-CA3-to-CA1 synapses.

This raises two key questions:

1. *To what extent does the molecular and morphological lateralization translate into a difference in plasticity between left-CA3-to-CA1 and right-CA3-to-CA1 synapses?*
2. *Is there a functional distinction between left and right CA3 in behaving animals as a consequence of the lateralization of CA3–CA1 synapses?*

Synaptic plasticity has been primarily studied through electrical stimulation of afferent fibers and intracellular or extracellular recording of excitatory postsynaptic potentials. Such studies have provided critical insights into mechanisms underlying the induction, expression, and maintenance of synaptic plasticity. However, the intermingling of distinct subpopulations of afferent fibers, as is the case with afferents from the left and right CA3 in the CA1 of the mouse hippocampus [1], means that postsynaptic potentials evoked by electrical stimulation will result from synaptic transmission at various synapses with potentially very different properties. An alternative approach is needed if the properties of these individual components are to be dissected. One solution, already mentioned above, is to transect fiber bundles carrying one set of inputs to the postsynaptic region of interest, thereby allowing the study of one or more sets of inputs in isolation from the transected fibers. Targeting precision can be further improved by restricting stimulation to a particular layer of a subfield. For example, while temporoammonic inputs from the entorhinal cortex arrive onto CA1 pyramidal cell dendrites in the *stratum lacunosum/moleculare*, those from the CA3 are restricted to the *stratum radiatum* and *stratum oriens* [1]. However, transection of one fiber bundle will rarely leave only one input intact in a central nervous system (CNS) circuit. For example, VHC transection, while removing contralateral hippocampal inputs into the CA1, leaves ipsilateral inputs from both the CA3 and CA2 as well as neuromodulatory dopaminergic and noradrenergic inputs into the *stratum radiatum* and *stratum* oriens of the CA1 intact. Another major caveat of this approach is that the prolonged removal of major inputs onto a set of neurons chronically reduces their overall activity. Such a reduction recruits homeostatic mechanisms that keep the overall firing rate of the postsynaptic neurons within a specific dynamic range. For example, prolonged reductions in neuronal firing cause neuron-wide increases in synaptic efficacy, a form of synaptic scaling [16]. Thus, fiber transection does not merely subtract one set of inputs from an otherwise unchanged repertoire of inputs, but additionally alters the properties of the remaining inputs. This confounds subsequent studies of synaptic function and especially of synaptic plasticity. These same limitations apply to manipulations in behaving animals aimed at isolating the contribution of CA3–CA1 synapses in one or the other hemisphere. Therefore, there is a need for tools that enable the interrogation of specific inputs within intact circuits if the physiological and behavioral function of these inputs is to be reliably characterized.

In the following sections, we first describe the basic principles of optogenetics and subsequently discuss how both optogenetic activation and silencing have been employed in conjunction with more traditional electrophysiological and behavioral methods to investigate the consequences of the molecular and morphological lateralization at CA3–CA1 synapses. In Section 3.2, we show that optogenetic activation can be used to isolate left-CA3-to-CA1 or right-CA3-to-CA1 inputs while keeping the rest of the circuitry intact, revealing a dramatic and previously undetected difference in the capacity for LTP between the two sets of synapses. In Section 3.3, we then describe the use of optogenetic silencing to achieve unprecedented temporal and spatial resolution when interrogating lateralization in the hippocampus of behaving animals. Finally, in Section 3.4, we outline how hypotheses regarding the development and function of this lateralization can be addressed in future studies by established and newly developed optogenetic approaches.

3.1 Principles of Optogenetics

Optogenetics involves the expression of a class of exogenous, rapidly activated, light-sensitive molecules under the control of cell-type-specific promoters to enable manipulation of neuronal activity with high spatiotemporal precision [17, 18]. Pioneered by Gero Miesenböck's laboratory in 2002, these genetically-targeted, light-sensitive molecules have become an invaluable tool in neuroscience, enabling gain or loss of function within defined neuronal populations on a timescale of milliseconds. The Miesenböck group first used hybrid photoreceptors part-derived from *Drosophila* and combined with G proteins [19] to enable robust light-induced depolarization of genetically-targeted vertebrate neurons. They followed this with another approach that combined genetically-targeted expression of a receptor-ion channel with high conductance, such as the vanilloid receptor 1 (transient receptor potential cation channel subfamily V member 1; TRPV1) or purinoreceptor 2X (P2X) channel, with application of a specific, photoactivatable agonist, enabling light-induced activation of these receptors [20, 21]. Meanwhile, the Isacoff and Kramer laboratories developed a third approach that included modifying endogenous ion channels, such as voltage-gated potassium channels [22] or ionotropic glutamate receptors [23], by attaching a gating molecule or agonist to the ion channel by a light-sensitive tether. A fourth approach, now most widely used in vertebrates, makes use of light-sensitive ion channels derived from algae and bacteria. The Deisseroth lab demonstrated in 2005 that the light-gated cation-selective channel Channelrhodopsin 2 (ChR2) can be expressed in vertebrate neurons and used to depolarize these neurons upon illumination with blue light [24]. Subsequently, many other microbial opsins were discovered or engineered using site-directed mutagenesis generating a large repertoire of optogenetic molecules with distinct light-sensitivity, conductance, kinetics, and expression patterns allowing a high level of flexibility in the types of neuronal manipulations achievable (for review, see [25, 26]).

ChR2 molecules can be expressed in the neuronal plasma membrane at the soma as well as dendrites and axon. It is therefore possible to activate particular subcompartments of neurons through focal light delivery. This has been especially useful for projection targeting, allowing direct activation of axons from genetically and anatomically defined neuronal populations. Furthermore, mutagenesis of ChR2 molecules has generated a battery of variants with distinct properties. For example, the expression of microbial ChR2 in mammalian systems is enhanced by replacing algal codons with mammalian codons, generating the so-called humanized ChR2 (hChR2; [27]). In addition, more targeted mutagenesis of ChR2 has been used to increase conductance (e.g. T159C), increase calcium permeability (e.g. L132C), achieve faster channel closure kinetics and hence improve temporal fidelity (e.g. E123T) or achieve slower channel closure kinetics and hence produce stable "steps" of depolarization that persist beyond initial light delivery (e.g. C128S; [18]). Several mutations can be combined to achieve hybrid functions. For example, the E123T/T159C hChR2 double mutant exhibits rapid kinetics without the concomitant decrease in photocurrents that is a feature of the E123T single mutant [28]. This "ultrafast" variant can therefore produce robust depolarizations that follow relatively high frequency trains of light stimulations with high fidelity [28]. Nevertheless, there is still an upper limit on the frequency of light pulses that can be reliably followed, which depends partly on the properties of the ChR2 variant and partly on the biophysical properties of the neurons in which they are expressed. The fastest currently available ChR2 variants begin to show unreliable activation at stimulation rates above 40 Hz in pyramidal neurons [28]. Thus, future modifications are needed to further improve the temporal fidelity of ChR2. In our hands, the hChR2(H134R) [5] and the hChR2(E123T/T159C) variants work well to control spiking in excitatory cells or their projections (see Section 3.2).

In addition to activation tools, a number of optogenetic silencing tools have been developed. Most notably, the yellow/green-light activated inwards chloride pump halorhodopsin from *Natronomonas pharaonis* (NpHR; [29]) and the *Halorubrum*-derived yellow/green-light activated proton pumps archaerhodopsin (Arch) and archaerhodopsinT (ArchT; [25, 30]) are capable of effective neuronal silencing through hyperpolarization. These hyperpolarizing tools have been used primarily to rapidly and reversibly silence neurons in order to assess the necessity of activity in defined neuronal populations for distinct behaviors and/or network states (e.g. [6, 31]). Mutagenesis of these proteins has generated a third generation of hyperpolarizing opsins capable of stronger effective photocurrents throughout several neuronal compartments. Through addition of an endoplasmic reticulum export sequence, and a neurite targeting sequence, third-generation NpHR3.0, Arch3.0, and ArchT3.0 exhibit increased targeting to the plasma membrane and enhanced expression in the dendrites and axons compared to their first-generation versions [25]. Thus, they may hold promise for silencing defined axonal projections, such as those made by the CA3 onto the CA1, and hence for assessing their necessity for memory processes. We used successfully eNpHR3.0 to silence spiking of CA3 neurons ([6]; see Section 3.3). However, in some experiments, especially when studying neural circuits in vivo, it may be desirable to optogenetically silence select projections to an area of interest while leaving the activity at the soma and other efferent projections unmodified (see Section 3.3).

A key power of optogenetics is the capacity to selectively target defined neuronal populations. Such targeting selectivity has been achieved through the generation of transgenic mouse lines that express opsins under the control of cell-type-specific promoters. For example, VGAT-ChR2(H134R)-EYFP and ChAT-ChR2(H134R)-EYFP mice exhibit ChR2 expression that is restricted to vesicular γ-aminobutyric acid (GABA) transporter (VGAT)-expressing GABAergic neurons or choline acetyltransferase (ChAT)-expressing cholinergic neurons respectively [18]. Stereotactic injection of viral vectors allows greater flexibility and spatial precision of opsin targeting. One approach is to generate a mouse line that expresses Cre-recombinase, instead of an opsin, under a cell-type-specific promoter. Subsequent stereotactic injection of a viral construct expressing the opsin of choice in a manner that is dependent on Cre-recombinase activity can allow spatial targeting of the opsin to a subset of neurons that express Cre. For example, combining CaMKIIα-Cre mice with stereotactic injection of a Cre-dependent ChR2 viral construct into either the left or right CA3 has allowed analysis of the properties of CA3–CA1 inputs from left or right CA3 pyramidal neurons [5]. An alternative approach to achieving a high level of spatial and cell-type selectivity is to use viral vectors that express opsins under the control of cell-type-specific promoters directly, without the need for transgenic animals. Constructs expressing opsins under the control of excitatory neuron-specific CaMKIIα promoter have been successfully packaged in adeno-associated virus (AAV)-based vectors [6, 25]. These can be used to target left or right CA3 neurons in wild-type animals [6]. However, such an approach is limited by the size of constructs that can be packaged into viral vectors, and other neuron-specific promoters, such as the parvalbumin promoter, are too large to include in a construct that fits into an adeno-associated virus. We used both aforementioned viral vector-based approaches to successfully target CA3 neurons in transgenic [5] and wild-type [6] animals.

Distinct viral vectors and serotypes have different tropisms for neurons, with some infecting cells more readily than others, and thus spreading differently once injected. It is therefore important to choose the appropriate serotype for the manipulation needed. For example, for the in vivo silencing of pyramidal neurons in the CA3, we chose the AAV5 serotype [6] since it spreads over a larger tissue volume, hence maximizing the percentage of targeted CA3 neurons [32, 33]. For ex vivo electrophysiology experiments, where maximizing virus spread is less critical, we used the more moderately spreading AAV2 serotype [5]. Furthermore, for some experiments, the spread of AAVs may be less tolerable, for example when the virus is likely to cross laminar borders or when illumination targeting is less feasible. Under these conditions, the pLenti virus, which spreads over a smaller volume, may be preferred [31].

To summarize, the recent application of optogenetic tools in mammalian systems has led to a revolution in the interrogation of neural circuits. The high spatiotemporal resolution and flexibility of optogenetic tools make them ideal for investigating the properties and functions of defined populations of neurons and inputs within intact neural circuits. In the following sections, we illustrate the use of optogenetic activation and silencing for investigating hippocampal lateralization.

3.2 Optogenetic Activation and Lateralization of CA3–CA1 Synaptic Physiology

The lateralization of synaptic composition and morphology at CA3–CA1 synapses raises the possibility that left-CA3-to-CA1 and right-CA3-to-CA1 synapses exhibit distinct physiological properties. In particular, given the importance of GluN2B subunits for LTP and the negative correlation between spine size and LTP capacity, it is possible to propose the following hypothesis:

Hypothesis: The small, GluN2B rich synapses made by the left CA3 onto CA1 neurons exhibit an enhanced capacity for LTP compared to those made by the right CA3.

Two potential caveats need to be addressed before this hypothesis can be tested. Firstly, the lateralization of GluN2B receptor composition is based primarily on data from VHC transected mice [3, 4]. As discussed before, VHC transection is likely to recruit compensatory changes at the synaptic level that may alter synaptic physiology. Indeed, synaptic GluN2B subunit levels are subject to tight control by prior activity [13, 34], and may thus be altered by the loss of input to CA1 neurons due to VHC transection and/or the resultant synaptic scaling at the intact inputs. A key priority, therefore, is to investigate the possible lateralization of GluN2B subunits at CA3–CA1 synapses and its physiological consequences within intact circuits. The second caveat relates to the mechanistic heterogeneity amongst different forms of LTP. In particular, the degree of block of LTP induction with GluN2B subunit-selective antagonists varies with the LTP induction protocol used [12]. A careful consideration of this heterogeneity is needed when designing and interpreting LTP experiments with left-CA3-to-CA1 and right-CA3-to-CA1 synapses. We discuss these caveats and how innovative optogenetic activation approaches are beginning to address them.

Optogenetics allows the isolation of defined inputs without the need for fiber transection. The combination of cell-type specificity, offered by expression of opsins under the control of cell-type-specific promoters, and anatomical targeting, through the stereotactic injection of viral expression vectors, allows defined neuronal populations and their distinct projections to be interrogated [18]. Expression of the blue-light activated cation permeable ChR2 allows subsequent light-evoked depolarization of neurons and/or axons. This optogenetic approach is particularly suitable for investigating differences between left-CA3-to-CA1 and right-CA3-to-CA1 synapses for a number of reasons:

1. Left and right CA3 neurons are sufficiently anatomically separated to enable selective stereotactic targeting of viral expression vectors ([1]; Figs. 1 and 3).
2. Projections of the CA3 to CA1 neurons are sufficiently separated from CA3 projections to other targets, and from the CA3 soma [1], to allow selective activation of CA3–CA1 axons with light sources such as lasers or LEDs (although antidromic spike propagation may reduce this specificity in vivo).
3. CA1 pyramidal neurons in mice receive converging intermingled inputs from both left and right CA3 pyramidal neurons [1], making the isolation of these two inputs practically impossible in an intact circuit with electrical stimulation, therefore necessitating optogenetic targeting.
4. The large density and laminar organization of CA3 axons in the CA1 enables robust and synchronous excitatory postsynaptic currents (EPSCs) and potentials (EPSPs) to be evoked, which enables reliable detection of changes in synaptic efficacy.
5. Transmission at CA3–CA1 synapses relies on release of small synaptic vesicles which, unlike the large dense core vesicles that mediate release of monoamines and neuropeptides, do not require sustained high frequencies of action potentials for evoked release [35]. This is important because the frequency with which action potentials can follow trains of light pulses is limited [18].

Together, these factors make it possible to use stereotactic viral targeting of opsin-expressing constructs (Fig. 3) and optogenetically evoke EPSCs and EPSPs at CA3–CA1 synapses that are comparable in size and kinetics to those evoked by conventional electrical stimulation (Figs. 4, 5, and 6).

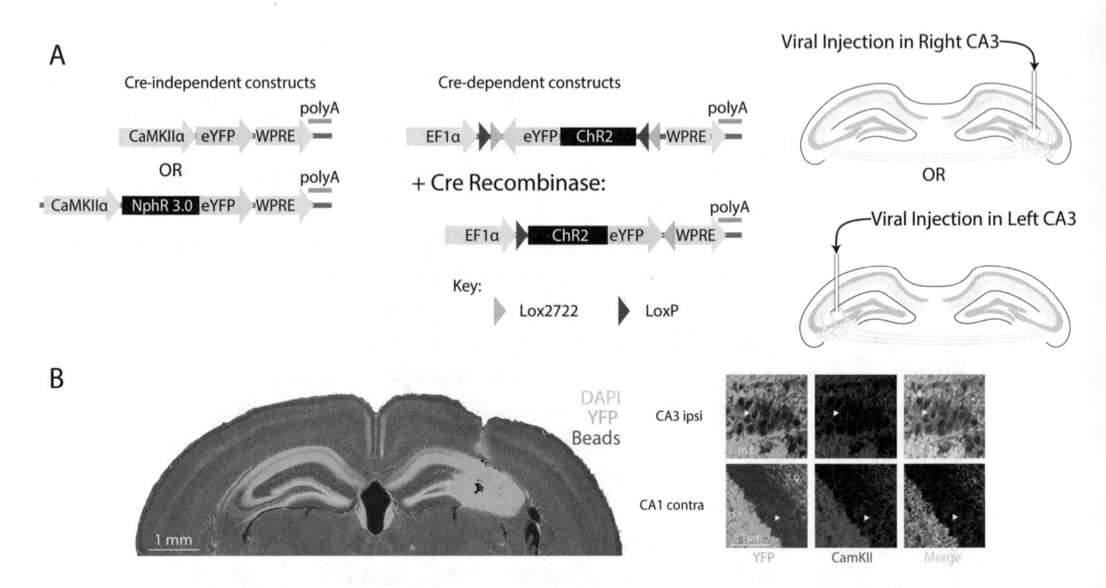

Fig. 3 Unilateral stereotactic targeting of viral constructs encoding opsins and matched controls to the CA3. (**a**) (*Left*) Schematics of gene constructs encoding enhanced yellow fluorescent protein (eYFP) or eYFP-fused to the third generation halorhodopsin from *Natronomonas pharaonis* (NphR3.0) under the control of the CaMKIIα promoter to restrict expression to excitatory neurons. The woodchuck hepatitis posttranscriptional regulatory element (WPRE) is added to enhance gene expression. (*Middle*) Gene construct encoding channelrhodopsin-2 (ChR2)-eYFP under the control of the EF1α promotor. When expressed, the Cre-recombinase enzyme catalyzes recombination at LoxP and Lox2722 sites resulting in the inversion of the construct which becomes in frame with the EF1α promoter and hence competent for expression. (*Right*) Adeno-associated viruses (AAV) containing one of these constructs are then injected into the left or right CA3 of wild-type C57BL/6 mice (Cre-independent constructs) or CaMKIIα-Cre mice (Cre-dependent constructs) using stereotactic apparatus. (**b**) (*Left*) An immunofluorescence image of a coronal slice from a mouse injected with an AAV containing a ChR2-eYFP construct. The viral injection site is marked by *red beads*. Expression of ChR2-eYFP, detected using an antibody against eYFP, is prominent in the CA3 and its projections to the ipsilateral and contralateral CA3 and CA1. (*Right*) Only pyramidal neurons (CaMKIIα expressing; arrowheads) in the ipsilateral CA3 region expressed ChR2-eYFP. *s.l.* stratum lucidum, *s.r.* stratum radiatum, *s.o.* stratum oriens. *Panel b* adapted from [5]

With optogenetics allowing the isolation of defined synapses, the next challenge is to investigate the receptor composition of these synapses. Whole-cell recording of CA1 pyramidal neurons in voltage-clamp mode allows isolation of AMPAR and NMDAR currents by two types of analyses. Firstly, a magnesium block of NMDARs at resting and hyperpolarized membrane potential results in a negligible contribution of NMDARs to EPSCs at holding voltage negative to −70 mV, allowing isolation of AMPAR currents at these holding voltages [7]. Secondly, the kinetics of NMDAR currents is slower than that of AMPAR currents [7]. Thus, at more depolarized membrane potentials, when both AMPARS and NMDARs contribute to EPSCs, the contribution of the latter can be quantified by recording the current after the AMPAR component has decayed but the NMDAR component

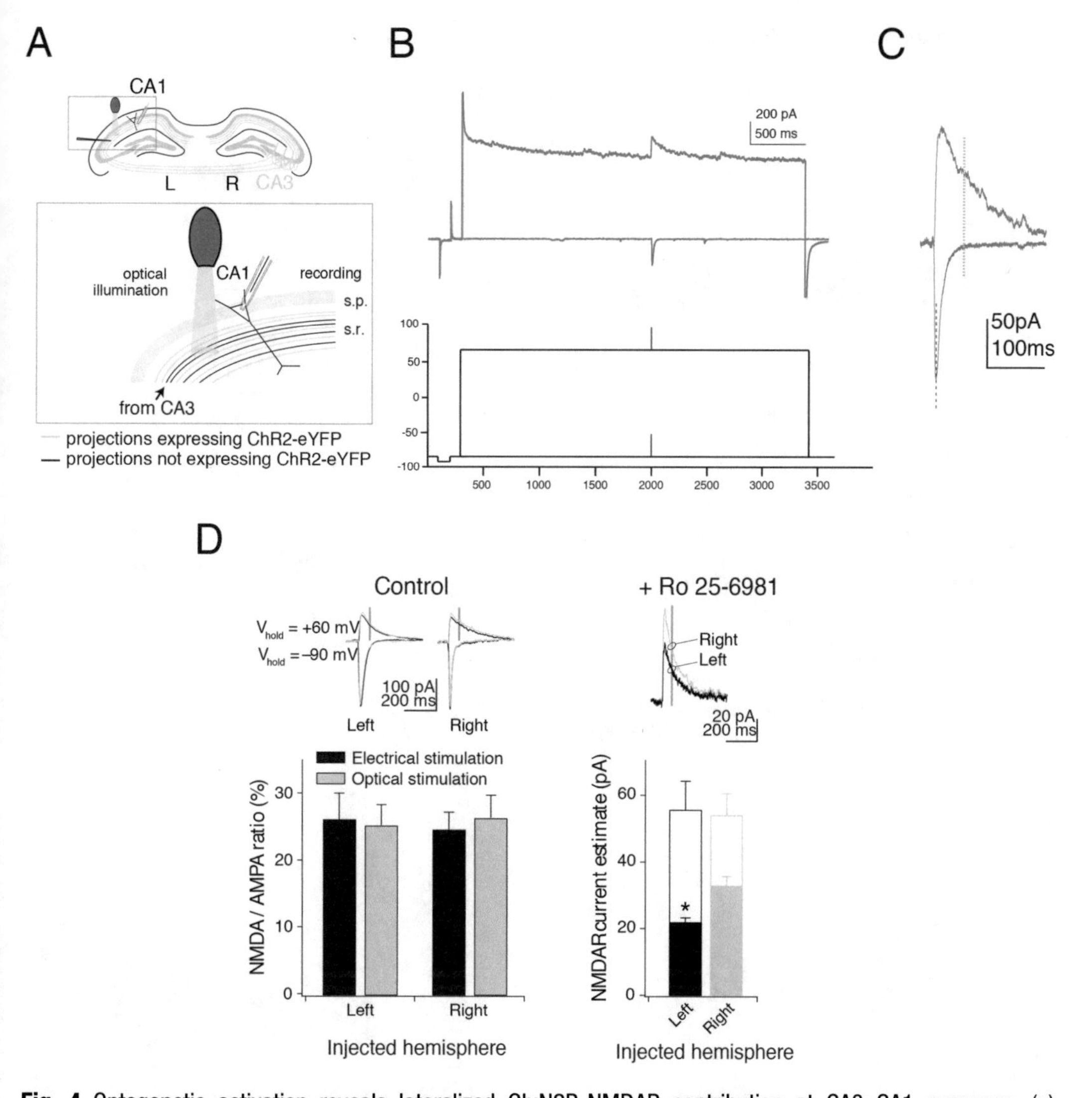

Fig. 4 Optogenetic activation reveals lateralized GluN2B-NMDAR contribution at CA3–CA1 synapses. (**a**) Optogenetic activation set up. A CA1 pyramidal neuron is patched in whole-cell patch mode and its responses monitored in voltage-clamp. A high-power laser is used to deliver 473 nm light to a focal point within the CA1 stratum radiatum in coronal hippocampal slices expressing ChR2 in CA3 projections (either Schaffer collateral or commissural) from either the left or right CA3. (**b**) The voltage-clamp recording protocol consists of alternating 3-s steps to −90 mV and 60 mV, returning to −90 mV in between. Each recorded sweep is 4 s, and there is 14 s between each stimulation (*red line*), which is delivered 2 s into each sweep to evoke an EPSC. In order to measure series resistance, a 5 mV test pulse of 100 ms duration is delivered at the start of each recording sweep. (**c**) Example EPSC traces. The AMPAR-mediated current is estimated using the peak amplitude at −90 mV (*black dashed line*) and the NMDAR-mediated current is estimated at a time window of 55–57 ms after stimulation (*grey dashed line*), at a time when the AMPAR-mediated component has decayed to less than 5 % of its peak value. (**d**) (*Left*) There was no difference in the overall NMDA/AMPA ratios between left and right for either electrical or optical stimulation (*dark gray box* indicates the time window for estimation of the NMDAR current). (*Right*) Selective block of GluN2B subunit-containing NMDARs with 0.5 μM Ro 25-6981 affected the NMDAR current evoked by left CA3 input more than that evoked by right CA3 input. *Open bars* indicate NMDAR current estimate in control, filled bars indicate remaining NMDAR current in the presence of 0.5 μM Ro 25-6981. Traces show representative optically evoked postsynaptic currents at +60 mV in the presence of 0.5 μM Ro 25-6981 for left and right CA3-injected animals. *Panel a* adapted from [6], *panel d* adapted from [5]

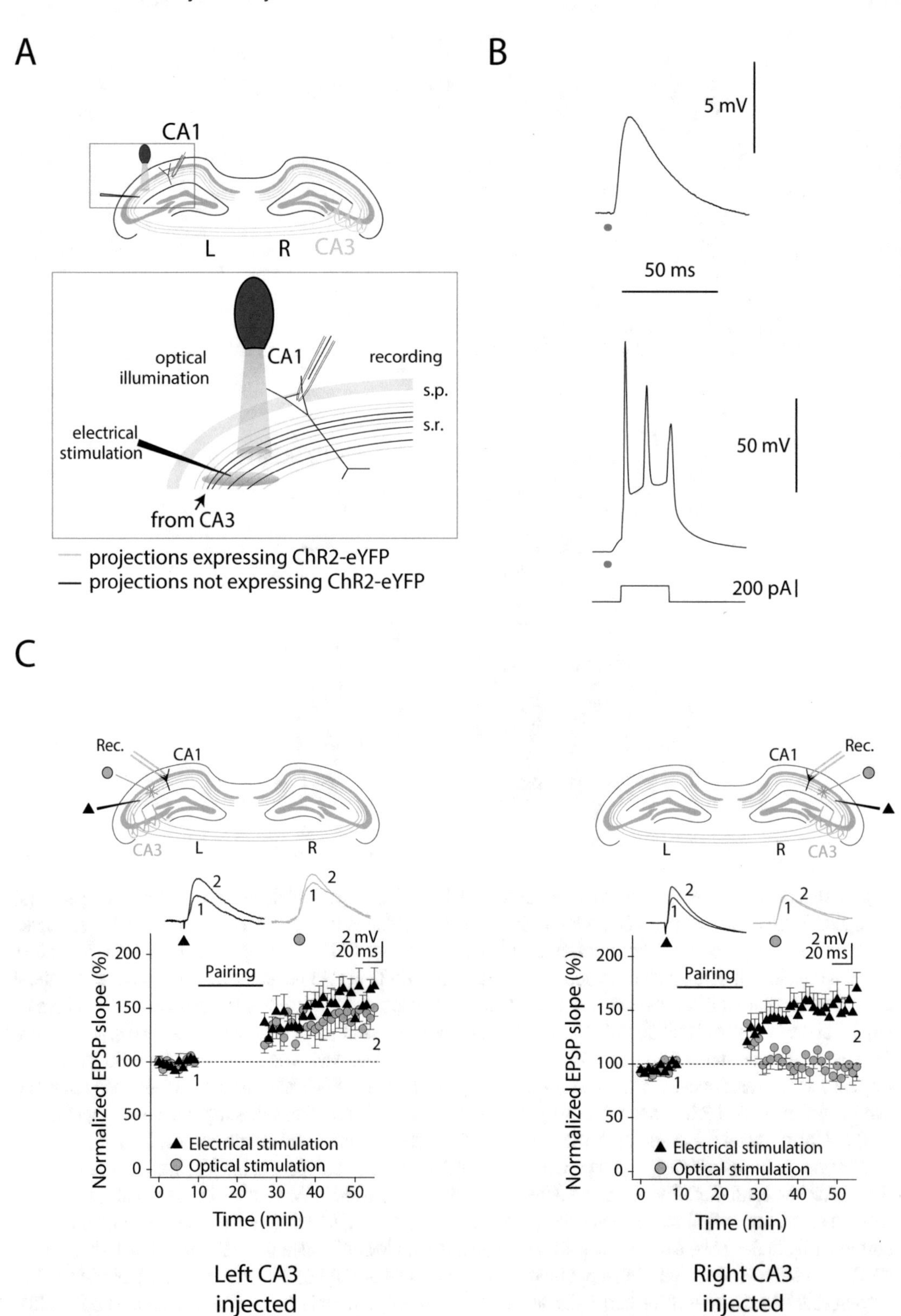

Fig. 5 Optogenetic activation reveals lateralization of t-LTP at CA3–CA1 synapses. (**a**) Optogenetic activation set up. A CA1 pyramidal neuron is patched in whole-cell mode and its responses monitored in current-clamp. A high power laser is used to deliver 473 nm light to a focal point within the CA1 stratum radiatum in coronal

persists (Fig. 4). This voltage clamp-based approach can be combined with increasingly more selective pharmacological tools that enable the isolation not only of AMPAR and NMDAR EPSCs but also of the contribution of particular subsets of these receptor types through subunit selective blockade or activation. A key example of this is the highly selective blockade of GluN2B subunit-containing NMDARs by ifenprodil and its derivatives [36]. By combining optogenetics, voltage-clamping, and selective pharmacological blockade, it is possible to investigate the functional glutamate receptor composition at defined synapses within relatively intact networks.

Kohl et al. employed the aforementioned combination of tools to investigate the NMDAR composition of left-CA3-to-CA1 and right-CA3-to-CA1 synapses ([5]; Fig. 4). The study made use of a transgenic mouse line expressing Cre-recombinase under the control of the CaMKIIα promoter. This, combined with an AAV2-packaged construct expressing ChR2 in a Cre-dependent manner under the control of the Elongation factor 1α promoter ensured ChR2 expression was restricted to excitatory projection neurons. Stereotactic injection of this construct into the left or right CA3 of adult mice enabled robust expression of ChR2 in CA3 pyramidal neurons and both their ipsilateral and contralateral axonal projections (Fig. 3). Whole-cell patch-clamp recordings from CA1 neurons from acute hippocampal slices expressing ChR2 in CA3 projections, and blue-light stimulation in the *stratum radiatum* of the CA1 using a laser enabled recording of AMPAR and NMDAR EPSCs from left-CA3-to-CA1 and right-CA3-to-CA1 synapses in isolation. The study found no left-right difference in the ratio of NMDAR and AMPAR EPSCs (N/A ratio), corroborating findings from VHC transected mice in which both the N/A ratio (when stimulating Schaffer collaterals; [3]) and the relative densities of NMDARs and AMPARs receptors were similar in the left or right hippocampus [4, 37]. Crucially, however, application of the ifenprodil derivative Ro 25-6981, to partially block GluN2B-

Fig. 5 (continued) hippocampal slices expressing ChR2 in CA3 projections from either the left or right CA3. Electrical stimulation is also delivered to the stratum radiatum of the CA1 at a separate set of fibers. (**b**) Protocol for inducing t-LTP. Optical stimulation at CA3 axons in the stratum radiatum of the CA1 induces excitatory postsynaptic potentials in CA1 pyramidal neurons (*top*). To induce t-LTP, single EPSPs are paired with a burst of three postsynaptic action potentials evoked by direct current injection into the patched pyramidal neuron (*bottom*). (**c**) Indiscriminate electrical stimulation (*triangles*) in the stratum radiatum produced robust t-LTP in CA1 pyramidal neurons. Selective optical stimulation (*circles*) of CA3 Schaffer collaterals (ipsilateral projections) and commissural fibers (contralateral projections; not shown) originating in the left hemisphere also both induced t-LTP. In contrast, optical stimulation of CA3 projections originating in the right hemisphere led to significantly less t-LTP than electrical stimulation. Insets show representative EPSPs at the indicated time points [1, 2]. *Panel a* adapted from [6], *panel c* adapted from [5].

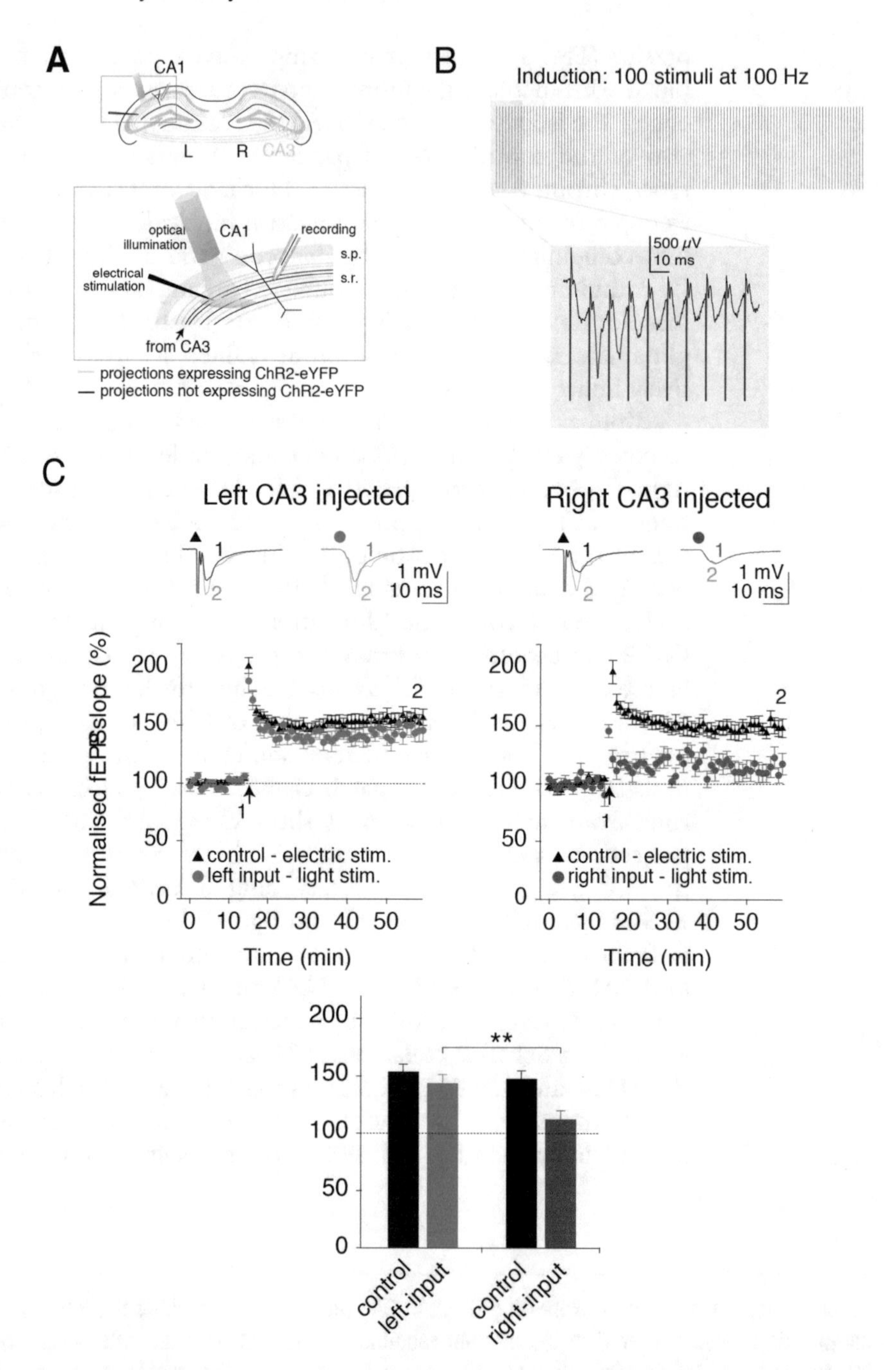

Fig. 6 Optogenetic activation reveals lateralization of HFS-LTP at CA3–CA1 synapses. (**a**) Projections from the CA3 in the injected hemisphere that express ChR2 are activated by optical stimulation in the stratum radiatum via blue (470 nm) light delivered from a high-power LED coupled to a 200 μm fiber. These projections are either hippocampal Schaffer collateral or commissural projections depending on whether the slice used is ipsilateral or contralateral to the injected hemisphere. Electrical stimulation recruits expressing and nonexpressing projections indiscriminately. The optical fiber and stimulation electrode are placed as closely as possible so that light and electrical stimulation recruit some of the same projections. Overlapping electrical and optical stimulation in the stratum radiatum enables the effect of high-frequency, electrical stimulation (HFS) to

NMDARs, caused a greater reduction of the NMDAR EPSC at left-CA3-to-CA1 compared to right-CA3-to-CA1 synapses ([5]; Fig. 4). Thus, the contribution of GluN2B-NMDARs to the total NMDAR current is greater for left-CA3-to-CA1 compared to right-CA3-to-CA1 synapses, confirming earlier findings in VHC transected mice [3].

The corroboration of the left-right asymmetry of GluN2B-NMDARs at CA3–CA1 synapses using optogenetic isolation of left-CA3-to-CA1 and right-CA3-to-CA1 inputs opens a window for investigating the consequences of this lateralization for synaptic plasticity. Given the difference in effectiveness of GluN2B subunit-selective NMDAR antagonists in blocking different forms of LTP, a suitable starting point is to investigate a highly GluN2B-sensitive form of LTP. One such form is timing-dependent LTP (tLTP), which can, for example, be induced by repeated pairings of an EPSP with a burst of three postsynaptic action potentials 5–10 ms later [38]. This stimulation pattern induces robust, slowly rising LTP (Fig. 5). Crucially, tLTP is completely abolished by partial blockade of GluN2B-NMDARs with Ro 25-6981 [5]. Furthermore, genetic knockout of *Grin2b*, the gene encoding the GluN2B subunit, or knock-in of a version of the *Grin2b* gene lacking the CaMKII binding sites on the GluN2B C-terminal tail replacing the wild-type *Grin2b* gene, abolishes tLTP [11, 12]. Together, these findings suggest that the induction of tLTP is strongly dependent on GluN2B-NMDARs. Given this strong dependence on GluN2B-NMDARs, Kohl et al. used optical stimulation of ChR2-expressing CA3 afferents in the *stratum radiatum* of the CA1 to assess the induction of tLTP at left-CA3-to-CA1 and right-CA3-to-CA1 synapses, in conjunction with more traditional electrical stimulation. Electrical stimulation, which recruits both Schaffer collateral and commissural CA3–CA1 inputs originating from the left and right CA3, induced identical tLTP in both the left and right CA1 (Fig. 4). However, optogenetically isolating CA3–CA1 inputs from one hemisphere revealed that while robust tLTP was induced at left-CA3-to-CA1 synapses, the same induction paradigm failed to induce tLTP at right-CA3-to-CA1 synapses

Fig. 6 (continued) be measured on selective optical stimulation. The fEPSP response is monitored by a glass recording pipette containing artificial cerebrospinal fluid (aCSF) in the stratum radiatum. (**b**) The induction protocol consists of a tetanus of 100 electrical stimulations at 100 Hz. The *gray inset* window shows an example field response to the first ten stimuli of the high-frequency induction protocol. (**c**) *Top*: HFS (*arrows*) produced robust LTP in the electrical pathway (*black triangles*), but the optical pathway (*circles*) only showed LTP when projections originated in the left CA3. *Insets* show representative fEPSPs at the indicated time points [1, 2]. *Bottom*: Significantly more LTP was observed in left-injected mice than in right-injected mice in the optical pathway. *Broken lines* represent baseline. Error bars represent s.e.m. $^{**}P < 0.01$. *Panel a* and *c* adapted from [6].

([5]; Fig. 4). This lateralization was apparent when considering CA3 afferents onto the CA1 both ipsilateral and contralateral to the ChR2-expressing CA3. Thus, the potential for tLTP at CA3–CA1 synapses depends strictly on the hemispheric origin of the CA3 afferents, and not the hemispheric location of the postsynaptic CA1 targets.

The left-right asymmetry of tLTP at CA3–CA1 synapses raises an important question: What is the relationship between the lateralization of GluN2B-NMDARs and that of LTP? A simple possibility is that the lateralization in tLTP is completely explained by the difference in the density of GluN2B subunits and their contribution to the total NMDAR EPSCS between left-CA3-to-CA1 and right-CA3-to-CA1 synapses. This would in turn predict that for less GluN2B sensitive forms of LTP, lateralization would be less apparent or completely absent. One form of LTP that is not blocked by GluN2B-selective NMDAR antagonists is high-frequency stimulation induced LTP (HFS-LTP). Induction of HFS-LTP, typically by presynaptic stimulation at 100 Hz for 1 s, produces a rapidly developing and long-lasting potentiation of synaptic transmission (Fig. 6). In stark contrast to tLTP, pharmacological blockade of GluN2B-NMDARs has no effect on the long-term component of such potentiation [39]. Furthermore, HFS-LTP is only partially reduced in *Grin2b* knockin mice lacking the CaMKII binding site [40]. HFS-LTP is therefore a suitable candidate to test whether the lateralization of LTP at CA3–CA1 synapses extends to paradigms with low sensitivity to selective interference with GluN2B subunits.

The optogenetic induction of HFS-LTP faces a major technical challenge; opsins are limited in their ability to follow high frequency trains of light stimuli for long durations [18]. Despite the molecular engineering of several “ultra-fast” ChR2 variants [28], the rapid kinetics needed for opsins to follow 100 Hz stimulation for 1 s remain elusive. To bypass this limitation, an alternative approach has been used. Shipton et al. injected an AAV5-packaged construct expressing the hChR2(E123T/T159C) variant under the direct control of the CaMKIIα promoter into the left or right CA3. HFS-LTP was induced in the *stratum radiatum* of the CA1 using conventional electrical stimulation while optogenetic activation of left or right CA3 afferents overlapping with those stimulated electrically allowed the effects of HFS-LTP induction on these distinct inputs to be probed ([6]; Fig. 6). Thus, unlike the assessment of tLTP by Kohl et al. [5], Shipton et al. [6] used optogenetics not to induce HFS-LTP, but to monitor left-CA3-to-CA1 or right-CA3-to-CA1 synapses, while HFS-LTP was induced through electrical stimulation. Surprisingly, the lateralization observed with tLTP was also seen with HFS-LTP; robust HFS-LTP was observed when monitoring left-CA3-to-CA1 synapses but not when monitoring right-CA3-to-CA1 synapses under the same conditions ([6]; Fig. 6). As with tLTP, it was the hemispheric origin of the CA3 afferents, rather than that of the CA1 targets, that dictated whether LTP was observed. Thus, lateralization of plasticity at CA3–CA1 synapses extends to LTP paradigms with low sensitivity to GluN2B antagonists.

The lateralization of CA3–CA1 plastic capacity across forms of LTP with distinct induction requirements suggests a fundamental distinction between left-CA3-to-CA1 and right-CA3-to-CA1 synapses [41]. Indeed, following the initial discovery of a lateralization in GluN2B-NMDARs [3], a number of other differences emerged between left-CA3-to-CA1 and right-CA3-to-CA1 synapses. The density of GluA1-containing AMPARs is higher at right-CA3-to-CA1 synapses [4]. In addition, right CA3 axons synapse onto larger, more mushroom-shaped CA1 boutons than those from the left CA3 [4]. Together, the lateralization in GluA1-AMPAR density and spine morphology suggests that right-CA3-to-CA1 synapses may represent previously potentiated synapses, a feature that would be expected to occlude the expression of subsequent LTP. Furthermore, immunoblotting of the synaptic fraction from the CA1 of VHC transected mice has suggested a larger density of metabotropic glutamate receptor 5 (mGluR5), a key modulator of some forms of LTP (e.g. [42]), at left-CA3-to-CA1 synapses [37]. Thus, it is likely that distinct molecular features at the level of LTP induction (GluN2B density), expression (GluA1 density and spine size) and modulation (mGluR5 density) cooperate to result in left-CA3-to-CA1 synapses are highly plastic while right-CA3-to-CA1 synapses being relatively stable across distinct induction paradigms.

The results described above were derived from recordings in the hippocampal slice preparation. The ease of synaptic recordings and drug delivery in such a preparation has made it highly popular among researchers studying CA3–CA1 synaptic transmission and plasticity. Furthermore, many of the findings derived from this preparation have been corroborated with in vivo recordings in intact animals. However, a number of studies have revealed that the slicing and incubation procedure gives rise to molecular and subcellular changes that alter synaptic physiology. For example, CA1 dendritic spine density increases significantly in slices compared to those in the perfusion-fixed hippocampus [43]. Similarly, the phosphorylation and activation states of signaling molecules such as CaMKII and protein kinase A and synaptic receptor subunits such as GluA1 are altered in hippocampal slices [44]. Such changes may alter the function and plasticity of CA3–CA1 synapses. Indeed, the capacity for LTD in adult CA3–CA1 synapses is highly sensitive to the exact slicing and incubation procedure [44]. These considerations should therefore encourage attempts to investigate lateralization of CA3–CA1 function in intact animals. A major strength of optogenetics is its applicability in vivo. Although light scattering is more pronounced in intact tissue, higher light power densities can be easily and safely achieved by high power lasers, allowing reliable activation of neurons and axons in vivo [18]. Corroboration of the findings from hippocampal slices in intact mice would serve to both strengthen existing results and allow more insight into forms of plasticity induced by naturalistic activity patterns and during behaviorally relevant network states.

To summarize, optogenetic activation of axonal projections enables an unprecedented level of spatiotemporal precision and flexibility in addressing synaptic lateralization. The power of this technique was demonstrated in three ways. First, through combining optogenetic activation of CA3–CA1 synapses with NMDAR/AMPAR current recordings and highly selective pharmacological agents, it was possible to directly assess functional receptor composition at defined synaptic inputs in living neurons. This has revealed a higher contribution of GluN2B-NMDARs at left-CA3-to-CA1 compared to right-CA3-to-CA1 synapses. Secondly, optogenetic projection stimulation can be used to specifically potentiate defined inputs, revealing a preferential induction of tLTP at left-CA3-to-CA1 synapses, and no tLTP induction at right-CA3-to-CA1 synapses. Finally, for situations where optogenetic activation with currently available opsins is incapable of directly inducing potentiation, as is the case with HFS-LTP, optogenetic activation can instead be used to monitor the impact of electrically induced potentiation on distinct inputs. This approach revealed that the left lateralization of LTP at CA3–CA1 synapses extends to HFS-LTP, a form of LTP with low sensitivity to GluN2B-selective NMDAR antagonists. Together, these findings suggest a fundamental lateralization in the capacity for LTP at CA3–CA1 synapses across distinct induction paradigms. However, corroboration of these results in vivo and investigations into other forms of plasticity (such as LTD) are necessary in order to understand the full extent of functional lateralization at CA3–CA1 synapses.

3.3 Optogenetic Silencing and Lateralization of CA3 Function During Hippocampus-Dependent Learning and Memory

The hippocampus is a key hub within the mammalian CNS [45]. Roles in memory processing and spatial navigation have made the hippocampus and its inputs and outputs the subject of intensive physiological and behavioral studies [46]. Consequently, more details concerning the functional roles of distinct hippocampal neurons are emerging. For example, the granule cells of the dentate gyrus, through their sparse firing and receipt of divergent inputs from the entorhinal cortex, likely perform a pattern separation process, orthogonalizing incoming information in order to minimize interference between neuronal ensembles prior to encoding these memories [47]. Conversely, the strongly recurrently connected CA3 pyramidal neurons may mediate pattern completion during memory retrieval, recapitulating the full memory ensemble from partial cues [47]. Together, distinct computations carried out within and between hippocampal subfields are thought to cooperate to mediate memory formation and retrieval.

While the basic hippocampal circuitry is similar between the hemispheres, the recently uncovered difference in synaptic composition, morphology, and plasticity between left-CA3-to-CA1 and right-CA3-to-CA1 synapses suggests that these synapses may be part of functionally distinct circuits. In particular, LTP has been strongly implicated in the formation of memories. The synaptic plasticity and memory hypothesis states that "*Activity-dependent synaptic plasticity is induced at appropriate synapses during memory formation, and is both necessary and sufficient for the information storage underlying the type of memory mediated by the brain area in which that plasticity is observed*" [48]. Several lines of evidence in rodents provide correlative support for this hypothesis. Pharmacological blockade or genetic knockout of NMDARs in the CA1, manipulations that block LTP induction at CA3–CA1 synapses, impairs performance on hippocampus-dependent spatial memory tasks ([49, 50], reviewed in [2]). Furthermore, LTP-like changes in synaptic efficacy and protein phosphorylation in the CA1, and occlusion of further LTP induction at CA3–CA1 synapses are seen with hippocampus-dependent learning (e.g. [51], reviewed in [2]). These findings are consistent with LTP at CA3–CA1 synapses playing a critical role in memory formation. Given the potential importance of LTP for memory formation, the lateralization in the capacity for LTP at CA3–CA1 synapses across distinct induction paradigms warrants the following hypothesis:

Hypothesis: The plastic left-CA3-to-CA1 synapses are preferentially involved in mediating long-term memory formation. Consequently, left, but not right, CA3 neurons are preferentially required for performance on hippocampus-dependent long-term memory tasks.

Previous studies of functional lateralization in the rodent hippocampus have produced mixed results. For example, Klur et al. used unilateral pharmacological inactivation of either the left or the right hippocampus during training on the Morris water maze (MWM; [52]). While learning of the MWM task was unaffected, performance on a retention probe trial was impaired specifically when left hippocampus was inactivated during the training period [52]. Conversely, unilateral lesions of the left or right hippocampus in mice caused a similarly small impairment in performance on the MWM, with no evidence of lateralization [53]. The discrepancy between these and other studies could reflect differences between mice and rats. Alternatively, the lack of lateralization seen in some studies could reflect the nature of the manipulations used. Prolonged manipulations, such as hippocampal lesions, could recruit compensatory synaptic and/or wiring changes that may impact performance on a given task. One dramatic demonstration of this principle was observed when comparing the effects of acute and prolonged silencing of CA1 neurons. While acute CA1 silencing resulted in a profound deficit in recall of remote contextual fear memories, prolonged (30 min) silencing of the CA1 produced no deficit [31]. This lack of an effect of prolonged CA1 silencing was accompanied by an enhanced recruitment of neurons in the anterior cingulate cortex, which may have compensated for the silenced CA1 neurons to mediate accurate memory retrieval [31]. Thus, acute, trial-limited synaptic silencing is highly desirable to avoid or minimize compensatory changes that alter the processing and/or routing of information through brain circuits.

The speed, reversibility and in vivo applicability of optogenetic tools make them of great value when acute manipulations are necessary. Making use of these advantages of optogenetics, Shipton et al. employed an optogenetic neuronal silencing approach as mice learnt a hippocampus-dependent long-term memory (LTM) task. A viral construct expressing the green light-activated chloride pump halorhodopsin (NpHR) under the control of the CaMKIIα promoter was injected into either the left or right CA3 (Fig. 3). This produced robust NpHR expression in CA3 pyramidal neurons. For light delivery, an optical cannula was implanted with its tip immediately above the dorsal CA3 in the injected hemisphere (Fig. 7). A separate group of mice were injected with a construct

Fig. 7 (continued) or eYFP only. An optical fiber contained in a metal cannula (*red arrow*) was inserted into the brain above the AAV-injected CA3 and secured to the skull using dental cement (*blue arrow*). *Bottom*: Green (532 nm) laser light was coupled to the inserted optical fiber and delivered to the CA3. (**b**) Illumination of CA3 neurons in anesthetized mice with green laser light for 30 s resulted in a reversible reduction in spontaneous spiking. The *graph* shows the normalized mean frequency of spiking. *Grey window* depicts one period of silencing. *Green bars* depict time with light on. Note the rebound activity observed at light off set [59]. (**c**) Elevated Y-maze task. *Left*: Mice were trained to find the rewarded arm (+), which remained in a fixed location relative to extra-maze cues, following release from one of the two start arms (S) allocated through a pseudorandom sequence. Mice entering the nonrewarded arm (–) were removed from the maze. Light was delivered throughout the trial. *Middle*: Mice received blocks of ten trials a day for 11 days and the number of correct arm entries was recorded each day. The performance of all four groups of mice improved over the course of testing, but to a different extent. On the final day of testing, the reward was given after the arm choice was made (post-choice baiting, P.C.B.). *Broken lines* represent chance performance of 50 %. Light delivery in mice injected with NpHR in the left CA3 (left-NpHR, *blue*) impairs their learning of reward location, but does not affect learning in right-NpHR mice (*red*) and control groups (left-eYFP: *black*, right-eYFP: *gray*). *Right*: The average performance on the penultimate day. *Broken lines* represent chance performance of 50 %. (**d**) Spontaneous alternation T-maze task. *Left*: Mice were placed in the start arm (S) and allowed to make a free choice of arm. Once they entered an arm, a barrier was put in place so that they explored the chosen arm for 30 s. Mice were then removed and placed immediately back in the start arm and the next arm choice was recorded. If mice chose the novel arm, this was counted as a spontaneous alternation. *Right*: The average rate of alternation in the light-on condition. *Broken line* represents chance performance of 50 %. Light delivery reduces performance of both right-NpHR and left-NpHR mice compared to their respective eYFP controls. $^{*}P<0.05$, $^{**}P<0.01$, $^{***}P<0.001$. Error bars represent s.e.m. *Panel b–d* adapted from [6]

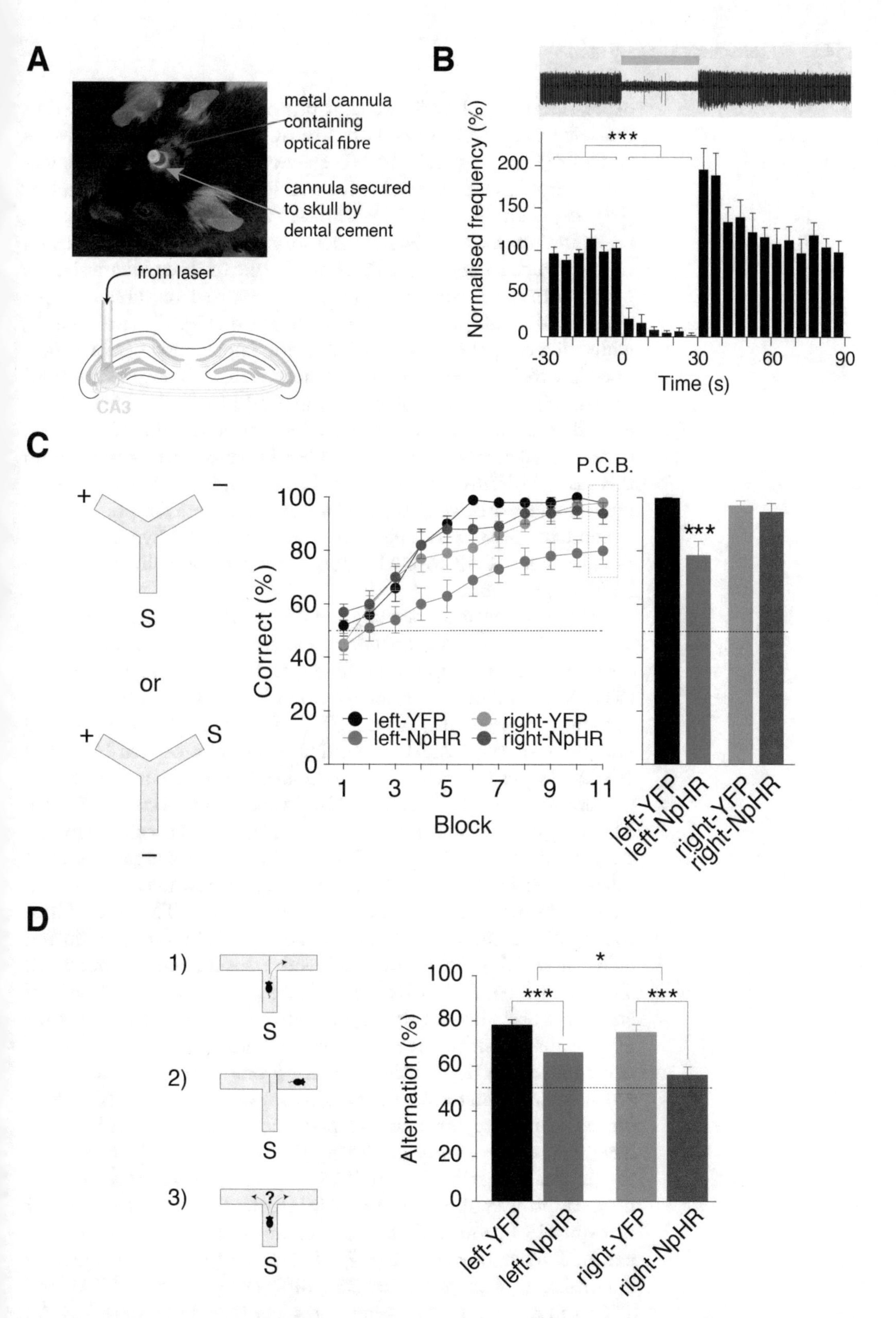

Fig. 7 Optogenetic silencing reveals lateralization of CA3 function during long-term, but not short-term, memory performance. (**a**) *Top*: Mice received unilateral CA3 injection of an AAV construct containing either eNpHR3.0-eYFP

expressing the yellow fluorescent protein (YFP) and implanted in the left or right CA3 (as with NpHR injected mice) to serve as controls (Figs. 3 and 7). NpHR achieves robust and acute neuronal hyperpolarization suitable for trial limited silencing of neurons ([6, 29, 54]; Fig. 7). The authors trained NpHR-expressing, implanted mice on a reference memory Y-maze task. This involved training mice to find a liquid reward in one of three identical arms using extra-maze cues for guidance, a task that requires multiple trials across multiple days for mice to learn (Fig. 7). Intriguingly, while silencing the right CA3 did not impair task performance, mice in which the left CA3 was silenced were significantly impaired in task acquisition compared to controls ([6]; Fig. 7). This did not extend to a hippocampus-independent visual discrimination task, as neither silencing left nor right CA3 impaired task acquisition in this case [6]. Thus, activity of left but not right CA3 neurons during trial performance is required for acquisition of a hippocampus-dependent LTM task. This is consistent with the hypothesis that the plastic left-CA3-to-CA1 synapses are preferentially involved in LTM acquisition.

While some evidence suggests that LTP at CA3–CA1 synapses mediates LTM acquisition, less is known about the mechanisms underlying hippocampus-dependent short-term memory (STM). Several lines of evidence point towards a mechanistic dissociation between LTM and STM. For example, while global knockout of the GluA1 or GluN2A subunits of AMPARs and NMDARs respectively impairs hippocampus-dependent STM, performance on the MWM, a LTM task, is unaffected [55, 56]. This may reflect the use of distinct circuits and/or cellular mechanisms by STM and LTM. This would in turn suggest that the relationship between synaptic lateralization and performance on STM tasks may differ from that observed with LTM tasks. Thus, to investigate the possible lateralization of CA3 function during STM tasks, Shipton et al. employed a spontaneous alternation T-maze task [6, 57]. Mice were placed at the base of a T-shaped enclosure and allowed to explore one of the two other arms for 30 s. Subsequently, mice were placed back at the base of the T and again given the choice of entering one of the remaining two arms. Mice are naturally inclined to choose the previously unexplored (alternate) arm in the majority of cases, provided they have formed a memory of visiting the first arm [57]. In contrast to the clear lateralization of CA3 function during LTM acquisition, Shipton et al. report that NpHR-mediated silencing of left or right CA3 produced an impairment in the spontaneous alternation T-maze task ([6], Fig. 7). This lack of a clear lateralization is consistent with a mechanistic difference between LTM and STM and may suggest that networks involving left and right CA3 each contribute unique computations that are necessary for STM performance. Alternatively, it may reflect a higher computational demand for STM, requiring much of the bilateral hippocampal circuitry for task performance. Further experiments will therefore be needed to dissect the mechanisms underlying the necessity for CA3 neurons bilaterally in STM.

Although optogenetic silencing of CA3 neurons with NpHR has been used successfully to address hippocampal function during learning, a number of methodological limitations need to be addressed.

1. Acute optogenetic manipulations provide a powerful way of rapidly interrogating neural circuits with minimal interference from compensatory changes. However, this strength can also be a weakness. The serial interconnectedness, parallel streams of processing, convergence of inputs, and divergence of outputs within neural networks can complicate the interpretation of acute manipulations in one region. This is because for a given node within a network, inputs from other nodes may be playing either an instructive or permissive role. Instructive nodes provide task relevant information to the node of interest for further processing. Other inputs may, however, simply supply a permissive input to the node of interest, for example by providing a "baseline" level of excitation or by allowing sufficient inhibition to maintain an appropriate excitation/inhibition balance that permits information transfer and/or processing from other nodes [58]. This becomes important when determining the importance of one neuronal population for relaying *information* necessary for a given task. Silencing a permissive node that does not participate in information processing associated with a task could impair performance of the task simply by disrupting the baseline activity of a downstream node. In the context of lateralization of CA3 function during memory processing, these considerations should encourage a direct assessment of the effects on neuronal information processing in the CA1 as a result of silencing left or right CA3 neurons. We discuss some ways this can be achieved in Section 3.4.
2. The use of NpHR, a chloride pump, means that in addition to the hyperpolarization produced by influx of chloride ions, the accumulation of intracellular chloride changes the reversal potential of $GABA_A$ receptors such that GABA becomes briefly depolarizing after NpHR activation [59]. This results in a pronounced rebound effect in NpHR expressing neurons following light cessation, whereby neuronal spiking transiently increases to levels significantly higher than the pre-light baseline before returning back to baseline levels ([59]; Fig. 7). Such an increase in excitation could potentially affect the activity of downstream neurons, which might contribute to the behavioral impairment seen following NpHR-mediated silencing in left CA3 neurons. It will therefore be important to investigate whether CA3 neuronal silencing with other hyperpolarizing tools that do not produce a rebound effect, such as

the outward proton pumps Arch and ArchT [30, 60], produces the same pattern of behavioral impairment as that seen with NpHR.

3. In addition to Schaffer collateral and commissural projections to the CA1, CA3 pyramidal neurons send projections to other targets including other CA3 pyramidal neurons and interneurons as well as neurons in the lateral septum [1]. A synapse selective manipulation is therefore necessary in order to make direct links between CA3–CA1 synapses and behavior. Both halorhodopsin and archaerhodopsin activation at presynaptic terminals can attenuate evoked transmitter release [61]. However, the exact nature of their effects on synaptic potentials is yet to be characterized.

To conclude, lateralization of plasticity at CA3–CA1 synapses suggests that left-CA3-to-CA1 and right-CA3-to-CA1 synapses may perform distinct functions. This would in turn predict a distinction between left and right CA3 neurons. The acute and highly selective neuronal silencing enabled by the green-light chloride pump NpHR has allowed this prediction to be tested directly. Silencing left but not right CA3 neurons impairs acquisition of a hippocampus-dependent LTM task. Thus, the left CA3, and its plastic inputs onto the CA1, could be part of a lateralized information processing channel participating in computations necessary for LTM during task performance. Interpretation of this result from optogenetic silencing in terms of information routing should, however, await more detailed interrogation of the exact effects of silencing left CA3 neurons on the activity of neuronal populations in the CA1 and elsewhere within the hippocampal circuitry.

3.4 Outstanding Questions and Methodological Ways of Addressing Them

The findings discussed in this chapter have illustrated how optogenetic activation and silencing are beginning to reveal intriguing insights into the lateralization of synaptic plasticity and memory processes within the hippocampus. Small left-right differences in receptor composition and morphology at CA3–CA1 synapses translate into dramatic differences in plastic capacity, and those distinct types of synapse may in turn participate in distinct computations during learning and memory. The simple and experimentally tractable nature of this lateralization has opened up a number of avenues for research, which we discuss below.

1. *Lateralization at CA3–CA1 synapses as a window into determinants of plastic capacity*

   The lateralization of plastic capacity suggests a fundamental molecular difference between left-CA3-to-CA1 and right-CA3-to-CA1 synapses. This in turn points towards a tightly regulated developmental programme dedicated to generating this lateralization. Investigating the mechanisms underlying the development of lateralization at CA3–CA1 synapses serves as a unique window into understanding the factors underlying plastic capacity at central synapses since it avoids confounding factors due to cross-region differences in plasticity mechanisms. Spine morphology and receptor composition evolve as an animal develops and are strongly modulated by the history of both presynaptic and postsynaptic neuronal activity [13, 34, 62, 63]. NMDARs in immature forebrain synapses of young animals are initially GluN2B rich but are gradually replaced by GluN2A-containing receptors in an activity-dependent manner as an animal develops into adulthood [13]. Furthermore, postsynaptic spine size and synaptic AMPAR density increase in an activity-dependent manner [63]. This suggests that left-CA3-to-CA1 synapses could be "trapped" in a less mature state relative to right-CA3-to-CA1 synapses, and indeed compared to the majority of forebrain synapses. Differences in the activity patterns of left and right CA3 neurons during development may have resulted in the preferential maturation of right but not left-CA3-to-CA1 synapses. Alternatively, or perhaps in addition, the sensitivity of receptor composition and spine morphology at left-CA3-to-CA1 and right-CA3-to-CA1 synapses to neuronal activity could be modulated by lateralized trans-synaptic factors. In order to address these possibilities, a number of approaches can be taken. We summarize some of these below:

(a) The activity-dependence of synaptic lateralization can be assessed through extracellular, multi-electrode recordings from left and right CA3 neurons in developing animals to investigate whether differences in activity patterns between left and right CA3 neurons exist. This could be complemented with optogenetic feedback silencing to equalize activity of left and right CA3 neurons and hence assess the *necessity* of any differences in overall neuronal firing rate for the development of the synaptic lateralization. Furthermore, optogenetic activation of CA3 neurons and/or their projections in isolated left and right hippocampi in a cell culture system could be used to investigate the *sufficiency* of distinct activity patterns for the development of synaptic lateralization.

(b) The involvement of trans-synaptic signaling in mediating the lateralization could be investigated by identifying candidate molecules to target from screens of lateralization in hippocampal gene expression (e.g. [64]), possibly restricted to CA3 pyramidal neurons, combined with genetic knockout studies that abolish the lateralization (e.g. [65]). Hemisphere-specific knockdown of messenger RNA (mRNA) encoding such candidate molecules from left or right CA3-pyramidal neurons could be used to assess their *necessity* in generating the lateralization. In addition, hemisphere-specific injection of viral constructs expressing candidate molecules to achieve ectopic expression in right or left CA3-pyramidal neurons could be used to assess the *sufficiency* of these molecules for generating a "left-CA3-to-CA1 synaptic" phenotype at right-CA3-to-CA1 synapses (or vice versa).

2. *Interrogation of synaptic plasticity and integration at the single synapse level*

   Activation of ChR2-expressing projections using conventional light delivery methods invariably recruits multiple synapses. While this has proved useful for interrogating the averaged plastic capacity of left-CA3-to-CA1 and right-CA3-to-CA1 synapses, an alternative approach is necessary if the properties of single synapses are to be assessed. Such precision is made possible by using ChR2 to directly depolarize individual presynaptic boutons, and hence induce transmitter release, when action potential propagation is pharmacologically blocked [66]. Direct depolarization of presynaptic boutons can be combined with existing electrophysiological and fluorescent imaging techniques to investigate synaptic plasticity at individual left-CA3-to-CA1 or right-CA3-to-CA1 synapses. This should help elucidate whether plastic capacity is uniform among left-CA3-to-CA1 synapses or whether further subdistinctions exist among these synapses. Furthermore, an application of this technique, termed subcellular ChR2-assisted circuit mapping (sCRACM), has previously been used to elucidate the dendritic distribution of defined inputs from thalamic nuclei and cortical regions onto target pyramidal neurons in the mouse barrel cortex [66]. An analogous application of sCRACM to map out the distribution of left-CA3-to-CA1 and right-CA3-to-CA1 synapses in the dendritic tree of CA1 neurons should pave the way for a clearer understanding of how these inputs are integrated at the single neuron level. Overall, functionally interrogating synaptic lateralization at the single synapse level should enable a more refined understanding of functional distinctions between and within populations of left-CA3-to-CA1 and right-CA3-to-CA1 synapses.

3. *What network computations are plastic left-CA3-to-CA1 synapses necessary for?*

   The unique requirement of left but not right CA3 neurons for LTM performance is consistent with a key prediction of the synaptic plasticity and memory hypothesis, namely, that plastic synapses, such as those made by the left CA3 onto the CA1, are necessary for memory storage [41, 48]. However, as discussed in Section 3.3, caution should be taken when interpreting findings from optogenetic silencing in terms of information routing. A key question therefore is precisely for what aspect of memory storage and/or processing, if any, are plastic left-CA3-to-CA1 synapses necessary?

Hippocampal LTM formation involves a number of mechanistically distinct but coordinated processes. As such, silencing left CA3 neurons could disrupt the encoding, consolidation and/or retrieval of learnt information [67]. Furthermore, the encoding process itself involves distinct subprocesses, such as spatial map formation and place-reward association [67, 68], either or both of which could be disrupted by silencing the left CA3. In order to identify the exact role of left CA3 neurons and their synapses onto the CA1 for performance on long-term memory tasks it will be necessary to interfere with CA3 neuronal activity and/or left-CA3-to-CA1 synaptic activity during distinct phases of task performance. This can involve interference with a distinct phase of each individual trial, for example silencing CA3 neurons only during reward consumption to test their role in reward-place association. In addition, in tasks that are learnt in distinct phases, silencing of left CA3 neurons during blocks of trials comprising a particular phase could enable a more detailed understanding of left CA3 involvement in learning. For example, hippocampus-dependent open-field and maze-based tasks often involve an exploratory phase whereby information regarding the spatial set-up of the context is available but information about reward location is not yet present. This is then followed by trials in which the reward is present at a specific location, and so reward-location associations can be made. The hippocampus has been proposed to play roles in both spatial map formation and associative learning [69, 70] and so it will be important to know which, if any, of these processes are dependent on left CA3 neuronal activity. Optogenetic silencing is well suited to such temporally restricted interference owing to its rapid onset and offset. Nevertheless, these tools are not without their caveats. For example, the rebound effect following NpHR activation (see Section 3.3) could limit its use for temporally restricted, within-trial interference because effects may extend beyond light cessation. The light-activated proton pumps Arch or ArchT may be more suitable for highly temporally restricted manipulations [30, 60]. This illustrates the general point that the choice of opsins for behavioral manipulation should be informed by the temporal profile of the physiological effects mediated by each tool, and the characteristics of the desired manipulation.

An approach complementary to understanding the role of left-CA3-to-CA1 synapses is to assess their necessity for hippocampal activity patterns. Pyramidal neurons in the CA3 and CA1 exhibit place cell firing; that is, they are preferentially active at a specific location, or place field, within an environment [71]. Together, hippocampal place cells can cover an entire environment and have, therefore, been strongly implicated in forming spatial maps of an environment [69]. Moreover, recent findings have indicated that learning of place-reward associations is accompanied by, and correlates with, changes in the firing rate and/or place field location of place cells [72, 73]. This learning modulation of place cell firing is especially apparent in the CA1. For example, learning to locate rewards on a cheeseboard maze is accompanied by shifts in place cell firing such that the number of place cells in the CA1 representing goal locations are significantly increased [73]. These changes could serve as a neural signature for place-reward associations. Coupling optogenetic silencing of left CA3 neurons, or left-CA3-to-CA1 synapses, with place cell recordings could provide mechanistic insights into their function. If this causes a specific disruption of learning-induced place cell reorganization, it would indicate that left CA3 neurons, and their outputs onto CA1 neurons, are involved in forming place-reward associations. Conversely, input from left CA3 neurons could be important for CA1 place cell formation and/or stability without necessarily taking part in associative learning [67]. A further possibility is that the temporal coherence of CA1 place cell populations, a property that correlates with memory performance [74], is critically dependent on input from the left CA3. Indeed, a recent study provided evidence for a critical role of inputs from the CA3 in the temporal coding of space within CA1 neuronal ensembles [75]. Thus, elucidating the effect of interfering with inputs from the left CA3 on CA1 place cell firing patterns during learning is essential for understanding the specific function of these left CA3 neurons and the highly plastic synapses they make in CA1.

4. *Hippocampal lateralization in humans*

   While optogenetic interference is only just beginning to bring some clarity into lateralization of the rodent hippocampus, lateralization of hippocampal function in humans is comparatively better established. Evidence for hippocampal lateralization in humans has come from studies employing virtual reality tasks on healthy volunteers and patients with unilateral hippocampal lesions, in combination with functional magnetic resonance imaging (fMRI; [76]). Early studies revealed that left hippocampal lesions preferentially impair episodic memories while those on the right affect spatial navigation [76]. While intriguing, such a dissociation poses a key question: What are the underlying differences in the computations performed by the left and right hippocampus? A number of subsequent studies have made steps in addressing this question. For example, an fMRI study has shown preferential activation of the left hippocampus when an egocentric strategy was used to solve a spatial navigation task, while allocentric navigation preferentially activated the right hippocampus [77]. Another fMRI study found that left but not right hippocampal activation reflects associative match-mismatch detection, whereby one of either spatial or temporal arrangements of current sensory stimuli is similar (match) but the other is different (mismatch) to the arrangements of previously encountered inputs [78]. This suggests that the left hippocampus may be preferentially involved in updating internal representations when incoming sensory input conveys novel information.

Despite the clear examples of hippocampal functional lateralization in humans, the mechanisms contributing to such lateralization are currently unclear. Lateralization of hippocampal computations in humans suggests a lateralization in underlying network function either within the hippocampal circuitry or at the level of its inputs. However, it is not clear whether the left-right differences in navigation strategies and match-mismatch detection reflect differences in the same fundamental computation, and underlying network processes, or are distinct. This uncertainty is in part due to the lack of clarity regarding the ubiquity of functional lateralization across mammalian species, and hence the lack of adequate, tractable animal models. It is tempting to speculate that the recently uncovered functional lateralization in mouse CA3 function could be related to one or more of the lateralizations discovered in humans. However, it is also possible that lateralization of hippocampal function could have evolved independently in rodents and humans. This uncertainty can be resolved by investigating whether some of the underlying hallmarks of rodent lateralization (e.g. the higher GluN2B density at left-CA3-to-CA1 synapses) are also seen in humans. Given the apparent absence of CA3 projections to the contralateral hemisphere in the posterior hippocampus of humans [1], immunohistochemical studies on human *post mortem* tissue might directly reveal subunit composition at left-CA3-to-CA1 and right-CA3-to-CA1 synapses. If rodent and human lateralization is indeed related, further research into the underlying molecular, network, and computational underpinnings of the rodent lateralization could have a direct impact on our understanding of human hippocampal lateralization. Conversely, if such lateralizations are fundamentally distinct, this would serve as a useful insight into the different strategies for functional lateralization used by hippocampal networks and their impact on the types of computations the hippocampus and its surrounding networks perform.

4 Conclusions

This chapter has focused on principles underlying the study of synaptic lateralization, using CA3–CA1 synapses as a case study. Optogenetics has provided a spatiotemporally highly precise means of investigating the properties and functions of defined neuronal and synaptic populations. The rapidly evolving repertoire of optogenetic tools, and their compatibility with various molecular, electrophysiological, and behavioral techniques has allowed previously intractable questions to be addressed. Throughout this chapter we have demonstrated distinct ways in which optogenetic activation of CA3–CA1 inputs, or silencing of CA3 neurons, has been employed to investigate the consequences of a molecular and subcellular lateralization at CA3–CA1 synapses. ChR2-mediated optogenetic activation was used both to induce LTP and to monitor the effects of electrically induced LTP on either left-CA3-to-CA1 or right-CA3-to-CA1 synapses. Results from these optogenetic activation studies revealed that left-CA3-to-CA1 synapses have a capacity for LTP that extends across distinct induction paradigms, while right-CA3-to-CA1 synapses are relatively stable. The importance of LTP for long-term memory suggests that plastic left-CA3-to-CA1 synapses are preferentially required for the formation of such memories. Consistent with this hypothesis, NpHR-mediated optogenetic silencing of left but not right CA3 neurons impairs performance on a hippocampus-dependent LTM task. Thus, the seemingly subtle left-right differences in CA3–CA1 synaptic receptor composition translate into dramatic differences in CA3–CA1 plastic capacity and CA3 neuronal function during learning. These findings make important first steps in investigating the mechanisms and consequences of synaptic lateralization. Advances in optogenetic tools and their combination with both new and established molecular, electrophysiological, and behavioral techniques promise to further our understanding of the mechanisms and consequences of synaptic lateralization not only at hippocampal CA3–CA1 synapses but throughout the brain.

Acknowledgments

We thank Dr. Olivia A. Shipton for many useful discussions. The authors' research is supported by the Biotechnology and Biological Sciences Research Council, UK.

References

1. Amaral D, Lavenex P (2007) Hippocampal neuroanatomy. In: Andersen C, Morris R, Amaral D, Bliss T, O'Keefe J (eds) The hippocampus book. Oxford University Press, Oxford, pp 4–33
2. Takeuchi T, Duszkiewicz AJ, Morris RG (2014) The synaptic plasticity and memory hypothesis: encoding, storage and persistence. Philos Trans R Soc Lond B Biol Sci 369(1633):20130288
3. Kawakami R, Shinohara Y, Kato Y, Sugiyama H, Shigemoto R, Ito I (2003) Asymmetrical allocation of NMDA receptor ε2 subunits in hippocampal circuitry. Science 300(5621): 990–994
4. Shinohara Y, Hirase H, Watanabe M, Itakura M, Takahashi M, Shigemoto R (2008) Left-right asymmetry of the hippocampal synapses with differential subunit allocation of glutamate receptors. Proc Natl Acad Sci U S A 105(49):19498–19503
5. Kohl MM, Shipton OA, Deacon RM, Rawlins JNP, Deisseroth K, Paulsen O (2011)

Hemisphere-specific optogenetic stimulation reveals left-right asymmetry of hippocampal plasticity. Nat Neurosci 14(11):1413–1415
6. Shipton OA, El-Gaby M, Apergis-Schoute J, Deisseroth K, Bannerman DM, Paulsen O, Kohl MM (2014) Left-right dissociation of hippocampal memory processes in mice. Proc Natl Acad Sci U S A 111(42):15238–15243
7. Dingledine R, Borges K, Bowie D, Traynelis SF (1999) The glutamate receptor ion channels. Pharmacol Rev 51(1):7–61
8. Huganir RL, Nicoll RA (2013) AMPARs and synaptic plasticity: the last 25 years. Neuron 80(3):704–717
9. Kasai H, Fukuda M, Watanabe S, Hayashi-Takagi A, Noguchi J (2010) Structural dynamics of dendritic spines in memory and cognition. Trends Neurosci 33(3):121–129
10. Hell JW (2014) CaMKII: claiming center stage in postsynaptic function and organization. Neuron 81(2):249–265
11. Barria A, Malinow R (2005) NMDA receptor subunit composition controls synaptic plasticity by regulating binding to CaMKII. Neuron 48(2):289–301
12. Shipton OA, Paulsen O (2014) GluN2A and GluN2B subunit-containing NMDA receptors in hippocampal plasticity. Philos Trans R Soc B Biol Sci 369(1633):20130163
13. Yashiro K, Philpot BD (2008) Regulation of NMDA receptor subunit expression and its implications for LTD, LTP, and metaplasticity. Neuropharmacology 55(7):1081–1094
14. Matsuzaki M, Honkura N, Ellis-Davies GC, Kasai H (2004) Structural basis of long-term potentiation in single dendritic spines. Nature 429(6993):761–766
15. Sobczyk A, Scheuss V, Svoboda K (2005) NMDA receptor subunit-dependent [Ca2+] signaling in individual hippocampal dendritic spines. J Neurosci 25:6037–6046
16. Turrigiano GG (2008) The self-tuning neuron: synaptic scaling of excitatory synapses. Cell 135(3):422–435
17. Miesenbock G, Kevrekidis IG (2005) Optical imaging and control of genetically designated neurons in functioning circuits. Annu Rev Neurosci 28:533–563
18. Yizhar O, Fenno LE, Davidson TJ, Mogri M, Deisseroth K (2011) Optogenetics in neural systems. Neuron 71(1):9–34
19. Zemelman BV, Lee GA, Ng M, Miesenböck G (2002) Selective photostimulation of genetically chARGed neurons. Neuron 33:15–22
20. Zemelman BV, Nesnas N, Lee GA, Miesenböck G (2003) Photochemical gating of heterologous ion channels: remote control over genetically designated populations of neurons. Proc Natl Acad Sci U S A 100:1352–1357
21. Lima SQ, Miesenböck G (2005) Remote control of behavior through genetically targeted photostimulation of neurons. Cell 121:141–152
22. Banghart M, Borges K, Isacoff E, Trauner D, Kramer RH (2004) Light-activated ion channels for remote control of neuronal firing. Nat Neurosci 7:1381–1386
23. Szobota S, Gorostiza P, Del Bene F, Wyart C, Fortin DL, Kolstad KD, Tulyathan O, Volgraf M, Numano R, Aaron HL, Scott EK, Kramer RH, Flannery J, Baier H, Trauner D, Isacoff EY (2007) Remote control of neuronal activity with a light-gated glutamate receptor. Neuron 54:535–545
24. Boyden ES, Zhang F, Bamberg E, Nagel G, Deisseroth K (2005) Millisecond-timescale, genetically targeted optical control of neural activity. Nat Neurosci 8:1263–1268
25. Mattis J, Tye KM, Ferenczi EA, Ramakrishnan C, O'Shea DJ, Prakash R, Gunaydin LA, Hyun M, Fenno LE, Gradinaru V, Yizhar O, Deisseroth K (2012) Principles for applying optogenetic tools derived from direct comparative analysis of microbial opsins. Nat Methods 9:159–172
26. Deisseroth K (2015) Optogenetics: 10 years of microbial opsins in neuroscience. Nat Neurosci 18:1213–1225
27. Zhang F, Wang LP, Boyden ES, Deisseroth K (2006) Channelrhodopsin-2 and optical control of excitable cells. Nat Methods 3(10):785–792
28. Gunaydin LA, Yizhar O, Berndt A, Sohal VS, Deisseroth K, Hegemann P (2010) Ultrafast optogenetic control. Nat Neurosci 13(3):387–392
29. Zhang F, Wang LP, Brauner M, Liewald JF, Kay K, Watzke N, Wood PG, Bamberg E, Nagel G, Gottschalk A, Deisseroth K (2007) Multimodal fast optical interrogation of neural circuitry. Nature 446(7136):633–639
30. Chow BY, Han X, Dobry AS, Qian X, Chuong AS, Li M, Henninger MA, Belfort GM, Lin Y, Monahan PE, Boyden ES (2010) High-performance genetically targetable optical neural silencing by light-driven proton pumps. Nature 463(7277):98–102
31. Goshen I, Brodsky M, Prakash R, Wallace J, Gradinaru V, Ramakrishnan C, Deisseroth K (2011) Dynamics of retrieval strategies for remote memories. Cell 147(3):678–689
32. Aschauer DF, Kreuz S, Rumpel S (2013) Analysis of transduction efficiency, tropism and axonal transport of AAV serotypes 1, 2,

5, 6, 8 and 9 in the mouse brain. PLoS ONE 8:e76310
33. Watakabe A, Ohtsuka M, Kinoshita M, Takaji M, Isa K, Mizukami H, Ozawa K, Isa T, Yamamori T (2015) Comparative analyses of adeno-associated viral vector serotypes 1, 2, 5, 8 and 9 in marmoset, mouse and macaque cerebral cortex. Neurosci Res 93:144–157
34. Lee MC, Yasuda R, Ehlers MD (2010) Metaplasticity at single glutamatergic synapses. Neuron 66(6):859–870
35. Zucker RS, Kullmann DM, Kaeser PS (2014) Release of neurotransmitters. In: Byrne J, Heidelberger R, Waxham MN (eds) From molecules to networks: an introduction to cellular and molecular neuroscience, 3rd edn. Academic, Cambridge, pp 443–488
36. Paoletti P, Neyton J (2007) NMDA receptor subunits: function and pharmacology. Curr Opin Pharmacol 7(1):39–47
37. Shinohara Y, Hirase H (2009) Size and receptor density of glutamatergic synapses: a viewpoint from left-right asymmetry of CA3–CA1 connections. Front Neuroanat 3:10
38. Pike FG, Meredith RM, Olding AW, Paulsen O (1999) Rapid report: postsynaptic bursting is essential for 'Hebbian' induction of associative long-term potentiation at excitatory synapses in rat hippocampus. J Physiol 518(Pt 2):571–576
39. Volianskis A, Bannister N, Collett VJ, Irvine MW, Monaghan DT, Fitzjohn S, Jensen MS, Jane DE, Collingridge GL (2012) Different NMDA receptor subtypes mediate induction of long-term potentiation and two forms of short-term potentiation at CA1 synapses in rat hippocampus in vitro. J Physiol 591:955–972
40. Halt AR, Dallapiazza RF, Zhou Y, Stein IS, Qian H, Juntti S, Wojcik S, Brose N, Silva AJ, Hell JW (2012) CaMKII binding to GluN2B is critical during memory consolidation. EMBO J 31:1203–1216
41. El-Gaby M, Shipton OA, Paulsen O (2015) Synaptic plasticity and memory: new insights from hippocampal left-right asymmetries. Neuroscientist 21(5):490–502
42. Kwag J, Paulsen O (2012) Gating of NMDA receptor-mediated hippocampal spike timing-dependent potentiation by mGluR5. Neuropharmacology 63(4):701–709
43. Kirov SA, Sorra KE, Harris KM (1999) Slices have more synapses than perfusion-fixed hippocampus from both young and mature rats. J Neurosci 19(8):2876–2886
44. Ho OH, Delgado JY, O'Dell TJ (2004) Phosphorylation of proteins involved in activity-dependent forms of synaptic plasticity is altered in hippocampal slices maintained in vitro. J Neurochem 91(6):1344–1357
45. Battaglia FP, Benchenane K, Sirota A, Pennartz CM, Wiener SI (2011) The hippocampus: hub of brain network communication for memory. Trends Cogn Sci 15(7):310–318
46. Andersen C, Morris R, Amaral D, Bliss T, O'Keefe J (eds) (2007) The hippocampus book. Oxford University Press, Oxford
47. Rolls ET (2015) Pattern separation, completion, and categorisation in the hippocampus and neocortex. Neurobiol Learn Mem. pii: S1074-7427(15)00129-X
48. Martin SJ, Grimwood PD, Morris RG (2000) Synaptic plasticity and memory: an evaluation of the hypothesis. Annu Rev Neurosci 23:649–671
49. Morris RG, Anderson E, Lynch GS, Baudry M (1986) Selective impairment of learning and blockade of long-term potentiation by an N-methyl-D-aspartate receptor antagonist, AP5. Nature 319(6056):774–776
50. Tsien JZ, Huerta PT, Tonegawa S (1996) The essential role of hippocampal CA1 NMDA receptor-dependent synaptic plasticity in spatial memory. Cell 87(7):1327–1338
51. Whitlock JR, Heynen AJ, Shuler MG, Bear MF (2006) Learning induces long-term potentiation in the hippocampus. Science 313(5790):1093–1097
52. Klur S, Muller C, Pereira de Vasconcelos A, Ballard T, Lopez J, Galani R, Certa U, Cassel JC (2009) Hippocampal-dependent spatial memory functions might be lateralized in rats: an approach combining gene expression profiling and reversible inactivation. Hippocampus 19:800–816
53. Gerlai RT, McNamara A, Williams S, Phillips HS (2002) Hippocampal dysfunction and behavioral deficit in the water maze in mice: an unresolved issue? Brain Res Bull 57:3–9
54. Gradinaru V, Zhang F, Ramakrishnan C, Mattis J, Prakash R, Diester I, Goshen I, Thompson KR, Deisseroth K (2010) Molecular and cellular approaches for diversifying and extending optogenetics. Cell 141:154–165
55. Reisel D, Bannerman DM, Schmitt WB, Deacon RM, Flint J, Borchardt T, Seeburg PH, Rawlins JN (2002) Spatial memory dissociations in mice lacking GluR1. Nat Neurosci 5:868–873
56. Bannerman DM, Niewoehner B, Lyon L, Romberg C, Schmitt WB, Taylor A, Sanderson DJ, Cottam J, Sprengel R, Seeburg PH, Köhr

G, Rawlins JN (2008) NMDA receptor subunit NR2A is required for rapidly acquired spatial working memory but not incremental spatial reference memory. J Neurosci 28:3623–3630
57. Deacon RM, Rawlins JN (2006) T-maze alternation in the rodent. Nat Protoc 1(1):7–12
58. Otchy TM, Wolff SB, Rhee JY, Pehlevan C, Kawai R, Kempf A, Gobes SM, Ölveczky BP (2015) Acute off-target effects of neural circuit manipulations. Nature 528(7582): 358–363
59. Raimondo JV, Kay L, Ellender TJ, Akerman CJ (2012) Optogenetic silencing strategies differ in their effects on inhibitory synaptic transmission. Nat Neurosci 15(8):1102–1104
60. Han X, Chow BY, Zhou H, Klapoetke NC, Chuong A, Rajimehr R, Yang A, Baratta MV, Winkle J, Desimone R, Boyden ES (2011) A high-light sensitivity optical neural silencer: development and application to optogenetic control of non-human primate cortex. Front Syst Neurosci 5:18
61. Mahn M, Prigge M, Ron S, Levy R, Yizhar O (2016) Biophysical constraints of optogenetic inhibition at presynaptic terminals. Nat Neurosci 19(4):554–556
62. Abraham WC (2008) Metaplasticity: tuning synapses and networks for plasticity. Nat Rev Neurosci 9(5):387
63. Hanse E, Seth H, Riebe I (2013) AMPA-silent synapses in brain development and pathology. Nat Rev Neurosci 14(12):839–850
64. Moskal JR, Kroes RA, Otto NJ, Rahimi O, Claiborne BJ (2006) Distinct patterns of gene expression in the left and right hippocampal formation of developing rats. Hippocampus 16(8):629–634
65. Kawahara A, Kurauchi S, Fukata Y, Martínez-Hernández J, Yagihashi T, Itadani Y, Sho R, Kajiyama T, Shinzato N, Narusuye K, Fukata M, Luján R, Shigemoto R, Ito I (2013) Neuronal major histocompatibility complex class I molecules are implicated in the generation of asymmetries in hippocampal circuitry. J Physiol 591(19):4777–4791
66. Petreanu L, Mao T, Sternson SM, Svoboda K (2009) The subcellular organization of neocortical excitatory connections. Nature 457(7233):1142–1145
67. Kandel ER, Dudai Y, Mayford MR (2014) The molecular and systems biology of memory. Cell 157(1):163–186
68. de Lavilléon G, Lacroix MM, Rondi-Reig L, Benchenane K (2015) Explicit memory creation during sleep demonstrates a causal role of place cells in navigation. Nat Neurosci 18(4):493–495
69. O'Keefe J, Nadel L (1978) The hippocampus as a cognitive map. Oxford University Press, Oxford
70. Eichenbaum H, Cohen NJ (2014) Can we reconcile the declarative memory and spatial navigation views on hippocampal function? Neuron 83(4):764–770
71. O'Keefe J, Dostrovsky J (1971) The hippocampus as a spatial map. Preliminary evidence from unit activity in the freely moving rat. Brain Res 34(1):171–175
72. Hollup SA, Molden S, Donnett JG, Moser MB, Moser EI (2001) Accumulation of hippocampal place fields at the goal location in an annular watermaze task. J Neurosci 21(5): 1635–1644
73. Dupret D, O'Neill J, Pleydell-Bouverie B, Csicsvari J (2010) The reorganization and reactivation of hippocampal maps predict spatial memory performance. Nat Neurosci 13(8):995–1002
74. Buzsáki G, Moser EI (2013) Memory, navigation and theta rhythm in the hippocampal-entorhinal system. Nat Neurosci 16(2): 130–138
75. Middleton SJ, McHugh TJ (2016) Silencing CA3 disrupts temporal coding in the CA1 ensemble. Nat Neurosci. doi:10.1038/nn.4311. [Epub ahead of print]
76. Burgess N, Maguire EA, O'Keefe J (2002) The human hippocampus and spatial and episodic memory. Neuron 35(4):625–641
77. Iglói K, Doeller CF, Berthoz A, Rondi-Reig L, Burgess N (2010) Lateralized human hippocampal activity predicts navigation based on sequence or place memory. Proc Natl Acad Sci U S A 107(32):14466–14471
78. Kumaran D, Maguire EA (2007) Match mismatch processes underlie human hippocampal responses to associative novelty. J Neurosci 27(32):8517–8524

Part III

Electroencephalographic, Imaging and Neuro-Stimulation Methods

Chapter 12

Transcranial Magnetic Stimulation

Luigi Cattaneo

Abstract

Transcranial magnetic stimulation is a technique that allows the induction of electrical current in the superficial brain tissue, by means of a rapidly changing magnetic field. It is a noninvasive technique which may be safely applied to awake and collaborating humans. The biological effects of transcranial magnetic stimulation can be classified as immediate, consisting of action potentials, and delayed, consisting of variably lasting changes in the excitability of neurons, outlasting stimulation itself. Accordingly, the impact of TMS on behavior can be generally categorized as "online" or "offline." TMS produces behavioral changes by manipulating the firing characteristics of neurons. As a consequence, TMS may be used to establish causal relationships between brain and behavior. The direct effects of TMS have a limited spatial distribution, in the order of 1–2 cm, thus making it an optimal tool for hemispheric localization of brain functions. TMS has been applied to the study of lateralization of brain functions in humans in multiple domains such as language, spatial attention, or executive functions.

Key words Magnetic stimulation, Behavior, Neurons, Action potentials, Focal stimulation

Abbreviations

CSP	Cortical silent period
DLPFC	Dorsolateral prefrontal cortex
EEG	Electroencephalography
EMG	Electromyography
ICF	Intracortical facilitation
ICI	Intracortical inhibition
ISI	Inter-stimulus interval
ISP	Ipsilateral silent period
MEP	Motor evoked potential
PAS	Paired associative stimulation
rTMS	Repetitive transcranial magnetic stimulation
TBS	Theta-burst stimulation
TES	Transcranial electrical stimulation
TMS	Transcranial magnetic stimulation

Lesley J. Rogers and Giorgio Vallortigara (eds.), *Lateralized Brain Functions: Methods in Human and Non-Human Species*, Neuromethods, vol. 122, DOI 10.1007/978-1-4939-6725-4_12, © Springer Science+Business Media LLC 2017

1 Background

1.1 Early History of Focal Brain Stimulation

Experimental focal stimulation of brain tissue has been a cornerstone in the study of brain–behavior relationships since the seminal experiment of Gustav Fritsch and Eduard Hitzig [1] who demonstrated that galvanic stimuli applied to a specific (anterior) portion of the cortex of a dog produced contralateral movements in the experimental animal. Fritsch and Hitzig's experiment allowed them to conclude that "one part of the brain is motor and another is not." In other words, motor functions are localized to a spatially restricted portion of the brain. The experiment demonstrated the validity of the hypothesis that cerebral functions are at least in part localized. The rationale behind the use of focal brain stimulation did not change much in the following 150 years. The two main assumptions behind its use are that (a) behavior is represented in the central nervous system according to topological rules forming a spatial map of different brain functions and (b) focal stimulation has indeed focal, i.e. spatially circumscribed, effects on the central nervous system. Focal brain stimulation is therefore a form of spatial sampling of the central nervous system, coupled with a measure of the effects of stimulation.

Since Fritsch and Hitzig's experiment, focal electrical brain stimulation in experimental animals has been, and still is, used extensively in brain–behavior research. In humans, the possibility of applying focal electrical stimulation to the brain has become available since the first half of the twentieth century. This was initially possible only with invasive methods, during neurosurgical procedures [2, 3]. Neurosurgeons started applying focal stimulation to the brain to obtain functional mapping of cortical functions in order to improve the outcome of surgery. The resulting collections of data were nevertheless astonishingly informative about the physiology of brain–behavior relations. The most salient example is the collection of maps of body part representations in the motor and somatosensory cortices of humans described by Wilder Penfield [4]. The author himself was able to summarize his findings on the somatotopy of body representations in perhaps one of the most famous icons in neuroscience: the motor and sensory homunculi. Invasive focal brain stimulation has several advantages, including extreme precision in spatial localization of the stimulus in the brain. It has, however, major drawbacks, the main one of which being that peri-surgical brain mapping cannot be planned ad hoc. Additionally, in some instances the underlying disorder may compromise or alter brain function to the point that it is no longer representative of physiological processes.

1.2 Noninvasive Brain Stimulation and the Advent of Transcranial Magnetic Stimulation (TMS)

Noninvasive electrical stimulation of human brain tissue is a methodological challenge. The skull provides a thick coating of electrical insulation from the outside, in that bone tissue offers a high resistance to the passage of electrical currents. It was not until 50 years after Penfield's works that the possibility of producing direct electrical stimulation of the human brain through an intact skull was described and applied. Transcranial electrical stimulation (TES) was made possible by the application of high-voltage, short-duration electrical pulses to the scalp. By this means in the early 1980s, transcranial stimulation of the corticospinal tract was achieved, thus starting the era of noninvasive brain stimulation [5, 6]. TES however had a problem: it was painful. Overcoming the skull's resistance required the use of electrical stimuli that activated massively the underlying scalp nociceptors. Shortly after, Tony Barker and his colleagues from Sheffield Hospital described transcranial magnetic stimulation (TMS). It was observed that application of a rapidly changing magnetic field to the scalp of an intact, awake human participant, produced movements in the limbs contralateral to the stimulated hemisphere [7]. Since then TMS has proved to be an outstanding tool for the study of brain–behavior relations in humans and its use has become widespread in the cognitive neurosciences.

2 The Physics of TMS

The general goal of TMS is to produce electrical stimulation of brain tissue. This objective is achieved by making an electrical current flow in a first conductor (the TMS coil), which produces a variable magnetic field around it, which in turn induces an electrical field in a second conductor (the brain) (Fig. 1). We will now review the series of physical phenomena that characterize TMS.

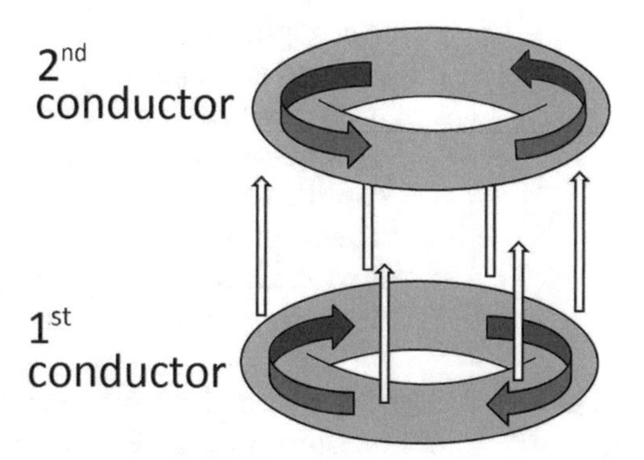

Fig. 1 Schematization of Faraday's law of electromagnetic induction: current flowing in the first conductor (the magnetic coil) produces a magnetic field (). The magnetic field in turn induces an electrical current in the second conductor (the brain)

2.1 Faraday's Law of Induction

In 1831 Faraday described a curious phenomenon in electrical circuits. When two circuits were placed in proximity to each other (but not in contact), the switching on of electrical current in one circuit induces a transient current flow in the nearby circuit. A similar phenomenon is produced by moving a magnet in the proximity of a circuit. This property of time-varying electrical currents was formalized in terms of Faraday's principle of electromagnetic induction. A time-varying electrical current in a circuit produces a magnetic field, the intensity of which is proportional to the intensity of current flow. The time-varying magnetic flux in turn induces an electrical field, i.e. a spatially distributed electromotive force, the amplitude of which is proportional to the rate of change over time of the magnetic field, according to the following formal relation:

$$\varepsilon = -N\frac{d\Phi}{dt}$$

where ε is the electromotive force, $d\Phi$ is the change in the magnetic flux and dt is the change in time. N refers to the number of coils in the circuit. The generated field of electromotive force does not yet correspond to a flow of electrical current. The relation between electromotive force and current flow is described by Ohm's law:

$$J = \sigma\varepsilon$$

where J is the current density, ε is the electromotive force and σ is the conductivity of the material in which the current flows. Transcranial magnetic stimulation operates very similarly to Faraday's original description of a circuit in which a current is switched on. Similarly, the peripheral portion of the magnetic stimulator consists in a coiled electrical circuit in which current is suddenly switched "on." The second conductor in which electrical current is induced is instead represented by brain tissue. Figure 2 summarizes the time course of the sequence of physical events that produce TMS. The direction of the induced magnetic field is perpendicular to that of the current in the first conductor. The direction of the induced current in the second conductor is perpendicular to the one of the magnetic flux. The result is that the induced current in the second conductor is parallel, but with opposite direction compared to the original current in the first conductor (Fig. 1).

2.2 Implications of Faraday's Law for TMS

There are several relevant implications of the sequence of physical events illustrated in Fig. 1 that are worth emphasizing. (a) The magnitude of the induced electromotive force is not a function of the magnitude of the original magnetic flux but rather a function

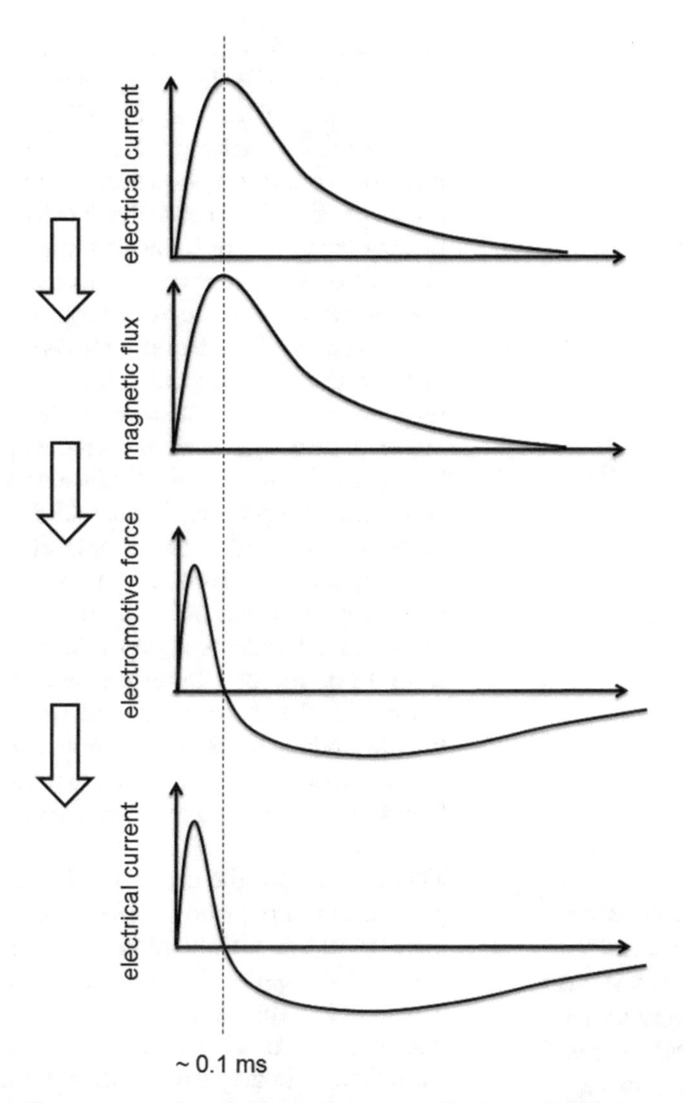

Fig. 2 The sequence of electromagnetic events at the basis of TMS. From *top* to *bottom*: the circuit in the stimulator is suddenly turned ON and the current flows in the magnetic coil. This gives rise to a magnetic flux that is proportional in amplitude to the intensity of electrical current in the coil. The magnetic field changing over time induces a field of electromotive force, the amplitude of which is proportional to the rate of change of the magnetic flux, i.e. to its first derivative. Finally, the electromotive filed is capable of powering an electrical current, with amplitude proportional to the amplitude of the electromotive force. However, according to Ohm's law, the amount of current in the second conductor depends strictly also on the conductivity properties of the conductor itself. A conductor with very low conductivity (such as, for example, air) will not allow any electrical current to flow in spite of the electromotive force being present. Notice that the relevant phenomena occur in a very short time period, i.e. in the first ~100 μs

of its rate of change over time. This implies that no matter how intense a magnetic field is, it is not capable of inducing transcranial electrical stimulation if it does not change rapidly over time. Commercially available TMS stimulators produce peak magnetic fields of around 2 T, however, it is not the absolute amplitude of the magnetic field what makes TMS effective in the process of electromagnetic induction, instead it is the rate of change of the magnetic field that produces the desired induction effects. (b) total time of exposure to the peak magnetic field is minimal, in the order of the tens of microseconds. This is a relevant datum for the assessment of the operator and subject's safety in terms of exposure to magnetic fields. (c) The decay of the magnetic flux is not linear but rather it decreases with the second power of the distance. This implies that doubling the distance between the first and the second conductor will produce a fourfold decrease of the induced current in the second conductor. Minimal changes in the distance between the scalp and the magnetic coil may result in considerable changes in the actual impact of TMS on brain tissue. (**d**) The induction of the electrical current is dependent upon the conductivity of the second conductor. Target materials with variable electrical resistance will therefore allow different intensities of current. The human body is spatially inhomogeneous in terms of conductivity. As a consequence, it is difficult to predict a priori where the current flow will be produced in brain tissue.

2.3 Spatial Characteristics of Electrical Stimulation of Brain Tissue: Focality, Anisotropy, and Depth of Stimulation

The rationale for the use of TMS in the cognitive neurosciences in general and in the study of lateralization in particular is that it produces focal brain stimulation, that is, spatially constrained stimulation of brain tissue. The issue of "where" TMS impacts brain tissue is of considerable interest, since it affects directly the design and interpretation of TMS studies. The spatial distribution of the electrical field in brain tissue depends upon several factors inherent to the geometry and electrical properties of the first and of the second conductor. The shape of the first conductor has considerable effect on the shape of the magnetic field. Commercially available stimulators are supplied with coils of two different shapes: the circular coil and the double-circular, or "figure of eight" or "butterfly" coil. The circular coil produces magnetic field of similar intensity in all points along its rim, which is therefore cylinder-shaped. The figure-of-eight coil consists of two partially superimposed circular coils, each of them producing a cylinder-shaped magnetic field. The two magnetic fields are superimposed in a small segment and therefore their intensity summates, thus generating local peak of magnetic field (Fig. 3), with an elongated cone-like shape. Commonly available figure-of-eight coils have single coil windings with diameters ranging between 35 and 70 mm. The region of peak focal stimulation in the second conductor is therefore oval-shaped with a larger diameter of around 2 cm and with current flowing in the direction

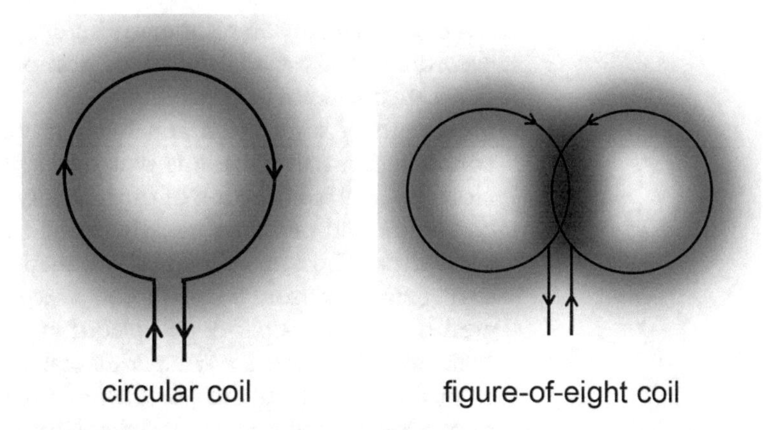

Fig. 3 Comparison of the amplitude of the magnetic field generated by the two most common types of coil. The field evoked by the circular coil is of the same intensity all around its rim. The field of the figure-of-eight coil has a central peak corresponding to the spatial overlapping of the two circular magnetic fields. The presence of the peak assures focality of stimulation

of the larger diameter of the oval. Figure-of-eight coils are the most commonly used coils in neurocognitive research, whereas circular coils find their field of application mainly in the clinical assessment of the corticospinal pathway.

The characteristics of the second conductor that determine the effects of TMS on neural tissue are far less predictable than those of the first conductor, outlined above. Indeed, to our purposes the second conductor is the human brain, a highly anisotropic material, as far as the induction and biological effects of current flow are concerned. Several factors make the effects of TMS on the brain directionally dependent (i.e. anisotropic). First of all, the conductivity of cephalic tissue varies in its different parts. Bony regions tend to distort the electrical flux and to concentrate it in proximity of foramina. Large collections of cerebrospinal fluid such as in the inter-hemispheric fissure or in correspondence of subarachnoid cisterns probably allow more current flow. The smearing of electrical flux on realistic cortical models has been consistently studied. However, to date, no reliable tool is commonly available to predict current flow on the cortical anatomy in routine TMS procedures. It is worth noting that pathological conditions can alter considerably the relative quantity and the geometry of cerebrospinal fluid in the brain and this is likely to alter our prediction of where the electrical flux will be localized in the brain. For example, chronic ischemic lesions result in large collection of cerebrospinal fluid, which produce dramatic changes in conductivity of brain tissue. A second factor that adds variability and unpredictability to the effects of TMS on brain tissue is that different portions of neurons respond differently to extracellular electrical currents and that the same part of a neuron responds differently to extracellular currents according to its spatial location. Section 3.1 describes in more detail such sources of variability.

Depth of stimulation is a frequently debated issue. The first consideration to be made is that the depth at which an effective electrical flow is a function of the intensity of the current in the first conductor and of distance of the second conductor from the first conductor. The same depth of stimulation can be achieved by a strong stimulus from a distant coil or from a weak stimulus delivered by a nearby coil. A coil placed at a fixed distance from the brain, as is the case in most experimental setups, will produce deeper or shallower electrical fluxes as a function of the intensity of stimulation. Commercially available stimulators at maximal intensity of their outputs probably produce intracranial electrical flows that reach well beyond the gray-white boundary of the more superficial gyri [8–10]. However, this information is of little interest, since in experimental conditions low-intensity stimulation is used (Sect. 3.2). In general, it should be noted that depth of stimulation is achieved at the cost of focality [11]. If we imagine the magnetic field produced by a figure-of-eight coil as a smoothed cone, which penetrates deeper into the brain tissue with increasing stimulation intensity, it becomes clear that reaching a deep structure bears the cost of stimulating a larger volume of brain tissue in the more superficial parts of the brain. A relevant consequence of the decrease of magnetic field with the distance from the coil is that, at low intensities, TMS has a strong bias in favor of the more superficial layers of the cortex, which will receive stronger stimuli compared to the deeper layers, which might receive no efficient stimulation at all.

2.4 The Engineering of a Magnetic Stimulator

A TMStimulator is made of a relatively small number of electrical components (Fig. 4). The core of the system is a capacitor, which stores a large amount of electromotive force. An electrically controlled switch, which allows current flow unidirectionally (a thyristor) is used to close the circuit and discharge the capacitors in the stimulating coils. The circuit as described produces a monophasic peak of current, which degrades exponentially over time. Stimulators that deliver similarly shaped pulse waveforms are termed *monophasic* stimulators. The addition to the circuit of a pathway that allows the energy to flow back into the capacitor changes radically the waveform of the electrical current in the coil. A negative phase (current flowing backwards) appears. Stimulators capable of delivering such pulse types are termed *biphasic* stimulators. There are several differences between monophasic and biphasic stimulation. First, from the point of view of the machinery, biphasic stimulation "recovers" most of the energy that flowed in the stimulating coil. The most relevant consequence of the backwards flow of current is that the stimulator is ready to discharge again in a shorter time compared to a monophasic stimulus. This sort of shorter "refractory period" allows higher frequency

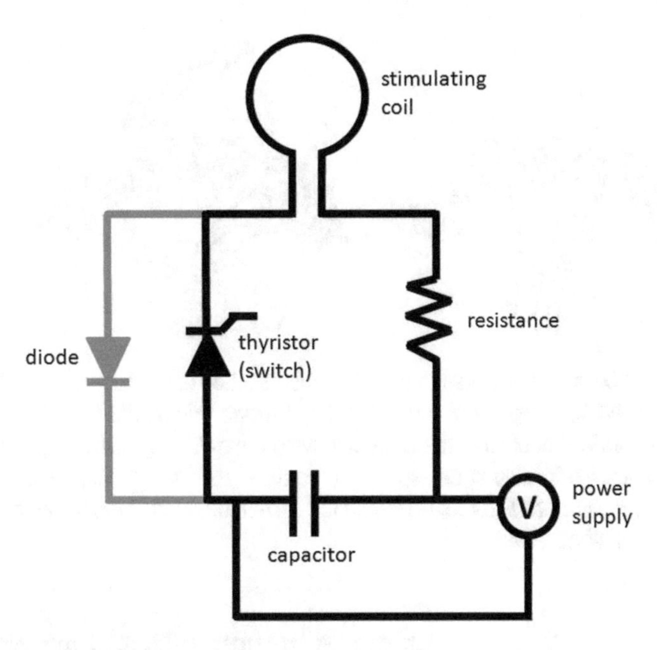

Fig. 4 The main components of a magnetic stimulator. The *black lines* in the equivalent circuit represent a monophasic stimulator. The addition of the grey components characterizes the biphasic stimulator

repetitive stimulation. Secondly, biphasic stimulators tend to be less prone to heating of the stimulating coil. Electricity flowing through the stimulating coil produces heat. Commercial stimulators are all endowed with a thermometer that switches off the stimulator whenever the coil temperature reaches 40–41 °C as a safety procedure to avoid discomfort or burning of the subject. From the point of view of brain tissue stimulation, biphasic currents yield potentially different biological effects, as shown empirically in the literature. It is therefore recommended to use monophasic or biphasic stimulators consistently within a single experiment. All devices are equipped with a system for the manual variation of stimulus intensity. Oddly, in all commercially available stimulators the stimulator output is expressed as a percentage of maximum output rather than as an absolute value of the magnetic field. Therefore the intensity regulator in all TMStimulators will be graded from 0% to 100%.

3 The Physiology of TMS

3.1 The Effect of TMS on Neuronal Membranes

The ultimate effect of TMS is an extracellular electrical flux. The neurons' cellular membranes, having excitable properties, are affected by the electrical current. The most prominent effect of TMS on neurons is the genesis of action potentials. Axons and axon-hillocks are the parts of a neuron with lowest threshold to electrical stimulation.

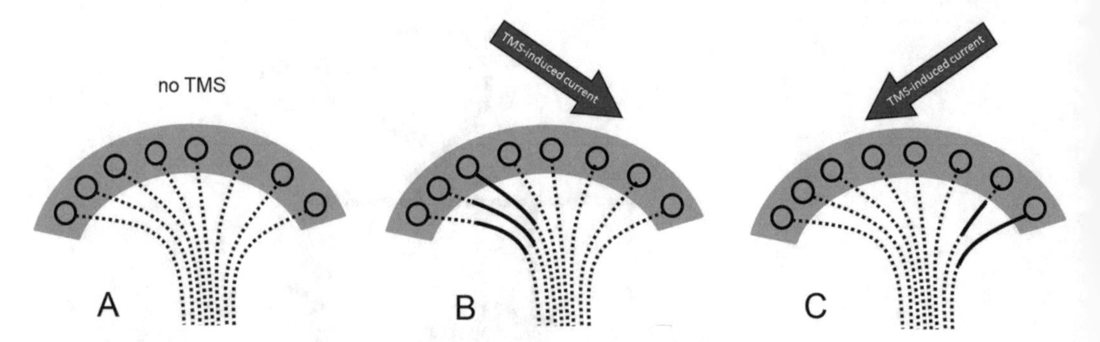

Fig. 5 (**a**) Schematic representation of a cross-section of a cortical gyrus. The *gray band* represents the cortex, the *dashed lines* represent the corticofugal axons originating from the cortical neurons (*circles*). (**b**) Low-intensity TMS is applied to the scalp overlying the gyrus with a left-right orientation of the current. The axons preferentially activated in this condition are those that have tracks parallel to the current and are highlighted by *continuous black lines*. (**c**) Changing the current orientation results in the activation of a completely different population of neurons

According to most influential models, at low stimulus intensities, such action potentials probably originate from the direct excitation of axons, rather than from the summation of dendritic/somatic potentials [12–23]. The orientation in space and the intrinsic geometrical properties of axons influence greatly the capacity of the TMS-induced electrical flux to excite them. Axons are preferentially depolarized by extracellular linear currents when they are oriented parallel to the flow of current. Additionally, axons seem to have a lower threshold to stimulation in correspondence of bends or branching. As a result, fiber orientation in the brain is a strong contributor to anisotropy of brain tissue with respect to TMS effects. This property is illustrated in Fig. 5. The practical consequence is that, at low stimulation intensities, TMS will excite different populations of neurons according to coil orientation, even if leaving invariant the point of stimulation. Changing the orientation of the TMS coil therefore results in sampling of different information from the brain, coded in different but spatially overlapping neuronal populations. Coil orientation should therefore be kept constant within- and possibly between-subjects in the context of an experiment. In addition, coil orientation should be systematically reported in the description of an experiment, to ensure replicability of the results.

More recent literature has investigated the effects of TMS on slices of neural tissue and brought further material in favor of the exceptional dependency of TMS on the geometrical features of the stimulated tissue. TMS applied to neuronal slices has proved to be particularly inefficient. However, careful geometrical arrangement of axon fibers in the slice was found to be of particular importance in shaping the susceptibility of neurons to TMS [24]. Similarly, it has been shown that intrinsic properties of the neuronal membrane, such as the length constant (a function of the axonal diameter) and the passive time constant (a function of the concentration and type of ion channels along the axon and axon hillock) are a main source of variability in the response of individual neurons to TMS [24]. Data from cellular cultures on thin slice have yielded additional information on the impact of TMS on neuronal membranes. Most authoritative models representing the effect of TMS on neuronal membranes imply that TMS acts preferentially on axons, at any given point. As a consequence the effects of TMS are thought to be spiking-dependent, i.e. they occur mainly as a consequence and in the form of action potentials. Some data however points in a different direction, providing evidence that at least in part TMS may act on somatic membranes as a first event, propagating next to the axon hillock [25].

3.2 The Effects of TMS on the Cortex as a Whole

Electrical stimulation of the cortex is an artificial condition, not resembling, nor mimicking any physiological process. In no physiological setting do the cortical neurons as a whole receive such a strong and synchronized depolarization. In addition, the populations of neurons that get excited by TMS are clustered from those neurons that are weakly (or not at all) stimulated, following geometrical and spatial biases rather than following functional rules. In other words TMS massively depolarizes groups of neurons mainly according to their spatial location in the cortex rather than according to the neurons' physiological role (Sect. 2.3). As a consequence, the effects of magnetic/electrical stimulation on cortical activity as a whole are fairly unpredictable on the basis of what we know of cortical physiology. The literature is rich in works that address the issue of what pattern of neural activity emerges at a population level, after a magnetic stimulus. It should be always kept in mind that such neuronal activity is potentially unrelated to physiological processes in the cortex. Therefore, the multiple accounts of how the cortex responds to TMS are best regarded as self-referential to the technique of TMS, rather than describing the way in which the neocortex actually works.

The first authors to characterize the cortical response to phasic electrical stimulation were Patton and Amassian who undertook a series of seminal experiments by stimulating the motor cortex of cats and recording the output of cortical stimulation from the axons of the pyramidal (corticospinal) tract [26]. The first striking finding was that the time-course of the appearance of action potentials along corticospinal axons was not monophasic, as one could have expected after direct stimulation of the axons of the pyramidal tract (Fig. 6a). Instead, the descending volleys were composed of several consecutive peaks. In other words, a single cortical stimulus produced descending volleys in consecutive waves rather than in a single wave (confront Fig. 6a and b). The interval between the waves was a striking 1.6 ms, corresponding to a frequency of around 600 Hz. The authors characterized the origin of the single

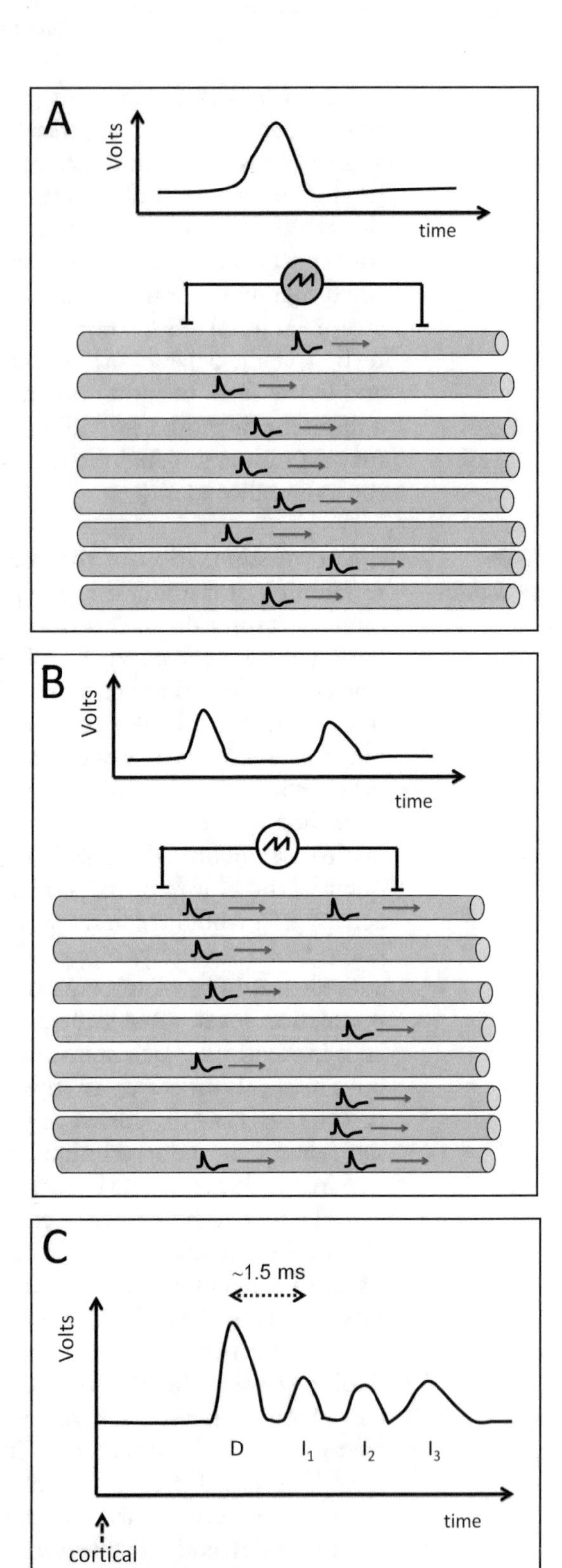

Fig. 6 (**a**) Temporal dispersion of action potentials along a bundle of axons following a monophasic distribution. The represents the appearance of the compound

waves in a series of control experiments. Ablation of the cortex abolished all but the first wave. This first wave was thought to be generated by direct excitation of the axons of pyramidal tract neurons at their exit from the cortex, therefore not requiring the integrity of the neuron's cell body. It was named thereafter "direct wave" or D-wave. The later waves were referred to as "indirect waves" or I-waves, because they were thought to be generated by trans-synaptic excitation of the pyramidal tract neurons by some other cortical neuron projecting an excitatory synapse upon them. Cooling of the cortex affected mainly the later I-waves, which were therefore thought to be produced by neurons in the more superficial layers of the cortex. Subsequent work in humans exploiting single-motor unit recordings and, more importantly, direct recording from the pyramidal tract in patients with implanted spinal stimulators, confirmed that the human motor cortex responds to TMS in a manner similar to the response of the feline motor cortex. D- and I-waves were demonstrated, with a frequency of discharge very similar to the cat's 600 Hz. Moreover, comparisons between TES and TMS showed that at low intensities TMS elicited preferentially I-waves rather than D-waves, probably due to the spatial inhomogeneity of TMS-induced electrical field, which is stronger in superficial cortical layers. The D- and I-wave model that was subsequently devised to explain the occurrence of the repetitive firing of neurons is illustrated in Fig. 7 [27].

D- and I-waves are not physiological entities, but rather are the epiphenomenon of cortical stimulation. However, the D- and I-wave model may teach us several important pieces of information. First of all it clarifies how it is possible that TMS can actually probe the excitability of populations of neurons. Changes in probability of firing of neurons are due mainly to changes of membrane potential in the dendrites and cell bodies and in changes of efficiency of the synapses projecting onto the neuron. If TMS elicited only D-waves, then changes in excitability of the corticospinal neurons would influence only minimally changes in probability of firing to the TMS pulse, because TMS would act at the level of the axon, downstream to most sites of neuronal plasticity. On the contrary, if TMS acts on neurons that project onto the corticospinal neurons, then the resulting action potential in the corticospinal axons will be dependent on changes of excitability of the synapse and of the dendritic and somatic membranes of the corticospinal neurons themselves. Trans-synaptic activation probably accounts for most of the trial-trial variability in TMS and is the single factor

Fig. 6 (continued) action potential, recorded by extracellular electrodes, resulting from the spatial summation of all single action potentials. (**b**) Schematic representation of a biphasic spatial distribution of action potentials, such as that observed in the pyramidal tract after cortical stimulation. (**c**) Schematic appearance of the D-wave and I-waves as recorded from the cervical corticospinal tract in humans

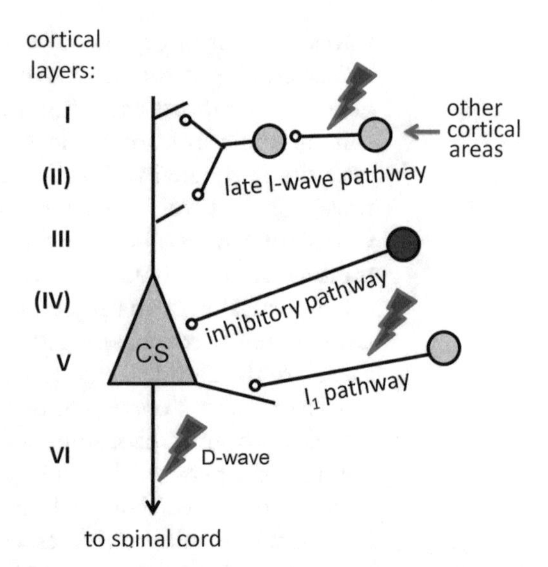

Fig. 7 Schematics of the sites of action of TMS (indicated by the *flash symbols*) in the genesis of the D-wave and of the I-waves. Note that, in the D-wave generation, the corticospinal (CS) axon is stimulated directly. On the contrary, the action potentials composing the I-waves are generated one or more synapses upstream of the CS neuron. This is thought to be the main factor contributing to the inter-peak interval between the different waves

that allows us to test the variability of the effects of TMS in different behavioral conditions. On the downside, the D- and I-wave model teaches us that TMS is far from being a uniform process; the net effects of TMS are likely to be the sum of the activation of several distinct populations of neurons, in a combination that is unlikely to be mimicking any physiological process. Importantly, both excitatory and inhibitory neurons participate in the net sum of the effects of TMS as demonstrated by paired-pulse paradigms (Sect. 4.2). The artificiality of the effects of TMS should always be kept in mind when interpreting the effects of TMS on behavior and the relation between the physiological and the behavioral effects of TMS.

3.3 TMS of the Motor Cortex

Even researchers who are interested in brain functions other than movement should possess some basic information on what are the effects of stimulation of the motor cortex. There are two reasons for this: first, the motor cortex has been the workbench of TMS since its original description in 1985 and, second, most safety parameters of TMS are referred to motor cortex measures of excitability. Most information about the physiological effects of TMS in humans are derived from studies of the motor cortex and then generalized to the remaining parts of the neocortex. The reason for this preference is that the motor cortex allows immediate objective evaluation of the output of magnetic stimulation, both at an immediate qualitative level through visual inspection of TMS-evoked movements and at a quantitative level by means of electromyography (EMG) of skeletal muscles. When we apply a magnetic coil to the central-lateral regions of a healthy participant's scalp and start delivering single stimuli of increasing intensity, at first we do not observe any apparent phenomena. However, at a given intensity we will start observing a single fast twitch of the upper limb contralateral to the stimulated hemi-scalp. Such twitches are time-locked to the stimulus, appearing almost simultaneously with TMS. They become of greater amplitude and tend to diffuse to proximal muscles with increasing stimulation intensity. The stimulus intensity at which the earliest motor responses appear is commonly referred to as the "motor threshold." The TMS-evoked twitch is the result of activation of the corticospinal pathway and subsequent excitation of spinal motor neurons, neuromuscular transmission, excitation of the muscular membrane and ultimately muscular contraction. EMG records the second to last of these stages, i.e. the depolarization of the cell membrane of skeletal muscle fibers. Surface EMG recordings consist of the sampling of the voltage difference between two surface electrodes, placed in proximity to the muscle of interest. These are commonly performed in a standard bipolar "belly-tendon" montage. The active electrode is placed on the belly of the muscle and the reference electrode is placed over the tendon of the muscle. The resulting recording consists of a waveform that is referred to as "motor evoked potential" (MEP). The onset latency of MEPs in the upper limb is of around 20 ms after the magnetic stimulus, of above 30 ms in the distal lower limb and of around 10 ms in the cranial region. It is worth noting that the mechanical events following the muscle action potential. The amplitude (or area) of the MEP is considered to be linearly correlated with the number of corticospinal neurons that have been activated in the motor cortex by the TMS pulse. A series of assumptions and approximations are required to establish such a relationship, as exemplified in Fig. 8. Given these approximations, the measure of the MEP amplitude is a good index of the excitability of the motor cortex.

3.4 The Effect of TMS on Neuronal Firing: Nonphysiological Versus Physiological Neural Activity

We have described how TMS directly affects neuronal firing; however, we have concluded that such neural activity is likely to be of little physiological significance. We will describe here the relation of TMS-induced neural activity with actual physiological neural activity. Also in this case the motor cortex is of great help. Easily recorded EMG signals can be treated as a direct index of cortical neuronal activity, thus allowing investigation of, albeit indirectly, the effects of TMS on neurons in humans, in vivo. If we ask a subject to perform a voluntary isometric contraction of his/her intrinsic hand muscles, resulting in an isometric finger grip and we

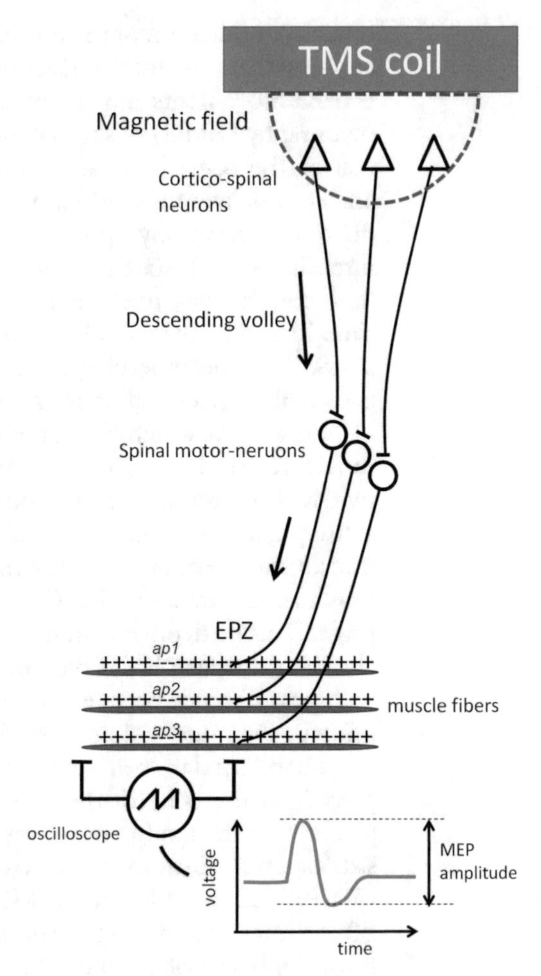

Fig. 8 Events linking TMS to the MEP. Several assumptions have to be met to linearly correlate the amplitude of the MEP to the number of cortico-spinal neurons that have been stimulated. (**a**) TMS is, electrically speaking, a very fast event. The duration of the effective current in the extracellular fluid of cortical neurons is in the order of the tens of microseconds. We assume that all biological events following electrical induction are as synchronized as TMS. However, as exemplified by the D- and I-wave model, activation of corticospinal neurons is far from being synchronous. (**b**) All corticospinal fibers and peripheral motor nerve axons conduct at the same speed, with no scattering over time of action potentials. In other words, we assume that action potentials that left the motor cortex simultaneously, reach the muscle fibers at the same time. (**c**) The single action potentials of individual muscle fibers (ap1-3) are "seen" as their spatial sum by the two surface electrodes. The amplitude of the compound surface potential is proportional to the number of active fibers only assuming that (1) single action potentials all start from the same point, i.e. that all neuromuscular junctions of all fibers are in a similar portion of the length of the fiber, the end-plate zone (EPZ). This is true of many muscles, including muscles of the hand but is not true, for example, of facial muscles. (2) Muscle fibers are parallel to each other, so that they can be recorded by the surface electrodes as a single large muscle fiber. Again, this is true for many muscles but not so in some muscles with complex geometry (for example in the tongue)

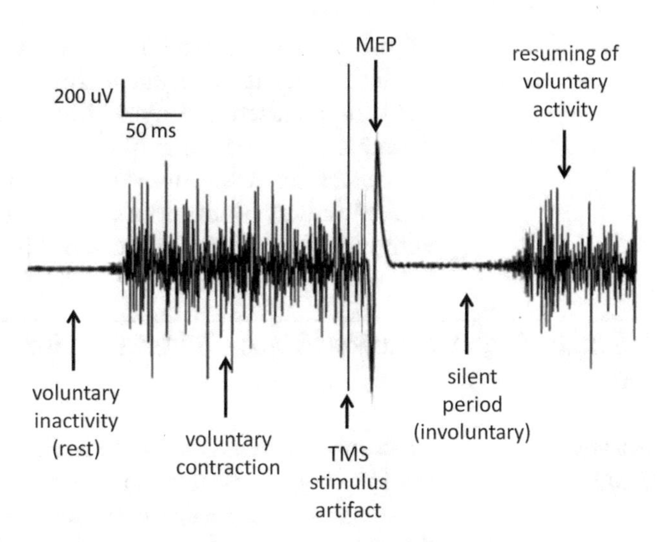

Fig. 9 The events occurring in EMG activity during voluntary contraction and TMS

record the EMG of the active muscles we will observe a series of voltage oscillations around the isoelectric line, corresponding to the spatial summation of the asynchronous activity of several different motor units. We can assume that this is paralleled in the motor cortex by frequent tonic discharge of several corticospinal neurons. TMS delivered to the motor cortex during such task produces two distinct phenomena: first a synchronized, high-amplitude discharge of the motor units (the MEP), occurring around 20 ms from TMS, which is immediately followed by a pause in voluntary activity, lasting for several tens of ms. This pause is referred to as "cortical silent period" (CSP) (Fig. 9). Interestingly, the MEP amplitude is strongly modulated by the amount of preceding voluntary activity: the stronger the contraction, the larger the MEP. The effects of TMS can therefore be classified into two large categories: an increase in neuronal firing, which is unrelated to the underlying physiological process and a subsequent decrease in the neuronal activity that is specifically related to the ongoing physiological process. In other words, TMS produces a burst of nonphysiological neuronal firing, together with inhibition of task-related physiological activity.

A few studies have investigated in vivo, in the visual cortex of anesthetized cats, what are the direct effects of TMS on neuronal firing patterns [28, 29] and they have shown complex interactions between stimulus intensity and the related effects on spontaneous and visually evoked firing. In one study [30] the neuronal activity evoked by visual stimuli was suppressed by TMS. On the contrary, spontaneous activity, unrelated to visual stimulation was increased. This duality between evoked (physiological) and spontaneous (nonphysiological) activity is strikingly similar to what is observed in the motor cortex. This excitation–inhibition pattern was confirmed using different techniques of in vivo optical imaging [31]. In a subsequent work, the effects of repetitive TMS on the stimulus–response curve of visual neurons to simple visual features were assessed. It was observed that the general decrease in visually evoked activity produced a narrowing of the neuron's tuning curve with increased selectivity for visual stimuli [32].

4 TMS in the Cognitive Neurosciences: Stimulating the Brain and Recording Behavior

4.1 The Impact of TMS on Behavior

TMS affects behavior. This is the main reason why TMS is a unique tool in research on brain–behavior relations. Contrary to most other noninvasive brain mapping techniques, such as functional magnetic resonance imaging, TMS adds information about causality in the brain–behavior relationship. A carefully controlled experimental design allows one to draw clear conclusions on the role of a given part of the brain in producing a certain type of behavior. This characteristic probably makes TMS and other neurostimulation techniques the single most valuable tools in cognitive neuroscience in humans. The quality of the impact of TMS on behavior has been thoroughly investigated in perceptual processes, mainly in the visual modality, probably because of the possibility to apply psychophysical analyses. Given the current available information, we are able to make reasonable predictions on the effects of TMS on perception. In other cognitive domains, the behavioral effects of TMS are less predictable and in many instances are evaluated and interpreted a posteriori. Since the earliest studies on vision, it has become clear that TMS application to the visual cortex can produce positive sensations, illusory perceptions, in the form of colorless flashes of light (phosphenes) [33]. At the same time TMS has been shown to disrupt conscious perception if delivered with appropriate timing, i.e. when the afferent volley along the visual pathways reaches the cortex [34]. The two phenomena, phosphenes and visual occlusion, were both intensity-dependent and seemingly occurred together [35]. The phenomenon of TMS-induced blinding led to the generalization that the impact of TMS on behavior was that of a loss-of-function, as clearly summarized in the term "virtual lesion"[36]. Several hypotheses have been put forward to explain the effect of TMS on perception in terms of changes in neural activity. Injection of neural noise, signal deletion, stochastic resonance, and activity-dependency have all been shown to be valid models to account for the effects of TMS on visual perception in different experimental conditions [37–41]. The resulting picture is not univocal and mandates further investigations of this topic.

The concept of virtual lesion is no longer accepted as pervasive to all aspects of TMS [42]. While it is true that in many cases TMS disrupts ongoing behavioral processes, it has become increasingly evident that the virtual lesion model cannot explain all the effects of TMS, which may be in the direction of a gain-of-function in many circumstances [43–48]. The effects of TMS on visual perception can switch from detrimental to facilitating according to different parameters. The best-known dimension that modulates the polarity of behavioral effects of TMS is the timing of the stimulus. High-frequency repetitive stimuli are generally detrimental to behavior when applied online but have been shown to improve some aspects of performance as an after-effect, at least in the motor cortex [49–51]. A second dimension that shapes the polarity of the behavioral effects of TMS is stimulus intensity: high stimulus intensities are more likely to impair perception; low stimulus intensities are more likely to improve perception in certain circumstances. The finding that TMS can result in behavioral gain of function is somewhat puzzling and at first glance might appear in contradiction with the possibility of inferring causality links between brain and behavior. Actually, the changes in behavior, regardless of them being a loss or a gain of function are supposedly specific to a given task, and are not observed in other domains of behavior. If this is the case, the finding that TMS affects specifically performance in a given task can be safely considered as evidence for an active involvement of the stimulated part of the brain in producing the task of interest.

The most intriguing of all the factors that may switch the TMS effects from inhibitory to facilitating is the so-called state-dependency of TMS. The fact that TMS effects are state-dependent is a trivial statement. It has been long known that identical magnetic stimuli produce different physiological and behavioral effects according to what task the stimulated subject is engaged in doing [52–55]. However, the term "state-dependent" TMS indicates an experimental procedure which is much more subtle than it actually suggests. The founding concept is that the pre-stimulus excitability of a given population of sensory neurons may be changed in a controlled way by means of consolidated psychophysical paradigms, in particular by inducing adaptation to a certain perceptual feature through repetitive exposure to that feature. The perceptual effects of TMS delivered to the cortical area that contains the adapted neurons boost the adapted perceptual category. In this way the effects of TMS can be selectively biased towards a single population of neurons (the adapted ones) in spite of the neurons being spatially overlapping with other neurons coding for different perceptual features (see [44] for a review on the topic). The original finding that started adaptation-TMS paradigms is particularly enlightening: TMS over the visual cortex induces a perceptual illusion, the phosphene, which is typically colorless. The inference from this finding could be that the early visual cortex is devoid of color representation. We know from physiology that this is not the case, the phosphene is colorless not because no neurons code for color, but rather because all neurons with tuning to each separate color are stimulated equally, and the net perceptual effect is of no particular color. In this case, the presence of spatially overlapping but functionally distinct neural populations renders the results of conventional TMS inconclusive in defining the functional aspects of the single neural populations. If we adapt the subject's vision to a color (e.g. red) for a minute or so, the resulting perceptual after-effect will be that of a green background. Stimulating the occipital cortex during this after-effect induces a phosphene that is surprisingly red. In controlled conditions TMS can reveal the existence of color coding in the occipital cortex.

The phenomenology of TMS adaptation is quite robust: repetitive behavioral processes produce adaptation-like after-effects, i.e. lesser sensitivity to the adapted process. If TMS is delivered during the after-effect, it boosts selectively the adapted process. Such paradigm has been replicated in several domains, from elementary visual features to more complex visual properties of stimuli, up to multimodal sensori-motor processes [56–60]. Figure 10 illustrates the expected results of a TMS-adaptation paradigm. Most of the ambiguity and the apparent contradictions between the expected

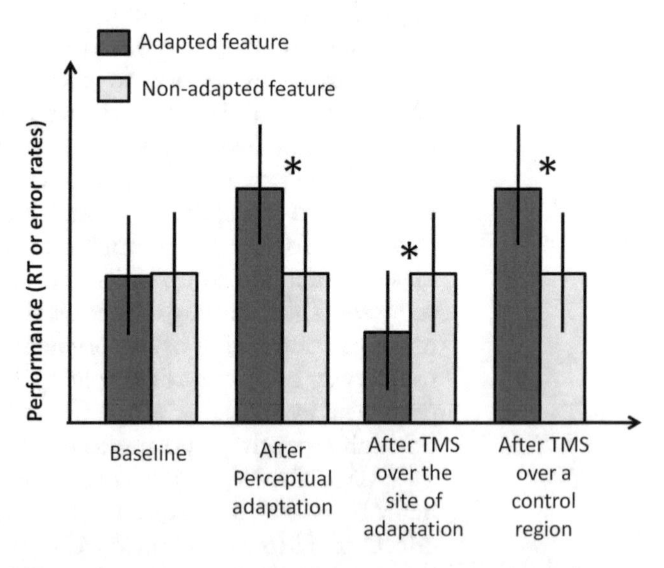

Fig. 10 Expected results of a TMS-adaptation paradigm. Adaptation introduces a perceptual cost in baseline conditions, indicated by the difference in performance between processing the adapted and the nonadapted stimuli. TMS over the critical brain area that contains adapted neurons reverses the cost of adaptation, making the processing of adapted features more efficient than that of the nonadapted ones

and the actual behavioral effects of TMS come from the faulty assumption that behavior and neural activity are simplistically "correlated." However appealing, we should resist the temptation to establish a direct connection between putative TMS-induced changes in neuronal activity (excitation or inhibition) and the polarity of the changes in behavior (facilitation or disruption). It is advisable to "dissociate the language of physiological effects from those of behavioral effects" [42]. In the current paucity of clear knowledge about the neuron-behavior relations it is probably preferable to consider and discuss the phenomenology of the effects of TMS on behavior without invoking potential underlying neural mechanisms.

4.2 The Toolbox of TMS: Single Pulses, Conditioned Pulses, and Repetitive Stimulation

The TMS toolbox comprises distinct patterns of stimulation that are conventional, well-characterized entities. The first and simplest of these tools is the single pulse. We have already reviewed what the effects of a single-pulse are: an initial increase in spontaneous neural activity and a subsequent inhibition of physiological neural activity. The behavioral effects of a single pulse are variable and dependent on timing, intensity, and the pre-stimulus state of the cortex (Sect. 4.1). The duration of the effects of a single pulse on behavior is of around 100 ms, as witnessed by the CSP following motor cortex stimulation.

Single pulses can be conditioned by other types of stimulation in classical conditioning stimulus/test stimulus paradigms in which the effects of the test TMS is evaluated alone or when preceded by the conditioning stimulus. Test stimuli are identical in both unconditioned and conditioned trials; therefore, if any difference emerges in the output of test stimuli between the two trials, it indicates functional connectivity between the target of conditioning TMS and the target of test-TMS (Fig. 11). The site of test TMS is in general the motor cortex. The site of the conditioning stimuli is anywhere it is thought that afferent projections to the motor cortex may originate. This includes cortical areas such as the premotor, parietal, and somatosensory cortices [61–66] as well as areas in the opposite hemisphere [67]. In case the conditioning stimulus is delivered by means of TMS, two coils are simultaneously placed over the scalp. This setting is commonly referred to as "twin-coil" or "dual-coil" technique. However, the conditioning stimulus may be applied to other portions of the nervous system such as peripheral sensory fibers as in the case of short-latency afferent inhibition [68] and in this case the stimulus type may be other than TMS. The time interval between the conditioning and the test stimuli (inter-stimulus interval—ISI) is usually set around the time required by a direct or quasi-direct (monosynaptic or oligosynaptic) pathway to travel from the conditioning to the test areas. For example, interactions between the premotor and motor cortex occur within the 5–7 ms ISI, between

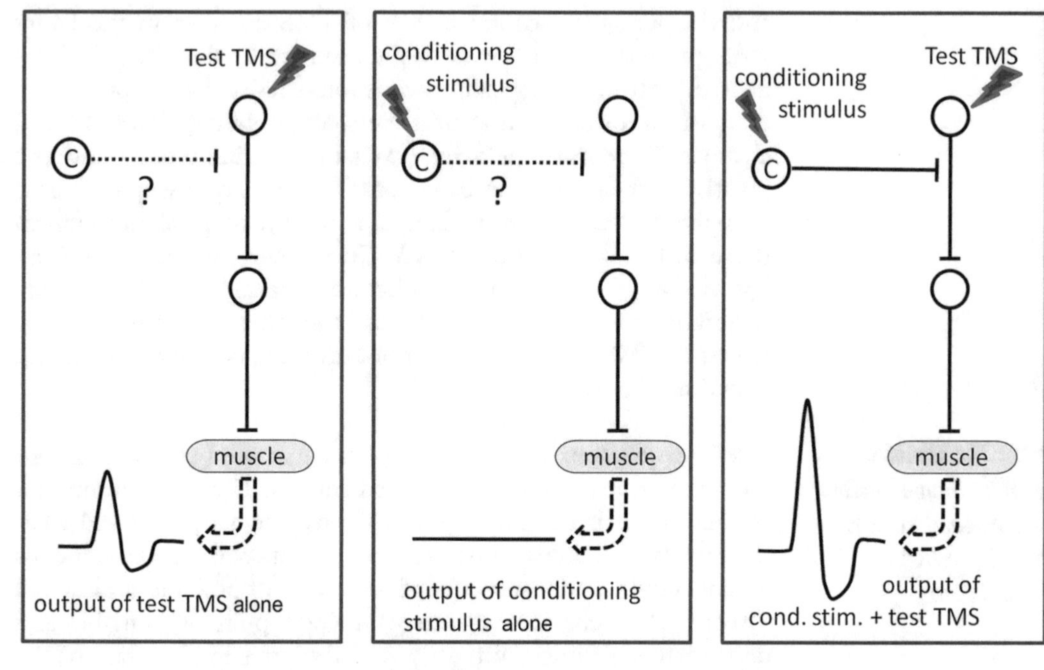

Fig. 11 Schematization of conditioned TMS paradigms. *Left*: TMS over the test area (the motor cortex) produces a known and measurable effect (the MEP and its amplitude). *Middle*: a given brain area, C, is thought to have weak connections to the motor cortex, but its stimulation does not produce any measurable output in the corticospinal system. *Right*: The output of test TMS is different (larger, in the current example) if test TMS is preceded by the conditioning stimulus. This finding confirms that the brain area C is functionally connected to the motor cortex

the ipsilateral and contralateral hemispheres occur at ISIs or above 10 ms. Interactions between a peripheral afferent stimulus from the upper limb and the sensorimotor cortex occurs at intervals at or above 20 ms. One peculiar form of conditioned TMS is that in which conditioning and test TMS are delivered through the same coil, on the same scalp position. In this case the interactions between different cellular populations within the same part of the cortex (generally the motor cortex) are probed. Changing the intensity of the conditioning stimulus compared to that of the test stimulus allows partly selective stimulation of different neuronal populations. The conventional paired-pulse paradigm consists of a low-intensity (submotor threshold) conditioning stimulus, followed by a high-intensity (above threshold) test stimulus [69]. Conditioning TMS at that intensity stimulates preferentially GABAergic inhibitory neurons. The resulting interaction between the two stimuli is complex. For inter-stimulus intervals (ISIs) between 1 and 4 ms a strong inhibitory effect is exerted by the conditioning stimulus on the test stimulus, a phenomenon referred to as intracortical inhibition (ICI). For ISIs higher than 7–10 ms a facilitatory interaction appears—intracortical facilitation (ICF).

The definition of repetitive TMS (rTMS) applies to any series of three or more magnetic stimuli. RTMS can employ single frequencies or patterns of frequencies. Single-frequency rTMS is generally classified into two main categories: low-frequency (0.2–1 Hz) and high frequency (>2 Hz) stimulation. Low-frequency rTMS is conventionally applied at 1 Hz. What is interesting about repetitive trains of TMS is that they induce cortical plasticity for a period that outlasts the train of stimuli itself. Such after-effects of rTMS are frequency-dependent, intensity-dependent, and duration-dependent. Observations in the motor cortex show that the after-effect of 1 Hz rTMS is that of a decrease in the cortical output to single TMS stimuli, i.e. a decrease in MEP amplitudes [70–74]. The After-effects of >5 Hz stimuli on the motor cortex are opposite, i.e. they lead to an increase in MEP size [75–78]. Following these findings, it has been argued that plasticity induced by low-frequency TMS results in impairment of cortical function and behavior, while high-frequency TMS would result in an increase in cortical activity and improvement of behavior. This generalization has proven to be a useful assumption in many instances but it often fails to explain all the behavioral effects of rTMS, which remain highly variable according to the site of application of stimulation. The duration of the after-effect is generally dependent on the duration of the stimulus train and is of comparable length, i.e. stimulating at 1 Hz for 15 min produces plastic changes in the cortex for another 15 min [74, 79]. Stimulus intensity correlates positively with both the duration and size of the after-effects [80, 81]. Patterned stimulation consists of repetitive TMS in which different frequencies are present in different patterns. The most popular example is theta-burst stimulation (TBS) [82, 83], consisting of short bursts of three TMS pulses at 50 Hz, repeated every 200 ms, i.e. with a frequency of 5 Hz, in the theta range. Continuous TBS over the motor cortex produces a long-lasting decrease in MEP amplitude, besides a decrease in performance in simple reaction times. It is worth noting that the relation between the train duration and the after-effect duration seem to be nonlinear: 2 min of TBS produce after-effects lasting for tens of minutes.

One last technique to be included in the toolbox of neurostimulation is paired-associative stimulation (PAS). This protocol has been demonstrated in the motor cortex and produces powerful and long-lasting plastic changes. It consists of repeated stimulation of a peripheral sensory nerve in the upper limb (usually the median nerve), paired and time-locked to single-pulse TMS of the motor cortex [84]. The polarity of the after-effect depends on timing of the two associated stimuli: the peripheral and the central one. The afferent volley reaches the motor cortex around 20 ms after median nerve stimulation. If TMS is delivered to the motor cortex 25 ms after median stimulation, as in the PAS25 protocol, it will be preceded by the afferent volley. If TMS is delivered 10 ms after median nerve stimulation, as in the PAS10 protocol, it will be followed by the afferent volley. The effects of the PAS25 protocol are those of potentiating cortical synapses, as shown by a stable increase in MEP size. The effects of the PAS10 protocol are a decrease in synaptic efficiency, as demonstrated by the subsequent decrease in MEP size.

4.3 Designing a TMS Experiment

TMS is a tool that does not produce any specific data by itself. In this respect TMS is erroneously assimilated and compared to other brain mapping techniques such as functional magnetic resonance imaging or electroencephalography (EEG) that record brain activity in specific domains and are therefore stand-alone instruments in brain imaging. On the contrary, TMS is necessarily coupled with a measure of some kind, ranging from simple recording techniques such as qualitative observation of behavior to complex instrumental measures such as electrophysiology of single cortical neurons. The first, obvious, step in designing a TMS experiment is to identify a measure of behavior or of neural activity that is of interest for the specific experimental hypothesis. Once this is done, it is possible to select one of several well-codified experimental paradigms: (1) cortical output measure, (2) offline TMS, (3) online TMS, and (4) state-dependent TMS (Fig. 12).

Measuring the cortical motor output is a simple technique that is applicable only to those parts of the cortex with a known and trial-by-trial measurable output. This restricts the field of operations to the primary motor cortex and to the primary sensory cortices. The aim of this experimental paradigm is to test whether a given behavioral condition (e.g. action observation) influences the excitability of the motor or sensory cortices, as measured from their output to TMS. Common examples include the effects on the motor cortex of speech processing, action observation, motor imagery, numerical processing, acoustic stimuli, or observation of pain [85–88], as well the effects on visual cortex excitability by attention allocation or numerical processing [89, 90]. In this paradigm, TMS is commonly used in an event-related way, i.e. TMS is delivered at time-locked intervals from the start of the behavioral condition. Event-related TMS allows exact chronometry of the neural process of interest, by varying the stimulus onset interval and identifying the time interval in which the behavioral task modulates cortical excitability (Fig. 12a). Stimulus onset delays may be varied between milliseconds to seconds, according to the specific experimental requirements. The major drawback of this paradigm is that it provides mainly correlative information rather than causal links between the stimulated area and behavior. However, the opportunity to assess the timing of neural processes can provide very useful information in support of hypothetic causal links [88, 91].

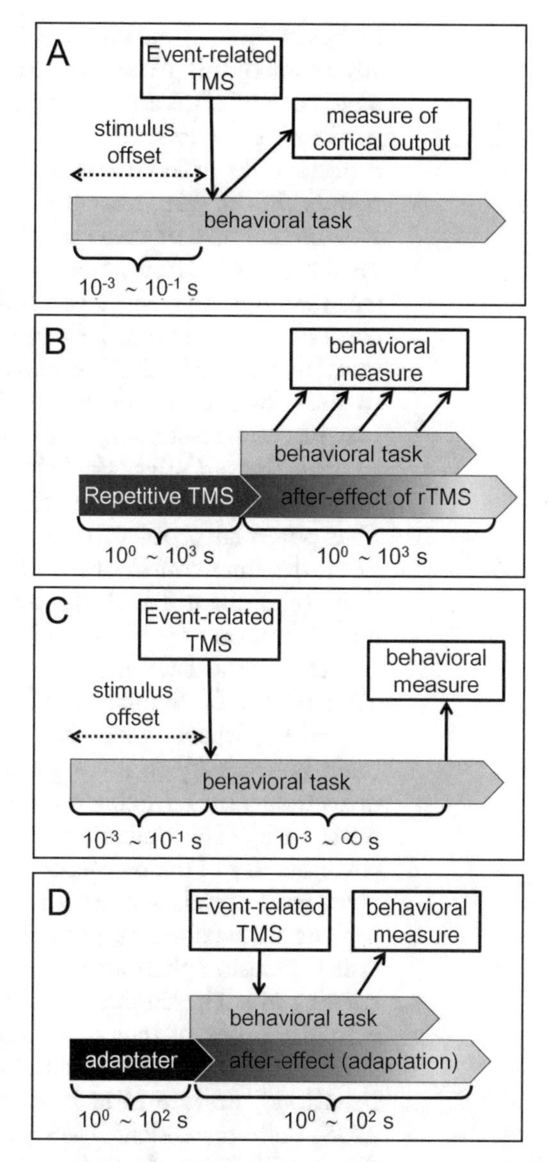

Fig. 12 Schematization of the different TMS protocols. (**a**) Cortical output measurement; (**b**) offline TMS; (**c**) online TMS; (**d**) state-dependent TMS

The offline TMS approach is based on the assumption that repetitive TMS has an after-effect outlasting the time of stimulation. In this paradigm, the experimental hypothesis consists in establishing whether the integrity of a given part of the brain is necessary to accomplish a given behavioral task. The task is performed during such after-effect and not during TMS itself. The time of rTMS varies according to protocols, ranging from seconds in high-frequency rTMS, to tens of seconds for TBS, to tens of minutes for 1 Hz-rTMS. Also the duration of the after-effect ranges in similar time intervals (see timings in Fig. 12b). The main advantage of this protocol is that several nonspecific effects (noise, pain, etc.) of TMS are not acting on the subject while he or she is performing the task. This is an obvious advantage, for example, in acoustic tasks or in attention allocation tasks. The main disadvantage is that it does not allow chronometry of neural processes. Possible sources of uncertainty are the non-univocal predictions on the polarity (facilitation or disruption) of the effects and the fact that the after-effect of rTMS wanes over time, with a time constant that is not entirely predictable; therefore behavior measured immediately after the end of TMS might not be comparable to that observed 5 min later. This latter disadvantage, however, may be counteracted in favor of clarity of the results. The demonstration that an effect on behavior decays over time after rTMS supports the idea that the effect was actually due to TMS itself. One potential problem intrinsic to the offline protocol is that, given the time-course of the after-effects of TMS, control conditions (e.g. sham TMS) can be delivered only in block designs, rather than on a trial-by-trial basis.

The online TMS approach is probably the most informative of TMS protocols. Similar to the offline TMS protocol, the experimental hypothesis is whether the integrity of a given part of the brain is necessary to accomplish a given behavioral task. TMS is delivered to the participant *during* the task, generally in an event-related way. The online protocol, being event-related allows for chronometry of the recruitment of the stimulated brain area in the given task. The classical example of the use of online TMS to establish the time-course of physiological processes in behavior is the visual occlusion phenomenon produced by TMS over the occipital cortex [34]. The timing of TMS depends on the task, but usually is in the order of tens of milliseconds to seconds. It should be stressed that TMS must be delivered online to the task, but the behavioral measure of task efficiency can occur at any time after TMS. State-dependent TMS has been explained in Sect. 4.1. In this case TMS is delivered in an event-related way during a behavioral task, similarly to online paradigms. However, in this protocol a period of repetitive behavior (the adapter period) may or may not precede the actual TMS. The finding of differential effects of TMS according to the preceding exposure to adapters allows one to suggest that the stimulated area contains neurons that are sensitive to the category that has been adapted.

4.4 The Problem of Control Conditions in TMS Experiments

TMS protocols require several levels of control conditions. These fall into two general categories: control tasks and control stimulations. Actually, to provide meaningful results, any experiment investigating the effects of TMS on behavior should include at least a control task and at least a control stimulation condition. These are usually mixed into a TASK*CONTROL factorial design.

The desired results, i.e. a specific effect of TMS over a specific brain area, on a specific task, will be represented by a TASK*CONTROL interaction, while any main effect is devoid of interest. As in most experimental approaches, control tasks are required to make sure that the effects of the experimental manipulation (in this case TMS) are specific to the task of interest. Control stimulation is more challenging, but necessary. Besides neural stimulation, TMS produces many phenomena that can potentially affect behavior and therefore are possible experimental confounds. These include: acoustic clicking noise; peripheral sensory nerve stimulation producing painful or nonpainful cutaneous sensations; peripheral motor nerve stimulation, causing muscular twitches and finally, less-specific factors such as head fixation or the weight of the coil on the head and neck [92]. An ideal control condition that accounts for all possible confounding variables is yet to be described. However, several possible options can be chosen according to different experimental requirements, namely, sham stimulation, effective stimulation at a different time and effective stimulation over different cortical areas. Sham stimulation refers to a procedure in which a TMS coil or a device externally looking like a TMS coil is positioned on the subject's scalp, producing sounds similar to those of effective TMS. Sham stimulation can be achieved by means of purpose-made sham coils. Alternatively, a standard TMS coil may be used by adding a spacer to the focal point so that the magnetic field reaching the scalp is no longer effective or by tilting the coil, thus turning away the focal point of the coil from the brain. Sham TMS controls well for acoustic stimulation, neck posture, coil weight, but does not control for electrical stimulation of peripheral sensory and motor nerves. Effective TMS can be used as a control if applied to a different cortical site. There are several drawbacks in this procedure: first of all choice of the control site, which should be, by definition, not involved in the task of interest. Second, locating TMS over a different scalp region changes the quality of the effects of peripheral nerve stimulation. For example, stimulating the inferior frontal cortex is often painful and produces collateral muscular twitches in cranial muscles, while stimulating the midline central region is completely painless and produces no muscular twitch. Often, as a control stimulation site, the "vertex" coordinate of the 10–20 system of EEG electrode placement is adopted, i.e. what is commonly referred to as the "Cz" coordinate. This is done under the bizarre assumption that the cortex under the vertex is unrelated to any task. The cortex under the Cz coordinate is likely to be between the lower-limb somatosensory area and the superior parietal lobule [93, 94]. Another popular control protocol is that of changing the timing of effective TMS. For example in a visual task, delivering TMS at or slightly before the onset of a visual stimulus should assure that stimulation has hit the brain well before the relevant sensory information reaches it.

This is a useful control condition in the "cortical output measure" paradigms. It may be less useful in online TMS protocols, because of potential delayed effects of TMS on behavior.

4.5 Dosing TMS—Safety

Individual variations in skull and cortex anatomy are the main source of variation in the amount of energy that is delivered to the brain by TMS. The distance between the scalp and the skull is the factor that most influences the capacity of TMS to reach the cortex. This is estimated to be on average 16 mm with a SD of 5 mm (see Table 4 of [93]). This measure however varies considerably in different scalp regions; it is maximal around the central-parietal midline regions and is minimal in the anterior frontal regions. Let us analyze the consequences of inter-individual variability of scalp–brain distance in terms of individual dosing of TMS intensity. Ina representative group of adults, the parietal region has an average scalp–brain distance of 17 mm, with a standard deviation of 2.6 mm [93]. This implies that in a strictly normally distributed sample of 20 individuals, 13 of them (corresponding to 68% of the population, distributed between +1SD and −1SD) will have a scalp–brain distance between 19.6 and 14.4 mm. The two subjects at the upper and lower extreme values will have scalp–brain distances of 20.9 and 13.1 mm. The minimum distance value expressed in relation to the maximum distance value is: 13.1/20.9 = 0.62, meaning that the minimum distance is 62% of the maximum distance. According to the law that describes the decay of magnetic fields in space, the magnetic field occurring at 62% of a given distance is 2.5 larger than the magnetic field at 100% of the distance. In conclusion, if the same stimulus intensity is applied to both subjects, the subject with the lowest scalp–brain distance will receive 250% more energy in his or her cortex compared to the subject with the highest scalp–brain distance. For this reason, it is never advisable to use fixed intensities of TMS, since the results would be extremely variable between subjects and, most importantly, in case of repetitive stimuli, the risk of delivering unsafe stimulation to a vulnerable portion of the population is very high.

The common way in which inter-subject variability in scalp–brain distance is taken into account is the measure of the motor threshold. This is the threshold intensity for activation of the corticospinal pathway with TMS of the motor cortex. There are several ways in which motor threshold can be assessed, the main distinction being between the active motor threshold, i.e. measured during voluntary contraction and the resting motor threshold (assessed at rest). The active motor threshold is systematically lower than the resting motor threshold [95]. The resting motor threshold is more variable in consecutive tests within the same individual, than the active motor threshold [96–101]. The corticospinal activity can be evaluated visually or by means of EMG recordings. In this case, it is said that the visual method tends to over-estimate the threshold intensity [102]. Specific guidelines have been published describing in detail the different modalities of threshold assessment [103, 104]. Once the individual motor threshold is assessed, the stimulation intensity to be used in a given experimental setup will be expressed as a function of motor threshold rather than as absolute values. Given that scalp–brain distances vary in a fairly predictable way within the same individual, between different scalp locations, adjustments to the stimulus intensity could be adopted to correct for this variability [105]. The use of motor threshold as a dosing measure is advisable also from the point of view of subject safety. It is beyond the scope of the present review to discuss safety issues of TMS, which are exhaustively described in two complementary, authoritative sets of guidelines [106, 107]. The two papers by Wassermann and Rossi et al. should be well-known to the TMS user before starting experimental brain stimulation. The authors identify some safety limits of TMS parameters, based on three distinct dimensions: stimulus frequency, stimulus intensity, and number of stimuli. The intensity of the stimuli in the safety curves is expressed as a function of individual motor threshold. As a consequence, it is not possible to know whether a stimulation being applied is safe, unless the motor threshold is known. To this purpose it should be always kept in mind that rTMS is a safe tool when used within safety limits. On the contrary, it can systematically and predictably produce adverse effects (epileptic seizures) when used outside published safety limits [108, 109].

5 TMS in the Study of Lateralization

Given its spatial resolution, TMS is a good tool for the study of brain–behavior relations if a strong claim is made on the focality of brain representations. On the other hand, the concept itself of lateralization is founded on spatial constraints within the brain (the right versus left distinction and the concept of homologous areas in the two hemispheres). Spatial distinction defining lateralization have very coarse spatial resolution, well within the possibilities of brain mapping with TMS, which therefore is an optimal tool for the study of the asymmetry of brain functions. Brain asymmetries have been found with TMS in the most diverse domains of cognition. Examples include the representation of space [110, 111]; semantic knowledge [112–116]; observed body and action representations [58, 117, 118], most components of language [119, 120]; episodic memory and working memory [121–123]. Some paradigmatic examples are presented in the following sections.

5.1 Left-Lateralization of Speech Articulation Versus Right-Lateralization of Prosody and Song

In 1851 Paul Broca described the appearance of aphasia in a patient following damage to the left frontal lobe, practically giving birth to modern neuropsychology [124]. More than a century later, TMS provided the means to produce a model of Broca's aphasia in healthy subjects, the so-called speech arrest phenomenon. This consists in a transient disruptive effect of online TMS on speech production, that can be obtained with high-frequency rTMS on the scalp region overlying the left inferior frontal gyrus [125]. Behaviorally it is a very variable phenomenon, ranging from a mild hesitation to complete aphemia [119, 126, 127]. The lateralization of the speech arrest effect is very robust, however, it is prone to producing a small but significant number of false localizations of language to the right hemisphere [128]. Speech production, the function that is impaired by the speech arrest procedure, can be dissociated by means of TMS from other language functions. In particular it is possible to obtain in a single healthy subject a dissociation between speech articulation, left-lateralized, and more automatic forms of utterance such as song production and formulaic speech, which tend to be right-lateralized [129]. Indeed, it was observed that left inferior frontal gyrus stimulation produced no interference on singing while having a heavy impact on reading aloud. On the contrary, right IFG stimulation produced a loss of melodic modulation [126]. A subsequent study replicated the single dissociation between speech and song in the left hemisphere but not in the right hemisphere [130].

5.2 Lateralized Effects of Frontal rTMS on Mood

RTMS produces after-effects on neuronal activity outlasting the period of stimulation. This characteristic has been exploited as a potential tool for treating neurological conditions in which abnormal activity of a population of cortical neurons is at the basis of the neurological or psychiatric symptoms. One medical condition that meets these requirements is unipolar mood disorder, or major depression. Dysfunctional neuronal activity in the prefrontal cortex has been associated with depression and rumination symptoms [131, 132], thus making it a possible target of repeated applications of rTMS. Several studies have now demonstrated the clinical usefulness of rTMS over the dorsolateral prefrontal cortex (DLPFC) on the mood of patients with unipolar depressive disorder [133–135]. Interestingly, the effects on mood of rTMS on the DLPFC are asymmetrical. Rapid-rate RTMS of the left DLPFC is effective in improving mood in unipolar patients, but slow-rate rTMS of the right hemisphere seems to be equally effective [136]. The offline effects of rapid rTMS are conventionally thought to be facilitatory, while the after-effects of slow rTMS are considered inhibitory (Sect. 4.2). The pattern of effects of unilateral TMS on mood seems to point at a hyperactive left frontal cortex and hypoactive right prefrontal cortex associated with depression, coherently with aetiopathological models of depression [131, 132]. Most authors warn against interpreting these results as evidence for a strict lateralization of mood. It is likely that the effects of TMS are in the sense of restoring a balance between collaborative activity in the two hemispheres, rather than identifying a hemisphere associated with good mood and a hemisphere associated with bad mood [137].

5.3 TMS to the Parietal Lobes Has Lateralized Effects on Numerical Abilities

The capacity of the brain to produce behavior that takes into account the numerosity of objects in the environment is shared by many animal species [138] and is of considerable interest in the human cognitive neuroscience. Functional neuroimaging in humans indicates a correlation between magnitudes and brain function in the posterior parietal cortex, in regions around the intraparietal sulcus (IPS) [139, 140]. Imaging studies failed to show clear lateralization of cortical activity, at odds with classical neuropsychological findings on patients with acquired acalculia, who suffer predominantly from left-hemisphere lesions [141]. Many TMS studies on numerical competences in humans have addressed the particular topic of lateralization, by having systematically two stimulation conditions: right and the left hemisphere stimulation, together with a sham stimulation condition. The vast majority of studies indicated that online TMS to the left posterior parietal cortex, near to the posterior portion of the IPS produces loss of function in numerical tasks. This effect is observed both when numerical quantities are represented as symbolic numerical notations or as nonsymbolic representation of different numerosity [142–146]. Similar effects have been found by online rTMS to the left parietal cortex, but not to the right parietal cortex on the spatial representation of ordinal sequences along the so-called "mental number line" [147, 148]. An effect of right parietal stimulation compared to left parietal stimulation was found only in a few instances on a selected type of numerical task [149, 150]. Bilateral and symmetric effects of parietal TMS on arithmetics have also been described [151]. There is a general methodological lesson to be learned from the paradigmatic case of TMS investigations of numerical capacities. Functional neuroimaging provides maps of cortical activity that tend to be overall bilateral, even in cognitive domains in which neuropsychology points at a clear-cut lateralization of brain functions as for example in visual face-processing [152, 153]. In this systematic dissociation between the effects of lesions and cortical activations, TMS tends to follow the pattern indicated by lesions, rather than the less conservative maps of brain function indicated by functional brain imaging.

5.4 A Pitfall in the Study of Hemispheric Specialization with TMS: Trans-Callosal Connectivity

The use of TMS in the study of hemispheric specialization suffers from a potential confounding effect caused by the transmission of TMS-induced excitation to the contralateral hemisphere through callosal connections. Since the first description of the spread of TMS activity at the whole brain level [154], it has become clear that the homologous region on the hemisphere contralateral to the stimulated one was activated very early (within tens of ms) and selectively by unilateral TMS. In other words, stimulating one part of a hemisphere produces immediate neural activity in the contralateral homologue region. This fast-track to the homologue regions of the opposite hemisphere is due to callosal fibers, which provide mainly homotopic interconnections between the hemispheres [155, 156], with only a few heterotopic connections. Also in this case, the motor cortex provides us with the neural substrate to investigate the physiological effects of TMS. If in a healthy subject at rest we record EMG activity from the intrinsic muscles of the right and left hand on two separate channels and we stimulate with TMS the right-hemisphere motor cortex, we will record an MEP in the left hand only. Stimulating the left hemisphere will produce an MEP in the right hand. These findings allow us to infer that the right hemisphere is dedicated to control of the left hand and the left hemisphere is dedicated to right hand movements. If we ask the subject to contract both hands and we stimulate the left hemisphere, we will record in the right hand a well-known sequence of MEP and CSP (Fig. 9). Surprisingly, in the left hand (ipsilateral to stimulation) we will record an interruption of voluntary activity, not preceded by a MEP, that has been named "ipsilateral silent period" (ISP) [67, 157–159]. This finding could lead us to the erroneous conclusion that both hands are represented in each hemisphere. On the contrary, the ISP is the quasi-exclusive product of inhibitory activity exerted by the stimulated hemisphere onto the contralateral hemisphere, which results ultimately in an interruption of voluntary activity in the hand ipsilateral to stimulation. This phenomenon is referred to as inter-hemispheric or transcallosal inhibition or sometimes as inter-hemispheric competition [67]. It is best studied and characterized by means of a conditioned, dual coil paradigm: in a subject at rest, test stimuli are applied to the motor cortex of one hemisphere, producing MEPs in the contralateral hand. Conditioning stimuli are applied to the motor cortex of the opposite hemisphere and they exert a powerful inhibition of the contralateral cortical output, when applied between 10 and 40 ms prior to the test stimulus. It is likely that most other cortical areas exert, when stimulated, significant effects on their contralateral homologue. In this respect, when addressing a lateralization issue it is probably best to include in all experimental designs the separate stimulation of both the hemisphere of interest and the contralateral homologue.

5.5 Conclusions

TMS is a powerful tool in lateralization research that allows sophisticated inference on the spatial localization and timing of neural processes producing a given behavior. Its main advantage is the possibility to produce significant effects on behavior and therefore to establish causal inferences. The use of TMS in experimental settings requires good knowledge of several physical and biological aspects of electromagnetic stimulation in order to produce repeatable and robust results. The main limitation of the use of TMS in the study of lateralization is the occurrence of neural effects of stimulation at regions distant from the stimulated one and in particular in the opposite hemisphere.

References

1. Fritsch GT, Hitzig E (1870) Ueber die elektrische Erregbarkeit des Grosshirns. In: Archiv für Anatomie, Physiologie und wissenschaftliche Medicin. G. Eichler, Berlin, pp 300–332
2. Cushing H (1909) A note upon the faradic stimulation of the post-central gyrus in conscious patients. Brain 32:44–53
3. Foerster O (1936) Motorische Felder und Bahnen. In: Bumke H, Foerster O (eds) Handbuch der Neurologie IV. Springer, Berlin, pp 49–56
4. Penfield W, Boldrey E (1937) Somatic motor and sensory representation in the cerebral cortex of man as studied by electrical stimulation. Brain 60:389–443
5. Marsden CD, Merton PA, Morton HB (1983) Direct electrical stimulation of corticospinal pathways through the intact scalp in human subjects. Adv Neurol 39:387–391
6. Merton PA et al (1982) Scope of a technique for electrical stimulation of human brain, spinal cord, and muscle. Lancet 2(8298):597–600
7. Barker AT, Jalinous R, Freeston IL (1985) Non-invasive magnetic stimulation of human motor cortex. Lancet 1(8437):1106–1107
8. Miranda PC (2013) Physics of effects of transcranial brain stimulation. Handb Clin Neurol 116:353–366
9. Roth Y et al (2007) Three-dimensional distribution of the electric field induced in the brain by transcranial magnetic stimulation using figure-8 and deep H-coils. J Clin Neurophysiol 24(1):31–38
10. Rudiak D, Marg E (1994) Finding the depth of magnetic brain stimulation: a re-evaluation. Electroencephalogr Clin Neurophysiol 93(5):358–371
11. Deng ZD, Lisanby SH, Peterchev AV (2013) Electric field depth-focality tradeoff in transcranial magnetic stimulation: simulation comparison of 50 coil designs. Brain Stimul 6(1):1–13
12. Basser PJ, Wijesinghe RS, Roth BJ (1992) The activating function for magnetic stimulation derived from a three-dimensional volume conductor model. IEEE Trans Biomed Eng 39(11):1207–1210
13. Basser PJ, Roth BJ (1991) Stimulation of a myelinated nerve axon by electromagnetic induction. Med Biol Eng Comput 29(3):261–268
14. Roth BJ, Basser PJ (1990) A model of the stimulation of a nerve fiber by electromagnetic induction. IEEE Trans Biomed Eng 37(6):588–597
15. Nagarajan SS, Durand DM, Warman EN (1993) Effects of induced electric fields on finite neuronal structures: a simulation study. IEEE Trans Biomed Eng 40(11):1175–1188
16. Silva S, Basser PJ, Miranda PC (2008) Elucidating the mechanisms and loci of neuronal excitation by transcranial magnetic stimulation using a finite element model of a cortical sulcus. Clin Neurophysiol 119(10):2405–2413
17. Abdeen MA, Stuchly MA (1994) Modeling of magnetic field stimulation of bent neurons. IEEE Trans Biomed Eng 41(11):1092–1095
18. Ravazzani P et al (1996) Magnetic stimulation of the nervous system: induced electric field in unbounded, semi-infinite, spherical, and cylindrical media. Ann Biomed Eng 24(5):606–616
19. Ruohonen J, Ravazzani P, Grandori F (1995) An analytical model to predict the electric field and excitation zones due to magnetic stimulation of peripheral nerves. IEEE Trans Biomed Eng 42(2):158–161
20. Hsu KH, Durand DM (2000) Prediction of neural excitation during magnetic stimulation using passive cable models. IEEE Trans Biomed Eng 47(4):463–471
21. Hsu KH, Nagarajan SS, Durand DM (2003) Analysis of efficiency of magnetic stimulation. IEEE Trans Biomed Eng 50(11):1276–1285
22. Rotem A, Moses E (2006) Magnetic stimulation of curved nerves. IEEE Trans Biomed Eng 53(3):414–420
23. Salvador R et al (2011) Determining which mechanisms lead to activation in the motor

cortex: a modeling study of transcranial magnetic stimulation using realistic stimulus waveforms and sulcal geometry. Clin Neurophysiol 122(4):748–758

24. Rotem A, Moses E (2008) Magnetic stimulation of one-dimensional neuronal cultures. Biophys J 94(12):5065–5078
25. Pashut T et al (2014) Patch-clamp recordings of rat neurons from acute brain slices of the somatosensory cortex during magnetic stimulation. Front Cell Neurosci 8:145
26. Patton HD, Amassian VE (1954) Single and multiple-unit analysis of cortical stage of pyramidal tract activation. J Neurophysiol 17(4): 345–363
27. Rusu CV et al (2014) A model of TMS-induced I-waves in motor cortex. Brain Stimul 7(3):401–414
28. Moliadze V et al (2005) Paired-pulse transcranial magnetic stimulation protocol applied to visual cortex of anaesthetized cat: effects on visually evoked single-unit activity. J Physiol 566(Pt 3):955–965
29. Moliadze V et al (2003) Effect of transcranial magnetic stimulation on single-unit activity in the cat primary visual cortex. J Physiol 553 (Pt 2):665–679
30. Allen EA et al (2007) Transcranial magnetic stimulation elicits coupled neural and hemodynamic consequences. Science 317(5846):1918–1921
31. Kozyrev V, Eysel UT, Jancke D (2014) Voltage-sensitive dye imaging of transcranial magnetic stimulation-induced intracortical dynamics. Proc Natl Acad Sci U S A 111(37):13553–13558
32. Kim T et al (2015) Transcranial magnetic stimulation changes response selectivity of neurons in the visual cortex. Brain Stimul 8(3):613–623
33. Meyer BU et al (1991) Magnetic stimuli applied over motor and visual cortex: influence of coil position and field polarity on motor responses, phosphenes, and eye movements. Electroencephalogr Clin Neurophysiol Suppl 43:121–134
34. Amassian VE et al (1989) Suppression of visual perception by magnetic coil stimulation of human occipital cortex. Electroencephalogr Clin Neurophysiol 74(6):458–462
35. Beckers G, Homberg V (1991) Impairment of visual perception and visual short term memory scanning by transcranial magnetic stimulation of occipital cortex. Exp Brain Res 87(2):421–432
36. Walsh V, Rushworth M (1999) A primer of magnetic stimulation as a tool for neuropsychology. Neuropsychologia 37(2):125–135
37. Ruzzoli M et al (2011) The effect of TMS on visual motion sensitivity: an increase in neural noise or a decrease in signal strength? J Neurophysiol 106(1):138–143
38. Harris JA, Clifford CW, Miniussi C (2008) The functional effect of transcranial magnetic stimulation: signal suppression or neural noise generation? J Cogn Neurosci 20(4):734–740
39. Ruzzoli M, Marzi CA, Miniussi C (2010) The neural mechanisms of the effects of transcranial magnetic stimulation on perception. J Neurophysiol 103(6):2982–2989
40. Schwarzkopf DS, Silvanto J, Rees G (2011) Stochastic resonance effects reveal the neural mechanisms of transcranial magnetic stimulation. J Neurosci 31(9):3143–3147
41. Perini F et al (2012) Occipital transcranial magnetic stimulation has an activity-dependent suppressive effect. J Neurosci 32(36):12361–12365
42. Miniussi C, Ruzzoli M, Walsh V (2010) The mechanism of transcranial magnetic stimulation in cognition. Cortex 46(1):128–130
43. Theoret H et al (2003) Exploring paradoxical functional facilitation with TMS. Suppl Clin Neurophysiol 56:211–219
44. Silvanto J, Pascual-Leone A (2008) State-dependency of transcranial magnetic stimulation. Brain Topogr 21(1):1–10
45. Abrahamyan A et al (2011) Improving visual sensitivity with subthreshold transcranial magnetic stimulation. J Neurosci 31(9): 3290–3294
46. Rahnev DA et al (2012) Direct injection of noise to the visual cortex decreases accuracy but increases decision confidence. J Neurophysiol 107(6):1556–1563
47. Abrahamyan A et al (2015) Low intensity TMS enhances perception of visual stimuli. Brain Stimul 8(6):1175–1182
48. Mulckhuyse M et al (2011) Enhanced visual perception with occipital transcranial magnetic stimulation. Eur J Neurosci 34(8): 1320–1325
49. Yozbatiran N et al (2009) Safety and behavioral effects of high-frequency repetitive transcranial magnetic stimulation in stroke. Stroke 40(1):309–312
50. Voss M et al (2007) An improvement in perception of self-generated tactile stimuli following theta-burst stimulation of primary motor cortex. Neuropsychologia 45(12):2712–2717

51. Nowak DA et al (2005) High-frequency repetitive transcranial magnetic stimulation over the hand area of the primary motor cortex disturbs predictive grip force scaling. Eur J Neurosci 22(9):2392–2396
52. Mazzocchio R et al (1994) Effect of tonic voluntary activity on the excitability of human motor cortex. J Physiol 474(2):261–267
53. Baker SN, Olivier E, Lemon RN (1995) Task-related variation in corticospinal output evoked by transcranial magnetic stimulation in the macaque monkey. J Physiol 488 (Pt 3):795–801
54. Massimini M et al (2005) Breakdown of cortical effective connectivity during sleep. Science 309(5744):2228–2232
55. Silvanto J et al (2008) Baseline cortical excitability determines whether TMS disrupts or facilitates behavior. J Neurophysiol 99(5): 2725–2730
56. Cattaneo Z et al (2008) Using state-dependency of transcranial magnetic stimulation (TMS) to investigate letter selectivity in the left posterior parietal cortex: a comparison of TMS-priming and TMS-adaptation paradigms. Eur J Neurosci 28(9):1924–1929
57. Silvanto J, Muggleton NG (2008) Testing the validity of the TMS state-dependency approach: targeting functionally distinct motion-selective neural populations in visual areas V1/V2 and V5/MT+. Neuroimage 40(4):1841–1848
58. Cattaneo L, Sandrini M, Schwarzbach J (2010) State-dependent TMS reveals a hierarchical representation of observed acts in the temporal, parietal, and premotor cortices. Cereb Cortex 20(9):2252–2258
59. Cattaneo L et al (2011) One's motor performance predictably modulates the understanding of others' actions through adaptation of premotor visuo-motor neurons. Soc Cogn Affect Neurosci 6(3):301–310
60. Jacquet PO, Avenanti A (2015) Perturbing the action observation network during perception and categorization of actions' goals and grips: state-dependency and virtual lesion TMS effects. Cereb Cortex 25(3): 598–608
61. Davare M et al (2009) Ventral premotor to primary motor cortical interactions during object-driven grasp in humans. Cortex 45(9):1050–1057
62. Baumer T et al (2009) Inhibitory and facilitatory connectivity from ventral premotor to primary motor cortex in healthy humans at rest—a bifocal TMS study. Clin Neurophysiol 120(9):1724–1731
63. Koch G et al (2008) Functional interplay between posterior parietal and ipsilateral motor cortex revealed by twin-coil transcranial magnetic stimulation during reach planning toward contralateral space. J Neurosci 28(23):5944–5953
64. Cattaneo L, Barchiesi G (2011) Transcranial magnetic mapping of the short-latency modulations of corticospinal activity from the ipsilateral hemisphere during rest. Front Neural Circuits 5:14
65. Maule F et al (2015) Haptic working memory for grasping: the role of the parietal operculum. Cereb Cortex 25:528–537
66. Parmigiani S, Barchiesi G, Cattaneo L (2015) The dorsal premotor cortex exerts a powerful and specific inhibitory effect on the ipsilateral corticofacial system: a dual-coil transcranial magnetic stimulation study. Exp Brain Res 233(11):3253–3260
67. Ferbert A et al (1992) Interhemispheric inhibition of the human motor cortex. J Physiol 453:525–546
68. Tokimura H et al (2000) Short latency inhibition of human hand motor cortex by somatosensory input from the hand. J Physiol 523 (Pt 2):503–513
69. Kujirai T et al (1993) Corticocortical inhibition in human motor cortex. J Physiol 471:501–519
70. Lee L et al (2003) Acute remapping within the motor system induced by low-frequency repetitive transcranial magnetic stimulation. J Neurosci 23(12):5308–5318
71. Gangitano M et al (2002) Modulation of input-output curves by low and high frequency repetitive transcranial magnetic stimulation of the motor cortex. Clin Neurophysiol 113(8):1249–1257
72. Touge T et al (2001) Are the after-effects of low-frequency rTMS on motor cortex excitability due to changes in the efficacy of cortical synapses? Clin Neurophysiol 112(11):2138–2145
73. Chen R et al (1997) Depression of motor cortex excitability by low-frequency transcranial magnetic stimulation. Neurology 48(5):1398–1403
74. Heide G, Witte OW, Ziemann U (2006) Physiology of modulation of motor cortex excitability by low-frequency suprathreshold repetitive transcranial magnetic stimulation. Exp Brain Res 171(1):26–34
75. Quartarone A et al (2005) Distinct changes in cortical and spinal excitability following high-frequency repetitive TMS to the human motor cortex. Exp Brain Res 161(1):114–124

76. Pascual-Leone A et al (1994) Responses to rapid-rate transcranial magnetic stimulation of the human motor cortex. Brain 117(Pt 4):847–858
77. Maeda F et al (2000) Interindividual variability of the modulatory effects of repetitive transcranial magnetic stimulation on cortical excitability. Exp Brain Res 133(4):425–430
78. Maeda F et al (2000) Modulation of corticospinal excitability by repetitive transcranial magnetic stimulation. Clin Neurophysiol 111(5):800–805
79. Peinemann A et al (2004) Long-lasting increase in corticospinal excitability after 1800 pulses of subthreshold 5 Hz repetitive TMS to the primary motor cortex. Clin Neurophysiol 115(7):1519–1526
80. Modugno N et al (2001) Motor cortex excitability following short trains of repetitive magnetic stimuli. Exp Brain Res 140(4):453–459
81. Wu T et al (2000) Lasting influence of repetitive transcranial magnetic stimulation on intracortical excitability in human subjects. Neurosci Lett 287(1):37–40
82. Di Lazzaro V et al (2005) Theta-burst repetitive transcranial magnetic stimulation suppresses specific excitatory circuits in the human motor cortex. J Physiol 565(Pt 3):945–950
83. Huang YZ et al (2005) Theta burst stimulation of the human motor cortex. Neuron 45(2):201–206
84. Stefan K et al (2000) Induction of plasticity in the human motor cortex by paired associative stimulation. Brain 123(Pt 3):572–584
85. Cattaneo L, Barchiesi G (2015) The auditory space in the motor system. Neuroscience 304:81–89
86. Avenanti A et al (2009) Freezing or escaping? Opposite modulations of empathic reactivity to the pain of others. Cortex 45(9):1072–1077
87. Glenberg AM et al (2008) Processing abstract language modulates motor system activity. Q J Exp Psychol (Hove) 61(6):905–919
88. Barchiesi G, Cattaneo L (2013) Early and late motor responses to action observation. Soc Cogn Affect Neurosci 8(6):711–719
89. Cattaneo Z et al (2009) The mental number line modulates visual cortical excitability. Neurosci Lett 462(3):253–256
90. Bestmann S et al (2007) Spatial attention changes excitability of human visual cortex to direct stimulation. Curr Biol 17(2):134–139
91. Ubaldi S, Barchiesi G, Cattaneo L (2015) Bottom-up and top-down visuomotor responses to action observation. Cereb Cortex 25(4):1032–1041
92. Rossi S et al (2007) A real electro-magnetic placebo (REMP) device for sham transcranial magnetic stimulation (TMS). Clin Neurophysiol 118(3):709–716
93. Okamoto M et al (2004) Three-dimensional probabilistic anatomical cranio-cerebral correlation via the international 10–20 system oriented for transcranial functional brain mapping. Neuroimage 21(1):99–111
94. Okamoto M, Dan I (2005) Automated cortical projection of head-surface locations for transcranial functional brain mapping. Neuroimage 26(1):18–28
95. Mills KR, Kimiskidis V (1996) Cortical and spinal mechanisms of facilitation to brain stimulation. Muscle Nerve 19(8):953–958
96. Koski L et al (2005) Normative data on changes in transcranial magnetic stimulation measures over a ten hour period. Clin Neurophysiol 116(9):2099–2109
97. Badawy RA et al (2012) Inter-session repeatability of cortical excitability measurements in patients with epilepsy. Epilepsy Res 98(2–3):182–186
98. Kimiskidis VK et al (2004) The repeatability of corticomotor threshold measurements. Neurophysiol Clin 34(6):259–266
99. Wassermann EM (2002) Variation in the response to transcranial magnetic brain stimulation in the general population. Clin Neurophysiol 113(7):1165–1171
100. Conforto AB et al (2004) Impact of coil position and electrophysiological monitoring on determination of motor thresholds to transcranial magnetic stimulation. Clin Neurophysiol 115(4):812–819
101. Mills KR, Nithi KA (1997) Corticomotor threshold to magnetic stimulation: normal values and repeatability. Muscle Nerve 20(5):570–576
102. Hanajima R et al (2007) Comparison of different methods for estimating motor threshold with transcranial magnetic stimulation. Clin Neurophysiol 118(9):2120–2122
103. Rossini PM et al (1994) Non-invasive electrical and magnetic stimulation of the brain, spinal cord and roots: basic principles and procedures for routine clinical application. Report of an IFCN committee. Electroencephalogr Clin Neurophysiol 91(2):79–92
104. Rossini PM et al (2015) Non-invasive electrical and magnetic stimulation of the brain, spinal cord, roots and peripheral nerves: basic principles and procedures for routine clinical and research application. An updated report from an I.F.C.N. Committee. Clin Neurophysiol 126(6):1071–1107

105. Stokes MG et al (2005) Simple metric for scaling motor threshold based on scalp-cortex distance: application to studies using transcranial magnetic stimulation. J Neurophysiol 94(6):4520–4527
106. Wassermann EM (1998) Risk and safety of repetitive transcranial magnetic stimulation: report and suggested guidelines from the International Workshop on the Safety of Repetitive Transcranial Magnetic Stimulation, June 5–7, 1996. Electroencephalogr Clin Neurophysiol 108(1):1–16
107. Rossi S et al (2009) Safety, ethical considerations, and application guidelines for the use of transcranial magnetic stimulation in clinical practice and research. Clin Neurophysiol 120(12):2008–2039
108. Kayser S et al (2013) Comparable seizure characteristics in magnetic seizure therapy and electroconvulsive therapy for major depression. Eur Neuropsychopharmacol 23(11):1541–1550
109. Lisanby SH et al (2003) Safety and feasibility of magnetic seizure therapy (MST) in major depression: randomized within-subject comparison with electroconvulsive therapy. Neuropsychopharmacology 28(10): 1852–1865
110. Fierro B et al (2001) Timing of right parietal and frontal cortex activity in visuo-spatial perception: a TMS study in normal individuals. Neuroreport 12(11):2605–2607
111. Bagattini C et al (2015) No causal effect of left hemisphere hyperactivity in the genesis of neglect-like behavior. Neuropsychologia 72:12–21
112. Bonni S et al (2015) Role of the anterior temporal lobes in semantic representations: paradoxical results of a cTBS study. Neuropsychologia 76:163–169
113. Gough PM, Nobre AC, Devlin JT (2005) Dissociating linguistic processes in the left inferior frontal cortex with transcranial magnetic stimulation. J Neurosci 25(35): 8010–8016
114. Devlin JT, Matthews PM, Rushworth MF (2003) Semantic processing in the left inferior prefrontal cortex: a combined functional magnetic resonance imaging and transcranial magnetic stimulation study. J Cogn Neurosci 15(1):71–84
115. Cattaneo Z et al (2010) The causal role of category-specific neuronal representations in the left ventral premotor cortex (PMv) in semantic processing. Neuroimage 49(3):2728–2734
116. Pobric G, Hamilton AF (2006) Action understanding requires the left inferior frontal cortex. Curr Biol 16(5):524–529
117. Aziz-Zadeh L et al (2002) Lateralization in motor facilitation during action observation: a TMS study. Exp Brain Res 144(1):127–131
118. Urgesi C et al (2007) Representation of body identity and body actions in extrastriate body area and ventral premotor cortex. Nat Neurosci 10(1):30–31
119. Aziz-Zadeh L et al (2005) Covert speech arrest induced by rTMS over both motor and nonmotor left hemisphere frontal sites. J Cogn Neurosci 17(6):928–938
120. Cattaneo L (2013) Language. Handb Clin Neurol 116:681–691
121. Rossi S et al (2006) Prefrontal and parietal cortex in human episodic memory: an interference study by repetitive transcranial magnetic stimulation. Eur J Neurosci 23(3):793–800
122. Kohler S et al (2004) Effects of left inferior prefrontal stimulation on episodic memory formation: a two-stage fMRI-rTMS study. J Cogn Neurosci 16(2):178–188
123. Innocenti I et al (2010) Event-related rTMS at encoding affects differently deep and shallow memory traces. Neuroimage 53(1):325–330
124. Broca P (1865) Sur le siège de la faculté du langage articulé. Bull Mém Soc Anthropol Paris 6(1):377–393
125. Pascual-Leone A, Gates JR, Dhuna A (1991) Induction of speech arrest and counting errors with rapid-rate transcranial magnetic stimulation. Neurology 41(5):697–702
126. Epstein CM et al (1999) Localization and characterization of speech arrest during transcranial magnetic stimulation. Clin Neurophysiol 110(6):1073–1079
127. Epstein CM et al (1996) Optimum stimulus parameters for lateralized suppression of speech with magnetic brain stimulation. Neurology 47(6):1590–1593
128. Epstein CM et al (2000) Repetitive transcranial magnetic stimulation does not replicate the Wada test. Neurology 55(7):1025–1027
129. Sidtis D, Canterucci G, Katsnelson D (2009) Effects of neurological damage on production of formulaic language. Clin Linguist Phon 23(4):270–284
130. Stewart L et al (2001) Transcranial magnetic stimulation produces speech arrest but not song arrest. Ann N Y Acad Sci 930:433–435
131. Soares JC, Mann JJ (1997) The anatomy of mood disorders—review of structural neuroimaging studies. Biol Psychiatry 41(1): 86–106
132. Drevets WC (2000) Neuroimaging studies of mood disorders. Biol Psychiatry 48(8): 813–829

133. Carpenter LL et al (2012) Transcranial magnetic stimulation (TMS) for major depression: a multisite, naturalistic, observational study of acute treatment outcomes in clinical practice. Depress Anxiety 29(7):587–596
134. O'Reardon JP et al (2007) Efficacy and safety of transcranial magnetic stimulation in the acute treatment of major depression: a multisite randomized controlled trial. Biol Psychiatry 62(11):1208–1216
135. Janicak PG, Dokucu ME (2015) Transcranial magnetic stimulation for the treatment of major depression. Neuropsychiatr Dis Treat 11:1549–1560
136. Chen J et al (2013) Left versus right repetitive transcranial magnetic stimulation in treating major depression: a meta-analysis of randomised controlled trials. Psychiatry Res 210(3):1260–1264
137. Schutter DJ (2009) Antidepressant efficacy of high-frequency transcranial magnetic stimulation over the left dorsolateral prefrontal cortex in double-blind sham-controlled designs: a meta-analysis. Psychol Med 39(1):65–75
138. Haun DB et al (2010) Origins of spatial, temporal and numerical cognition: Insights from comparative psychology. Trends Cogn Sci 14(12):552–560
139. Pinel P et al (2001) Modulation of parietal activation by semantic distance in a number comparison task. Neuroimage 14(5):1013–1026
140. Piazza M et al (2007) A magnitude code common to numerosities and number symbols in human intraparietal cortex. Neuron 53(2):293–305
141. Gerstmann J (1940) Syndrome of finger agnosia, disorientation for right and left, agraphia and acalculia—local diagnostic value. Arch Neurol Psychiatry 44(2):398–408
142. Dormal V, Andres M, Pesenti M (2008) Dissociation of numerosity and duration processing in the left intraparietal sulcus: a transcranial magnetic stimulation study. Cortex 44(4):462–469
143. Cappelletti M et al (2007) rTMS over the intraparietal sulcus disrupts numerosity processing. Exp Brain Res 179(4):631–642
144. Knops A et al (2006) On the functional role of human parietal cortex in number processing: How gender mediates the impact of a 'virtual lesion' induced by rTMS. Neuropsychologia 44(12):2270–2283
145. Sandrini M, Rossini PM, Miniussi C (2004) The differential involvement of inferior parietal lobule in number comparison: a rTMS study. Neuropsychologia 42(14):1902–1909
146. Sasanguie D, Gobel SM, Reynvoet B (2013) Left parietal TMS disturbs priming between symbolic and non-symbolic number representations. Neuropsychologia 51(8):1528–1533
147. Gobel S, Walsh V, Rushworth MF (2001) The mental number line and the human angular gyrus. Neuroimage 14(6):1278–1289
148. Gobel SM et al (2006) Parietal rTMS distorts the mental number line: simulating 'spatial' neglect in healthy subjects. Neuropsychologia 44(6):860–868
149. Cohen Kadosh R et al (2007) Virtual dyscalculia induced by parietal-lobe TMS impairs automatic magnitude processing. Curr Biol 17(8):689–693
150. Cohen Kadosh R, Bien N, Sack AT (2012) Automatic and intentional number processing both rely on intact right parietal cortex: a combined FMRI and neuronavigated TMS study. Front Hum Neurosci 6:2
151. Andres M et al (2011) Role of distinct parietal areas in arithmetic: an fMRI-guided TMS study. Neuroimage 54(4):3048–3056
152. McCarthy G et al (1997) Face-specific processing in the human fusiform gyrus. J Cogn Neurosci 9(5):605–610
153. Damasio AR, Damasio H, Van Hoesen GW (1982) Prosopagnosia: anatomic basis and behavioral mechanisms. Neurology 32(4):331–341
154. Ilmoniemi RJ et al (1997) Neuronal responses to magnetic stimulation reveal cortical reactivity and connectivity. Neuroreport 8(16):3537–3540
155. Hofer S, Frahm J (2006) Topography of the human corpus callosum revisited—comprehensive fiber tractography using diffusion tensor magnetic resonance imaging. Neuroimage 32(3):989–994
156. Wahl M et al (2007) Human motor corpus callosum: topography, somatotopy, and link between microstructure and function. J Neurosci 27(45):12132–12138
157. Wassermann EM et al (1991) Effects of transcranial magnetic stimulation on ipsilateral muscles. Neurology 41(11):1795–1799
158. Giovannelli F et al (2009) Modulation of interhemispheric inhibition by volitional motor activity: an ipsilateral silent period study. J Physiol 587(Pt 22):5393–5410
159. Meyer BU, Roricht S, Woiciechowsky C (1998) Topography of fibers in the human corpus callosum mediating interhemispheric inhibition between the motor cortices. Ann Neurol 43(3):360–369

Chapter 13

Electroencephalographic Asymmetries in Human Cognition

Veronica Mazza and Silvia Pagano

Abstract

Electroencephalographic (EEG) measures provide high-quality information on the time course of brain activity during stimulus processing. This chapter highlights the contribution of EEG to the study of human asymmetries in cognition. In the first part, we discuss the main methodological issues related to EEG recording, and describe a distinction between Event-Related Potential (ERP) responses and induced frequency activity. In the second part, we review some key findings on human lateralization in perception, language, and emotion. We conclude with some recent attempts to combine EEG with other neuroimaging techniques in the study of lateralized cognitive functions.

Key words Electroencephalography, EEG, Event-related potentials, ERPs, Oscillations, Electrodes, Montage, Early perceptual asymmetries, Emotion, Language, Multimodal imaging

1 Introduction

The study of human asymmetries in cognitive functions has ancient origins. Thanks to the advancements in neuroimaging techniques we can now provide a direct visualization of the brain correlates of lateralized cognition. This chapter explores the main contribution of Electroencephalography (EEG) to this issue. What type of brain signals can be measured using EEG? Do EEG measures capture the various hemispheric specializations of human cognition? We address these and other questions below. The chapter is divided into four sections. In the first section, we discuss some of the main methodological aspects to consider when planning an EEG study. The second section presents some crucial advantages, as well as some disadvantages, of EEG with respect to the other neuroimaging methods; here we will also describe a distinction between Event-Related Potential (ERP) responses and induced frequency oscillations. In the third section we review some key EEG findings on human lateralization, as well as briefly provide accounts of the functional role of some lateralized functions found in perception, language, and emotion. Finally, we present a description of the

Lesley J. Rogers and Giorgio Vallortigara (eds.), *Lateralized Brain Functions: Methods in Human and Non-Human Species*, Neuromethods, vol. 122, DOI 10.1007/978-1-4939-6725-4_13, © Springer Science+Business Media LLC 2017

recent attempts to combine EEG with other neuroimaging or neurostimulation techniques in order to reach the full complexity of the brain dynamics underlying human lateralization.

2 Methodological Aspects

What methodological aspects need to be considered when conducting an EEG study of human asymmetries? There are no large differences between running an EEG experiment on cognition and running an EEG experiment investigating lateralized cognitive functions in humans. However, there are some aspects that become particularly relevant for studying EEG activity related to lateralized functions. Below we summarize the key aspects related to the setup of an EEG experiment, and highlight some of the peculiarities related to the recordings of asymmetric EEG activity.

2.1 Electrodes and Montage

The electrodes used for EEG studies in cognitive neuroscience are made of various materials, but currently the most - used is Ag/AgCl [1] (Fig. 1). Two important aspects to consider are related to the number of electrodes and to their location on the scalp. In the past decade the number of electrodes used in EEG studies has increased considerably. A large number of electrodes (e.g. 64 electrodes) is useful when researchers are interested in source localization, namely when they want to estimate the anatomical brain counterpart of a specific neural EEG response recorded on the scalp. While EEG is not the ideal technique for studying localization, researchers have developed methods to estimate source localization (e.g., [2–4]). For all these methods, having a relatively large number of electrodes is useful to guarantee good sampling of the signal (e.g., see [5]).

Another reason for having a large number of electrodes is to get greater precision in artifact correction. One method for dealing with the various types of electrical artifacts produced by participants (e.g. eye movements, eye blinks) is to correct the signal by isolating the components related to the artifacts (e.g., via *Independent Component Analysis*, ICA, [6]) and subtract them from the continuous signal. Having a relatively large number of electrodes is useful for computing ICA correctly.

EEG caps with pre-assigned electrode locations (Figs. 1 and 2) have allowed researchers to record from a relatively large number of electrodes, and to reduce dramatically the time for the preparation phase. The specific selection of the electrode positions largely depends on the type of cognitive functions investigated. For example, in studying lateralized phenomena in visual perception one may want to prioritize the posterior parts of the scalp (where the signal is expected to be largest, given that visual areas are located in posterior areas). When studying asymmetric cognitive

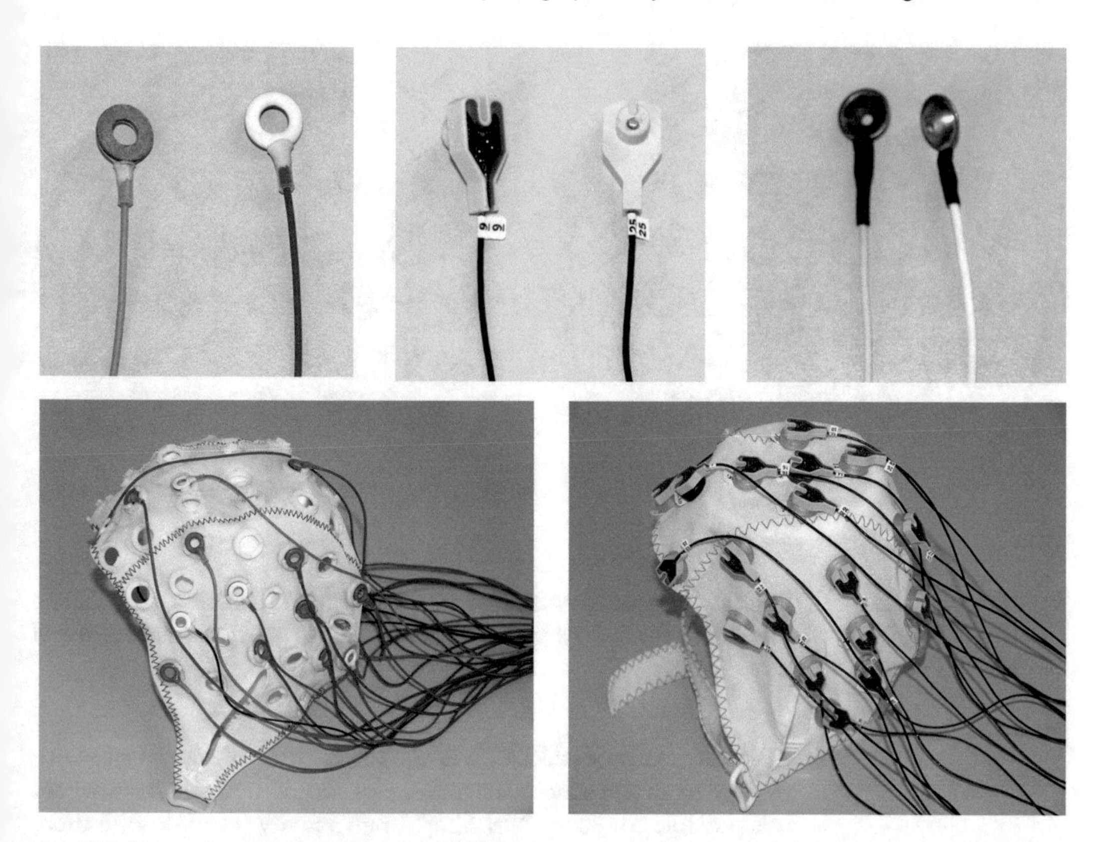

Fig. 1 (*Top*) Example of electrodes used in EEG experiments (*left*: two passive Ag/AgCl ring electrodes; *middle*: two active electrodes; *right*: two passive cup electrodes). (*Bottom*) *Left*: EEG cap with pre-mounted passive electrodes. *Right*: EEG cap with an active electrode system

functions lateral electrodes (i.e., on the left and right parts of the scalp) should be used. However, the general consensus within the EEG community is that the measurements should be taken from multiple locations, as this allows for a better individuation of artifacts and of their contribution in the generation of an effect (see [1]; Fig. 2).

2.2 Reference and Filters

One of the methodological choices one has to make when using EEG is related to the reference electrode. This is an intrinsic aspect of all EEG experiments, due to the fact that differential amplifiers are used to record EEG. For this reason, the EEG signal recorded never represents the activation recorded from a single electrode; instead, what is measured is the difference between the activity recorded from one electrode and the activity recorded from a "reference" electrode (and actually, things are slightly more complicated, as this difference is then subtracted from the activation recorded from another, "ground" electrode, see [7]).

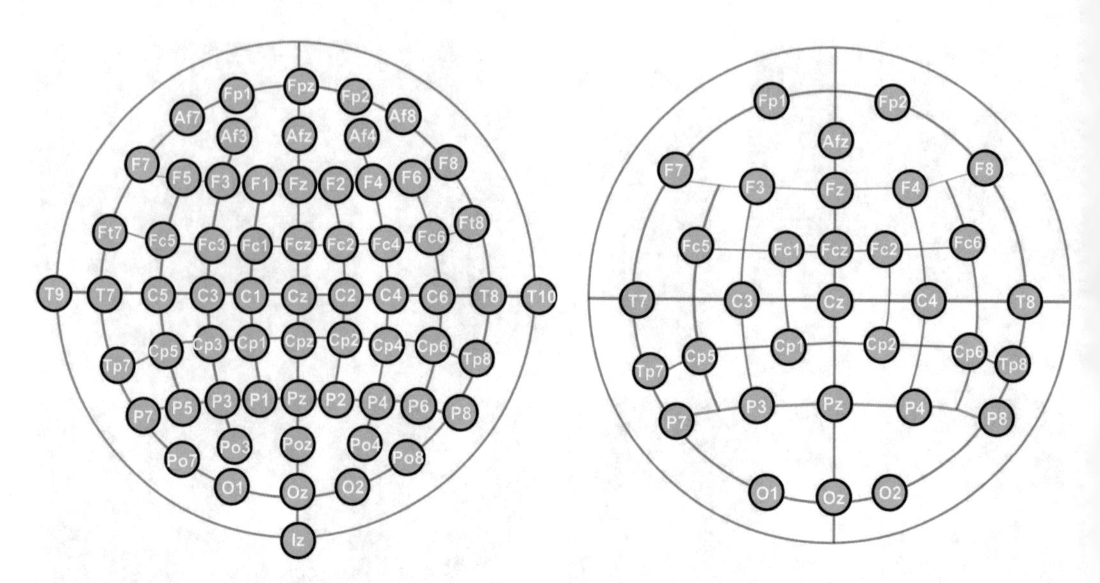

Fig. 2 Electrode placement. The most common system is called 10–20 System [149], because electrodes are placed at 10% or 20% distance. Other systems have been proposed recently, to account for the increase of the number of electrodes (e.g., see [150])

The reference electrode is supposed to be a neutral site with respect to all the other electrodes, and should be a site from where it is possible to record "baseline" brain activity (even though there is no site that is truly neutral [7]). Most of the experiments use a single reference for all the other electrodes (unipolar recordings). The reference can vary from the vertex (electrode Cz) to the linked mastoids (the bones behind the ears) or the tip of the nose, depending on the choice made by the majority of studies for a specific topic. Experts in the field (e.g. [8]) recommend choosing a reference electrode that is comfortable for the participants (e.g., that is not distracting). As ERPs may change as a function of the reference site, a further recommendation is to use the same reference electrode used by other investigators in a specific research field. This greatly facilitates comparisons across different studies.

Finally, a critical aspect to take into consideration for the evaluation of asymmetric functions relates to the lateralization of the reference. Since the focus here is to test for any left-right difference in hemispheric activity, one should ensure that an "unbiased" reference is used, namely that there is no a priori spatial imbalance in the reference activity. Therefore, a good strategy may be to use the combined activity of both mastoids (or earlobes) as reference. This reference is frequently used in studies investigating the contralateral versus ipsilateral activation of the two hemispheres with respect to the side of presentation of a visual target. Due to possible electrical shunting, the use of linked mastoids (or earlobes) during recording is usually discouraged [1, 7, 8]. Indeed, some studies have also shown an important reduction of some kind of asymmetries with linked earlobes or mastoids compared to other reference sites (e.g., [9–11]). To avoid this, experts recommend the use of one mastoid/earlobe as active reference, and the offline re-referencing to the average activity of both sites [8]. Almost all software for data analysis includes this (re-referencing) function. Another option available to researchers is to compute an offline common average reference, namely the re-referencing to the average of all the electrodes used in a study. There are several articles to consult for more details on this issue (e.g. [7, 12]).

Filters are another aspect that deserves attention when planning an EEG experiment. It is beyond the scope of this chapter to provide an extensive description on filters (see [13, 14] for detailed methodological descriptions). Here we will only point out that filtering is an essential part of an EEG experiment, in the effort to reduce noise level and to have a high-fidelity reproduction of the signal of interest. There are two classes of filters. Analog filters are applied during the recording phase on the continuous signal, and are typically employed for preventing aliasing phenomena (namely, the misrepresentation of original high frequencies due to an under sampling of the signal). Because filters can severely distort the original brain signal, researchers try to minimize the use of analog filters, and rely on off-line digital filtering on the recorded signal.

2.3 Artifacts

A crucial step in EEG data processing is the rejection/correction of artifacts. Artifacts can significantly distort the EEG signal of interest because they are relatively large (and larger than the signal of interest), and/or because they are temporally correlated with the signal. We can divide artifacts in two main classes: external versus internal noise with respect to the participant.

The most frequent source of noise *external* to the participant is represented by the electrical environment (either 50 or 60 Hz, depending on the country) and instruments. While the noise related to the electrical environment could be easily reduced or eliminated by having electrically shielded booths for recording, in fact most instruments (e.g. pc monitors, loudspeakers) that are present inside the (shielded) booths are electrically operated. These aspects may create the presence of a periodic artificial 50 (or 60) Hz activity in the EEG measurements, especially in the case of electrodes with poor connection. Instruments can also produce transient rather than periodic noise. For instance, the refresh rate of CTR monitors produces a sharp activity in the EEG signal, and this is problematic because it is time-locked to the onset of the stimulus.

There are several approaches one could take to reduce or eliminate the electrical noise, from using either analog or digital filters on the signal to having electrically shielded cages for the instruments close to the participants, in addition to (or as an alternative for) a shielded booth (for a discussion, see [8]). The last approach seems particularly promising, as it allows the experimenter to obtain good quality data already in the acquisition phase. Another source of external noise artifact is represented by poorly connected electrodes, which cause different kinds of distortion in the EEG signal. Fortunately, an easy solution for this consists in replacing the bad electrode as soon as one notices the distortion in the measurements.

The other class of artifacts is represented by electrical noise *internally* produced by the volunteers during the experiment (Fig. 3). Internal noise can vary from individual to individual, and could be so severe as to lead to the exclusion of the entire set of data for a subject. The most common artifacts visible in EEG are related to eye and muscle activity.

During the experiment participants tend to make eye movements and blinks, even when they are explicitly instructed to refrain from doing so (at least during specific time intervals, i.e., during

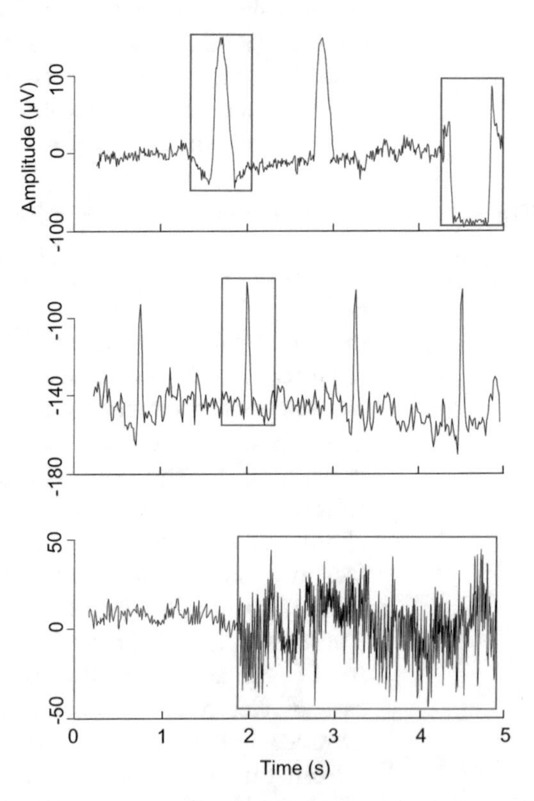

Fig. 3 Examples of "internal" artifacts. *Top*: artifact produced by eye blinks (*left*) and horizontal eye movements (*right*) visible at frontal electrodes. The typical square shape of horizontal eye artifact reflects the large movement of the eye to reach the periphery of the visual field. *Middle*: artifact produced by cardiac activity at a posterior electrode. The artifact captures the typical shape of the cardiac activity and its cyclic repetition with a frequency around 1 Hz. *Bottom*: muscular artifact produced by mouth contractions visible from temporal electrodes

stimulus presentation). The artifacts created by eye movements and blinks are a serious concern for investigators, as they tend to be relatively large and often time-locked to the stimulus presentation (Fig. 3). As these artifacts are mainly visible in the anterior parts of the scalp, eye-related activity becomes of relevance particularly for research investigating asymmetric activity in frontal/anterior areas of the scalp [9].

Another common source of artifacts in EEG is represented by muscle activity (Fig. 3). This type of activity (mainly caused by tension in the forehead and at the temples) is also mostly recorded from frontal and temporal scalp sites, and thus can be a source of concern when investigating asymmetric functions, such as language and emotion.

There are various ways in which artifacts can be handled, the most successful of which would be to avoid the acquisition of artifacts altogether. The experimenter usually asks participants to reproduce some artifacts (e.g., eye movements) prior to the start of the experimental session, and then "trains" them on how not to produce such artifacts. Subsequently, subjects are instructed to fixate a central point and to minimize any movement as much as possible during a block of trials. While this procedure is effective for most subjects, it is practically impossible to have EEG measurements that do not contain any artifacts. There are several offline procedures available that can be used to reduce/eliminate the impact of artifacts on the acquired signal. These can be sorted in two main classes.

One class of procedures aims at rejecting all the trials containing a specific artifact [8]. Trial *rejection* can be computed on the basis of a threshold value (e.g., ±75 μV for blink rejection), or on the difference between maximum versus minimum values (peak-to-peak value). Artifact rejection can be very effective (although the rejection criterion may vary from individual to individual). However, the main risk of this technique is that the experimenter is left with too few good trials for each condition of interest. Since the EEG signal strongly depends on the number of repetitions of a specific condition, these procedures may not be applicable for all experiments and/or populations because of the typically small size of the signal of interest with respect to the noise. For instance, while studying perceptual functions in aging we have realized that it is virtually impossible to use artifact rejection procedures for dealing with eye movement artifacts, since older people tend to move their eyes quite often, and certainly more often than young people. This could become even more serious when clinical populations are used. As a consequence, applying artifact rejection would reduce the number of good trials to an unacceptably low level for distinguishing signal from noise.

An alternative solution to artifact rejection is correcting for the artifact [15]. Artifact *correction* procedures are based on algorithms that isolate the noise related to a specific noncerebral activity and subtract this from the entire signal. Several methods have been developed for this purpose. The early techniques consisted of algorithms that would estimate (and afterwards, remove) the propagation of the artifact via linear regression [16]. More recently, several researchers have relied on correction methods based on ICA, particularly for the rejection of ocular artifacts. ICA is a data-driven method that isolates a set of statistically independent components, including those related to artifacts [6]. The estimated artifact-related components are then subtracted from the other channels, ideally leading to an "artifact-free" EEG signal [17]. The obvious advantage of these techniques is that there is no data loss, and if correctly applied to the data can lead to a clean signal. As mentioned above, this could be essential for studies with special populations (e.g. children, old people, and patients). However, one important caveat is that ICA requires visual inspection to ensure that the component isolated by the method as an artifact actually represents the artifact of interest. Therefore, some expertise is needed to ensure optimal use of the method. More generally, if correction methods are poorly applied, it becomes very difficult to understand what has been removed from the signal, and therefore to avoid distortion of the original cerebral activity. Details on artifact rejection and correction can be found in [8, 15, 18].

2.4 Subjects

A final point specifically related to the study of EEG asymmetries is about participants. Individual differences can be a thorny issue for studying lateralized functions. Indeed while the relevance of some factors is well established, such the inclusion of only right-handers in research focusing on language processing, the impact of other features related to individual variability is less clear. For instance, the participants' age may be a critical factor to consider when exploring asymmetric processing. Indeed, there is accumulating evidence that aging importantly reduces hemispheric asymmetries in typically lateralized functions such as language [19]. More generally, there seem to be trait-driven differences in brain asymmetries within the human population (see [11]).

3 Types of EEG Signals

Why would one rely on EEG to study lateralized/asymmetric functions? In other words, what are the advantages of using EEG for understanding the brain activity associated with cognitive functions? And what type of brain signals is measured through EEG? Below we will describe the merits (and limits) of EEG with respect to other neuroimaging methods, and we will discuss the difference between evoked signals and induced frequency oscillations.

3.1 Advantages and Disadvantages of the EEG Technique

EEG measurements of the brain activity during the execution of a task offer several advantages in the study of cognition, and therefore of asymmetric cognitive functions, too. On the practical side, the equipment and instruments required to set up an EEG lab, as well as the overall maintenance costs, are relatively low with respect to other neuroimaging techniques. There are also several methodological advantages in using EEG.

First, EEG measures represent a direct and instantaneous measure of the neuronal activity, and thus do not require any interpretation of how a certain physiological parameter is related to brain activity. This is substantially different from techniques such as fMRI (functional Magnetic Resonance Imaging) where the signal recorded (Blood Oxygen Level Dependent signal) is not a direct representation of the activity of the neurons, but represents the hemodynamic effects of neural activity. Together with magnetoencephalographic (MEG) measures, EEG measures provide a unique window into the temporal dynamics of the cortical activity associated with a cognitive function. These techniques are currently the only noninvasive methods with millisecond precision. This aspect makes EEG a well-suited technique for addressing questions concerning the timing of a specific event or phenomenon (e.g. "Does object recognition occur early on during scene analysis?"), but not when the question concerns the specific brain areas where an effect occurs. The spatial resolution of EEG is probably its weakest point, due to the fact that a spatial pattern of activity recorded from the scalp can potentially be generated by an infinite number of neural sources (this is the so-called inverse solution problem, [20]). EEG has complementary strengths and weaknesses to fMRI, which has an optimal spatial resolution but limited temporal precision. Over the past decade, researchers have increasingly focused on multimodal imaging, combining different imaging techniques [21]. In the last part of the chapter we describe multimodal imaging in studies of cognitive asymmetries.

Technological advances have substantially reduced the size of the EEG amplifiers and have developed electrodes that reduce electrical noise even without environmental shielding (these sensors are called "active" electrodes, as opposed to the "passive" electrode systems traditionally used in research, see [22]; Fig. 1), making EEG a portable technique for brain recording. Portability is definitely an advantage as it makes EEG research well suited for different contexts (e.g. hospitals, outdoor environments), and across different populations (from new-born infants to immobilized patients). Additionally, portability has made EEG the elite technique for multimodal imaging, as it is easy to integrate with fMRI, MEG, and neurostimulation methods such as Transcranial Magnetic Stimulation (TMS) [23]. EEG provides a continuous measure of brain activity from the onset of an event of interest, thereby allowing the investigation of implicit processing of a stimulus in the absence of an overt response to it. This is another good feature when testing individuals who are not able to communicate (for instance, because they have lost this ability, as in the case of coma, see [24]). More generally, the ability to acquire a continuous measure of neural processing allows the investigation of the moment-by-moment neural activations during fast stimulation (such as word presentation in sentence reading).

On the less bright side, EEG studies cannot directly provide information on causality, as they provide only correlational links between a specific pattern of neural activity and the implementation of a cognitive function. This aspect is common to all the other neuroimaging techniques and should always be taken into account in interpreting neuroimaging data (but see [25] for a different approach to causality).

We have already mentioned that spatial resolution is not one of the strengths of EEG measures, and that EEG is not well suited for questions primarily concerned with the anatomical counterpart of a cognitive phenomenon (although several methods have been proposed to estimate the EEG sources, see [26] for a review). Finally, one important disadvantage of EEG measures is that the neural activity is recorded from a distance (i.e., from the scalp), and hence the signal is typically tiny and smaller with respect to sources of noise. For this reason, a fairly large number of trials per condition are needed in order to obtain reliable measures. Because the average number of trials needed may change as a function of the sensory modality and of the specific cognitive function under study, we recommend the readers review the literature of interest for their study and use the average number of trials adopted by other researchers in the field [27].

Even when a relatively large number of trials is used, it is hard to exclude the possibility that the signal, for instance any asymmetric EEG signals, is not visible because it is too tiny with respect to the noisy EEG background, or because the overall asymmetric effect visible in behavioral data is the result of relatively weak asymmetric processing occurring at multiple stages of processing. More generally, EEG asymmetries could be reduced by volume conduction, an aspect that is intrinsic to EEG recordings.

3.2 Evoked Versus Induced Activity

In cognitive neuroscience, EEG activity is usually recorded with reference to the occurrence of a specific event—a stimulus or a response initiated by the participant—cognitive neuroscientists are rarely interested in purely spontaneous EEG patterns. A general distinction can be made between signals related to neural activity strictly in phase with the critical stimulation, and signals that reflect activity related to the critical stimulation but that are not strictly in phase with the onset of the stimulation [28, 29] (see Fig. 4).

The term "*evoked activity*" refers to the signals that are time-locked and phase-locked to the stimulus onset, and therefore are

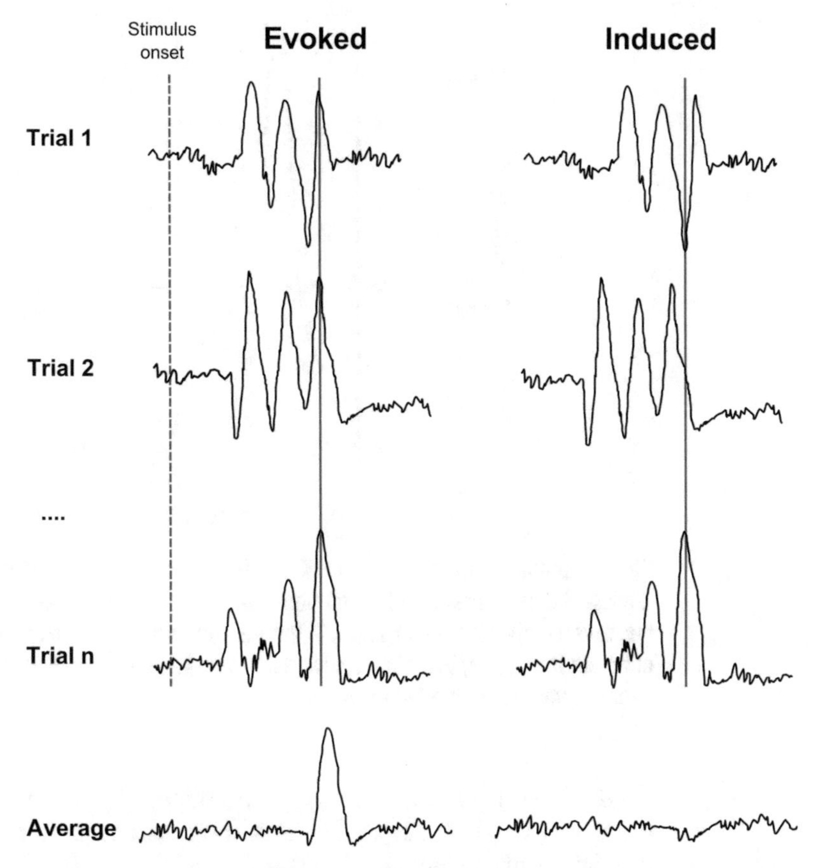

Fig. 4 Difference between evoked and induced activity. *Left*: evoked activity elicited by a response that is both phase- and time-locked to the stimulus onset. The phase of the third peak is always the same in all trials. When averaging these trials the consistent activity on the third peak remains visible while the other peaks are averaged out. The evoked activity is typically visible in event-related potential (ERP) analysis. *Right*: induced activity elicited by responses that are time-locked, but not phase-locked, to the stimulus onset. All peaks reach their maximum amplitude in different moments in time, and are therefore not in phase with stimulus onset. When averaging this activity there is no visible peak. The induced activity can be visualized only using time-frequency analyses

visible when averaging together trials for a given condition. While evoked activity can be measured in terms of frequency or amplitude values, the best known example of evoked EEG is represented by Event-Related Potentials (ERPs), which represent the mean amplitude variations for a certain set of frequencies (usually, in the 0.01–40 Hz range) that are time-locked with respect to the onset of a critical event. Since these are the result of an averaging procedure, ERPs additionally represent that portion of the EEG signal that is phase-locked to the stimulus onset across trials.

Since the beginning of use of ERPs to study cognition (e.g. starting from the 1960s), there has been an explosion of articles using these measures to study almost all cognitive functions one could consider. It is beyond the scope of this chapter to provide a

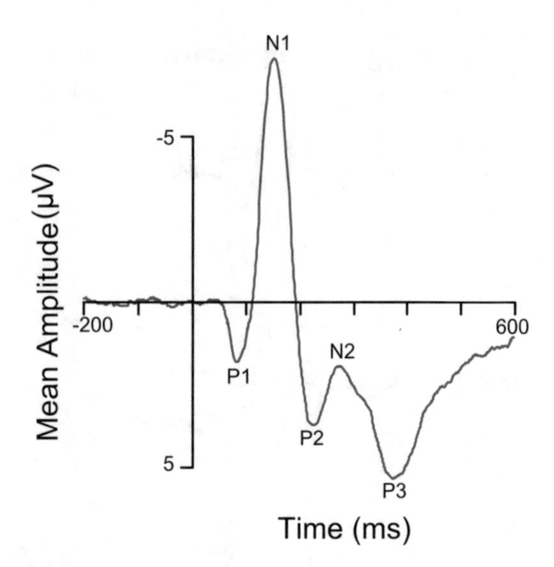

Fig. 5 Example of a typical visual ERP time-locked to stimulus onset. This figure depicts the three first positive components—P1, P2, and P3—and the first two negative components—N1 and N2 (the N2 component is a negative deflection although the activity is in the positive quadrant). As in traditional EEG recordings, negative values are plotted upwards

detailed description of the ERP componentry (for a detail discussion, see [8, 20, 27, 29]). We limit the description to some of the more relevant elements.

ERPs are usually evaluated in terms of their presence or absence with respect to a critical stimulus/event, their amplitude changes between two or more experimental conditions, and/or in terms of latency (either of the ERP onset or of the peak) across conditions. While ERPs are typically measured as average responses across many trials, recent attempts to extract meaningful signals on a single-trial basis have opened new methods for studying ERPs [30, 31]. An ERP response is typically defined by a letter and a number. The letter (P or N) stands for the polarity of the response (either positive or negative); an exception is represented by the visual C1, which is the very first component visible in simple stimulation tasks that varies its polarity as a function of the position (lower versus upper) of the visual stimulus [32]. The number indicates the ordinal position of the response (e.g., P1 indicates the first positive ERP visible on the average waveform, see Fig. 5). Alternatively, researchers have used a number to indicate the peak latency of the ERP response (e.g. P100 to indicate a positive ERP with a peak latency of 100 ms).

According to some previous classifications, early components (up to approximately 200 ms) are purely modulated by exogenous factors, such as physical stimulus parameters, and are impenetrable to more "cognitive" factors. For instance, the visual P1 is sensitive to variation in stimulus contrast or luminance [8]. In contrast, late ERPs (e.g., the well-known P300 response) would reflect high-level cognitive states because they are modulated by endogenous factors (such as expectancy and target relevance), but not sensory parameters. Nevertheless, it is now clear that even late ERPs can to some extent be modulated by physical parameters, and, in turn, some relatively early components are modulated by high-level "cognitive" factors. For instance, seminal work on visual processing has indicated that attention to a specific location in space modulates visual stages as early as the ones reflected in the P1 component [33].

While most of the ERPs represent the normal brain response to a stimulus (for example, all relevant visual stimuli elicit a P1, N1, and P3 components), some are visible only in some specific conditions. This is the case for the so-called Mismatch Negativity (MMN), a component that is visible in the 150–300 ms post-stimulus onset as a difference between the response to a series of frequent stimuli and the response elicited by rare and infrequent items [34]. Similarly, some ERPs are elicited only by stimuli in a specific domain. For instance, this is the case of the P600, a response measured from centro-parietal electrodes in response to syntactic violation in language processing [35].

Finally, according to an old convention the polarity in the visualization of the ERPs is inverted, so that positive values are plotted downward (Fig. 5). Although the origin of this convention is unclear and has been a contentious issue [8], it is unlikely that this tendency will be inverted, and it is important to keep this in mind to avoid confusion in reading and interpreting findings from other researchers.

A different way to look at the neural activity related to a cognitive function is to consider those EEG signals that are not strictly in phase with the onset of the stimulation but are nonetheless elicited by it. The term "***induced activity***" denotes exactly this (Fig. 4). This class of brain activity in association with cognitive events has attracted widespread interest in the last couple of decades. More precisely, most of the research on this class of responses has focused on frequency oscillations and their modulation over time. There are several methods one could use to extract oscillations of a specific frequency, the most popular of which has so far been wavelet analysis (for a recent review, see [36]). The analysis extracts the oscillatory signals by relying on the construction of a set of signals based on a prototypical wavelet. The advantage of using this analysis with respect to other methods, such as the tradition Fast Fourier Transform, is that it is possible to retain information about the time course of the frequency modulation with respect to the event of interest [28] (see Fig. 6).

As in the case of ERP research, there is no consensus about the functional role of the specific frequency ranges, and there are

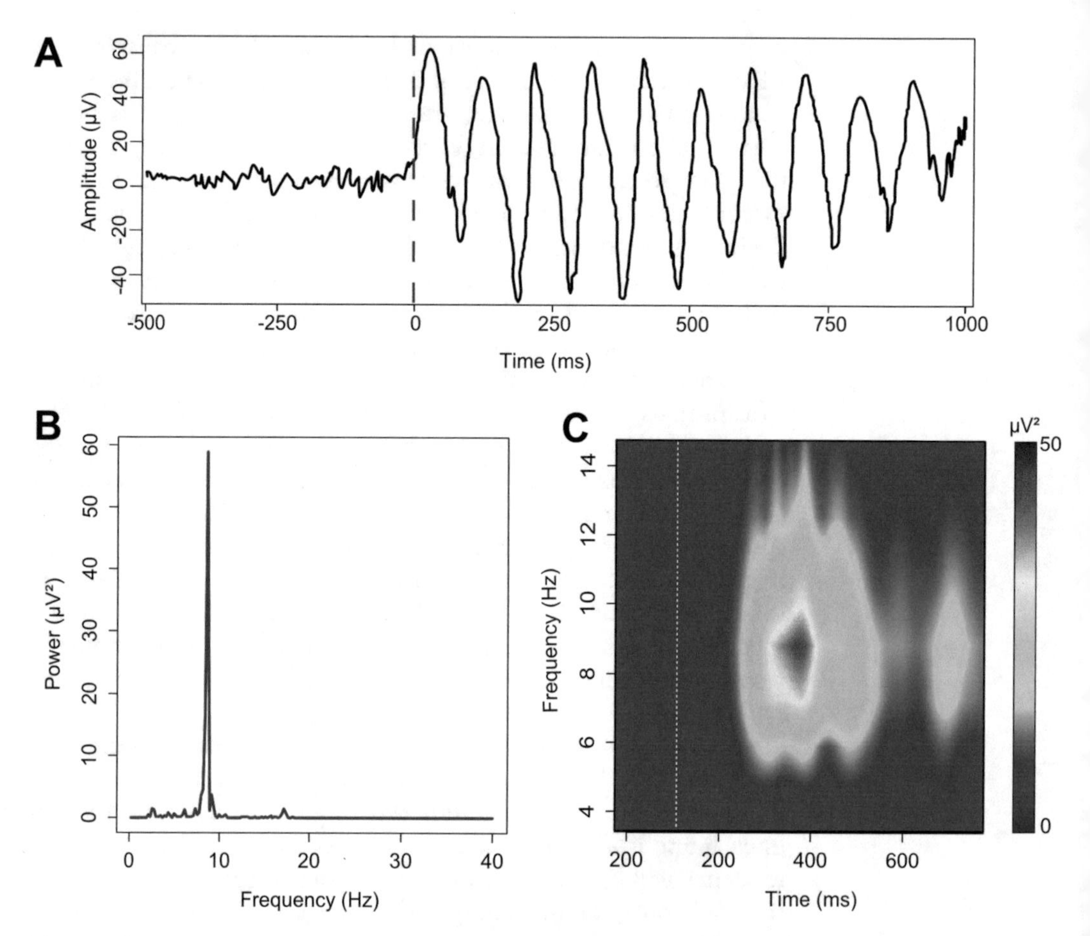

Fig. 6 (**a**) EEG signal at a posterior electrode filtered between 5 and 15 Hz in order to enhance the alpha activity. (**b**) A spectrogram showing the Fourier transform of the same EEG signal. The *y* axis represents the power of the oscillation; the *x* axis represents the various frequencies from 0 to 40 Hz. The alpha activity is prominent in the signal considered and for this reason we observe a high peak in correspondence of 8 Hz. (**c**) Wavelet decomposition of the EEG signal represented in *panel a*. The spectrogram shows the temporal evolution of the frequencies over time. The *red spot* marks the highest power in the data (around 8 Hz)

entire books dedicated to this issue (e.g. [37]). For this reason, the description we provide below should by no means be considered as exhaustive.

Over the last 15 years, researchers have become more and more interested in identifying oscillation patterns linked to various cognitive contexts (for a recent review, see [38]; but for a more cautious view on the usefulness of oscillations in cognition, see [39]). In particular, delta (1–3 Hz), theta (4–7 Hz), alpha (8–13 Hz), beta (14–30 Hz), and gamma (30–90 Hz) waves are the major sets of frequency oscillations. Higher frequencies (e.g., frequencies higher than 90 Hz) have recently been studied [40], although these seem to be more visible through MEG measures. Slow oscillations are typically associated with large networks in the brain, while high frequencies are visible in focal cortical areas. There are different measurements to evaluate oscillations, ranging from amplitude/power changes to phase locking and coherence between sites and/or across the various frequencies.

Perhaps the most studied frequency is represented by alpha (Fig. 6). Oscillations in the alpha range (8–13 Hz) are observed during the execution of a variety of tasks, and for this reason have been interpreted as the "brain signature" for several cognitive functions. According to one interpretation, alpha oscillations reflect the setting of the brain to specific states. For instance, a common finding in studies on visual attention with lateral relevant objects (namely, objects presented in the left or right visual field) is a decrease of the alpha level (usually measured in terms of Event-Related Desynchronization, ERD, see [41]) in posterior areas contralateral to the object side. The opposite pattern (i.e., increase of alpha activity, measured as Event-Related Synchronization, ERS) is measured at the ipsilateral sites [42, 43]. According to some investigators, the ERD in alpha waves may reflect the release of inhibition of the areas involved in the processing of the relevant element, so that processing is facilitated. In contrast, the (sometimes simultaneous, sometimes sequential) ERS for the opposite hemisphere would represent the suppression of the representation of the non-relevant objects [44, 45]. However, there is no general consensus on the precise function of alpha oscillations. For instance, variation in the alpha *phase* (rather than amplitude) may have a direct role in the mechanism of attention and consciousness. More specifically, some accounts have proposed that synchronization between alpha and other oscillations (such as gamma and beta) is the key factor for the generation of coherent representations in conscious perception and memory [46].

Another rhythmic set of oscillations that has been the object of intense research in cognitive neuroscience is represented by the gamma range (with a center frequency between 30 and 90 Hz). Some studies provide accounts of the gamma in terms of a set of mechanisms that allows for the "binding" of visual features into a coherent and unitary percept (e.g., [47]). More generally, like alpha waveforms, modulations in gamma oscillations have been found in studies addressing diverse cognitive functions, such as attention, working memory, and conscious processing [38]. For this reason, recent studies are now focusing on establishing how alpha and gamma oscillations act together for the generation of object representations [48, 49].

Gamma oscillations are strongly modulated by stimulus physical properties, such as size and eccentricity; therefore, studies comparing conditions with different sensory parameters may result in failure to find reliable effects on gamma [28]. Another issue concerning gamma frequencies is their link with the ocular phenomena known as microsaccades. Microsaccades produce electrical artifacts visible in high frequencies [50]. For this reason, some researchers have cast doubt on the reliability of gamma to study the activity of the brain [51].

4 EEG Studies of Human Asymmetries

Humans show various degrees of left-right specialization for some cognitive domains. For instance, both behavioral observations and neuropsychological syndromes underscore some aspects of spatial perception; object representation and language are lateralized to either the left or right hemisphere. Do EEG measures capture this type of specialization? If so, how? And what is the functional role of these asymmetries?

Given that lateralized functions are defined as those functions that rely more on the operation of one of the two hemispheres, one should expect an asymmetric pattern of neural activation at some point during stimulus processing. This may take the form of a difference in the signal strength (e.g. amplitude), or a difference in activation latency between the two hemispheres [52]. As for the first aspect, sometimes researchers have used an index of asymmetric strength. The Asymmetry (or Laterality) index is computed by considering a combination of the left and right hemisphere activations (e.g. Left-Right/Left + Right, see [9]; or Left-Right/average (Left and Right), see [42]), which allows one to have in one measure an idea of the direction and strength of any asymmetry.

Below we will discuss some of the key EEG findings on asymmetric functions. In particular, we will describe studies on lateralized activations in perception/attention, in language, and in emotion. We will also provide a brief description of some interpretations on the functional role of these asymmetries. Unfortunately, researchers investigating EEG lateralization rarely discussed the evolutionary aspect of asymmetries. For a discussion on this issue, the interested reader should consult the specialized literature.

4.1 Perception and Attention

4.1.1 Space

The left and right portions of the visual field are not treated equally by the brain. For instance, in the well-known neuropsychological syndrome of neglect there is an altered spatial processing for one hemifield (usually the left one) relative to the other. ERP studies on hemineglect patients have demonstrated that the first stages of stimulus processing are preserved (i.e., up to 130 ms post-stimulus, see [53]), while the rightward bias becomes apparent only in mid-latency components (i.e. around 150 ms). The spatial alteration is also visible for the auditory modality around the same time interval [54]; the magnitude of the ERP effect also correlates with the severity of the neglect [55].

Healthy, neurologically intact individuals exhibit a phenomenon known as "pseudo-neglect", wherein an advantage in the processing of elements in the *left* rather than right visual field is observed [56]. In the past decade, ERP and oscillation studies have provided compelling support for the involvement of right posterior (parieto-occipital) areas in the generation of this asymmetric processing, and, crucially, they have revealed the time course of such asymmetry.

In line with studies on neglect (see above), it has been shown that the leftward bias in a line bisection task does not correlate with the earliest responses visible in ERPs (i.e., the P1 response) but rather with an effect occurring at approximately 170 ms (N1, a marker for ventral stream processing) and lasting for hundreds of millisecond thereafter [57, 58]. The right hemisphere advantage is visible in two different ways. First, the amplitude of the ERP responses is larger in the right scalp electrodes for the line bisection task relative to a control condition. Second, this differential activation starts earlier (approximately 40 ms) than the activation over the left parieto-occipital areas.

This spatial leftward bias in healthy individuals can be found in a variety of tasks and is associated with spontaneous fluctuations in the degree of activation of the right hemisphere. Indeed, studies on induced oscillations and pre-stimulus neural asymmetries have shown that the spatial leftward bias is linked to different modulation of the pre-stimulus alpha level activity of posterior scalp areas [59]. The degree of right-left pre-stimulus alpha asymmetry additionally predicts individual variation in response speed.

The magnitude of the pseudo-neglect phenomenon can be reduced by several factors. Some concern sensory and perceptual parameters. For instance, line length is significantly correlated with modulations of lateralized post-stimulus N1 effect [57], as well as with a reduction of pseudo-neglect as visible from accuracy. In addition, the same N1 effect for the right occipital and parietal regions is attenuated for high relative to low perceptual loads [60].

In the frequency domain, increasing the perceptual load of the task reduces the pseudo-neglect in behavioral performance, and additionally correlates with a reduction of the right alpha dominance seen in posterior electrode sites post-stimulus onset [61]. Other factors that reduce the predominance of the right hemisphere for perceptual/attention processing are more generally related to cognitive effort and fatigue. It has been found that increasing the time-on-task factor (thus, presumably inducing fatigue) can attenuate the pre-stimulus lateralized right alpha level, and leads to a behavioral reduction of the left spatial bias [59]. Recently, it has also been shown that the pre-stimulus alpha activity is additionally biased towards the lower portion of the hemifield, and is predictive of faster responses to targets in the lower versus upper hemifield [62]. Overall, these studies point out that the spatial bias for the left hemifield is flexibly modulated by several factors, and that may interact with a vertical preference as well.

Spatial leftward asymmetries are usually interpreted in terms of the dominance of the right hemisphere in governing spatial attention [63, 64]. The extant EEG findings on the reduction of pseudo-neglect provide an account for the decrease of the leftward bias in terms of a reduction of the right hemisphere's dominance in attention control rather than an increased activity of the left hemisphere towards the right visual space (e.g. [61]).

The studies described above suggest the presence of spatial asymmetries in attentional distribution. Nevertheless, it is important to note that such asymmetries are not extremely large. Indeed, especially for visual stimuli, processing is predominantly "symmetrical" and distributed in the two hemispheres. For example, neuroimaging studies have indicated that each fronto-parietal network directs resources toward the contralateral hemifield in a rather "symmetrical" way [65, 66]. Along these lines, ERP studies have indicated that attention selection of lateralized (single and multiple) relevant objects is reflected in a negative posterior contralateral response occurring at approximately 200 ms from stimulus onset (N2pc) [67–69]. Consistently, EEG investigations on oscillations typically report symmetrical contralateral versus ipsilateral variation of alpha magnitude, with respect to the target side [42, 49].

4.1.2 Objects

The posterior areas of the two hemispheres are not only differentially sensitive to spatial aspects but also seem to be differentially involved in other aspects of visual stimulus processing. On the assumption that there is a hierarchical organization of the visual input according to global versus local elements (e.g. rooms are usually made of various pieces of furniture.; see [70]), it has consistently been shown that the right posterior areas encode the global aspects of a scene (in our examples, rooms) while the left posterior areas are preferentially activated by local aspects (e.g., the pieces of furniture in the example above). ERP studies with Navon-like stimuli have additionally indicated that while early stages of processing (e.g., the P1 response) are not modulated by global versus local processing differentially in the two hemispheres, later responses (N2, a response occurring at approximately 260 ms post-stimulus onset) are larger for local details in the left than in the right hemisphere [71–73]. This however occurs only when participants have to discriminate between local and global configurations, namely in a divided attention condition [71].

One functional interpretation of this global versus local asymmetry points to differential properties of the visual system for processing spatial frequencies. Mirroring the results on local versus global configurations, some studies have found a left-right asymmetry for high versus low spatial frequencies, with the right hemisphere exhibiting a preference for stimuli with low spatial frequencies and the left hemisphere favoring high spatial frequencies [74, 75]. Moreover, eliminating low but not high spatial frequencies reduces the right-hemisphere bias for global elements visible at approximately 200 ms post-stimulus onset [74, 75]. This suggests the operation of a relatively high-level perceptual mechanism that takes into account the spatial frequencies and the global-local distinction, as well as their interaction (for a theoretical account on asymmetries in spatial frequency processing, see [76]).

Studies on oscillations indicate that spontaneous pre-stimulus deactivation of the right versus left posterior areas (as visible through event-related synchronization in the alpha range, see [41]) correlate with lower performance for global and local configurations respectively [77]. In particular, larger amplitudes of the right alphas (likely indicating deactivation) are associated to faster responses for local features, whereas the opposite is measured for global features.

4.1.3 "Special Objects": EEG Asymmetries in Face Processing

Given their social relevance, faces represent a peculiar type of object. A well-known fact coming from several areas of research (brain imaging, neuropsychology) is that face processing triggers the right more than the left posterior hemisphere [78, 79]. Pioneering ERP studies with faces have complemented this information by providing evidence of a relatively early locus of face recognition. Indeed, several studies have found a face-specific modulation of the N170 response in the right hemisphere, indicating a larger N170 for human faces compared to animal faces or other inanimate stimuli (e.g., [80]). In addition to the N170 amplitude modulation for face compared to nonface stimuli, researchers have found a modulation of the N170 as a function of face orientation, with upright faces eliciting smaller or earlier amplitudes [81]. The fact that N170 is not affected by emotional face expression or face familiarity seems to suggest an interpretation of N170 in terms of structural face coding [79]. Thus, these results support the view that face perceptual encoding occurs relatively early and provides some foundation for other face recognition stages.

Recent attempts at single-trial analyses of face processing have provided interesting findings [82, 83]. For instance, it has been shown that the subjective uncertainty relative to face presentation is not due to variation in the efficiency of face-specific encoding (as visible from the analysis on average data), but rather reflects trial-by-trial fluctuations in the temporal activation of face-specific encoding [83].

Finally, studies on oscillations have found a frequency counterpart of the N170 effect in the gamma range (see [84]). Specifically, induced gamma band was found to be larger and earlier for faces than other objects. To our knowledge however, no evidence for *asymmetric* oscillations in face processing has been provided yet [85].

4.2 Language

Ever since the first neurological evidence of a major involvement of the left hemisphere in speech production, language has been considered as a function pertaining exclusively to the left hemisphere (for a review, see [86]). In the last decades however, an increasing number of studies on healthy individuals have revealed that both hemispheres, albeit with different roles, are involved in language processing/comprehension. Below we will briefly consider some of the key findings coming from research on language.

4.2.1 (Visual) Word Recognition

There are several levels of word representation, from the orthographic (i.e. correct writing of words) to the semantic one. One of the earliest stages sensitive to printed words (as opposed to nonwords, such as symbols) is measured in a left N1 response, peaking at around 150–170 ms [87]. Since the sensitivity consists in a larger response for words as compared to nonwords (such as symbols), researchers have proposed that this ERP response reflects the orthographic level of word processing. Moreover, due to the neuro-anatomical pattern found by MEG recordings [88], this ERP effect could reflect a function similar to that attributed to the Visual Word Form Area found in fMRI studies [89].

4.2.2 Semantic Processing

The paragraph above highlights the predominance of the left hemisphere for word processing. However, growing evidence supports the view that there is no absolute prerogative of the left hemisphere for all aspects of language processing. Thus, a better way to frame questions on asymmetries in language is to ask what language aspects are processed by the left versus right hemispheres. For instance, a distinction has been proposed concerning the type of contextual information that each hemisphere uses during language comprehension [90].

One of the traditional paradigms for studying hemispheric specialization in semantic processing is the Visual Half field presentation method, wherein two different stimuli (typically words) are presented to the left and right hemisphere, respectively [91]. By combining this method with ERP measures, and in particular with a late response occurring at approximately 400 ms post-stimulus and associated with the detection of semantic context violation (i.e. the N400 component), it has been demonstrated that the right hemisphere is also sensitive to semantic violations. However, only the left hemisphere discriminates between unexpected elements belonging to the same category and unexpected elements from unexpected categories (e.g., [92]). On the basis of this and other findings, it has been suggested that the right hemisphere processing may reflect a general mechanism that integrates the incoming information with contextual information. In contrast, processing in the left hemisphere would reflect the operation of a "predictive" mechanism that detects violation based on prediction rather than the actual context (see [90, 93]).

Overall, the N400 is usually recorded from *bilateral* scalp sites, and is seen in a variety of situations where semantic or world-knowledge is violated [94]. Studies attempting to localize the neural source of the N400 response have indicated a role for several structures in the left hemisphere, from the anterior temporal areas [95] to the Left Inferior Prefrontal Cortex [94], in the generation of the N400 (see [93]).

4.2.3 Syntactic Processing

ERP studies on sentence comprehension in the past 20 years have revealed the occurrence of a negative response (often) lateralized to the left hemisphere in the context of syntactic processing. This response has a latency of approximately 300 ms post-stimulus onset and is distributed mainly over the left anterior part of the scalp (Left Anterior Negativity, LAN, [96]). The LAN (300–500 ms) has been associated to detection of morphosyntactic violations in a variety of contexts (i.e., gender agreement, number agreement), and would represent the first component of a biphasic pattern in syntactic processing (the second component is a late bilateral activation visible from centro-parietal scalp areas, the P600, [97]). The relative overlap in timing between the LAN and the N400 response described above seems to suggest that at least some syntactic and semantic aspects of word processing take place in parallel (for a different view, see [96]). Although the LAN has now been reported in several languages, there is currently no consensus on the exact interpretation of the functional meaning of this response, as well as on its role in language comprehension [96, 98, 99].

An intriguing aspect emerging from language research concerns individual differences in the degree of hemispheric specialization. It is well known that language asymmetries are correlated with manual preference, so that the left-hemisphere dominance for language is mainly visible in right-handers compared to left-handers [100]. However, recent research seems to indicate that some aspects of this dominance even in right-handers are differentially modulated by genetic factors such as familial sinistrality. For instance, a recent study by Lee and Federmeier [101] indicates that, in contrast with right-handers without familial sinistrality, right-handed individuals with left-handed family members process syntactic violations with both hemispheres.

Finally, one of the most exciting lines of research in language deals with developmental studies. Thanks to the methodological advancements in noninvasive brain activity measurements, and in particular EEG, it is now possible to directly look at the infant brain activity in a variety of experimental settings. Recent research on language comprehension in infants has successfully proven a certain degree of linguistic competence in these very young individuals. For instance, an early left ERP response to syntactic violations is visible already in infants of approximately 24 months [102] (for a recent review see [103]).

In the past decades there have been attempts to study semantic and syntactic processing by means of oscillation analyses, in addition to ERPs. For instance, some evidence has been offered for a frequency segregation of language processing, with higher frequency modulations (in the gamma range) associated with semantic processing and mid-range frequencies (beta) more responsive to syntactic processing (e.g., [104]). However, to date evidence for *lateralized* oscillations in this context is scanty.

4.2.4 Speech Processing

A feature of language processing that consistently shows a lateralized pattern of activity is (auditory) speech processing. According to some theories (e.g., [105, 106]) the left lateralization for speech processing emerges from a specific specialization for temporal information that is critical for, but not strictly driven by, linguistic input. For instance, it has been proposed that the left hemisphere is specialized for rapid serial processing of auditory information, which represents a fundamental analysis in speech processing, whereas the right hemisphere preferentially reacts to spectral patterns, thus favoring music perception [106]. ERP studies [107] on the Mismatch Negativity component (MMN) have provided compelling support to this view. As mentioned above, the MMN is a large response arising between 150 and 300 ms, typically associated to the detection of an unexpected and infrequent sound among frequent identical sounds. Over the years, findings have indicated that the MMN can be used as an index for phoneme and consonant-vowel discrimination [108]. What seems to distinguish the nonlinguistic versus linguistic deviance detection is the lateralization of the MMN: whereas both hemispheres are active in the processing of the acoustic features of sounds, there is a left-hemisphere predominance for speech-related deviant sounds (visible in terms of amplitude enhancement of the MMN). A similar ability has been found in infants and even in newborns [109].

Researchers have recently advanced a compelling proposal centered on the role of neural oscillations in auditory processing in general, and in language in particular [105]. The hypothesis argues for an oscillation-based decoding of some language features, and assumes that high (gamma) frequencies dominate in the left hemisphere, while low (theta) frequencies dominate in the right auditory areas. Some EEG-fMRI studies confirm this assumption by indicating that spontaneous variation in the gamma power correlates with the activity of the left auditory cortex, while a right-hemisphere correlation is measured for theta power [110].

4.3 Emotion

It is a well-known fact that expression and processing of emotional contents are lateralized functions in the brain. Seminal work on brain lesions and patients consistently found that damages to the left hemisphere led to depressive symptoms while damages to the right hemisphere led to euphoric states [111, 112], suggesting that each hemisphere processes emotion differently. In recent years, EEG has proven to be an effective tool to study lateralization in the field of emotion. In particular, we can distinguish between oscillatory and ERP measures of emotional functions, as well as between studies that focus on trait affective styles and those focusing on emotional states. Trait affective styles refer to the stable traits of personality that define how a person deals with emotional situations and contents in different environmental conditions (such as the individual tendency to approach or to withdraw from positive or negative situations). Studies on emotional states focus on changes occurring to an emotional experience in a certain moment in time that applies only to a specific environmental condition [113]. Below we will highlight some of the most interesting findings on lateralized measures of emotions using EEG methods.

ERP research on emotional processing has focused mainly on induced emotional states using two different components, the late positive potential (LPP) and the early posterior negativity (EPN). LPP is a positive component visible around 470 ms post-stimulus onset, and it is associated to the processing of emotional stimuli presented in an incongruent context [114, 115]. Studies on LPP indicate larger LPP amplitude in the right than left hemisphere when negative stimuli are presented in a positive context [115, 116]. When participants evaluate good versus bad socially relevant concepts a larger right frontal LPP is observed for concepts rated bad, whereas a greater left frontal LPP is observed for concepts rated good [117]. In both adults [118] and children [119] the LPP is modulated by emotional significance and by the context in which information is presented. Therefore, it is considered as an indicant of attentional deployments towards emotional stimuli that trigger fundamental motivational systems [120].

The EPN is an attentional component arising around 150 ms post-stimulus, and is related to the selective processing of positive and negative emotional stimuli as compared to neutral ones [121]. EPN is observed in tasks in which either emotional pictures or words are presented and is considered as a relatively early measure of evaluation of the emotional content of a stimulus [122]. EPN shows lateralization according to the valence of the stimuli involved. For instance, larger EPN amplitudes have been observed in the left versus right hemisphere when processing negative as compared to neutral nouns [123] and adjectives [124]. In a task in which an emotional or neutral background picture is presented while participants are engaged in an attentional task the right EPN is larger during the presentation of emotional than neutral pictures [125]. This suggests that this component can process emotional content even implicitly.

Finally, ERP studies have found that components associated to attentional processing such as P2, N2, and P3 may have a role in the processing of emotional content. For instance, greater right parietal N2 and P3 are elicited by the processing of negative and positive arousing pictures in a passive viewing task [126]. Larger P2 and P3 amplitudes in the right hemisphere are visible in association to threatening words during an emotional Stroop task [127], supporting the notion that emotional content is processed differently in the right and left hemisphere according to its valence.

One of the most important oscillation measures in the field of emotional functions is the frontal alpha asymmetry at rest [128–130]. High levels of left frontal alpha are associated with a trait predisposition for approaching new situations, whereas the increase in right frontal alpha leads to withdrawal-related responses [128, 129, 131, 132]. Many studies focus on individual differences in trait affective styles and on how these styles can influence the response to emotional content and the expression of emotions [130, 133–135]. For instance, a study investigated how trait predisposition, measured by frontal alpha asymmetry, influences how humans respond to emotional film clips with fearful, happy, or neutral content [130]. The degree of frontal alpha lateralization predicted how participants would react to emotional content: larger left frontal alpha at rest was associated to a more positive response to happy movies, and larger right frontal alpha was associated to more negative responses to fearful movies. Active manipulation of left or right frontal resting alpha activity using neurofeedback can even change the way participants respond to emotional movies in the direction of the manipulation [133], suggesting a causal role of frontal alpha asymmetry in generating approach-like or withdrawal-like reactions to emotional videos. Frontal alpha asymmetry additionally predicts the style of social approach and expression of emotions, with higher right frontal alpha associated with a higher degree of shyness and higher left frontal alpha associated to a higher degree of sociability in children [136] and adults [134, 135].

Other studies have focused on how asymmetric frontal brain activity can be associated with state measures of emotional responses. Some studies have shown that the probability of feeling angry after an insult correlates with the level of left frontal alpha [137, 138], and that the level of sympathy towards the insulter has been linked to the increase or decrease of left frontal alpha [139].

5 Combination with Other Techniques

The recent technical advancements in the field of neuroimaging have made it possible to combine two or more techniques to observe the brain activity in vivo during the execution of cognitive tasks. For example, it is now possible to measure combined EEG-fMRI signals to exploit both the excellent time resolution of EEG and the excellent spatial resolution of fMRI to investigate spatiotemporal brain dynamics (see [140]). It is also possible to obtain concurrent TMS-EEG measurements to search for the causal involvement of brain areas in the generation of the EEG signal (for reviews of the method see [141, 142]). Below we will highlight some technical aspects related to the combination of fMRI and TMS with EEG, and some relevant findings in the field of lateralization.

The combination of different imaging modalities comes with some technical requirements that depend on the nature of the signals involved. First, the electrodes used to record EEG must be compatible with the other signal. For instance, in EEG-fMRI experiments, one needs an fMRI-compatible EEG set. Standard EEG equipment is usually composed by ferromagnetic material that, if introduced in the MR room, may be attracted by the static magnetic field and cause harm to the participant. Similarly, to record concurrent TMS-EEG one needs a set of TMS-compatible electrodes that do not overheat during the stimulation, and a TMS-compatible EEG amplifier that is not saturated by the large electric signal induced by TMS pulses.

Second, once the signal is recorded, one needs to deal with artifacts that are related to the functioning of fMRI or TMS and that affect profoundly the shape and amplitude of the EEG signal. For instance, in an EEG-fMRI experiment the magnetic gradient of the MR that repeatedly switches on and off causes a high frequency noise in the EEG signal (Fig. 7). Another important source of EEG noise when measuring concurrent EEG-fMRI is the cardioballistic artifact that is related to the pulsatile movement of the head and of the EEG electrodes in the static magnetic field, due to the blood pulse wave underneath the electrodes.

When recording concurrent EEG-TMS, the pulses produced by TMS elicit large electromagnetic artifacts that have to be removed if the EEG signal of interest is temporally close to the TMS pulse (Fig. 7). Other important artifacts are related to muscle activity triggered by TMS both on the head and on the eyes, as they may affect greatly the EEG signal and are usually time-locked to the TMS pulse. All these artifacts add up to the normal EEG artifacts described in the first section of the chapter, and must be carefully considered when combining EEG with other techniques.

Although the combination of EEG with TMS and fMRI is a great technical challenge, it is also an interesting tool to study lateralization of some cognitive functions. For example, the combination of EEG and TMS can show causal links between the deactivation of some brain areas, the related EEG signal and behavioral outcome. Studies on visual perception have indicated that TMS-induced interference before stimulus presentation leads to a large synchronization of alpha rhythm and to a decreased capacity to detect visual stimuli only when TMS is delivered over the right

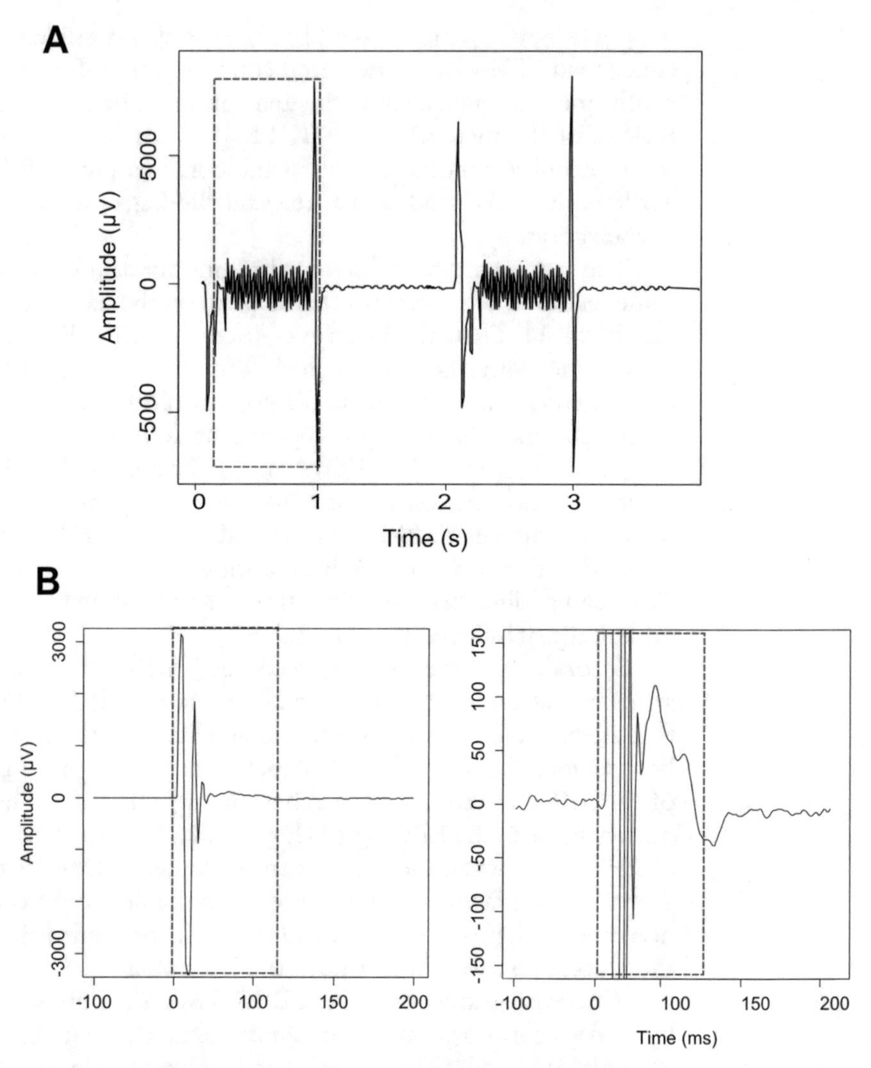

Fig. 7 (**a**) EEG activity at electrode Fp1 during fMRI acquisition. In *red* is highlighted the artifact produced by the magnetic fields during acquisition. Note the high amplitude of the scale with respect to a standard EEG. Data were taken from a public tutorial (http://fsl.fmrib.ox.ac.uk/eeglab/fmribplugin/#tutorial). (**b**) EEG activity during TMS stimulation on electrode Fc1. *Left*: artifact produced by the TMS pulse in the range of −3000/3000 μV. *Right*: the same EEG signal is zoomed to visualize the second part of the TMS artifact, which represents a TMS-induced muscle twitching in the range of −50/100 μV. Data were taken from a public tutorial (http://www.fieldtriptoolbox.org/tutorial)

parietal cortex, suggesting that attentional selectivity may depend particularly on the functioning of the right hemisphere [143]. Combined EEG-TMS can be used successfully to highlight lateralized differences in how we process the semantic content of words [144], in unconscious states such as sleep [145] or to study the causal role of brain areas in lateralized cognitive functions such as emotion and mood expression [119].

The combination between fMRI and EEG can help achieving a proper understanding of the spatiotemporal dynamics of the neural activity underlying the execution of a certain task. For example, combined EEG-fMRI has been used to study lateralization in language processing and word production. Rowan et al. [146] found that lateralization in ERPs is present only during verb generation, while fMRI shows lateralized activity across different language tasks. Combined EEG-fMRI has been used also to investigate attentional processing [147] and to elucidate the interplay between different frequency bands of EEG during working memory tasks [148]. In particular, the study by Michels et al. [148] found that during a delayed match-to-sample task both the BOLD signal in the left dorsolateral prefrontal cortex and the gamma activity increase as a function of task load. This suggests that fast neuronal synchrony is strongly required by high task demands in the brain regions that support task execution.

References

1. Picton TW, Bentin S, Berg P, Donchin E, Hillyard SA, Johnson R, Taylor MJ (2000) Guidelines for using human event-related potentials to study cognition: recording standards and publication criteria. Psychophysiology 37(02):127–152
2. Acar ZA, Makeig S (2013) Effects of forward model errors on EEG source localization. Brain Topogr 26(3):378–396
3. Michel CM, Murray MM, Lantz G, Gonzalez S, Spinelli L, de Peralta RG (2004) EEG source imaging. Clin Neurophysiol 115(10):2195–2222
4. Nunez PL, Srinivasan R (2006) Electric fields of the brain: the neurophysics of EEG. Oxford University Press, New York
5. Lantz G, de Peralta RG, Spinelli L, Seeck M, Michel CM (2003) Epileptic source localization with high density EEG: how many electrodes are needed? Clin Neurophysiol 114(1):63–69
6. Onton J, Westerfield M, Townsend J, Makeig S (2006) Imaging human EEG dynamics using independent component analysis. Neurosci Biobehav Rev 30(6):808–822
7. Nunez PL, Srinivasan R, Westdorp AF, Wijesinghe RS, Tucker DM, Silberstein RB, Cadusch PJ (1997) EEG coherency: I. Statistics, reference electrode, volume conduction, Laplacians, cortical imaging, and interpretation at multiple scales. Electroencephalogr Clin Neurophysiol 103(5):499–515
8. Luck SJ (2005) An introduction to the event-related potential technique. MIT Press, Cambridge, MA
9. Davidson RJ (1988) EEG measures of cerebral asymmetry: conceptual and methodological issues. Int J Neurosci 39(1–2):71–89
10. Van Petten C, Kutas M (1988) The use of event-related potentials in the study of brain asymmetries. Int J Neurosci 39(1–2):91–99
11. Hagemann D (2004) Individual differences in anterior EEG asymmetry: methodological problems and solutions. Biol Psychol 67(1):157–182
12. Bertrand O, Perrin F, Pernier J (1985) A theoretical justification of the average reference in topographic evoked potential studies. Electroencephalogr Clin Neurophysiol 62(6):462–464
13. Cook EW, Miller GA (1992) Digital filtering: background and tutorial for psychophysiologists. Psychophysiology 29(3):350–367
14. Edgar JC, Stewart JL, Miller GA (2005) Digital filtering in EEG/ERP research event-related potentials. In: Handy TC (ed) Event-related potentials: a methods handbook. MIT Press, Cambridge, MA, pp 85–113
15. Talsma D, Woldorff MG (2005) Methods for the estimation and removal of artifacts and overlap. In: Handy TC (ed) Event-related potentials: a methods handbook. MIT Press, Cambridge, MA, pp 115–148
16. Gratton G, Coles MG, Donchin E (1983) A new method for off-line removal of ocular

artifact. Electroencephalogr Clin Neurophysiol 55(4):468–484

17. Jung TP, Humphries C, Lee TW, Makeig S, McKeown MJ, Iragui V, Sejnowski TJ (1998) Extended ICA removes artifacts from electroencephalographic recordings. Adv Neural Inf Process Syst, 10894–900
18. Fisch BJ (1991) Spehlmann's EEG primer. Elsevier, Amsterdam
19. Cabeza R (2002) Hemispheric asymmetry reduction in older adults: the HAROLD model. Psychol Aging 17(1):85–100
20. Sanei S, Chambers JA (2013) EEG signal processing. John Wiley & Sons, New York
21. Debener S, Ullsperger M, Siegel M, Engel AK (2006) Single-trial EEG-fMRI reveals the dynamics of cognitive function. Trends Cogn Sci 10(12):558–563
22. Kappenman ES, Luck SJ (2010) The effects of electrode impedance on data quality and statistical significance in ERP recordings. Psychophysiology 47(5):888–904
23. Thut G, Miniussi C (2009) New insights into rhythmic brain activity from TMS-EEG studies. Trends Cogn Sci 13(4):182–189
24. Bekinschtein TA, Dehaene S, Rohaut B, Tadel F, Cohen L, Naccache L (2009) Neural signature of the conscious processing of auditory regularities. Proc Natl Acad Sci U S A 106(5):1672–1677
25. Herrmann CS, Strüber D, Helfrich RF, Engel AK (2016) EEG oscillations: from correlation to causality. Int J Psychophysiol 103:12–21
26. Custo A, Vulliemoz S, Grouiller F, Van De Ville D, Michel C (2014) EEG source imaging of brain states using spatiotemporal regression. Neuroimage 96:106–116
27. Woodman GF (2010) A brief introduction to the use of event-related potentials in studies of perception and attention. Atten Percept Psychophys 72(8):2031–2046
28. Herrmann CS, Grigutsch M, Busch NA (2005). EEG oscillations and wavelet analysis. In: Event-related potentials: a methods handbook. T.C. Handy, Cambridge, MA. MIT Press
29. Schomer DL, Da Silva FL, T.C. Handy, Cambridge, MA (2012) Niedermeyer's electroencephalography: basic principles, clinical applications, and related fields. Lippincott Williams & Wilkins, Philadelphia, PA
30. Delorme A, Makeig S (2004) EEGLAB: an open source toolbox for analysis of single-trial EEG dynamics including independent component analysis. J Neurosci Methods 134(1):9–21
31. Quian Quiroga R, Garcia H (2003) Single-trial event-related potentials with wavelet denoising. Clin Neurophysiol 114:376–390
32. Clark VP, Fan S, Hillyard SA (1995) Identification of early visually evoked potential generators by retinotopic and topographic analyses. Hum Brain Mapp 2:170–187
33. Hillyard SA, Vogel EK, Luck SJ (1998) Sensory gain control (amplification) as a mechanism of selective attention: electrophysiological and neuroimaging evidence. Philos Trans R Soc B Biol Sci 353(1373):1257–1270
34. Näätänen R, Paavilainen P, Rinne T, Alho K (2007) The mismatch negativity (MMN) in basic research of central auditory processing: a review. Clin Neurophysiol 118(12):2544–2590
35. Hagoort P, Brown CM (2000) ERP effects of listening to speech compared to reading: the P600/SPS to syntactic violations in spoken sentences and rapid serial visual presentation. Neuropsychologia 38(11):1531–1549
36. Gross J (2014) Analytical methods and experimental approaches for electrophysiological studies of brain oscillations. J Neurosci Methods 228:57–66
37. Buzsaki G (2006) Rhythms of the brain. Oxford University Press, Oxford
38. da Silva FL (2013) EEG and MEG: relevance to neuroscience. Neuron 80(5):1112–1128
39. Sejnowski TJ, Paulsen O (2006) Network oscillations: emerging computational principles. J Neurosci 26(6):1673–1676
40. Gobbelé R, Buchner H, Curio G (1998) High-frequency (600 Hz) SEP activities originating in the subcortical and cortical human somatosensory system. Electroencephalogr Clin Neurophysiol 108(2):182–189
41. Pfurtscheller G, Da Silva FL (1999) Event-related EEG/MEG synchronization and desynchronization: basic principles. Clin Neurophysiol 110(11):1842–1857
42. Thut G, Nietzel A, Brandt SA, Pascual-Leone A (2006) α-Band electroencephalographic activity over occipital cortex indexes visuospatial attention bias and predicts visual target detection. J Neurosci 26(37):9494–9502
43. Worden MS, Foxe JJ, Wang N, Simpson GV (2000) Anticipatory biasing of visuospatial attention indexed by retinotopically specific-band electroencephalography increases over occipital cortex. J Neurosci 20:1–6
44. Klimesch W, Sauseng P, Hanslmayr S (2007) EEG alpha oscillations: the inhibition–timing hypothesis. Brain Res Rev 53(1):63–88
45. Hanslmayr S, Aslan A, Staudigl T, Klimesch W, Herrmann CS, Bäuml KH (2007) Prestimulus oscillations predict visual perception performance between and within subjects. Neuroimage 37(4):1465–1473

46. Palva S, Palva JM (2007) New vistas for α-frequency band oscillations. Trends Neurosci 30(4):150–158
47. Tallon-Baudry C, Bertrand O (1999) Oscillatory gamma activity in humans and its role in object representation. Trends Cogn Sci 3(4):151–162
48. Jensen O, Gips B, Bergmann TO, Bonnefond M (2014) Temporal coding organized by coupled alpha and gamma oscillations prioritize visual processing. Trends Neurosci 37:357–369
49. Sauseng P, Klimesch W, Heise KF, Gruber WR, Holz E, Karim AA, Hummel FC (2009) Brain oscillatory substrates of visual short-term memory capacity. Curr Biol 19(21):1846–1852
50. Yuval-Greenberg S, Tomer O, Keren AS, Nelken I, Deouell LY (2008) Transient induced gamma-band response in EEG as a manifestation of miniature saccades. Neuron 58(3):429–441
51. Ray S, Maunsell JH (2015) Do gamma oscillations play a role in cerebral cortex? Trends Cogn Sci 19(2):78–85
52. Brancucci A (2010) Electroencephalographic and magnetoencephalographic indices of hemispheric asymmetry. In: Hugdahl K, Westerhausen R (eds) The two halves of the brain: information processing in the two hemispheres. MIT Press, Cambridge, MA, pp 211–250
53. Di Russo F, Aprile T, Spitoni G, Spinelli D (2008) Impaired visual processing of contralesional stimuli in neglect patients: a visual-evoked potential study. Brain 131:842–854
54. Deouell LY, Bentin S, Soroker N (2000) Electrophysiological evidence for an early (pre-attentive) information processing deficit in patients with right hemisphere damage and unilateral neglect. Brain 123:353–365
55. Tarkka IM, Luukkainen-Markkula R, Pitkänen K, Hämäläinen H (2011) Alterations in visual and auditory processing in hemispatial neglect: an evoked potential follow-up study. Int J Psychophysiol 79(2):272–279
56. Bowers D, Heilman KM (1980) Pseudoneglect: effects of hemispace on a tactile line bisection task. Neuropsychologia 18:491–498
57. Benwell CS, Harvey M, Thut G (2014) On the neural origin of pseudoneglect: EEG-correlates of shifts in line bisection performance with manipulation of line length. Neuroimage 86:370–380
58. Foxe JJ, McCourt ME, Javitt DC (2003) Right hemisphere control of visuospatial attention: line-bisection judgments evaluated with high-density electrical mapping and source analysis. Neuroimage 19(3):710–726
59. Newman DP, O'Connell RG, Bellgrove MA (2013) Linking time-on-task, spatial bias and hemispheric activation asymmetry: a neural correlate of rightward attention drift. Neuropsychologia 51:1215–1223
60. O'Connell RG, Schneider D, Hester R, Mattingley JB, Bellgrove MA (2011) Attentional load asymmetrically affects early electrophysiological indices of visual orienting. Cereb Cortex 21:1056–1065
61. Perez A, Peers PV, Valdes-Sosa M, Galan L, Garcia L, Martinez-Montes E (2009) Hemispheric modulations of alpha-band power reflect the rightward shift in attention induced by enhanced attentional load. Neuropsychologia 47:41–49
62. Loughnane GM, Shanley JP, Lalor EC, O'Connell RG (2015) Behavioral and electrophysiological evidence of opposing lateral visuospatial asymmetries in the upper and lower visual fields. Cortex 63:220–231
63. Corbetta M, Shulman GL (2002) Control of goal-directed and stimulus-driven attention in the brain. Nat Rev Neurosci 3(3):201–215
64. Mesulam M (1981) A cortical network for directed attention and unilateral neglect. Ann Neurol 10(4):309–325
65. Hopfinger JB, Buonocore MH, Mangun GR (2000) The neural mechanisms of top-down attentional control. Nat Neurosci 3(3):284–291
66. Kastner S, Pinsk MA, De Weerd P, Desimone R, Ungerleider LG (1999) Increased activity in human visual cortex during directed attention in the absence of visual stimulation. Neuron 22(4):751–761
67. Eimer M (1996) The N2pc component as an indicator of attentional selectivity. Electroencephalogr Clin Neurophysiol 99(3):225–234
68. Luck SJ, Hillyard SA (1994) Electrophysiological correlates of feature analysis during visual search. Psychophysiology 31(3):291–308
69. Mazza V, Turatto M, Caramazza A (2009) Attention selection, distractor suppression and N2pc. Cortex 45(7):879–890
70. Navon D (1977) Forest before trees: the precedence of global features in visual perception. Cogn Psychol 9(3):353–383

71. Heinze HJ, Hinrichs H, Scholz M, Burchert W, Mangun GR (1998) Neural mechanisms of global and local processing: a combined PET and ERP study. J Cogn Neurosci 10(4):485–498
72. Flevaris AV, Martínez A, Hillyard SA (2014) Attending to global versus local stimulus features modulates neural processing of low versus high spatial frequencies: an analysis with event-related brain potentials. Front Psychol 5:277
73. Proverbio AM, Minniti A, Zani A (1998) Electrophysiological evidence of a perceptual precedence of global vs. local visual information. Cogn Brain Res 6(4):321–334
74. Boeschoten MA, Kemner C, Kenemans JL, Engeland HV (2005) The relationship between local and global processing and the processing of high and low spatial frequencies studied by event-related potentials and source modeling. Cogn Brain Res 24:228–236
75. Han S, Yund EW, Woods DL (2003) An ERP study of the global precedence effect: the role of spatial frequency. Clin Neurophysiol 114:1850–1865. doi:10.1016/S1388-2457(03)00196-2
76. Robertson LC, Ivry R (2000) Hemispheric asymmetries: attention to visual and auditory primitives. Curr Dir Psychol Sci 9:59–63
77. Volberg G, Kliegl K, Hansimayr S, Greenlee MW (2009) EEG alpha oscillations in the preparation for global and local processing predicts behavioral performance. Hum Brain Mapp 30:2173–2183
78. Gainotti G (2013) Laterality effects in normal subjects' recognition of familiar faces, voices and names. Perceptual and representational components. Neuropsychologia 51(7):1151–1160
79. Palermo R, Rhodes G (2007) Are you always on my mind? A review of how face perception and attention interact. Neuropsychologia 45(1):75–92
80. Bentin S, Allison T, Puce A, Perez E, McCarthy G (1996) Electrophysiological studies of face perception in humans. J Cogn Neurosci 8(6):551–565
81. Rossion B, Gauthier I, Tarr MJ, Despland P, Bruyer R, Linotte S, Crommelinck M (2000) The N170 occipito-temporal component is delayed and enhanced to inverted faces but not to inverted objects: an electrophysiological account of face-specific processes in the human brain. Neuroreport 11(1):69–72
82. De Vos M, Thorne JD, Yovel G, Debener S (2012) Let's face it, from trial to trial: comparing procedures for N170 single-trial estimation. Neuroimage 63(3):1196–1202
83. Navajas J, Ahmadi M, Quiroga RQ (2013) Uncovering the mechanisms of conscious face perception: a single-trial study of the N170 responses. J Neurosci 33(4):1337–1343
84. Zion-Golumbic E, Bentin S (2007) Dissociated neural mechanisms for face detection and configural encoding: evidence from N170 and induced gamma-band oscillation effects. Cereb Cortex 17(8):1741–1749
85. Güntekin B, Başar E (2014) A review of brain oscillations in perception of faces and emotional pictures. Neuropsychologia 58:33–51
86. Dronkers NF, Plaisant O, Iba-Zizen MT, Cabanis EA (2007) Paul Broca's historic cases: high resolution MR imaging of the brains of Leborgne and Lelong. Brain 130(5):1432–1441
87. Bentin S, Mouchetant-Rostaing Y, Giard MH, Echallier JF, Pernier J (1999) ERP manifestations of processing printed words at different psycholinguistic levels: time course and scalp distribution. J Cogn Neurosci 11(3): 235–260
88. Tarkiainen A, Helenius P, Hansen PC, Cornelissen PL, Salmelin R (1999) Dynamics of letter string perception in the human occipitotemporal cortex. Brain 122(11): 2119–2132
89. McCandliss BD, Cohen L, Dehaene S (2003) The visual word form area: expertise for reading in the fusiform gyrus. Trends Cogn Sci 7(7):293–299
90. Federmeier KD (2007) Thinking ahead: the role and roots of prediction in language comprehension. Psychophysiology 44(4):491–505
91. Chiarello C, Whitaker HA (1988) Lateralization of lexical processes in the normal brain: a review of visual half-field research. In: Contemporary reviews in neuropsychology. Springer New York, pp 36–76
92. Federmeier KD, Kutas M (1999) Right words and left words: electrophysiological evidence for hemispheric differences in meaning processing. Cogn Brain Res 8(3):373–392
93. Kutas M, Federmeier KD (2011) Thirty years and counting: Finding meaning in the N400 component of the event related brain potential (ERP). Annu Rev Psychol 62:621
94. Hagoort P, Hald L, Bastiaansen M, Petersson KM (2004) Integration of word meaning and world knowledge in language comprehension. Science 304(5669):438–441
95. Nobre AC, Mccarthy G (1995) Language-related field potentials in the anterior-medial temporal lobe: II. Effects of word type and semantic priming. J Neurosci 15(2): 1090–1098

96. Friederici AD (2002) Towards a neural basis of auditory sentence processing. Trends Cogn Sci 6(2):78–84
97. Osterhout L, Holcomb PJ (1992) Event-related brain potentials elicited by syntactic anomaly. J Mem Lang 31(6):785–806
98. Molinaro N, Barber HA, Carreiras M (2011) Grammatical agreement processing in reading: ERP findings and future directions. Cortex 47(8):908–930
99. Steinhauer K, Drury JE (2012) On the early left-anterior negativity (ELAN) in syntax studies. Brain Lang 120(2):135–162
100. Knecht S, Dräger B, Deppe M, Bobe L, Lohmann H, Flöel A, Henningsen H (2000) Handedness and hemispheric language dominance in healthy humans. Brain 123(12):2512–2518
101. Lee CL, Federmeier KD (2015) It's all in the family brain asymmetry and syntactic processing of word class. Psychol Sci 26:997–1005. doi:10.1177/0956797615575743
102. Bernal S, Dehaene-Lambertz G, Millotte S, Christophe A (2010) Two-year-olds compute syntactic structure on-line. Dev Sci 13(1): 69–76
103. Dehaene-Lambertz G, Spelke ES (2015) The infancy of the human brain. Neuron 88(1): 93–109
104. Bastiaansen MCM, Hagoort P (2015) Frequency-based segregation of syntactic and semantic unification during online sentence level language comprehension. J Cogn Neurosci 27(11):2095–2107
105. Giraud AL, Poeppel D (2012) Cortical oscillations and speech processing: emerging computational principles and operations. Nat Neurosci 15(4):511–517
106. Zatorre RJ, Belin P, Penhune VB (2002) Structure and function of auditory cortex: music and speech. Trends Cogn Sci 6(1):37–46
107. Garrido MI, Kilner JM, Stephan KE, Friston KJ (2009) The mismatch negativity: a review of underlying mechanisms. Clin Neurophysiol 120(3):453–463
108. Näätänen R (2001) The perception of speech sounds by the human brain as reflected by the mismatch negativity (MMN) and its magnetic equivalent (MMNm). Psychophysiology 38(01):1–21
109. Dehaene-Lambertz G, Gliga T (2004) Common neural basis for phoneme processing in infants and adults. J Cogn Neurosci 16(8):1375–1387
110. Giraud AL, Kleinschmidt A, Poeppel D, Lund TE, Frackowiak RS, Laufs H (2007) Endogenous cortical rhythms determine cerebral specialization for speech perception and production. Neuron 56(6):1127–1134
111. Goldstein K (1939) The Organism: A Holistic Approach to Biology Derived from Pathological Data in Man. New York: American Book Company.
112. Perria L, Rosandini G, Rossi GF (1961) Determination of side of cerebral dominance with amobarbital. Arch Neurol 4(2):173–181
113. Cattell, R. B., & Scheier, I. H. (1961). The meaning and measurement of neuroticism and anxiety. New York: Ronald Press.
114. Cacioppo JT, Crites SL, Gardner WL, Berntson GG (1994) Bioelectrical echoes from evaluative categorizations: I. A late positive brain potential that varies as a function of trait negativity and extremity. J Pers Soc Psychol 67(1):115
115. Cacioppo JT, Crites SL, Gardner WL (1996) Attitudes to the right: evaluative processing is associated with lateralized late positive event-related brain potentials. Pers Soc Psychol Bull 22(12):1205–1219
116. Ito TA, Larsen JT, Smith NK, Cacioppo JT (1998) Negative information weighs more heavily on the brain: the negativity bias in evaluative categorizations. J Pers Soc Psychol 75(4):887
117. Cunningham WA, Espinet SD, DeYoung CG, Zelazo PD (2005) Attitudes to the right-and left: frontal ERP asymmetries associated with stimulus valence and processing goals. Neuroimage 28(4):827–834
118. Schupp HT, Cuthbert BN, Bradley MM, Cacioppo JT, Ito T, Lang PJ (2000) Affective picture processing: the late positive potential is modulated by motivational relevance. Psychophysiology 37(2):257–261
119. Dennis TA, Hajcak G (2009) The late positive potential: a neurophysiological marker for emotion regulation in children. J Child Psychol Psychiatry 50(11):1373–1383
120. Lang PJ, Bradley MM, Cuthbert BN (1997) International affective picture system (IAPS): technical manual and affective ratings. NIMH Center for the Study of Emotion and Attention, pp 39–58
121. Schupp HT, Junghöfer M, Weike AI, Hamm AO (2003) Attention and emotion: an ERP analysis of facilitated emotional stimulus processing. Neuroreport 14(8):1107–1110

122. Junghöfer M, Bradley MM, Elbert TR, Lang PJ (2001) Fleeting images: a new look at early emotion discrimination. Psychophysiology 38(02):175–178
123. Frühholz S, Jellinghaus A, Herrmann M (2011) Time course of implicit processing and explicit processing of emotional faces and emotional words. Biol Psychol 87(2):265–274
124. Herbert BM, Pollatos O, Schandry R (2007) Interoceptive sensitivity and emotion processing: an EEG study. Int J Psychophysiol 65(3):214–227
125. Bode S, Bennett D, Stahl J, Murawski C (2014) Distributed patterns of event-related potentials predict subsequent ratings of abstract stimulus attributes. PLoS One 9(10):e109070. doi:10.1371/journal.pone.0109070
126. Kayser J, Tenke C, Nordby H, Hammerborg D, Hugdahl K, Erdmann G (1997) Event-related potential (ERP) asymmetries to emotional stimuli in a visual half-field paradigm. Psychophysiology 34(4):414–426
127. Thomas LA, De Bellis MD, Graham R, LaBar KS (2007) Development of emotional facial recognition in late childhood and adolescence. Dev Sci 10(5):547–558
128. Harmon-Jones E, Allen JJ (1997) Behavioral activation sensitivity and resting frontal EEG asymmetry: covariation of putative indicators related to risk for mood disorders. J Abnorm Psychol 106(1):159
129. Harmon-Jones E, Allen JJ (1998) Anger and frontal brain activity: EEG asymmetry consistent with approach motivation despite negative affective valence. J Pers Soc Psychol 74(5):1310
130. Wheeler RE, Davidson RJ, Tomarken AJ (1993) Frontal brain asymmetry and emotional reactivity: a biological substrate of affective style. Psychophysiology 30(1):82–89
131. Sutton SK, Davidson RJ (1997) Prefrontal brain asymmetry: a biological substrate of the behavioral approach and inhibition systems. Psychol Sci 8(3):204–210
132. Wiedemann G, Pauli P, Dengler W, Lutzenberger W, Birbaumer N, Buchkremer G (1999) Frontal brain asymmetry as a biological substrate of emotions in patients with panic disorders. Arch Gen Psychiatry 56(1):78–84
133. Allen JJ, Harmon-Jones E, Cavender JH (2001) Manipulation of frontal EEG asymmetry through biofeedback alters self-reported emotional responses and facial EMG. Psychophysiology 38(04):685–693
134. Schmidt LA (1999) Frontal brain electrical activity in shyness and sociability. Psychol Sci 10(4):316–320
135. Schmidt LA, Fox NA (1994) Patterns of cortical electrophysiology and autonomic activity in adults' shyness and sociability. Biol Psychol 38(2):183–198
136. Fox NA, Rubin KH, Calkins SD, Marshall TR, Coplan RJ, Porges SW, Stewart S (1995) Frontal activation asymmetry and social competence at four years of age. Child Dev 66(6):1770–1784
137. Harmon-Jones E, Sigelman J (2001) State anger and prefrontal brain activity: evidence that insult-related relative left-prefrontal activation is associated with experienced anger and aggression. J Pers Soc Psychol 80(5):797
138. Harmon-Jones E, Sigelman J, Bohlig A, Harmon-Jones C (2003) Anger, coping, and frontal cortical activity: the effect of coping potential on anger-induced left frontal activity. Cognit Emot 17(1):1–24
139. Harmon-Jones E, Vaughn-Scott K, Mohr S, Sigelman J, Harmon-Jones C (2004) The effect of manipulated sympathy and anger on left and right frontal cortical activity. Emotion 4(1):95
140. Huster RJ, Debener S, Eichele T, Herrmann CS (2012) Methods for simultaneous EEG-fMRI: an introductory review. J Neurosci 32(18):6053–6060
141. Ilmoniemi RJ, Kičić D (2010) Methodology for combined TMS and EEG. Brain Topogr 22(4):233–248
142. Thut G, Pascual-Leone A (2010) Integrating TMS with EEG: how and what for? Brain Topogr 22(4):215–218
143. Capotosto P, Babiloni C, Romani GL, Corbetta M (2012) Differential contribution of right and left parietal cortex to the control of spatial attention: a simultaneous EEG–rTMS study. Cereb Cortex 22(2):446–454
144. Fuggetta G, Rizzo S, Pobric G, Lavidor M, Walsh V (2009) Functional representation of living and nonliving domains across the cerebral hemispheres: a combined event-related potential/transcranial magnetic stimulation study. J Cogn Neurosci 21(2):403–414

145. Stamm M, Aru J, Bachmann T (2011) Right-frontal slow negative potentials evoked by occipital TMS are reduced in NREM sleep. Neurosci Lett 493(3):116–121
146. Rowan A, Liégeois F, Vargha-Khadem F, Gadian D, Connelly A, Baldeweg T (2004) Cortical lateralization during verb generation: a combined ERP and fMRI study. Neuroimage 22(2):665–675
147. Bledowski C, Prvulovic D, Goebel R, Zanella FE, Linden DE (2004) Attentional systems in target and distractor processing: a combined ERP and fMRI study. Neuroimage 22(2): 530–540
148. Michels L, Bucher K, Lüchinger R, Klaver P, Martin E, Jeanmonod D, Brandeis D (2010) Simultaneous EEG-fMRI during a working memory task: modulations in low and high frequency bands. PLoS One 5(4):e10298
149. American Electroencephalographic Society (1994) Guidelines for standard electrode position nomenclature. J Clin Neurophysiol 11:111–113
150. Oostenveld R, Praamstra P (2001) The five percent electrode system for high-resolution EEG and ERP measurements. Clin Neurophysiol 112(4):713–719

Chapter 14

Noninvasive Imaging Technologies in Primates

William D. Hopkins and Kimberley A. Phillips

Abstract

The study of neuroanatomical and functional asymmetries has been the topic of considerable scientific debate and research. While early research primarily focused on neuropsychological investigations of clinical populations and analysis of postmortem materials, with the advent of noninvasive neuroimaging, it has afforded many advantages for comparative studies in primates. Here, we describe the various methods that have been used to quantify neuroanatomical and functional asymmetries in nonhuman primates and the results that have emerged from these studies. We further discuss the limitations of some of these methods and offer suggestions for future research.

Key words Asymmetry, Neuroanatomy, Functional asymmetries, Primates, Magnetic resonance imaging (MRI), Positron emission tomography (PET), Resting-state functional MRI (R-fMRI), Diffusion tensor imaging (DTI)

1 Introduction

Primate brain evolution has been the topic of considerable empirical investigation in the neurosciences for more than 150 years [1, 2]. A central focus of many comparative studies in the anthropological field has been to characterize allometric variation in different dimensions of cortical organization in different primates species and of making inferences about evolutionary changes that reflect human specific adaptations, for example, language, tool manufacture and use, and intelligence [3, 4]. Until 20–25 years ago, many comparative studies in cortical organization relied on collections of postmortem brains as the primary database from which to make a variety of different anatomical measures, such as sulcus morphology, overall and region specific volumes and surface areas [5–14]. It is clear that many comparative studies of cortical organization, such as those focusing on cellular or molecular organization, require postmortem material; however, studies on morphology have benefitted recently with the advent of in vivo modern imaging technologies. Specifically, in the past 25 years, advances in both

Lesley J. Rogers and Giorgio Vallortigara (eds.), *Lateralized Brain Functions: Methods in Human and Non-Human Species*, Neuromethods, vol. 122, DOI 10.1007/978-1-4939-6725-4_14, © Springer Science+Business Media LLC 2017

structural and functional in vivo imaging technologies have allowed considerable advancement in our understanding of primate brain volume with respect to gray and white matter volume, gyrification, cortical thickness, and a variety of regions of interest [15].

In this chapter, we describe the application and use of in vivo imaging technologies to the study of cortical asymmetries, or left-right differences, in different dimensions of cortical organization including morphology, function, and connectivity. Though interest in structural and functional differences in neuroanatomy and morphology in the human and nonhuman primate can be dated back to the early publications of Dax, Broca, Wernicke, and Cunningham, it has not been until recently that more systematic and statistically powerful studies on cerebral asymmetries have been conducted in human and nonhuman primates. We believe that studies using in vivo imaging technologies have three important strengths over traditional morphological studies using postmortem materials. First, in vivo imaging can be done in many more subjects and one does not need to euthanize or wait until death to obtain quantitative measures of the brain. For instance, most studies of the length of sulci or size of surface areas obtained from postmortem brains have typically had relatively small sample sizes with limited statistical power to test for differences between ages or sexes [16–18]. Additionally, obtaining measures of cortical organization, including asymmetry, in living animals allows the anatomical measures to be correlated with different behaviors or traits without a lengthy delay between collection of the two types of data. A significant pragmatic benefit of in vivo imaging technologies, at least for some methods, is that multiple regions can be quantified because the images are virtually cut and visualized in all three planes. For example, in the same brain, a scientist can measure the planum temporale in the cortical plane and, subsequently, take the same brain and measure the inferior frontal gyrus in the transverse plane. If using postmortem material, a decision regarding the plane in which the tissue will be cut must be made for a given study and this sectioning in procedure will not always allow measurement of other brain areas that require visualization in a different plane. Finally, the variety of different imaging methods allows collection of multiple dimensions of cortical organization to be obtained in the same subjects. For example, anatomical, white matter connectivity and functional connectivity can be obtained in same scan session from the same subject. This allows the measurement of asymmetry at the morphological, connective, and functional levels of analysis, which is quite powerful. In this chapter, we describe different in vivo imaging methods for the measurement of asymmetries in morphology and connectivity in human and nonhuman primates.

2 Morphological and Structural Asymmetries

2.1 Structural MRI and Region of Interest Approaches

Perhaps the most common approach to the measurement of cortical asymmetries has come from studies of regional variation in volumes of different regions measured from magnetic resonance images (MRI). In the typical region of interest (ROI) MRI study using free or commercially available software, the pointer on their computer mouse can be used to define the boundaries of a region of interest (sometimes referred to as an object map or mask) [19–24]. For example, we have previously used a ROI method for quantifying the motor-hand area of the precentral gyrus [25], posterior superior temporal gyrus [26] and inferior frontal gyrus of the chimpanzee (*Pan troglodytes*) brain for comparison with equivalent regions in humans [27, 28]. ROI measures of different regional volumes have also been linked to different behavioral traits including handedness [29–33], social behavior and personality [34], and social cognition [26, 35]. An important aspect of all ROI methods is the establishment of reliable landmarks (i.e., sulci or other cortical features) for defining the region of interest.

High resolution MRI scans acquired in 3D are incredibly useful because one can obtain excellent contrast in gray matter, white matter, and cerebrospinal fluid (CSF). Further, most available software programs used for ROI tracing allow for several important functions. First, specific regions can be visualized in multiple planes simultaneously, thereby increasing the precision of defining the area. Second, because the contrast between gray and white matter is fairly robust, depending on the scanner strength and imaging protocol, several available software programs (i.e., FSL, SPM) can automatically segment the MRI T1-scan into individual gray and white matter volumes. Third, when drawing an ROI and saving the object map or mask, these can be applied to the segmented gray and white matter volumes, thereby allowing independent measures of the tissue composition comprising a given region of interest (see Fig. 1). For example, the object map for the inferior frontal gyrus (IFG) applied to a segmented gray and white matter volume is shown in Fig. 1. The distinction between gray and white matter within a specific region of interest is by no means trivial because how they relate to different subject characteristics or behavioral phenotypes may reflect different levels of neurological properties. Gray matter reflects the localized neurons within a specific region while the white matter values reflect the degree of myelination within the region, which indirectly indicate the level of potential connectivity within the region. It is possible, and indeed likely, that asymmetries might be present in one aspect of cortical organization but not another. For instance, Hopkins et al. [36] examined variation in volumetric and lateralized differences in the motor-hand

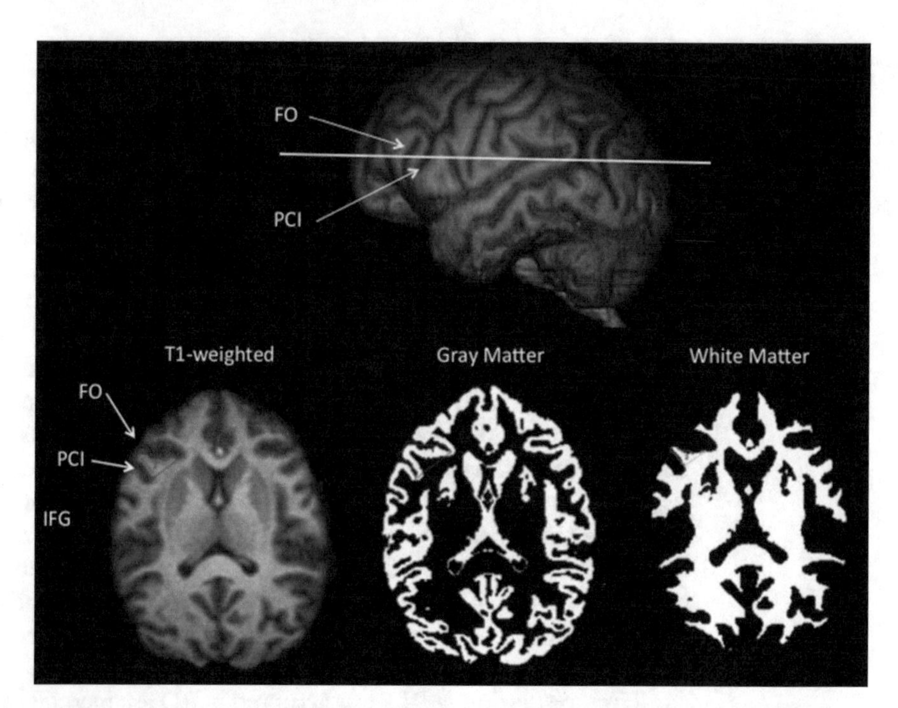

Fig. 1 Example of a region-of-interest tracing. *Upper panel*: 3D figure of a chimpanzee MRI scan. *FO* fronto-orbital sulcus, *PCI* precentral inferior sulcus. *Horizontal line* represents the axial plane at which the brain was cut. *Lower panel*: Shows tracing the inferior frontal gyrus (IFG) (on a single slice) with the entire gyrus traced in *red* and white matter traced in *green*. Application of the objects maps applied to segmented gray and white matter

area and IFG between chimpanzees that learned to throw and those that did not and found significant differences in white but not gray matter. We have also used ROI methods combined with segmentation to quantify gray matter variation in the posterior superior temporal gyrus (p_STG) in vervet monkeys (*Chlorocebus aethiops sabaeus*), rhesus monkeys (*Macaca mulatta*), bonnet monkeys (*Macaca radiata*), and chimpanzees [37]. The p_STG overlaps with the planum temporale, a region of the human brain that has been consistently shown to be left lateralized [38, 39]. Shown in Fig. 2 are the mean percentage of total gray matter volume belonging to the p_STG in each species and hemisphere. There are two important results from this study. First with respect to asymmetry, only chimpanzees showed a significant population bias (leftward). Second, the percentage of total gray matter within the p_STG in much lower in chimpanzees compared to the Old World monkeys.

2.2 Sulci Surface Area, Depth, and Gray Matter Thickness

Historically, the most commonly used measure of brain asymmetry in primates is the linear measurement of sulci length. Sulci length measures have been obtained by direct measurement on postmortem brains, from photographs, endocasts, and MRI scans

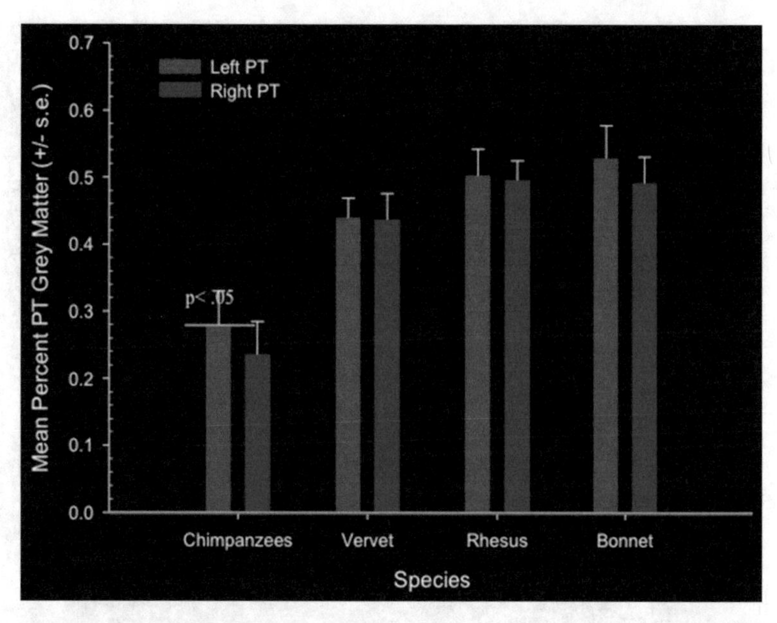

Fig. 2 Mean percentage of whole gray matter volume in the left (*red*) and right (*blue*) hemisphere posterior superior temporal gyrus (PT) in chimpanzees, vervet monkeys, rhesus monkeys, and bonnet monkeys. Image reprinted from Lyn et al. [37]

[6, 8, 17, 40–45]. For instance, in one of the early systematic studies of the sylvian fissure, Yeni-Komshian and Benson [17] measured the length of this sulcus in humans, chimpanzees, and rhesus monkeys and reported significant leftward biases in humans and chimpanzees but not rhesus monkeys. One limitation of measuring solely the hull length of the sulcus (this is the outer contour of the sulcus) is that it does not capture the depth of the fold, and thereby only indirectly measures the entire surface area.

There are several recently developed software programs that extract cortical folds from in vivo and postmortem MRI scans and these provide more comprehensive approaches to the measurement of asymmetries in sulci. One influential program is called BrainVisa (BV) that has an automatic pipeline processing procedure for extracting the surface area, maximum depth, mean depth, hull length, gray matter thickness, and the sulci width from MRI scans (see Figs. 3 and 4). BV also has several add-on tools that provide automatic quantification of overall gyrification and estimates cortical thickness. In nonhuman primates, data produced from BV analyses have been used to characterize age-related changes in cortical folding [46, 47], heritability in cortical organization [48–50], comparative differences in central sulcus shape [51], and quantification of the pli-de-passage in humans and chimpanzees [52, 53]. Recently, Bogart et al. [54] examined interhemispheric differences in surface area and mean depth in

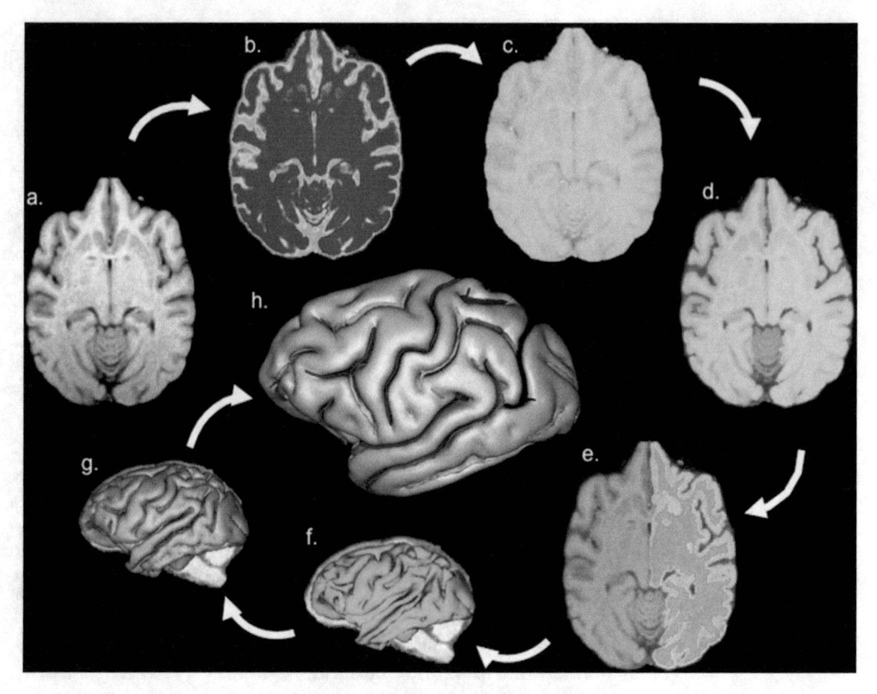

Fig. 3 Pipeline steps in BrainVisa (see Hopkins et al. 2010 for description)

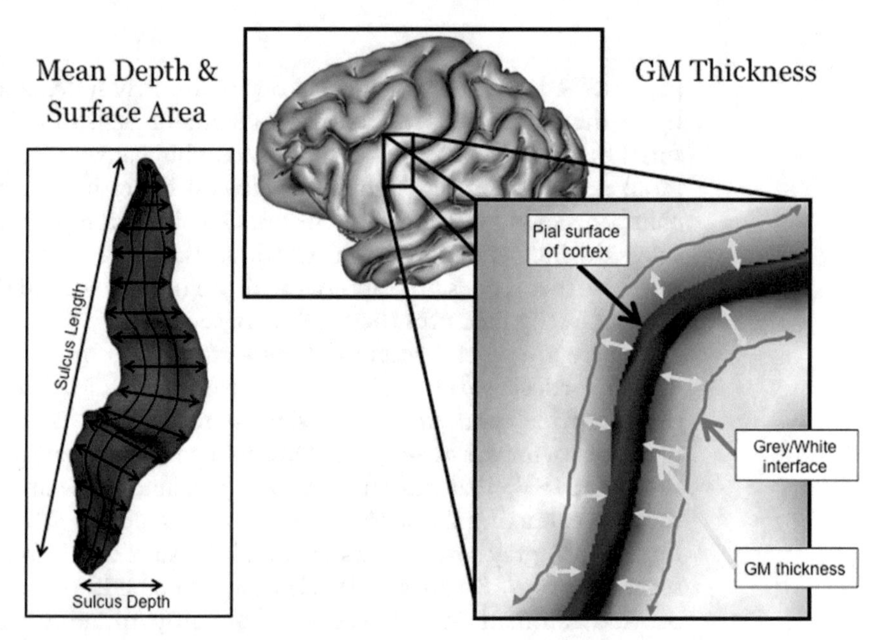

Fig. 4 Different measures obtained from BrainVisa. *Left*: Surface area and mean depth. *Right*: Calculation of gray matter thickness for cortical fold in BrainVisa

primary cortical sulci in chimpanzees (11 sulci) in rhesus monkeys (7 sulci). For the chimpanzee surface areas, significant leftward asymmetries of the fronto-orbital, superior frontal, sylvian fissure, and inferior post-central sulcus were found. For

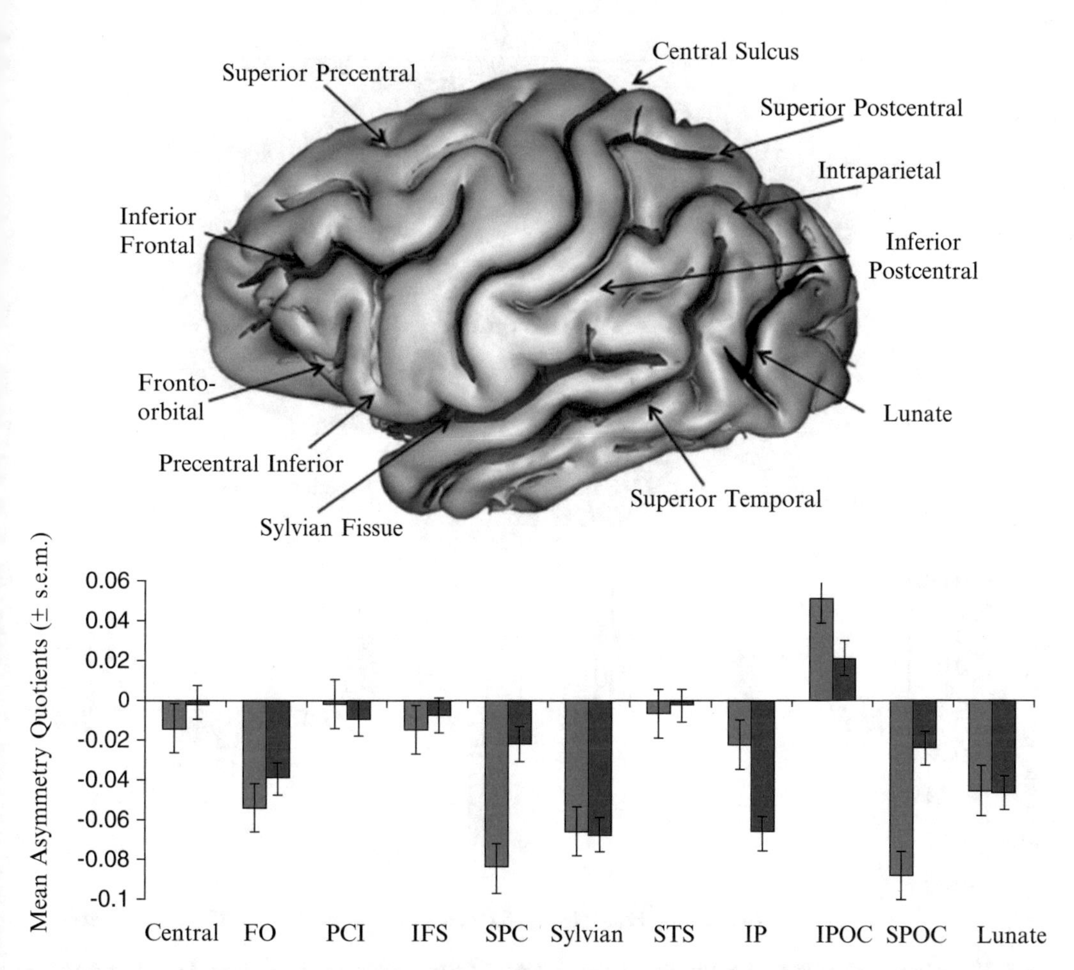

Fig. 5 *Upper panel*: 3D rendering of chimpanzee brain with sulci extracted and labeled. *Lower panel*: Mean asymmetry quotient (AQ) scores (±s.e.) for surface areas (*green*) and mean depths (*red*) for each sulcus

chimpanzee mean sulci depths, significant leftward asymmetries were found for the fronto-orbital, superior frontal, sylvian fissure, intraparietal, and lunate sulci (see Fig. 5). For the surface areas of rhesus monkeys, a significant rightward asymmetry was found for the superior temporal sulcus and no population-level biases were found for the mean depth measures (see Fig. 6).

2.3 Voxel-Based Morphometry

Rather than focus solely on ROI measures of brain asymmetry, recent methodological advancements now allow for whole-brain analysis of interhemispheric differences in gray or white matter on a voxel-by-voxel basis using voxel-based morphometry (VBM) [55]. Procedurally, from the standpoint of left-right differences, VBM approaches attempt to map individual brains into a common stereotaxic space, thereby aligning all subjects on a voxel-by-voxel basis in the lateral-medial (x), anterior-posterior (y), and dorsal-ventral (z) planes. After the brains are mapped into the coordinate

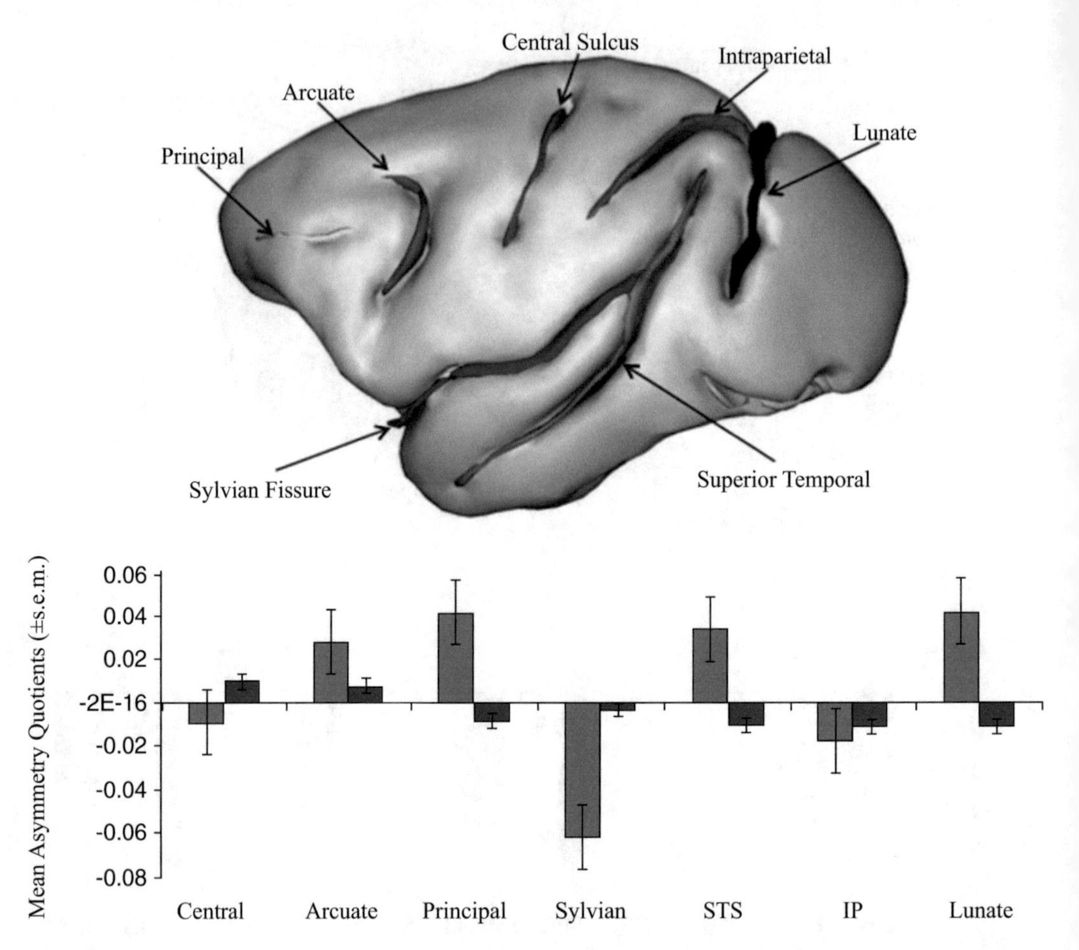

Fig. 6 *Upper panel*: 3D rendering of rhesus monkey brain with sulci extracted and labeled. *Lower panel*: Mean asymmetry quotient (AQ) scores (±s.e.) for surface areas (*green*) and mean depths (*red*) for each sulcus

space, the images are segmented into gray or white matter (most studies are done on gray matter) and typically the partial volumes are then saved for subsequent analysis. In the partial volumes, the values at each voxel represent the probability that it belongs to either gray (or white matter) and vary between 0% and 100% and these are essentially the dependent measures. Because individual brains vary in the amount of morphing that is necessary to place them into template space, the voxel probability values can be multiplied by the warping dimensions (called the Jacobian) to create estimates of volume at each voxel. Once all these steps have been accomplished, essentially one-sample *t*-tests are conducted on each voxel against a hypothetical value of zero (i.e., no difference in voxel intensity); the ones that differ significantly can be identified. It should be pointed out that because there are literally thousands of voxels in the brain (depending on imaging resolution and brain size), alpha must be adjusted for multiple comparisons to

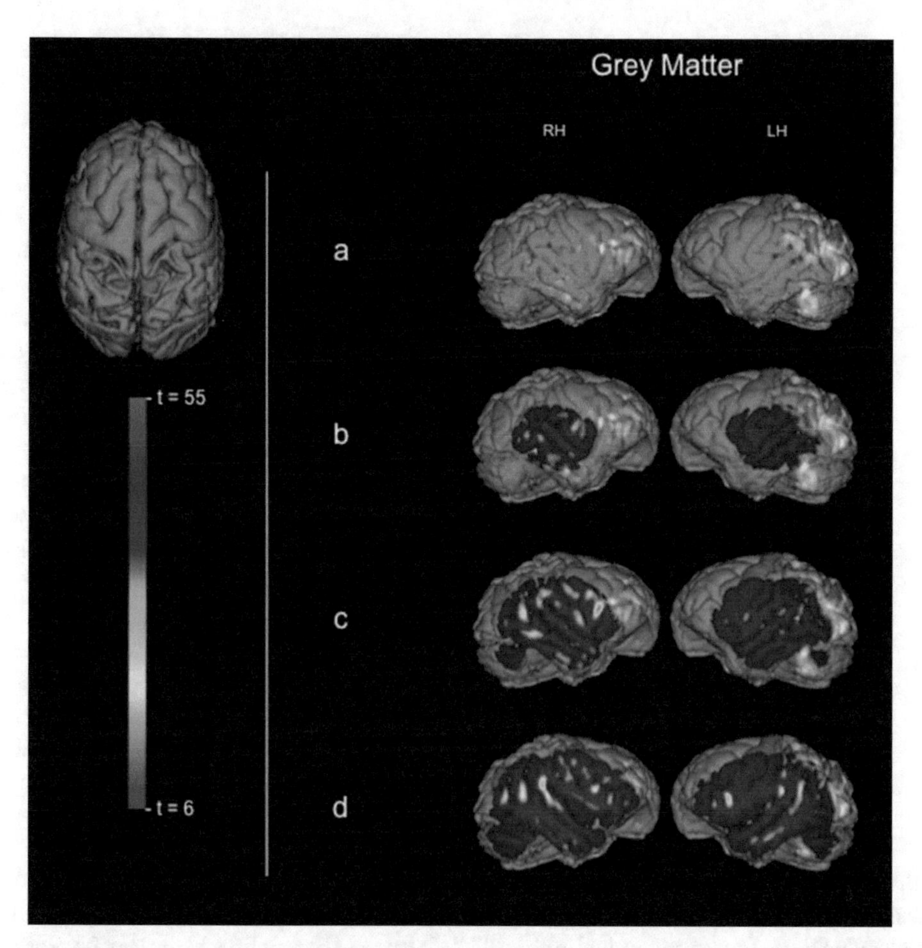

Fig. 7 (**a**–**d**) Left-right differences in gray matter location in chimpanzees as revealed by VBM at various lateral to medial planes. Reprinted from Hopkins et al. [56].

rule out potential Type I errors. To date, there are only a few studies on asymmetry using VBM in nonhuman primates [56, 57]. A 3D rendering of differences in the left and right hemisphere differences in gray matter in a sample of chimpanzees is shown in Fig. 7.

Though VBM is very powerful, there are a few challenges that merit discussion. First, a key feature of VBM is the creation of a species-specific and study-specific template for the analyses. In short, the successful registration of the images into a representative template brain for the species is important. Second, in the creation of the template, inherent asymmetries present in many or all individuals will be expressed in the template and some have suggested that to test for asymmetries, there is a need for the creation of a symmetric template. Third, many of the existing commercially available software programs for VBM analyses (e.g., FSL, SPM) are primarily designed for use with human brains. Thus, modification to existing scripts and changes to certain registration and segmentation parameters may be necessary for their successful use in nonhuman primates. For instance, for VBM and related analyses, the first step requires that the skull be removed from the image and the most common software tool, BET, does a very good job of automatically doing this with human and chimpanzee brains but, in our experience, it encounters troubles when applied to smaller brained species such as rhesus monkeys, baboons, or capuchin monkeys. Thus, manual segmentation or developing species-specific protocols may be needed in these cases. Fourth, segmentation of the T1-weighted MRI scans into gray matter, white matter, and CSF is an important step in VBM or related analyses. Many factors can influence segmentation including scanner magnet strength, signal to noise ratio, and inhomogeneities in the signal within the scan. Default values used in many algorithms in commonly used programs like FSL or SPM may not work best with all primate brains so it is important to evaluate the effectiveness of the segmentation process. Finally, inferential statistics can be performed on the VBM volumes such as tests for sex or handedness differences but if they are done on the template registered volumes, to some extent one might be limiting their ability to detect differences. This is because, in the registration process, the images are homogenized, in that, the goal is to remove individual variation. However, it is exactly this variation that one might hope to capture with certain types of tests.

2.4 Combining Structural and Probabilistic Methods

As many comparative neuroanatomists know, one of the challenges when using region-of-interest methods is characterizing the anatomical and sulcal features that define a region because there are considerable individual differences in sulcal morphology within and between species. For instance, in human brains, within the pars opercularis, a portion of Broca's area, some subjects show a small dimple within the gyrus but others do not and, it turns out that whether humans show a leftward asymmetry in this region depends on presence of the dimple [58, 59]. Likewise, in chimpanzees, the precentral inferior (PCI) sulcus is used to define the posterior border of the inferior frontal gyrus (IFG), which is the homolog to the pars opercularis in the human brain [28]. The PCI sulcus bifurcates in some individuals, thus whether the anterior or posterior limb should be considered the proper border for the IFG is not always clear [60, 61]. One approach that has been used is to create probabilistic morphological or cytoarchitectonic maps [62–66]. With respect to morphology, the basic method involves registering each individual MRI scan to a common template brain, thereby placing each individual brain into the same stereotaxic space. Once this is done, individual ROI masks are drawn on each registered brain. When each mask has been drawn, they can all be superimposed on each other, which displays the degree of overlap in the size of each mask across all subjects. The overlapping masks can then be thresholded at different levels to reflect the percentage of overlapping regions that are shared across all subjects. Higher thresholds will increasingly constrain the masks to those locations of the ROI that are most common to all subjects. Subsequently, rather than use individual ROIs drawn on native space MRI scans to quantify a region of interest, the probabilistic maps can be applied to segmented gray or white matter volumes to estimate the proportion of each tissue type that is found within the masks [67].

For instance, we have traced the IFG in the axial plane in a sample of 66 chimpanzees scanned at 3 T. The individual brains were registered to an average chimpanzee template brain in order to place them in the same stereotaxic space. We outlined the IFG in the left and right hemisphere of each brain using the fronto-orbital sulcus as the anterior border and the most anterior limb of precentral inferior sulcus as the posterior border. We then create a volume of the overlapping IFG masks for the left and right hemispheres and thresholded at 30%, 50% and 70%, reflecting the percentage of overlap in IFG voxels present in the mask within the sample (see Fig. 8). The volume of the 30%, 50%, and 70% masks were reduced with increasing threshold and this makes sense because, with increasing threshold values, one is constraining the most extreme voxels within the sample. We then applied the 30%, 50%, and 70% left and right hemisphere masks to the segmented gray matter volumes of each subject and computed the volume of gray matter voxels found within each masks. To adjust for overall gray matter brain volume, the volume of gray matter within each mask and hemisphere was divided by the total gray matter volume within the hemisphere and multiplied by 100. We then compared the interhemispheric differences in gray matter percentage. Shown in Fig. 9 are the mean asymmetry scores for chimpanzees at each threshold level. As can be seen, at lower thresholds with the lowest overlap, essentially no interhemispheric differences are found; however, for the highest threshold level (i.e., the regions that are most overlapping across subjects), a significant left hemisphere bias emerges.

A similar approach can be used to create probabilistic maps based not on morphology, but on cytoarchitectonics [62, 68]. For instance, Schenker et al. [62] quantified Brodmann areas 44 and 45, the constituent regions comprising Broca's area, in a sample of 12 chimpanzee postmortem brains. Before the brains were sectioned, they were scanned postmortem and registered to a common stereotaxic space. Following sectioning, masks reflecting the distribution of BA44 and BA45 cells for each individual brain were created on each subject's postmortem MRI scan. Because the postmortem MRI scans are aligned in the same space, the individual masks can be overlaid on each other and thresholded at different levels, creating probabilistic cytoarchitectonic maps. These cytoarchitectonic maps can then be applied to functional or anatomical

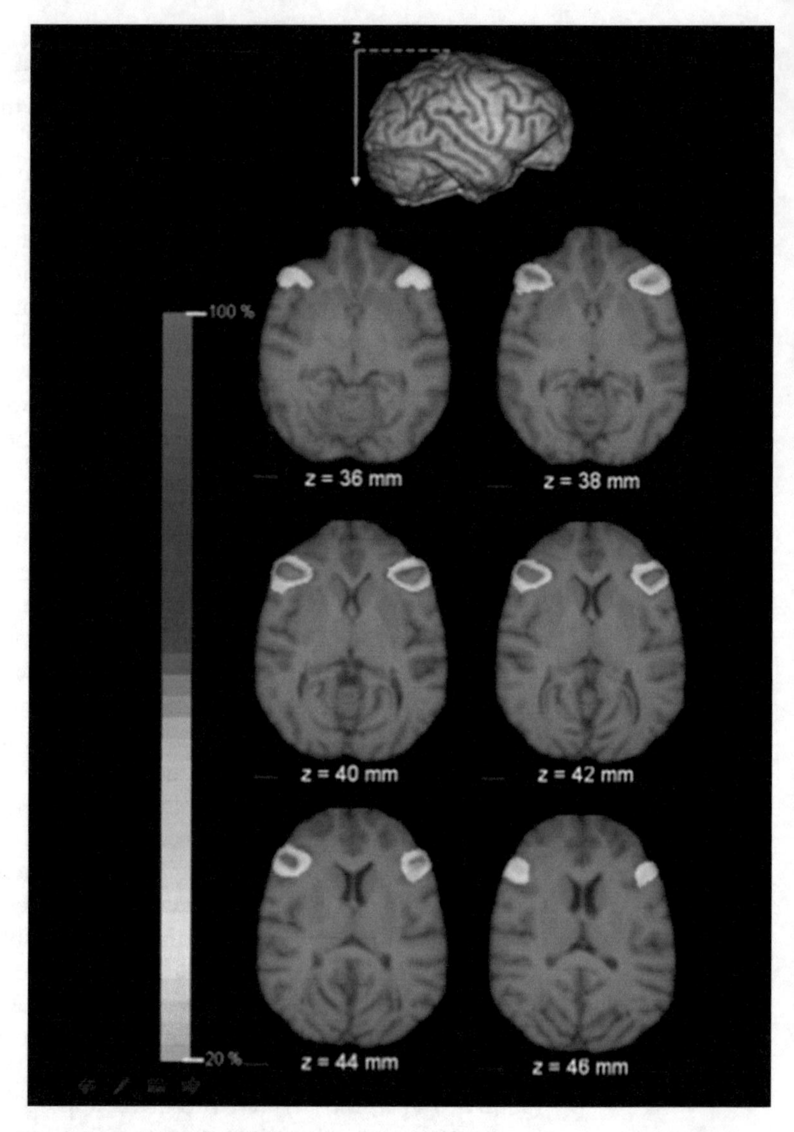

Fig. 8 Example probabilistic map of the inferior frontal gyrus in chimpanzees. The *different colors* reflect the extent of overlap in the voxels across subjects, which vary from 20 % (*orange*) to 100 % (*purple*)

images to quantify changes in metabolic and physiological activity within the region of interest [64].

2.5 Shape Asymmetries (Petalia)

The human brain exhibits a right-frontal, left-occipital shape or torque asymmetry (sometimes called a petalia asymmetry) [69]. Although the mechanism and functional significance underlying this asymmetry remains unclear [70, 71], whether nonhuman primates exhibit similar patterns has been the topic of several studies. Early worked attempted to quantify petalia asymmetries by observing the shape of the brain from endocasts created from the internal

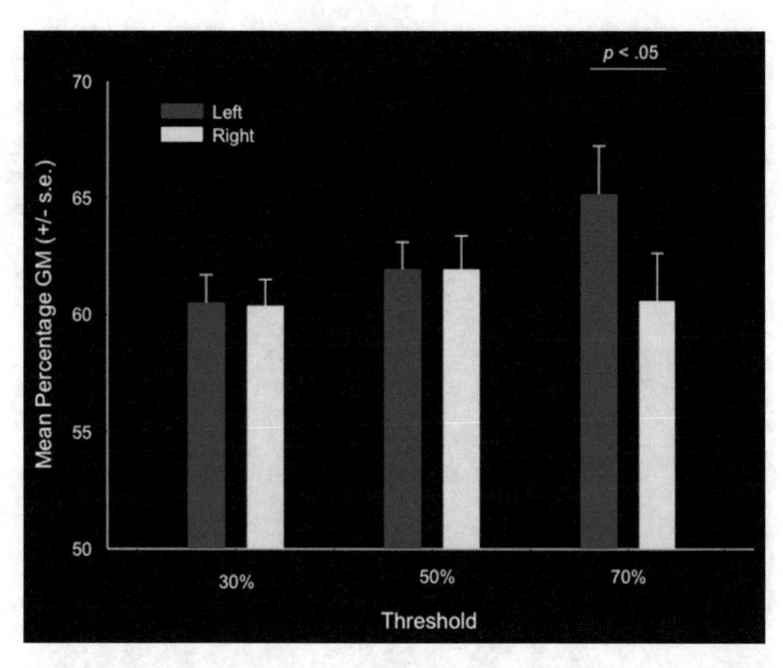

Fig. 9 Percentage of left (*blue*) and right (*yellow*) hemisphere gray matter at 30 %, 50 %, and 70 % overlap in the IFG probabilistic maps

cavity of the skull [8, 12, 72, 73]. More recently, with the advent of 3D computerized tomography (CT), the ability to characterize brain asymmetries from skulls has advanced significantly. In a novel study, Balzeau and Gilissen [74], quantified the petalia asymmetries of a sample of bonobo (*Pan paniscus*), chimpanzees, and gorilla (*Gorilla gorilla*) skulls. Though there was some species variation, there was a small but significant right frontal, left occipital asymmetry in all three species that was similar to results found in humans [75].

In addition to the use of CT, others have quantified the torque asymmetries using ROI and VBM approaches in human and, to a lesser extent, nonhuman primate brains [25, 76–80]. For instance, Hopkins et al. [56] reported a right frontal, left occipital gray matter asymmetry based on a VBM analysis in a sample of chimpanzees. Similarly, Pilcher et al. [25] quantified the volume of left and right frontal and occipital poles in a sample of chimpanzees and similarly found a right frontal, left occipital torque asymmetry. We have recently combined these two methods to assess consistency in torque asymmetries in chimpanzees. In a sample of 91 chimpanzee MRI scans, to assess the volume of the ROIs, the MRI scans were aligned in the sagittal, coronal, and axial planes using standard anatomical landmarks and cut into 1 mm coronal slices using multiplanar reformatting software (ANALYZE 11.0). The first slice anterior to the genu of the corpus callosum was identified and served as the posterior border of the frontal region. The sagittal midpoint

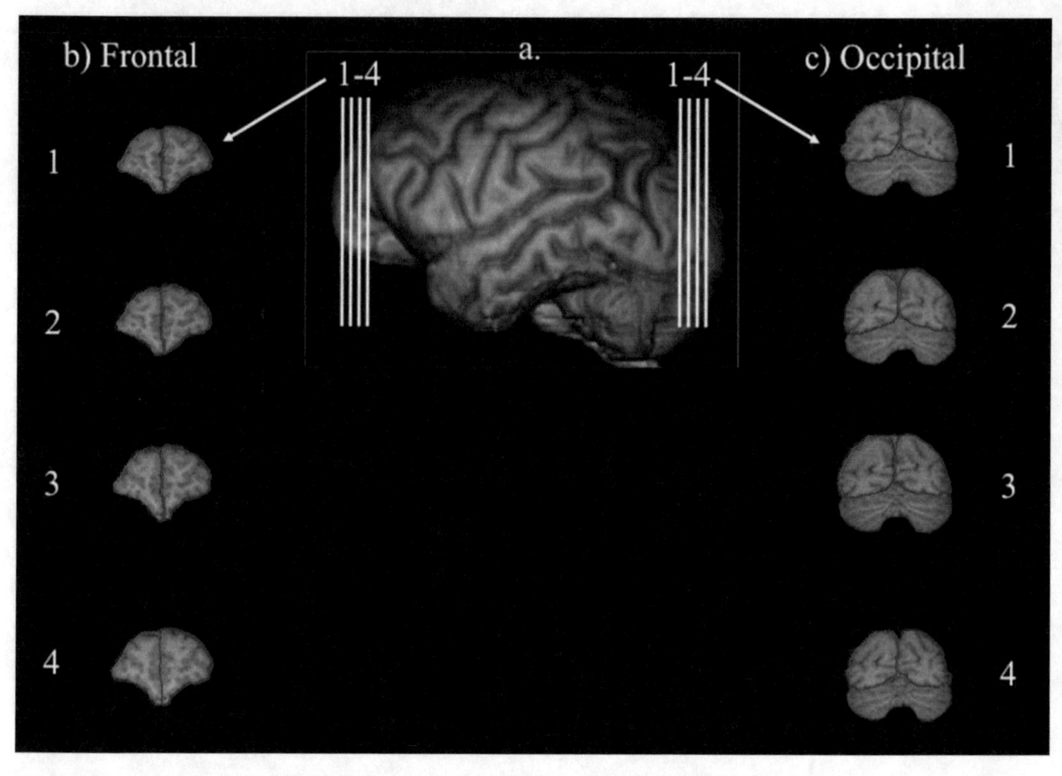

Fig. 10 (**a**) 3D rendering of a chimpanzee brain illustrating the sections measured from the anterior and posterior poles. (**b**) Coronal views and example tracing of the left (*blue*) and right (*red*) frontal lobe areas. (**c**) Coronal views and example tracing of the left (*blue*) and right (*red*) occipital areas.

between the posterior border of the corpus callosum (splenium) and the lateral surface of the occipital pole approximated the anterior border of the posterior region (see Fig. 10). Using these two borders, 1 mm successive coronal slices were obtained until reaching the end points of the frontal and occipital poles, respectively (see Fig. 10). Each 1 mm coronal slice was traced using a freehand tool in ANALYZE 11.0. Within the frontal and occipital regions, the total number of 1 mm slices was divided in half to derive separate measures of the anterior and posterior regions. Frontal 1 and 2 corresponded to the anterior and posterior regions within the frontal pole. Posterior 1 and 2 corresponded to the anterior and posterior regions of the occipital region (although more than occipital lobe tissue was included in these regions). Sub-cortical structures were not included in the cerebral volumetric measures. Two independent raters, blind to the handedness of the apes and the hypotheses of the study, measured the torque in ten randomly selected specimens and the coefficients of agreement were .981 and .994 for the left and right hemispheres, respectively.

For the VBM method, each individual MRI scan was coregistered to a chimpanzee template described above using three-dimensional voxel registration with a linear transformation (Analyze

Table 1
Mean asymmetry scores (±s.e.) for combined ROI and VBM approach to measure of torque asymmetry in chimpanzees

	Region			
	Frontal 1	**Frontal 2**	**Occipital 1**	**Occipital 2**
ROI	.039	.027	−.036	−.082
s.e.	.012	.006	.010	.022
VBM	42.64	25.84	−40.56	−22.95
s.e.	16.40	16.11	16.67	19.80

11.0, Mayo Clinic). The MRI scans were then segmented into gray, white, and CSF tissue using FSL (Analysis Group, FMRIB, Oxford, UK) [81] and low-pass filtered with a 5 mm kernel. For this study, we only used the segmented gray matter (GM) volumes. Subsequently, each individual subject's segmented GM volume was then flipped 180° in the left-right axis to create a mirror-image brain volume. Each of these flipped volumes was then re-registered to the chimpanzee template. For each ape, the mirror image GM volume was then subtracted from the normally-oriented volume to create a "subtracted" or "reflected" volume. We created four object maps that were drawn on the chimpanzee template using the same landmarks that had been used in the ROI analysis (see above). The object maps corresponded to the two frontal and two posterior regions that had been quantified in the ROI analysis. Because we were using the subtracted volumes, the objects were applied to the right hemisphere in order to derive a single quantitative measure of asymmetry. The total number of positive and negative voxels was summed within each object. Because the objects were placed on the right hemisphere, positive values reflected right hemisphere biases in gray matter and negative values reflected left hemisphere biases.

For both the ROI $F(3, 267) = 11.52$, $p < .001$and VBM $F(3, 189) = 6.88$, $p < .001$ approaches, we found significant differences (see Table 1). For the ROI method, significant population-level rightward asymmetries were found for frontal regions 1 $t(90) = 3.29$, $p < .001$ and 2 $t(90) = 4.42$, $p < .001$ and significant population-level leftward biases were found for the posterior regions 1 $t(90) = -3.06$, $p < .001$ and 2 $t(09) = -2.83$, $p < .01$. For the VBM analysis, population-level rightward asymmetries were found for frontal region 1 $t(64) = 2.83$, $p < .01$ and population-level leftward biases were found for posterior region 1 $t(64) = 3.34$, $p < .01$. A trend toward leftward asymmetry was found for the posterior region 2 $t(64) = 1.90$, $p < .06$. Pearson Product Moment correlations were performed between the four cerebral torque asymmetry measures for both the ROI and VBM methods. To minimize Type I error due to multiple comparisons, we averaged the two frontal and two occipital measures for each technique into single measures of asymmetry and correlated the resulting values. For both the frontal ($r=.526$, df=89, $p<.001$) and occipital regions ($r=.430$, df=89, $p<.001$), significant positive correlations were found between the AQ measures derived from the ROI volumetric and VBM techniques.

3 Anatomical and Functional Connectivity

The ROI, measurement of sulcal length, and VBM approaches discussed above are all techniques largely used to identify structural brain asymmetries in overall volume or gray matter. Now we turn our attention to techniques used to investigate asymmetries in white matter connectivity, functional connectivity, and functional activation.

3.1 Diffusion Tensor Imaging (DTI)

DTI allows for the in vivo study of structural connectivity, or the anatomical connections between brain regions. Diffusion MRI measures the displacement of water molecules; in white matter water displacement is anisotropic, with water diffusion faster along white matter fibers rather than perpendicular to these fibers. Diffusion anisotropy measures the difference between these two directions of water motion. Combining diffusion anisotropy and the principle diffusion characteristics, tractography programs (such as FSL) are able to estimate the white matter connectivity patterns [82]. In addition to tractography, several quantitative measures can be obtained from DTI that are of considerable usefulness to researchers. Fractional anisotropy (FA), the normalized standard deviation of the diffusivities, is one of the most commonly reported measures of white matter architecture [83]. While FA is not a direct measure of myelination, it is believed to reflect anatomical features of white matter such as fiber density and myelination [82]. Mean diffusivity (MD), the average of the eigenvalues, is another quantitative measure obtained through DTI and provides information about tissue structure. Some believe MD may allow a better understanding of how the diffusion tensor is changing [82]. Many studies are beginning to report both of these quantitative measures.

Only a few studies have used DTI to investigate lateralization of white matter pathways in nonhuman primates, and those that have been conducted tend to focus on questions concerning evolution of the fiber tract pathways or neural substrates underlying handedness and tool use. Rilling et al. [84] used diffusion imaging to investigate the evolution of arcuate fasciculus, a white matter tract involved in human language, in humans, chimpanzees, and rhesus macaques. The arcuate fasciculus was found to have strong connections to the medial temporal gyrus and the inferior temporal gyrus in humans but not chimpanzees or macaques. These gyri are typically included in Wernicke's area, which suggests that new connections were made in the human lineage to support the demands of language. More recently, in a follow up study with a larger sample size, Rilling et al. [85] examined asymmetries in the dorsal and ventral white matter language pathways in a sample of 26 female humans and chimpanzees. For both species, significant population-level leftward asymmetries were found in the dorsal but not ventral pathways; however, within each hemisphere, the ratio in size of the dorsal to ventral pathways was significantly greater in humans. Similarly, Hecht et al. [86] compared the superior longitudinal fasciculus (SLF) of humans and chimpanzees. The SLF provides frontoparietal connectivity, and a branch of the SLF (SLF III) is related to representation of action sequences and goals, such as in complex tool use. Humans were shown to have a right-lateralized SLF III whereas chimpanzees did not show this lateralization. Additionally, in humans but not chimpanzees, the SLF III showed an asymmetry in the frontal terminations to the inferior frontal gyrus with greater connectivity in the right hemisphere.

Li et al. [87] investigated asymmetry in the corticospinal pathway (CST) and handedness in female chimpanzees. Their sample of chimpanzees showed a left hemisphere bias in FA and a right hemisphere bias in MD of the corticospinal tract, but these were not correlated with handedness. Li et al. [87] also found that the terminal location of the CST pathways was more posterior in the left compared to right hemisphere.

Phillips et al. [88] examined performance asymmetries in tool use in both male and female chimpanzees and related this performance to organization of the corpus callosum (CC) as informed by regional FA values. In their study, the strength of hand preference was positively correlated with FA of the anterior-most region of the CC. Thus, the authors suggested that more strongly lateralized chimpanzees had greater interhemispheric communication across the prefrontal cortex. Furthermore, asymmetry in performance was correlated with FA in the region of the CC, which connects temporal, parietal and occipital cortices, with more strongly lateralized individuals having greater FA in this region.

Iturria-Medina et al. [89] used diffusion-weighted imaging to compare asymmetries in structural network properties in humans and one nonhuman primate, a rhesus macaque. Using graph theory framework, the authors reported that in both primate species the right hemisphere was more efficient and interconnected than the left hemisphere. The left hemisphere was described as possessing more "indispensable regions" for whole-brain networks—that is, left hemisphere pathways were considered to be crucial to cognitively demanding specific processes such as language, whereas the right hemisphere pathways were considered to be more specific for general processing and integration. Overall, these studies indicate that characteristic of white matter integrity and pathways, as informed by DTI, contribute to the complexity of our understanding of the relationships between behavioral laterality and brain organization and function.

Several fiber tractography algorithms exist for the analysis of DTI data. Several preprocessing steps are required including motion correction and eddy current correction. Seed-based analysis, including probabilistic tractography from seed regions [90, 91] using FSL's FMRIB's Diffusion Toolbox (FDT) software package, can then be used. In this analysis, an index of connectivity is assigned for each brain voxel; this index represents the number of generated paths that pass through the voxel from the seed region. Tract-based statistical (TBSS) is another technique that can be used to analyze parameters about pathways, performing voxel-wise comparisons, such as comparing the FA in right and left hemisphere pathways. In TBSS, all subjects' FA data are projected onto a mean FA tract skeleton, and then voxelwise cross-subject statistics are performed [92]. Another analysis technique is graph analysis, which can be used to calculate an estimation of the efficiency of a network [89]. This analysis assigns each brain voxel a probability of connection with each seed region.

3.2 Resting-State fMRI (R-fMRI)

Functional MRI (fMRI) is an indirect measure of brain activity. Blood-oxygen-level-dependent (BOLD) contrast imaging is commonly used; it is based on the assumption that changes in regional blood flow and metabolism are linked to changes in regional neuronal activity. fMRI typically uses a task-based or stimulus-driven paradigm so that processing can be related to patterns of brain activity, as quantified by the percent signal change in the BOLD response in specific regions of the brain. As is implied, in order to conduct an fMRI experiment, the subject typically must be alert, awake, and interactive in the magnet.

Functional connectivity refers to the temporal dependency between spatially remote neurophysiological events [93, 94], or how brain regions work together. Even at "rest", the brain is continually engaged in information processing and functional connectivity among multiple brain regions [95–97]. Resting-state fMRI (R-fMRI) is the application of the fMRI technique to the "at rest" brain. R-fMRI is a relatively new method used to evaluate the functional connectivity of the brain, which is inferred from spontaneous BOLD signal fluctuations arising from low frequency (<0.1 Hz) brain activity. Spatially distinct regions with synchronized brain activity comprise resting-state networks. Task-activated functional networks are strongly correlated with nontask activated, or "at rest" networks [98]. Thus, these networks of the brain continue to process information and retain functional connectivity in the resting state. This technique can be used to study functional connectivity and lateralized connectivity networks in nonhuman primates.

In contrast to conventional fMRI, the R-fMRI technique does not require the subject to be awake, alert, or performing a task while in the magnet. Perhaps surprisingly, as this technique clearly has the potential to provide tremendous insight and understanding of comparative brain function, R-fMRI experiments on nonhuman primates are relatively scarce and the majority of these investigations have been conducted on macaques [99–104]. Comparison of functional connectivity networks in humans and macaques has identified six networks (sensorimotor, executive control, auditory, visual, default mode, and attention) that are markedly conserved across the two species [100, 102, 105, 106].

Functional connectivity networks have also been studied in New World monkeys and great apes [99, 104, 107]. Belcher et al. [99] investigated R-fMRI functional connectivity networks in common marmosets (*Callithrix jacchus*) that were habituated to restraint after 3 weeks of daily training. Distinct large-scale networks that corresponded to networks frequently described in humans and anesthetized Old World monkeys were identified. Additionally, as anesthesia was not used, the subjects were truly awake while undergoing scanning—thereby coming the closest to modeling the conditions under which a human R-fMRI is obtained—and potential confounds from anesthesia were avoided (see below for a discussion of anesthesia-related confounds).

To date, only one investigation has specifically addressed lateralized functional connectivity networks in nonhuman primates. Wey et al. [104] examined these networks in humans, chimpanzees, baboons (*Papio hamadryas*), and capuchin monkeys (*Cebus apella*). The results revealed that all species displayed strikingly similar functional right and left-lateralized networks involving the frontal and parietal lobes (see Fig. 11). However, only humans had a left-lateralized network that involved connectivity of the frontal and parietal cortices. In nonhuman primates these networks were split into separate lateralized networks. In humans, the left lateralized inter-lobar fronto-parietal connectivity is associated with higher aspects of cognition including attention and language. While the nonhuman primates in this study were lightly anesthetized (which may confound the hemodynamic response and functional connectivity networks, see discussion below), this is the only study to date to investigate lateralized functional networks in primates.

There are two key methodological considerations of which one needs to be aware—subject-related confounds and the effects of anesthesia on connectivity networks. Subject-related confounds include motion artifacts, and the effects of cardiac and respiratory

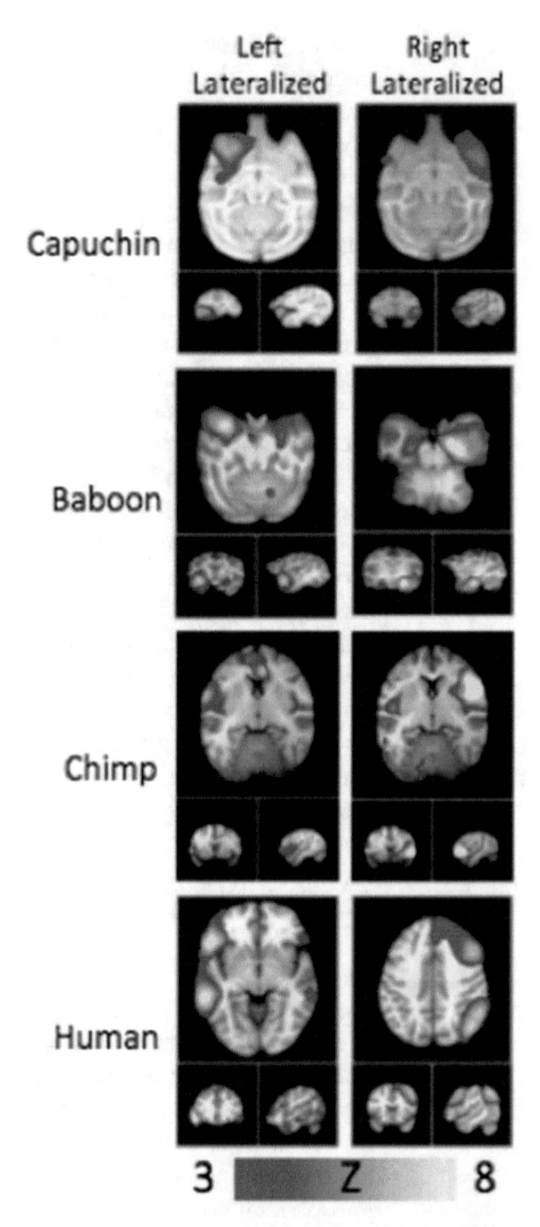

Fig. 11 Unilateral intrinsic connectivity networks across primate species. Left- and right-lateralized ICNs that correspond top-down cognitive functionality. In humans, the left lateralized fronto-parietal network is associated with speech and language processing, while the right lateralized fronto-parietal network is associated with reasoning, attention, inhibition, and working memory. In contrast to bilateral ICNs, these networks are primarily confined to a single node in non-human primates, while activity is multinodal and spans the frontal and parietal lobes in humans. Reprinted from Wey et al. [104]

noise. Cardiac and respiratory processes are frequently recorded during acquisition of R-fMRI, and estimates of how these fluctuations influence the BOLD response can be modeled. However, if these physiological recordings are not able to be obtained during scan acquisition, several methods exist to estimate and remove the confounds. These methods are standard preprocessing techniques in all fMRI data analyses, regardless of whether physiological recordings and head restraints (to reduce motion) have been used. Noise removal methods include a bandpass filter of the data, which will remove fluctuations outside the frequency range of interest, so that these do not affect the analysis. Typically frequencies less than 0.01 Hz and greater than .1 Hz are filtered out [108]. Motion correction is also a standard preprocessing technique used in both R-fMRI and fMRI analyses. For a thorough discussion of these and other subject-related confounds and techniques for removal, the reader is referred to Murphy et al. [108] and Teichert et al. [109].

Acquiring R-fMRI from restrained, alert, and truly awake animals is ideal as it most closely models conditions under which human R-fMRI are acquired. Several research groups have successfully habituated marmosets and Old world monkeys to restraint and the noise of a magnet for acquisition of fMRI [98, 99, 107, 110–112]. However, this is not always feasible. Thus, it is important to understand the specific effects different anesthesia agents and doses have on functional connectivity networks in primates.

Hutchison et al. investigated different isoflurane levels on distributed connectivity networks in macaques [100]. Dose-dependent effects on interhemispheric cortical functional connectivity networks were found at moderate and high dosages. Stable functional connectivity networks were present at isoflurane dosages between 1.0% and 1.5%. In a study of marmosets, compared activity in the somatosensory pathway and connectivity networks acquired from awake marmosets with those acquired while under propofol anesthesia [107]. A mild electrical stimulation of the median nerve was administered to assess the integrity of the somatosensory pathway. In awake marmosets, a robust BOLD response was seen throughout the somatosensory pathway, including the thalamus and somatosensory cortex. The BOLD response was detected in this pathway in anesthetized marmosets, but under both low and high doses of propofol there was significant attenuation of the response in all regions. Awake marmosets had a stronger amplitude of the fMRI response compared to anesthetized marmosets. The effects of the propofol anesthesia on connectivity networks within the somatosensory cortex was region specific. Intrahemispheric connectivity between primary somatosensory cortex (SI) and secondary somatosensory cortex (SII) was reduced with high propofol dose, but interhemispheric connectivity of SI, and interhemispheric connectivity of SII was not reduced.

While there are several approaches used in the analysis of R-fMRI data, two are most commonly used: a seed-based approach (e.g. Vincent et al. [102]) and independent components analysis (ICA). In the seed-based approach, ROIs are selected and the average BOLD time course of voxels within those ROIs are correlated with each other and with the time courses of all the other voxels in the brain. For example, if a researcher is interested in the functional connectivity of the left planum temporale, a resting-state time-series analysis of that selected seed region would be correlated with the time-series of all other voxels, therein creating a functional connectivity map. The resulting analysis would indicate which brain voxels show similar time-series and thus are functionally connected. This is a model-dependent method [113], as the experimenter has to make a priori decisions about which regions to analyze.

Independent components analysis (ICA) analyses functional connectivity networks in a data-driven fashion, which avoids potential bias from a priori hypotheses [105]. ICA is thus a model-free method for the analysis for R-fMRI data. In probabilistic ICA, the entire group data is concatenated, forming a single string of data. Group-wide 4D fMRI data are then decomposed into spatial and temporal components based on the relevant time course. These spatial maps are then thresholded and registered to a standard template brain (which will likely need to be made in-house for nonhuman primate species). The process results in the output of a spatial map that represents functional connectivity as well as the output of the relevant time course. A key consideration in analysis of R-fMRI data is the applied number of components in the ICA, as the outcome depends strongly on the applied number of components, which may modify the networks that are identified [114].

The use of nonhuman primates in resting-state investigations has tremendous translational potential [115]. The application of R-fMRI to understanding how the brain is altered in neurological disease has the potential to increase the translation of fMRI into clinical care. One example of this advantage concerns R-fMRI circumventing task-related confounds. Such confounds may arise if subjects are unable to perform various cognitive or behavioral tasks. If this were to occur, R-fMRI could circumvent this confound, as resting-state networks are correlated with task-related functional networks.

3.3 Resting-State and Functional Imaging Using Positron Emission Tomography (PET)

As noted above, several laboratories have developed fMRI in Old and New World monkeys for the purposes of examining neurofunctional correlates of different sensory, perceptual, and cognitive processes. An alternative functional imaging approach used with nonhuman primates has been positron emission tomography (PET). Like fMRI, PET is a method for visualizing neural activity that primary works by tagging radioactive isotopes to normally existing, active biological chemicals. For example, flurodeoxyglucose (FDG) is a radioisotope that attaches to glucose, which is used by the brain when engaged in different processes. In the typical PET paradigm, subjects are initially dosed with a radioactive ligand and subsequently engaged in some type of behavioral task during the uptake period. At the end of the uptake period, which can vary depending on the ligand, the subjects are usually anesthesized and then scanned in the PET machine. The PET machine detects the release of the positrons that are released by the breakdown of the radionuclide, which can be detected by the camera and aligned to the spatial coordinates of the brain.

PET has been used in both monkeys and chimpanzees to measure both resting-state metabolic activity and functional activation. Rilling and colleagues [116] gave eight humans and five chimpanzees FDG and measured uptake while the subjects were at rest (i.e., not engaged in any specific tasks). Humans and chimpanzees showed similar resting-state activation, though some small differences were noted. Notably, both species showed large, bilateral activation of the medial prefrontal cortex and cingulate cortex. Bilateral activation was also found in dorsolateral and rostrolateral cortex, though this effect appeared more left lateralized in humans compared to the chimpanzees. In contrast to humans and chimpanzees, Kojima et al. [117] examined resting state in three monkeys using O^{15} as the ligand and found very few regions of activation at rest in macaque monkeys and none of these regions overlapped with the data from humans and chimpanzees. Notwithstanding, Kokjima et al. [117] did find that when the monkeys were given different cognitive tasks, the relative amounts of activation in medial cortical regions were significantly higher in the at-rest compared to activation condition, suggesting that these regions are part of the default system. Similarly, in chimpanzees, Barks et al. [118] compared PET activation in four chimpanzees during a social discrimination task and when at rest and also found deactivation of cortical midline regions, consistent with other evidence for a default network.

Parr et al. [119] compared PET activation in five chimpanzees when performing a matching-to-sample task with two different types of stimuli, faces, and objects. With respect to face processing, a number of bilateral cortical regions were activated during this task and there were very few asymmetries. The most robust was in the right ventromedial orbitofrontal cortex. Taglialatela et al. [120] imaged three chimpanzees while producing manual gestures and producing oro-facial movements during a communication task. As in the Parr et al. [119] study, a number of bilateral regions were activated in the cortex. With respect to laterality, leftward asymmetries were found for the pre- and post-central gyri, left inferior frontal gyrus, caudate/putamen, orbital gyrus, parahippocampus, and inferior temporal gyrus. Rightward asymmetries were found for the middle temporal gyrus, superior parietal and frontal gyri, and right middle frontal gyrus. In contrast to communication production, Taglialatela et al. [121] also quantified activation in three chimpanzees using PET in response to two different classes of chimpanzee vocalizations, including broadcast and proximal calls. As a baseline condition, the chimpanzees were presented with chimpanzees sounds played backwards. Proximal calls elicited much more robust patterns of cortical activation than distal calls;

however, for both calls, predominantly rightward asymmetries were found, particularly throughout the middle and superior temporal gyri.

Finally, there are two studies using PET to visualize auditory perception in macaque monkeys. Gil-da-Costa et al. [122] recorded PET activity in response to two classes of vocalizations in macaques, screams and coos in three rhesus monkeys. For the baseline condition, PET activation was recorded in response to nonbiological sounds. Overall, significant activation was found in the ventral premotor and posterior superior temporal gyri in all three monkeys but no consistent laterality effects were found across subjects. Finally, Poremba et al. [123] measured PET activation in response to different classes of auditory stimuli, including species-specific vocalizations, in eight neurologically intact macaque monkeys. In terms of laterality, they found that rightward asymmetries in the superior portion of the temporal lobe but leftward asymmetries in the temporal pole for responding to species-specific vocalizations. These patterns of asymmetry were not found for other classes of auditory stimuli.

4 Summary

In conclusion, there are a number of noninvasive imaging technologies that can be used with nonhuman primates to characterize anatomical and functional asymmetries. ROI and VBM are the most reasonable methods for quantifying morphological asymmetries while DTI and related methods offer great potential for measuring laterality in white matter connectivity. Resting-state fMRI provides the opportunity to quantify asymmetries in functional connectivity while fMRI and PET imaging are powerful methods for assessing real-time functional asymmetries. With this said, the use of these technologies has not been fully developed in nonhuman primates, at least with respect to studies with large sample sizes. Indeed, for R-fMRI, fMRI, and PET, for pragmatic reasons, the sample sizes are very small and any definitive claims of population-level asymmetry based on the extant findings warrants considerable caution.

Despite the appeal of using noninvasive imaging technologies, there are challenges in implementing their use with nonhuman primates. Notably, a majority of the software developed for use with different types of images (i.e., anatomical or functional) has been developed for use with human subjects. Thus, there is some modification of the software that is necessary for it to work effectively with nonhuman primate brains. This is not an insurmountable problem but attempts to use existing programs without a working knowledge of what the different software operations perform will lead to problems. Thankfully, there are an increasing number of scientists who are using the available software with nonhuman primates and many have made their code available or described in great detail what software modifications were necessary to make the software tools functional for their specific species of interest.

Advances in technology have been instrumental in fostering scientists' ability to characterize individual and phylogenetic differences in cortical organization since the conception of the field of neuroscience. A mere 40 years ago, structural MRI scanning was in its infancy and we have witnessed considerable advancement in the application of MRI and other imaging technologies since that time. We expect that in the next 40 years, even better methods and technologies will be developed that will further our understanding of cortical organization, including anatomical and functional asymmetries.

Acknowledgement

This research was supported in part by NIH grants NS-42867, NS-73134, HD-38051, and HD-56232 to W.D.H. and NS-70717 to K.A.P.

References

1. Sherwood CC, Baurernfeind AL, Bianchi S, Raghanti MA, Hof PR (2012) Human brain evolution writ large and small. In: Hofman MA, Falk D (eds) Progress in brain research. Elsevier, New York
2. Deacon TW (1997) The symbolic species: the coevolution of language and the brain. W. W. Norton and Company, New York
3. Rilling JK (2006) Human and non-human primate brains: are they allometrically scaled versions of the same design? Evol Anthropol 15(2):65–77
4. Schoenemann PT (2006) Evolution of size and functional areas of the human brain. Annu Rev Anthropol 35:379–406
5. Heilbronner PL, Holloway RL (1988) Anatomical brain asymmetries in New World and Old World monkeys. Stages of temporal lobe development in primate evolution. Am J Phys Anthropol 76:39–48
6. Heilbronner PL, Holloway RL (1989) Anatomical brain asymmetry in monkeys: Frontal, temporoparietal, and limbic cortex in *Macaca*. Am J Phys Anthropol 80:203–211
7. Falk D, Cheverud J, Vannier MW, Conroy GC (1986) Advanced computer-graphics technology reveals cortical asymmetry in endocasts of rhesus-monkeys. Folia Primatol 46:98–103
8. Falk D, Hildebolt C, Cheverud J, Vannier M, Helmkamp RC, Konigsberg L (1990) Cortical asymmetries in the frontal lobe of rhesus monkeys (*Macaca mulatta*). Brain Res 512:40–45
9. LeMay M (1976) Morphological cerebral asymmetries of modern man, fossil man and nonhuman primates. Ann N Y Acad Sci 280:349–366
10. LeMay M (1977) Asymmetries of the skull and handedness. J Neurol Sci 32:243–253
11. LeMay M (1982) Morphological aspects of human brain asymmetry: an evolutionary perspective. Trends Neurosci 5:273–275
12. LeMay M (1985) Asymmetries of the brains and skulls of nonhuman primates. In: Glick SD (ed) Cerebral lateralization in nonhuman species. Academic, New York, pp 223–245
13. LeMay M, Geschwind N (1975) Hemispheric differences in the brains of great apes. Brain Behav Evol 11:48–52
14. Zilles K, Palomero-Gallagher N, Amunts K (2013) Development of cortical folding during evolution and ontogeny. Trends Neurosci 36(5):275–284
15. Hecht E, Stout D (2015) Techniques for studying brain structure and function. In: Bruner E (ed) Human paleoneurology. Springer International Publishing, Switzerland

16. Gannon PJ, Kheck N, Hof PR (2008) Leftward interhemispheric asymmetry of macaque monkey temporal lobe language area homolog is evident at the cytoarchitectural, but not gross anatomic level. Brain Res 1199:62–73
17. Yeni-Komshian G, Benson D (1976) Anatomical study of cerebral asymmetry in the temporal lobe of humans, chimpanzees and monkeys. Science 192:387–389
18. Gannon PJ, Holloway RL, Broadfield DC, Braun AR (1998) Asymmetry of chimpanzee Planum Temporale: humanlike pattern of Wernicke's language area homolog. Science 279:220–222
19. Semendeferi K, Damasio H (2000) The brain and its main anatomical subdivisions in living hominids using magnetic resonance imaging. J Hum Evol 38:317–332
20. Semendeferi K, Damasio H, Frank R, Van Hoesen GW (1997) The evolution of the frontal lobes: a volumetric analysis based on three dimensional reconstructions of the magnetic resonance scans of human and ape brains. J Hum Evol 32:375–388
21. Semendeferi K, Lu A, Schenker NM, Damasio H (2002) Humans and great apes share a large frontal cortex. Nat Neurosci 5(3):272–276
22. McBride T, Arnold SE, Gur RC (1999) A comparative volumetric analysis of the prefrontal cortex in human and baboon MRI. Brain Behav Evol 54:159–166
23. Rilling JK, Insel TR (1998) Evolution of the cerebellum in primates: differences in relative volume among monkeys, apes and humans. Brain Behav Evol 52:308–314
24. Fears SC, Scheibel K, Abaryan Z, Lee C, Service SK, Jorgensen MJ, Fairbanks LA, Cantor RM, Freimer NB, Woods RP (2011) Anatomic brain asymmetry in vervet monkeys. PLos One 6(12):e28243
25. Pilcher D, Hammock L, Hopkins WD (2001) Cerebral volume asymmetries in non-human primates as revealed by magnetic resonance imaging. Laterality 6:165–180
26. Hopkins WD, Misiura M, Reamer LA, Schaeffer JA, Mareno MC, Schapiro SJ (2014) Poor receptive joint attention skills are associated with atypical grey matter asymmetry in the posterior superior temporal gyrus of chimpanzees (*Pan troglodytes*). Front Cogn 5(7):1–8
27. Cantalupo C, Hopkins WD (2001) Asymmetric Broca's area in great apes. Nature 414:505
28. Keller SS, Roberts N, Hopkins WD (2009) A comparative magnetic resonance imaging study of the anatomy, variability and asymmetry of Broca's area in the human and chimpanzee brain. J Neurosci 29:14607–14616
29. Phillips K, Sherwood CC (2005) Primary motor cortex asymmetry correlates with handedness in capuchin monkeys (*Cebus apella*). Behav Neurosci 119:1701–1704
30. Hopkins WD, Cantalupo C (2004) Handedness in chimpanzees is associated with asymmetries in the primary motor but not with homologous language areas. Behav Neurosci 118:1176–1183
31. Hopkins WD, Russell JL, Cantalupo C (2007) Neuroanatomical correlates of handedness for tool use in chimpanzees (*Pan troglodytes*): implication for theories on the evolution of language. Psychol Sci 18(11):971–977
32. Cantalupo C, Freeman HD, Rodes W, Hopkins WD (2008) Handedness for tool use correlates with cerebellar asymmetries in chimpanzees (*Pan troglodytes*). Behav Neurosci 122:191–198
33. Taglialatela JP, Cantalupo C, Hopkins WD (2006) Gesture handedness predicts asymmetry in the chimpanzee inferior frontal gyrus. Neuroreport 17(9):923–927
34. Blatchley B, Hopkins WD (2010) Subgenual cingulate cortex and personality in chimpanzees (*Pan troglodytes*). Cognitive, Affective, & Behavioral Neuroscience 10(3):414–421
35. Hopkins WD, Taglialatela JP (2012) Initiation of joint attention is associated with morphometric variation in the anterior cingulate cortex of chimpanzees (*Pan troglodytes*). Am J Primatol 75(5):441–449
36. Hopkins WD, Russell JL, Schaeffer JA (2012) The neural and cognitive correlates of aimed throwing in chimpanzees: a magnetic resonance image and behavioural study on a unique form of social tool use. Philos Trans R Soc B Biol Sci 367(1585):37–47
37. Lyn HL, Pierre P, Bennett AJ, Fears SC, Woods RP, Hopkins WD (2011) Planum temporale grey matter asymmetries in chimpanzees (*Pan troglodytes*), vervet (*Chlorocebus aethiops sabaeus*), rhesus (*Macaca mulatta*) and bonnet (*Macaca radiata*) monkeys. Neuropsychologia 49:2004–2012
38. Shapleske J, Rossell SL, Woodruff PW, David AS (1999) The planum temporale: a systematic, quantitative review of its structural, functional and clinical significance. Brain Res Rev 29:26–49
39. Josse G, Mazoyer B, Crivello F, Tzourio-Mazoyer N (2003) Left planum temporale: an anatomical marker of left hemispheric

specialization for language comprehension. Cogn Brain Res 18:1–14

40. Hopkins WD, Pilcher DL, MacGregor L (2000) Sylvian fissure length asymmetries in primates revisited: a comparative MRI study. Brain Behav Evol 56:293–299
41. Cantalupo C, Pilcher D, Hopkins WD (2003) Are planum temporale and sylvian fissure asymmetries directly related? A MRI study in great apes. Neuropsychologia 41:1975–1981
42. Liu ST, Phillips KA (2009) Sylvian fissure asymmetries in capuchin monkeys (*Cebus apella*). Laterality 14(3):217–227
43. Ide A, Rodriguez E, Zaidel E, Aboitiz F (1996) Bifurcation patterns in the human sylvian fissure: hemispheric and sex differences. Cereb Cortex 6(5):717–725
44. Gilissen E (1992) The neocortical sulci of the capuchin monkey (*Cebus*): evidence for asymmetry in the sylvian sulcus and comparison with other primates. C R Acad Sci III 314:165–170
45. Gilissen E (2001) Structural symmetries and asymmetries in human and chimpanzee brains. In: Falk D, Gibson KR (eds) Evolutionary anatomy of the primate cerebral cortex. Cambridge University Press, Cambridge, pp 187–215
46. Kochunov PV, Mangin JF, Coyle T, Lancaster JL, Thompson P, Riviere D, Cointepas Y, Regis J, Schlosser A, Royall DR, Zilles K, Mazziotta J, Toga AW, Fox PT (2005) Age-related morphology trends in cortical sulci. Hum Brain Mapp 26(3):210–220
47. Autrey MM, Reame LA, Mareno MC, Sherwood CC, Herndon JG, Preuss TM, Schapiro SJ, Hopkins WD (2014) Age-related effects in the neocortical organization of chimpanzees: gray and white matter volume, cortical thickness, and gyrification. Neuroimage 101:59–67
48. Kochunov PV, Glahn DC, Fox PT, Lancaster JL, Saleem KS, Shelledy W, Zilles K, Thompson PM, Coulon O, Mangin JF, Blangero J, Rogers J (2010) Genetics of primary cerebral gyrification: heritability of length, depth and area of primary sulci in an extended pedigree of *Papio* baboons. Neuroimage 53(3):1126–1134
49. Rogers J, Kochunov PV, Lancaster JL, Sheeledy W, Glahn D, Blangero J, Fox PT (2007) Heritability of brain volume, surface area and shape: an MRI study in an extended pedigree of baboons. Hum Brain Mapp 28:576–583
50. Rogers J, Kochunov PV, Zilles K, Shelledy W, Lancaster JL, Thompson P, Duggirala R, Blangero J, Fox PT, Glahn DC (2010) On the genetic architecture of cortical folding and brain volume in primates. Neuroimage 53:1103–1108
51. Hopkins WD, Meguerditchian A, Coulon O, Bogart SL, Mangin JF, Sherwood CC, Grabowski MW, Bennett AJ, Pierre PJ, Fears SC, Woods RP, Hof PR, Vauclair J (2014) Evolution of the central sulcus morphology in primates. Brain Behav Evol 84:1930
52. Hopkins WD, Coulon O, Mangin JF (2010) Observer-independent characterization of sulcal landmarks and depth asymmetry in the central sulcus of the chimpanzee brain. Neuroscience 171:544–551
53. Cykowski MD, Coulon O, Kochunov PV, Amunts K, Lancaster JL, Laird AR, Glahn C, Fox PT (2008) The central sulcus: an observer-independent characterization of sulcal landmarks and depth asymmetry. Cereb Cortex 18:1999–2009
54. Bogart SL, Mangin JF, Schapiro SJ, Reamer L, Bennett AJ, Pierre PJ, Hopkins WD (2012) Cortical sulci asymmetries in chimpanzees and macaques: a new look at an old idea. Neuroimage 61:533–541
55. Kurth F, Gaser C, Luders E (2015) A 12-step user guide for analyzing voxel-wise gray matter asymmetries in statistical parametric mapping. Nat Protoc 10(2):293–304
56. Hopkins WD, Taglialatela JP, Meguerditchian A, Nir T, Schenker NM, Sherwood CC (2008) Gray matter asymmetries in chimpanzees as revealed by voxel-based morphometry. Neuroimage 42(2):491–497
57. Hopkins WD, Taglialatela JP, Nir T, Schenker NM, Sherwood CC (2010) A voxel-based morphometry analysis of white matter asymmetries in chimpanzees (*Pan troglodytes*). Brain Behav Evol 76(2):93–100
58. Keller SS, Crow TJ, Foundas AL, Amunts K, Roberts N (2009) Broca's area: nomenclature, anatomy, typology and asymmetry. Brain Lang 109:29–48
59. Keller SS, Highley JR, Garcia-Finana M, Sluming V, Rezaie R, Roberts N (2007) Sulcal variability, stereological measurement and asymmetry of Broca's area on MR images. J Anat 211:534–555
60. Keller SS, Deppe M, Herbin M, Gilissen E (2012) Variabilty and asymmetry of the suclal contours defining Broca's area homologue in the chimpanzee brain. J Comp Neurol 520:1165–1180
61. Sherwood CC, Broadfield DC, Holloway RL, Gannon PJ, Hof PR (2003) Variability of Broca's area homologue in great apes:

implication for language evolution. Anat Rec 217A:276–285

62. Schenker NM, Hopkins WD, Spocter MA, Garrison AR, Stimpson CD, Erwin JM, Hof PR, Sherwood CC (2010) Broca's area homologue in chimpanzees (*Pan troglodytes*): probabilistic mapping, asymmetry, and comparison to humans. Cereb Cortex 20:730–742
63. Amunts K, Schleicher A, Bürgel U, Mohlberg H, Uylings HB, Zilles K (1999) Broca's region revisited: cytoarchitecture and intersubject variability. J Comp Neurol 412(2):319–341
64. Horwitz B, Amunts K, Bhattacharyya R, PAtkin D, Jeffries K, Zilles K, Braun AR (1999) Activation of Broca's area during the production of spoken and signed language: a combined cytoarchitectonic mapping and PET analysis. Neuropsychologia 41(14):1868–1876
65. Paus T, Tomaiuolo F, Otaky N, MacDonald D, Petrides M, Atllas J, Morris R, Evans AC (1996) Human cingulate and paracingulate sulci: attern, variabilty, asymmetry and probabilstic map. Cereb Cortex 6:207–214
66. Tomaiuolo F, MacDonald JD, Caramanos Z, Posner G, Chiavaras M, Evans AC, Petrides M (1999) Morphology, morphometry and probability mapping of the pars opercularis of the inferior frontal gyrus: an *in vivo* MRI analysis. Eur J Neurosci 11:3033–3046
67. Hopkins WD, Taglialatela JP (2011) The role of Broca's area in socio-communicative processes in chimpanzees. In: Ferrari P, de Waal F (eds) The primate mind: built to connect with other minds. Harvard University Press, Cambridge
68. Spocter MA, Hopkins WD, Garrison AR, Stimpson CD, Erwin JM, Hof PR, Sherwood CS (2010) Wernicke's area homolog in chimpanzees (*Pan troglodytes*): probabilstic mapping, asymmetry and comparison with humans. Proc R Soc B Biol Sci 277: 2165–2174
69. Toga AW, Thompson M (2003) Mapping brain asymmetry. Nature 4:37–48
70. Singh M, Nagashima M, TInoue Y (2004) Anatomical variations of occpital bone impressions for dural venous sinuses around the torcular Herophili, with special reference to the consideration of clinical significance. Surg Radiol Anat 26:480–487
71. Williams NA, Close JP, Giouzeli M, Crow TJ (2006) Accelerated evolution of *Protocadherin 11X/Y*: a candidate gene-pair for cerebral asymmetry and language. Am J Med Genet B Neuropsychiatr Genet 141B:623–633
72. Cain DP, Wada JA (1979) An anatomical asymmetry in the baboon brain. Brain Behav Evol 16:222–226
73. Holloway RL, De La Coste-Lareymondie MC (1982) Brain endocast asymmetry in pongids and hominids: Some preliminary findings on the paleontology of cerebral dominance. Am J Phys Anthropol 58:101–110
74. Balzeau A, Gilissen E (2010) Endocranial shape asymmetries in *Pan panuscus*, *Pan troglodytes*, and *Gorilla gorilla* assessed via skull based landmark analysis. J Hum Evol 59:54–69
75. Balzeau A, Holloway RL, Grimaud-Herve D (2012) Variations and asymmetries in regional brain surface in the genus Homo. J Hum Evol 62:696–706
76. Barrick TR, Mackay CE, Prima S, Maes F, Vandermeulen D, Crow TJ, Roberts N (2005) Automatic analysis of cerebral asymmetry: an exploratory study of the relationship between torque and planum temporale asymmetry. Neuroimage 24:678–691
77. Narr KL, Bilder RM, Luders E, Thompson PM, Woods RP, Robinson D, Szeszko PR, Dimtcheva T, Gurbani M, Toga AW (2007) Asymmetries of cortical shape: effects of handedness, sex and schizophrenia. Neuroimage 34:939–948
78. Luders E, Gaser C, Jancke L, Schlaug G (2004) A voxel-based approach to gray matter asymmetries. Neuroimage 22:656–664
79. Watkins KE, Paus T, Lerch JP, Zijdenbos A, Collins DL, Neelin P, Taylor J, Worsley KJ, Evans AC (2001) Structural asymmetries in the human brain: a voxel-based statistical analysis of 142 MRI scans. Cereb Cortex 11:868–877
80. Phillips KA, Sherwood CS (2007) Cerebral petalias and their relationship to handedness in capuchin monkeys (*Cebus apella*). Neuropsychologia 45:2398–2401
81. Smith SM, Jenkinson M, Woolrich MW, Beckmann CF, Behrens TEJ, Johansen-Berg H, Bannister PR, De Luca M, Drobniak I, Flitney DE, Niazy R, Saunders J, Vickers J, Zhang Y, De Stafano N, Brady JM, Matthews PM (2004) Advances in functional and structural MR image analysis and implementation of FSL. Neuroimage 23(S1):208–219
82. Alexander AL, Lee JE, Lazar M, Field AS (2007) Diffusion tensor imaging of the brain. Neurotherapeutics 4(3):316–329
83. Basser PJ, Pierpaoli C (1996) Microstructural and physiological features of tissues elucidated by quantitative-diffusion-tensor MRI. J Magn Reson B 111:209–219

84. Rilling JK, Glasser MF, Preuss TM, Ma X, Zhang X, Zhao T, Hu X, Behrens T (2008) The evolution of the arcuate fasciculus revealed with comparative DTI. Nat Neurosci 11:426–428
85. Rilling JK, Glasser MF, Jbabdi S, Andersson J, Preuss TM (2012) Continuity, divergence and the evolution of brain language pathways. Front Evol Neurosci 3(11):1–6
86. Hecht EE, Gutman DA, Bradley BA, Preuss TM, Stout D (2015) Virtual dissection and comparative connectivity of the superior longitudinal fasciculus in chimpanzees and humans. Neuroimage 108:124–137
87. Li L, Preuss TM, Rilling JK, Hopkins WD, Glasser MF, Kumar B, Nana R, Zhang X, Hu X (2009) Chimpanzee pre-central corticospinal system asymmetry and handedness: a diffusion magnetic resonance imaging study. PLos One 5(9):e12886
88. Phillips KA, Schaeffer J, Barrett E, Hopkins WD (2013) Performance asymmetries in tool use are associated with corpus callosum integrity in chimpanzees (*Pan troglodytes*): a diffusion tensor imaging study. Behav Neurosci 127(1):106–113. doi:10.1037/a0031089
89. Iturria-Medina Y, Fernández AP, Morris DM, Canales-Rodríguez EJ, Haroon HA, Pentón LG, Augath M, García LG, Logothetis N, Parker GJM, Melie-García L (2011) Brain hemispheric structural efficiency and interconnectivity rightward asymmetry in human and nonhuman primates. Cereb Cortex 21:56–67
90. Behrens TEJ, Woolrich MW, Jenkinson M, Johansen-Berg H, Nunes RG, Clare S, Matthews PM, Brady JM, Smith SM (2003) Characterization and propagation of uncertainty in diffusion-weighted MR imaging. Magn Reson Med 50(5):1077–1088
91. Behrens TEJ, Johansen-Berg H, Woolrich MW, Smith SM, Wheeler-Kingshott CAM, Boulby PA, Barker GJ, Sillery EL, Sheehan K, Ciccarelli O, Thompson AJ, Brady JM, Mathews PM (2003) Non-invasive mapping of connections between human thalamus and cortex using diffusion imaging. Nat Neurosci 6(7):750–757
92. Smith SM, Jenkinson M, Johansen-Berg H, Rueckert D, Nichols TE, MacKay CE, Watkins KE, Ciccarelli O, Cader MZ, Mathews PM, Behrens TEJ (2006) Tract-based spatial statistics: voxelwise analysis of multi-subject diffusion data. Neuroimage 31:1487–1505
93. Biswal BB, Yetkin FZ, Haughton VM, Hyde JS (1995) Functional connectivity in the motor cortex of resting human brain using echo-planar MRI. Magn Reson Med 34:537–541
94. Friston KJ, Frith CD, Liddle PF, Frackowiak RSJ (1993) Functional connectivity: the principal-component analysis of large (PET) data sets. J Cereb Blood Flow Metab 13:5–14
95. Buckner RL, Vincent JL (2007) Unrest at rest: default activity and spontaneous network correlations. Neuroimage 102:1091–1096
96. Greicius MD, Krasnow B, Reiss AL, Menon V (2003) Functional connectivity in the resting brain: a network analysis of the default mode hypothesis. Proc Natl Acad Sci U S A 100(1):253–258
97. Lowe MJ, Dzemidzic M, Lurito JT, Mathews VP, Phillips MD (2000) Correlations in low-frequency BOLD fluctuations reflect cortico-cortical connections. Neuroimage 12(5):582–587
98. Smith SM, Fox PT, Miller KL, Glahn DC, Fox PM, Mackay CE, Fillippini N, Watkins KE, Toro R, Laird AR, Beckman CF (2009) Correspondence of the brain's functional architecture during activation and rest. Proc Natl Acad Sci U S A 106(31):13040–13045
99. Belcher AM, Yen CC, Stepp H, Gu H, Lu H, Yang Y, Silva AC, Stein EA (2013) Large-scale networks in the awake, truly resting marmoset monkey. J Neurosci 33: 16796–16804
100. Hutchison RM, Leung LS, Mirsaattari SM, Gati JS, Menon RS, Everling S (2011) Resting-state networks in the macaque at 7T. Neuroimage 56:1546–1555
101. Hutchison RW, Culham JC, Flanagan JR, Everling S, Gallivan JP (2015) Functional subdivisions of medial parieto-occipital cortex in humans and nonhuman primates using resting-state fMRI. Neuroimage 116:10–29
102. Vincent JL, Patel GH, Fox MD, Synder AZ, Baker JT, Van Essen DC, Zempel JM, Synder LH, Corbetta M, Raichle ME (2007) Intrinsic functional architecture in the anaesthetized monkey brain. Nature 447:83–86
103. Mantini D, Gerits A, Nelissen K, Durand J-B, Joly O, Simone L, Sawamura H, Wardak C, Orban GA, Buckner RL, Vanduffel W (2011) Default mode of brain function in monkeys. J Neurosci 31:12954–12962
104. Wey HY, Phillips KA, McKay DR, Laird AR, Kochunov P, Davis MD, Glahn DC, Duong TQ, Fox PT (2014) Multi-region hemispheric

specialization differentiates human from non-human primate brain function. Brain Struct Funct 219:2187–2194

105. Beckmann CF, DeLuca M, Devlin JT, Smith SM (2005) Investigations into resting-state connectivity using independent components analysis. Philos Trans R Soc Lond B Biol Sci 360:1001–1013

106. Damoiseaux JS, Rombouts SARB, Barkhof F, Scheltens P, Stam CJ, Smith SM, Beckmann CF (2006) Consistent resting-state networks across healthy subjects. Proc Natl Acad Sci U S A 103(37):13848–13853

107. Liu JV, Hirano Y, Nascimento GC, Stefanovic B, Leopold DA, Silva AC (2013) fMRI in the awake marmoset: somatosensory-evoked responses, functional connectivity, and comparison with propofol anesthesia. Neuroimage 78:186–195

108. Murphy K, Birn RM, Bandettini PA (2013) Resting-state fMRI confounds and cleanup. Neuroimage 80:349–359

109. Teichert T, Grinband J, Hirsch J, Ferrera VP (2010) Effects of heartbeat and respiration on macaque fMRI: implications for functional connectivity. Neuropsychologia 48: 1886–1894

110. Goense JB, Whittingstall K, Logothetis NK (2010) Functional magnetic resonance imaging of awake behaving macaques. Methods 50:178–188

111. Andersen AH, Zhang Z, Barber T, Rayens WS, Zhang J, Grondin R, Hardy P, Gerhardt GA, Gash DM (2002) Functional MRI studies in awake rhesus monkeys: methodological and analytical strategies. Methods 118:141–152

112. Logothetis NK, Guggenberger H, Peled S, Pauls J (1999) Functional imaging of the monkey brain. Nat Neurosci 2:555–562

113. van den Heuvel MP, Hilleke E, Hulshoff Pol HE (2010) Exploring the brain network: a review on resting-state fMRI functional connectivity. Eur Neuropsychopharmacol 20:519–534

114. Margulies DS, Bottger J, Long X, Lv Y, Kelly C, Schafer A, Goldhahn D, Abbushi A, Milham MP, Lohmann G, Villringer A (2010) Resting developments: a review of fMRI post-processing methodologies for spontaneous brain activity. MAGMA 23:289–307

115. Hutchison RW, Everling S (2012) Monkey in the middle: why non-human primates are needed to bridge the gap in resting-state investigations. Front Neuroanat 6:20

116. Rilling JK, Barks SK, Parr LA, Preuss TM, Faber TL, Pagnoni G, Bremmer JD, Votaw JR (2007) A comparison of resting-state brain activity in humans and chimpanzees. Proc Natl Acad Sci U S A 104(43):17146–17151

117. Kojima T, Onoe H, Hikosaka K, Tsutsui KI, Tsukada H, Watanabe M (2009) Default mode of brain activity demonstrated by positron emission tomography imaging in awake monkeys: higher rest-related than working memory-related activity in medial cortical areas. J Neurosci 29(46):14463–14471

118. Barks SK, Parr LA, Rilling JK (2015) The default mode network in chimpanzees (*Pan troglodytes*) is similar to that of humans. Cereb Cortex 25:538–544

119. Parr LA, Hecht E, Barks SK, Preuss TM, Votaw JR (2009) Face processing in the chimpanzee brain. Curr Biol 19:50–53

120. Taglialatela JP, Russell JL, Schaeffer JA, Hopkins WD (2008) Communicative signaling activates "Broca's" homolog in chimpanzees. Curr Biol 18:343–348

121. Taglialatela JP, Russell JL, Schaeffer JA, Hopkins WD (2009) Visualizing vocal perception in the chimpanzee brain. Cereb Cortex 19(5):1151–1157

122. Gil-da-Costa R, Martin A, Lopes MA, Munoz M, Fritz JB, Braun AR (2006) Species-specific calls activate homologs of Broca's and Wernicke's areas in the macaque. Nat Neurosci 9(8):1064–1070

123. Poremba A, Malloy M, Saunders RC, Carson RE, Herscovitch P, Mishkin M (2004) Species-specific calls evoke asymmetric activity in the monkey's temporal poles. Nature 427:448–451

Chapter 15

Imaging Techniques in Insects

Marco Paoli, Mara Andrione, and Albrecht Haase

Abstract

The present chapter describes how to apply optical neuroimaging to study brain lateralization in insects. It provides two complete protocols, one for in vivo imaging to obtain information on functional lateralization, and one on histochemical techniques to study morphological asymmetries. Both sections start with the animal preparation, and illustrate the different possibilities for brain tissue labeling. Then, imaging techniques are presented, concentrating on wide-field fluorescence microscopy, confocal, and two-photon laser scanning microscopy. After some remarks on the main methods for data analysis, studies on functional and morphological lateralization in insects are reviewed.

Key words Lateralization, Calcium imaging, Two-photon microscopy, Honeybee, Drosophila

1 Introduction

The discovery of brain lateralization in insects [1, 2] opened up a promising new test ground for one of the most exciting questions in evolutionary neuroscience: the origin of brain asymmetry [3–5]. While first results were reported from behavioral studies (for a review see [6]), subsequently, attention focussed also on the underlying mechanisms within the insect brain. The search for correlations of the behavioral findings with anatomical asymmetries was started via optical imaging, which had already offered several techniques for visualizing brain structure and morphology. Initially, brain functions had been mainly studied using electrophysiology [7, 8], which offered the most direct access to neuronal activity. However, although the strength of this method is the direct measurement of local activity, it does not allow detection of precise response pattern distribution. Therefore, scientists started to develop functional imaging of insect brains, which succeeded in the early 1990s [9] and became a standard tool shortly after [10, 11]. Initial studies on functional lateralization, whose sensitivity and resolution were still relatively low, were unable to reveal significant effects [12]. With the recent advances in 3D in vivo imaging technology and molecular biology, functional

Lesley J. Rogers and Giorgio Vallortigara (eds.), *Lateralized Brain Functions: Methods in Human and Non-Human Species*, Neuromethods, vol. 122, DOI 10.1007/978-1-4939-6725-4_15, © Springer Science+Business Media LLC 2017

imaging has overcome these limitations. Today, optical imaging represents a promising method for studying both morphological and functional lateralization [13]. The aim of this chapter is to provide a guideline through the multitude of available methods of sample preparation, image acquisition, and data post-processing. We have concentrated on their implementation to research on lateralization in the honeybee *Apis mellifera* and the fruit fly *Drosophila melanogaster*, but most of the techniques can be easily adapted to study other insect species.

2 In Vivo Imaging

The antennal lobe (AL) is the primary olfactory center of the insect brain, and the functional equivalent of the vertebrate olfactory bulb. Due to its primary role in sensory coding and perception, and to its favorable position in terms of optical accessibility, the AL represents one of the most common anatomical models to investigate functional brain lateralization in invertebrates. Antennal lobe in vivo imaging can be performed using a similar protocol in different insect species. In both honeybees and fruit flies, olfactory information is conveyed via olfactory receptor neurons (ORNs) to the AL, with ORNs of the same receptor family (i.e. expressing the same odor receptor protein) converging onto a single glomerulus, AL anatomical, and functional unit. The honeybee AL consists of about 160 roughly spherical glomeruli of diameter $d \approx 50$ μm [14], while the fruit fly AL consists of about 54 glomeruli of $d \approx 20$ μm [15]. Glomeruli are densely packed into a bi-dimensional layer to form the shell of the antennal lobe, which reaches an overall $d \approx 400$ μm in the honeybee, and $d \approx 80$ μm in the fruit fly. Local interneurons (LINs) synapse among glomeruli within each AL, and from each glomerulus, projection neurons (PNs) relay the processed olfactory signal to central brain areas, such as the mushroom bodies (MBs) and the lateral horn (LH).

2.1 Preparation

2.1.1 Honeybee

To obtain fluorescently stained antennal lobes for calcium imaging analysis in the honeybee, two main approaches might be followed. The first one involves staining all cell types present in the AL, and measuring the overall activity. The second one aims to measure the activity of a selected cellular population, the AL projection neurons. In the first case, a cell-permeable calcium sensor is applied as bath staining above the whole brain surface. With the second approach, a membrane-impermeable calcium indicator is injected onto the antennal protocerebral tracts, where it will be absorbed by the projection neurons' axonal branches. Within 4–6 h, the dye will diffuse retrogradely to the PNs' cell bodies and dendrites. The advantage of selective probe-injection is the specificity of the fluorescent signal collected, and the higher contrast. Disadvantages are the long preparation time and the lower success rate.

Common to both methods are the preparative steps preceding the staining phase. The methods described here consist of adaptations of the procedure developed for the first time in 1998 by Galizia and collaborators [12]. Briefly, bees are collected via a transparent plexiglass pyramid at the hive entrance during late morning or early afternoon, when they are leaving the hive to forage. This allows collection of mainly forager bees, which are usually preferred over younger/male bees for olfaction studies. Bees are subsequently brought indoors, and transferred into a BugDorm cage, where they can be left undisturbed for several hours or a few days. In this case, the bees should be kept in the dark, food should be provided (i.e. honey, or sucrose solution), and the environment should be enriched with fragments of wax from the combs, in order to provide exposure to a familiar odor. Immediately before starting the staining procedure, the bees are transferred from the box into small plastic tubes with perforated lids.

To immobilize the animal for handling, the plastic tube is placed on ice or at −20 °C for 3–4 min, until the bee stops moving. It is important to avoid unnecessarily long low-temperature incubation, because excessive permanence of the bee at low temperatures markedly affects the bee's survival. Also, the time necessary and sufficient to immobilize a bee without causing damage fluctuates greatly according to the season and the physiological state of the individual animal. Once the bee has been immobilized, it is fixed inside a custom-made plexiglass mount (Fig. 1) by gently

Fig. 1 Animal preparation using a plexiglass mounting stage (**a**). After the bee's neck has been inserted into the thin stage fissure (**b**, **c**), the head is fixed with a thin plastic foil (**d**), and sealed to the mounting stage with hard wax (**e**). The head is further immobilized with soft dental wax (**f**)

sliding its neck through a thin stage fissure, so that the head is accessible, and antennae and mouthparts are directed towards the front side of the mounting stage. This operation is carried out with large rounded-end entomological forceps, in order to prevent the bee from being damaged. Immediately after, a small rectangular piece of plastic foil ($\approx 7 \times 4$ mm) is placed onto the stage behind the bee's head, thereby applying a slight pressure to push the neck forward. The foil is attached to the mounting stage with hard wax (i.e. Siladent, Deiberit 502) melted with the hot tip of a soldering iron (60 °C). It is important to avoid touching the bee's head with the hot wax, as its temperature is high enough to damage the animal. The plastic cover prevents the insect from escaping the mount, but head movements have to be further limited. For this reason, once the hard wax has solidified, a piece of soft dental wax (Kerr, Soft Boxing Wax Sticks) is shaped to cover the mounting stage surface all around the bee's head. The soft dental wax becomes malleable through warming it by hand, and it can be handled comfortably using a wooden toothpick.

The next step consists of opening the head cuticle and exposing the brain. This will allow administration of the dye either by non-specific bath-application, or by localized injection. In both cases, it is essential to preserve the animal's antennae from any damage by keeping them facing forward. This can be achieved by means of two insect pins skewered into the soft wax, and gently pushing the antennae forward. The cuticle is then incised three-times with a scalpel to obtain a window bordered by the antenna sockets on its anterior side, by the ocelli on its posterior side, and by one of the compound eyes on the lateral side. On the side without a cut, a third insect pin is placed into the wax to act as a vertical fulcrum and keep the cuticle window open. Afterwards, glands and tracheae covering the neural tissue are pushed aside or removed with precision forceps to reveal the whole area of interest: the antennal lobe, in the case of bath staining, or the injection point (*i.e.* the intersection between the lateral and medial antenna-protocerebral tracts; Fig. 2) in the case of PNs' backfilling.

In the case of bath staining, the brain is incubated for about 1 h with a drop of dissolved dye at the appropriate concentration. During this time, the cuticle window should be closed and sealed with Eicosan (easily melted using a soldering iron at 40 °C) in order to prevent the brain from drying. An alternative method is a semi-in vivo approach [16], in which the head is cut off and incubated in a solution containing the dye of choice, but keeping the antennae dry. This method might be preferable in the cases in which movements are a major concern, and long imaging times are not required. For backfilling of PNs, injection of the dextran-conjugated dye is made via tapered glass capillaries (*i.e.* WPI Patch Clamp Glass, OD 1.65 mm/ID 1.1 mm), previously pulled to a tip size of few micrometers. Capillaries are then loaded by covering their tip surface with small amounts of the crystallized dye. This is achieved by

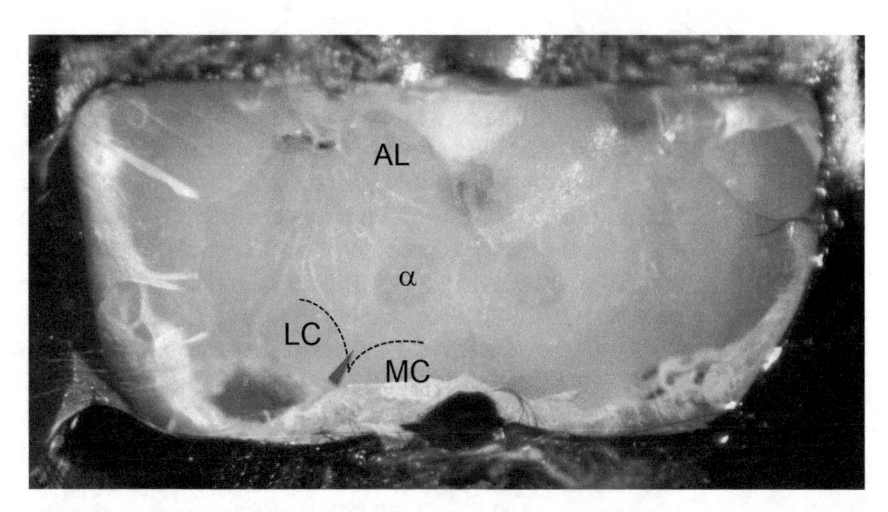

Fig. 2 Backfilling of projection neurons. The proper injection site for the dye is at the intersection between the lateral and medial antenno-protocerebral tracts (*red arrowhead*). The site can be recognized as a small indent between medial (MC) and lateral calices (LC) of the mushroom bodies, which border the posterior part of the brain. The α-lobe (α) may also be used as a reference

adding single drops (<0.5 μl) of distilled water to a few crumbs of dye on a glass slide. When the water starts evaporating, the electrode tip is rolled in the viscous solution until it is covered with a blob of fluorescent dye. Importantly, the solution should not be too liquid, to prevent it from being sucked into the capillary. The loaded tip is then inserted into the tissue at the described location (Fig. 2) [17] and kept in place for a few seconds, allowing the dye to dissolve. The injection should be as perpendicular as possible to the brain surface. In order to study the AL lateralization, it might be useful to perform injections on both sides, to obtain bilaterally stained ALs. After the injection, the cuticle window is closed and sealed with three drops of liquid eicosane. The antennae are freed and the bee is fed on a 50% (v/v) sucrose solution. Bees are kept in the dark at room temperature before imaging.

On the following day (or 1 h after in case of bath staining incubation), and immediately before imaging, animals are prepared in a manner allowing clear optical access to the brain, and to suppress all possible movements. This becomes even more critical when working in two-photon microscopy, since a displacement of a few micrometers in any direction will cause large artifacts. Therefore, the antennae are again gently moved forward-facing using insect pins. Then, the antennal joints both at the socket and between scrape and pedicel may be further fixed with a small drop of eicosane, to keep them parallel to each other and distant enough from the mouthparts, to avoid wetting.

The cuticle window is then re-opened with a scalpel, and the removal of glands and tracheae is completed to expose the antennal lobe. Optionally, to further reduce brain movements, a cut into the cuticle 1 mm above the mouthparts allows pulling out of the esophagus with precision forceps. However, this should be done with considerable care as it may severely reduce viability. A piece of plastic foil matching the shape of the mount and the bee's head is placed between the antennae and cuticle window, and it is fixed with melted hard wax to the plexiglass mount to isolate the antennae from the imaging area. Care has to be taken to avoid the plastic foil from interfering with the movements of the microscope's objective. Finally, the regions around the cuticle window are sealed with a bi-component silicone elastomer (Kwik-Sil, WPI) to form a water-proof environment for the physiological solution (Ringer solution: 130 mM NaCl, 6 mM KCl, 4 mM $MgCl_2$, 5 mM $CaCl_2$, 160 mM sucrose, 25 mM glucose, 10 mM HEPES, pH 6.7, 500 mOsmol [18]). Alternatively, the whole brain can be covered in transparent Kwik-Sil. This has the advantage of a more stable brain preparation, and avoids introducing excessive fluid in the animal's head. On the other hand, this method is not suitable for pharmacological experiments since access to the brain for drug administration is blocked by the silicon. As a last step, a piece of foam gently pressed against the bee's thorax and abdomen can further reduce body movements, which may cause pumping motions within the head.

2.1.2 Fruit Fly

The preparation of *Drosophila melanogaster* for in vivo functional imaging is particularly delicate due to the small size of the animal and, therefore, of its brain. Bearing this in mind, the preparation in itself closely resembles the protocol adopted for honeybee handling described above. This preparation can be adapted to wide-field fluorescence, confocal, and two-photon imaging.

For an optimal preparation, flies of an age between 4 and 15 days old are preferred, unless otherwise required by the experimental question. At younger age, in fact, the head cuticle is still very soft and hence difficult to cut. In addition, females are generally preferred because they are bigger in size, and, therefore, easier to manipulate. To facilitate animal handling, flies are briefly immobilized (<1 min) in an ice-cooled petri dish, and gently transported to a custom-made plexiglass mount, where a thin copper plate, cut in the middle, had been previously installed. While still immobilized, the fly is slid by its neck through the slit across the copper plate, where the head is held in place by a small drop of colophony resin applied between the eyes and the copper plate. In this way, the head of the fruit fly remains still, a necessary condition for the preparation, while the body and legs are free to move. At this stage, the antennal plate is carefully pulled forward with surgical forceps, and a thin metal wire (0.015 mm thick) is gently inserted along the frontal suture to keep the antennae in a forward-facing position, and the antennal plate slightly detached from the head. A small piece of polyethylene foil, with a $\approx$0.5 mm wide square hole in the middle is then placed above the head in such a way that eyes and antennae are covered by the foil, but the cuticle of the head behind the antennae remains accessible. It is recommended that the viability of the animal is checked after this step. The head is then covered by a thin layer of two-component silicone (Kwik-Sil, WPI) in order to seal the fissures between the polyethylene foil and the head. This last step is crucial to prevent the preparation from leaking. With the help of a sharp sapphire blade, a window is cut through the cuticle to expose the brain, which is rapidly covered with a drop of Ringer solution [18]. Eventually, for a clearer view of the antennal lobes or of the brain structures of interest, tracheae and glands obstructing the view can be removed [19, 20].

A successful operation will be characterized by a very lively animal, whose legs will be actively moving during the functional imaging procedure, and until the end of the experiment. This will inevitably result in small movements of the fly's brain during image acquisition, which, if excessive, may have to be treated with movement-correction post-processing filters. Also, compared to bigger animals—such as the honeybee—the fruit fly is more delicate and less resistant to the described manipulation. As a consequence, it is more difficult to image the same animal for a long time, or to expose the animal to chemical treatments or stimulus conditioning between subsequent imaging sessions. These practical drawbacks are partly compensated by the advantages of working with an animal model for which a wide array of genetic tools are available. The possibility of expressing a given functional indicator in virtually any cellular population of its central nervous system avoids potential damage induced by invasive probe delivery, and allows detailed investigations of each elemental unit of the fly's brain, from the single cell to the entire network.

2.2 Functional Indicators

In vivo imaging in the honeybee and in *Drosophila* allows investigation of different aspects of neural activity. In the first section, we will introduce those functional organic probes that are most suitable for studying neural dynamics and lateralization in the olfactory circuit of the honeybee. In the next section, genetically encoded probes that are most relevant for functional imaging in the fruit fly will be described.

2.2.1 Fluorescent Probes

Voltage-sensitive dyes (VSDs) are membrane-binding chromophores that quickly modify their absorption/emission fluorescence intensity in response to electric field changes caused by modifications in the membrane potential. Hence, optical imaging using VSDs represents a direct measure of neuronal activity, allowing insights into fast cellular dynamics (i.e. action potentials and sub-threshold potentials). It allows monitoring of membrane potentials even in locations where direct electrical recording would be barely feasible, such as in thin neurites and dendritic spines.

Also, like other optical methods, it allows one to follow the dynamics of large subsets of neurons, or even entire brains, simultaneously. However, this method has a few important limitations. The main drawback is the low signal-to-noise ratio [21]. In the case of nongenetically encoded VSDs, one of the most common methods of probe delivery is bath application. Bath-applied voltage sensitive dyes will interact nonspecifically with any cell membrane in the preparation (bulk staining), so that various sources of fluorescence fluctuations may mask the relevant signals. Furthermore, even reducing the nonspecific signal by intracellular delivery of the functional reporter, the relatively low amount of membrane available for the dye to interact with, compared to the total cellular volume available i.e. for cytoplasmic calcium reporters, greatly limits the detectability of the fluorescence changes [22].

For these reasons, VSD-based methodology has not been broadly applied yet. Relevant publications regarding honeybee neuroimaging include two studies from Galizia and collaborators [16, 23], who used the voltage sensitive dye RH795 in bulk staining. RH795 belongs to the class of the fast VSDs, and has excitation/emission wavelengths of 530/712 nm. In the mentioned experiments, it was necessary to average over 8–16 odor stimulation responses in order to reach an acceptable signal-to-noise ratio. The dye was helpful in revealing important insights on the phasic-tonic behavior of different glomeruli in response to different odors. Anyhow, because of the bulk staining approach adopted in that study, it was not possible to disentangle the contributions of different cell types (PNs, ORNs, LINs) from the observed voltage signal variations.

In insects, retrograde staining of specific neuronal populations with VSDs has not yet been implemented, although this is in principle possible, at least with some hydrophobic VSDs, as shown by Wenner et al. [24] in the chick embryo. Voltage sensitive dyes retrograde staining of PNs would in principle allow a better understanding of the temporal dynamics of odor responses within the chosen cell population, and strongly improve the temporal resolutions in this kind of studies. Also, in the near future, a new generation of functional dyes, developed from fluorine substitutions of existing hemicyanine dyes, and providing a wider spectrum of excitation/emission wavelengths, may improve two-photon implementations of voltage imaging, and its combination with calcium imaging [25].

As far as the use of calcium sensitive dyes is concerned, there is an extensive literature regarding the functional study of the honeybee olfactory pathway, and the AL in particular. The calcium imaging technique is based on the visualization by fluorescence microscopy of intracellular calcium changes, representing an indirect measure of neuronal activity [26]. Indeed, it has been shown that action potentials can be studied in terms of the voltage-dependent intracellular calcium increase that they produce [27]. The method relies on the use of fluorescent dyes with a calcium-sensitive absorption/emission spectrum, developed in the '80s [28], and based on the structures of calcium chelators such as EGTA and BAPTA. Such dyes have the intrinsic property of modulating their excitation/emission spectra upon interaction with calcium, and can therefore be used to monitor neuronal activity in vivo. Interestingly, some calcium sensitive dyes—ratiometric dyes—allow measurement of intracellular calcium concentration with high accuracy by exploiting the properties of their absorbance spectrum. Ratiometric calcium imaging requires two measurements taken in rapid sequence: the first one at the excitation wavelength showing the maximum ratio between fluorescence intensity of calcium-bound and calcium-free states (i.e. 340 nm in Fura-2), the second one at the excitation wavelength showing a minimum ratio between the two values (i.e. 380 nm in Fura-2). The emitted fluorescence is measured in both cases at the maximum of the emission peak (510 nm in Fura-2). The ratio between the two measured values (F_{340nm}/F_{380nm}) allows estimation of the absolute calcium concentration, independently of artifacts due to uneven dye distribution among the cells. This sequential imaging technique is well suited for widefield microscopy, whereas laser-scanning techniques do not allow alteration of the excitation wavelength with the required speed [26].

The protocol for performing functional calcium imaging analysis in the honeybee was first established by Galizia and collaborators in 1997 [16], who stained the brain through bath application of Calcium Green-1 AM and Calcium Green-2 AM, in a semi in vivo preparation. One year later, the protocol was implemented for in vivo studies [12]. Thanks to functional calcium imaging—which greatly improved signal-to-noise ratio with respect to VSDs—the olfactory code could soon be characterized in terms of spatio-temporal patterns of activation of the AL glomeruli.

The first data on olfactory coding and olfactory circuit plasticity were collected via calcium imaging using the bath-staining technique. Dyes of choice for these preparations were cell-permeable acetoxymethyl (AM) ester forms of chromophores, such as Calcium Green-1 and Calcium Green-2. The AM coating makes the dye electrically neutral, and allows it to cross the lipid bilayer of the cell membrane. Once inside the cell, the AM groups are quickly removed by endogenous esterases, trapping the chromophore (now electrically charged) inside the cytoplasm. The signal collected in the AL through the bath staining procedure is a complex nonspecific signal, in which the ORNs' activity contribution predominates [18], but it is mixed with that one of other neuronal populations, and of glial cells.

When PNs backfilling was finally introduced in honeybee olfaction studies [17], for the first time, a signal originating from a defined cellular population could be studied optically (PNs' staining procedure is described in detail in Sect. 2.1.1). The signals obtained through the two methods (bulk staining and backfilling of PNs) are correlated [17]. However, it is preferable to choose the second approach over bath staining, even if the preparation is more complex, since the collection of specific calcium signals makes it easier to draw relevant conclusions about the network properties and its plasticity. Nowadays, most functional imaging studies conducted in the honeybee brain are performed using the Fura-2 fluorescent reporter, and cell population specificity is obtained through selective PNs staining. The ratiometric properties of the dye are exploited in the case of single-photon excitation by sequential excitation at 340/380 nm. Instead, in two-photon microscopy, a single excitation wavelength at ≈800 nm is used. Maximum excitation wavelengths for one and two-photon transitions regarding the most used functional probes can be found in Table 1.

2.2.2 Genetically Encoded Indicators

A drawback of the use of organic calcium-sensitive dyes is the invasiveness of the technique. In fact, they have to be introduced into the target neurons by "forced uptake", as described in Sect. 2.1.1, or through a patch pipette. Alternatively, it is possible to employ membrane permeable dyes for a less invasive application, but at the expenses of signal specificity [16]. The use of *Drosophila melanogaster* as a model organism allows one to circumvent the issue of invasiveness via in situ expression of genetically encoded functional indicators [29, 30]. During the last two decades, in fact, genetically encoded fluorescent reporters have been developed and applied to different animal models to track different aspects of neuronal activity, i.e. intracellular calcium concentration [31, 32], membrane potential [33, 34], synaptic vesicle cycling [35], or chloride concentration [36, 37]. In the following paragraphs, we will briefly introduce the molecular basis of genetically encoded calcium and voltage indicators.

The first generation of genetically encoded calcium indicators (GECI) relies on a Förster resonance energy transfer (FRET) effect between two green fluorescent protein (GFP) variants, one acting as an electron-donor, the other as an electron acceptor, linked together by a calcium-binding module based on the binding domain of calmodulin. In the presence of calcium, the two fluorescent proteins undergo conformational changes reducing their intermolecular distance, and energy transfer from the donor to the acceptor protein takes place. If the donor chromophore, usually a blue variant of GFP, is continuously excited, an increase in intracellular calcium can be detected by a decrease in emission intensity from the donor and an increase in emission intensity from the acceptor molecule. This is true in the case of Cameleon-like calcium indicators [31]. A second generation of calcium sensors was developed based on a single GFP-like protein coupled to a calcium-binding

Table 1
Most common probes in fluorescent imaging

Fluorescent probe	Single-photon excitation wavelength (nm)	Two-photon excitation wavelength (nm)	Dichroic cut-off wavelength (nm)	Emission wavelength (nm)
Alexa350	343	690	415	441
Hoechst33258	352	690	415	455
DAPI	358	690	415	463
Lucifer yellow	428	–	470	536
GFP	488	900–1000	495	507
RH795	490	1040	580	688
Alexa488	490	985	495	525
FITC	490	947	495	525
Rhodamin-123	507	913	515	529
Cy-3	552	1032	565	570
Alexa 546	556	1028	565	573
Alexa555	553	–	565	568
TRITC	557	–	565	576
Alexa594	590	1040	600	617
Alexa647	650	–	660	665
fura-2	340	810	416	510
indo-1	338	700	444	475[a] (400[b])
fluo-3	506	810	514	526
fluo-4	494	810	505	516
Calcium Green-1	507	820	514	529
GCaMP-6	488	940	495	512

In the *gray sections* are reported the most common calcium-sensitive probes. The maximum single-photon and two-photon excitation wavelength, the emission wavelengths [76–79] ([a]calcium-free, [b]calcium-bound), and an appropriate cut-off wavelength for the dichroic filter to separate emission from excitation light are given (data from Chroma Technology and Thermo Fisher Scientific)

domain. This is the case of the widely used GCaMP indicators [32]. In GCaMPs, a GFP protein has been circularized, and new C- and N-terminals have been introduced and linked to a calmodulin sequence and to a calmodulin binding peptide. The interaction with calcium ions induces a conformational change of the protein, which enhances its fluorescence emission. Both types of chromophores have been refined over the years to improve various parameters (i.e. sensibility, dynamic range, brightness), and they have been used extensively to study the neuronal activity within the mammalian and insect brain [38, 39].

A more direct way of monitoring neuronal activity is to measure changes in membrane potential via genetically encoded voltage indicators (GEVIs). Notably, activity measurements conducted with voltage sensitive probes reflect more faithfully the neuronal dynamics, providing information on both action potentials and sub-threshold activity, which do not generate relevant changes the intracellular calcium concentration. However, the short duration of voltage changes demands faster kinetics and higher sensitivity for imaging, and these physical constraints have consistently slowed down the development of genetically encoded voltage indicators. Moreover, voltage biosensors are necessarily associated with the plasma membrane, strongly reducing the active imaging volume and worsening the signal-to-noise ratio. Of the different voltage-sensitive probes developed during the last two decades, most were generated by coupling a fluorescent reporter to a voltage-gated ion-channel protein (i.e. FlaSh [33], SPARC [40]), or to its voltage-sensitive domain (VSFP 1 [34]). A second generation of indicators has been developed based on the voltage-sensitive domain of a phosphatase of *Ciona intestinalis* (Ci-VSP [41]). Various probes have been developed based on the Ci-VSP domain [30] with improved kinetics and signal dynamics. The family includes VSFP2.3 and VSFP butterfly 1.2, both successfully expressed in the mouse cortex and used to monitor neuronal activity in vivo. Notably, in order to make the detection of sub-threshold potentials possible, and in order to capture signal kinetics at higher temporal resolution, voltage sensors' sensitivity still needs to be improved.

In conclusion, GEVIs main advantage remains in providing broader coverage of neuronal activity, including sub-threshold potential variations. Conversely, action potentials (with their large calcium transients) are better detected by GECIs, which show slower kinetics but a more favorable dynamic range. Therefore, both voltage and calcium sensors should be used to capture complementary aspects of neuronal dynamics. However, we expect that further improvements of genetically encoded voltage indicators will be able to produce a more universal and reliable tool for direct monitoring of neuronal activity.

2.3 Stimulus Generator

Functional imaging analyses of brain lateralization have been conducted, comparing the olfactory representations in the left and right antennal lobes of the honeybee [42]. For this kind of study, special attention needs to be dedicated to the odor delivery apparatus. In particular, it needs: (1) to produce a stimulus of a consistent intensity across several trials, and across a given time period, (2) to ensure a fast onset and offset of the stimulus, and (3) to avoid biases in the geometric properties of the resulting air flow, which could transmit lateralized information to the animal.

Several set-ups have been developed for this purpose. A simple odor delivery method relies on the programmable Stimulus Air Controller CS-55 (Syntech). This device is an autonomous apparatus designed to deliver controlled olfactory stimulation. By means of independent pumps, the device produces a continuous flow (carrier flow, generally of clean background air) directed towards the animal, and a "stimulus flow" that converges into the main one at the selected time point, and with the chosen intensity and duration. A third blank channel, of the same flow intensity as the stimulus one, is added to the carrier and is switched off during stimulation to keep the total air-flow constant, thus avoiding mechanical biases in olfactory conditioning or perception experiments. The CS-55 allows some basic customizations such as flow rate regulation and pulse timing by means of an adjustable timer. However, this set-up is limited in terms of temporal precision and signal reproducibility. In fact, in this apparatus, odors are generally placed on filter papers inside Pasteur pipettes, and tend to evaporate differently according to local temperature and air exposure. Therefore, the signal may vary considerably across trials. Also, it does not provide the possibility of performing multiple odor stimulation either in parallel or in sequence [43].

An alternative custom-made apparatus, that was first developed by Szyszka and colleagues [44], and later replicated by our research group (unpublished, see Fig. 3), allows achievement of higher signal reproducibility, greater temporal resolution, and the possibility of performing multiple-odor (or mixture) stimulation [45, 46]. In this type of device, the switch between background airflow and olfactory stimulus is tightly controlled by high-precision solenoid valves (Lee Products Ltd.), providing a temporal resolution of the

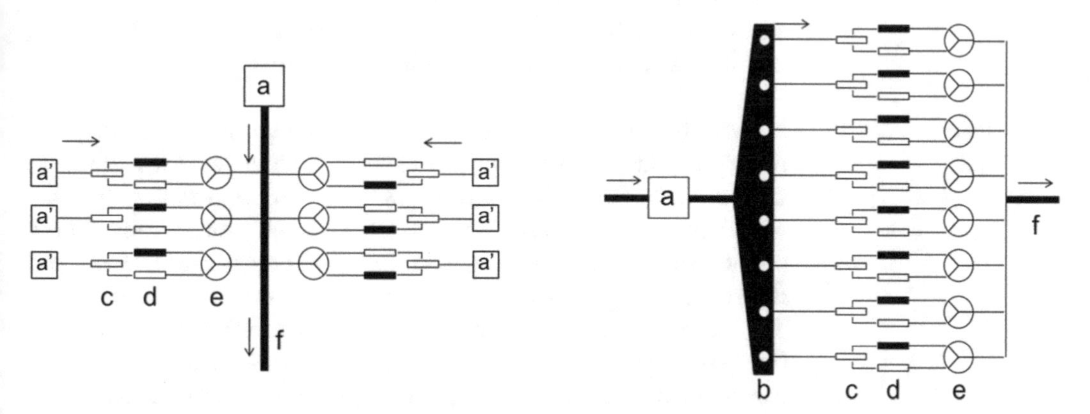

Fig. 3 Scheme of two versions of an odor delivery device. On the left, according to the project of Szyszka et al. [44], six independent channels, each comprising a blank (*white*) and an odor-containing vial (*black*) (**d**), and gated by a three-way solenoid valve (**e**), converge into a main air carrier. Airflows are independently generated for the carrier (**a**, ≈3000 mL/min) and for the odor/blank vials' channels (**a′**, ≈300 mL/min). On the right, a layout of the apparatus developed in our laboratory: a single airflow (**a**) is split in eight valve-gated channels (**b**). All channels converge then in a final connector piece (**f**)

stimulus onset/offset in the order of few milliseconds. Stimulus reproducibility was enhanced by loading mineral oil-diluted fragrances in Teflon-sealed glass vials. This expedient prevents odors from evaporating, and helps to maintain the stimulus intensity constant across a far greater number of experimental sessions. However, a characterization of odor signals generated by this so-called "olfactometer" should be performed at regular intervals, by means of a photoionization detector, in order to ensure stimulus reproducibility. A further advantage of such custom-built stimulus delivery apparatus is the possibility of combining multiple independent odor channels (i.e. 6–8 channels) into the same device, and to control them independently via computer. This permits acquisition of stimulus-elicited responses from several odors in the same animal/antennal preparation in a fast and automatized way. Also, the high temporal resolution of individual channels allows study of how the olfactory circuit deals with synchronous and asynchronous odor mixtures [47].

If the aim of the experimenter is to collect odor responses across a great number of odor samples and concentrations, and a high temporal precision is not required, an alternative option is to use a computer-controlled autosampler, such as the commercially available PAL System (CTC Analytics AG). This system is built for high-throughput data collection and can automatically sample tens of odors at once [48]. By means of a syringe, the PAL System collects the selected volume of headspace from a sealed glass vial, and injects it into a constant flow of clear air with the programmed injection speed. After each injection, the autosampler syringe is rinsed with nitrogen and washed with pentane to minimize the presence of residual odorant between subsequent measurements. Finally, a dedicated software allows easy manipulation of injection parameters (i.e. injected volume, injection time, number of pulses, and injection speed), as well as the number of analyzed samples, and intensity of the syringe cleaning procedure [48].

2.4 Image Acquisition

Nowadays, the methods for insect brain imaging are almost completely based on fluorescence microscopy. Fluorescent probes, as described in Sects. 2.2 and 3.1.3, are used to mark either the whole brain, specific regions, or selected cell populations. These probes are excited at their excitation wavelength, and the emitted fluorescence signal is collected at their peak emission wavelength. The basic hardware required for this procedure is a light source (i.e. gas discharge lamp, light emitting diodes, or lasers), a set of wavelength-selective filters, microscope optics, and a detector (i.e. camera, photomultiplier, or avalanche photodiode). Which types of each of these components are required depends on the imaging modality of choice. In the following sections, we will describe the different options available, together with their advantages and disadvantages.

2.4.1 Widefield Microscopy

Pros and Cons

Widefield microscopy is the classical implementation of fluorescence microscopy, where the whole field of view is excited at once via a wide field light source, and spatial information is obtained by detecting the fluorescence light through a 2D array of detectors. Advantages of this approach are contained costs, simplicity, and robustness. Nowadays, all kinds of implementations are commercially available, and these setups are almost maintenance-free and easy to use. Limitations are due to the fact that the fluorescence excitation is not restricted to a single focal plane, so that signal contributions from off-focus excited regions limit axial resolution. Moreover, the off-focus excitation causes a large sample volume to continuously absorb light, which enhances both photo-damage and photo-bleaching effects. Finally, most fluorescent markers are excited by blue or green light. These wavelengths are strongly scattered within the tissue, thus limiting the imaging depth. However, limitations in resolutions do not always constitute a disadvantage, especially if aiming at imaging structures that are bigger than the resolution limit. In fact, in this way the contribution of motion artifacts is strongly reduced, allowing one to obtain a more stable measure of the average activity of the brain area of interest.

Light Sources

Typical light sources in widefield fluorescence imaging are broadband gas discharge lamps, the excitation bandwidth of which is limited by the appropriate color filter set. The most common ones are mercury vapor lamps or xenon arc lamps. Their working principle is very similar, except for dimensions and enclosed gases. Mercury lamps produce, on top of a continuous spectrum, several characteristic peaks around wavelengths of 365, 400, 440, 546, and 580 nm. Xenon lamps have a more uniform intensity profile across the entire visible spectrum. Therefore, the choice of the light source and the fluorescent probe should be connected. If the excitation bands of the markers overlap with the mercury lines, a mercury lamp might be the better choice; otherwise the uniformly distributed spectrum of the xenon lamp represents a more universal solution. The authors use a 120 W mercury-vapor lamp with a lifespan of 2000 h (X-Cite 120Q).

Optics

The optics are based on a standard epifluorescence microscope, meaning that the sample is illuminated from above, and the signal collected with the same objective in the backward direction. This optical conformation requires separation of the induced fluorescent signal from the far more intense backscattered excitation light. This can be achieved with the introduction of dichroic mirrors and additional blocking filters in the detection arm. The necessary filters are commonly available in packages optimized for the used fluorescent markers. Such filter sets contain (1) an excitation filter along the illumination pathway that limits the bandwidth of the white light source to the excitation window of the dye, (2) a dichroic mirror that separates the elicited fluorescence from the excitation light, and (3) an emission filter along the detection pathway, which additionally blocks all the light outside of the emission window of the used marker. Filter combinations for the most frequently used dyes can be found in Table 1.

The main optical component is the objective. In order to choose the correct objective, one should know beforehand:

1. What magnification is required for the planned experiments
2. What is the field of view that needs to be covered
3. If any immersion liquid will be used (i.e. water, oil)
4. In which spectral range (excitation/emission) the analysis will be performed

Apart from these parameters, an optimal objective should offer the highest possible numerical aperture with respect to the available budget, since this guarantees the highest possible light-collection angle and the best resolution. In widefield microscopy experiments on functional lateralization in honeybees, a 5× objective was used to image both antennal lobes simultaneously, and a 40×, NA 0.6 air objective for high resolution functional imaging of single antennal lobes [12].

Finally, ocular lenses are needed for live observations of the sample, and a detector is required for image recording. There are several implementations for splitting the fluorescent signal between the eyepieces and the detector. The most commonly adopted is a switchable mirror. Also, semi-transparent mirrors, or transparent camera chips, are available, but both cause partial signal loss during data acquisition.

Detectors

Since widefield microscopy is a scan-less technique, spatial resolution is created by the detector. As far as detectors are concerned, a variety of different technologies and features are available. The most common ones are CCD (Charge-Coupled Device) cameras, but CMOS (Complementary Metal Oxide Silicon) cameras have recently become more and more popular. While CCD is a more mature technology, high speed and low costs mostly favor CMOS cameras. When choosing the best possible detector, the following questions should be posed beforehand:

1. What spatial resolution is required (pixel size and number)?

The camera should limit neither the microscope's resolution, nor its field of view. Given the total optical resolution of the camera, one should verify that the optical resolution limit of about 250 nm will be sampled sufficiently over several pixels under the high resolution configuration.

2. What temporal resolution is required?

Acquisition frame-rate will dictate the temporal resolution of the imaging data. Depending on the scientific question, the temporal resolution of the detector should match either the timescale at which signal changes take place or, at least, the timescale of the stimulus duration. This should permit acquisition of a sufficient number of frames for basic data averaging.

3. What is the expected fluorescence intensity range within a single image?

The dynamic range of the camera chip and the bit-depth of the collected signal need to be high enough to cover the different intensity levels of the biological sample. A 12 bit resolution is the absolute minimum, but, ideally, a 16 bit resolution chip is to be preferred.

4. Is there a need to resolve extremely low signals?

In this context, the quantum efficiency of the camera chip comes into play (i.e. the probability of detecting single photons). In this regard, it is important not to focus on the advertised peak quantum efficiency (QE), but rather at the QE in the spectral region that is going to be imaged. Cameras tend to show a drastic decrease of this efficiency in the blue, but modern phosphors-based coatings allow pushing of the QE up to 70–80%. In low-intensity imaging, additional noise sources must be reduced. One possible strategy to overcome such limitations consists in cooling the CCD chip, which reduces the thermal noise, and is available in combination with several commercial detectors.

Experimental Control

The issue of experimental control should not be underestimated. Imaging functional lateralization is a complex process, and many experimental parameters, inside and outside the microscope, need to be controlled precisely. Therefore, a versatile microscope software should meet the following criteria:

1. It should allow automatized acquisition in the most flexible way.
2. Parameters of light source and camera should be controllable automatically, and should ideally be saveable in configuration files.
3. The main computer should have signal output and input ports to synchronize additional devices with the acquisition process. A trigger input should allow remote starting of the process, and a frame clock output should allow synchronization of accessory devices (i.e. stimulators) with single image acquisitions.
4. The image file format should be as universal as possible, and should be readable by programs that are not from the microscope

software provider. This will simplify advanced image processing, which always needs to be performed using external programs (Matlab, Python, R, ImageJ, etc.).

2.4.2 Two-Photon Microscopy

Pros and Cons

Two-photon excitation laser scanning microscopy was first used in 1990 [49] and, 25 years later, optical neuroimaging would be unimaginable without it. The fact that the two-photon absorption is limited to the high intensity region at the focal point has several important consequences. In the first place, it provides intrinsic three-dimensional resolution, which means that by focusing the light onto a certain point within the sample, the generated fluorescent signal can be detected without spatially resolving it. Also, as a consequence, photo-damage and photo-bleaching are markedly reduced, whereas in the case of single-photon excitation a large portion of the sample is continuously exposed to light, thus suffering photo-damage. Finally, the two-photon excitation resonances are in the near infrared, so that the light penetrates much more deeply into the sample than the corresponding blue light used for single-photon excitation. This allows imaging down to 0.5 mm of depth, granting optical access to most parts of many insect brains.

The main disadvantage of this imaging set-up is its high cost. To achieve peak intensities as required for two-photon excitation, ultra-short pulsed lasers are needed. Although new technologies are entering the market, reliable lasers providing these pulses are still expensive. Also, since it is a laser-scanning-based technique, further optics and optomechanics are involved, which add to the cost and cause the need for regular maintenance. Also, the main advantage of the method, its high axial resolution, has a negative side effect, due to the fact that any kind of movement artifact within that resolution regime is detected, and greatly affects the quality of specific signals. Hence, great care has to be taken to shield the whole setup from vibrations and to prepare the animals in a way that minimizes biological movements. Finally, the two-photon elicited signals are of very low intensity, and this should be compensated by using high efficiency detectors. For this reason, the experimental stage needs to be carefully isolated from any external light sources.

Light Source

In order to excite efficiently two-photon fluorescence in a living sample, the highest possible peak-power, which determines the excitation probability, in combination with the lowest possible average power, which is responsible for photo-damage, is needed. For this reason, adequate light sources are ultra-short pulsed lasers with a pulse length not exceeding a few hundred femtoseconds (10^{-15} s). The classical solution is a Titanium:Sapphire laser (our laboratory is equipped with a Mai Tai Deep See HP, Spectra-Physics). Besides its short pulse width of 100 fs, it can be tuned between 690 and 1040 nm, which allows two-photon excitation of a large variety of dyes. An alternative is the single wavelength Nd:YLF laser, which has the drawback of being fixed to a single wavelength of 1047 nm. This disadvantage can be overcome by combining the laser with an optical parametric oscillator (OPO), which converts the starting wavelength into an arbitrary wavelength by means of nonlinear crystals. Recently, interesting new solutions have entered the market, such as erbium- or ytterbium-doped fiber lasers, which promise to be more compact, more robust, and less expensive than conventional lasers. Regarding the necessary laser power, assuming 100 fs pulse length, an average laser power of 10–20 mW is required at the sample, in order to image with reasonable contrast also in deep areas of the tissue.

The mentioned pulse length is a further critical aspect. The shorter the pulses are, the less average power is needed. However, short pulses will undergo dispersion along the optical path, especially through the objective, causing the pulses to spread. An important feature of the system is therefore dispersion pre-compensation, which is often already integrated in commercial laser systems. Otherwise, it should be added to the setup, since it greatly increases the fluorescence signal collected at a certain excitation power. Some commercial systems allow calibration and saving compensation curves for different wavelengths and objectives, thus granting a flexible use of the set-up while maintaining a maximum fluorescence signal. Fast intensity modulation of the laser source is also needed to quickly move between different excitation wavelengths, and it is usually performed with a Pockels cell.

Microscope Configuration

Two-photon microscopes are nowadays available from a large number of providers, and with countless options to choose among. Some parameters are dictated by the biological sample geometry. For instance, when imaging a whole insect brain, an upright microscope (i.e. objective above the sample) in epifluorescence configuration (the same objective that excites the sample, also collects the fluorescence signal) is needed. A choice that needs to be made is whether the sample can be positioned on a movable stage, or if the microscope should be movable with respect to a fixed sample.

Scanning

A fundamental choice is the lateral scanning modality, which usually falls among one of three options: a pair of directly controllable galvanometric mirrors (one of which can be driven resonantly), acousto-optical modulators (AOMs), or spatial light modulators (SLMs). All these scanning techniques have different advantages and disadvantages. While the singly addressable mirrors are very flexible regarding scanning traces and ranges, they are the slowest of all available options. Resonant mirrors allow fast 2D scanning, but without any control of the scanning patterns. The freely addressable and almost instantaneous switchable AOMs represent a technique allowing so-called random-access microscopy, where several points can be imaged almost instantaneously with fast repetition rates. Spatial light modulators are a flexible solution that allows generation of arbitrary intensity patterns via phase-modulation in the conjugate plane, thus permitting the simultaneous excitation of different regions in a "scan-less" fashion. In the scan-less mode, spatial resolution has to be provided by the detector.

While galvanometric mirrors are the standard option, AOMs should be considered if transient dynamics need to be resolved, and if the molecular markers in use provide a time resolution that matches the one of the scanning process. Spatial light modulation is the newest technique and it is only about to be introduced into commercial products. A flexible solution could be switching between different scanning modes. The authors' set-up allows switching between freely and resonantly scanning Galvo-mirrors.

Axial scanning of the focal plane can be achieved by motorized scanning of the objective position. Alternatively, piezo-electric scanners can very much increase scanning speed, although with a limited scanning range (100–200 μm). Fast axial scanning can also be achieved via AOMs and SLMs, but those techniques are still in the development phase.

Using a galvanometric setup, functional images can be acquired along one dimension, by repeatedly scanning single (custom defined) traces, with scanning rates up to 140 Hz for 256 pixels traces. Resonant scanning allows for 2D frame rates of 60 Hz for 256 × 256 pixels images, or 15 kHz for straight 1D line-scans. By adding the piezo-driven axial scanner, a full 3D functional image with 256 × 256 × 10 voxels can be achieved at 5 Hz sampling rate.

Objective

The most important choice regarding the optics is the objective. Here, the same criteria hold true as in Sect. 2.4.1 for widefield microscopy. Here, a high numerical aperture is even more crucial, because the two-photon excited signal has a low absolute intensity and, therefore, as much light as possible should be collected. An additional aspect of two-photon microscopy is that the objectives should be able to correct chromatic aberrations over a wide range of wavelengths (*i.e.* from infrared to blue), because the excitation wavelength can reach up to 1040 nm, and fluorescence light is collected in the visible region of the spectrum down to 400 nm. Most of these objectives are water-immersion, which makes them compatible with in vivo brain imaging, and reaching numerical apertures up to 1.3. The authors use a 20× water-immersion objective (NA = 1.0, Olympus) to image single honeybee antennal lobes. To image both hemispheres simultaneously a 10× objective is used (NA = 0.6, Olympus). In Drosophila, a 20× objective allows contemporary imaging of the entire brain, and a 40× is best suited for AL imaging.

Filters

In two-photon microscopy, color filters are needed for four different purposes: (1) to separate the fluorescence from the backscattered excitation light, (2) to split fluorescence between different detection channels, (3) to limit the bandwidth of the single channels, and (4) to protect the sensitive detectors against residual laser light. In particular:

1. The dichroic mirror, which reflects the fluorescent light towards the detector, is a high-pass filter, with a cut-off wavelength just below the lowest excitation wavelength of the laser. Using a Ti:Sa laser whose tuning range starts from 690 nm, a cut-off wavelength of 660 nm is an appropriate choice.
2. A standard way of separating the elicited fluorescent signal is to split it into a green and a red channel. This can be achieved with another high-pass filter. The authors limit the detection windows by two band-pass filters: one centered at 525 nm, with 70 nm bandwidth, and a second one centered at 607 nm, with 45 nm bandwidth.
3. Finally, the detectors should have another band-pass filter in front of them to block residual laser light centered at 530 nm with 120 nm bandwidth.

Detector

Two-photon microscopy is a laser scanning technique, in which the detector does not need to acquire spatial information but should maximize fluorescent signal collection. Therefore, the criteria for choosing a detector are all aimed at increasing the photon count number, while limiting background noise. Available options in this regard are photomultiplier tubes (PMTs), or avalanche photodiodes (APDs). In the case of special applications, such as scan-less microscopy, high quantum efficiency CCD cameras may be used. If fluorescence lifetime information is needed, time-gated detectors or single photon counting detectors are also available.

Photomultipliers are the most frequently used devices, with high quantum efficiency and very low noise. Their main drawback is their bulky size, and the high voltage (1000 V) they require, which cannot be switched on and off at high rates. Avalanche photodiodes can compete with the PMTs with regard to quantum efficiency, but they have a relatively small active area, which requires, for example, de-scanning the beam. The authors use photomultiplier tubes from Hamamatsu.

Experimental Control

Also for two-photon microscopy, all points highlighted in Sect. 2.4.1 apply. However, it should be considered that in this configuration the number of parameters to control is greater (see Fig. 4), i.e. laser intensity, scanning trace, scanning time. Therefore, the software interface should be laid out clearly. In particular:

1. The system should be as open as possible, to allow integrating and synchronizing other components of the setup with the image acquisition apparatus.

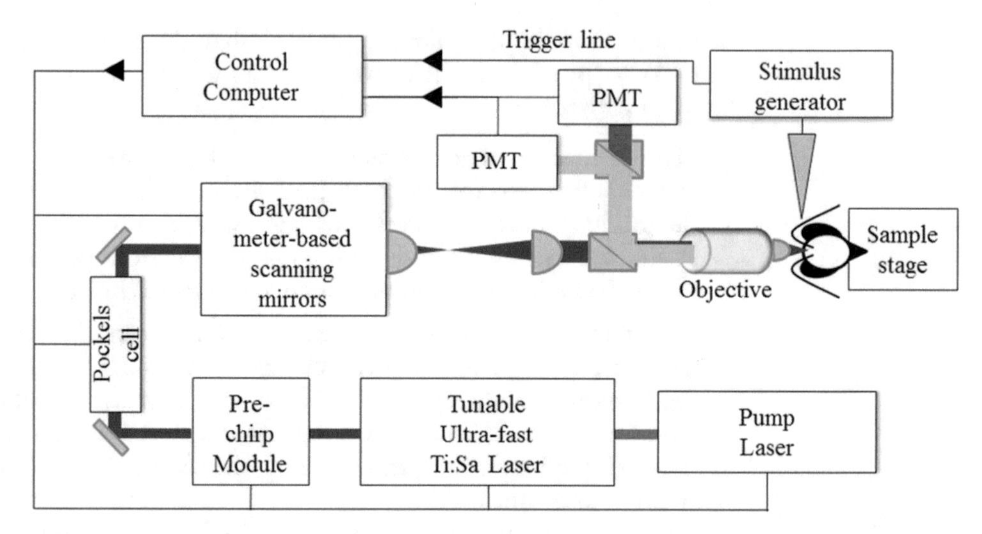

Fig. 4 Schematic setup for functional imaging: a Ti:Sa Laser provides ultrashort pulsed light. A Pockels cell controls the beam intensity. Transversal scanning is achieved by two galvanometric mirrors. Light is then focussed onto the sample via an objective with high numerical aperture. Fluorescence is collected in epi-configuration by the same objective, separated from the backscattered excitation light by a dichroic mirror, and further filtered before being sent to two photomultiplier detectors. A computer controls all elements of the microscopy setup and synchronizes the acquisition with external processes, such as odor stimulus generation (reprinted from [43] with kind permission of Società Italiana di Fisica)

2. Acquisition modes should be as flexible as possible, allowing choice of the dimensionality and directions of the scans. The system should allow one to limit the scanning to regions of interest, and to freely choose exposure times and sampling rates.
3. The availability of different scanning modalities—i.e. line-scans, spiral-scans, or custom defined traces—is of great advantage.

Laboratory Infrastructure

Three issues we have already mentioned make a two-photon imaging experiment much more demanding than a normal wide-field microscopy experiment. The corresponding requirements to the laboratory environment are the following:

1. The general intensity of the fluorescence signal is low and requires sensitive detectors. This also implies that the experimental stage must be completely shielded from external light sources, so a lightproof housing has to be built around the microscope. Simply switching off room light will not be enough, since the light from a computer monitor already saturates the detectors.
2. The sensitivity to mechanical movements, due to the high spatial resolution, requires decoupling of the microscope from the oscillations of the laboratory floor. Highly recommendable is an actively air-damped optical table.

3. The high laser intensity required for two-photon excitation demands a great care regarding laser safety. The utilized lasers are of class IV, and a complete shielding of the laser beam along the entire optical path is highly recommendable.

2.5 Data Analysis

2.5.1 Raw Data Correction

Time-dependent data is usually acquired in fluorescence microscopy in the form of a series of 2D images. Another approach that can be used in laser scanning techniques (such as two-photon microscopy), and that can greatly enhance time-resolution, is reducing the image to a single curve of arbitrary shape, which is then repeatedly scanned by the laser. This second approach produces a time series of 1D lines, which can be collectively visualized as a 2D image. Notably, in functional in vivo imaging analysis, the measure of interest is the stimulus-related change in fluorescence, with respect to a baseline signal. Therefore, the relative change in fluorescence $\Delta F/F$ has to be calculated. As a reference, a baseline F is needed, which can be calculated by averaging over several frames from the pre-stimulus phase.

In widefield and two-photon microscopy, the main signal processing operation is the signal baseline subtraction. Indeed, especially during prolonged experiments, drifts of the fluorescence signal over time can greatly affect the observed fluorescence. There are essentially two approaches that correct for this effect. The first is a dynamic approach, in which a low pass filter, used to average out all activity-induced fluctuations, is applied to the data. This provides a dynamic baseline, which is afterwards subtracted from the data. The second is a static approach, more laborious but often more precise. In this case, the experiment is fractionated into windows encompassing the single stimulus and its pre- and post-stimulus periods. Then, a linear interpolation is performed between pre-stimulus and post-stimulus periods, and is subsequently subtracted from the data in the window. A simplified version of this would be to reset the baseline just before every stimulus at the start of the experiment, by averaging over a few pre-stimulus frames, and subtracting this value from the whole stimulus period.

A second important aspect is the correction for photobleaching. After a certain number of excitation and emission cycles, fluorescent dyes will end up in a nonfluorescent state. The bleaching rates are dependent on the kind of fluorophores in use, excitation power, and exposure time. In widefield and confocal fluorescence microscopy, it is an important issue, because large parts of the sample are continuously illuminated. A correction for the signal decay due to bleaching can be achieved by fitting an exponential function to the data, involving only the pre- and post-stimulus period, or by running an experiment without a stimulus to use as baseline. In two-photon microscopy, due to the limited excitation volume, bleaching effects are usually negligible. In the case of imaging sessions lasting over several minutes, or at very high intensities, bleaching might become relevant, and should be corrected in the same way as described above.

Further corrections of these data may be needed when artifacts caused by biological movements are present. In conventional fluorescence imaging, these are mostly transversal motions (to the optical axis), since axial motions are usually averaged out due to the limited axial resolution. These corrections can be performed by registering consecutive frames via affine transformations. In two-photon microscopy combined with line-scan recording, transversal motion can be corrected by means of the same concept of registration, but through overlapping 1D lines, which can be accomplished by mean of snake algorithms. However, the major issue in two-photon microscopy are movements along the optical axis. These movements do not cause shifts in the signal, but simply a bright/dark modulation, which is hardly distinguishable from an activity signal. The best strategy to overcome such a drawback is to look for correlations of these modulations over many glomeruli that are specific for these motion artifacts. Correction against high frequency vibrations, instead, can be achieved by low-pass filtering temporal and spatial data.

2.5.2 Functional Data Analysis

For a higher level of data processing, signals have to be associated to single functional units. This can be done by manually selecting regions of interest following the glomerular structures in the images. While two-photon microscopy, thanks to its intrinsically high spatial resolution, provides relatively good contrast between single glomeruli, in widefield imaging, boundaries among anatomical structures are often unclear. Therefore, after the functional imaging experimental session, it can be helpful to acquire additional high-resolution images, which can be used for morphological reconstruction and for assigning functional data to specific anatomical structures. Another way to isolate independent functional units is to calculate glomerular boundaries by analyzing cross-correlations between the functional data of neighboring pixels. These methods are implemented, for example, in [50].

When glomerular signals are extracted, it is noticeable how the activity of each glomerulus changes depending on the stimulus presented. In coincidence with specific olfactory stimuli, some glomeruli are excited, showing calcium or voltage increase, others are inhibited (calcium or voltage decrease), and many glomeruli remain unaffected. Also, the glomerular response dynamics vary between phasic, phasic-tonic, and tonic [23]. Therefore, a comparative analysis of odor codes becomes more precise the more glomerular responses it contains. The analysis of these codes is, hence, a multidimensional problem and requires methods that are able to take into consideration all the available data.

Principal Component Analysis

One way to reduce the dimensionality of the problem without losing much information is to rely on principal component analysis (PCA), developed in multivariate statistics. Principal component analysis transforms a number of possibly interlaced factors into a reduced set of linearly uncorrelated variables, the principal components (PCs). The dimensions of the new transformed coordinate space are formally still the same, but now single components are ordered with regard to their contribution in describing the system's variance, and all are orthogonal to each other, so that there is no redundancy in information anymore. The first component has the highest influence on the variance, the second the next highest, and so on. In systems such as a neuronal network, where single nodes are strongly correlated, the contribution of the principal components decays rapidly with their order. This allows reduction of the dimensionality of the problem by restricting the analysis on the first two or three PCs, and still describing most of the system's dynamics. These principal components are then adopted as the new coordinate system for describing the activity signal observed, so that odor-specific patterns of response in the AL can be plotted as a function in two or three Cartesian coordinates of space. In this way, it is possible to easily visualize and compare the neural representation of different odors. For the computation of PCs, mathematical softwares, in which algorithms are already implemented, can be used (i.e. Matlab). Using PCA, different types of datasets might be explored. For instance, it may be of interest to observe how the representations of various odors diverge over the course of an odor stimulus pulse, and which is the moment of maximal separation between the odors. For this analysis, the odor response code for each odorant is plotted throughout different time points over the course of the odor stimulation, as a function of the principle components (as an example, see Fig. 5).

Euclidean Distances

Alternatively, Euclidean distances can be used to compare multidimensional odor codes. When handling odor response data from a large subset of glomeruli, an analysis via multiple comparisons has the drawback of increasing the family-wise error rate and, as a result, differences between single glomeruli are still barely intuitive. The highest possible compression of the information regarding the similarity between two odors can be achieved via their Euclidean distance represented by a single number. Briefly, for calculation of Euclidean distances, the intensity of response within each glomerulus is considered as a single dimension, and the full odor code is represented by a vector in an n-dimensional space, where n is the number of considered glomeruli. The difference between two odor codes can then be described simply by the distance between the two vectors within the n-dimensional space. This parameter is calculated as the sum of all the distances between

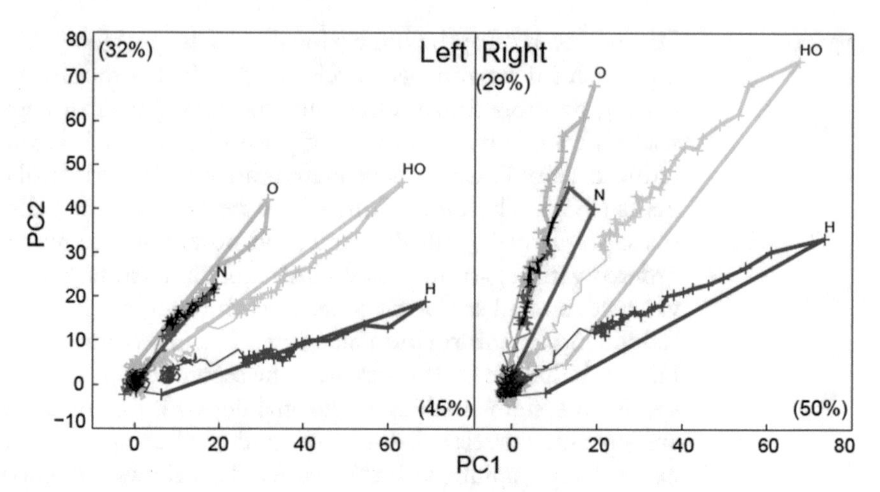

Fig. 5 Trajectories of the odor responses in all glomeruli of left ALs (426 glomeruli) and right ALs (455 glomeruli) calculated using PCA. *Crosses* represent values for subsequently sampled frames during odor stimulation (125 ms inter-frame interval). Higher distances between activation vectors in the right AL means greater odor distinguishability (reprinted from [42] with permission from The Royal Society)

each two corresponding vector components, meaning the differences in response in the single glomeruli. When *xi* and *yi* are the responses relative to odor *x* and odor *y* within glomerulus *i*, the Euclidean distance *dx,y* is given by

$$d_{x,y} = \sqrt{\sum_{i=1}^{n} (x_i - y_i)^2}$$

where summation over the squared differences prevents the summands from canceling each other. This allows calculation of an average distance between two odors, by using for each glomerulus its average activity signal ($\Delta F/F$) in a relevant period within the odor pulse. To describe the response dynamics, Euclidean distances can also be calculated at various time points in the experiment, to monitor how differences between odors are building up in the AL. In general, the Euclidean distance is the best quantitative way of analyzing the distinguishability of two odors in the AL. It has also been demonstrated that such an estimate of distance correlates well with the perceived similarity/dissimilarity of the same two odors in a behavioral paradigm [51].

2.6 Results and Perspectives

2.6.1 Honeybee *Apis mellifera*

The first results of comparative imaging of odor codes in insects were published by Galizia et al. in 1998 [12]. The authors studied odor coding in honeybees via wide-field microscopy with calcium-sensitive markers. They used bath application of Calcium Green-2-AM, which is able to penetrate cell membranes and thus stains all cells in the AL. The resulting odor-induced fluorescence changes are therefore composed of all calcium changes in olfactory receptor neurons, local interneurons, projection neurons, and glial cells.

For imaging, a conventional fluorescent microscope was used, equipped with a 40×, NA 0.6 LD air objective, and fluorescence changes were recorded with a cooled CCD camera (Photometrics CH250A). The odor response pattern was acquired with a spatial resolution of 51 × 51 pixels (binned on chip from 512 × 512 pixels) over a field of view of 250 × 250 μm, which allowed imaging the superficial layer of single antennal lobes. Morphological information on the glomerular structure was obtained by counterstaining the bee brain with the membrane permeable dye RH795, and by imaging the whole brain with a 5× objective. Statistical tests on the functional response pattern of left against right antennal lobe to 13 odor compounds in 30 glomeruli showed no significant asymmetries between sides ($n = 25$ bees).

This lack of asymmetry was, however, not confirmed more recently by Rigosi et al. [42], using a similar optical setup: a wide-field fluorescence microscope Olympus X-50WI with a water immersion objective (20×, NA = 0.95, Olympus) and a CCD camera (Imago QE, Till Photonics). The spatial resolution was 172 × 130 pixels over a field of view of 300 μm. The changes in intracellular calcium concentration of the projection neurons, backfilled with the cell-impermeant Fura-2-dextran, were recorded. The signals were temporally resolved with a 10 Hz frame-rate, which allowed a more detailed analysis of the signal dynamics. This analysis was based on 33 bees, with 8–20 identified glomeruli per individual. The quantification of differences in odor representations between brain sides was based on the measure of the Euclidean distance within the multi-glomerular coding space. This approach allowed detection of a significant lateralization within the odor response patterns. The code showed a better distinguishability of odor stimuli within the right antennal lobe. Figure 5 shows the bilateral odor coding dynamics projected into two dimensions by a principal components analysis. Single curves represent the activation pattern elicited by each odor during and after the stimulus delivery. The distance between the single activation vectors is a measure of the difference between the odor coding patterns.

2.6.2 Fruit Fly Drosophila melanogaster

Louis et al. [52] found an asymmetry in chemotaxis in *D. melanogaster* larvae, i.e. a difference in performance between animals with just left or right functioning olfactory receptor neurons (ORNs). They indeed created larvae expressing a single functional olfactory receptor Or42a or Or1a either bilaterally, only at the left or right side, or at neither side. Besides the finding that the overall accuracy of navigation in odor gradients is enhanced by bilateral sensory input, they showed that a right-functional ORN-animal performed significantly better than the corresponding left-functional animal. Behavioral experiments were performed blindly with respect to the expression type, which was determined after each experiment by removing the anterior tips of larvae, fixing the brain, and imaging it in confocal microscopy (see Sect. 3.2.3). Nonfunctional ORNs were tagged with RFT, functional ORNs with GFP. Larvae were imaged in 3D. The resulting z-stacks were then projected onto a single 2D image to evaluate the overall GFP expression, which allowed identification of the functional ORNs. The observed hypersensitivity shown in the right side was explained by differential gene expression levels, suggesting that this lateralization is an innate feature of the insect brain.

2.6.3 Perspectives

While the observed functional effects represent proof for the existence of some degree of lateralization in the antennal lobes of insect brains, the possibilities that are offered by modern fluorescence imaging techniques are far from being fully exploited. So far, odor response signals are averaged over all projection neurons within a single glomerulus, and only the intensity of the calcium signal is considered as a measure for the odor representation. In vivo application of two-photon microscopy offers a spatial resolution that allows one to resolve single neurons and subcellular structures, and a temporal resolution that may resolve single action potentials. This will permit decomposition of the odor code into single neuron spiking responses, to further investigate from where lateralization in the AL is originating. Furthermore, the optical access offered by two-photon microscopy allows investigation of subsurface brain regions, such as honeybee PNs belonging to the T2–T4 tracts, which have not been studied yet regarding possible functional asymmetries.

3 Ex Vivo Imaging

3.1 Preparation: Methods for Immunohistochemistry

Structure accessibility and staining specificity limit in vivo imaging analysis. Study of the nervous system via classical and immunohistochemical techniques has always been adopted to complement in vivo optical and electrophysiological recordings, as well as behavioral experiments. In the following paragraphs, we will introduce the main histological methods used to investigate lateralization in insects' central nervous systems, from tissue handling, preparation, and staining, to image acquisition, and we will discuss the main issues of data analysis.

3.1.1 Tissue Handling

Careful brain dissection and proper tissue handling are crucial steps for any anatomical and histological analysis. Insect brain tissue should be dissected with the appropriate care in phosphate buffer saline (PBS), or insect Ringer solution [53]. The operation should be conducted at 4 °C, in order to delay as much as possible the degradation processes triggered by anoxia and stressful conditions. Therefore, the whole procedure should be performed on an ice-cooled stage. After a quick dissection phase, the brain tissue is immediately placed in the fixative solution, commonly a 4% solution of paraformaldehyde (PFA) in PBS [53]. Alternatively, the whole head capsule (or the entire animal) may be fixed prior to brain dissection by overnight incubation in fixative solution [54]. This step facilitates the dissection process and limits the extent of manipulation-induced tissue damage. Moreover, fixing the tissue within its natural anatomical environment helps to conserve its original structure [15]. A wide range of fixation conditions can be found in the literature, varying in PFA concentration, duration, temperature, and possible additives (i.e. glutaraldehyde) [53–56]. A standard protocol for fixation of the honeybee brain consists of a 12/24-h incubation of the tissue in 4% PFA in PBS at 4 °C under slow shaking on a bench rocker. For the fruit fly's smaller brain, 1 h in the same conditions is sufficient [57]. Afterwards, the tissue can be stored at 4 °C in PBS indefinitely.

3.1.2 Tissue Labeling

Brain labeling, and subsequent image analysis, can follow two distinct approaches: the fixed tissue can be analyzed in the whole mount (i.e. the brain structure is conserved intact), or it can be included in paraffin for subsequent slicing. The first approach, normally discarded for bigger-sized brains, becomes very efficient when working with miniature brains such as insect brains. In fact, their limited thickness may make them accessible to both antibody staining and optical imaging. This approach becomes particularly effective when coupled with confocal or two-photon microscopy, as both optical techniques provide high 3D resolution. Alternatively, paraffin-embedded tissue can be cut into 5–20 μm-thick sections. The latter method is more laborious than the former one, but it has several advantages. First of all, working on thin sections eliminates the need for a confocal or two-photon microscope for the imaging process. Secondly, both fluorescent and nonfluorescent markers can be employed for labeling, allowing the use of bright-field microscopy. Also, the binding rate of some antibodies to their antigen improves in these conditions, due to a higher level of exposure of the epitopes. Finally, neighboring sections can be stained with different markers, allowing a virtual multiple-staining approach, not feasible when working with a whole mount preparation.

In both cases, prior to immunolabeling, a permeabilization treatment is required to allow the primary antibody to penetrate the tissue and reach the target antigen. Tissue permeabilization can be achieved by brief exposure to a detergent, which will partly dissolve cellular membranes making the various cell compartments accessible to labeling. Detergent concentration and duration of the treatment are variable and need to be optimized case by case. Protocols vary from a 10 min incubation with 0.1–0.2% of Triton-X100 in PBS, to a higher concentration (0.5% in PBS) and longer tissue exposure time (30–60 min) when milder substances such as Tween 20 are used. Before primary antibody incubation, all accessible nonspecific epitopes should be blocked to prevent nonspecific binding of the antibody. This can be obtained by incubating the tissue in a 5–10% solution of normal serum in optimized conditions of duration and temperature (i.e. a time window ranging from 30 min at room temperature to 24 h at 4 °C, based on the protocol specific for each antibody). Normal serum is rich in antibodies that will bind to endogenous immunoglobulins, thus preventing the nonspecific binding of both primary and secondary antibodies used in the assay. Serums from different animals are commercially available. However, it is of paramount importance not to use a serum from the same species, in which the primary antibody was raised, because in this case the secondary antibody would recognize and bind the antibodies present in the blocking solution along with the primary antibodies bound to the target antigen. After blocking, it is recommended to perform sufficient washing (i.e. three washes of 20 min on the shaker) to remove excess blocking solution that may impair optimal detection of the target epitope.

After the permeabilization and blocking steps, the tissue is ready for immunolabeling. This can be done either directly or indirectly. Direct immunolabeling is achieved by using a labeled antibody, which binds directly to the target protein. Alternatively, indirect immunolabeling can be performed, by using two types of antibodies. The first one (primary antibody) it is not labeled and recognizes the target epitope, while the second one (secondary antibody) is labeled and specifically targets the primary antibody. This method has two major advantages with respect to direct labeling. The first one is that it amplifies the signal, because several secondary antibodies can bind to a single primary antibody. The second consequence is that with this approach it is not necessary to label each different primary antibody, but it is possible to use commercial secondary antibodies already labeled with the desired probe. In fact, the same secondary antibody will interact with any specific primary antibody of a given species. For these reasons, indirect immunolabeling is generally the standard approach adopted in immunohistochemistry. Briefly, after the permeabilization and blocking steps, the tissue is exposed to the primary antibody. This phase is highly dependent on the antibody used, both in terms of concentration (from 10^{-2} to 10^{-4}; indications are usually provided by manufacturers) and incubation time (from a few hours to several days). Whereas the antibody concentration needs to be tested for each batch of product, a 24 h incubation period at 4 °C is usually appropriate. After washing the tissue to eliminate the excess primary antibody, the labeling procedure continues with a more standard exposure to the secondary antibody (usually 1–2 h at the suggested concentration). Afterwards, excess secondary antibody is rinsed off in a few quick washes (i.e. 3×20 min washing).

If secondary antibody was fluorescently labeled, the brain tissue is then ready for visualization. Instead, if it was cross-linked to an enzyme requiring a chromogenic substrate for bright-field microscopy visualization, the tissue is now ready to be exposed to the substrate for signal developing. These two different labeling approaches are described briefly in the following paragraphs.

3.1.3 Probes

Fluorescent Probes

Two categories of probes are available for immunolabeling. As discussed in the previous paragraphs, immunohistochemistry can be conducted with both fluorescence and bright-field microscopy, and the probes will need to be selected according to the choice of imaging set-up.

Fluorescence microscopes are now readily available in the majority of research facilities, and use of fluorescent imaging for specific visualization of cellular compartments has become standard practice. For this purpose, antibodies can be cross-linked by a functional group to a fluorescent dye, and will indirectly (in the most common case, i.e. of fluorescent secondary antibodies) bind to a target structure. A large choice of fluorescent probes has been developed and they are currently available from different manufacturers, in order to match a wide range of excitation wavelengths, from UV to red light (see Table 1). An appropriate combination of excitation, dichroic, and emission filters on the imaging apparatus will easily allow one to perform multiple labeling on the same tissue without signal overlapping [53, 58, 59]. One of the main concerns, when working with fluorescent dyes, is signal stability upon light exposure. In fact, prolonged or repetitive excitation of a fluorescent molecule will inevitably lead to signal bleaching. However, with respect to the first organic dyes (i.e. Fluorescein/FITC, or Rhodamine), the more recently developed ones (i.e. Alexa, or Cy families) show a much higher photostability [60]. Moreover, state-of-art fluorescent probes yield a much brighter signal due to higher resistance to self-quenching effects (i.e. the situation in which marker molecules form complexes which are not excitable anymore). In fact, i.e. FITC conjugates quench more rapidly as more fluorophores are added to the same antibody. On the contrary, *i.e.* Alexa probes have a higher resistance to self-quenching, giving rise to a more brilliant signal and allowing use of less conjugate during the procedure, thus reducing background/nonspecific fluorescence. Fluorescent labeling is intrinsically sensitive to light exposure. For this reason, all procedures involving fluorescent probes should be carried out in the dark or under illumination that does not excite the markers (i.e. yellow-red). In this respect, it is also advisable to protect the sample during the secondary labeling step by keeping it inside a dark box, or in a dedicated room. As far as sample storage is concerned, fluorescent staining is suitable for medium term storage if kept at 4 °C (or at −20 °C) in the dark. However, samples prepared more than a few months apart should not be compared directly.

Nonfluorescent Probes

Alternatively, immunohistochemistry can be conducted in bright-field microscopy, by employing secondary antibodies cross-linked to enzymes such as horseradish peroxidase (HRP) or alkaline phosphatase (AP). These enzymes are not directly detectable, but, in the appropriate conditions, they catalyze the conversion of a chromogenic substrate into a colored precipitate. For instance, in presence of hydrogen peroxide applied as an oxidizing agent, HRP oxidizes diaminobenzidine (DAB), turning it into a dark-brown precipitate, which will indirectly indicate the localization of the antibody-labeled structure. In a similar way, AP catalyses the hydrolysis of a phosphate group from the substrate molecule of choice, resulting in a colored product. Importantly, chromogenic substrates are intrinsically sensitive to oxidation, and brain tissue is rich in peroxidases that can quickly oxidize the substrate producing a strong background signal. Hence, before incubation with enzyme-labeled secondary antibody, tissue treatment with saturating amount of hydrogen peroxide (0.3% H_2O_2 in TBS or PBS for 15 min) results in irreversible inactivation of endogenous peroxidases and consequently causes much reduction of the nonspecific background signal.

For both enzymes, various chromogenic substrates have been made available, allowing HRP and AP (or other enzymes) to generate reaction products with different chromatic properties (i.e. product colors vary from intense black to brown, blue, and red of different shades). This permits one to conduct multiple labeling experiments on the same tissue, and to pair any labeling with the counterstaining of choice. Nonetheless, when dealing with spatially close or overlapping signals, it must be considered that this method performs less well than fluorescent probe labeling in terms of signal discrimination and colocalization analysis [61]. Importantly, differing from fluorescent reporters, signal intensity is not a meaningful estimator of the amount of the conjugates present in the sample, but can be modulated by adjusting the developing time (i.e. the period in which the enzyme operates on the substrate). This may vary from a few seconds to several minutes according to tissue properties, labeling efficiency, enzyme and substrate quality. For this reason, it is not possible to quantify the concentration of the target structure based on signal intensity. An advantage associated with nonfluorescent label is that the colored product of the oxidative or dephosphorilative enzymes is permanent and stable in time, and is unaffected by extensive light exposure. In fact, immunolabeled sections can be conveniently stored at room temperature indefinitely, with no effect on staining quality. Finally, the use of nonfluorescent probes can be coupled with several classical methods of histological staining, developed during the pre-fluorescence imaging era, and still routinely used due to their simplicity and the reliability of the developing process. Histochemical counterstaining can be used to conveniently highlight subcellular compartments (i.e. nucleus, cytoplasm, mitochondria, or cellular membranes), or tissue structures, such as densely myelinated areas or amyloid deposits (for an extensive review on the most common reagents for histological counterstaining, see [61]).

3.2 Image Acquisition

Imaging methods for fixed tissues are technically less demanding compared to in vivo imaging analysis. They range from classical histological studies based on bright-field microscopy, to fluorescence-based methods. Besides widefield fluorescence microscopy, which provides high transversal but low axial resolution, there are two competing techniques for 3D fluorescent imaging: confocal microscopy and two-photon microscopy. The main goal of this section is to briefly introduce bright-field microscopy, and to compare the two 3D fluorescence microscopy techniques, regarding their benefits in anatomical imaging of insect brains.

3.2.1 Bright-Field Microscopy

Bright-field microscopy is the simplest of all forms of microscopy. The specimen is illuminated from below, if transparent, or from the side (or above), if nontransparent. While this method is still used widely for animal preparation and histological analyses, quantitative measurements are nowadays mostly based on fluorescence microscopy. Nevertheless, it is worth mentioning that, before buying a bright-field microscope to perform insect brain imaging, the following points should be considered:

Magnification Range and Working Distance

The magnification can be changed within a certain range by adjusting the microscope tube length, i.e. the distance between the objective and the ocular lenses. One should ascertain that the chosen magnification range allows a full view of the sample mount, and that it allows zooming into specific brain centers to identify structures of interest. Also, whenever manipulation of the sample is supposed to happen under the microscope, a comfortable working distance, which is the distance between sample and objective, is needed (10 cm are a recommendable working distance).

Illumination

Again, in the case of manipulation of the sample under the microscope, a light source with the smallest possible angle against the optical axis, and with a high flexibility regarding illumination direction, should be chosen, in order to avoid shadows due to hands or instruments during dissection and manipulations. Traditionally, flexible light guides were the best solution, nowadays, rings of LED, can be mounted around the objective and they provide a valid alternative. If fluorescent samples are prepared under the bright-field microscope, an illumination out of the excitation window of the markers should be used, to prevent the sample from bleaching. Therefore, high pass color filters can be used, that transmit only yellow-red light.

Camera

In bright-field microscopy, image or video recording is often needed. Therefore, a camera should be part of the set-up. While the easiest and cheapest solution is a camera mounted onto one of the eye pieces, it can be useful to record via camera and, simultaneously, monitor or manipulate the sample via the eye pieces. In this case, one should consider buying a trinocular microscope with a camera port, through which the signal can be split between camera and binoculars. Otherwise, it is possible to employ a transparent CCD camera, to be integrated into the optical path. The latter is more flexible, but also more expensive.

3.2.2 Widefield Fluorescence Microscopy

The most common technique for ex vivo fluorescence imaging, is widefield fluorescence microscopy. Thanks to its accessible costs, this is the method of choice for analysis of tissue slices, which usually do not need 3D resolution. Regarding the design of the imaging set-up, the points discussed in Sect. 2.4.1 remain valid, apart from the fact that the sample is now static, so no time resolution is needed. In Table 1, some of the most used fluorescent probes are listed, together with their spectral properties, i.e. excitation and emission wavelengths, as well as the cut-off wavelengths suggested to separate the fluorescence from the excitation light. As soon as samples become thicker (>10 μm), the need to suppress off-focus fluorescence to obtain a reasonable image contrast, will strongly favor confocal or two-photon techniques.

3.2.3 Confocal Microscopy

Confocal microscopy is an extension of the widefield fluorescence method, in which, by post-selection via a pinhole, only light from a single focal plane is allowed to reach the detector. The pinhole is placed in the confocal plane, i.e. the conjugated plane to the focus in the detector pathway of the microscope. Adjusting the pinhole size, it is possible to select how strict the blocking of off-focus light should be. This will control the axial resolution, but also the intensity of the detected signal. Due to the pinhole, confocal microscopy allows imaging of only a single pixel at the time. Two-dimensional imaging is achieved by laser scanning excitation. Three-dimensional scanning is achieved by moving the objective perpendicularly to the sample mount. This implies that many components of the microscope will be identical or similar to those which are needed for the two-photon microscopy. This has already been covered in Sect. 2.4.2, therefore, only the differences will be highlighted in the following paragraphs.

Pros and Cons

Confocal microscopy is the most common technique adopted to investigate the morphology of thick samples. Its clear advantage over widefield imaging is the additional axial resolution, which allows 3D imaging. Its advantage over two-photon microscopy is the relatively lower costs, due to the fact that laser sources do not need to be short-pulsed, as the excitation is here based on single photon transition as in conventional fluorescence microscopy. However, that is also one of the disadvantages of this method. Although light is collected only from the focal plane, absorption happens in the whole light cone, producing strong photo-bleaching effects, which needs to be considered if samples are supposed to be imaged repeatedly. Image post-processing can be required to correct for this effect. Especially in multi-beam confocal microscopy, bleaching is a serious issue, since major portions of the sample are illuminated continuously. This has to be taken into account when choosing an adequate light source. A second drawback of confocal microscopy is the fact that the single-photon transitions are excited in the visible range, mostly in the blue or green, while the same transitions would be excited by infrared light in two-photon microscopy. Optical properties of brain tissue are strongly wavelength-dependent [62], and this causes the penetration depth to be at least twice as high in two-photon microscopy (300–500 μm in two-photon imaging of untreated tissue [63]), compared to single-photon excitation.

Light Source

The light source is the component that differs substantially between confocal and two-photon microscopy, since in the former single photon excitation of the fluorophores is needed. However, since the power density of commercial lamps is not high enough, focussed laser beams are required. To cover the whole spectral window needed for the most frequently used markers, several laser sources are generally combined within one setup. The necessary wavelengths will, once again, be dictated by the fluorophores to be used (see Table 1). Typically, five emission wavelengths are needed to excite efficiently almost all commercially available probes. Standard solutions combine Helium-cadmium Lasers (emitting at 325, 353, and 442 nm), Argon-ion lasers (488 and 514 nm), Argon-Krypton mixed-gas lasers (488, 568, and 647 nm), and Helium-Neon laser (543 and 633 nm). Until now these gas lasers have been the standard solution but in the future they will most certainly be replaced by laser diodes and solid-state-lasers, which are more stable, more compact, and easier to cool. A different class of confocal microscopes use multiple-beam systems. Such set-ups are often equipped with arc lamps from widefield fluorescence microscopy (Sect. 2.4.1.2) to reduce photo-damage.

Scanning

The 2D image formation is based on laser scanning, and it is used in the same way as in two-photon microscopy. Hence, the same considerations as in Sect. 2.4.2 hold true. Multiple-beam scanning confocal microscopes are based on a different mechanism. At any time, excitation is directed not towards a single point, but on a spinning disk, which has an array of pinholes and microlenses illuminating simultaneously an array of points. This mechanism greatly increases the scan rate.

Objective

As in other fluorescence imaging techniques, the objective is the most important optical element for the imaging quality. Issues that need to be considered before purchasing objectives are the same as in widefield microscopy, Sect. 2.4.1.3. In confocal microscopy, it is even more important to maximize the numerical aperture of the lens, because the more the laser beam is focussed, the more efficiently the post-selection process of the emitted signal from the focal plane works. The optical pathway beyond the objective, in confocal microscopy, differs slightly from that used in two-photon microscopy. In fact, the fluorescence collected in epi-configuration is sent back via the scanning unit to be de-scanned, and is then separated from the backscattered excitation light via the appropriate filters. An additional element that we have already mentioned is the adjustable pinhole in the conjugated plane to the focus, which permits axial selection of the signal before it reaches the detectors. The adjustable size of the pinhole allows one to balance the generated signal between resolution and intensity. The size is measured in Airy units ($1\ AU = 0.8\lambda/2NA$), a unit related to the diffraction limit (thus, dependent on wavelength λ and numerical aperture NA). A pinhole of 1 AU ($\approx 0.2\ \mu m$ for $\lambda = 488$ nm and $NA = 1.0$) usually provides the best signal-to-noise ratio. If the signal is sufficiently intense, the pinhole can be further closed to ≈ 0.5 AU to improve spatial resolution.

Filters

Also in confocal microscopy, the optical configuration is dependent on the fluorophores, and on the light sources used to excite them. The general set-up consists of an excitation filter limiting the bandwidth of the excitation source, a dichroic filter that separates the fluorescence from the backscattered excitation light, and emission filters, which select the spectral detection window. If simultaneous imaging on more than one detection channel is performed, an additional beam-splitter and separate emission filters are needed. Important in confocal, as well as in widefield fluorescence microscopy, is that the dichroic filter splitting the induced fluorescence from the excitation light must be a very steep lowpass filter, because excitation and emission wavelengths are much closer than in two-photon microscopy, and residual excitation light produces a strong background signal.

Detectors

Since single beam confocal microscopy is a laser scanning technique, the detector does not have to provide spatial resolution. However, since post-selection of the signals from the focal plane strongly reduces the fluorescence intensity, in this case also high sensitivity detectors are necessary. Photomultipliers are the most common solution, although avalanche photodiodes can be a more compact alternative, if the beam can be well focussed on the smaller active area. In multiple beam scanning confocal microscopes, the signal has to be spatially resolved by an array detector, such as a CCD or a CMOS camera. High efficiency/low noise models including cooling are of advantage.

3.2.4 Two-Photon Microscopy

Brain Imaging In Situ

The anatomical imaging of antennal lobes without brain extraction has the advantage of avoiding artifacts due to tissue isolation, fixation, and dehydration [64]. This procedure is not common, but can be useful when minimum structure alterations due to tissue fixation must be avoided. For reasons of enhanced penetration depth, in this case it is recommended the use of two-photon microscopy. The imaging system is equivalent to the in vivo case, described in Sect. 2.4.2, and filters have to be adapted to the fluorescent markers. In studies measuring glomerular volume, i.e., RH795 is used [55], which is excited best at 1040 nm in two-photon microscopy (490 nm in the single-photon case), and has its spectral peak of fluorescence at 690 nm.

Extracted Brain

When operating with extracted and fixed brain samples, the increased penetration depth of two-photon microscopy can be easily compensated in one-photon confocal configuration by optical clearing [65]. An efficient preparation for AL morphological imaging is, for example, synaptic staining with antibodies, such as anti-synapsin, raised against synaptic proteins (Sect. 3.1.2) and linked to Alexa546 fluorophore, which is efficiently excited via two-photons at 800 nm and emits fluorescence around 570 nm.

3.3 Data Analysis

3.3.1 Image Segmentation

One of the main approaches in ex vivo analysis for investigating brain asymmetries in insects is volume reconstruction of anatomical structures, based on image segmentation. In image processing, the term segmentation indicates the process of partitioning a digital image into multiple regions. Its aim is to reduce the information contained in an image into something easier to analyze, usually by defining clear boundaries of identifiable structures, or by setting a threshold that will automatically segregate meaningful signals from background. When applied to a stack of images, the boundaries defined by the segmentation process can be used to reconstruct a three-dimensional model of the structures of interest, generally in combination with several smoothing/sharpening and interpolating functions provided by any reconstruction software. This analytical approach is best exploited in confocal or two-photon microscopy, both offering higher axial resolution (and imaging depth) than conventional microscopy.

For a reliable 3D reconstruction, it is crucial to dispose of a carefully prepared tissue sample, in which the structure that will be imaged is very well preserved and evenly stained. Synaptic markers such as anti-synapsin or, for *Drosophila*, anti-nc82 antibodies, are preferentially used because they produce homogeneous staining throughout the neural tissue, and enhance anatomical structures and boundaries [15, 53, 59]. Images should be acquired at high resolution and in stacks with a relatively small step size along the optical axis. Clearly, these parameters are linked to the structure under investigation, and they will need to be adjusted accordingly.

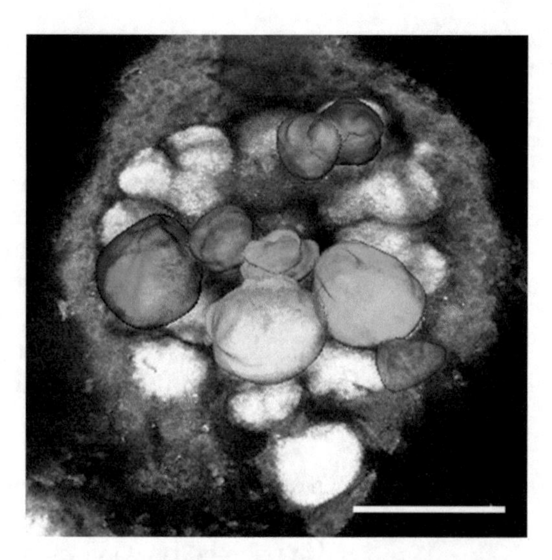

Fig. 6 Focal plane of a honeybee antennal lobe labeled with an anti-synapsin antibody, imaged in a two-photon microscope with a 20× water immersion objective. Image segmentation across the entire z-stack and volume reconstruction of a subset of glomeruli (identified by different colors) was conducted with the AMIRA 3D imaging software, and the reconstructed regions were overlapped to the optical section. Scale bar = 100 μm

Due to its accessibility, and to the interest in insects' olfactory circuit for the study of brain lateralization, one of the most studied neuropils is the antennal lobe. In honeybee, the antennal lobes are spherical structures of about 400 μm in diameter. The single AL can be correctly reconstructed by imaging with a 20× water immersion objective, and acquiring images at 512 × 512 pixels resolution, in steps of 3 μm along the anteroposterior axis [12, 55] (Fig. 6). In the fruit fly, the system is much smaller, and good results are achieved by imaging with a 40× water immersion objective in steps of 1 μm along the anteroposterior axis [15].

Even if this approach is better suited for whole mount preparation coupled to high-resolution fluorescence microscopy, a first attempt to study lateralization in the honeybee antennal lobes was performed on paraffin-embedded 5 μm-thick slices. In that case, the tissue was stained with Luxol fast blue, a type of histochemical staining that sharply highlights the boundaries of AL glomeruli. Volumes were estimated using Cavalieri's direct estimator of volume, a stereological technique that allows volumes of arbitrarily shaped objects to be estimated from a set of randomly oriented, parallel histological sections [56].

Two major issues have been encountered in performing fine measurements of volume lateralization. The first problem concerns the quality of 3D reconstruction, and it is strictly dependent on the process of segmentation, on the quality of the specimen, and on the quality of the acquired images. Segmentation can be done automatically, with a threshold or a watershed algorithm, but the quality of the staining, as well as the presence of disturbing structures (i.e. shadows, tracheae), may markedly alter the reliability and comparability of the results. Alternatively, due to the common presence of imperfection in biological preparations, it is possible to conduct the segmentation manually, by selecting the boundaries of the anatomical structure of interest in the three orthogonal planes, then using a wrapping algorithm to interpolate the full 3D surface. One solution to identify poor segmentation quality is to vary slightly the starting parameters of the reconstruction procedure. If the outcome varies too much, the image quality is probably insufficient, and such a dataset should be discarded.

The second issue concerns the high variability of measured volumes across individuals. Volume differences from left to right side in a lateralized structure may be consistent, but also relatively small if compared to inter-individual variability. For instance, the absolute volumes measured on the individual glomeruli of the honeybee AL vary significantly among different individuals, both due to real anatomical inter-individual differences and to unequal quality of the staining. Therefore, it is recommendable to directly compare volumetric data from the left and right side of the brain within a single animal. Such left/right difference can be quantified by a lateralization index $L = V_R / (V_R + V_L)$ where V_R and V_L denote the right and left volume, respectively [55]. Lateralization index L ranges from 0 to 1 around the symmetry point 0.5. This will allow comparison of the degree of lateralization among different individuals, without the influence of inter-individual variability in size and in histochemical preparation.

3.3.2 Quantification of Objects

Brain asymmetries can be quantified by object counting, where "objects" can be any biological structure present in multiple copies in both the left and right brain hemisphere (i.e. receptor neurons, synapses, neuropils, mushroom bodies microglomeruli). To the best of our knowledge, this approach has been used widely to identify brain differences across animals of different developmental stages [66] or subjected to different conditioning procedures [58], but it has not yet been adopted to studying anatomical correlates of brain lateralization. This method is suited for both whole brain and sliced brain preparations, stained with both fluorescent and nonfluorescent probes. Hourcade et al. [58] have investigated the density of microglomeruli (MGs) in the honeybee mushroom bodies, by double staining those structures with anti-synapsin antibodies (pre-synaptic sites) and fluorescently labeled phalloidin (post-synaptic dendritic spines), and quantifying them. The bee brain was previously sliced in 5 μm sections, and two optical sections containing the medial lips and collars of the two medial calyces were considered in each animal for the analysis. Then, the same circular area was drawn in each of the regions of interest, lip and collar of the medial calices of the mushroom bodies, and MGs were individually counted in each circle. This operation can be performed in several regions of interest and across multiple individuals, in order to obtain a reliable quantification of the number, or density, of the microglomerular structures. A similar approach was used to measure how size-related division of labor in polymorphic leaf-cutting ant workers is reflected in the number of MGs in olfactory calyx subregions [66]. In this study, the same synapsin/phalloidin double staining was performed. To quantify positive boutons within selected volume samples in the lip, the innermost part of the right medial calyx was scanned at high spatial resolution and up to a depth of 10 μm, at intervals of 0.5 μm.

Object counting in both 2D areas and 3D volumes can be carried out by commercial or custom-made functions of image processing software (i.e. Amira, ImageJ), as well as manually. The first approach has the clear advantage of allowing quick high-throughput analyses of several samples in a nonbiased manner. However, imperfection of the preparation and unequal staining may result in processing problems and detection of artifacts. For this reason, in the majority of the cases, a manual counting approach is still preferred. In this case, the approach usually consists of selecting multiple regions of interest inside the analyzed structure. The counts are then averaged across the different ROIs, separately for each individual, and a mean value is then calculated for each group (i.e. different treatment, age, brain compartment). The total counting can be then be used to extrapolate the number of objects in the entire area or volume of the investigated structure.

3.3.3 Fluorescence Intensity Measurements

A third quantitative approach that can be used to compare bilateral brain structures is the analysis of fluorescence intensity. This method is used widely in immunohistochemical analysis to retrieve information on the effects of a manipulation (from the behavioral to the molecular level). Notably, this method is based on the fact that the intensity of a nonsaturated fluorescent signal is directly correlated with the amount of antigens available for labeling, representing an indirect measure of the amount of the target molecule present in the sample.

To our knowledge, this approach, which has been routinely used to address other biological questions [66–68], has not yet been applied to the study of brain lateralization. When adopting this methodology one has to consider carefully that the preparation of different samples is highly variable in terms of background autofluorescence (as well as in immunostaining intensity). For this reason, signal quantification is generally measured in relation to an intensity of reference rather than as an absolute value, i.e. relative to the mean fluorescence of the analyzed structure or relative to the background signal. When the research hypothesis is probing the relative change between two structures after a certain treatment, however, the difference in fluorescent signal can be optimally quantified as the ratio between the signal intensity of the two structures, before and after the treatment [69]. We believe this last approach to be very powerful in highlighting subtle differences linked to brain lateralization. For our purpose, an approach that measures directly the ratio between left and right structures (such as a certain glomerulus, or the microglomeruli of a specific mushroom body area) will enhance any small but consistent differences in expression of a given epitope. On the contrary, an approach in which the fluorescence intensities of the investigated structures are normalized against a reference value prior to a direct comparison could flatten a potential difference between signal intensities.

3.4 Results and Perspectives

3.4.1 Honeybee *Apis mellifera*

Standard Atlas of the Bee Brain

The basis for all anatomical lateralization studies in insects brains are standard atlases, which allow, by means of specific landmarks, identification and comparison of specific functional centers. In the case of the honeybee, *Apis mellifera*, the first 3D model of an antennal lobe was constructed by Flanagan and Mercer in 1989 [70] via phase-contrast bright-field microscopy. Brains were fixed, dehydrated, paraffin-embedded, and cut into 10 μm-thick serial slices. Retrograde injection of cobalt chloride was used to impale and stain cells. Histological sections were then digitized and 3D-reconstructed.

A first high resolution digital atlas of the antennal lobe was created by Galizia and colleagues in 1999 [14] using confocal microscopy. The ORNs axons were traced via neurobiotin, and stained with Cy3-conjugated streptavidin, injected via the antennal nerves. Brains were then extracted, formaldehyde-fixed, and imaged using a Leica TCS 4D with an LD 16× oil-immersion objective (NA = 0.5). As a light source, the 568 nm line of an ArKr 150 mW laser was used and fluorescence was separated from the excitation light via a LP 590 filter. The labeling resulted in clear fluorescent staining of the glomeruli, and the use of confocal microscopy allowed the entire AL to be scanned as a whole-mount, with no need to slice the preparation. Thus, resulting optical slices were perfectly aligned and the AL could be directly reconstructed and segmented by manual tracing of the glomerular borders (Fig. 7). The resulting atlas provides a clear three-dimensional grid for the glomerular positions with improved resolution (2.4 μm axial slice distance and no loss of information between them, so that no glomeruli are missing). Thanks to its digital form, it can be re-sliced at any angle. Brandt and colleagues created the first common spatial reference map of the entire honeybee brain [71] via confocal microscopy. Preparations and imaging techniques were identical to those of Galizia et al. [14], except for the fact that a Leica HC PL Apo ×10/0.4 dry objective had to be used to increase the field of view. Single image stacks were recombined, segmented, and reconstructed using AMIRA 3D Software. The standard brain atlas was created as an average-shape atlas of 20 individual bee brains, which were normalized and registered via affine transformations (Fig. 8).

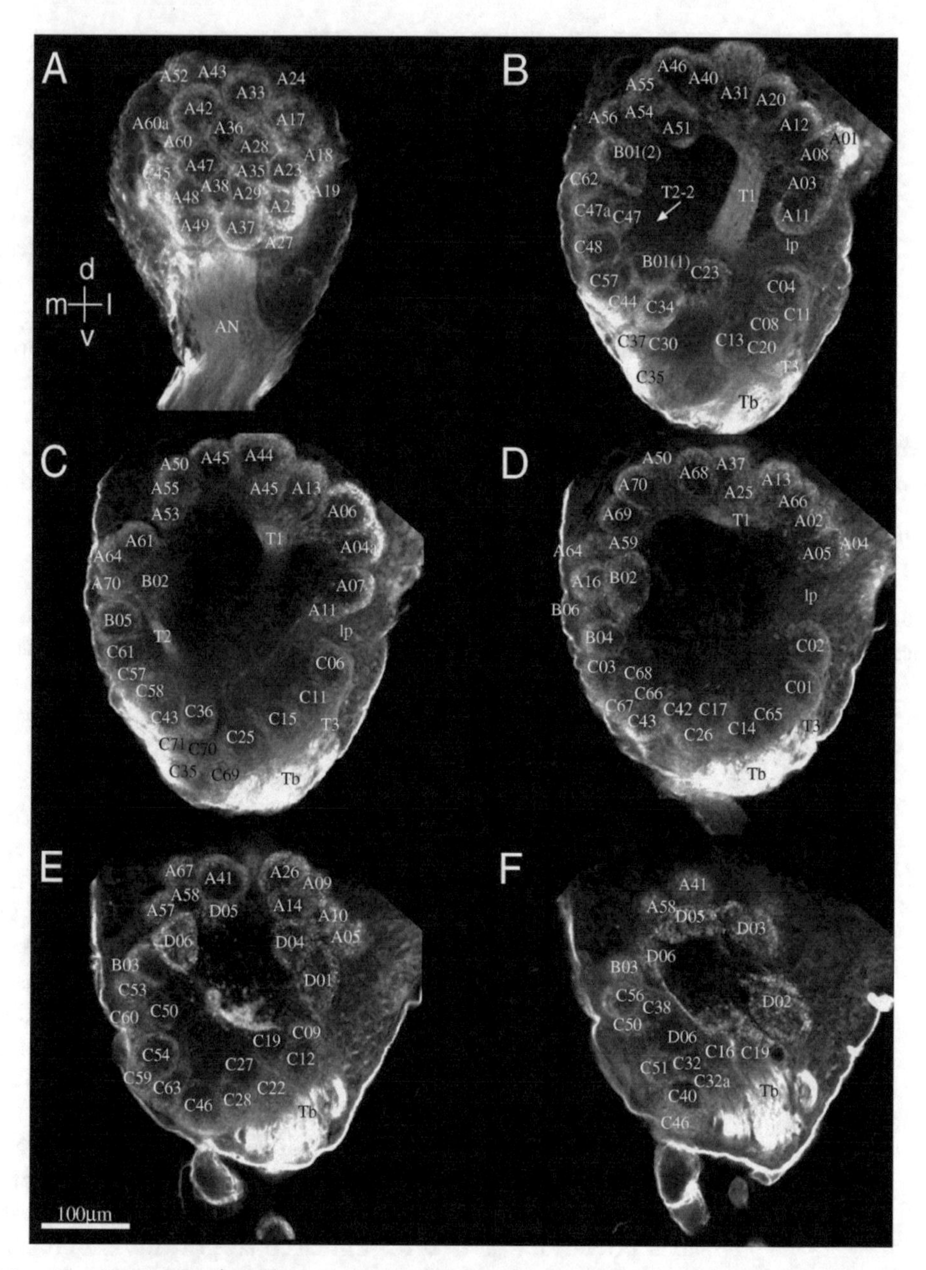

Fig. 7 A series of six confocal optical horizontal sections (**a**) 26 μm, (**b**) 78 μm, (**c**) 104 μm, (**d**) 130 μm, (**e**) 169 μm, (**f**) 195 μm) of the left AL of *Apis mellifera* in which ORNs' axons were fluorescently stained. Individual glomeruli are labeled (reprinted with kind permission from Springer Science + Business Media [14])

Comparative Volume Measurements

Winnington et al. [56] published the first results on comparative volume measurements in the honeybee brain using bright-field microscopy. Brains were extracted, fixed, paraffin-embedded, and sectioned into 5 μm-thick slices. Slices were stained either with Luxol fast blue/cresyl violet, haematoxylin, and eosin, or by cobalt chloride backfilling of antennal sensory afferents and Timm's

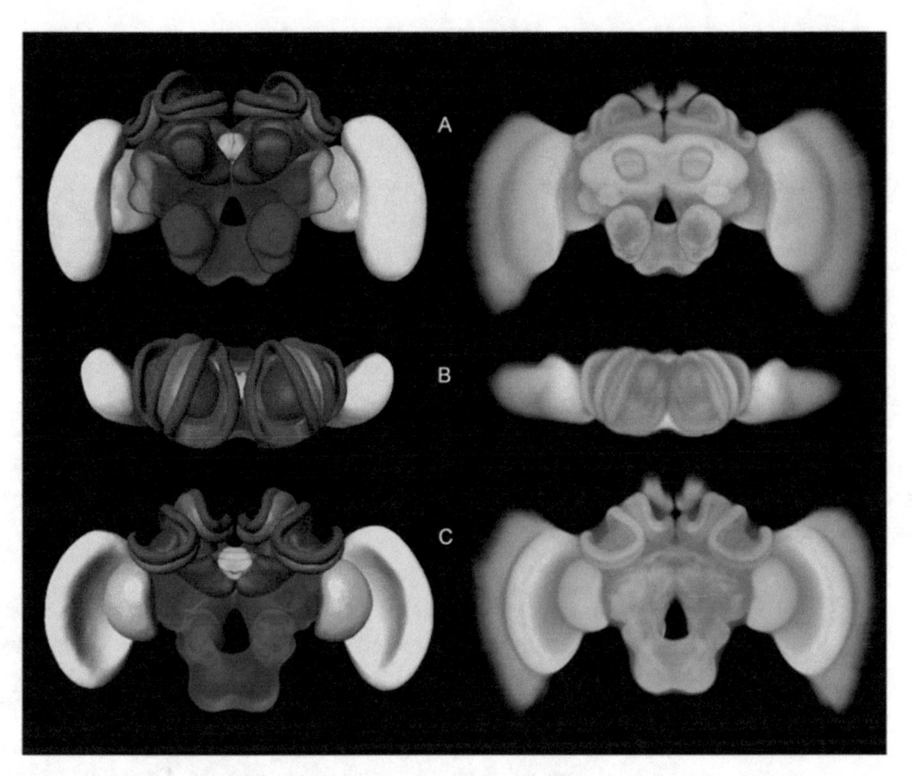

Fig. 8 The Honeybee Standard Brain. The *left panel* shows in grayscale the surface reconstruction of honeybee brain, the *right panel* is a volume rendering of the corresponding average *gray image*. (**a**) Frontal view, (**b**) top view. (**c**) Posterior view. Reprinted from [71] with kind permission from John Wiley and Sons

intensification. For quantitative size measurements, a transparent acetate sheet with a square grid of known size was placed on a magnified image of the brain region of interest. The images were acquired with an Olympus BHS system microscope equipped with a Panasonic WV-CL500 video camera. A count was made of all grid intersections, or points, lying within the cross-sectional area of the structure of interest. The volume of the structure was calculated by multiplying this area measure by slice thickness. Although this work did not concentrate on a bilateral comparison, it found no significant difference between right and left antennal lobe volumes in 4-day-old ($n=6$), 6-day-old ($n=4$), or 28-day-old ($n=6$) bees.

A first study explicitly devoted to the search for volumetric lateralization was performed by Rigosi et al. in 2011 [55], using two-photon microscopy. The fluorescent marker RH795 was bath-applied to the brain in situ. Imaging was performed before dissection and fixation using an Ultima IV two-photon microscope (Prairie Technologies, Bruker) in combination with an ultra-short pulsed laser (Mai Tai Deep See HP, Spectra-Physics). The dye was excited at 1040 nm. An Olympus 40×, NA = 0.8 water-immersion objective provided a field of view of approximately 300 μm. The average laser power used was 10 mW at the sample. Volumetric

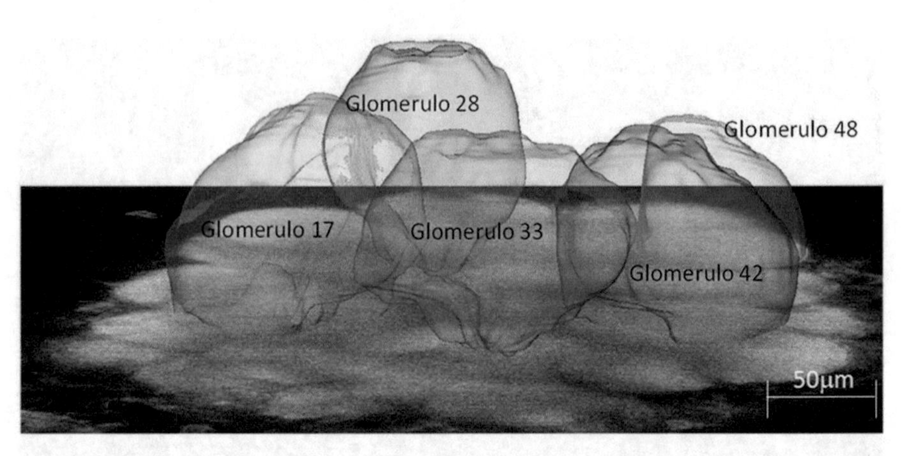

Fig. 9 Single image of the left antennal lobe of *Apis mellifera* at an imaging depth of approximately 80 μm, superimposed with the reconstructed 3D images of five glomeruli (reprinted from [55], copyright 2011, with permission from Elsevier)

measurements were obtained by collecting image stacks in steps of 3 μm along the anteroposterior axis. The imaging depth was limited to 150 μm by the diffusion depth of the bath-applied dye. After 3D image reconstruction and segmentation (Fig. 9), no significant asymmetry in glomerular volumes was found in a subset of five glomeruli averaged over nine animals. The standard error of the lateralization index suggests that a possible volumetric difference would have to be below 5 %.

3.4.2 Fruit Fly Drosophila melanogaster

Standard Atlas of the Antennal Lobe

The first atlas of the *Drosophila* antennal lobe was produced by Stocker et al. in 1983 [72], by means of a bright-field microscope. Samples were prepared in vivo via local injections of cobalt chloride into the antennae. After orthograde diffusion of the dye into the antennal lobe, samples were fixed, paraffin-embedded, and cut into 15 μm serial sections. After imaging more than 400 cobalt stained preparations, 19 glomeruli could be identified regarding their location, shape, and size in each antennal lobe. By staining only single antennas, glomeruli were characterized with regard to their ipsi- or bilateral input. Centring injection in flagellar regions bearing only a single type of sensillum, it allowed description of patterns of glomerular projections, which suggested that individual glomeruli were forming functional units.

The first full 3D brain atlas was created by Laissue and colleagues in 1999 [73], using confocal microscopy. Fly brains were fixed and dissected, and nc82 antibody labeling was performed in whole mount preparation. Images were taken at three images per μm density with 512 × 512 pixel resolution, resulting in an average of 140 images per AL. Eight representative brains were processed for 3D analysis and reference brain reconstruction. This study confirmed the presence of the previously described glomeruli, and revealed the existence of eight new ones. Also, the higher staining quality and imaging resolution allowed better discrimination of glomerular boundaries and to observe glomerular compartments.

A systematic study of the expression of different olfactory receptor types was conducted by Couto et al. in 2005 [74]. This analysis confirmed previous anatomical results and provided further insight into anatomical and functional organization of the AL. By constructing a set of mCD8-GFP reporter lines for all 62 odorant receptor promoters, the authors were able to localize reliably and identify all Or-expressing neurons and to confirm the general validity of the "one receptor-one glomerulus" principle. By exploiting the GAL4 enhancer-trap system, Tanaka and colleagues [75] were then able to further improve the understanding of the AL organization, by identifying 29 neuronal types associated to this neuropil. Morphological analysis of the neuronal arborisations revealed novel information about the organization of the local inter-glomerular network, providing a base for further morpho-functional studies of odor processing.

Dissection and fixation procedures cause strong volumetric and geometrical modifications of the fly brain, resulting in unpredictable deformation of the AL glomeruli, and making it impossible to compare directly in vitro and in vivo images of the brains. Hence, Grabe and colleagues [15] recently produced an in vivo digital 3D atlas of the *Drosophila* antennal lobe. Transgenic flies expressing the red fluorescent protein DsRed directly fused to the presynaptic protein n-synaptobrevin, under the control of the pan-neuronal promoter elav, allowed genetic staining of the antennal lobes in the living animal. Optical sectioning in vivo was performed with a confocal microscopy, equipped with a 40× water-immersion objective. Individual glomeruli were reconstructed using the segmentation software AMIRA. This approach provided a highly reliable topology of the antennal lobe in vivo, a very valuable tool for glomerular identification during functional imaging analysis.

Brain Asymmetries

One of the few studies on fly brain lateralization was conducted by Pasqual and colleagues, who investigated the relationship between brain asymmetry and long-term memory [57]. They described an unknown asymmetrical structure in the *Drosophila* brain, positioned in the right hemisphere, and hence named asymmetrical body. This structure was revealed by anti-fasciclinII (FasII) antibody staining, together with a Bodian counterstaining used to visualize surrounding nerve fibers. The analysis was conducted on paraffin-embedded brain samples, sectioned into 7 μm frontal sections and inspected using bright-field microscopy. Subsequently, other fruit flies were trained for associative memory formation and tested for short term (3 h), and long-term (4 days) memory retention. Good-learner and bad-learner flies were systematically stained with anti-fasII antibody, and imaged via confocal microscopy. With this method, the authors found that such FasII-expressing

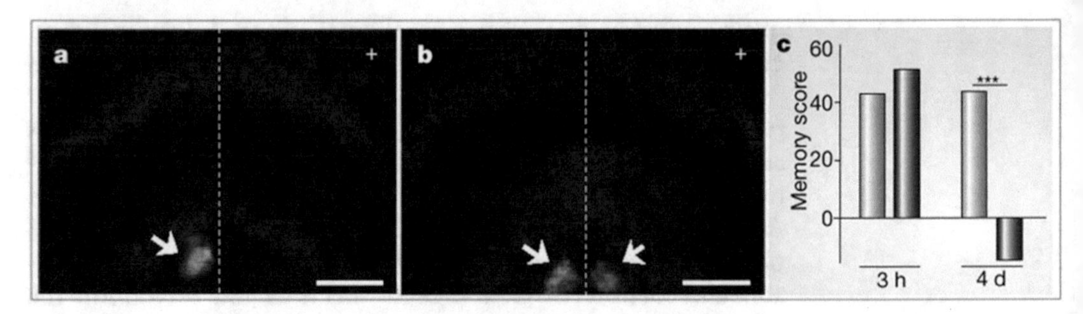

Fig. 10 Single confocal section showing the FasII-labeled asymmetrical body (*arrow*) in the majority of wild-type flies (**a**), while in few flies the brain is symmetrical, presenting a double structure (**b**), scale bar 20 μm. (**c**) Learning performances measured 3 h and 4 days after conditioning, showing that brain lateralization strongly correlates with long-term memory. Reprinted by permission from Macmillan Publishers Ltd: Nature [57], copyright (2005)

structure is bilateral in 7.6% of the wild-type population (Fig. 10). Moreover, they observed that short-term memory performance was independent of the presence of the asymmetrical body. By contrast, 4 days after the associative training, memory scores in the flies with the asymmetrical brain were the same as in the short-term, whereas flies with the symmetrical brain showed no signs of a long-term memory formation.

3.4.3 Perspectives

Although a morphological asymmetry is apparent in fruit flies, in honeybees no anatomical lateralization has so far been detected. However, relevant studies have been comparing brain regions on rather large scales. Thanks to modern fluorescence microscopy techniques, such as confocal and two-photon imaging, the search for anatomical correlates of functional lateralization in the honeybee antennal lobes can be brought to another level, that of single neurons, or even subcellular features. Single neuron staining will make it possible to quantitatively evaluate neuronal branching, while synaptic staining will provide measures for the synaptic density in different neuropils. This will mean continuation of the search for anatomical correlates of behavioral and functional lateralization at a new scale.

References

1. Letzkus P, Ribi W WA, Wood JT et al (2006) Lateralization of olfaction in the honeybee Apis mellifera. Curr Biol 16:1471–1476. doi:10.1016/j.cub.2006.05.060
2. Kells AR, Goulson D (2001) Evidence for handedness in bumblebees. J Insect Behav 14:47–55
3. Vallortigara G, Rogers LJ, Bisazza A (1999) Possible evolutionary origins of cognitive brain lateralization. Brain Res Brain Res Rev 30:164–175
4. Rogers LJ, Vallortigara G, Andrew RJ (2013) Divided brains: the biology and behaviour of brain asymmetries. Cambridge University Press, New York
5. Frasnelli E (2013) Brain and behavioral lateralization in invertebrates. Front Psychol 4:939
6. Frasnelli E, Vallortigara G, Rogers LJ (2012) Left-right asymmetries of behaviour and nervous system in invertebrates. Neurosci Biobehav Rev 36:1273–1291. doi:10.1016/j.neubiorev.2012.02.006

7. Hodgkin AL, Huxley AF (1945) Resting and action potentials in single nerve fibres. J Physiol 104:176–195
8. Matsumoto SG, Hildebrand JG (1981) Olfactory mechanisms in the Moth Manduca sexta: response characteristics and morphology of central neurons in the antennal lobes. Proc R Soc B Biol Sci 213:249–277
9. Lieke EE (1993) Optical recording of neuronal activity in the insect central nervous system: odorant coding by the antennal lobes of honeybees. Eur J Neurosci 5:49–55. doi:10.1111/j.1460-9568.1993.tb00204.x
10. Joerges J, Küttner A, Galizia CG, Menzel R (1997) Representations of odours and odour mixtures visualized in the honeybee brain. Nature 387:285–288. doi:10.1038/387285a0
11. Galizia CG, Sachse S, Rappert A, Menzel R (1999) The glomerular code for odor representation is species specific in the honeybee Apis mellifera. Nat Neurosci 2:473–478. doi:10.1038/8144
12. Galizia CG, Nägler K, Hölldobler B, Menzel R (1998) Odour coding is bilaterally symmetrical in the antennal lobes of honeybees (Apis mellifera). Eur J Neurosci 10:2964–2974
13. Haase A, Rigosi E, Frasnelli E et al (2011) A multimodal approach for tracing lateralisation along the olfactory pathway in the honeybee through electrophysiological recordings, morpho-functional imaging, and behavioural studies. Eur Biophys J 40:1247–1258. doi:10.1007/s00249-011-0748-6
14. Galizia CG, McIlwrath SL, Menzel R (1999) A digital three-dimensional atlas of the honeybee antennal lobe based on optical sections acquired by confocal microscopy. Cell Tissue Res 295:383–394. doi:10.1007/s004410051245
15. Grabe V, Strutz A, Baschwitz A et al (2015) Digital in vivo 3D atlas of the antennal lobe of Drosophila melanogaster. J Comp Neurol 523:530–544. doi:10.1002/cne.23697
16. Galizia CG, Joerges J, Küttner A et al (1997) A semi-in-vivo preparation for optical recording of the insect brain. J Neurosci Methods 76:61–69. doi:10.1016/S0165-0270(97)00080-0
17. Sachse S, Galizia CG (2002) Role of inhibition for temporal and spatial odor representation in olfactory output neurons: a calcium imaging study. J Neurophysiol 87:1106–1117
18. Galizia GC, Vetter RS (2004) Methods in insect sensory neuroscience. Adv Insect Sens Neurosci. doi:10.1201/9781420039429
19. Stökl J, Strutz A, Dafni A et al (2010) A deceptive pollination system targeting drosophilids through olfactory mimicry of yeast. Curr Biol 20:1846–1852. doi:10.1016/j.cub.2010.09.033
20. Silbering AF, Bell R, Galizia CG, Benton R (2012) Calcium imaging of odor-evoked responses in the Drosophila antennal lobe. J Vis Exp 1–10. doi:10.3791/2976
21. Zochowski M, Wachowiak M, Falk CX et al (2000) Imaging membrane potential with voltage-sensitive dyes. Biol Bull 198:1–21
22. Homma R, Baker BJ, Jin L et al (2009) Wide-field and two-photon imaging of brain activity with voltage- and calcium-sensitive dyes. Philos Trans R Soc Lond B Biol Sci 364:2453–2467. doi:10.1098/rstb.2009.0084
23. Galizia CGC, Küttner A, Joerges J, Menzel R (2000) Odour representation in honeybee olfactory glomeruli shows slow temporal dynamics: an optical recording study using a voltage-sensitive dye. J Insect Physiol 46:877–886. doi:10.1016/S0022-1910(99)00194-8
24. Wenner P, Tsau Y, Cohen LB et al (1996) Voltage-sensitive dye recording using retrogradely transported dye in the chicken spinal cord: staining and signal characteristics. J Neurosci Methods 70:111–120
25. Yan P, Acker CD, Zhou W-L et al (2012) Palette of fluorinated voltage-sensitive hemicyanine dyes. Proc Natl Acad Sci U S A 109(50):20443–20448. doi:10.1073/pnas.1214850109
26. Grienberger C, Konnerth A (2012) Imaging calcium in neurons. Neuron 73:862–885. doi:10.1016/j.neuron.2012.02.011
27. Galizia CG, Kimmerle B (2004) Physiological and morphological characterization of honeybee olfactory neurons combining electrophysiology, calcium imaging and confocal microscopy. J Comp Physiol A 190:21–38. doi:10.1007/s00359-003-0469-0
28. Takahashi A, Camacho P, Lechleiter JD, Herman B (1999) Measurement of intracellular calcium. Physiol Rev 79:1089–1125
29. Knöpfel T (2012) Genetically encoded optical indicators for the analysis of neuronal circuits. Nat Rev Neurosci 13:687–700. doi:10.1038/nrn3293
30. Broussard GJ, Liang R, Tian L (2014) Monitoring activity in neural circuits with genetically encoded indicators. Front Mol Neurosci 7:97. doi:10.3389/fnmol.2014.00097
31. Miyawaki A, Llopis J, Heim R et al (1997) Fluorescent indicators for Ca^{2+} based on green fluorescent proteins and calmodulin. Nature 388:882–887. doi:10.1038/42264
32. Nakai J, Ohkura M, Imoto K (2001) A high signal-to-noise Ca(2+) probe composed of a single green fluorescent protein. Nat Biotechnol 19:137–141. doi:10.1038/84397
33. Siegel MS, Isacoff EY (1997) A genetically encoded optical probe of membrane voltage.

Neuron 19:735–741. doi:10.1016/S0896-6273(00)80955-1
34. Sakai R, Repunte-Canonigo V, Raj CD, Knöpfel T (2001) Design and characterization of a DNA-encoded, voltage-sensitive fluorescent protein. Eur J Neurosci 13:2314–2318
35. Miesenböck G, De Angelis DA, Rothman JE (1998) Visualizing secretion and synaptic transmission with pH-sensitive green fluorescent proteins. Nature 394:192–195
36. Kuner T, Augustine GJ (2000) A genetically encoded ratiometric indicator for chloride: capturing chloride transients in cultured hippocampal neurons. Neuron 27:447–459. doi:10.1016/S0896-6273(00)00056-8
37. Arosio D, Ricci F, Marchetti L et al (2010) Simultaneous intracellular chloride and pH measurements using a GFP-based sensor. Nat Methods 7:516–518. doi:10.1038/nmeth.1471
38. Reiff DF (2005) In vivo performance of genetically encoded indicators of neural activity in flies. J Neurosci 25:4766–4778. doi:10.1523/JNEUROSCI.4900-04.2005
39. Hendel T, Mank M, Schnell B et al (2008) Fluorescence changes of genetic calcium indicators and OGB-1 correlated with neural activity and calcium in vivo and in vitro. J Neurosci 28:7399–7411. doi:10.1523/JNEUROSCI.1038-08.2008
40. Ataka K, Pieribone VA (2002) A genetically targetable fluorescent probe of channel gating with rapid kinetics. Biophys J 82:509–516. doi:10.1016/S0006-3495(02)75415-5
41. Murata Y, Iwasaki H, Sasaki M et al (2005) Phosphoinositide phosphatase activity coupled to an intrinsic voltage sensor. Nature 435:1239–1243
42. Rigosi E, Haase A, Rath L et al (2015) Asymmetric neural coding revealed by in vivo calcium imaging in the honey bee brain. Proc R Soc B Biol Sci 282:20142571. doi:10.1098/rspb.2014.2571
43. Haase A, Rigosi E, Trona F et al (2010) In-vivo two-photon imaging of the honey bee antennal lobe. Biomed Opt Express 2:131–138. doi:10.1364/BOE.1.000131
44. Szyszka P, Demmler C, Oemisch M et al (2011) Mind the gap: olfactory trace conditioning in honeybees. J Neurosci 31:7229–7239. doi:10.1523/JNEUROSCI.6668-10.2011
45. Paoli M., A. Anesi, R. Antolini, G. Guella, G. Vallortigara, and A. Haase (2016) Differential odour coding of isotopomers in the honeybee brain. Sci Rep, 6:21893
46. Paoli M, Weisz N, Antolini R, Haase A. (2016) Spatially resolved time-frequency analysis of odour coding in the insect antennal lobe. Eur J Neurosci. 2016 Sep;44(6):2387–95. doi:10.1111/ejn.13344
47. Stierle JS, Galizia CG, Szyszka P (2013) Millisecond stimulus onset-asynchrony enhances information about components in an odor mixture. J Neurosci 33:6060–6069. doi:10.1523/JNEUROSCI.5838-12.2013
48. Strauch M, Lüdke A, Münch D et al (2014) More than apples and oranges—detecting cancer with a fruit fly's antenna. Sci Rep 4:1–9. doi:10.1038/srep03576
49. Denk W, Strickler J, Webb W (1990) Two-photon laser scanning fluorescence microscopy. Science 248:73–76. doi:10.1126/science.2321027
50. Locatelli FF, Fernandez PC, Villareal F et al (2013) Nonassociative plasticity alters competitive interactions among mixture components in early olfactory processing. Eur J Neurosci 37:63–79. doi:10.1111/ejn.12021
51. Guerrieri F, Schubert M, Sandoz J-C, Giurfa M (2005) Perceptual and neural olfactory similarity in honeybees. PLoS Biol 3:e60. doi:10.1371/journal.pbio.0030060
52. Louis M, Huber T, Benton R et al (2008) Bilateral olfactory sensory input enhances chemotaxis behavior. Nat Neurosci 11:187–199. doi:10.1038/nn2031
53. Sinakevitch IT, Smith AN, Locatelli F et al (2013) Apis mellifera octopamine receptor 1 (AmOA1) expression in antennal lobe networks of the honey bee (Apis mellifera) and fruit fly (Drosophila melanogaster). Front Syst Neurosci 7:70. doi:10.3389/fnsys.2013.00070
54. Hourcade B, Perisse E, Devaud J-M, Sandoz J-C (2009) Long-term memory shapes the primary olfactory center of an insect brain. Learn Mem 16:607–615. doi:10.1101/lm.1445609
55. Rigosi E, Frasnelli E, Vinegoni C et al (2011) Searching for anatomical correlates of olfactory lateralization in the honeybee antennal lobes: a morphological and behavioural study. Behav Brain Res 221:290–294. doi:10.1016/j.bbr.2011.03.015
56. Winnington AP, Napper RM, Mercer AR (1996) Structural plasticity of identified glomeruli in the antennal lobes of the adult worker honey bee. J Comp Neurol 365:479–490. doi:10.1002/(SICI)1096-9861(19960212)365:3<479::AID-CNE10>3.0.CO;2-M
57. Pascual A, Huang K, Neveu J, Préat T (2004) Neuroanatomy: brain asymmetry and long-term memory. Nature 427:605–606. doi:10.1038/427605a
58. Hourcade B, Muenz TS, Sandoz J-C et al (2010) Long-term memory leads to synaptic reorganization in the mushroom bodies: a memory trace in the insect brain? J Neurosci 30:6461–6465. doi:10.1523/JNEUROSCI.0841-10.2010

59. Gaudry Q, Hong EJ, Kain J et al (2013) Asymmetric neurotransmitter release enables rapid odour lateralization in Drosophila. Nature 493:424–428. doi:10.1038/nature11747
60. Panchuk-Voloshina N, Haugland RP, Bishop-Stewart J et al (1999) Alexa dyes, a series of new fluorescent dyes that yield exceptionally bright, photostable conjugates. Institute of Physics and Engeneering in Medicine. J Histochem Cytochem 47:1179–1188. doi:10.1177/002215549904700910
61. Elsam J (2015) Histological and histochemical methods: theory and practice, 5th edn. Biotech Histochem 1. Institute of Physics and Engeneering in Medicine
62. Jacques SL (2013) Optical properties of biological tissues: a review. Phys Med Biol 58:R37–R61. doi:10.1088/0031-9155/58/11/R37
63. Haase A (2011) Simultaneous morphological and functional imaging of the honeybee's brain by two-photon microscopy. Nuovo Cim C 34:1–10. doi:10.1393/ncc/i2011-10960-4
64. Bucher D, Scholz M, Stetter M et al (2000) Correction methods for three-dimensional reconstructions from confocal images: I. Tissue shrinking and axial scaling. J Neurosci Methods 100:135–143. doi:10.1016/S0165-0270(00)00245-4
65. Hama H, Kurokawa H, Kawano H et al (2011) Scale: a chemical approach for fluorescence imaging and reconstruction of transparent mouse brain. Nat Neurosci 14:1481–1488. doi:10.1038/nn.2928
66. Groh C, Kelber C, Grübel K, Rössler W (2014) Density of mushroom body synaptic complexes limits intraspecies brain miniaturization in highly polymorphic leaf-cutting ant workers. Proc Biol Sci 281:20140432. doi:10.1098/rspb.2014.0432
67. Danielson E, Lee SH (2014) SynPAnal: software for rapid quantification of the density and intensity of protein puncta from fluorescence microscopy images of neurons. PLoS One 9:e115298. doi:10.1371/journal.pone.0115298
68. Sturt BL, Bamber BA (2012) Automated quantification of synaptic fluorescence in C. elegans. J Vis Exp. doi:10.3791/4090
69. Berdnik D, Chihara T, Couto A, Luo L (2006) Wiring stability of the adult Drosophila olfactory circuit after lesion. J Neurosci 26:3367–3376. doi:10.1523/JNEUROSCI.4941-05.2006
70. Flanagan D, Mercer AR (1989) An atlas and 3-D reconstruction of the antennal lobes in the worker honey bee, Apis mellifera L. (Hymenoptera: Apidae). Int J Insect Morphol Embryol 18:145–159. doi:10.1016/0020-7322(89)90023-8
71. Brandt R, Rohlfing T, Rybak J et al (2005) Three-dimensional average-shape atlas of the honeybee brain and its applications. J Comp Neurol 492:1–19. doi:10.1002/cne.20644
72. Stocker RF, Singh RN, Schorderet M, Siddiqi O (1983) Projection patterns of different types of antennal sensilla in the antennal glomeruli of Drosophila melanogaster. Cell Tissue Res 232:237–248. doi:10.1007/BF00213783
73. Laissue PP, Reiter C, Hiesinger PR et al (1999) Three–dimensional reconstruction of the antennal lobe in Drosophila melanogaster. J Comp Neurol 405:543–552, doi:10.1002/(SICI)1096-9861(19990322)405:4<543::AID-CNE7>3.0.CO;2-A [pii]
74. Couto A, Alenius M, Dickson BJ (2005) Molecular, anatomical, and functional organization of the Drosophila olfactory system. Curr Biol 15:1535–1547. doi:10.1016/j.cub.2005.07.034
75. Tanaka NK, Endo K, Ito K (2012) Organization of antennal lobe-associated neurons in adult Drosophila melanogaster brain. J Comp Neurol 520:4067–4130. doi:10.1002/cne.23142
76. Bestvater F, Spiess E, Stobrawa G et al (2002) Two-photon fluorescence absorption and emission spectra of dyes relevant for cell imaging. J Microsc 208:108–115. doi:10.1046/j.1365-2818.2002.01074.x
77. Albota MA, Xu C, Webb WW (1998) Two-photon fluorescence excitation cross sections of biomolecular probes from 690 to 960 nm. Appl Opt 37:7352. doi:10.1364/AO.37.007352
78. Wokosin DL, Loughrey CM, Smith GL (2004) Characterization of a range of fura dyes with two-photon excitation. Biophys J 86:1726–1738. doi:10.1016/S0006-3495(04)74241-1
79. Svoboda K, Yasuda R (2006) Principles of two-photon excitation microscopy and its applications to neuroscience. Neuron 50:823–839. doi:10.1016/j.neuron.2006.05.019

Part IV

Genetic Techniques

Chapter 16

Genetics of Human Handedness and Laterality

Silvia Paracchini and Tom Scerri

Abstract

Handedness is the most evident lateralized trait in humans. A weak (~25%) genetic component has been consistently reported for hand preference across independent studies. Genomic technologies have made rapid progress in increasing throughput and resolution of genetic mapping but their success is largely dependent on the availability of large sample sizes. Hand preference in humans is easily determined and available for tens of thousands of individuals for which genomic data are also available. Yet, strong genetic candidates for hand preference have not been proposed yet. Genetics analyses using quantitative measures of handedness have demonstrated that genes involved in establishing left/right structural laterality play a role in controlling behavioral laterality. Therefore it is crucial how handedness is measured and it is clear that other genetic determinants remain to be discovered. The combination of detailed phenotypes with high throughput technologies promises to advance, in the near future, our understanding of handedness genetics and, in turn, the relevant biology. These discoveries will contribute to addressing one of the long-standing questions in laterality research: what makes some people left-handed?

Key words Human genetics, GWAS, Quantitative genetics, Handedness, Brain asymmetries

1 Handedness and Laterality

Humans display different types of laterality. At a structural level visceral organs are asymmetric in shape and position; for example the heart is normally on the left and the liver on the right. The two brain hemispheres are not symmetrical in structure and function. A typical lateralized function of the brain is language. Language dominance normally resides in the left hemisphere [1]. But, undoubtedly the most obvious and visible left-right asymmetry in humans is hand preference, with the vast majority of people having a clear preference to perform most tasks with one hand.

World-wide the majority of people are right-handed. This population bias implies an evolutionary advantage raising the question of why left-handedness is maintained. Several theories have been proposed to explain this phenomenon, suggesting the bias is regulated by evolutionary forces and results from the interactions between functionally asymmetric individuals [2]. Handedness

Lesley J. Rogers and Giorgio Vallortigara (eds.), *Lateralized Brain Functions: Methods in Human and Non-Human Species*, Neuromethods, vol. 122, DOI 10.1007/978-1-4939-6725-4_16, © Springer Science+Business Media LLC 2017

presents a correlation, although imperfect, with language lateralization. Language dominance in the left hemisphere is observed for over 90 % of right-handers but only in 70 % of left-handers [3]. Since language is considered by most to be a feature unique to humans, this weak correlation has contributed to the high interest in handedness. Atypical asymmetries have been observed in different neurodevelopmental traits such as schizophrenia or dyslexia. Understanding the molecular mechanisms of asymmetries may help dissecting the cause/effect relationship between laterality and disorders [4].

Studying the genetics of human handedness will provide a relatively accessible tool to investigate complex phenomena such as cerebral asymmetries, complex neurological disorders, and human evolution. Recent progress in genomics technology is providing unprecedented platforms for studying genetically determined traits.

2 Handedness Genetics

Before studying the genetics of a particular human trait, it is important to establish whether a genetic component really does exist and that the observed phenotypic variation is not solely due to environmental factors.

Different lines of evidence support genetic determination, at least partially, of handedness (reviewed by [5]). Ultrasound scanning has shown that 85 % of fetuses preferentially move their right hand at 10 weeks of gestation reliably predicting the hand preference that they will develop later in life and hence suggesting an early determination of hand preference [6]. A higher correlation of handedness between parents and their children is observed in biological—compared to adoptive-families [7]. Twin studies are the most common approach used to estimate the genetic contribution explaining the variation of a trait across individuals, referred to as **heritability**. This method compares identical (monozygotic, or MZ) twins with fraternal (dizygotic, or DZ) twins to distinguish genetic and environmental contributions for a particular phenotype. Both MZ and DZ twins normally share a highly similar environment during their in utero gestation, birth, and upbringing. However, MZ twins share the same genome (Box 1), while DZ twins are generally no more genetically similar than non-twin siblings. If a trait is determined by a genetic component we expect to see MZ twins more similar, or concordant, for that trait compared to DZ twins. Concordance rates in MZ compared to DZ twins provide an estimate of the heritability of such trait (for a more detailed and theoretical explanation see [8]). Heritability close to zero indicates little or no genetic contributions while a figure close to one suggests that a particular trait is completely under genetic control.

Heritability for hand preference is not high, but independent studies have consistently reported a genetic contribution of about 25% [9]. This figure indicates that about one quarter of what determines human handedness can be assigned to genetic variation alone. For comparison the heritability of human height is estimated to be approximately 80% [10]. Genetic control for behavioral laterality has been reported for other species. For example, in fish (poeciliid fish, *Girardinus falcatus*) a test that reproduces the choice of turning direction when approaching a predator has an estimated heritability greater than 50% [11].

3 Is There a Gene for Handedness?

Hand preference frequency and heritability estimate fit nicely with single gene theories, such as the right shift model [12] and the 'dextral/chance' model [13]. However, to date no gene determining hand preference has been convincingly reported and it is safe to say, that if such a gene existed, we would have had a very good chance of finding it by now. The identification of specific genes or genetic variants underlying a trait is largely driven by the available technologies (Fig. 1). A key element to increase the chances of identifying genes contributing to a specific trait is the availability of large samples. Given hand preference is easily measured and left-handedness is relatively common, obtaining samples large enough for genetic screenings is not particularly challenging. In fact, it is remarkable that, in spite of large samples being available, a convincing genetic mechanism controlling hand preference has not yet been described. As we will discuss later, different strategies might be necessary to map hand preference genes, starting from different definitions of handedness.

4 Linkage Analysis and Candidate Gene Studies

In the past, the challenges involved in generating genotype data were the main limiting factors for genetic research. Researchers have therefore prioritized fractions of the genome using the so-called candidate gene approach, which selects genes based on their known function or for residing in chromosomal regions selected by linkage analysis. The principle of linkage analysis involves identifying **genetic markers** (see Box 2), that co-segregate with a phenotype, such as handedness, within a large family or a collection of nuclear families. A useful genetic marker is any location within the genome that shows substantial genetic variability and can be defined as polymorphic; this could be simple **single nucleotide polymorphisms** (**SNPs**) or more complicated repeated segments

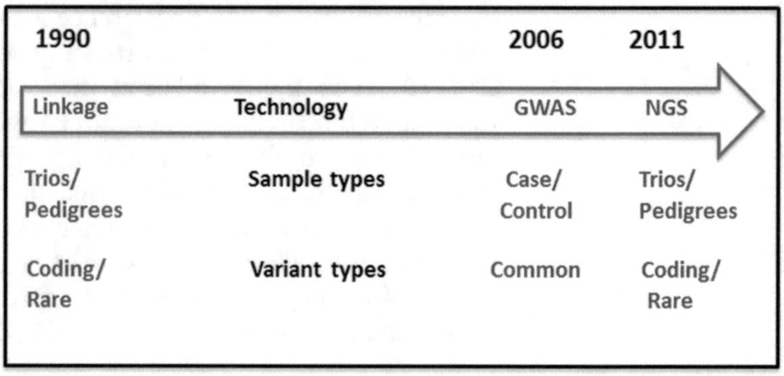

1990		2006	2011
Linkage	**Technology**	GWAS	NGS
Trios/ Pedigrees	**Sample types**	Case/ Control	Trios/ Pedigrees
Coding/ Rare	**Variant types**	Common	Coding/ Rare

Fig. 1 Human genetics is a technology-driven field. The technology has shaped the type of genetic analysis in terms of statistical approaches and types of genetic variants that could be mapped. Linkage analysis, traditionally, is a low-resolution approach relying on tracing inheritance patterns of highly polymorphic markers, such as microsatellites, across families. It has been applied successfully to map highly penetrant and rare variants in collections of trios or large pedigrees, with specific and often severe phenotypes. Linkage has been successful to identify many genes for monogenetic diseases for which mutations in one gene reliably predict a clinical condition. Scanning effectively single nucleotide polymorphisms (SNPs) at high density across the genome has led to the development of GWASs. Analysis of hundreds of thousands of markers across large samples of cases and controls has identified common variants associated to many complex and common disorders. These variants explain a very small proportion of the heritability underlying complex traits. It is now possible to re-sequence the genome to find rare and highly penetrant variants which might also contribute to complex traits. Linkage and GWAS approaches have been used to map the genetic component of handedness

known as **microsatellites**. Microsatellites are usually the marker of choice for linkage analysis, whereas SNPs are employed for association studies. The power of linkage analysis is determined by several parameters, notably the size and the number of the families, the number of individuals with and without a particular phenotype, the type and number of genetic markers tested, the **penetrance** of the underlying genetic etiology, and the underlying genetic model (e.g. **recessive** or **dominant**). A large pedigree ideal for linkage analysis would normally present the phenotype of interest in multiple family members at each generation (see also Fig. 7). This pattern of inheritance would be suggestive of a dominant model, where a single mutation in one copy of a gene is sufficient to cause the phenotype in that family, normally by disrupting gene function. In contrast, under a recessive model, two mutations (identical or different) in the same gene are required to lead to the phenotype. Normal functioning requires both copies of a gene under a dominant model, whilst single nonmutated gene is sufficient under a recessive model. It is estimated that all of us are carriers of several pathogenic mutations within different genes, but they do not lead to any clinical consequence due to the other functioning copy in our genomes.

Linkage analysis for highly penetrant mutations is followed by re-sequencing of the genes within "linked" chromosomal segments, to identify the genetic variant that might disrupt gene function and cause the phenotype.

Linkage analysis for multifactorial traits or conditions is normally carried out in a large number of families. In this case, the hypothesis is that the linked chromosomal segments carry variants in the same gene contributing to the same phenotype across unrelated individuals. Such variant is expected to have a small **effect size. Association studies** (as opposed to re-sequencing) targeting the genes within the linked chromosomal fragment normally represent the following step (Fig. 2). The basic principle of an association study is to test whether **allele** frequencies of a tested marker are significantly different between categorical groups (e.g. cases and controls, short and tall people, or right-handers and left-handers). Association studies conducted for quantitative measures (e.g. height) usually test whether alleles are associated with a shift from the mean of a normal distribution.

Genetic studies of language impairment provide good examples for both approaches. Linkage analysis in the KE family, presenting a large number of individuals affected by severe dyspraxia, identified a linked region on chromosome 7 [14]. Subsequent screening of this region led to the identification of a mutation within the *FOXP2* gene [15]. Dyspraxia is a rare and severe difficulty in articulating language due to difficulties in fine orofacial movements of form. Specific language impairment (SLI) is a more common form of language disorder and, while it recurs in families, it is not observed with patterns consistent with a dominant model such as in the KE family. Linkage analysis of SLI in a collection of nuclear families has led to the identification of several linked chromosomal segments (for a review see [16]). Genes within these segments were prioritized for further screening based on their characteristics, such as a known role in neurodevelopment or specific expression patterns in the brain, as in a typical candidate gene approach. These studies have led to the identification of the genes *CMIP* and *ATP2C2* [17]. The *CNTNAP2* gene was also implicated in SLI using a different candidate gene strategy [18]. *CNTNAP2* was selected because it was observed to have a direct interaction with the protein product of the *FOXP2* gene.

Similar approaches have been applied to identify genes influencing hand preference but with less success. Linkage for hand preference has been reported at 10q26 [19] and 12q21-23 [20] but these findings have not been independently replicated and candidate genes within these regions have not been reported. A recent

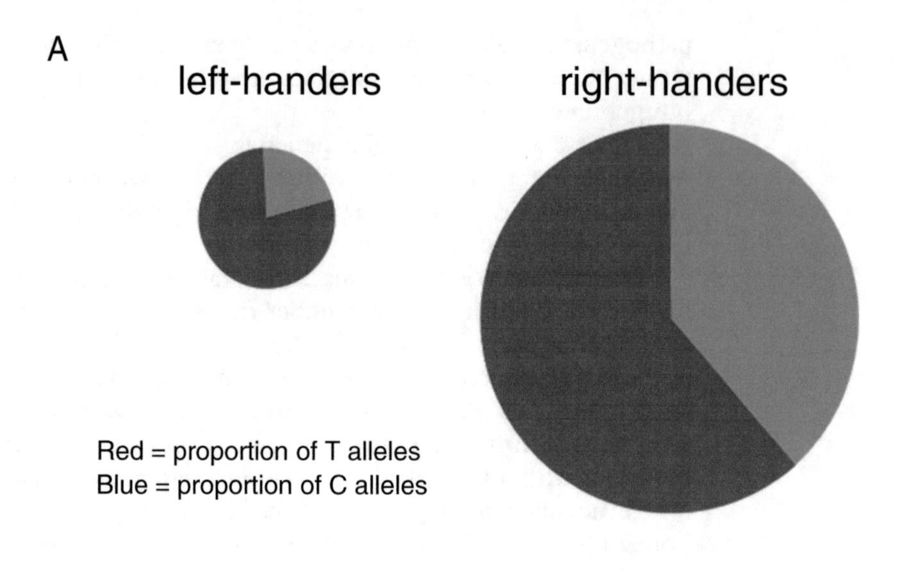

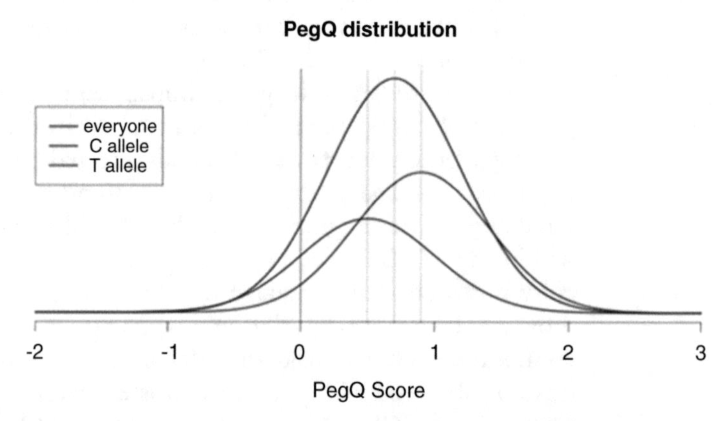

Fig. 2 Principles of GWASs for handedness: (**a**) Categorical GWASs compare allele frequency of millions of markers across the genome in left- and right-handers. In this hypothetical example, the alleles (C and T) of one SNP show substantial difference between the two groups. We see that the frequency of allele C is approximately 20 % in the left-handers compared to 40 % in the right-handers (represented by the blue slices of the pie charts). Typically the number of left-handers is about 1/10 of the right-handers. (**b**) A quantitative GWAS for relative hand skill (or PegQ) tests whether any allele is associated with a different phenotypic distribution compared to the general population (*black line*). PegQ is normally distributed with a positive mean, because most people will be better with their right hand. The null hypothesis is that both alleles of a SNP show a phenotypic distribution with the same mean. A marker is associated with a phenotype when the phenotypic distributions observed for the two alleles show shifts of the mean compared to the entire population. In this case the T allele is associated with a stronger lateralization towards relative stronger skills with the right hand while the C allele is associated to a mean shifted towards the left. The *dotted vertical lines* align with the mean of the three distributions

linkage study on 37 Dutch families also did not highlight any significant findings [21]. Linkage analysis in large pedigrees with left-handed individuals has not led to any significant results. If such families were identified today, they would be analyzed directly with deep sequencing strategies (i.e. Next-generation sequencing (NGS), see below and Fig. 7). Indeed, researchers are actively recruiting such large families with a predominance of left-handed individuals for genetic analysis (e.g. a project led by the Max Planck Institute for Psycholinguistics http://www.mpi.nl/departments/language-and-genetics/projects/brain-and-behavioral-asymmetries/genetics-of-handedness).

One candidate gene study looked at a sample of 27 nuclear families with at least two left-handed children. Genes (*DNAHC1*, *DNAHC2*, *DNAHC6*, *DNAHC8*, *DNAHC13*, *LRD*, and *NODAL*) were selected on the basis of their known function in controlling the development of the left-right anatomical axis, but no significant findings were reported [22].

The *APOE* gene, better known for conferring a strong risk for Alzheimer's disease [23], was also selected as a candidate for handedness because of its implication in a range of cognitive functions [24]. Some *APOE* alleles were reported at a higher frequency in left-handers but the findings were not subsequently replicated [25, 26].

More support has instead been provided for a role of the X chromosome. Combining the reports of linkage on chromosome X [27, 28] and the hypothesis that prenatal exposure to testosterone would influence laterality, a candidate gene study reported association with hand-preference at the androgen receptor (*AR*) gene in a sample of 783 individuals [29]. The genetic marker used for association was a microsatellite that encoded a polymorphic polyglutamine tract (a varying number of glutamine amino acids in the protein product), and could be quantified by measuring the number of CAG repeats in the *AR* gene. The study reported that left-handedness was associated with number of CAG repeats. Females and males were more likely to be left-handed if they had greater or smaller number of repeats, respectively. The number of repeats also correlated, positively in males and negatively in females, with testosterone levels. Therefore, the authors concluded that lower testosterone levels are related to left-handedness, both in males and females. To some extent, further support for a role of the *AR* CAG repeats in handedness was reported but with different trends [30]. Longer CAG repeats were observed in mixed-handers compared to right- and left-handers who did not show significant difference. One possible difference across the two studies was that instead of classifying handedness as a binary category [29], the replication studies measured handedness along a scale which allowed identifying the mix-handers [30] (see below discussion about phenotype definition). It remains a possibility that these findings represent false positives.

5 Genome-Wide Association Studies (GWASs)

The key conclusion we can derive from the lack of a strong candidate for hand preference is that we are not dealing with a monogenetic trait but rather with a complex phenotype influenced by multiple genes, each probably contributing a small effect. These kinds of genetic factors are very challenging to be identified with linkage analysis.

Technological advances have allowed the genetic mapping of complex traits by overcoming the challenges of generating genotype data (Fig. 1). Different chips are now available to query human DNA samples for hundreds of thousands of genetic variants in a single experiment at an affordable cost. Genotypes can then be analyzed for **genome-wide association studies (GWASs)**, which in recent years have seen an exponential growth [31]. GWASs apply the principle of association studies to the entire genome, so that allele frequencies of millions of SNPs are compared in cases and controls or right- and left-handers (Fig. 3). A marker showing a statistically significant difference between the two groups indicates a genetic association for the trait that distinguishes the groups. Both the generation of data and their analysis has now become routine and follows standard pipelines (e.g. see [32] for a description of statistical analysis). In addition to the genotype data directly genotyped (typically up to one million markers generated on a single chip), it is also possible to infer, or impute, genotypes at other markers. This is achieved through a probabilistic process by utilizing human genome reference sequences from large-scale projects that have annotated the genomes of different populations (i.e. the HapMap project http://hapmap.ncbi.nlm.nih.gov/ or, more recently, the 1000 Genomes project http://www.1000genomes.org/). An awareness of ethnicity in association studies is important to avoid false positive association signals. Genetic markers have different allelic frequencies across populations, therefore comparing cases and controls derived from different populations might lead to false associations simply because of an ethnicity effect, known as **population stratification**. In some instances this can be controlled for by using principle component analysis (PCA). The main limiting factor for GWASs is the availability of adequately large samples and the quality of the phenotype measurements, but the ideal design requires wider consideration (Table 1). Large samples are needed to have the statistical power to counterbalance millions of tests. Good quality phenotypes are equally important. The debate is still open on whether it is better to run analysis on extremely large samples characterized with crude phenotype or to focus the analysis on more detailed phenotypes, which however can be extremely laborious and expensive to be collected. Furthermore, what works for one trait might not apply to another, as it could well be the case for handedness.

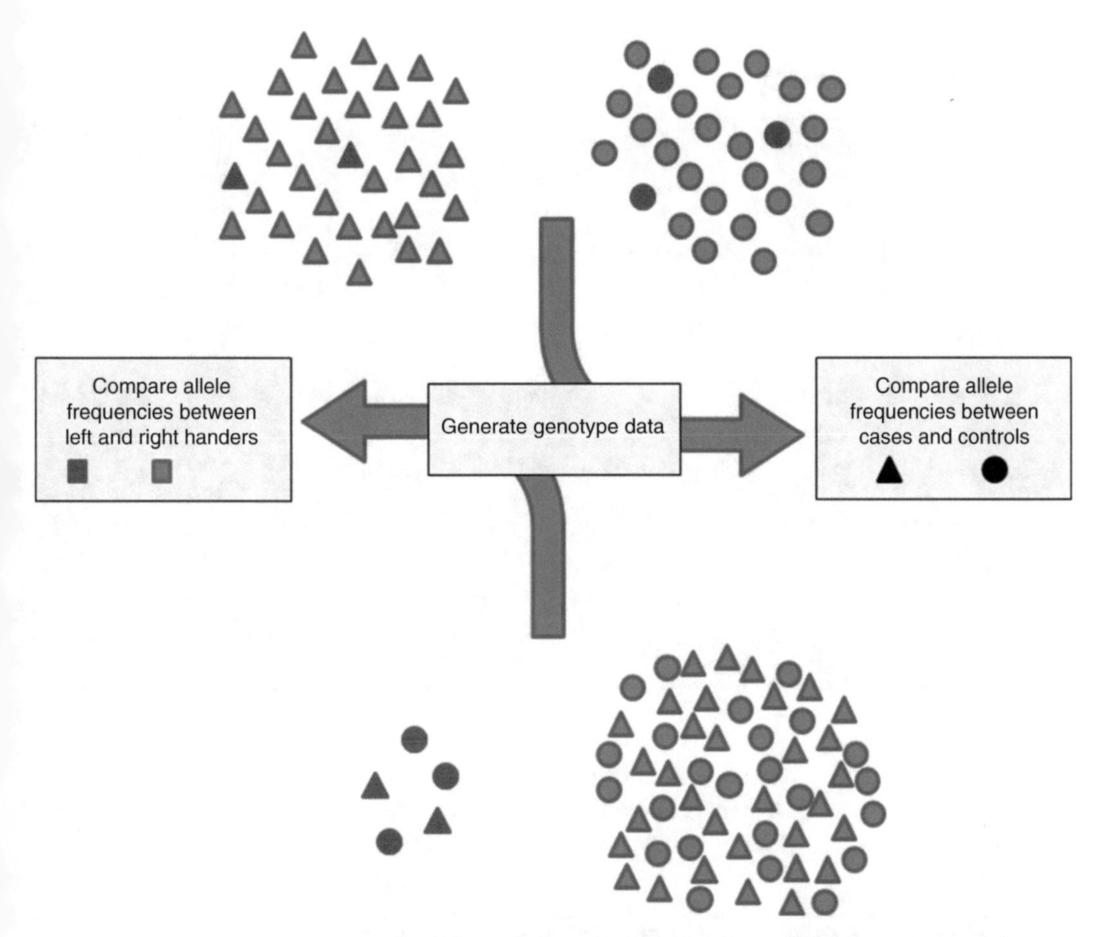

Fig. 3 Re-analysis of GWAS data. DNA samples from cases (*triangles*) and controls (*circles*) selected for a disease are used to generate hundreds of thousands to millions of genotype data to conduct genome-wide association analysis. Allele frequency of each genotyped marker is compared across cases and controls to detect statistically significant differences suggestive of genetic associations. The same genotype data can be re-analyzed by regrouping the study participants as left- (*red*) or right (*blue*)-handers. This is possible because hand preference is routinely collected as a simple tick-box questionnaire in large clinical cohorts

The few GWASs published so for hand preference have reported no significant associations [33, 34]. This adds further weight to the hypothesis that it is highly unlikely that a single gene can control handedness, whereas polygenetic models are more appropriate [35]. Hundreds of thousands of patients have now been analyzed in GWASs for the most diverse complex traits, and these studies have resulted in finding many genes underlying common diseases. Hand preference data is available in a large proportion of these cohorts, as it is collected as a standard tick-box question on most clinical questionnaires. Therefore, by re-analyzing the genotype data for a left/right hand preference rather than a case/control definition it is possible to rerun effectively a GWAS study for handedness (Fig. 3). In spite of the availability of large

Table 1
Quick guide for a GWAS design

Factor	Comments
Sample size	Large sample sizes (2000 < *N* < no upper limit) to have sufficient power
Matching cases and controls	Ethnicity, sex, and age should ideally be matched between groups
Covariates	Collect data on other measures that could be relevant for a particular trait, such as smoking habits for lung cancer, or sugar intake for diabetes, or IQ for reading ability
Technology	Choose an array with a dense number of genetic markers (*N* > 700,000) and relevant to the ethnicity of studied cohorts
Imputation	Should be performed in the analysis to improve resolution
Statistical analysis	Data QC and association analyses can be performed according to standard pipeline
Phenotype	An open debate: should we use easily collected phenotypes (e.g. tick box on questionnaire) in large samples or invest in collecting high-quality measures?
Test	Categorical or quantitative, allelic, genotypic, dominant or recessive, logistical or linear regression

samples (tens of thousands), candidate genes for hand preference have not been reported yet. But is hand preference the most appropriate phenotype for genetic studies?

6 Handedness: A Categorical or Quantitative Phenotype?

So far we have treated handedness as a categorical trait: left or right. This is readily established by asking which hand an individual prefers for writing. However, handedness is more accurately defined by degree and not category (Fig. 4). The Edinburgh inventory [36], or modified versions of it, asks which hand is preferred for a number of tasks. A laterality quotient from the scores gives a J-shaped distribution, with most people using either the right or left hand for the majority of tasks and few people showing mixed use of both hands (Fig. 4b). The questionnaire measures the direction of handedness (left or right) and the degree of handedness (how much one hand is preferred). There are no fixed criteria to define systematically this degree. A variation of the questionnaire, i.e. the Crovitz–Zener Handedness Inventory [37], was used in the replication study for the *AR* association discussed above [30]. A particular test, the pegboard test (Fig. 5), assesses fine motor skills but provides also a quantitative measure of dexterity [38]. It measures the time taken to complete a task with the left (L) and right (R) hand. $\text{PegQ} = [2(\text{L} - \text{R})/(\text{L} + \text{R})]$ measures relative hand skill, indicating how much one individual is better with one hand,

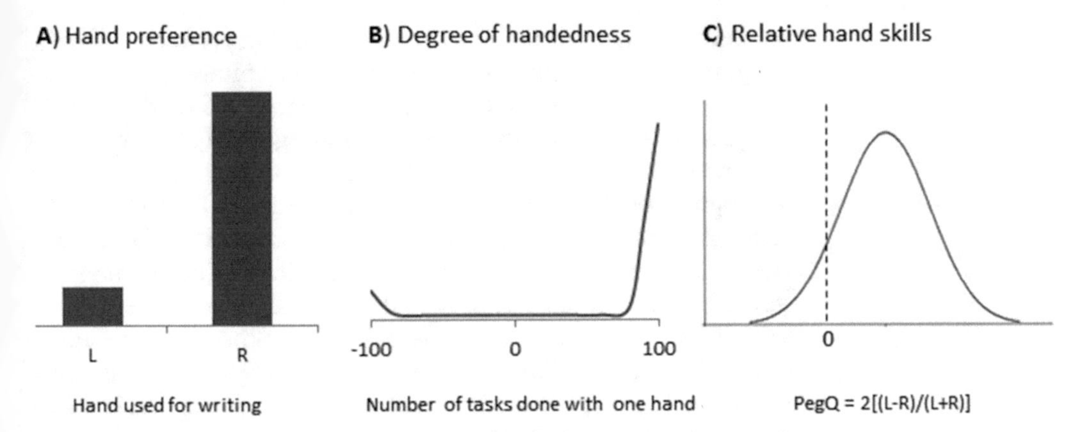

Fig. 4 How to measure handedness: (**a**) Hand preference, measured by asking which hand is used for writing, results in two groups with a ratio of about one left (L) to nine right (R)-handers. (**b**) The Edinburgh Inventory measures the number of tasks performed with one hand on a list of typically 15–20 items (positive and negative index indicates right or left hand preference respectively). (**c**) PegQ is a measure derived by the pegboard task and leads to a normally distributed measure describing how much one individual is better with one hand versus the other. It measures relative hand skills by comparing the time employed to complete the task with the left hand (L) versus the right (R) hand

Fig. 5 The pegboard. During the pegboard task children move ten pegs from a row of ten holes to another with the left (L) and right (R) hands alternated. The time taken to complete the task with each hand provides a fine motor control measure. The relative difference in time, PegQ = [2(L – R)/(L + R)], is a measure of lateralization and shows how much an individual is better with one hand versus the other (Fig. 4c). PegQ strongly correlates with hand preference. The picture is a courtesy of Priti Kashyap

or is lateralized. PegQ correlates with hand preference [39], is normally distributed and has a positive mean, as expected, because most individuals are faster with their right hand (Fig. 4c). Heritability estimates are higher for measures of degree of handedness (up to 67%) [40] compared to hand preference (25%), suggesting they are more suitable for genetic studies.

The PegQ measure has been used for linkage and candidate gene analyses in a limited number of studies. Linkage with PegQ has been conducted in a cohort originally selected for dyslexia. The strongest signal was detected on chromosome 2p12-q11 [27, 41]. A few candidate genes within this segment were then selected for association analysis on the basis of an established pattern of gene expression in the brain. A **haplotype** upstream of the *LRRTM1* gene showed association when paternally inherited [42]. Although this finding does not replicate in independent cohorts unaffected with dyslexia, the same haplotype was also associated with schizophrenia when paternally inherited [42–44].

Other studies have found associations with handedness when looking in cohorts selected for clinical diagnosis. For example, analysis of a cohort selected for bipolar disorder showed a variant in the *COMT* gene (Val158Met) associated with relative hand skill, measured by the time required to make dots in a series of circles [45]. The Val158Met variant was selected for a candidate gene approach because of its association is reported in a range of psychiatric disorders including bipolar disorder, schizophrenia, and ADHD.

These latter findings raise the question whether ascertainment bias might be crucial for gene identification. Quantitative laterality measures are more systematically assessed in these groups of patients. Therefore, the results could also be due to the availability of larger datasets in patients with neurodevelopmental and psychiatric disorders. However, these observations would support the idea that a link between laterality and neurological disorders, hypothesized for a long time, does exist [5].

7 The PCSK6 Association

Indeed, the identification of the only gene (*PCSK6*) that reached genome-wide statistical significance (P-value $< 5 \times 10^{-8}$) depended on dyslexia diagnosis ascertainment criteria and the use of the PegQ measure [39, 46]. The total sample size ($N = 728$) was relatively small for a GWAS, but the association was identified in three independent cohorts all selected for dyslexia. Furthermore, a completely independent study reported association for degree of handedness with a **VNTR** (variable number tandem repeat) marker at the same *PCSK6* locus (Fig. 6) [47]. What makes *PCSK6* an extremely attractive candidate for handedness is its known biological function. PCSK6 activates the NODAL pathway, one of the key players in setting up left/right body asymmetries [48]. In the early phases of development, the embryo forms a cavity, known as the node, in which motile cilia generate a flow through a

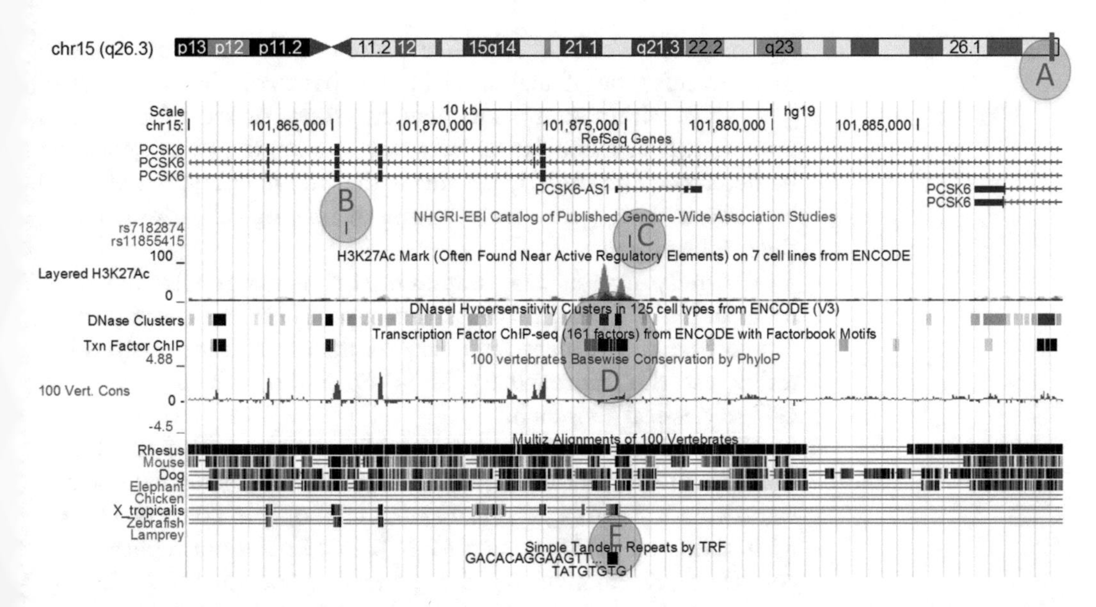

Fig. 6 The PCSK6 region associated with handedness. A screenshot from the UCSC Genome Browser (www.genome.ucsc.edu). The browser allows the user to zoom-in to different parts of the human genome. Here, we have zoomed into a section of *PCSK6*. The *blue horizontal lines* show the extent of *PCSK6*, with the chevrons indicating the introns and the *dark blue blocks* indicating the exons of the gene. Towards the bottom of the figure, the *black horizontal patches* show the level of DNA conservation between human and other species; more *black* indicates more conservation. A few elements are highlighted by *pink circles*; (**A**) The location of PCSK6 on chromosome 15. (**B**, **C**) The SNPs rs7162874 and rs11855415 associated with relative hand skill. (**D**) A region of DNA implicated in transcriptional regulation. (**E**) The VNTR associated with degree of handedness

unidirectional rotation. This flow is detected by nonmotile mechanosensory cilia, transducing the signal into the asymmetrical expression of NODAL on the left side of the embryo to set up the structural left-right asymmetries. The Nodal pathway is very conserved and has been found to control structural asymmetries across many species including chirality in snails [49]. Nodal has also been found to be relevant for behavioral asymmetries in zebrafish. The zebrafish *fsi* (frequent situs inversus) line displays reversal of body, some neuronal asymmetries, and behavioral phenotypes, like eye preference and direction of prey approach [50]. The *fsi* phenotype supports the idea, suggested by the *PCSK6* association with handedness, that the Nodal pathway, known to control structural asymmetries, may also contribute to behavioral laterality.

8 Pathway Analysis

Pathway analysis (or **gene set analysis, GSA**) is a promising approach to assess the combined effect of multiple genes, rather than single marker/traits associations, by re-analyzing GWAS data. While there is substantial consensus on the validity of such approaches, we still do not have commonly accepted pipelines to perform this type of analysis [51]. The basic principle is to aggregate SNPs within an a priori defined group, or set, of genes and test whether any association between a trait and a set of genes can be detected. The underlying idea is that the phenotype is affected by variants across genes contributing to a biological pathway rather than isolated genes.

Pathway analysis has the potential to generate false positives, for example by testing many gene groups. Therefore it is important not to base the primary discovery results on pathway analysis but rather use this approach to follow up findings from an original GWAS classical approach.

For example, GSA was used to follow up the GWAS that led to the PCSK6 association [39]. *PCSK6* was the only gene that reached statistical significance, and it was detected only in cohorts selected for a definition of dyslexia ($N=728$). *PCSK6* did not replicate in a cohort representative of the general population ($N=2666$). However, the strongest association with PegQ in this latter cohort was detected in proximity of the *GPC3* gene on the X chromosome. Mutations in *GPC3* cause heart and lung asymmetry defects in mice [52] and therefore, similarly to *PCSK6*, implicated the biology that controls structural body asymmetries in handedness. Accordingly, GSA was conducted by testing specifically genes known to control structural asymmetries. A set of genes was derived from the mouse literature for their role in leading to laterality defects. Re-analysis of the GWAS data with a GSA approach showed an over-representation of genes known to cause left/right asymmetry phenotypes. In particular, three phenotypes showed association with PegQ both in the general population and in the dyslexia cohorts: heterotaxia (the right/left transposition of thoracic and/or abdominal organs); *situs inversus* (a reversal of organ asymmetry); and double outlet right ventricle (a heart asymmetry defect). Among the top associated genes in the GSA analysis conducted in the dyslexia cohort were *MNS1*, *RFX3*, *PKD2*, *GLI3*. Disruption of *Mns1*, *Rfx3*, or *Pkd2* in mice causes *situs inversus* [53–55]. *Rfx3* regulates the expression of *Gli3*, which is required for the formation of the corpus callosum, a bundle of fibers connecting the left and right hemispheres [56]. Interestingly, *PCSK6* is highly expressed in the corpus callosum in the adult brain.

Therefore the results further supported the initial hypothesis indicated by the *PCSK6* association, suggesting that shared biology between structural and behavioral asymmetries.

This hypothesis is contrasted by the observation that, individuals with *situs inversus* do not show an increased chance of being left-handed [57], implying different mechanisms at the basis of body and brain asymmetry [58]. *Situs inversus* is a rare condition and very few individuals have been analyzed up to now. But very likely, the genetics of handedness is far more complex than being explained by a few genes determining early patterning of asymmetries. In fact, the effect sizes of *PCSK6* and other asymmetry genes are very small explaining a minimal fraction of variability in handedness measures. Much larger GWASs for PegQ, and ideally for other handedness-related phenotypes, will be required to identify additional genetic factors.

9 Next-Generation Sequencing

One of the latest technologies available to geneticists today is known as next-generation sequencing (NGS). Before NGS technology, genetic sequencing was routinely performed by a process known as Sanger technology, which was relatively slow, expensive, and had a low output. Sanger sequencing is a very targeted process that normally sequences a single short and precise fragment of the genome. With NGS technology, we can sequence the 70 Mb of human exonic DNA (known as whole exome sequencing; WES) or the entire 3500 Mb human genome (known as whole genome sequencing; WGS) with relative ease and cost-effectiveness. However, NGS is a stochastic process whereby millions of random DNA fragments are sequenced and so this requires intensive computational processing of the raw data to assemble them together using complex bioinformatics pipelines. Table 2 lists a selection of tools routinely used in bioinformatics analysis providing a flavor of the variety of steps and required and the complexity involved in interpreting this type of data. Common guidelines are emerging to ensure best practise and homogeneity across studies. One of the most established and commonly used procedures is the GATK protocol (https://www.broadinstitute.org/gatk/). The final output of NGS consists of stacks of DNA sequence, or reads, and the level of stacking is known as the depth, or coverage. The greater the depth, the more reliable the consensus sequence of the sample analyzed, and the greater the likelihood of identifying a mutation. Typically, a depth of ×30 or greater is sufficient to reliably detect mutations.

NGS has accelerated the identification of mutations that cause inherited diseases and has now been implemented in large scale projects [59] and publically accessible databases: http://exac.broadinstitute.org/. The novelty or rarity of any variants discovered with NGS can be determined by looking them up in databases of 1000s of individuals that have already been sequenced, such as in the 1000 Genomes Project (1KGP) (http://www.1000genomes.org/) database or the NHLBI Exome Sequencing Project's (ESP) Exome Variant Server (http://evs.gs.washington.edu/EVS/). Rare or novel genetic variants might have no effect on any phenotype, therefore a suite of special computer programs have been

Table 2
Commonly used bioinformatics tools for analyzing and interpreting genetic data

Category	Description	Name	Website
Genome browsers and genetic variants databases	These are the front ends for vast amounts of genetic data, such as genes and variants, enabling them to be visualized and downloaded	UCSC	genome.ucsc.edu
		Ensembl	ensembl.org/
		NCBI	ncbi.nlm.nih.gov/gene/
		1000 Genomes	1000genomes.org/
		HapMap	hapmap.ncbi.nlm.nih.gov/
		DGV	dgv.tcag.ca/dgv/app/home
		EVS	evs.gs.washington.edu/EVS/
		IGV	broadinstitute.org/igv/
NGS tools	These tools primarily (but not exclusively) QC, clean, process, manipulate, recalibrate, analysis, variant detect and/or output NGS sequencing data in one form or another	fastQC	bioinformatics.babraham.ac.uk/projects/fastqc/
		Novoalign	novocraft.com/
		Samtools	samtools.sourceforge.net/
		Picard	broadinstitute.github.io/picard/
		GATK	broadinstitute.org/gatk/
		BWA	bio-bwa.sourceforge.net/
		bedtools	bedtools.readthedocs.org/en/latest/
		vcftools	vcftools.sourceforge.net/
Variant analysis	Variants often require annotating, e.g. to give them allele frequencies, or provide predictions for their pathogenicity	SIFT	sift.bii.a-star.edu.sg/
		Polyphen2	genetics.bwh.harvard.edu/pph2/
		Annovar	openbioinformatics.org/annovar/
		SeattleSeq	snp.gs.washington.edu/SeattleSeqAnnotation141/
Variant processing, quality controls and analysis	Variants, such as SNPs and microsatellites, can be imputed, cleaned, transformed, tested for linkage or association with traits such as handedness. Also, estimates of linkage **disequilibrium** (**LD**) can be made. Tests for inbreeding can be performed	PLINK	pngu.mgh.harvard.edu/~purcell/plink/
		PLINK2	cog-genomics.org/plink2/
		MERLIN	csg.sph.umich.edu//abecasis/Merlin/
		QTDT	csg.sph.umich.edu/abecasis/QTDT/
		Haploview	broadinstitute.org/scientific-community/science/programs/medical-and-population-genetics/haploview/haploview
		IMPUTE2	mathgen.stats.ox.ac.uk/impute/impute_v2.html
		SHAPEIT	shapeit.fr/
		FEstim	genestat.cephb.fr/software/index.php/FEstim
Networks and pathways	Create gene sets, and visualize interactions between genes	STRING	string-db.org/
		DEPICT	www.broadinstitute.org/depict
		MAGENTA	broadinstitute.org/mpg/magenta/
		KEGG pathways	genome.jp/kegg/pathway.html
		Go Ontologies	geneontology.org/

(continued)

Table 2 (continued)

Category	Description	Name	Website
Disease and phenotypes related to genetics	Databases that have linked together genetic associations with phenotypes (diseases and traits)	OMIM DECIPHER ClinVar	omim.org/ decipher.sanger.ac.uk/ ncbi.nlm.nih.gov/clinvar/
CNV analysis	CNV detection primarily from SNP chip data, but also other	PennCNV	openbioinformatics.org/penncnv/

developed to computationally gauge pathogenicity or biological functionality (e.g. ANNOVAR [60]). The GATK website provides best practise guidelines for this analysis as well. In the case of handedness, NGS could be theoretically useful to study families with preponderance of left-handed individuals. The first NGS study for handedness was conducted in a large consanguineous family from Pakistan [61]. Consanguinity increases the chances expressing recessive **recessive** genetic traits and disorders because individuals are more likely to carry two identical copies of rare mutations. Large families have the advantage to allow segregation analysis by observing how a mutation co-segregates with a particular trait across different family members (Fig. 7). Nevertheless, WES in this family did not lead to any clear candidate contributing to a nonright-handed status.

Even potentially functional variants at protein level cannot be immediately implicated in a disease or trait such as left-handedness, and they will need to be investigated for their potential functional effect. For example, finding variants in genes known to influence the asymmetry of the body plan or known to interact with established candidates, such as *PCSK6* (Fig. 8), could be a starting point for interrogation.

10 From Genetic Studies to Biological Function

Indeed the identification of a genetic variant, either associated to a phenotype or identified by NGS, is only the first step in understanding its biological relevance. The location of such variant usually points to a gene. The key question is then how genetic variation at that specific site might impact gene function. Rare and coding variants identified by NGS are expected to significantly impact the protein coded by that particular gene by introducing, for example,

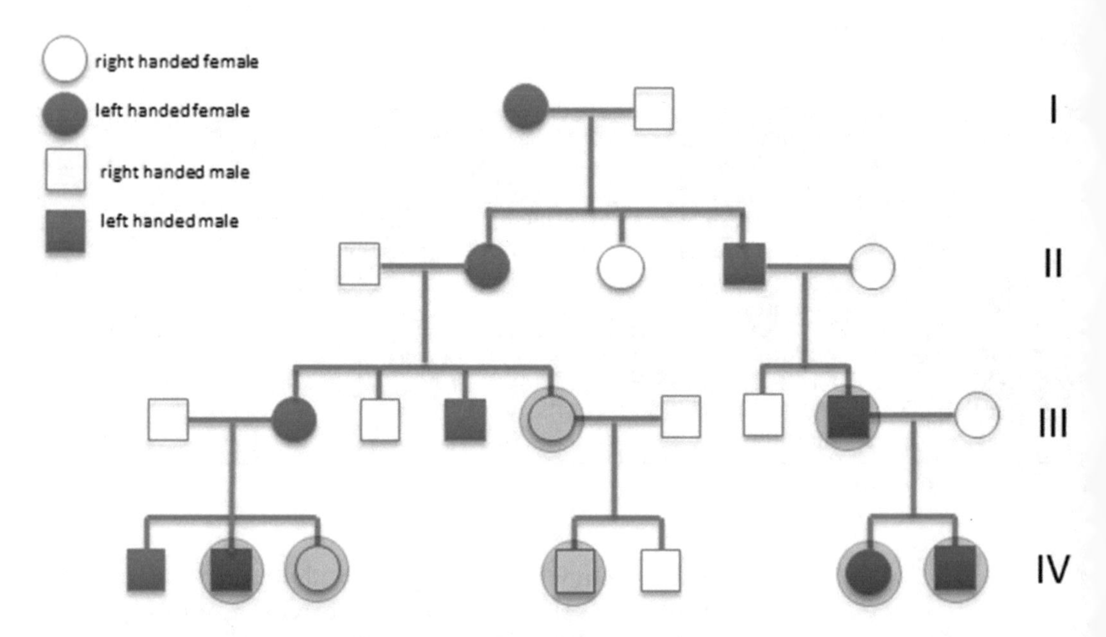

Fig. 7 A pedigree with genetic information and a preponderance of left handed individuals. In this family, the left-hand preference (L) is passed down through each generation in a dominant genetic manner, suggesting a single variant lead to left-handedness. Seven individuals (highlighted by *red circles*) were selected for whole-exome sequencing (WES); four left-handed and three right-handed. Normally, DNA is available for individuals of the most recent generations. If DNA availability and cost were not an issue, all family members could undergo WES. The goal of the analysis is to identify a rare variant shared by the left-handers but not the right-handers. Candidate variants will then be analyzed in the remainder available DNA samples

an amino acid change or by truncating the protein. Genetic variants identified by association studies are not easily interpreted for their functional consequences as they usually do not impact directly a protein sequence. Instead, it has been shown that a significant proportion of GWAS associations occur at regulatory sequences [62, 63], i.e. region of the genome that controls when a gene should be activated (i.e. expressed or upregulated) or silenced (i.e. inhibited or downregulated). This is consistent with the idea that common variants contribute to trait variabilities through very subtle effects. Different molecular mechanisms can mediate these processes. A common scenario is that the two alleles of a SNP located within a **promoter region** yield different DNA protein binding affinities for transcription factors required to regulate gene expression. This mechanism has been described for genetic variants associated with dyslexia [64]. Most of the time such SNPs affect the expression of nearby genes (*cis* effects) but they can also alter the function of a regulatory region acting on a long range (*trans* effect). To help interpret GWAS results for these types of effects, large-scale studies have generated databases of SNPs found to be associated with gene expression level. These types of data are

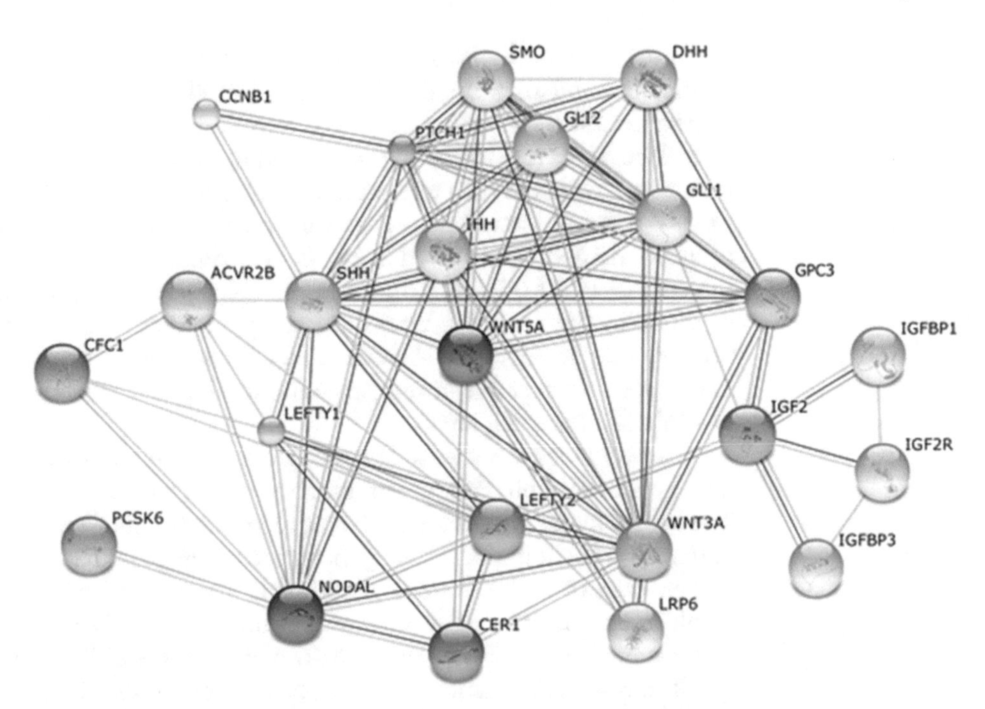

Fig. 8 Computational prediction of PCSK6 networks. Computer algorithms are able to pull together vast amounts of biological data. Here, evidence is taken from numerous databases to produce a network of known and predicted interactions between genes. The different colored lines represent different facets of evidence: *olive*—text-mining: the co-occurrence of protein names in a published reports suggests they are involved in a similar pathway; *pink*—experimental data has directly linked proteins together; *blue*—evidence obtained from a number of database linking together proteins; *violet*—homology between two proteins suggests they are functionally related; *black*—coexpression of genes suggests they are coregulated and so are functionally linked. For example, from text-mining and database searches we can see that *PCSK6* has been linked to *NODAL*, and in turn *NODAL* has been linked to both *LEFTY1*, *LEFTY2,* and experimental data also links *NODAL* to *GPC3* via *WNT3A*. These connections can be used to create gene sets that may be tested for enrichment of their genes in the top signals of GWASs. The graph was created with STRING (http://string-db.org)

known as eQTL (expression quantitative trait loci) databases and are derived by using gene expression levels as the phenotype in GWAS, using particular cell types or tissues derived from many individuals. One limitation of eQTLs is that data are collected in limited tissues, mainly in adult lymphoblastic cells (white blood cells) because they are accessible from blood. However, eQTLs might have a very tight temporal/spatial regulation. Therefore, it is possible that a genuine eQTL effect of a GWAS association might not be present in a database because expression data might not have been collected in the relevant tissue or developmental stage. To partly address this issue the GTEx project (http://www.gtex-portal.org/home/) is aiming to generate eQTL post-mortem data in up to 43 tissues derived from 900 donors. The database includes various brain tissues. Other large studies have looked specifically at gene expression during brain development [65].

Interestingly, the *PCSK6* associations cluster in proximity of a predicted regulatory region likely to affect the transcription of shorter *PCSK6* transcripts (Fig. 6). Specifically, one of the top associated SNPs (rs11855415) alters binding affinity of a transcription factor, which in turn might affect the expression of a particular *PCSK6* transcript in an allele-specific manner [66]. However, none of the *PCSK6* SNPs are listed in eQLT databases. Considering that the role of PCSK6 in activating the NODAL pathway is in place precisely in the very first weeks of embryonic development, it is possible that any genuine allele-specific effects on gene expression might not be detectable in later stages of development and more so in adult tissues or cell lines.

These kinds of analyses can provide a mechanism to explain how genetic and phenotypic variations correlate with one another. The subsequent, and probably the more important challenge, is to understand the function of the genes affected by genetic variation. In the case of *PCSK6*, gene function is known and indeed its involvement in left/right asymmetry further corroborates the genetic association. However, PCSK6 might have a different role specifically affecting handedness that has not been described yet.

While a specific function can be assigned to many of the approximately 20,000 human genes, for a good proportion of them it remains unknown or not fully explained. For example, the three most commonly studied candidate genes for dyslexia susceptibility (*KIAA0319*, *DCDC2*, and *DYX1C1*) have been implicated in brain development [67], but their function is not exactly understood. Only recently, functional studies in cellular and animal models are revealing a role in ciliogenesis and cilia function. As we discussed earlier, cilia are crucial in controlling the establishment of left/right asymmetries through the Nodal pathway. Defective cilia lead to ciliopathies, a class of disorders often characterized by laterality defects. Interestingly *DYX1C1* and *DCDC2* have been found mutated in patients with ciliopathies. These observations provide a hypothesis to interpret the dyslexia-specific *PCSK6* association. It is possible that complex gene by gene interactions, or **epistatic** effects, between the genes determining laterality interact with genes involved in brain development and relevant to dyslexia [5]. This interpretation is still a speculation, but it exemplifies the process on how we move from statistical genetic associations to the understanding of the biology underlying complex traits. In other words genetic studies generate hypothesis but gene characterization in biological models is necessary to understand function.

11 Genetics and Brain Asymmetries

Research in handedness genetics is being driven strongly by the possibility of opening new pathways in our understanding of brain asymmetries. The human brain is asymmetric both structurally (the two hemispheres are not the mirror images of one another) and functionally. The left hemisphere controls the right limbs and vice-versa and as discussed above, handedness displays a weak correlation with language dominance lateralization. A twin study has found an intriguing pattern of heritability for handedness and brain measures [68]. By studying 72 MZ twins and 67 DZ twins, they found that the cerebral lobar volumes of the right hemispheres were under greater genetic control (i.e. more heritable). Furthermore, by sub-categorizing the DZ and MZ twins into right–right concordant, right–left discordant and left–left concordant pairs, they further showed that the heritability estimates for cerebral lobar volume were stronger in the right–right categories of twins.

One way to address what genes might be responsible for controlling brain lateralization has been to compare gene expression patterns across the two hemispheres. An early study looked specifically at how gene expression differed between the two hemispheres during embryogenesis [69]. They studied human left and right hemispheres at ages 12–19 weeks, and reported differences in expression for numerous genes between the two hemispheres. However, numerous other larger studies have been unable to find significant differences [70–72]. One of these studies [72] assessed gene expression using both traditional microarrays mRNA-seq (mRNA sequencing by NGS). mRNA-seq can unbiasedly detect any transcripts regardless of a previous knowledge of gene sequencing. For example it could detect alternatively spliced isoforms not described in public repositories that may be specific to particular tissues or developmental stages. This has resulted in a single interesting finding—a bilateral differential expression of the *SHANK3* gene [72]. While this result was detected by mRNA-seq, it could not be replicated in their microarray dataset, and therefore the possibility of it being an artifact cannot be ruled out. However, *SHANK3* is of particular interest due to its association with speech and language disorders [73, 74].

Others studies have tested genetic variants for association with hemispheric differences of brain structures or volume. For example, a deletion within the dyslexia candidate gene *DCDC2* was reported to be associated with pronounced volumetric changes in the left hemisphere [75]. This study however used a small sample size ($N = 56$). A much larger meta-analysis GWAS looking at over 3000 individuals found no genome-wide associations for brain structure laterality, but did not test the *DCDC2* deletion [76].

To illustrate in a direct manner how genetic variation can play a role in neuronal development in a specific brain hemisphere, recent studies utilizing NGS technology have begun to investigate the role of somatic DNA mutations in focal brain malformations. Somatic mutations in one hemisphere alone have been identified when gross neuronal malformations were also detected in the same hemisphere. In these cases, DNA mutations that were generated in a single cell after the establishment of the two hemispheres, is then propagated to the daughter brain cells which remained localized in the same hemisphere. While these studies do not explain the biology of brain asymmetries, they demonstrate the role of single genes and single mutations in establishing pathogenic unilateral defects [77, 78].

12 Conservation Analysis

Genome comparisons across species (Fig. 6) might also offer interesting insight into the function of candidate genes. By taking advantage of the conservation of interspecies (orthologous) and intra-species (paralogous) DNA sequences, we can predict the effect of newly discovered mutations in humans. For example, if a candidate mutation for a human trait lies in an area of the genome that is highly conserved between species, then this is strong evidence that the mutation will have a biological effect. However, if there is little conservation in this region, then it suggests that mutations in this part of the genome would be tolerated or benign. The principle of sequence conservation is applied in the quest of what make us human. By comparing mutation rate along the genome it is possible to identify regions that exhibit significantly higher variability in humans specifically [79]. The *FOXP2* gene provides a clear example of what we can learn from conservation analysis. *FOXP2* encodes one of the most highly conserved proteins among vertebrates, yet two amino acid substitutions in exon 7 have been fixed on the human lineage after the split from chimpanzees. Furthermore, re-sequencing of a region adjacent to exon 7 showed an allele frequency spectrum, consistent with positive selection events.

The data suggests that subsequent to the human–chimpanzee split, there have been two successive evolutionary events at *FOXP2* [80]. The mutations in exon 7 occurred prior to the human-Neandertal split (~400,000 years ago), and changed FOXP2 function. Modeling in mice of this genetic variant confirmed an impact on brain development [81]. A later event (~200,000 years ago) at *FOXP2* noncoding sequences probably affected gene expression regulation.

This approach is therefore particularly relevant when applied to lateralized traits to understand how they emerged and which role they played in human evolution. The advances in technology both in their throughput and affordability will allow the generation of high-quality data for an increasing number of species, including from the remains of different hominid lines (such as Neanderthals), that might contribute to this type of analyses.

13 Concluding Remarks and Future Directions

The advances in genomic technologies and bioinformatics tools pose virtually no limit in how many genomes can be sequenced. In theory, the genetic component of any trait can be dissected following standard and well-established human genetic procedures, for which the key element is considered to be the availability of large samples. Yet, sequencing capacity and large sample sizes need to be coupled with careful study designs taking into account what we have learnt so far about handedness genetics. In fact, the most intuitive definition of handedness, left and right, might not be the most suitable classification for genetic studies. How we assess handedness, for example with the use of the pegboard test, might be the crucial factor to identify genes. Therefore, the potential of GWASs has not been exhausted yet and analysis in larger cohorts characterized with quantitative laterality measures promise to unveil further common genetics factors for handedness. Quantitative assessment of handedness, however, is not straightforward, and collection of good quality data in adequately powered samples requires long-term and collaborative efforts. The pegboard has often been collected for different reasons (e.g. testing fine motor skills) rather than elucidating handedness genetics. It is therefore possible that other measures might be better for this purpose if designed specifically to assess handedness. This is an exciting time to study handedness genetics; we have the technology for large-scale screenings and we have learnt a lot from their application on other traits. Paradoxically, while handedness appeared an easy trait to be assessed, we might not have developed yet the most suitable test. Furthermore, genetic data alone will only generate hypotheses, but the elucidation of handedness biology will require functional characterization in biological models and the integration of different approaches such as cognitive neuroscience, cell biology, neuroimaging, and comparative genomics.

14 Box Elements

14.1 Box 1. The Human Genome

The human genome consists of roughly 3.5 Gb (gigabase pairs) of nucleotides (A, C, G, and Ts) distributed amongst 23 pairs of linear chromosomes and hundreds of identical small circular mitochondrial chromosomes. Each chromosome is a molecule of deoxyribonucleic acid (DNA). Humans are diploid organisms; the 23 pairs of chromosomes make 46 chromosomes in total, with each member of a pair inherited either maternally or paternally. Of the 23 pairs, 22 are autosomal and one pair is represented by the sex chromosomes, XX in females or XY in males. A very small fraction of the entire genome constitutes the genes, which are estimated to be about 20,000 in total. The DNA of genes can be divided into introns and exons. The exons of genes code for proteins and account for only 2% (or about 70 Mb) of the human genome. Between two randomly chosen human genomes there could be hundreds of thousands of differences known as genetic variants (see Box 2), which contribute in making each of us unique. Each cell of one individual has the same DNA sequence, with few exceptions (e.g. red blood cells and germ cells) but cell types clearly look and function differently. Gene regulation is one of the factors leading to diverse cell types and tissue differentiation. DNA regulatory sequences, for example promoter regions found upstream of genes, control this process. DNA variant in these sequences can therefore affect gene expression regulation.

14.2 Box 2. Genetic Variants

On any given chromosome, a genetic marker is a highly specific location that shows variability. Variants could be germline, meaning they are inherited from a parent, or they could be somatic, meaning they have spontaneously arisen and may be present in only group of cells or in a tissue depending at which stage of development they occurred. A simple bi-allelic variant comes in just two states; for example a point variant may have just two alleles such C and T. As humans are diploid, they may carry two copies of the same allele (known as homozygosity), one on each autosomal chromosome (e.g. CC or TT), or otherwise two different alleles (known as heterozygosity, i.e. CT). When both alleles are detectable within a population they are referred to as polymorphic. Single nucleotide polymorphisms (SNPs) are the markers used for GWASs. Small insertion/deletion (INDEL), which typically consists of changes of 1–5 bases of DNA can also be polymorphic. A different type of polymorphism is represented by number of repeats. Microsatellites, which refers to the repeat of few nucleotides (e.g. CAG) has until recently been the marker of choice for linkage studies. Variable number of tandems repeats (VNTRs) refer to repeats of longer sequences. **Copy number variants (CNVs)**, like INDELS, consist of insertions or deletion of DNA but on a much grander scale. CNVs are normally genotyped in "states" corresponding to how many copies are present. A state of zero (0) means both copies are absent. A state of one (1) means the deletion of one chromosomal copy, and states of three (3), four (4) and upwards indicate duplications (i.e. one, two, and upwards). CNVs are commonly found in the population, but they may also be unique events directly affecting a phenotype of interest. CNVs can be analyzed as single events or for their cumulative effect on the genome (CNV burden analysis).

14.3 Box 3. Glossary

Allele A specific variant of a genetic marker. For example, the two alleles of the bi-allelic SNP rs12345 are C and T.

Association study Statistic method that tests whether alleles frequencies differ between two groups of individuals (usually diseased subjects and healthy controls; categorical association) or display a different distribution of a score within a population (quantitative association).

Candidate gene study An association study testing specific genes on the basis of an a priori hypothesis.

*Dominant genetic model*A single mutation is able to cause a disease.

Effect size (of genetic variants) Often measured as an odds ratio or beta coefficient, it provides the magnitude and direction of effect of an allele on a phenotype.

Epistasis Interaction of alleles from different loci (cis or trans) that may influence a phenotype in a manner differently to the individual alleles alone.

Genetic marker A variable site in the human.

Genotype The aggregate of alleles from both chromosomes of a one marker. A biallelic marker (e.g. with the A and T alleles) can have three genotype (AA, AT, and TT).

Germline variant Any variant that is passed down from ancestors to progeny.

GSA Gene set analysis; a computational approach that tests whether an a priori defined set of genes shows an over-representation of association with a particular phenotype than expected by chance.

GWAS Genome-wide association study; an association study testing the entire genome.

Haplotype The allelic combination of different markers.

Heritability The proportion of phenotypic variability that can be attributed to genetic variability.

Linkage studies Two genetic variants located nearby to each other on a chromosome, are inherited together. By testing segregation of known markers with a phenotype it is possible to identify portions of chromosome that are likely to carry genetic variants responsible for the phenotype.

LD Linkage disequilibrium is a measure of correlation between the alleles of two different genetic markers. Markers on the same chromosome and relatively close (e.g. within 10 kb), may be in strong LD.

Microsatellite A short DNA sequence, such as CAG, that may be repeated a varied number of times on different chromosomes.

Monogenic condition A trait or disorder caused by a mutation in a single gene, for example cystic fibrosis or Huntington's disease.

NGS Next-generation sequencing, a technology that can re-sequence either the coding portion (WES) or the entire genome (WGS). Useful to detect rare or novel variants.

Penetrance The probability of an individual manifesting a disease or phenotype, given a particular genotype.

Phenotype A characteristic or trait, such as a disease, disorder, or attribute like handedness.

Population stratification Cases and controls of a particular study are not of a similar ethnicity, thereby potentially leading to false positive signals in an association study.

Promoter region A region often upstream of a gene that regulate the expression of a gene by recruiting proteins that drive transcription.

Recessive genetic model A gene requiring two mutations, one on each chromosome, before a phenotype is observed. The mutation may be the same, e.g. as a result of inbreeding, or affect the gene with two separate variants.

SNP Single nucleotide polymorphism is a biallelic marker consisting of a single base difference and with an appreciable frequency in a population (e.g. >5 %).

Somatic mutations These mutations arise in any cell of the human body and are not passed through the next generations.

Twin studies Studies comparing identical to fraternal twins to estimate the genetic contribution to a trait.

WES Whole exome sequencing. NGS technology, targeting exons.

WGS Whole genome sequencing. NGS technology targeting the entire genome.

Acknowledgements

Silvia Paracchini is a Royal Society University Research Fellow. The authors would like to thank Dr. William Brandler for useful comments on the manuscript of this chapter.

References

1. Berker EA, Berker AH, Smith A (1986) Translation of Broca's 1865 report. Localization of speech in the third left frontal convolution. Arch Neurol 43(10):1065–1072
2. Ghirlanda S, Vallortigara G (2004) The evolution of brain lateralization: a game-theoretical analysis of population structure. Proc Biol Sci 271(1541):853–857
3. Knecht S, Drager B, Deppe M, Bobe L, Lohmann H, Floel A, Ringelstein EB, Henningsen H (2000) Handedness and hemispheric language dominance in healthy humans. Brain 123(Pt 12):2512–2518
4. Bishop DV (2013) Cerebral asymmetry and language development: cause, correlate, or consequence? Science 340(6138):1230531
5. Brandler WM, Paracchini S (2014) The genetic relationship between handedness and neurodevelopmental disorders. Trends Mol Med 20(2):83–90
6. Hepper PG (2013) The developmental origins of laterality: fetal handedness. Dev Psychobiol 55(6):588–595
7. Carter-Saltzman L (1980) Biological and sociocultural effects on handedness: comparison between biological and adoptive families. Science 209(4462):1263–1265
8. Bishop DV (2015) The interface between genetics and psychology: lessons from developmental dyslexia. Proc Biol Sci 282(1806): 20143139
9. Medland SE, Duffy DL, Wright MJ, Geffen GM, Hay DA, Levy F, van-Beijsterveldt CE, Willemsen G, Townsend GC, White V, Hewitt AW, Mackey DA, Bailey JM, Slutske WS, Nyholt DR, Treloar SA, Martin NG, Boomsma DI (2009) Genetic influences on handedness: data from 25,732 Australian and Dutch twin families. Neuropsychologia 47(2): 330–337

10. Macgregor S, Cornes BK, Martin NG, Visscher PM (2006) Bias, precision and heritability of self-reported and clinically measured height in Australian twins. Hum Genet 120(4):571–580
11. Bisazza A, Facchin L, Vallortigara G (2000) Heritability of lateralization in fish: concordance of right-left asymmetry between parents and offspring. Neuropsychologia 38(7):907–912
12. Annett M (2002) Non-right-handedness and schizophrenia. Br J Psychiatry 181:349–350
13. McManus IC (1985) Handedness, language dominance and aphasia: a genetic model. Psychol Med Monogr Suppl 8:1–40
14. Fisher SE, Vargha-Khadem F, Watkins KE, Monaco AP, Pembrey ME (1998) Localisation of a gene implicated in a severe speech and language disorder. Nat Genet 18(2):168–170
15. Lai CS, Fisher SE, Hurst JA, Vargha-Khadem F, Monaco AP (2001) A forkhead-domain gene is mutated in a severe speech and language disorder. Nature 413(6855):519–523
16. Newbury DF, Monaco AP, Paracchini S (2014) Reading and language disorders: the importance of both quantity and quality. Genes (Basel) 5(2):285–309
17. Newbury DF, Winchester L, Addis L, Paracchini S, Buckingham LL, Clark A, Cohen W, Cowie H, Dworzynski K, Everitt A, Goodyer IM, Hennessy E, Kindley AD, Miller LL, Nasir J, O'Hare A, Shaw D, Simkin Z, Simonoff E, Slonims V, Watson J, Ragoussis J, Fisher SE, Seckl JR, Helms PJ, Bolton PF, Pickles A, Conti-Ramsden G, Baird G, Bishop DV, Monaco AP (2009) CMIP and ATP2C2 modulate phonological short-term memory in language impairment. Am J Hum Genet 85(2):264–272
18. Vernes SC, Newbury DF, Abrahams BS, Winchester L, Nicod J, Groszer M, Alarcon M, Oliver PL, Davies KE, Geschwind DH, Monaco AP, Fisher SE (2008) A functional genetic link between distinct developmental language disorders. N Engl J Med 359(22):2337–2345
19. Van Agtmael T, Forrest SM, Del-Favero J, Van Broeckhoven C, Williamson R (2003) Parametric and nonparametric genome scan analyses for human handedness. Eur J Hum Genet 11(10):779–783
20. Warren DM, Stern M, Duggirala R, Dyer TD, Almasy L (2006) Heritability and linkage analysis of hand, foot, and eye preference in Mexican Americans. Laterality 11(6): 508–524
21. Somers M, Ophoff RA, Aukes MF, Cantor RM, Boks MP, Dauwan M, de Visser KL, Kahn RS, Sommer IE (2015) Linkage analysis in a Dutch population isolate shows no major gene for left-handedness or atypical language lateralization. J Neurosci 35(23):8730–8736
22. Van Agtmael T, Forrest SM, Williamson R (2002) Parametric and non-parametric linkage analysis of several candidate regions for genes for human handedness. Eur J Hum Genet 10(10):623–630
23. Corder EH, Saunders AM, Strittmatter WJ, Schmechel DE, Gaskell PC, Small GW, Roses AD, Haines JL, Pericak-Vance MA (1993) Gene dose of apolipoprotein E type 4 allele and the risk of Alzheimer's disease in late onset families. Science 261(5123):921–923
24. Bloss CS, Delis DC, Salmon DP, Bondi MW (2010) APOE genotype is associated with left-handedness and visuospatial skills in children. Neurobiol Aging 31(5):787–795
25. Piper BJ, Yasen AL, Taylor AE, Ruiz JR, Gaynor JW, Dayger CA, Gonzalez-Gross M, Kwon OD, Nilsson LG, Day IN, Raber J, Miller JK (2013) Non-replication of an association of Apolipoprotein E2 with sinistrality. Laterality 18(2):251–261
26. Hubacek JA, Piper BJ, Pikhart H, Peasey A, Kubinova R, Bobak M (2013) Lack of an association between left-handedness and APOE polymorphism in a large sample of adults: results of the Czech HAPIEE study. Laterality 18(5):513–519
27. Francks C, Fisher SE, MacPhie IL, Richardson AJ, Marlow AJ, Stein JF, Monaco AP (2002) A genomewide linkage screen for relative hand skill in sibling pairs. Am J Hum Genet 70(3):800–805
28. Laval SH, Dann JC, Butler RJ, Loftus J, Rue J, Leask SJ, Bass N, Comazzi M, Vita A, Nanko S, Shaw S, Peterson P, Shields G, Smith AB, Stewart J, DeLisi LE, Crow TJ (1998) Evidence for linkage to psychosis and cerebral asymmetry (relative hand skill) on the X chromosome. Am J Med Genet 81(5):420–427
29. Medland SE, Duffy DL, Spurdle AB, Wright MJ, Geffen GM, Montgomery GW, Martin NG (2005) Opposite effects of androgen receptor CAG repeat length on increased risk of left-handedness in males and females. Behav Genet 35(6):735–744
30. Hampson E, Sankar JS (2012) Hand preference in humans is associated with testosterone levels and androgen receptor gene polymorphism. Neuropsychologia 50(8):2018–2025
31. Welter D, MacArthur J, Morales J, Burdett T, Hall P, Junkins H, Klemm A, Flicek P, Manolio T, Hindorff L, Parkinson H (2014) The NHGRI GWAS Catalog, a curated resource of SNP-trait associations. Nucleic Acids Res 42(database issue):D1001–D1006

32. Clarke GM, Anderson CA, Pettersson FH, Cardon LR, Morris AP, Zondervan KT (2011) Basic statistical analysis in genetic case-control studies. Nat Protoc 6(2):121–133
33. Eriksson N, Macpherson JM, Tung JY, Hon LS, Naughton B, Saxonov S, Avey L, Wojcicki A, Pe'er I, Mountain J (2010) Web-based, participant-driven studies yield novel genetic associations for common traits. PLoS Genet 6(6):e1000993
34. Armour JA, Davison A, McManus IC (2014) Genome-wide association study of handedness excludes simple genetic models. Heredity 112(3):221–225
35. McManus IC, Davison A, Armour JA (2013) Multilocus genetic models of handedness closely resemble single-locus models in explaining family data and are compatible with genome-wide association studies. Ann N Y Acad Sci 1288:48–58
36. Oldfield RC (1971) The assessment and analysis of handedness: the Edinburgh inventory. Neuropsychologia 9(1):97–113
37. Crovitz HF, Zener K (1962) A group-test for assessing hand- and eye-dominance. Am J Psychol 75:271–276
38. Annett M (1985) Left, right, hand and brain: the right shift theory. Psychology Press, London
39. Brandler WM, Morris AP, Evans DM, Scerri TS, Kemp JP, Timpson NJ, Pourcain BS, Smith GD, Ring SM, Stein J, Monaco AP, Talcott JB, Fisher SE, Webber C, Paracchini S (2013) Common variants in left/right asymmetry genes and pathways are associated with relative hand skill. PLoS Genet 9(9):e1003751
40. Lien YJ, Chen WJ, Hsiao PC, Tsuang HC (2015) Estimation of heritability for varied indexes of handedness. Laterality 20:469–482
41. Francks C, Fisher SE, Marlow AJ, MacPhie IL, Taylor KE, Richardson AJ, Stein JF, Monaco AP (2003) Familial and genetic effects on motor coordination, laterality, and reading-related cognition. Am J Psychiatry 160(11):1970–1977
42. Francks C, Maegawa S, Lauren J, Abrahams BS, Velayos-Baeza A, Medland SE, Colella S, Groszer M, McAuley EZ, Caffrey TM, Timmusk T, Pruunsild P, Koppel I, Lind PA, Matsumoto-Itaba N, Nicod J, Xiong L, Joober R, Enard W, Krinsky B, Nanba E, Richardson AJ, Riley BP, Martin NG, Strittmatter SM, Moller HJ, Rujescu D, Clair DS, Muglia P, Roos JL, Fisher SE, Wade-Martins R, Rouleau GA, Stein JF, Karayiorgou M, Geschwind DH, Ragoussis J, Kendler KS, Airaksinen MS, Oshimura M, DeLisi LE, Monaco AP (2007) LRRTM1 on chromosome 2p12 is a maternally suppressed gene that is associated paternally with handedness and schizophrenia. Mol Psychiatry 12(12):1129–1139, 1057
43. Ludwig KU, Mattheisen M, Muhleisen TW, Roeske D, Schmal C, Breuer R, Schulte-Korne G, Muller-Myhsok B, Nothen MM, Hoffmann P, Rietschel M, Cichon S (2009) Supporting evidence for LRRTM1 imprinting effects in schizophrenia. Mol Psychiatry 14(8):743–745
44. McManus C, Nicholls M, Vallortigara G (2009) Editorial commentary: is LRRTM1 the gene for handedness? Laterality 14(1):1–2
45. Savitz J, van der Merwe L, Solms M, Ramesar R (2007) Lateralization of hand skill in bipolar affective disorder. Genes Brain Behav 6(8):698–705
46. Scerri TS, Brandler WM, Paracchini S, Morris AP, Ring SM, Richardson AJ, Talcott JB, Stein J, Monaco AP (2011) PCSK6 is associated with handedness in individuals with dyslexia. Hum Mol Genet 20(3):608–614
47. Arning L, Ocklenburg S, Schulz S, Ness V, Gerding WM, Hengstler JG, Falkenstein M, Epplen JT, Gunturkun O, Beste C (2013) VNTR polymorphism is associated with degree of handedness but not direction of handedness. PLoS One 8(6):e67251
48. Hirokawa N, Tanaka Y, Okada Y, Takeda S (2006) Nodal flow and the generation of left-right asymmetry. Cell 125(1):33–45
49. Grande C, Patel NH (2009) Nodal signalling is involved in left-right asymmetry in snails. Nature 457(7232):1007–1011
50. Barth KA, Miklosi A, Watkins J, Bianco IH, Wilson SW, Andrew RJ (2005) fsi zebrafish show concordant reversal of laterality of viscera, neuroanatomy, and a subset of behavioral responses. Curr Biol 15(9):844–850
51. Mooney MA, Nigg JT, McWeeney SK, Wilmot B (2014) Functional and genomic context in pathway analysis of GWAS data. Trends Genet 30(9):390–400
52. Ng A, Wong M, Viviano B, Erlich JM, Alba G, Pflederer C, Jay PY, Saunders S (2009) Loss of glypican-3 function causes growth factor-dependent defects in cardiac and coronary vascular development. Dev Biol 335(1):208–215
53. Zhou J, Yang F, Leu NA, Wang PJ (2012) MNS1 is essential for spermiogenesis and motile ciliary functions in mice. PLoS Genet 8(3):e1002516
54. Bonnafe E, Touka M, AitLounis A, Baas D, Barras E, Ucla C, Moreau A, Flamant F, Dubruille R, Couble P, Collignon J, Durand B, Reith W (2004) The transcription factor RFX3 directs nodal cilium development and left-right asymmetry specification. Mol Cell Biol 24(10):4417–4427

55. Pennekamp P, Karcher C, Fischer A, Schweickert A, Skryabin B, Horst J, Blum M, Dworniczak B (2002) The ion channel polycystin-2 is required for left-right axis determination in mice. Curr Biol 12(11):938–943
56. Benadiba C, Magnani D, Niquille M, Morle L, Valloton D, Nawabi H, Ait-Lounis A, Otsmane B, Reith W, Theil T, Hornung JP, Lebrand C, Durand B (2012) The ciliogenic transcription factor RFX3 regulates early midline distribution of guidepost neurons required for corpus callosum development. PLoS Genet 8(3):e1002606
57. McManus IC, Martin N, Stubbings GF, Chung EM, Mitchison HM (2004) Handedness and situs inversus in primary ciliary dyskinesia. Proc Biol Sci 271(1557):2579–2582
58. Sun T, Walsh CA (2006) Molecular approaches to brain asymmetry and handedness. Nat Rev Neurosci 7(8):655–662
59. Deciphering Developmental Disorders S (2015) Large-scale discovery of novel genetic causes of developmental disorders. Nature 519(7542):223–228
60. Wang K, Li M, Hakonarson H (2010) ANNOVAR: functional annotation of genetic variants from high-throughput sequencing data. Nucleic Acids Res 38(16):e164
61. Kavaklioglu T, Ajmal M, Hameed A, Francks C (2015) Whole exome sequencing for handedness in a large and highly consanguineous family., Neuropsychologia
62. Nicolae DL, Gamazon E, Zhang W, Duan S, Dolan ME, Cox NJ (2010) Trait-associated SNPs are more likely to be eQTLs: annotation to enhance discovery from GWAS. PLoS Genet 6(4):e1000888
63. Bernstein BE, Birney E, Dunham I, Green ED, Gunter C, Snyder M (2012) An integrated encyclopedia of DNA elements in the human genome. Nature 489(7414):57–74
64. Dennis MY, Paracchini S, Scerri TS, Prokunina-Olsson L, Knight JC, Wade-Martins R, Coggill P, Beck S, Green ED, Monaco AP (2009) A common variant associated with dyslexia reduces expression of the KIAA0319 gene. PLoS Genet 5(3):e1000436
65. Kang HJ, Kawasawa YI, Cheng F, Zhu Y, Xu X, Li M, Sousa AM, Pletikos M, Meyer KA, Sedmak G, Guennel T, Shin Y, Johnson MB, Krsnik Z, Mayer S, Fertuzinhos S, Umlauf S, Lisgo SN, Vortmeyer A, Weinberger DR, Mane S, Hyde TM, Huttner A, Reimers M, Kleinman JE, Sestan N (2011) Spatio-temporal transcriptome of the human brain. Nature 478(7370):483–489
66. Shore R, Covill L, Pettigrew KA, Brandler WM, Diaz R, Xu Y, Tello J, Talcott J, Newbury DF, Stein J, Monaco AP, Paracchini S (2016) The handedness-associated PCSK6 locus spans a bidirectional promoter regulating novel transcripts. Hum Mol Genet 25(9):1771–1779
67. Paracchini S, Scerri T, Monaco AP (2007) The genetic lexicon of dyslexia. Annu Rev Genomics Hum Genet 8:57–79
68. Geschwind DH, Miller BL, DeCarli C, Carmelli D (2002) Heritability of lobar brain volumes in twins supports genetic models of cerebral laterality and handedness. Proc Natl Acad Sci U S A 99(5):3176–3181
69. Sun T, Patoine C, Abu-Khalil A, Visvader J, Sum E, Cherry TJ, Orkin SH, Geschwind DH, Walsh CA (2005) Early asymmetry of gene transcription in embryonic human left and right cerebral cortex. Science 308(5729):1794–1798
70. Johnson MB, Kawasawa YI, Mason CE, Krsnik Z, Coppola G, Bogdanovic D, Geschwind DH, Mane SM, State MW, Sestan N (2009) Functional and evolutionary insights into human brain development through global transcriptome analysis. Neuron 62(4):494–509
71. Lambert N, Lambot MA, Bilheu A, Albert V, Englert Y, Libert F, Noel JC, Sotiriou C, Holloway AK, Pollard KS, Detours V, Vanderhaeghen P (2011) Genes expressed in specific areas of the human fetal cerebral cortex display distinct patterns of evolution. PLoS One 6(3):e17753
72. Pletikos M, Sousa AM, Sedmak G, Meyer KA, Zhu Y, Cheng F, Li M, Kawasawa YI, Sestan N (2014) Temporal specification and bilaterality of human neocortical topographic gene expression. Neuron 81(2):321–332
73. Durand CM, Betancur C, Boeckers TM, Bockmann J, Chaste P, Fauchereau F, Nygren G, Rastam M, Gillberg IC, Anckarsater H, Sponheim E, Goubran-Botros H, Delorme R, Chabane N, Mouren-Simeoni MC, de Mas P, Bieth E, Roge B, Heron D, Burglen L, Gillberg C, Leboyer M, Bourgeron T (2007) Mutations in the gene encoding the synaptic scaffolding protein SHANK3 are associated with autism spectrum disorders. Nat Genet 39(1):25–27
74. Waga C, Okamoto N, Ondo Y, Fukumura-Kato R, Goto Y, Kohsaka S, Uchino S (2011) Novel variants of the SHANK3 gene in Japanese autistic patients with severe delayed speech development. Psychiatr Genet 21(4):208–211
75. Meda SA, Gelernter J, Gruen JR, Calhoun VD, Meng H, Cope NA, Pearlson GD (2008) Polymorphism of DCDC2 reveals differences in cortical morphology of healthy individuals—a preliminary voxel based morphometry study. Brain Imaging Behav 2(1):21–26

76. Guadalupe T, Zwiers MP, Teumer A, Wittfeld K, Vasquez AA, Hoogman M, Hagoort P, Fernandez G, Buitelaar J, Hegenscheid K, Volzke H, Franke B, Fisher SE, Grabe HJ, Francks C (2014) Measurement and genetics of human subcortical and hippocampal asymmetries in large datasets. Hum Brain Mapp 35(7):3277–3289
77. Leventer RJ, Scerri T, Marsh AP, Pope K, Gillies G, Maixner W, MacGregor D, Harvey AS, Delatycki MB, Amor DJ, Crino P, Bahlo M, Lockhart PJ (2015) Hemispheric cortical dysplasia secondary to a mosaic somatic mutation in MTOR. Neurology 84(20):2029–2032
78. Lee JH, Huynh M, Silhavy JL, Kim S, Dixon-Salazar T, Heiberg A, Scott E, Bafna V, Hill KJ, Collazo A, Funari V, Russ C, Gabriel SB, Mathern GW, Gleeson JG (2012) De novo somatic mutations in components of the PI3K-AKT3-mTOR pathway cause hemimegalencephaly. Nat Genet 44(8):941–945
79. Paabo S (2014) The human condition—a molecular approach. Cell 157(1):216–226
80. Ayub Q, Yngvadottir B, Chen Y, Xue Y, Hu M, Vernes SC, Fisher SE, Tyler-Smith C (2013) FOXP2 targets show evidence of positive selection in European populations. Am J Hum Genet 92(5):696–706
81. Enard W, Gehre S, Hammerschmidt K, Holter SM, Blass T, Somel M, Bruckner MK, Schreiweis C, Winter C, Sohr R, Becker L, Wiebe V, Nickel B, Giger T, Muller U, Groszer M, Adler T, Aguilar A, Bolle I, Calzada-Wack J, Dalke C, Ehrhardt N, Favor J, Fuchs H, Gailus-Durner V, Hans W, Holzlwimmer G, Javaheri A, Kalaydjiev S, Kallnik M, Kling E, Kunder S, Mossbrugger I, Naton B, Racz I, Rathkolb B, Rozman J, Schrewe A, Busch DH, Graw J, Ivandic B, Klingenspor M, Klopstock T, Ollert M, Quintanilla-Martinez L, Schulz H, Wolf E, Wurst W, Zimmer A, Fisher SE, Morgenstern R, Arendt T, de Angelis MH, Fischer J, Schwarz J, Paabo S (2009) A humanized version of Foxp2 affects cortico-basal ganglia circuits in mice. Cell 137(5):961–971

Chapter 17

Genetic and Transgenic Approaches to Study Zebrafish Brain Asymmetry and Lateralized Behavior

Erik R. Duboué and Marnie E. Halpern

Abstract

In just over 25 years, the zebrafish has emerged as a valuable model to discover genes controlling vertebrate development, owing to its amenability to mutagenesis and unbiased phenotype-based screens. The transparency of early developmental stages enables abnormal morphology to be readily detected in many organs and tissues types, including the central nervous system. The same attribute makes embryonic and larval zebrafish well suited to cataloging spatial patterns of gene expression and for time-lapse imaging of fluorescently labeled cells. The application of mobile transposable elements expedites the production of transgenic animals, thereby advancing techniques to visualize neuronal populations and their axonal projections, to selectively destroy them, or to manipulate their synaptic activity. Neural activity of specific regions or throughout the brain of a live, behaving larva can be monitored by transgenic strategies capitalizing on genetically encoded calcium indicators. This chapter will outline how such genetic and transgenic approaches are being applied to study left-right (L-R) asymmetry of the zebrafish brain, to learn how and where it develops, and examine its impact on neural processing and lateralized behaviors.

Key words Left-right brain asymmetry, Epithalamus, Habenula, Fear/anxiety

1 Epithalamic Left-Right Asymmetry: Setting the Stage for Molecular Genetic Approaches

Upon preparing a brain atlas for the frog *Rana esculenta* [1], Italian neuroanatomist Milena Kemali rediscovered pronounced L-R differences in the habenular nuclei of the dorsal diencephalon that were previously observed in other amphibian and fish species [2–4]. In fact, newts with *situs inversus*, including reversed habenular asymmetry, had been collected from the wild as well as generated through experimental manipulations [5]. The epithalamus, derived from the dorsal diencephalon, includes the bilateral habenulae and the prominent commissure that runs between them, the pineal complex and the stria medullaris bundle of afferent fibers to the habenulae. The habenulae consist of separate dorsal (dHb) and ventral (vHb) nuclei (designated as medial and lateral in mammals)

Lesley J. Rogers and Giorgio Vallortigara (eds.), *Lateralized Brain Functions: Methods in Human and Non-Human Species*, Neuromethods, vol. 122, DOI 10.1007/978-1-4939-6725-4_17, © Springer Science+Business Media LLC 2017

and only the dorsal pair appear asymmetric. The pineal complex varies widely in vertebrates; the pineal organ often has an accessory structure, which in some species of fish is the unilaterally positioned parapineal. A comprehensive survey of the literature [6] affirmed the presence of L-R differences in epithalamic structures throughout the vertebrate lineage, from primitive fish to mammals. Using several *Rana* species, Kemali and colleagues determined that the left and right dHb differ in their size, organization, neuropil density, neurochemistry, and cytoarchitecture [7–12]. However, an appreciation of how these differences arise, and their impact on habenular function, was limited by the technical approaches available at the time.

2 Conserved Left-Right Signaling Pathway Functions in the Developing Zebrafish Brain

A renewed interest in diencephalic asymmetry came from the unexpected finding that the same signaling pathway underlying the placement and directional morphology of the visceral organs in vertebrate embryos is also transiently activated on the left side of the zebrafish anterior neural tube. Genes encoding the secreted Nodal peptide, a member of the Transforming Growth Factor beta (TGF-β) superfamily, are unilaterally expressed along the vertebrate body axis and are important for the rightward looping of the heart tube, anticlockwise coiling of the gut, and positioning of the pancreas, gallbladder, and liver within the body cavity (refer to [13]). Three *nodal-related* (*ndr*) genes were identified in the zebrafish genome, two of which, *ndr2* and *southpaw* (*spaw*), are transcribed along the left flank of the embryo in the lateral plate mesoderm (LPM) [14, 15], in a similar pattern first described for the chick embryo [16]. Additionally, commencing at ~20 hours post fertilization (h) *ndr2* and other members of the Nodal signaling pathway are transiently expressed on the left side of the dorsal diencephalon (i.e., the presumptive epithalamus). The zebrafish *lefty1* (*lft1*) homologue (previously referred to as *antivin*), which encodes a related TGF-β family member that functions as a Nodal antagonist, is also transcribed on the left side of the neural tube [17, 18] in a small domain that was erroneously thought to correspond to the left habenula [18]. Subsequent double labeling analyses demonstrated that left-sided gene expression is actually located within the presumptive pineal organ, the pineal anlage [19, 20].

The signaling cascade is further conserved in the left pineal anlage as the Nodal-regulated gene *paired-like homeodomain 2* (*pitx2*) is also transiently expressed in this region [19–21]. The *ndr2*, *lft1*, and *pitx2* transcripts are not entirely coextensive but, rather, are located in partially overlapping domains within the left pineal anlage. Curiously, the *one eyed pinhead* (*oep*) gene (renamed *teratocarcinoma-derived growth factor 1* [*tdgf1*]), encoding an essential component of the Nodal receptor complex [22], is expressed bilaterally in the same region of the developing diencephalon [19], indicating that both sides of the brain can respond to Nodal signals.

The precise function of transient Nodal activity in the left pineal anlage is still not completely resolved as it is necessary to uncouple it from the role of Nodal signaling present earlier in development or in the left lateral plate mesoderm. One way this has been partly addressed was to restore receptivity to Nodal signals in the *oep* mutant up until gastrulation, but not when the Ndr2 pathway is activated on the left side during late somitogenesis [19, 20]. These and other experiments indicate that activation of the Nodal pathway in the developing zebrafish diencephalon is not required for the generation of asymmetry but, rather, sets the directionality of this asymmetry. This function, however, may be unique to the teleostean brain, as a recent study of agnathan and cartilaginous fish reports that Nodal signaling plays an essential role in the determination of dHb asymmetry, not merely its directionality [23].

3 Sequential Appearance of Morphological and Molecular Left-Right Differences

Comprehensive screening of collections of cDNA clones for their temporal and tissue-specific patterns of expression by whole-mount in situ hybridization in zebrafish embryos and larvae [24] has yielded valuable "markers" to track anatomical and molecular asymmetries in the dorsal diencephalon. Approximately 6 h after the expression of Nodal pathway members in the left diencephalon, a small cluster of cells emerges to the left of the pineal anlage to give rise to the parapineal organ [25, 26]. Formation of the asymmetrically positioned parapineal was initially monitored by expression of genes shared with the pineal anlage such as those encoding the transcription factors Otx5 (Orthodenticle homolog 5), Noto (Notochord homeobox formerly known as Floating head) and Foxd3 (Forkhead box D3) [25–28]. Subsequently, the pineal anlage undergoes dramatic morphological changes to produce the photoreceptive pineal organ and its axonal stalk, which in the majority of adult zebrafish emanates from the roof of the brain, just left of the diencephalic midline [20]. In contrast, the parapineal persists as a small cluster of cells in the left side of the brain at the base of the pineal stalk and, in some species of fish such as medaka, is eventually incorporated within the left habenula [29].

Molecular evidence of habenular asymmetry is first detected by expression of the *potassium channel tetramerisation domain containing 12.1* gene [*kctd12.1*, formerly known as *leftover* (*lov*)], whose transcripts appear on the left side of the diencephalon at approximately 38–40 h, in a strip of cells closely apposed to the newly arising parapineal [28]. The *kctd12.1* gene is later transcribed in an expanded domain in the left dHb compared to the right, in a pattern presumed to correspond to a lateral subnucleus that is significantly larger in the left dHb than the right [30]. Conversely, two related genes, *kctd 12.2* (renamed from *right on*) and *kctd8* (renamed from *dexter*), are more extensively expressed in the right habenula than in the left [31]. Transgenic lines that label asymmetric and complementary regions of the dHb have also been taken to represent lateral (dHbL) and medial (dHbM) subnuclei (refer to Sect. 5.2). In larval zebrafish, the medial region is slightly larger in the right dHb than the left, while the lateral one is considerably larger on the left and largely corresponds to the asymmetric domain of *kctd12.1* expression [28, 30].

Many other genes have been found to be differentially expressed in the dHb (for comprehensive list refer to [32]), including those of a duplicated cholinergic gene locus (CGLb) that encodes proteins important for the synthesis of the neurotransmitter acetylcholine (i.e., Choline O-acetyltransferase b, ChATb) and for packaging it into synaptic vesicles (i.e., Solute carrier family 18 [vesicular acetylcholine transporter], member 3b, Slc18a3b) (refer to [33]). In larval zebrafish, *chatb* and *slc18a3b* transcripts are present in both vHb, but in the dHb they are enriched in the right nucleus [34].

The inequality in size, and hence gene expression, between the medial and lateral subregions of the left and right dHb has been attributed to a difference in the timing of proliferation of habenular progenitor cells [35]. An abundance of early proliferating progenitors on the left side of the diencephalon contribute to the dHbL, whereas later dividing progenitors generate the dHbM. Although this L-R asymmetry in neurogenesis seems dependent on Nodal signaling, it does not require an intact parapineal [36]. The Notch pathway is also involved, as hyperactivation of Notch signaling inhibits neurogenesis and the differentiation of early born dHbL neurons, resulting in right isomerized dHb. Conversely, mutants defective in Notch signaling show premature and excessive neurogenesis, leading to an enlarged dHbL and a reduced dHbM, equivalent to left isomerized dHb nuclei [35]. Genetic screens have also identified other factors that regulate habenular progenitors to generate either dHbM or dHbL neurons (refer to Sect. 4.3).

By the fourth day of development, the left dHb is larger in volume and, as in *Rana* species [37, 38], contains markedly dense clusters of neuronal processes or neuropil compared to the right [19, 28]. The Kctd12.1 and 12.2 proteins appear to regulate the extent of neuropil elaboration in the zebrafish dHb, presumably through their suppression of Ulk2 (Unc-51 like autophagy activating kinase 2), a kinase that is known to promote axonal growth by stimulating trafficking of endosomes at the ends of neurites [39].

Also established by 4 days are the efferent connections of dHb neurons [30, 31], whose axons project through the fasciculus retroflexus (FR) fiber bundle to terminate onto an unpaired target, the interpeduncular nucleus (IPN), in the midline of the ventral midbrain (refer to [40]). In rodents, the IPN is comprised of subnuclear regions that are distinguished largely on the basis of anatomical features and neurotransmitter distribution ([41–43] and refer to [44]). The IPN of adult zebrafish, which has not been well characterized, consists of independent dorsal (dIPN), intermediate (iIPN), and ventral (vIPN) regions that are structurally recognizable but likely divisible into smaller functional units [32, 45].

Habenular connections with the IPN in larval and adult zebrafish were initially traced by immunolabeling and lipophilic dyes, and later in transgenic lines that label axons with fluorescent proteins (refer to Sect. 5.2). Antisera produced against the Kctd12.1 and 12.2 proteins differentially label neurons in the dHb, in a pattern similar to that of transcripts, and also label efferent axons throughout the length of the FR to their synaptic terminals at the IPN. Kctd12.1 immunoreactive axons innervate the entire dorsoventral extent of the IPN, whereas Kctd12.2 positive axons predominantly terminate at the vIPN, a pattern confirmed by labeling of the left and right dHb with differently colored dyes [31, 46]. However, in parallel studies, it was concluded from dye-labeling experiments that dHb projections form a "laterotopic" map with the target, whereby most left dHb neurons innervate the dIPN and right dHb ones the vIPN [30]. As discussed below (refer to Sect. 5.2), the analysis of transgenic lines that selectively label neurons in the dHbL or dHbM subregions and their axonal projections, has helped to reconcile these seemingly conflicting interpretations of the dHb-IPN connectivity map.

Differential dorsoventral targeting of the IPN by the paired dHb can be explained, in part, by the unilateral expression of the *neuropilin 1a* (*nrp1a*) gene in neurons of the left nucleus [46]. Nrp1a is a transmembrane glycoprotein and coreceptor for members of the secreted Class III Semaphorin (Sema3) family of axon guidance molecules, including Sema3D, which is synthesized by cells along the FR trajectory to the IPN [46]. Injection of antisense morpholinos (MOs) directed to either the *nrp1a* or *sema3D* gene did not affect axonal outgrowth from the left dHb or guidance through the FR but, instead, inhibited growth cone extension and branching at the dorsal IPN. The results suggest that Sema3D/Nrp1a signaling serves as an attractant for left habenular innervation of the dIPN, and is just one of many guidance cues that likely direct Hb-IPN connectivity.

Labeling of single neurons by focal electroporation of plasmid DNA containing the green fluorescent protein (*gfp*) gene revealed another asymmetric feature of the zebrafish dHb, at the level of the stereotypic morphology of their axonal endings [47].

Early histological staining techniques on brains from various vertebrate species demonstrated that habenular axons extend in a unique S-shaped pattern, traversing the midline multiple times as they spiral around the IPN (e.g., [48, 49]). The axonal terminals of neurons individually labeled in the dHb of larval zebrafish have a similar trajectory, but adopt very different morphologies depending on which side of the brain they originate from [47]. Most neurons from the left dHb show highly branched axonal terminals that encircle and extend over a broad territory of the target. In contrast, axons emanating from most right dHb neurons have fewer branches, a flattened profile, and innervate a more confined IPN region [47].

The vast majority of individuals (>95 %) in wild-type strains of zebrafish activate the Nodal pathway in the left diencephalon, show a left positioned parapineal and habenular nuclei with L-R specific neuroanatomical, molecular, and connectivity properties. Two hypotheses could account for the corresponding directionality of parapineal position and dHb asymmetry: First, as with Nodal signaling in the left LPM that organizes visceral asymmetries, an early global signal to the neural tube could set up left-right polarity across the entire tissue. Alternatively, and the model most consistent with the experimental evidence (refer to Sect. 5.1) is that a small bias, such as the presence or absence of the parapineal, influences the identity of the adjacent habenula and its neural connections.

4 Identification of Genes Controlling Epithalamic Asymmetry

Insights into how L-R asymmetry emerges in the developing zebrafish brain and the underlying cellular mechanisms have come from testing candidate genes or from the isolation of mutations and their corresponding affected genes. Many of the implicated genes and genetic pathways have been recently reviewed [50]. Initially, mutations already known to affect other developmental processes were secondarily determined to alter L-R patterning of the heart, viscera, and diencephalon. Then, in a "reverse genetics" approach, mutations or antisense MOs were targeted to genes with presumed functions, owing to their confirmed roles in visceral organ L-R asymmetry in other vertebrates. Most significantly, and an important advantage of the zebrafish model, is the ability to discover unappreciated genetic pathways or cellular functions through unbiased screening of embryos descended from fish mutagenized by the DNA alkylating agent N-ethyl-N-nitrosourea (ENU [51]). Once asymmetries could be reliably recognized by the morphology of developing organs or through patterns of gene expression, "forward genetic" screens were performed to identify those embryos that phenotypically deviated from wild-type.

4.1 Fortuitous Discovery of Mutations Affecting L-R Patterning

The entry point for genetic analyses was the discovery that ENU-induced *cyclops* mutations, which cause a loss of cells in the ventral midline of the brain and lack of medial floor plate cells along the length of the ventral neural tube [52], reside in the zebrafish *nodal*-related gene *ndr2* [53, 54]. Owing to the known role of Nodal in other vertebrates, it was then recognized that homozygous *cyclops* mutants also have laterality defects, including bilateral rather than left-sided diencephalic expression of Nodal pathway members [53, 54]. Embryos mutant for the *nodal*-related *spaw* gene also show an increase in bilateral as well as L-R reversed expression in the brain [55]. By contrast, in mutants lacking diencephalic function of the nodal coreceptor Oep [19, 20], transcripts for *ndr2*, *lft1* and *pitx2* are not detected in the pineal anlage, affirming an autoregulatory requirement for Nodal signaling.

Mutations that block formation of the notochord, the midline mesoderm underlying the neural tube, were also found to secondarily affect directional asymmetry of the brain. In zebrafish embryos with defective notochord development, Nodal is expressed bilaterally in the pineal anlage [19, 54, 56], resulting in L-R randomization of epithalamic asymmetry in the mutant population. The notochord may be required as a barrier to the passage of Nodal factors from the left to the right side of the embryo or to preserve midline expression of Lefty, which normally restricts Nodal signaling by disrupting binding to its receptor [17]. Consistent with this model, MO-mediated depletion of Lefty1 also causes bilateral expression of *ndr2* and *pitx2* in the dorsal diencephalon, presumably through the inappropriate spread of Nodal signals [57].

Whether expression of Nodal pathway genes is bilateral or absent in the pineal anlage, the outcome is the same; approximately half the population shows a L-R reversal in parapineal position and habenular identity [25, 28, 58]. Such findings reveal a role for Nodal signaling in regulating gene expression on the left side of the brain and setting the direction of epithalamic asymmetry. Another interesting observation from the analysis of mutant and MO injected embryos is that the direction of morphological asymmetry does not necessarily correspond between the brain and the visceral organs [54, 56, 58, 59], suggesting that the Nodal pathway normally helps coordinate L-R situs throughout the developing organism.

4.2 Targeted Disruption of Genes with Suspected Roles in Brain Asymmetry

Considerable progress has been made on deciphering key mechanisms in the establishment of the vertebrate L-R body axis and identifying the genetic and cellular components that are involved (refer to [60]). These studies have provided many genes that are candidates for regulating asymmetry of the brain as well.

Although there is still some controversy as to the timing and nature of the initial symmetry-breaking event [61, 62], it is well established that in the developing embryos of zebrafish, frog, and mouse, a ciliated epithelial tissue, the so-called L-R organizer (LRO, refer to [63, 64]), plays an essential role. The atypical monocilia of the LRO are arranged in a polarized fashion and are motile, with their rotation generating a directional fluid flow. The direction of the flow is thought to be perceived and transmitted by nearby mechanosensory cells with immotile cilia through mechanisms involving Ca^{++} flux ([65] and refer to [66]). The result is the transfer of asymmetric signals from the LRO to neighboring cells, with expression of genes encoding Nodal antagonists on the right and activation of the Nodal pathway on the left (refer to [60]).

In zebrafish, the LRO is a transient structure called Kupffer's vesicle (KV), which is derived from a specialized group of cells (the dorsal forerunners) that migrate posteriorly from the dorsal margin of the gastrulating embryo to coalesce into a ciliated epithelial vesicle in the tailbud [67, 68]. Some of the epithelial cells possess motile cilia that generate flow within the fluid filled vesicle, whereas others have immotile cilia [69, 70] that appear to sense fluid flow and initiate calcium oscillations [71]. The net result is enhanced expression of a Nodal antagonist (DAN domain family, member 5 formerly known as Charon) in cells to the right of KV (refer to [70, 72]) and the transmission of signals to the LPM on the left. As in the chick node [73], generation of these L-R differences involves an asymmetry in extracellular Ca^{++} in addition to the action of Notch and other signaling pathways (e.g., [74–77]). The critical factor that induces L-R asymmetry in the anterior neural tube is encoded by the *spaw* gene, which itself is not transcribed in the zebrafish brain [14]. Expression of *spaw* is activated unilaterally in the posterior left LPM and, through transcriptional autoregulation and cleavage of the proprotein, production of this Nodal-related factor is propagated and its range of action expanded anteriorly along the length of the LPM [78].

Accordingly, mutations in any genes important for the formation of KV or for the generation, polarity or motility of its cilia, for mechanosensation and initiation of asymmetric gene expression in cells that surround KV, or for induction and propagation of Spaw in the left LPM, are expected to perturb L-R asymmetry of the brain. Indeed, a ciliary dynein gene (*dynein, axonemal, heavy chain 9*; previously known as *left-right dynein-related1*) selectively expressed in the KV [79] and the *polycystic kidney disease 2* gene (*pktd2*), which encodes a Ca^{++}-activated cation channel important for left-sided restriction and propagation of Spaw [80, 81] are just two of the many genes validated thus far. Mutations alter the normally left-sided pattern of gene expression in the pineal anlage and, when assayed, randomize the direction of epithalamic asymmetry within the affected population. Additional candidates have been netted by targeting zebrafish homologues of human disease genes, those responsible for congenital heart abnormalities due to heterotaxia and for ciliopathies that lead to structural or physiological malfunctioning of organ systems (e.g., [82–84]).

4.3 Implicating New Genes Through Mutagenesis and Molecular Screens

Although it is gratifying that genes with known roles in setting up the L-R body axis also affect asymmetry in the developing zebrafish brain, there are many questions that remain. Notably, how is the Spaw signal transmitted to the anterior dorsal neural tube to induce activation of *ndr2*, *lft1*, and *pitx2* on its left side, what controls the formation and asymmetric positioning of the parapineal, and how does the parapineal direct the neighboring habenula to develop differently? Forward genetic screens as well as reverse genetic strategies are beginning to answer these questions, and reveal unanticipated new players.

In a standard three generation genetic screen, families of fish are derived from mutagenized adult males and then siblings mated to recover homozygous mutants among their progeny [51]. Mutations that affect L-R asymmetry have been identified by careful examination of embryos for morphological criteria such as directional looping of the heart or gut, positioning of visceral organs or formation of pronephric cysts ([85–87] and refer to [88]).

To specifically isolate mutations that affect asymmetry of the brain, third generation larvae have also been screened by RNA in situ hybridization for patterns of gene expression, such as the L-R distribution of *kctd12.1* transcripts in the dHb. The first mutant to be identified in this manner showed nearly equivalent *kctd12.1* expression in both dHb, but in a spatial pattern more characteristic of the right nucleus (i.e., right isomerized). The mutation was mapped and the affected gene determined to encode the T-box 2b (Tbx2b) transcription factor, which itself is synthesized in the pineal anlage [26]. The primary defect of mutants appears to be the misspecification of parapineal cells; they are fewer in number and fail to migrate properly to the left side of the brain [26]. As a consequence, the left dHb is not appropriately specified.

A mutation in the *fibroblast growth factor 8a* (*fgf8a*) gene was also found to disrupt the leftward migration of parapineal progenitors and cause right isomerization of the dHb [89]. Local application of a bead soaked in this factor was sufficient to restore cell migration even in the absence of Nodal signaling. The action of Fgf8a may occur even earlier on neuronal specification because, in its absence, cells of the anterior pineal anlage that would normally give rise to parapineal precursors become cone photoreceptors instead [90]. A genetic screen for abnormal parapineal development also demonstrated a role for the Mediator complex, an essential coordinator of transcriptional regulation, in the maintenance of pineal photoreceptors and projection neurons. In mutants for the Mediator complex subunit Med12, the parapineal fails to mature, habenular neurogenesis is deficient, and *tbx2b* transcripts and FGF signaling are both reduced [91].

Pitx2c, an isoform of the Pix2 transcription factor regulated by Nodal signaling and the only variant of this protein produced in the left pineal anlage, was also shown to control the number of cells that contribute to the parapineal [92]. Unexpectedly, loss of

Pitx2 function results in an increase rather than a decrease in parapineal cells and right isomerization of the dHb. L-R asymmetry can be restored if the enlarged parapineal is partially destroyed [92].

Altogether, the analyses of mutations in candidate or newly identified genes indicate that disrupting the development of the parapineal, either by altering its size, composition, or location, secondarily affects L-R asymmetry of the dHb, providing further support for the parapineal's instructive role.

Screens for abnormal gene expression in the dHb demonstrate that the Wnt/β-catenin pathway is also involved at multiple steps in the development of diencephalic asymmetries, from regulating expression of Nodal pathways members in the neural tube to the generation of habenular precursors and the specification of dHb subregions. A preexisting mutation in the *axin1* gene, which encodes an inhibitory protein that targets cytoplasmic β-catenin for degradation, was initially found to activate expression of Nodal pathway genes bilaterally in the pineal anlage [59]. Surprisingly, bilateral expression still occurs in the absence of Spaw. This finding is consistent with a model in which Axin1, through inhibition of Wnt signaling, normally acts to repress *ndr2*, *lft1*, and *pitx2* gene expression in the diencephalon, a repression that is alleviated on the left side through the unilateral activity of Spaw [59, 93]. However, in contrast, to other mutants with bilateral diencephalic expression, and that later show L-R randomization of epithalamic asymmetry, *axin1* mutants develop with symmetric right isomerized habenulae. Conversely, zebrafish embryos bearing mutated alleles of *transcription factor 7-like 2* (*tcf7l2*), a gene that encodes a transcriptional activator responsive to Wnt signaling, develop with symmetric dHb that are left isomerized [94]. Increased or decreased Wnt signaling, therefore, results in bilaterally symmetric dHb of opposite identity, and in the presence of an intact and normally positioned parapineal. Additional experiments to manipulate levels of Wnt/β-catenin signaling [94] indicate that it must be precisely modulated during neurogenesis, at a time when bipotential habenular progenitors shift from producing dHbL neurons to dHbM ones. The dependence of dHb organization and asymmetry on the regulation of neurogenesis is further supported by the recovery of a mutation in the *sec61a-like 1* (*sec61al1*) gene, which affects the integrity of the neuroepithelium that generates habenular neurons and the timing of proliferation, consequently expanding the dHbL at the expense of the dHbM [95].

Wnt signaling is not only essential for specifying the identity of newly born dHb neurons, but also for regulating the development of habenular progenitors. Mutations in the *wntless* (*wls*) gene, which encodes a protein important for intracellular transport and secretion of Wnts, results in small dHb, owing to a reduced number of progenitor cells [96]. Additionally, both *wls* and *tcf7l2* mutants lack the ventral habenular nuclei [96, 97]. The recovery and analyses of mutations affecting a variety of genes in the Wnt/β-catenin pathway confirm its complex and reiterative actions in habenular development, which may only be fully deciphered when individual functions are determined for the numerous Wnt proteins present in the developing zebrafish embryo.

A complementary approach to a mutagenesis screen is a molecular one, or reverse genetic screen, in which candidate genes are identified on the basis of their differential expression, then subsequently assessed for function using MOs or other targeted techniques for gene inactivation. The *dishevelled associated activator of morphogenesis 1a* (*daam1a*) gene, isolated from a screen for transcripts that are differentially enriched between the left and right halves of the zebrafish brain, was verified by in situ hybridization to be expressed to a greater extent in the left dHb than the right [98]. MO-mediated depletion of Daam1a protein reduced the growth of habenular dendrites and efferent axons, presumably by altering the stability of cytoskeletal elements. Molecular strategies, therefore, revealed that two novel components, the kinase Alk2 and Daam1a, a member of the Formin family that regulate actin assembly, have synergistic roles in the formation of the asymmetric neuropil bundles of the dHb [39, 98].

With a few exceptions, the majority of zebrafish mutations that perturb development or L-R asymmetry of the epithalamus are pleiotropic, meaning that they affect additional tissues essential for viability or other regions of the brain, which complicates the interpretation of behavioral assays. For this reason, transgenic methods that allow more precise manipulation of only neuronal populations of interest have become a preferred approach for assessing dHb function and the impact of their L-R differences on behavior.

5 Transgenic Tools to Probe Epithalamic L-R Asymmetry

A significant advance for the zebrafish model was the adoption of transposable elements for efficient integration of exogenous DNA into the genome. In particular, transposition using the Tol2 element, isolated from the Japanese medaka fish (*Oryzias latipes*) results in single copy insertions, an increased frequency of germline transmission, and rapid generation of zebrafish transgenic lines [99]. Because zebrafish lack this transposon, injection of constructs flanked by Tol2 terminal sequences, along with mRNA encoding the corresponding transposase, into one-cell stage embryos is a highly effective method for producing transgenic individuals. With Tol2 transposition, it is possible to integrate even large genomic regions (e.g., ~100 kb) contained within zebrafish bacterial artificial chromosomes (BACs) [100]. The advantage of a BAC is that it often possesses the requisite DNA regulatory elements for directing transcription of genes expressed in cells of interest. By introducing the *gfp* gene into a transcribed sequence within the BAC [101], its expression can be controlled by endogenous regulatory elements, permitting in vivo labeling.

Alternatively, engineering BACs by inserting genes that encode either the yeast transcription factor Gal4 or the bacterial Cre recombinase offers added flexibility. Where and when produced, Gal4 binds to a particular upstream activating sequence (UAS) to regulate transcription of an adjacent gene [102]. Cre mediates recombination between two loxP sites (i.e., "floxed" allele) and, through deletion of intervening DNA sequence, can promote activation of gene expression (refer to [103]). Such so-called binary systems allow one transgenic line (i.e., Gal4 or Cre "driver") to regulate the expression of multiple responder lines that bear the appropriate UAS or loxP sequences. Integration into gene loci with known cell type-specific expression is one way of producing driver lines. Another is to screen for desired patterns of labeling in individuals recovered by genome-wide enhancer traps, whereby unidentified regulatory sequences control the expression of transgenes that have integrated nearby [104–107]. Responder lines can be "reporters" of cellular or subcellular structure through transcriptional activation of fluorescent protein genes or "effectors" whose gene products modify cellular functions. A variety of transgenic effectors have been successfully applied to study the function of neuronal populations in zebrafish, including methods to kill cells such as the bacterial *nfsB* gene, which encodes a nitroreductase that converts the drug metronidazole into a cytotoxin [108] or KillerRed, a genetically encoded photosensitizer that generates an excess of reactive oxygen species [109], methods to block neural transmission by disrupting neurotransmitter release through botulinum neurotoxin [110] or tetanus toxin light chain [111] or optogenetic tools such as channelrhodopsin, halorhodopsin and Arch3 [108, 112, 113], that function as light-gated ion channels or pumps to activate or suppress neuronal activity selectively.

Although transgenic tools offer powerful and versatile approaches, it is necessary to be aware of potential problems that could arise with their use. First, not all transgenes accurately recapitulate the expression of endogenous genes, typically when requisite regulatory DNA sequences are lacking to activate or repress transcription completely. For reporter lines, there is a temporal delay before fluorescent proteins are properly folded and accumulate at high enough levels for visualization. Progressive silencing of integrated transgenes can also cause variability in expression, particularly when multicopy UAS sites are used to regulate adjacent genes. These sequences, which contain CpG dinucleotides as part of the Gal4 binding site, are prone to DNA methylation and transcriptional repression [114, 115]. Transgenes capitalizing on alternative binary regulatory systems, such as the Q system (QF/QUAS) of *Neurospora crassa* or the bacterial tryptophan repressor

(TrpR/tUAS) appear more resistant to silencing from one zebrafish generation to the next [116, 117], and their broader adoption could ensure greater reproducibility in functional experiments that rely on their efficacy. The inducible TetON system offers the added advantages of reversibility and temporal control of transgene expression [118].

5.1 Monitoring Parapineal Formation and Position

Transgenic zebrafish, in which selective cell populations are labeled by expression of genes encoding fluorescent proteins, are valuable in vivo reporters of diencephalic asymmetries. They have proven useful not only in developmental studies, but also for recognizing larvae with L-R reversed or symmetric brains prior to behavioral testing.

Time-lapse confocal imaging of the pineal anlage, which is labeled in *Tg(foxd3:GFP)* embryos, showed that the parapineal forms from the sequential, leftward migration of a chain of cells (~10–15), followed by their compaction [26]. Tracing of individual cells, after focal activation of caged fluorescein in the *Tg(flh:GFP)* transgenic background, demonstrated that cells from both sides of the anterior pineal anlagen, not just from the *ndr2* expressing population on the left, contribute to the parapineal [25]. The ability to visualize the emerging cells that form the parapineal permits their extirpation via laser ablation at a stage prior to the appearance of habenular L-R differences. This experimental approach provided strong evidence that the parapineal normally plays an instructive role in directing the left dHb to adopt distinct properties; in its absence, both habenulae develop symmetrically with an identity characteristic of the right dHb [25, 28].

The unilateral position of the fluorescently labeled parapineal also serves as an indicator of the directionality of brain asymmetry in live animals. As noted above, injection of translation blocking antisense MOs into one-cell stage embryos [14] or a mutation in the *spaw* gene [55] results in randomization of brain asymmetry. Thus, injecting *spaw* MOs into *Tg(foxd3:GFP)* embryos generates nearly equal numbers of individuals with a parapineal on the left (L_{pp}) or right (R_{pp}) [58]. Embryos can be readily sorted by the position of the fluorescently labeled parapineal for later use in behavioral assays (refer to Sect. 6.1).

5.2 Habenular Organization, Neuronal Identity, and Connectivity

Transgenic lines that label the dHb or their subregions were obtained by targeting known genetic loci on the basis of their expression patterns [30, 119, 120]. Random gene or enhancer trap screens have yielded additional lines including habenula-specific Gal4 drivers (e.g., [106, 107, 121, 122]). These can be useful tools for examining habenular structure. For example, L-R differences in neuropil organization of the dHb can be visualized in vivo using a transgenic line produced by introducing DNA sequences that encode membrane-tagged GFP into a gene (*guanine nucleotide binding protein (G protein), gamma 8* [*gng8*]) that is selectively expressed in the habenular region at 5 days of development [123]. In *TgBAC(gng8:Eco.NfsB-2A-CAAX-GFP)* larvae, the expanded dense neurophil of the left dHb is distinguished from that of the right dHb, which consists of three smaller, separate clusters. From the combinatorial expression patterns of *kctd-related* genes, the dHb were proposed to contain six molecularly distinct domains at the larval stage that differed between the left and right nucleus [31]. As noted above, transgenes have been used to designate asymmetric subnuclear regions. Notably, the dHb were subdivided into medial and lateral subnuclei, delineated by the non-overlapping labeling of the *Tg(pou4f1-hsp70l:GFP)* [previously named *Tg(brn3a-hsp70:GFP)*] [30] and *Tg(nptx2:Gal4-VP16)* [119] lines, respectively. This regional designation of the dHb coincides with the mHb of rodents that contain separate cholinergic and substance P neuronal populations [124], only these are asymmetrically distributed in zebrafish. In the larval brain, the right dHbM corresponds with the prominent *chatb* and *slc18a3b* coexpressing neuronal population [32, 34]. Optogenetic activation of dHb neurons together with electrophysiological recording of target neurons confirmed that, as in mammals, the Hb-IPN pathway is indeed cholinergic [34]. Peptidergic neurons have been detected within the adult dHbL including a group that expresses *tachykinin 1* (*tac1*), which codes for a precursor protein post-translationally processed into Substance P and other small peptides. A somatostatinergic cluster that is larger in the right dHb was also recently discovered [32].

Despite the adoption of the dHbL and dHbM designation, it is important to validate that patterns of transgene labeling truly represent subnuclear organization. One issue is that there is incomplete overlap of labeling with neuronal populations defined by neurotransmitter phenotype [32]. For example, *Tg(pou4f1-hsp70l:GFP)* is taken to demarcate a discrete dHbM subnucleus, but the endogenous *pou4f1* gene is transcribed throughout the dorsal as well as the ventral habenulae [35]. Moreover, larvae from the *Tg(nptx2:Gal4-VP16)* driver line tend to exhibit variable expression depending on the UAS reporter line under its control [32]. Irrespective of these caveats, transgenic lines have been used effectively to correlate neuronal regions of the dHb with connectivity and function (refer to Sect. 6.1).

The L-R identity of the zebrafish dHb dictates how connections are made with other brain regions and transgenic reporters have begun to illuminate some of these afferent and efferent projections. The parapineal is usually positioned to the left of the pineal anlage and labeling of its neurons by *Tg(foxd3:GFP)* corroborates that they innervate only the left dHb, where their axons terminate in a stereotypic arborization pattern [25]. By contrast, when the parapineal is situated on the right side of the brain the pattern of innervation becomes highly variable. Parapineal axons can be truncated, have extended and more elaborate branches or, in some cases, project bilaterally to innervate both the left and right dHb [125]. It is not yet understood how these variations in axonal morphology correlate with altered behavior (refer to Sect. 6.2), but the results suggest that reversal of directionality does not apply to all features of the asymmetric epithalamus.

Fortuitously, a small population of neurons that project only to the right dHb was identified from a transgenic line constructed with DNA sequences upstream of the *LIM homeobox 2a* (*lhx2a*) gene and a yellow fluorescent protein reporter gene [126]. In *Tg(lhx2a-YFP)* larvae, a labeled subset of mitral cells on both sides of the olfactory bulb project their axons bilaterally through the telencephalon to the dHb; however, both the ipsilateral and contralateral projections terminate only at the right nucleus. Synaptic endings are restricted to a small region where a gene of undetermined function, *family with sequence similarity 84, member B* (*fam84b*), is unilaterally expressed [123]. Colabeling of the membranes of habenular neurons and their dendritic processes using *TgBAC(gng8:Eco.NfsB-2A-CAAX-GFP)* further associated the olfactory axon terminals with the medial cluster of dense neuropil in the right dHb [123]. The function of the asymmetric *Tg(lhx2a-YFP)* olfactory projection to the right dHb or of the *fam84b*-expressing target cells is unknown; however, as might be expected, sidedness of both features is L-R reversed in the dHb of R_{pp} larvae [123, 127]. Surprisingly, in the absence of just the parapineal, axons from the bilateral mitral cells now terminate and form synapses with both the left and right dHb [123]. Loss of *nrp1a* expression [46] and gain of *fam84b*-expressing cells in the left dHb [123] accompanies this ectopic innervation; however, experiments using MO against these genes (unpublished observations) indicate that other molecules must serve as the cues that guide olfactory neurons to the right dHb.

With respect to efferent connections, *Tg(pou4f1-hsp70l:GFP)* and *Tg(nptx2:Gal4-VP16)* provided labeling of dHbM and dHbL axonal processes, respectively, and their complementary patterns of IPN innervation. The overly simplistic model of a laterotopic map in which left dHb neurons innervate the dIPN and right dHb neurons the i/vIPN [30, 47] was revised with the realization that medial and lateral neurons on both sides of the brain project to the same general regions of the IPN. A coherent picture has thus emerged: Neurons from a large left and much smaller right dHbL project their axons to the dIPN. Neurons in the left and right dHbM, are more equivalent in number and innervate the intermediate and ventral IPN regions ([119] and refer to [128]).

As with pre-synaptic input, perturbations of L-R asymmetry of the dHb predictably alter IPN connectivity. Consistent with earlier dye labeling and immunolabeling experiments [30, 31], the origin of IPN afferents is L-R reversed in R_{pp} larvae and adults:

Tg(pou4f1-hsp70l:GFP) labeled axons from the left and right dHbM predominantly terminate at the dIPN instead of the i/vIPN [30]. In individuals with the parapineal removed, the majority of efferents from the bilaterally symmetric dHb project to the i/vIPN [31, 47], consistent with the chiefly dHbM identity of their neurons. Importantly, this manipulation affecting less than 20 cells, demonstrates how a small bias in the developing brain not only alters the identity of adjacent neurons, but also those of distant, pre- and post-synaptic partners, and presumably other downstream members of the neural circuitry.

Transgenic lines that illuminate the targets of the zebrafish IPN have not yet been described; however, labeling by focal application of retrograde and anterograde tracers demonstrated that the griseum centrale receives efferents from the dIPN and neurons from the vIPN project to the median Raphe, regions thought to be homologous to those of the mammalian brain that modify behaviors provoked by fear/anxiety ([119]; refer to [128]).

6 Functional Correlates to Habenular Left-Right Asymmetry

Differences in the neuroanatomy, neuronal identity, and connectivity between the left and right sides of the nervous system have been found through the animal kingdom [129–134], yet in only a few cases is their relevance to behavior understood. The Hb-IPN pathway of zebrafish is a particularly advantageous model for determining the influence of brain asymmetry on behavior, owing to the feasibility of manipulating L-R differences using genetic and cellular approaches (refer to Fig. 1). How such changes might influence behavioral lateralization, those functions already attributed to the habenular nuclei, or their differential neuronal activity, are questions that are being actively pursued.

6.1 Behavioral Lateralization in Fish: A Role for the Hb-IPN System?

There is considerable evidence for lateralized behaviors in fish, meaning that individuals can exhibit a directional bias in their response to physical challenges or specific stimuli. Whether there is any involvement of the Hb-IPN pathway and, in particular, a clear correlation between asymmetry of the dHb and the lateralized responses of zebrafish is still under debate.

An early demonstration of lateralized behavior was the direction that male mosquitofish (*Gambusia holbrooki*) swam around a clear barrier to reach a group of conspecific females (termed the "detour test"; Fig. 2a; [135]). Adult males showed a significant bias in turning to the left of the barrier when swimming to the target. Males did not exhibit a L-R bias when the barrier was either U-shaped or opaque, thereby visually obscuring the females, or the target females were replaced with males [135]. Goal directed behavior in mosquitofish therefore seems biased to the left.

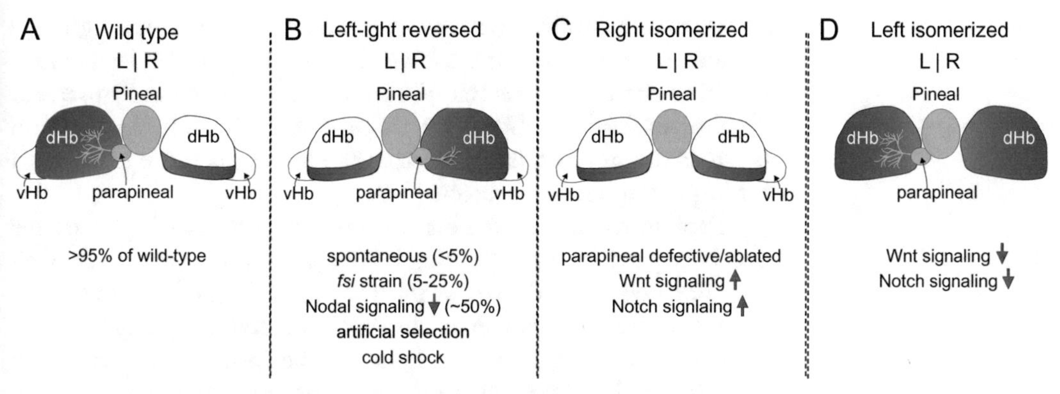

Fig. 1 Approaches to produce zebrafish with altered dHb L-R asymmetry. (**a–d**) Schematic dorsal view of the epithalamic region with the pineal complex in *green* and dorsal (dHb) expression of *kctd12.1* in *blue*. (**a**) In the vast majority of wild-type larvae, the parapineal forms to the left of the pineal anlage, and its neurons innervate the left dHb in a stereotypic arborization pattern. Expression of *kctd12.1* is expanded in the left dHb compared to the right. (**b**) In a small percentage of wild type, the parapineal, develops on the right side and its neurons show highly variable innervation of the dHb (a representative truncated arborization pattern is depicted) [125]. Expression of kctd12.1 in the dHb is L-R reversed. The proportion of individuals with L-R reversed epithalamic asymmetry can be increased by disruption of Nodal signaling [14, 19, 20, 55], reducing temperature during early development [127], artificial selection [146, 147], or in the fsi zebrafish strain [148], and by many mutations (described in the text). (c) Removal of the parapineal [25, 28] or its abnormal development [90, 91], or increased activation of the Wnt [59] or Notch signaling pathways [35] produces larvae with right isomerized dHb. (d) Reducing Notch [35] or Wnt signaling, either genetically [94] or pharmacologically [127], results in larvae with left isomerized dHb and loss of the ventral habenulae (vHb)

The detour test was also used to measure turning preference in a related species, *G. falcatus*, in response to a "dummy" predator [136]. At the population level, adults showed a significant preference to swim towards the left, although individuals were found that predominantly turned to the right or had no bias at all. Those mosquitofish showing a significant preference to turn left also tended to inspect the perceived predator with the right eye. On the other hand, individuals turning rightward more frequently exhibited increased viewing with the left eye. Fish without a clear bias did not demonstrate a strong eye preference. The results suggest that directionality in turning behavior is correlated with preferential eye use.

Zebrafish also show a turning bias in the detour test and a preference in eye use [137], but whether it is left or right biased depends on the visual stimulus presented. At the population level, adult zebrafish prefer to view an empty space with their left eye; however, when confronted with a more complex environment, they favor the right one [138]. Exposure to a novel object, such as a marble when food is anticipated, initially elicits right eye viewing and biting behavior. The frequency of right eye use and the number of bites both declined during subsequent trials [139]. When other species of fish were presented, the fighting fish *Betta splendens* or a familiar social species, zebrafish favored the right eye to view the former and the left eye for the latter [138]. These studies led to the hypothesis that the right eye is preferentially used for viewing unfamiliar objects or those that elicit a strong reaction (such as the fighting fish), whereas the left eye is more often used for inspecting familiar objects [140].

Paradigms for assessing bias in motor responses and eye use have also been devised for larval stages (Fig. 2b and c). Larval zebrafish have been challenged in a "swim-way" consisting of multiple compartments that are connected by small, central openings they can swim through and in which light can be individually controlled (Fig. 2b; [141]). Typically, zebrafish larvae are attracted to light and avoid darkness [142]. In one version of the swim-way test a single larva was placed in the first compartment with the light on. The light was gradually turned off and turned on in the second compartment. This pattern was repeated in succession until the larva either turned into a specific compartment or reached the end of the swim-way without turning. The behavior of wild-type larvae was somewhat variable, with approximately two thirds showing a strong tendency to turn left and the remainder having no bias [141]. Zebrafish larvae are also known to become active if lights are rapidly turned off instead of gradually dimmed, demonstrating locomotor behavior akin to a startle-response [142]. When lights were rapidly turned off in the swim-way test, they significantly preferred to turn to the right [141]. Past experience also may affect turning behavior. In a modified version of the swim-way, where black stripes were added to the right or left sides of the compartment, larvae tended to turn away from the stripes. However, if they had been raised in containers with similar stripes, a preference for turning either towards or away from the stripes was not observed [141].

The "mirror test," is a commonly used assay for measuring visual lateralization. Originally applied to adult teleosts [143], the testing apparatus is comprised of a rectangular tank with the long walls made of mirrors and the shorter walls of opaque plastic (Fig. 2c). A variety of teleost species show a bias for viewing their mirror reflection with the left eye [143]. Zebrafish larvae also seem to have a significant preference for left eye viewing [144], although

Fig. 2 (continued) or fearful [125] behaviors. (**d**) In a conditioned fear assay, adult zebrafish are trained to associate a red light (conditioned stimulus) with an electric shock (unconditioned stimulus). After training, zebrafish display freezing behavior when only the red light is presented (adapted from [119]). (**e**) The novel tank assay measures innate fear/anxiety in adult zebrafish. More time spent in the bottom half of a novel tank is taken as a measure of increased anxiety (adapted from [157]). (**f**) The apparatus for the confined box test is approximately twice the length of an adult zebrafish. It is fitted with two escape holes that are open upon removal of a lid, and the amount of time it takes to emerge is scored (adapted from [125]). The testing chambers are not drawn to scale as their size depends on whether larval or adult fish are being assayed

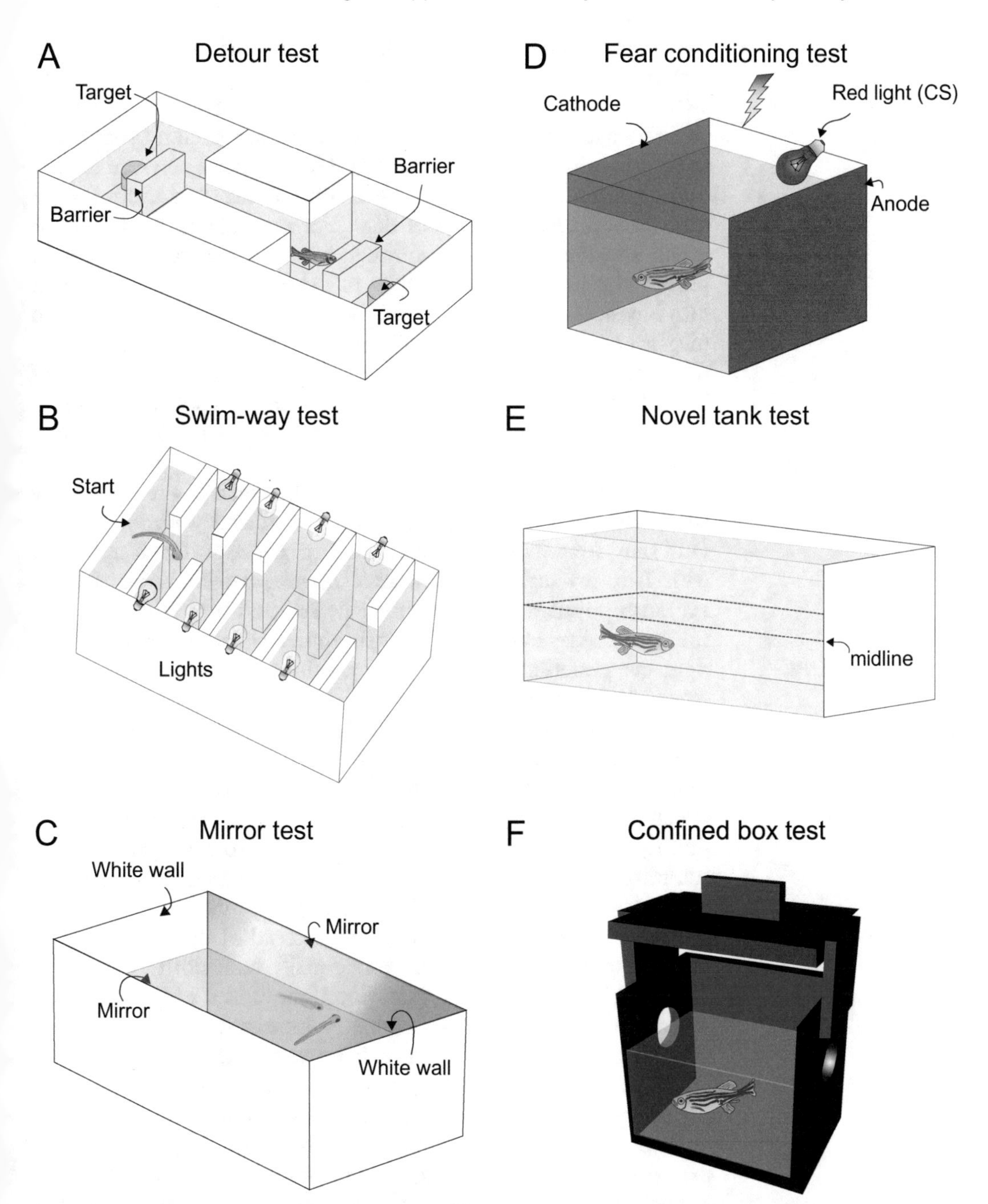

Fig. 2 Assays to measure behavioral lateralization and fear/anxiety. (**a**) The detour test consists of a target blocked by a barrier. The direction adult fish swim around the barrier to reach the target is recorded (adapted from [135]). (**b**) The apparatus for the swim-way test consists of multiple, individually lit compartments connected by a small passage. In response to changes in light conditions, a larva will either turn left or right into a compartment or swim through to the last compartment (adapted from [141]). (**c**) For both larval and adult zebrafish, the mirror test has been used to measure preference in eye use upon viewing their reflection (adapted from [143, 144]). The apparatus is a rectangular tank with mirrors as the long walls and opaque white plastic as the short ones. In others studies, interaction with the mirrors has revealed aggressive [149]

the degree depends on a number of factors including the age and genetic background of the subjects and the test interval [144, 145]. An independent study, however, did not find a significant difference in left or right eye preference in approaching the mirror in the majority of larvae tested [58]. Discrepancies between studies may be accounted for, in part, by nonstandardized methods for the analysis and interpretation of data. For example, in some experiments all time points were summed for left versus right eye use in mirror viewing [58, 144] while, in others, results were grouped into bins according to the angle of a larva's body axis relative to the mirror [145]. In addition, early reports suggesting a left eye bias relied on visual inspection of video records and manual tracking of larval behavior [144, 145]. Automated tracking software in place of manual annotation [58] could be more consistent, thereby enhancing reproducibility.

Efforts to explore the relationship between brain asymmetry and behavioral laterality have also produced varying results. Different strategies have been used to generate larval zebrafish with L-R reversals of epithalamic asymmetry. In some studies, zebrafish populations were artificially selected by testing larvae in the mirror assay and raising only those that showed a strong left or right eye preference. After five generations of artificial selection, a strain exhibiting more frequent right eye use (termed TLRE) had a higher proportion of individuals with L-R reversed *kctd12.1* expression in the dHb compared to the unselected background they were derived from [146]. Subsequently, the TLRE line was selected further to enrich for R_{pp} individuals, scored in vivo using the *Tg(foxd3:gfp)* line [147]. When adults from this strain were submitted to a variation of the mirror test, an octagonal arena with mirrors fitted to each of the eight-walls, the L_{pp} group had a significant bias in right eye use. However, instead of increased left eye use, R_{pp} adults viewed the mirrors equally with both eyes [147]. This result suggests that behavioral responses are not merely reversed when epithalamic asymmetry appears to be.

Another study used larvae from the *frequent-situs-inversus* (*fsi*) strain, a background of unknown genetic identity in which the parapineal is located to the right of the pineal (R_{pp}) in as high as 25% of the population [148; see also Chap. 21]. Larvae were pre-sorted according to parapineal position and later submitted to a series of behavioral tests. In the biting test, L_{pp} adults viewed the marble with the right eye before approaching, whereas R_{pp} siblings more often used the left [148]. During the mirror test, a pronounced change in eye use occurred between the fourth and fifth minute of the 5-min testing period: L_{pp} larvae shifted from left to right eye viewing and R_{pp} larvae also shifted but in the opposite direction [148]. In the swim-way test, L_{pp} and R_{pp} larvae showed variable, nonsignificant differences in turning behavior similar to what had been observed for wild-type larvae [141]. The overall conclusion was that some, but not all, lateralized behaviors are influenced by epithalamic asymmetry [148].

A third strategy generated L-R reversals of epithalamic asymmetry experimentally, through injections of the *spaw* antisense MO into one-cell stage *Tg(foxd3:gfp)* embryos. L_{pp} and R_{pp} larvae were pre-sorted and tested for turning direction during a startle response or upon presentation of a lateralized stimulus, and for eye use in the mirror test. Significant differences were not found between the two groups in any of these assays [58], further supporting that the direction of neuroanatomical asymmetry does not neatly correspond with the sidedness of lateralized behaviors.

6.2 Lateralized dHb Function in Modulation of Fear and Anxiety

When performing what were considered to be assays for lateralized behaviors, R_{pp} larvae and adults were discovered to manifest unexpected behaviors indicative of increased fear/anxiety [58, 125]. Compared to L_{pp} siblings, they showed a delay in the onset of swimming and reduced exploration of the testing arena in the mirror test, staying in close proximity to the walls. Abnormal behavior was triggered by their reflection, because when the mirrors were replaced with opaque walls, R_{pp} individuals showed normal exploratory behavior [125]. Interestingly, for African cichlids, the mirror image provokes aggressive behaviors and is interpreted as a fearful stimulus [149].

The Hb-IPN pathway is known to modulate fear and anxiety in mammals (refer to [150]). In an early study, rats that had undergone bilateral lesions of the FR manifested signs of increased anxiety, such as fewer entries into the open arms of an elevated plus maze and reduced exploration in the open field test [151]. Similar lesions to the FR raised levels of plasma corticosterone, the rodent equivalent of fish and human cortisol [152]. In mice, the dorsal region of the medial Hb (dMHb) is thought to regulate fearful behavior and the ventral region (vMHb) to modulate anxiety [153]. Following lesions to the bed nucleus of the anterior commissure whose neurons specifically project to the dMHb, mice were shown to increase freezing behavior in a fear-conditioning paradigm. Lesions to the triangular septum, which sends efferent projections to the vMHb, reduced exploratory behavior in the open field assay and elevated plus maze [153].

Fear conditioning has also been applied to juvenile and adult zebrafish, through which they learn to associate a conditioned cue (CS), such as the flash of a red light, with an unconditioned stimulus (US) like a mild electric shock [119, 154]. Typically, juveniles (3–5 weeks old) associate the two stimuli and actively avoid the electric shock by moving to the opposite side of the test tank. After training, presentation of the CS alone elicits similar behavior. However, transgenic zebrafish expressing KillerRed or TeTxLC under the control of an enhancer trap line, which labels (but is not specific to) dHb neurons did not exhibit avoidance behavior in response to the CS. Instead, they remained immobile for a prolonged period, displaying behavior akin to freezing [154]. These findings suggest that the dHb are necessary for active avoidance of a fearful stimulus.

Transgenic techniques permit selective manipulations of dHb subregions. For example, by crossing transgenic lines *Tg(nptx2:Gal4-VP16)* and *Tg(UAS:nfsB-mCherry)* or *Tg(UAS-TeTxLC)* to either ablate or inactivate neurons, a role for the lateral subregion of the dHb in modulating experience-dependent fear/anxiety was established [119]. Following training in a conditioned fear assay (Fig. 2d), anticipation of a cued electric shock caused adult zebrafish to increase their speed of swimming in an acknowledged fear response. Before conditioning, control adults and those with silenced or ablated dHbL responded to electric shock comparably, with increased swimming velocity. After conditioning, however, adults with ablated or silenced dHbL covered less distance and exhibited an approximately fivefold increase in the duration of freezing relative to controls with an intact dHbL [119]. That the dorsal region of the medial Hb of rodents [153] and the dHbL of zebrafish [119] both modulate fear responses by suppressing freezing behavior is indicative of a functional homology between them [155]. Furthermore, the pronounced size difference in the domain of *Tg(nptx2:Gal4-VP16)* expression between the left and right dHb further suggests that it is the left subnucleus that is predominantly responsible for calibrating the response to experience-dependent fear in adult zebrafish.

Altering their L-R asymmetry can directly test the differential impact of the zebrafish dHb on fear-induced behaviors. A commonly used assay for measuring innate fear/anxiety in adult zebrafish is the "novel tank" test (Fig. 2e). When first introduced into an unfamiliar tank, adult zebrafish initially spend more time in the bottom half but over time, will explore the top half to a greater degree [156, 157]. Prior exposure to drugs that are known to reduce anxiety in humans (anxiolytic agents) increases the duration of time that a zebrafish swims in the top half, whereas compounds that promote distress (anxiogenic agents) reduce it [157, 158]. Thus, the proportion of time spent in the bottom portion of the tank is taken as a measure of anxiety. In this assay, R_{pp} adults showed increased bottom dwelling relative to L_{pp} controls, although prior treatment with the anxiolytic drug buspirone, enhanced the exploratory behavior of both groups [125].

An additional behavioral assay, the "confined box test" was devised to corroborate that R_{pp} zebrafish have different behavioral responses than their L_{pp} siblings [125]. Single adults were placed in a small, rectangular box approximately twice their body length, and fitted with two exit holes blocked by a lid. Removal of the lid exposed the holes to a large, brightly illuminated tank and the time it took adults to emerge from the box was scored (Fig. 2f). Relative to L_{pp} controls, R_{pp} adults showed a significant delay in exiting. Some fish protruded just their head from the exit hole, remaining in this position for an extended period of time. This seemingly anxious response was never observed in controls.

A direct physiological readout of anxiety is the level of whole body cortisol [156, 159, 160]. Methods to measure cortisol levels using enzyme linked immunosorbent assays (ELISA) have been standardized for adult zebrafish [156]. When temporarily removed from water for a 4-min period (i.e., hypoxic stress), whole body cortisol levels were significantly elevated in R_{pp} adults relative to L_{pp} siblings [125], indicative of enhanced anxiety.

How L-R reversal of brain asymmetry causes physiological and behavioral changes is unclear. From analyses of molecular markers [6, 31], labeling of habenular efferents using lipophilic dyes [30, 31, 46, 161] and labeling habenular afferents by transgenic reporters [123], it was assumed that R_{pp} zebrafish are a complete reversal of the more typical form. This assumption, however, was reevaluated following behavioral studies of larval and adult zebrafish with bilaterally symmetric dHb (i.e., right isomerized) caused by parapineal ablation [125]. These individuals explored less of the test tank and took a longer time to initiate swimming relative to controls when tested in the mirror assay. However, in the absence of mirrors, both groups behaved similarly. These and other experiments indicate that right isomerized and R_{pp} zebrafish share aberrant behavioral traits, supporting that L-R reversed epithalamic asymmetry represents an abnormal condition.

6.3 Differential Neuronal Responses in the Left and Right dHb

A continuing challenge is to correlate neuronal populations of the dHb with behavioral responses attributed to their L-R asymmetry and function. One promising approach is the implementation of genetically encoded calcium indicators (GECI) that allow monitoring of neuronal activity in live, behaving animals. Fluorescence based reporters, such as GCaMP, are powerful tools for assessing neural activation in vivo [162]. GCaMP consists of three functional units: a circularly permuted green fluorescent protein (cpGFP), a Calmodulin (CaM) subunit, and an M13 linker domain derived from the myosin light chain kinase. In the absence of calcium, the cpGFP is in a folded state and fluoresces at a very low level. Firing of an action potential is accompanied by a large influx of calcium ions that, through binding to CaM, causes a conformational change in the cpGFP and a substantial increase in fluorescence. Many versions of GCaMP have been engineered in an effort to optimize the signal to noise ratio and on and off response times [162–167].

Transgenic zebrafish larvae and juveniles expressing GCaMP throughout the brain are particularly well suited for functional imaging of neuronal activity. By 5 days of development, the larval brain is roughly 800 μm long × 400 μm wide × 300 μm thick, which permits imaging of neuronal responses throughout its entirety and with single cell resolution [168–171]. On account of their location on the dorsal surface of the diencephalon, dHb neurons are readily accessible for recording of calcium transients using confocal microscopy.

Exposure to light or odorants activates neurons in the dHb, as assessed by increased GCaMP fluorescence [127, 172–174]. Light-induced fictive swimming activity, which is inferred from extracellular potentials recorded from motor neurons [175], is accompanied by an increase in calcium transients in the dHb [172]. Light can also elicit lateralized responses. In 4-day-old larvae, red light (625 nm) predominantly activates neurons in the left dHb, whereas neurons in the right dHb are preferentially activated by odorants derived from dried food dissolved in water [127]. In larval zebrafish with L-R reversed dHb asymmetry, this pattern is reversed, with light-evoked calcium transients largely detected in the right dHb and food odors activating neurons in the left. In larvae with right isomerized dHb, the light cue did not increase neural activity and bilateral activation was observed in response to odors. The opposite pattern applied to larvae with left isomerized dHb. The habenular response to light is thought to require retinal input because calcium transients are not increased in dHb neurons following light exposure in homozygous mutants for the *retinal homeobox gene 3* gene, that fail to form retinas [127].

The lateralized response to odors appears more complex. Exposure of juvenile zebrafish (21–28 day old) to amino acids, bile, and nucleic acids also activates more neurons in the right dHb, in spatially distinct sets [173]. Thus, subgroups of neurons in the right dHb may be selective for different odorants. The bile salt, glychochenodeoxycholic acid (GCDA) evokes dose-dependent responses in dHbM that could have behavioral significance [174]. Low concentrations of GCDA are attractive to zebrafish, while high concentrations are repulsive [174]. Bile salt is often used as a positive social cue by teleost species [176–178], but is toxic at high concentrations and can lead to tissue damage in the liver [179] and mitochondrial breakdown [180]. Therefore, graded neuronal activity in the right dHb in response to GCDA may permit attraction to bile salts at low concentrations, but avoidance at higher ones [174].

The relevance of differential processing of light and odor by the dHb is unclear. Mapping pre-synaptic input to habenular neurons may prove informative. Neural connections between the dHb and the visual system have not been described. As discussed in Sect. 5.2, bilateral projections from a subset of olfactory mitral cells, identified by *Tg(lhx2a-YFP)* labeling, specifically innervate only the right dHb [126]. It has been asserted that chondroitin sulfate, an aversive component of extracts from fish skin, might activate this mitral cell population to elicit fearful behaviors mediated by dHb neurons [181, 182]. Although odorants derived from skin extracts and chondroitin sulfate induce robust expression of the neural activity marker *fosab* (formerly known as *c-fos*) in the olfactory bulb, it does not colocalize to *Tg(lhx2a-YFP)* labeled cells [123]. Moreover, in studies where neuronal activity was measured by Ca^{++} signaling, delivery of skin extract, or chondroitin sulfate activated a subset of right dHb neurons, but not in the vicinity of the *Tg(lhx2a-YFP)* labeled axon terminals [173].

7 Future Directions

In a relatively short period of time, the zebrafish dorsal diencephalon has become a useful template for probing how L-R asymmetry is established in the brain, the underlying molecular mechanisms and, ultimately, the consequences on neuronal activity and behavior. Genetic and cellular approaches afforded by this vertebrate model, such as the ability to alter epithalamic asymmetry in predictable ways or to perform whole brain imaging of neuronal activation, also make it an especially powerful one.

Forward and reverse genetics, and accumulating knowledge of the development of visceral organ asymmetry, have contributed to establishing a useful framework to determine how asymmetry might arise in a vertebrate brain. To date, the only region of the developing zebrafish brain found with overt L-R differences is the epithalamus. However, even a seemingly small perturbation such as removal of the parapineal, alters dHb identity, afferent innervation, and efferent connections, demonstrating that a localized asymmetry can exert widespread changes in the brain.

Progress has been made in exploring the relationship between dHb asymmetry and behavior, yet little is known about the neuronal subtypes that execute the responses or the intervening neural pathways. For example, neurons in the dHb show selectivity for different sensory modalities such as light and odors [127, 173, 174], but at the cellular level, their molecular identity or specific pre- and postsynaptic partners are unknown. Nor do we fully appreciate the effect of such lateralization on behavioral responses. Furthermore, how does differential dHb input to subregions of the IPN and presumably to their diverse neuronal subtypes activate or inhibit the dorsal raphe and griseum centrale to modulate targets further downstream?

We are beginning to have a greater understanding of the neuronal identities and organization of the dHb, but considerably more work is required to classify all of the neuronal cell types and to obtain a more refined connectivity map between the dHb and IPN. Transcriptional profiling of isolated brain regions will provide new molecular markers for cataloguing neurons and for tracing axonal projections with greater precision. Sparked by recent advances, optogenetic control of activity will enable the functional assessment of selective neuronal populations and imaging of calcium transients will correlate their responses to sensory stimuli. State-of-the-art in vivo targeting of the cell-type-specific neuronal genes using Crispr/Cas9 technology ([183] and refer to [184]) will greatly accelerate the production of transgenic tools to pursue such studies. More sophisticated strategies involving intersectional gene expression will likely be required to restrict manipulations to only the desired neuronal subtypes. The zebrafish larval brain undergoes a dramatic morphogenetic transformation and growth to achieve the adult form [185, 186], which is accompanied by changes in the degree of dHb L-R asymmetry and the appearance of new neuronal cell types [32]. A rigorous investigation into the nature of these changes and their consequence on IPN connectivity and behavior in adult zebrafish is also warranted. An exciting new study on the fighting behavior of adult males suggests that differential activity between the dHbM and dHbL subregions corresponds with whether a fish will be a winner or a loser [187]. It will be of great interest to uncover the precise neurons that mediate such extreme outcomes.

As it is now possible to image neuronal activity in almost every neuron in the larval zebrafish brain at single cell resolution [168, 170], presenting stimuli or provoking fictive behaviors in combination with whole brain calcium imaging could reveal the entire neural network that acts upon and through which the left and right dHb differentially exert their effects. Applying an unconditioned electric shock to head immobilized larvae, for instance, enables a fearful stimulus to be correlated with changes in neuronal activity in the dHb and elsewhere in the brain (Duboué and Halpern, unpublished). Alternative methods to calcium imaging provide the means to analyze neuronal activation in freely behaving animals of all ages and minimize the need for complex microscopy and computational strategies [188]. For example, depolarization stimulates the Ras-ERK pathway; specifically, calcium influx into neurons results in ERK phosphorylation. Thus, comparative immunolabeling of phosphorylated and total ERK identifies those neurons that have recently become activated [189–192].

The focus of this chapter has been on the zebrafish, largely because of the genetic tools that are available. A rapidly amassing number of genes have been shown to have an impact on dHb development, either controlling the degree or directionality of L-R asymmetry. However, many classes of fish, amphibians, and reptiles show L-R differences in epithalamic morphology ([23, 161, 193] and refer to [6]) that could have arisen from variations on the same genetic themes or from novel mechanisms. Comparative studies are essential and, in the near future, these should also be facilitated by Crispr/Cas9 genome editing, to inactivate functions or introduce new ones into candidate genes.

A pressing question is whether what is learned from the fish epithalamus will be relevant for understanding L-R specializations of the mammalian brain. In mice, while functions such as maternal recognition of ultrasonic calls from pups [194], fear learning from observing others in distress [195] and fear conditioning [196], seem to be lateralized, neuroanatomical differences are not obvious and only a few molecular ones have been reported (e.g., [197, 198]). L-R differences in the rodent brain may be quite subtle, as evidenced by studies on the hippocampus where CA3 pyramidal neurons synapse onto the dendritic spines of CA1 neurons. Whether the innervating fibers emanate from CA3 neurons on the left or right side of the brain strongly influences the type of synapse that is formed [199]. CA1 dendritic spines receiving axon terminals from left CA3 neurons are thin, their synapses are smaller and contain the glutamate NMDA receptor subunit NR2B. In contrast, CA1 dendrites receiving input from right CA3 neurons show enlarged, mushroom-shaped spines with synapses enriched for the glutamate AMPA receptor subunit GluR1. Axonal terminals of left and right CA3 neurons must be specialized to direct the form of CA1 dendritic spines and their glutamate receptor profiles in distinct ways, but how such properties are established developmentally is currently unknown. Intriguingly, in mutants with L-R reversed visceral asymmetry, these hippocampal features become symmetric (i.e., right isomerized) resulting in defective learning and memory [200].

In the human cortex, where L-R reversal of neuroanatomical asymmetries in the temporal lobe is not necessarily coupled with a hemispheric shift in language lateralization [201–203], the correlation of functional specializations with underlying neuronal pathways may be a daunting task. As noted long ago by the science historian Jane Oppenheimer [204], "The most unfathomable asymmetry of all is that of our brains and minds—at a far profounder level than the simple structural asymmetry of the amphibian habenulae." From the neuroanatomical, molecular, and functional studies described in this Chapter, the dHb of zebrafish seem far from simple. Yet even with a more complete picture of their neuronal populations and lateralized activities, we are still a far way from knowing whether hemisphere-specific tasks in the human brain arose from or rely on similar processes.

References

1. Kemali M, Braitenberg V (1969) Atlas of the frog's brain. Springer Verlag, Heidelberg
2. Braitenberg V, Kemali M (1970) Exceptions to bilateral symmetry in the epithalamus of lower vertebrates. J Comp Neurol 138(2):137–146. doi:10.1002/cne.901380203
3. Frontera JG (1952) A study of the anuran diencephalon. J Comp Neurol 96(1):1–69
4. von Woellwarth C (1950) Experimentelle Untersuchungen über den Situs inversus der Eingeweide und der Habenula des Zwischenhirns bei Amphibien. Wilhelm Roux' Arch 144:178–256
5. Wehrmaker A (1969) Right-left asymmetry and situs inversus in Triturus alpestria. Wilhelm Roux Arch Entwickl Mech Org 163:1–32

6. Concha ML, Wilson SW (2001) Asymmetry in the epithalamus of vertebrates. J Anat 199(Pt 1–2):63–84
7. Kemali M (1976) The dense core of vertebrate central nervous system synapses revealed by potassium permanganate fixation as formed by small membrane-bounded vesicles. Neurosci Lett 2(2):67–71
8. Kemali M, Guglielmotti V (1977) An electron microscope observation of the right and the two left portions of the habenular nuclei of the frog. J Comp Neurol 176(2):133–148. doi:10.1002/cne.901760202
9. Kemali M, Guglielmotti V (1984) The distribution of substance P in the habenulo-interpeduncular system of the frog shown by an immunohistochemical method. Arch Ital Biol 122(4):269–280
10. Kemali M, Sada E (1973) Myelinated cell bodies in the habenular nuclei of the frog. Brain Res 54:355–359
11. Kemali M, Sada E (1974) Histology and ultrastructure of the dorsal habenular nuclei of the frog. Z Mikrosk Anat Forsch 88(1):167–176
12. Vota-Pinardi U, Kemali M (1990) Neuroelectrophysiology of the morphologically asymmetric habenulae of the frog. Comp Biochem Physiol A Comp Physiol 96(3):421–424
13. Varlet I, Robertson EJ (1997) Left-right asymmetry in vertebrates. Curr Opin Genet Dev 7(4):519–523
14. Long S, Ahmad N, Rebagliati M (2003) The zebrafish nodal-related gene southpaw is required for visceral and diencephalic left-right asymmetry. Development 130(11):2303–2316
15. Rebagliati MR, Toyama R, Fricke C, Haffter P, Dawid IB (1998) Zebrafish nodal-related genes are implicated in axial patterning and establishing left-right asymmetry. Dev Biol 199(2):261–272
16. Levin M, Johnson RL, Stern CD, Kuehn M, Tabin C (1995) A molecular pathway determining left-right asymmetry in chick embryogenesis. Cell 82(5):803–814
17. Bisgrove BW, Essner JJ, Yost HJ (1999) Regulation of midline development by antagonism of lefty and nodal signaling. Development 126(14):3253–3262
18. Thisse C, Thisse B (1999) Antivin, a novel and divergent member of the TGFbeta superfamily, negatively regulates mesoderm induction. Development 126(2):229–240
19. Concha ML, Burdine RD, Russell C, Schier AF, Wilson SW (2000) A nodal signaling pathway regulates the laterality of neuroanatomical asymmetries in the zebrafish forebrain. Neuron 28(2):399–409
20. Liang JO, Etheridge A, Hantsoo L, Rubinstein AL, Nowak SJ, Izpisua Belmonte JC, Halpern ME (2000) Asymmetric nodal signaling in the zebrafish diencephalon positions the pineal organ. Development 127(23):5101–5112
21. Essner JJ, Branford WW, Zhang J, Yost HJ (2000) Mesendoderm and left-right brain, heart and gut development are differentially regulated by pitx2 isoforms. Development 127(5):1081–1093
22. Gritsman K, Zhang J, Cheng S, Heckscher E, Talbot WS, Schier AF (1999) The EGF-CFC protein one-eyed pinhead is essential for nodal signaling. Cell 97(1):121–132
23. Lagadec R, Laguerre L, Menuet A, Amara A, Rocancourt C, Pericard P, Godard BG, Celina Rodicio M, Rodriguez-Moldes I, Mayeur H, Rougemont Q, Mazan S, Boutet A (2015) The ancestral role of nodal signalling in breaking L/R symmetry in the vertebrate forebrain. Nat Commun 6:6686. doi:10.1038/ncomms7686
24. Thisse C, Thisse B (2008) High-resolution in situ hybridization to whole-mount zebrafish embryos. Nat Protoc 3(1):59–69. doi:10.1038/nprot.2007.514
25. Concha ML, Russell C, Regan JC, Tawk M, Sidi S, Gilmour DT, Kapsimali M, Sumoy L, Goldstone K, Amaya E, Kimelman D, Nicolson T, Grunder S, Gomperts M, Clarke JD, Wilson SW (2003) Local tissue interactions across the dorsal midline of the forebrain establish CNS laterality. Neuron 39(3):423–438
26. Snelson CD, Santhakumar K, Halpern ME, Gamse JT (2008) Tbx2b is required for the development of the parapineal organ. Development 135 (9):1693–702.
27. Gamse JT, Shen YC, Thisse C, Thisse B, Raymond PA, Halpern ME, Liang JO (2002) Otx5 regulates genes that show circadian expression in the zebrafish pineal complex. Nat Genet 30(1):117–121. doi:10.1038/ng793
28. Gamse JT, Thisse C, Thisse B, Halpern ME (2003) The parapineal mediates left-right asymmetry in the zebrafish diencephalon. Development 130(6):1059–1068
29. Ishikawa Y, Inohaya K, Yamamoto N, Maruyama K, Yoshimoto M, Iigo M, Oishi T, Kudo A, Ito H (2015) The parapineal is incorporated into the habenula during ontogenesis in the medaka fish. Brain Behav Evol 85(4):257–270. doi:10.1159/000431249

30. Aizawa H, Bianco IH, Hamaoka T, Miyashita T, Uemura O, Concha ML, Russell C, Wilson SW, Okamoto H (2005) Laterotopic representation of left-right information onto the dorso-ventral axis of a zebrafish midbrain target nucleus. Curr Biol 15(3):238–243. doi:10.1016/j.cub.2005.01.014
31. Gamse JT, Kuan YS, Macurak M, Brosamle C, Thisse B, Thisse C, Halpern ME (2005) Directional asymmetry of the zebrafish epithalamus guides dorsoventral innervation of the midbrain target. Development 132(21):4869–4881. doi:10.1242/dev.02046
32. deCarvalho TN, Subedi A, Rock J, Harfe BD, Thisse C, Thisse B, Halpern ME, Hong E (2014) Neurotransmitter map of the asymmetric dorsal habenular nuclei of zebrafish. Genesis 52(6):636–655. doi:10.1002/dvg.22785
33. Eiden LE (1998) The cholinergic gene locus. J Neurochem 70(6):2227–2240
34. Hong E, Santhakumar K, Akitake CA, Ahn SJ, Thisse C, Thisse B, Wyart C, Mangin JM, Halpern ME (2013) Cholinergic left-right asymmetry in the habenulo-interpeduncular pathway. Proc Natl Acad Sci U S A 110(52):21171–21176. doi:10.1073/pnas.1319566110
35. Aizawa H, Goto M, Sato T, Okamoto H (2007) Temporally regulated asymmetric neurogenesis causes left-right difference in the zebrafish habenular structures. Dev Cell 12(1):87–98. doi:10.1016/j.devcel.2006.10.004
36. Roussigne M, Bianco IH, Wilson SW, Blader P (2009) Nodal signalling imposes left-right asymmetry upon neurogenesis in the habenular nuclei. Development 136(9):1549–1557. doi:10.1242/dev.034793
37. Aizawa H, Amo R, Okamoto H (2011) Phylogeny and ontogeny of the habenular structure. Front Neurosci 5:138. doi:10.3389/fnins.2011.00138
38. Kemali M, Guglielmotti V, Fiorino L (1990) The asymmetry of the habenular nuclei of female and male frogs in spring and in winter. Brain Res 517(1–2):251–255
39. Taylor RW, Qi JY, Talaga AK, Ma TP, Pan L, Bartholomew CR, Klionsky DJ, Moens CB, Gamse JT (2011) Asymmetric inhibition of Ulk2 causes left-right differences in habenular neuropil formation. J Neurosci 31(27):9869–9878. doi:10.1523/JNEUROSCI.0435-11.2011
40. Sutherland RJ (1982) The dorsal diencephalic conduction system: a review of the anatomy and functions of the habenular complex. Neurosci Biobehav Rev 6(1):1–13
41. Groenewegen HJ, Ahlenius S, Haber SN, Kowall NW, Nauta WJ (1986) Cytoarchitecture, fiber connections, and some histochemical aspects of the interpeduncular nucleus in the rat. J Comp Neurol 249(1):65–102. doi:10.1002/cne.902490107
42. Hamill GS, Lenn NJ (1984) The subnuclear organization of the rat interpeduncular nucleus: a light and electron microscopic study. J Comp Neurol 222(3):396–408. doi:10.1002/cne.902220307
43. Hamill GS, Olschowka JA, Lenn NJ, Jacobowitz DM (1984) The subnuclear distribution of substance P, cholecystokinin, vasoactive intestinal peptide, somatostatin, leu-enkephalin, dopamine-beta-hydroxylase, and serotonin in the rat interpeduncular nucleus. J Comp Neurol 226(4):580–596. doi:10.1002/cne.902260410
44. Morley BJ (1986) The interpeduncular nucleus. Int Rev Neurobiol 28:157–182
45. Tomizawa K, Katayama H, Nakayasu H (2001) A novel monoclonal antibody recognizes a previously unknown subdivision of the habenulo-interpeduncular system in zebrafish. Brain Res 901(1–2):117–127
46. Kuan YS, Yu HH, Moens CB, Halpern ME (2007) Neuropilin asymmetry mediates a left-right difference in habenular connectivity. Development 134(5):857–865. doi:10.1242/dev.02791
47. Bianco IH, Carl M, Russell C, Clarke JD, Wilson SW (2008) Brain asymmetry is encoded at the level of axon terminal morphology. Neural Dev 3:9. doi:10.1186/1749-8104-3-9
48. Cajal SRY (1911) Histologie du systeme nerveux: de l'homme & des vertebres, vol 2, 2nd edn. Consejo Superior de Investigaciones Científicas, Instituto Ramón y Cajal, 1972, Madrid, Spain
49. Herrick JC (1948) The brain of the tiger salamander: Ambystoma tigrinum. University of Chicago Press, Chicago, IL
50. Duboc V, Dufourcq P, Blader P, Roussigne M (2015) Asymmetry of the brain: development and implications. Annu Rev Genet 49:647–672. doi:10.1146/annurev-genet-112414-055322
51. Haffter P, Granato M, Brand M, Mullins MC, Hammerschmidt M, Kane DA, Odenthal J, van Eeden FJ, Jiang YJ, Heisenberg CP, Kelsh RN, Furutani-Seiki M, Vogelsang E, Beuchle D, Schach U, Fabian C, Nusslein-Volhard C (1996) The identification of genes with unique and essential functions in the development of the zebrafish, Danio rerio. Development 123:1–36

52. Hatta K, Kimmel CB, Ho RK, Walker C (1991) The cyclops mutation blocks specification of the floor plate of the zebrafish central nervous system. Nature 350(6316):339–341. doi:10.1038/350339a0
53. Rebagliati MR, Toyama R, Haffter P, Dawid IB (1998) Cyclops encodes a nodal-related factor involved in midline signaling. Proc Natl Acad Sci U S A 95(17):9932–9937
54. Sampath K, Rubinstein AL, Cheng AM, Liang JO, Fekany K, Solnica-Krezel L, Korzh V, Halpern ME, Wright CV (1998) Induction of the zebrafish ventral brain and floorplate requires cyclops/nodal signalling. Nature 395(6698):185–189. doi:10.1038/26020
55. Noel ES, Verhoeven M, Lagendijk AK, Tessadori F, Smith K, Choorapoikayil S, den Hertog J, Bakkers J (2013) A Nodal-independent and tissue-intrinsic mechanism controls heart-looping chirality. Nat Commun 4:2754. doi:10.1038/ncomms3754
56. Bisgrove BW, Essner JJ, Yost HJ (2000) Multiple pathways in the midline regulate concordant brain, heart and gut left-right asymmetry. Development 127(16):3567–3579
57. Feldman B, Concha ML, Saude L, Parsons MJ, Adams RJ, Wilson SW, Stemple DL (2002) Lefty antagonism of Squint is essential for normal gastrulation. Curr Biol 12(24):2129–2135
58. Facchin L, Burgess HA, Siddiqi M, Granato M, Halpern ME (2009) Determining the function of zebrafish epithalamic asymmetry. Philos Trans R Soc Lond B Biol Sci 364(1519):1021–1032. doi:10.1098/rstb.2008.0234
59. Carl M, Bianco IH, Bajoghli B, Aghaallaei N, Czerny T, Wilson SW (2007) Wnt/Axin1/beta-catenin signaling regulates asymmetric nodal activation, elaboration, and concordance of CNS asymmetries. Neuron 55(3):393–405. doi:10.1016/j.neuron.2007.07.007
60. Hamada H, Tam PP (2014) Mechanisms of left-right asymmetry and patterning: driver, mediator and responder. F1000Prime Rep 6:110. doi:10.12703/P6-110
61. Blum M, Schweickert A, Vick P, Wright CV, Danilchik MV (2014) Symmetry breakage in the vertebrate embryo: when does it happen and how does it work? Dev Biol 393(1):109–123. doi:10.1016/j.ydbio.2014.06.014
62. Vandenberg LN, Levin M (2010) Far from solved: a perspective on what we know about early mechanisms of left-right asymmetry. Dev Dyn 239(12):3131–3146. doi:10.1002/dvdy.22450
63. Amack JD (2014) Salient features of the ciliated organ of asymmetry. Bioarchitecture 4(1):6–15. doi:10.4161/bioa.28014
64. Blum M, Feistel K, Thumberger T, Schweickert A (2014) The evolution and conservation of left-right patterning mechanisms. Development 141(8):1603–1613. doi:10.1242/dev.100560
65. McGrath J, Somlo S, Makova S, Tian X, Brueckner M (2003) Two populations of node monocilia initiate left-right asymmetry in the mouse. Cell 114(1):61–73
66. Yoshiba S, Hamada H (2014) Roles of cilia, fluid flow, and Ca^{2+} signaling in breaking of left-right symmetry. Trends Genet 30(1):10–17. doi:10.1016/j.tig.2013.09.001
67. Cooper MS, D'Amico LA (1996) A cluster of noninvoluting endocytic cells at the margin of the zebrafish blastoderm marks the site of embryonic shield formation. Dev Biol 180(1):184–198. doi:10.1006/dbio.1996.0294
68. Melby AE, Warga RM, Kimmel CB (1996) Specification of cell fates at the dorsal margin of the zebrafish gastrula. Development 122(7):2225–2237
69. Kramer-Zucker AG, Olale F, Haycraft CJ, Yoder BK, Schier AF, Drummond IA (2005) Cilia-driven fluid flow in the zebrafish pronephros, brain and Kupffer's vesicle is required for normal organogenesis. Development 132(8):1907–1921. doi:10.1242/dev.01772
70. Sampaio P, Ferreira RR, Guerrero A, Pintado P, Tavares B, Amaro J, Smith AA, Montenegro-Johnson T, Smith DJ, Lopes SS (2014) Left-right organizer flow dynamics: how much cilia activity reliably yields laterality? Dev Cell 29(6):716–728. doi:10.1016/j.devcel.2014.04.030
71. Yuan S, Zhao L, Brueckner M, Sun Z (2015) Intraciliary calcium oscillations initiate vertebrate left-right asymmetry. Curr Biol 25(5):556–567. doi:10.1016/j.cub.2014.12.051
72. Matsui T, Bessho Y (2012) Left-right asymmetry in zebrafish. Cell Mol Life Sci 69(18):3069–3077. doi:10.1007/s00018-012-0985-6
73. Raya A, Kawakami Y, Rodriguez-Esteban C, Ibanes M, Rasskin-Gutman D, Rodriguez-Leon J, Buscher D, Feijo JA, Izpisua Belmonte JC (2004) Notch activity acts as a sensor for extracellular calcium during vertebrate left-right determination. Nature 427(6970):121–128. doi:10.1038/nature02190
74. Lenhart KF, Lin SY, Titus TA, Postlethwait JH, Burdine RD (2011) Two additional midline barriers function with midline lefty1 expression to maintain asymmetric Nodal signaling during left-right axis specification in zebrafish. Development 138(20):4405–4410. doi:10.1242/dev.071092

75. Peterson AG, Wang X, Yost HJ (2013) Dvr1 transfers left-right asymmetric signals from Kupffer's vesicle to lateral plate mesoderm in zebrafish. Dev Biol 382(1):198–208. doi:10.1016/j.ydbio.2013.06.011
76. Sarmah B, Latimer AJ, Appel B, Wente SR (2005) Inositol polyphosphates regulate zebrafish left-right asymmetry. Dev Cell 9(1):133–145. doi:10.1016/j.devcel.2005.05.002
77. Schilling TF, Concordet JP, Ingham PW (1999) Regulation of left-right asymmetries in the zebrafish by Shh and BMP4. Dev Biol 210(2):277–287. doi:10.1006/dbio.1999.9214
78. Tessadori F, Noel ES, Rens EG, Magliozzi R, Evers-van Gogh IJ, Guardavaccaro D, Merks RM, Bakkers J (2015) Nodal signaling range is regulated by proprotein convertase-mediated maturation. Dev Cell 32(5):631–639. doi:10.1016/j.devcel.2014.12.014
79. Essner JJ, Amack JD, Nyholm MK, Harris EB, Yost HJ (2005) Kupffer's vesicle is a ciliated organ of asymmetry in the zebrafish embryo that initiates left-right development of the brain, heart and gut. Development 132(6):1247–1260. doi:10.1242/dev.01663
80. Bisgrove BW, Snarr BS, Emrazian A, Yost HJ (2005) Polaris and Polycystin-2 in dorsal forerunner cells and Kupffer's vesicle are required for specification of the zebrafish left-right axis. Dev Biol 287(2):274–288. doi:10.1016/j.ydbio.2005.08.047
81. Schottenfeld J, Sullivan-Brown J, Burdine RD (2007) Zebrafish curly up encodes a Pkd2 ortholog that restricts left-side-specific expression of southpaw. Development 134(8):1605–1615. doi:10.1242/dev.02827
82. Guimier A, Gabriel GC, Bajolle F, Tsang M, Liu H, Noll A, Schwartz M, El Malti R, Smith LD, Klena NT, Jimenez G, Miller NA, Oufadem M, Moreau de Bellaing A, Yagi H, Saunders CJ, Baker CN, Di Filippo S, Peterson KA, Thiffault I, Bole-Feysot C, Cooley LD, Farrow EG, Masson C, Schoen P, Deleuze JF, Nitschke P, Lyonnet S, de Pontual L, Murray SA, Bonnet D, Kingsmore SF, Amiel J, Bouvagnet P, Lo CW, Gordon CT (2015) MMP21 is mutated in human heterotaxy and is required for normal left-right asymmetry in vertebrates. Nat Genet 47(11):1260–1263. doi:10.1038/ng.3376
83. Perles Z, Moon S, Ta-Shma A, Yaacov B, Francescatto L, Edvardson S, Rein AJ, Elpeleg O, Katsanis N (2015) A human laterality disorder caused by a homozygous deleterious mutation in MMP21. J Med Genet 52(12):840–847. doi:10.1136/jmedgenet-2015-103336
84. Zhao C, Malicki J (2007) Genetic defects of pronephric cilia in zebrafish. Mech Dev 124(7–8):605–616. doi:10.1016/j.mod.2007.04.004
85. Chen JN, van Eeden FJ, Warren KS, Chin A, Nusslein-Volhard C, Haffter P, Fishman MC (1997) Left-right pattern of cardiac BMP4 may drive asymmetry of the heart in zebrafish. Development 124(21):4373–4382
86. Hochgreb-Hagele T, Yin C, Koo DE, Bronner ME, Stainier DY (2013) Laminin beta1a controls distinct steps during the establishment of digestive organ laterality. Development 140(13):2734–2745. doi:10.1242/dev.097618
87. Smith KA, Noel E, Thurlings I, Rehmann H, Chocron S, Bakkers J (2011) Bmp and nodal independently regulate lefty1 expression to maintain unilateral nodal activity during left-right axis specification in zebrafish. PLoS Genet 7(9):e1002289. doi:10.1371/journal.pgen.1002289
88. Drummond IA (2005) Kidney development and disease in the zebrafish. J Am Soc Nephrol 16(2):299–304. doi:10.1681/ASN.2004090754
89. Regan JC, Concha ML, Roussigne M, Russell C, Wilson SW (2009) An Fgf8-dependent bistable cell migratory event establishes CNS asymmetry. Neuron 61(1):27–34. doi:10.1016/j.neuron.2008.11.030
90. Clanton JA, Hope KD, Gamse JT (2013) Fgf signaling governs cell fate in the zebrafish pineal complex. Development 140(2):323–332. doi:10.1242/dev.083709
91. Wu SY, de Borsetti NH, Bain EJ, Bulow CR, Gamse JT (2014) Mediator subunit 12 coordinates intrinsic and extrinsic control of epithalamic development. Dev Biol 385(1):13–22. doi:10.1016/j.ydbio.2013.10.023
92. Garric L, Ronsin B, Roussigne M, Booton S, Gamse JT, Dufourcq P, Blader P (2014) Pitx2c ensures habenular asymmetry by restricting parapineal cell number. Development 141(7):1572–1579. doi:10.1242/dev.100305
93. Inbal A, Kim SH, Shin J, Solnica-Krezel L (2007) Six3 represses nodal activity to establish early brain asymmetry in zebrafish. Neuron 55(3):407–415. doi:10.1016/j.neuron.2007.06.037
94. Husken U, Stickney HL, Gestri G, Bianco IH, Faro A, Young RM, Roussigne M, Hawkins TA, Beretta CA, Brinkmann I, Paolini A, Jacinto R, Albadri S, Dreosti E, Tsalavouta M, Schwarz Q, Cavodeassi F, Barth AK, Wen L, Zhang B, Blader P, Yaksi E, Poggi L, Zigman M, Lin S, Wilson SW, Carl M (2014) Tcf7l2 is required for left-right

asymmetric differentiation of habenular neurons. Curr Biol 24(19):2217–2227. doi:10.1016/j.cub.2014.08.006
95. Doll CA, Burkart JT, Hope KD, Halpern ME, Gamse JT (2011) Subnuclear development of the zebrafish habenular nuclei requires ER translocon function. Dev Biol 360(1):44–57. doi:10.1016/j.ydbio.2011.09.003
96. Kuan YS, Roberson S, Akitake CM, Fortuno L, Gamse J, Moens C, Halpern ME (2015) Distinct requirements for Wntless in habenular development. Dev Biol 406(2):117–128. doi:10.1016/j.ydbio.2015.06.006
97. Beretta CA, Dross N, Bankhead P, Carl M (2013) The ventral habenulae of zebrafish develop in prosomere 2 dependent on Tcf7l2 function. Neural Dev 8:19. doi:10.1186/1749-8104-8-19
98. Colombo A, Palma K, Armijo L, Mione M, Signore IA, Morales C, Guerrero N, Meynard MM, Perez R, Suazo J, Marcelain K, Briones L, Hartel S, Wilson SW, Concha ML (2013) Daam1a mediates asymmetric habenular morphogenesis by regulating dendritic and axonal outgrowth. Development 140(19):3997–4007. doi:10.1242/dev.091934
99. Kawakami K, Shima A, Kawakami N (2000) Identification of a functional transposase of the Tol2 element, an Ac-like element from the Japanese medaka fish, and its transposition in the zebrafish germ lineage. Proc Natl Acad Sci U S A 97(21):11403–11408. doi:10.1073/pnas.97.21.11403
100. Suster ML, Sumiyama K, Kawakami K (2009) Transposon-mediated BAC transgenesis in zebrafish and mice. BMC Genomics 10:477. doi:10.1186/1471-2164-10-477
101. Jessen JR, Meng A, McFarlane RJ, Paw BH, Zon LI, Smith GR, Lin S (1998) Modification of bacterial artificial chromosomes through chi-stimulated homologous recombination and its application in zebrafish transgenesis. Proc Natl Acad Sci U S A 95(9):5121–5126
102. Scheer N, Campos-Ortega JA (1999) Use of the Gal4-UAS technique for targeted gene expression in the zebrafish. Mech Dev 80(2):153–158
103. Sauer B (1998) Inducible gene targeting in mice using the Cre/lox system. Methods 14(4):381–392. doi:10.1006/meth.1998.0593
104. Balciunas D, Davidson AE, Sivasubbu S, Hermanson SB, Welle Z, Ekker SC (2004) Enhancer trapping in zebrafish using the Sleeping Beauty transposon. BMC Genomics 5(1):62. doi:10.1186/1471-2164-5-62
105. Kawakami K, Takeda H, Kawakami N, Kobayashi M, Matsuda N, Mishina M (2004) A transposon-mediated gene trap approach identifies developmentally regulated genes in zebrafish. Dev Cell 7(1):133–144. doi:10.1016/j.devcel.2004.06.005
106. Parinov S, Kondrichin I, Korzh V, Emelyanov A (2004) Tol2 transposon-mediated enhancer trap to identify developmentally regulated zebrafish genes in vivo. Dev Dyn 231(2):449–459. doi:10.1002/dvdy.20157
107. Scott EK, Mason L, Arrenberg AB, Ziv L, Gosse NJ, Xiao T, Chi NC, Asakawa K, Kawakami K, Baier H (2007) Targeting neural circuitry in zebrafish using GAL4 enhancer trapping. Nat Methods 4(4):323–326. doi:10.1038/nmeth1033
108. Bergeron SA, Carrier N, Li GH, Ahn S, Burgess HA (2015) Gsx1 expression defines neurons required for prepulse inhibition. Mol Psychiatry 20(8):974–985. doi:10.1038/mp.2014.106
109. Teh C, Chudakov DM, Poon KL, Mamedov IZ, Sek JY, Shidlovsky K, Lukyanov S, Korzh V (2010) Optogenetic in vivo cell manipulation in KillerRed-expressing zebrafish transgenics. BMC Dev Biol 10:110. doi:10.1186/1471-213X-10-110
110. Auer TO, Xiao T, Bercier V, Gebhardt C, Duroure K, Concordet JP, Wyart C, Suster M, Kawakami K, Wittbrodt J, Baier H, Del Bene F (2015) Deletion of a kinesin I motor unmasks a mechanism of homeostatic branching control by neurotrophin-3. eLife 4. doi:10.7554/eLife.05061
111. Asakawa K, Suster ML, Mizusawa K, Nagayoshi S, Kotani T, Urasaki A, Kishimoto Y, Hibi M, Kawakami K (2008) Genetic dissection of neural circuits by Tol2 transposon-mediated Gal4 gene and enhancer trapping in zebrafish. Proc Natl Acad Sci U S A 105(4):1255–1260. doi:10.1073/pnas.0704963105
112. Arrenberg AB, Del Bene F, Baier H (2009) Optical control of zebrafish behavior with halorhodopsin. Proc Natl Acad Sci U S A 106(42):17968–17973. doi:10.1073/pnas.0906252106
113. Douglass AD, Kraves S, Deisseroth K, Schier AF, Engert F (2008) Escape behavior elicited by single, channelrhodopsin-2-evoked spikes in zebrafish somatosensory neurons. Curr Biol 18(15):1133–1137. doi:10.1016/j.cub.2008.06.077
114. Akitake CM, Macurak M, Halpern ME, Goll MG (2011) Transgenerational analysis of transcriptional silencing in zebrafish. Dev Biol 352(2):191–201. doi:10.1016/j.ydbio.2011.01.002
115. Goll MG, Anderson R, Stainier DY, Spradling AC, Halpern ME (2009) Transcriptional

silencing and reactivation in transgenic zebrafish. Genetics 182(3):747–755. doi:10.1534/genetics.109.102079

116. Subedi A, Macurak M, Gee ST, Monge E, Goll MG, Potter CJ, Parsons MJ, Halpern ME (2014) Adoption of the Q transcriptional regulatory system for zebrafish transgenesis. Methods 66(3):433–440. doi:10.1016/j.ymeth.2013.06.012
117. Suli A, Guler AD, Raible DW, Kimelman D (2014) A targeted gene expression system using the tryptophan repressor in zebrafish shows no silencing in subsequent generations. Development 141(5):1167–1174. doi:10.1242/dev.100057
118. Knopf F, Schnabel K, Haase C, Pfeifer K, Anastassiadis K, Weidinger G (2010) Dually inducible TetON systems for tissue-specific conditional gene expression in zebrafish. Proc Natl Acad Sci U S A 107(46):19933–19938. doi:10.1073/pnas.1007799107
119. Agetsuma M, Aizawa H, Aoki T, Nakayama R, Takahoko M, Goto M, Sassa T, Amo R, Shiraki T, Kawakami K, Hosoya T, Higashijima S, Okamoto H (2010) The habenula is crucial for experience-dependent modification of fear responses in zebrafish. Nat Neurosci 13(11):1354–1356. doi:10.1038/nn.2654
120. Chen YC, Cheng CH, Chen GD, Hung CC, Yang CH, Hwang SP, Kawakami K, Wu BK, Huang CJ (2009) Recapitulation of zebrafish sncga expression pattern and labeling the habenular complex in transgenic zebrafish using green fluorescent protein reporter gene. Dev Dyn 238(3):746–754. doi:10.1002/dvdy.21877
121. Distel M, Wullimann MF, Koster RW (2009) Optimized Gal4 genetics for permanent gene expression mapping in zebrafish. Proc Natl Acad Sci U S A 106(32):13365–13370. doi:10.1073/pnas.0903060106
122. Marquart GD, Tabor KM, Brown M, Strykowski JL, Varshney GK, LaFave MC, Mueller T, Burgess SM, Higashijima S, Burgess HA (2015) A 3D searchable database of transgenic zebrafish Gal4 and Cre lines for functional neuroanatomy studies. Front Neural Circuits 9:78. doi:10.3389/fncir.2015.00078
123. deCarvalho TN, Akitake CM, Thisse C, Thisse B, Halpern ME (2013) Aversive cues fail to activate fos expression in the asymmetric olfactory-habenula pathway of zebrafish. Front Neural Circuits 7:98. doi:10.3389/fncir.2013.00098
124. Cuello AC, Emson PC, Paxinos G, Jessell T (1978) Substance P containing and cholinergic projections from the habenula. Brain Res 149(2):413–429
125. Facchin L, Duboue ER, Halpern ME (2015) Disruption of epithalamic left-right asymmetry increases anxiety in zebrafish. J Neurosci 35(48):15847–15859. doi:10.1523/JNEUROSCI.2593-15.2015
126. Miyasaka N, Morimoto K, Tsubokawa T, Higashijima S, Okamoto H, Yoshihara Y (2009) From the olfactory bulb to higher brain centers: genetic visualization of secondary olfactory pathways in zebrafish. J Neurosci 29(15):4756–4767. doi:10.1523/JNEUROSCI.0118-09.2009
127. Dreosti E, Vendrell Llopis N, Carl M, Yaksi E, Wilson SW (2014) Left-right asymmetry is required for the habenulae to respond to both visual and olfactory stimuli. Curr Biol 24(4):440–445. doi:10.1016/j.cub.2014.01.016
128. Okamoto H, Agetsuma M, Aizawa H (2012) Genetic dissection of the zebrafish habenula, a possible switching board for selection of behavioral strategy to cope with fear and anxiety. Dev Neurobiol 72(3):386–394. doi:10.1002/dneu.20913
129. Deng C, Rogers LJ (1997) Differential contributions of the two visual pathways to functional lateralization in chicks. Behav Brain Res 87(2):173–182
130. Geschwind N, Levitsky W (1968) Human brain: left-right asymmetries in temporal speech region. Science 161(3837):186–187
131. Pascual A, Huang KL, Neveu J, Preat T (2004) Neuroanatomy: brain asymmetry and long-term memory. Nature 427(6975):605–606. doi:10.1038/427605a
132. Sarin S, O'Meara MM, Flowers EB, Antonio C, Poole RJ, Didiano D, Johnston RJ Jr, Chang S, Narula S, Hobert O (2007) Genetic screens for Caenorhabditis elegans mutants defective in left/right asymmetric neuronal fate specification. Genetics 176(4):2109–2130. doi:10.1534/genetics.107.075648
133. Wes PD, Bargmann CI (2001) C. elegans odour discrimination requires asymmetric diversity in olfactory neurons. Nature 410(6829):698–701. doi:10.1038/35070581
134. Young EJ, Williams CL (2010) Valence dependent asymmetric release of norepinephrine in the basolateral amygdala. Behav Neurosci 124(5):633–644. doi:10.1037/a0020885
135. Bisazza A, Pignatti R, Vallortigara G (1997) Laterality in detour behaviour: interspecific variation in poeciliid fish. Anim Behav 54(5):1273–1281
136. Facchin L, Bisazza A, Vallortigara G (1999) What causes lateralization of detour behavior

in fish? Evidence for asymmetries in eye use. Behav Brain Res 103(2):229–234
137. Bisazza A, Cantalupo C, Capocchiano M, Vallortigara G (2000) Population lateralisation and social behaviour: a study with 16 species of fish. Laterality 5(3):269–284. doi:10.1080/713754381
138. Miklosi A, Andrew RJ, Savage H (1997) Behavioural lateralisation of the tetrapod type in the zebrafish (Brachydanio rerio). Physiol Behav 63(1):127–135
139. Miklosi A, Andrew RJ (1999) Right eye use associated with decision to bite in zebrafish. Behav Brain Res 105(2):199–205
140. Vallortigara G, Rogers LJ (2005) Survival with an asymmetrical brain: advantages and disadvantages of cerebral lateralization. Behav Brain Sci 28(4):575–589. doi:10.1017/S0140525X05000105, discussion 589–633
141. Watkins J, Miklosi A, Andrew RJ (2004) Early asymmetries in the behaviour of zebrafish larvae. Behav Brain Res 151(1–2):177–183. doi:10.1016/j.bbr.2003.08.012
142. Burgess HA, Granato M (2007) Modulation of locomotor activity in larval zebrafish during light adaptation. J Exp Biol 210(Pt 14):2526–2539. doi:10.1242/jeb.003939
143. Sovrano VA, Rainoldi C, Bisazza A, Vallortigara G (1999) Roots of brain specializations: preferential left-eye use during mirror-image inspection in six species of teleost fish. Behav Brain Res 106(1–2):175–180
144. Sovrano VA, Andrew RJ (2006) Eye use during viewing a reflection: behavioural lateralisation in zebrafish larvae. Behav Brain Res 167(2):226–231. doi:10.1016/j.bbr.2005.09.021
145. Andrew RJ, Dharmaretnam M, Gyori B, Miklosi A, Watkins JA, Sovrano VA (2009) Precise endogenous control of involvement of right and left visual structures in assessment by zebrafish. Behav Brain Res 196(1):99–105. doi:10.1016/j.bbr.2008.07.034
146. Facchin L, Argenton F, Bisazza A (2009) Lines of Danio rerio selected for opposite behavioural lateralization show differences in anatomical left-right asymmetries. Behav Brain Res 197(1):157–165. doi:10.1016/j.bbr.2008.08.033
147. Dadda M, Domenichini A, Piffer L, Argenton F, Bisazza A (2010) Early differences in epithalamic left-right asymmetry influence lateralization and personality of adult zebrafish. Behav Brain Res 206(2):208–215. doi:10.1016/j.bbr.2009.09.019
148. Barth KA, Miklosi A, Watkins J, Bianco IH, Wilson SW, Andrew RJ (2005) fsi zebrafish show concordant reversal of laterality of viscera, neuroanatomy, and a subset of behavioral responses. Curr Biol 15(9):844–850. doi:10.1016/j.cub.2005.03.047
149. Desjardins JK, Fernald RD (2010) What do fish make of mirror images? Biol Lett 6(6):744–747. doi:10.1098/rsbl.2010.0247
150. Viswanath H, Carter AQ, Baldwin PR, Molfese DL, Salas R (2013) The medial habenula: still neglected. Front Hum Neurosci 7:931. doi:10.3389/fnhum.2013.00931
151. Murphy CA, DiCamillo AM, Haun F, Murray M (1996) Lesion of the habenular efferent pathway produces anxiety and locomotor hyperactivity in rats: a comparison of the effects of neonatal and adult lesions. Behav Brain Res 81(1–2):43–52
152. Murray M, Murphy CA, Ross LL, Haun F (1994) The role of the habenula-interpeduncular pathway in modulating levels of circulating adrenal hormones. Restor Neurol Neurosci 6(4):301–307. doi:10.3233/RNN-1994-6406
153. Yamaguchi T, Danjo T, Pastan I, Hikida T, Nakanishi S (2013) Distinct roles of segregated transmission of the septo-habenular pathway in anxiety and fear. Neuron 78(3):537–544. doi:10.1016/j.neuron.2013.02.035
154. Lee A, Mathuru AS, Teh C, Kibat C, Korzh V, Penney TB, Jesuthasan S (2010) The habenula prevents helpless behavior in larval zebrafish. Curr Biol 20(24):2211–2216. doi:10.1016/j.cub.2010.11.025
155. Okamoto H, Aizawa H (2013) Fear and anxiety regulation by conserved affective circuits. Neuron 78(3):411–413. doi:10.1016/j.neuron.2013.04.031
156. Cachat J, Stewart A, Grossman L, Gaikwad S, Kadri F, Chung KM, Wu N, Wong K, Roy S, Suciu C, Goodspeed J, Elegante M, Bartels B, Elkhayat S, Tien D, Tan J, Denmark A, Gilder T, Kyzar E, Dileo J, Frank K, Chang K, Utterback E, Hart P, Kalueff AV (2010) Measuring behavioral and endocrine responses to novelty stress in adult zebrafish. Nat Protoc 5(11):1786–1799. doi:10.1038/nprot.2010.140
157. Levin ED, Bencan Z, Cerutti DT (2007) Anxiolytic effects of nicotine in zebrafish. Physiol Behav 90(1):54–58. doi:10.1016/j.physbeh.2006.08.026
158. Bencan Z, Sledge D, Levin ED (2009) Buspirone, chlordiazepoxide and diazepam effects in a zebrafish model of anxiety. Pharmacol Biochem Behav 94(1):75–80. doi:10.1016/j.pbb.2009.07.009

159. Barcellos L, Ritter F, Kreutz L, Quevedo R, Bolognesi da Silva L, Bedin A, Finco J, Cericato L (2007) Whole-body cortisol increases after direct and visual contact with a predator in zebrafish, Danio rerio. Aquaculture 272:774
160. Selye H (1976) Stress in health and disease. Butterworths, Boston
161. Kuan YS, Gamse JT, Schreiber AM, Halpern ME (2007) Selective asymmetry in a conserved forebrain to midbrain projection. J Exp Zool B Mol Dev Evol 308(5):669–678. doi:10.1002/jez.b.21184
162. Nakai J, Ohkura M, Imoto K (2001) A high signal-to-noise Ca(2+) probe composed of a single green fluorescent protein. Nat Biotechnol 19(2):137–141. doi:10.1038/84397
163. Akerboom J, Chen TW, Wardill TJ, Tian L, Marvin JS, Mutlu S, Calderon NC, Esposti F, Borghuis BG, Sun XR, Gordus A, Orger MB, Portugues R, Engert F, Macklin JJ, Filosa A, Aggarwal A, Kerr RA, Takagi R, Kracun S, Shigetomi E, Khakh BS, Baier H, Lagnado L, Wang SS, Bargmann CI, Kimmel BE, Jayaraman V, Svoboda K, Kim DS, Schreiter ER, Looger LL (2012) Optimization of a GCaMP calcium indicator for neural activity imaging. J Neurosci 32(40):13819–13840. doi:10.1523/JNEUROSCI.2601-12.2012
164. Chen TW, Wardill TJ, Sun Y, Pulver SR, Renninger SL, Baohan A, Schreiter ER, Kerr RA, Orger MB, Jayaraman V, Looger LL, Svoboda K, Kim DS (2013) Ultrasensitive fluorescent proteins for imaging neuronal activity. Nature 499(7458):295–300. doi:10.1038/nature12354
165. Muto A, Ohkura M, Abe G, Nakai J, Kawakami K (2013) Real-time visualization of neuronal activity during perception. Curr Biol 23(4):307–311. doi:10.1016/j.cub.2012.12.040
166. Tallini YN, Ohkura M, Choi BR, Ji G, Imoto K, Doran R, Lee J, Plan P, Wilson J, Xin HB, Sanbe A, Gulick J, Mathai J, Robbins J, Salama G, Nakai J, Kotlikoff MI (2006) Imaging cellular signals in the heart in vivo: cardiac expression of the high-signal Ca^{2+} indicator GCaMP2. Proc Natl Acad Sci U S A 103(12):4753–4758. doi:10.1073/pnas.0509378103
167. Tian L, Hires SA, Mao T, Huber D, Chiappe ME, Chalasani SH, Petreanu L, Akerboom J, McKinney SA, Schreiter ER, Bargmann CI, Jayaraman V, Svoboda K, Looger LL (2009) Imaging neural activity in worms, flies and mice with improved GCaMP calcium indicators. Nat Methods 6(12):875–881. doi:10.1038/nmeth.1398
168. Ahrens MB, Li JM, Orger MB, Robson DN, Schier AF, Engert F, Portugues R (2012) Brain-wide neuronal dynamics during motor adaptation in zebrafish. Nature 485(7399):471–477. doi:10.1038/nature11057
169. Fetcho JR, Cox KJ, O'Malley DM (1997) Imaging neural activity with single cell resolution in an intact, behaving vertebrate. Biol Bull 192(1):150–153
170. Portugues R, Feierstein CE, Engert F, Orger MB (2014) Whole-brain activity maps reveal stereotyped, distributed networks for visuomotor behavior. Neuron 81(6):1328–1343. doi:10.1016/j.neuron.2014.01.019
171. Renninger SL, Orger MB (2013) Two-photon imaging of neural population activity in zebrafish. Methods 62(3):255–267. doi:10.1016/j.ymeth.2013.05.016
172. Ahrens MB, Huang KH, Narayan S, Mensh BD, Engert F (2013) Two-photon calcium imaging during fictive navigation in virtual environments. Front Neural Circuits 7:104. doi:10.3389/fncir.2013.00104
173. Jetti SK, Vendrell-Llopis N, Yaksi E (2014) Spontaneous activity governs olfactory representations in spatially organized habenular microcircuits. Curr Biol 24(4):434–439. doi:10.1016/j.cub.2014.01.015
174. Krishnan S, Mathuru AS, Kibat C, Rahman M, Lupton CE, Stewart J, Claridge-Chang A, Yen SC, Jesuthasan S (2014) The right dorsal habenula limits attraction to an odor in zebrafish. Curr Biol 24(11):1167–1175. doi:10.1016/j.cub.2014.03.073
175. Masino MA, Fetcho JR (2005) Fictive swimming motor patterns in wild type and mutant larval zebrafish. J Neurophysiol 93(6):3177–3188. doi:10.1152/jn.01248.2004
176. Hellstrom T, Doving KB (1986) Chemoreception of taurocholate in anosmic and sham-operated cod, Gadus morhua. Behav Brain Res 21(2):155–162
177. Jones KA, Hara TJ (1985) Behavioural responses of fishes to chemical cues: results from a new bioassay. J Fish Biol 27:495–504
178. Koide T, Miyasaka N, Morimoto K, Asakawa K, Urasaki A, Kawakami K, Yoshihara Y (2009) Olfactory neural circuitry for attraction to amino acids revealed by transposon-mediated gene trap approach in zebrafish. Proc Natl Acad Sci U S A 106(24):9884–9889. doi:10.1073/pnas.0900470106
179. Rust C, Wild N, Bernt C, Vennegeerts T, Wimmer R, Beuers U (2009) Bile acid-induced apoptosis in hepatocytes is caspase-6-dependent. J Biol Chem 284(5):2908–2916. doi:10.1074/jbc.M804585200

180. Palmeira CM, Rolo AP (2004) Mitochondrially-mediated toxicity of bile acids. Toxicology 203(1–3):1–15. doi:10.1016/j.tox.2004.06.001
181. Concha ML, Bianco IH, Wilson SW (2012) Encoding asymmetry within neural circuits. Nat Rev Neurosci 13(12):832–843. doi:10.1038/nrn3371
182. Mathuru AS, Jesuthasan S (2013) The medial habenula as a regulator of anxiety in adult zebrafish. Front Neural Circuits 7:99. doi:10.3389/fncir.2013.00099
183. Kimura Y, Hisano Y, Kawahara A, Higashijima S (2014) Efficient generation of knock-in transgenic zebrafish carrying reporter/driver genes by CRISPR/Cas9-mediated genome engineering. Sci Rep 4:6545. doi:10.1038/srep06545
184. Auer TO, Del Bene F (2014) CRISPR/Cas9 and TALEN-mediated knock-in approaches in zebrafish. Methods 69(2):142–150. doi:10.1016/j.ymeth.2014.03.027
185. Amo R, Aizawa H, Takahoko M, Kobayashi M, Takahashi R, Aoki T, Okamoto H (2010) Identification of the zebrafish ventral habenula as a homolog of the mammalian lateral habenula. J Neurosci 30(4):1566–1574. doi:10.1523/JNEUROSCI.3690-09.2010
186. Folgueira M, Bayley P, Navratilova P, Becker TS, Wilson SW, Clarke JD (2012) Morphogenesis underlying the development of the everted teleost telencephalon. Neural Dev 7:32. doi:10.1186/1749-8104-7-32
187. Chou MY, Amo R, Kinoshita M, Cherng BW, Shimazaki H, Agetsuma M, Shiraki T, Aoki T, Takahoko M, Yamazaki M, Higashijima S, Okamoto H (2016) Social conflict resolution regulated by two dorsal habenular subregions in zebrafish. Science 352(6281):87–90. doi:10.1126/science.aac9508
188. Randlett O, Wee CL, Naumann EA, Nnaemeka O, Schoppik D, Fitzgerald JE, Portugues R, Lacoste AM, Riegler C, Engert F, Schier AF (2015) Whole-brain activity mapping onto a zebrafish brain atlas. Nat Methods 12(11):1039–1046. doi:10.1038/nmeth.3581
189. Cancedda L, Putignano E, Impey S, Maffei L, Ratto GM, Pizzorusso T (2003) Patterned vision causes CRE-mediated gene expression in the visual cortex through PKA and ERK. J Neurosci 23(18):7012–7020
190. Hussain A, Saraiva LR, Ferrero DM, Ahuja G, Krishna VS, Liberles SD, Korsching SI (2013) High-affinity olfactory receptor for the death-associated odor cadaverine. Proc Natl Acad Sci U S A 110(48):19579–19584. doi:10.1073/pnas.1318596110
191. Itoh M, Yamamoto T, Nakajima Y, Hatta K (2014) Multistepped optogenetics connects neurons and behavior. Curr Biol 24(24):R1155–R1156. doi:10.1016/j.cub.2014.10.065
192. Ji RR, Baba H, Brenner GJ, Woolf CJ (1999) Nociceptive-specific activation of ERK in spinal neurons contributes to pain hypersensitivity. Nat Neurosci 2(12):1114–1119. doi:10.1038/16040
193. Signore IA, Guerrero N, Loosli F, Colombo A, Villalon A, Wittbrodt J, Concha ML (2009) Zebrafish and medaka: model organisms for a comparative developmental approach of brain asymmetry. Philos Trans R Soc Lond B Biol Sci 364(1519):991–1003. doi:10.1098/rstb.2008.0260
194. Ehret G (1987) Left hemisphere advantage in the mouse brain for recognizing ultrasonic communication calls. Nature 325(6101):249–251. doi:10.1038/325249a0
195. Kim S, Matyas F, Lee S, Acsady L, Shin HS (2012) Lateralization of observational fear learning at the cortical but not thalamic level in mice. Proc Natl Acad Sci U S A 109(38):15497–15501. doi:10.1073/pnas.1213903109
196. Young EJ, Williams CL (2013) Differential activation of amygdala Arc expression by positive and negatively valenced emotional learning conditions. Front Behav Neurosci 7:191. doi:10.3389/fnbeh.2013.00191
197. Marlin BJ, Mitre M, D'Amour JA, Chao MV, Froemke RC (2015) Oxytocin enables maternal behaviour by balancing cortical inhibition. Nature 520(7548):499–504. doi:10.1038/nature14402
198. Sun T, Patoine C, Abu-Khalil A, Visvader J, Sum E, Cherry TJ, Orkin SH, Geschwind DH, Walsh CA (2005) Early asymmetry of gene transcription in embryonic human left and right cerebral cortex. Science 308(5729):1794–1798. doi:10.1126/science.1110324
199. Kawakami R, Dobi A, Shigemoto R, Ito I (2008) Right isomerism of the brain in inversus viscerum mutant mice. PLoS One 3(4):e1945. doi:10.1371/journal.pone.0001945
200. Goto K, Kurashima R, Gokan H, Inoue N, Ito I, Watanabe S (2010) Left-right asymmetry defect in the hippocampal circuitry impairs spatial learning and working memory in iv mice. PLoS One 5(11):e15468. doi:10.1371/journal.pone.0015468
201. Ihara A, Hirata M, Fujimaki N, Goto T, Umekawa Y, Fujita N, Terazono Y, Matani A, Wei Q, Yoshimine T, Yorifuji S, Murata T (2010)

Neuroimaging study on brain asymmetries in situs inversus totalis. J Neurol Sci 288(1–2): 72–78. doi:10.1016/j.jns.2009.10.002

202. Kennedy DN, O'Craven KM, Ticho BS, Goldstein AM, Makris N, Henson JW (1999) Structural and functional brain asymmetries in human situs inversus totalis. Neurology 53(6):1260–1265

203. Tanaka S, Kanzaki R, Yoshibayashi M, Kamiya T, Sugishita M (1999) Dichotic listening in patients with situs inversus: brain asymmetry and situs asymmetry. Neuropsychologia 37(7):869–874

204. Oppenheimer JM (1974) Asymmetry revisited. Am Zool 14(3):867–879

Chapter 18

Methods to Study Nervous System Laterality in the *Caenorhabditis elegans* Model System

Berta Vidal and Oliver Hobert

Abstract

The anatomically and genetically very accessible nervous system of the nematode *Caenorhabditis elegans*, composed of a total of 302 neurons in the hermaphrodite, displays a number of striking neuronal lateralities which come largely in two forms: unilateral neurons found only on one side of the nervous system and functional differences in otherwise bilaterally symmetric neuron pairs. Two recent reviews have described in detail the genetic mechanisms that specify the most prominent sensory lateralities in two bilaterally symmetric sensory neuron classes in *C. elegans*. In this Neuromethods chapter, we provide a general overview of the specific methods and opportunities that exist in *C. elegans* to identify lateralities, to decipher their functional relevance and to dissect the genetic control mechanisms that establish these lateralities. These specific advantages include (a) the ability to identify and visualize neuronal lateralities on the anatomical, gene expression, and neuronal activity level with single cell resolution; (b) the ability to assign function to lateralized neurons using behavioral analysis and genetic manipulation of neuronal activity; (c) the ability to conduct genetic screens for mutants that disrupt lateralities, thereby deciphering the genetic patterning mechanisms that instruct neuronal lateralities.

Key words *Caenorhabditis elegans*, Functional laterality, Sensory receptor, Fluorescent reporter, Genetic screen, Whole genome sequencing

1 Introduction

Asymmetries in brain anatomy and function have been described across the animal kingdom [1–7]. However, anatomical asymmetries are often defined only at very gross anatomical levels and their functional relevance is usually poorly understood. Conversely, the molecular and cellular bases of known functional lateralities have not been well mapped out. The situation is remarkably distinct in the invertebrate model system *C. elegans* [6]. The nervous system of this nematode is largely bilaterally symmetric but displays well mapped anatomical asymmetries [6, 8]. Apart from its anatomical asymmetries, there are a number of well-defined functional lateralities in the *C. elegans* nervous system, which, as in vertebrate nervous systems, are superimposed onto anatomically symmetric structures. A

Lesley J. Rogers and Giorgio Vallortigara (eds.), *Lateralized Brain Functions: Methods in Human and Non-Human Species*, Neuromethods, vol. 122, DOI 10.1007/978-1-4939-6725-4_18, © Springer Science+Business Media LLC 2017

number of reviews have described these anatomical and functional asymmetries in detail [6, 9, 10]. We will rather focus here on the description of the very specific experimental and methodological advantages of the *C. elegans* that predestine this system for studies on how nervous system asymmetries are genetically controlled. We group these advantages into six sections and briefly describe how these experimental advantages have been exploited so far. We hope that a precise description of these advantages will motivate researchers to join in the exciting venture of studying nervous system lateralities in *C. elegans.*

2 Methodological Advantage #1: Anatomical Description of Neuronal Lateralities

The exquisitely well-described anatomy of the *C. elegans* nervous system, composed of 302 neurons that fall into 118 distinct anatomical classes [8], has put this model system into a unique position in the animal kingdom. In no other system can one examine the extent of morphological symmetries and asymmetries in the nervous system at comparable levels of resolution and comprehensiveness. This detailed knowledge is a result of an electron micrographical reconstruction of the entire nervous system [8] (all summarized in WormAtlas [11]).

As discussed in a comprehensive review a number of years ago [6], a look at the gross anatomy of neuronal soma arrangement in the main head and tail ganglia and the positioning of the axonal/dendritic tracks shows that the nervous system of *C. elegans* is largely bilaterally symmetric (Fig. 1). The electron microscopical reconstruction of the *C. elegans* nervous system demonstrates overall symmetry on impressively fine-grained anatomical levels: Most neurons in the head and tail ganglia come as bilaterally symmetric neuron pairs with strikingly similar patterns of axo/dendritic patterning and synaptic connectivity (Fig. 1).

The most obvious deviation from symmetry is the ventral nerve cord, which is populated by a string of unpaired motor neurons and axonal tracts that run within the cord from either the anterior or posterior end. Curiously, most of these axonal tracts are positioned along the right side of the animal, with "right" and "left" being defined by an ectodermal ridge structure that separates the axonal tracts. Other nematodes do not show such asymmetry [6] and therefore we will not discuss ventral nerve cord asymmetries further.

Fig. 1 (continued) *otIs356[rab-3p::2xNLS::GFP]* and *otIs392[eat-4p6::GFP, ttx-3::dsRed]*. Anterior is to the left, posterior is to the right, all pictures are dorsoventral views. (**b**) Representative sampling of neurons from the head ganglia of *C. elegans*. Most head neurons come as bilaterally symmetric pairs, a few come as radially symmetric groups that are composed of bilaterally symmetric pairs (e.g., the URY neuron class shown here) and some are unilateral neurons (e.g., the RIS and AVL neurons shown here). All images shown here are from www.wormatlas.org [11]

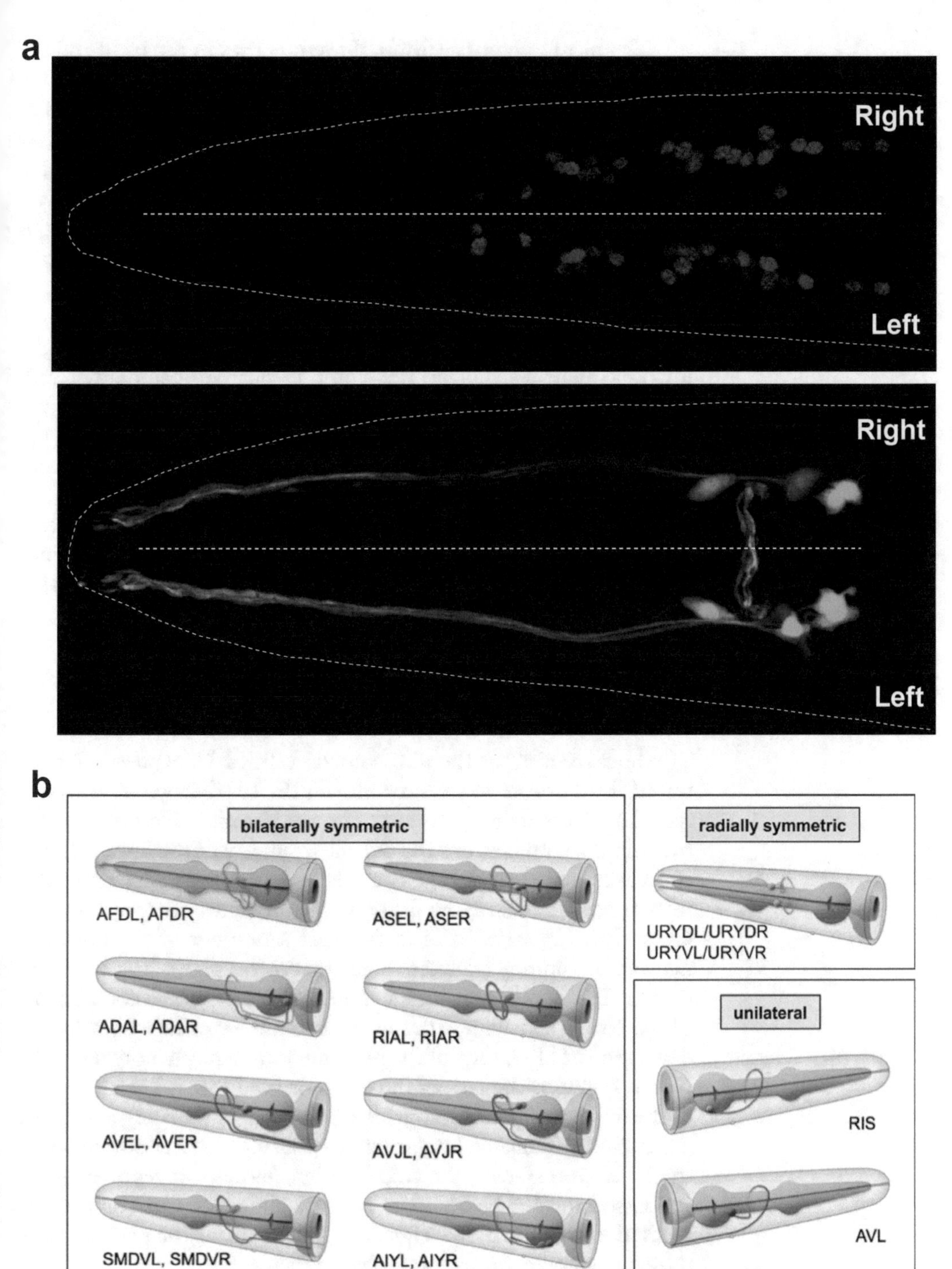

Fig. 1 The *C. elegans* nervous system is mostly bilaterally symmetric. (**a**) Representative pictures of neuronal bilateral symmetry in the head of the worm. The *top panel* is a middle section showing expression of a nuclear localized panneuronal reporter (*rab-3*); the *bottom panel* shows cytoplasmic expression of a promoter fragment of the vesicular glutamate transporter gene (*eat-4*), which is expressed in five pairs of bilaterally symmetric neurons. Images were taken on a confocal microscope. Genotypes of imaged strains are:

The most striking anatomical asymmetries in the head and tail ganglia of the worm are the presence of unilateral neurons, i.e. neurons only present on one side of the animal. These few unilateral neurons—some examples shown in Fig. 1b—stand in contrast to the largely bilaterally symmetric neuron pairs in the head and tail ganglia. Of these unilateral neurons, the GABAergic RIS and AVL neurons are the best characterized. The AVL motor neuron is involved in defecation behavior [12] and the RIS interneuron is involved in controlling sleep-like behavior, possibly via secretion of as yet unknown neuropeptidergic substances [13].

In addition to unilateral neurons, there are a number of single, unpaired neurons that do not come in bilateral pairs either but, rather than being unilateral, they are located essentially on the midline of the animal. One of these neurons, ALA, is, like the above-mentioned RIS neuron, also involved in controlling sleep-like behavior [14].

The existence of unilateral neurons poses a simple question—how are neurons generated just on one side of the animal, but not the other? This question brings us to the problem of cell lineage, discussed in the next section.

3 Methodological Advantage #2: Cell Lineage History

The lineage of all of the 958 somatic cells of *C. elegans* has been traced from zygote to mature animal [15, 16]. Many bilateral neurons have bilaterally symmetric lineage histories. However in the case of unilateral neurons the overall symmetry of the lineage has to be broken and their lineage contralateral homologs adopt very distinct fates (Fig. 2a). How do these distinct fates become adopted? Clear clues are based on the fact that a number of the unilateral neurons are lineally related to a cell that forms part of the excretory system. Early studies on lineage mutants have identified components of the Notch signaling pathway, a paradigm for cellular signaling in development [17]. One of the prominent mutant phenotypes of the Notch system is the lack of the excretory cell that is lineally related to the unilateral neurons, suggesting that Notch signaling affects the generation of unilateral neurons [18]. Using another advantage of the *C. elegans* system, the ease of using fluorescent reporter genes (discussed further below), has recently allowed us to confirm the prediction that the Notch signaling system affects the generation of these unilateral neurons (unpublished data).

How does a Notch signal induce two, otherwise bilaterally symmetric lineage branches to execute a different fate? The symmetry-breaking event that specifies the fate of the lineage branch that will give rise to the excretory cell and unilateral neurons is a cell–cell interaction (the so-called 4th Notch interaction) between the descendants of the mesodermal MSap blastomere, which express the LAG-2 delta-like notch ligand and the ABplpapp blastomere which

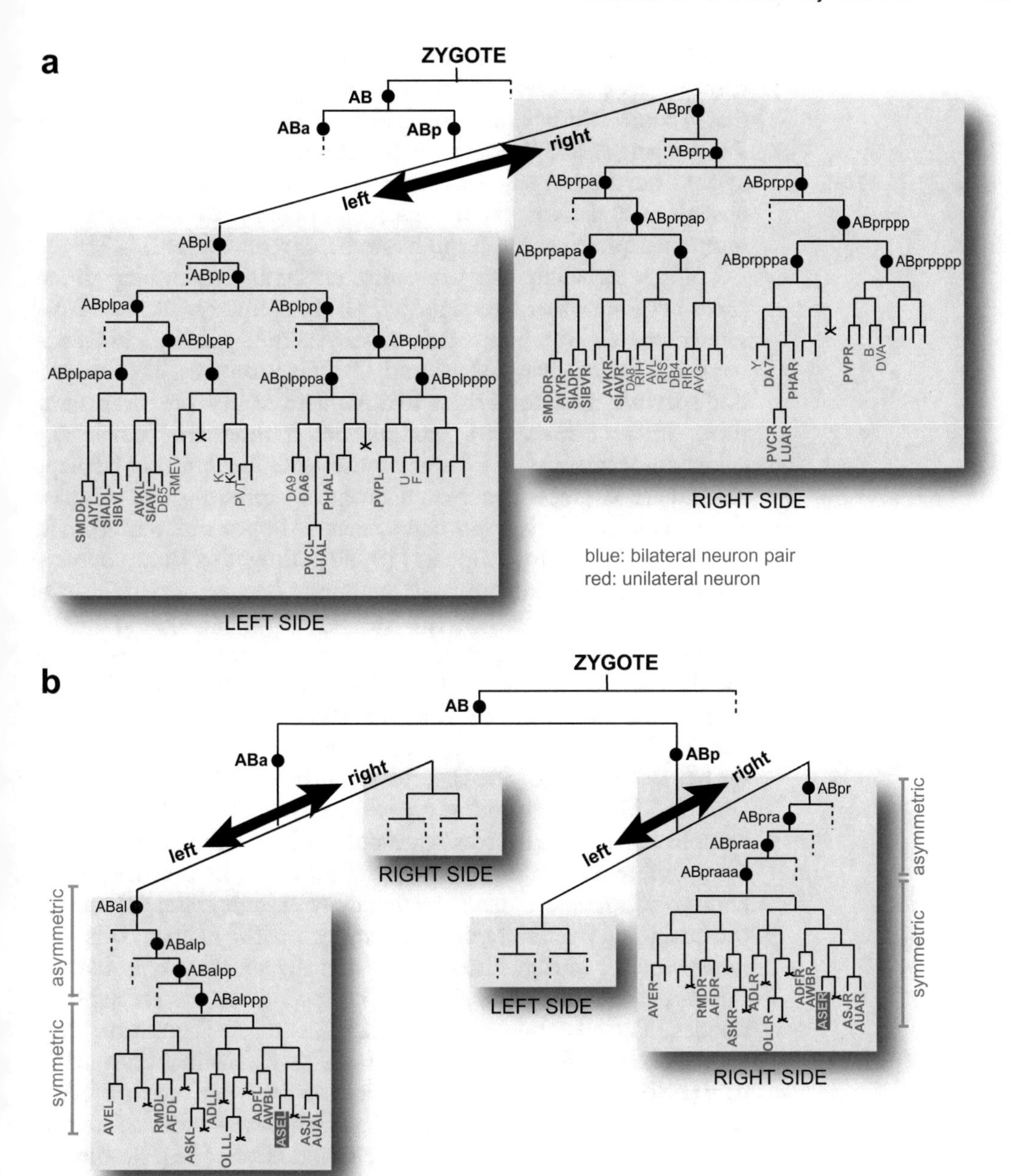

Fig. 2 Lineage history and neuronal symmetry. Lineage histories have been elucidated by Sulston et al. [16]. "a/p/l/r" indicate cell divisions along the anterior/posterior/left/right axis. (**a**) Example of lineage branches with a bilaterally symmetric history that arises from an early division of the ABp blastomere across the left/right axis. These two branches contain bilaterally symmetric neuron pairs, indicated in *blue*, but they also contain unilateral neurons (three- or four-letter code), or unilateral neuroblasts (that divide postembryonically; single-letter code), indicated in *red*. (**b**) Lineage branches that produce bilaterally symmetric neurons but derive from distinct blastomeres and only begin to share a bilaterally symmetric cleavage pattern after gastrulation. The branch shown here contains 11 neuron pairs that do not share a bilaterally symmetric lineage history. The ASE neuron pair is shaded in *gray*. This image has been reproduced with slight revisions from [9]

expresses the Notch receptor [19–21]. *lag-2* expression in MSap descendants is determined by an anterior-posterior cell fate decision that is controlled by noncanonical Wnt signaling [22, 23]. The *lag-2* expressing MSap descendants (signaling cells) are asymmetrically located on the left side of the embryo due to a skewed anterior-posterior cell division axis that places them in close proximity to the receptor-expressing ABplpapp blastomere (receiving cell) [22].

Notch signaling also has earlier effects in establishing differences between other left/right pairs of blastomeres. At the 12-cell stage there is a Notch interaction between the signaling MS blastomere and the receiving ABara and ABalp blastomeres. This interaction specifies the fate of these two AB derived blastomeres making them different from their contralateral counterparts (ABala and ABarp respectively) [24]. Later on the LAG-2 expressing ABalapp blastomere contacts the Notch receptor-expressing cell ABplaaa and this interaction is important to make ABplaaa different from its contralateral homolog ABpraaa [19, 20]. Altogether these consecutive Notch inductions break the symmetry between left-right pairs of blastomeres and diversify the fates across the left-right axis.

Another intriguing revelation of cell lineage histories is that the almost 100 bilateral neuron pairs fall into two different categories: About half of the neuron pairs show bilateral symmetric cell lineage history (Fig. 2a), while the other half of bilaterally symmetric neuron pairs do not share a symmetric lineage history [16] (Fig. 2b). This is intriguing for two reasons: (1) it opens the still completely non-understood question of how neurons that come from different lineage branches and locations in the developing embryo adopt an eventually bilaterally symmetric state; (2) it raises the intriguing possibility that bilateral symmetric neuron pairs with nonsymmetric lineage histories may actually not be as symmetric as they appear. In one particular case, this prediction has indeed been fulfilled. The bilaterally symmetric ASEL and ASER gustatory neurons are indeed functionally lateralized, have a distinct lineage history (ASEL is derived from ABa, while ASER is derived from ABp) and the basis for the functional lateralization was indeed found in the distinct lineage history of these two neurons [25]. Studies on ASEL/R lateralization were triggered by the initial discovery of asymmetric gene expressionin the two ASE neurons, which brings us to the next section.

Fig. 3 (continued) into the fragments by the primers. Primers A and B amplify the genomic region of interest (amplicon #1). Note that the fusion can be made to either the coding sequence of the gene (i.e., translational fusion) or only to the promoter of the gene (i.e., transcriptional fusion). Primer B adds a 24 bp overlap in frame to the fluorophore coding region. Primers C and D amplify the reporter gene (e.g., GFP) and 3′ UTR (amplicon #2). Primers A* and D* are used to fuse amplicon #1 and amplicon #2 (*red line* indicates 24 bp sequence overlap). The resulting fusion product (amplicon #3) can be directly injected into *C. elegans* and after 6 days approximately you can obtain stable transgenic lines (at the F2 generation). For more specific details on the generation of fluorescent reporters refer to [54]

4 Methodological Advantage #3: Reporter Gene Technology Reveals Expression Asymmetries in Bilaterally Symmetric Neurons

C. elegans is easily amenable to transgenesis which enabled Martin Chalfie in the mid-90s to establish *gfp* reporter gene technology in this organism [26]. Ever since its establishment, fluorescent protein reporter gene technology has been extensively used to define gene expression patterns (Fig. 3). Two studies that quickly

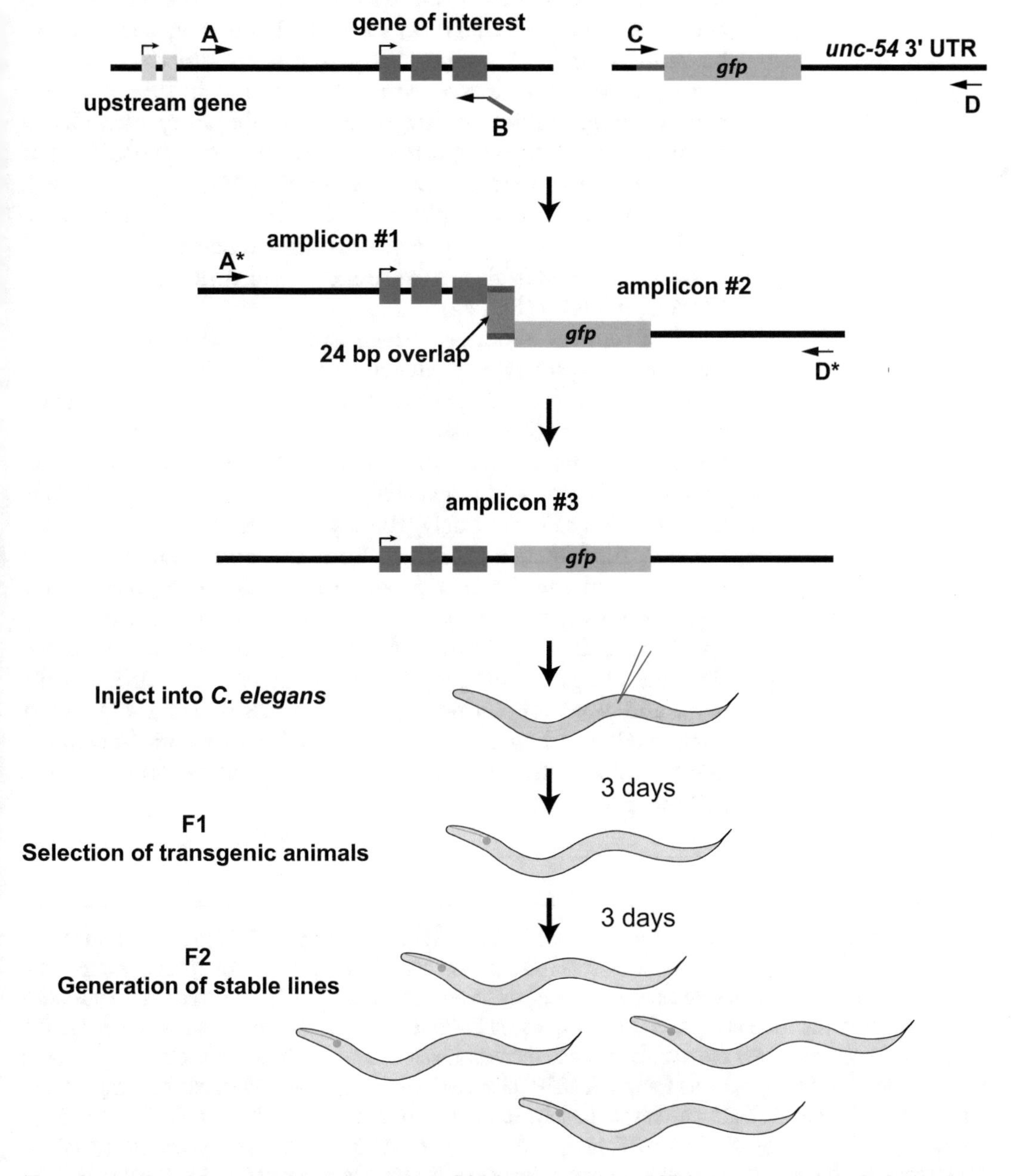

Fig. 3 Generation of fluorescent reporters using a PCR fusion approach. PCR fusions generate a single amplicon from two fragments that contain a small overlapping region of sequence homology (~24 bp). The overlap is engineered

followed the establishment of *gfp* reporter gene technology examined the expression patterns of putative sensory receptors [27, 28] and these two studies revealed completely unanticipated patterns of left/right asymmetric gene expression. In the course of studying the expression patterns of putative olfactory receptors, the Bargmann laboratory found pervasive bilaterally symmetric expression of many olfactory neuron pairs [27]. One striking exception was a putative olfactory receptor, *str-2*, that was found to be expressed in a left/right asymmetric manner in the otherwise bilaterally symmetric olfactory neuron pair AWC left (AWCL) and AWC right (AWCR) (Fig. 4a) [29]. Expression was found to be "antisymmetric," a term that describes the stochastic expression of a specific feature on either the left or right side of an animal. Ensuing studies by the Bargmann laboratory identified a number of additional G-protein coupled receptors (GPCRs) that show an antisymmetric expression in the AWC neurons [30]. Expression of these antisymmetric GPCRs is strictly correlated. For example, *str-2* is always expressed in one neuron ("AWC on" cell), while *srsx-3* is always expressed in the contralateral neuron ("AWC off cell") (Fig. 4a).

The Garbers laboratory, which had a long-standing interest in studying receptor-type guanylyl cyclases (rGCs) in vertebrates found in an expression pattern analysis of *C. elegans* rGC genes that three rGCs are expressed in a left/right asymmetric manner in the gustatory neuron pair ASEL and ASER [28]. Our own laboratory found several additional rGCs to be expressed asymmetrically in ASEL/R [31]. Strikingly, the asymmetric expression of these rGCs is fundamentally distinct from the antisymmetric GPCR expression in the AWC neurons—the rGCs are expressed in a directionally asymmetric manner, a term that describes the stereotyped expression of a feature exclusively in one side of the animal. For example, *gcy-6* and *gcy-7* are always expressed in ASEL, while *gcy-4* and *gcy-5* are always expressed by ASER (Fig. 4b). Apart from rGC-encoding genes, a number of neuropeptide-encoding genes have also been found to be asymmetrically expressed in the ASE neurons [32].

Fig. 4 (continued) which is symmetrically expressed in both ASE (*arrowheads*) and AWC (*arrows*) neuron pairs. Genotypes of imaged strains are: *kyls408[str-2p::dsRed, srsx-3p::GFP, elt-2p::GFP]* and *otls263[ceh-36p2::TagRFP]*. (**b**) Representative pictures showing directional asymmetric gene expression in the gustatory ASE neuron pair. The rGC gene *gcy-5* is only expressed in ASER while *gcy-7* is only expressed in ASEL. The *right panels* show examples of mutants in which the expression pattern has been symmetrized. *die-1* (zinc finger transcription factor) mutants show a "2 ASER" phenotype while *cog-1* (homeodomain transcription factor) mutants show a "2 ASEL" phenotype. Genotypes of imaged strains are: *ntls1[gcy-5p::GFP, lin-15(+)]*, *otls3[gcy-7p::GFP, lin-15(+)]*, *die-1(ot26);ntls1*, and *cog-1(ot28);otls3*. Images were taken on a confocal microscope. All pictures are dorsoventral views, anterior is to the top and posterior is to the bottom

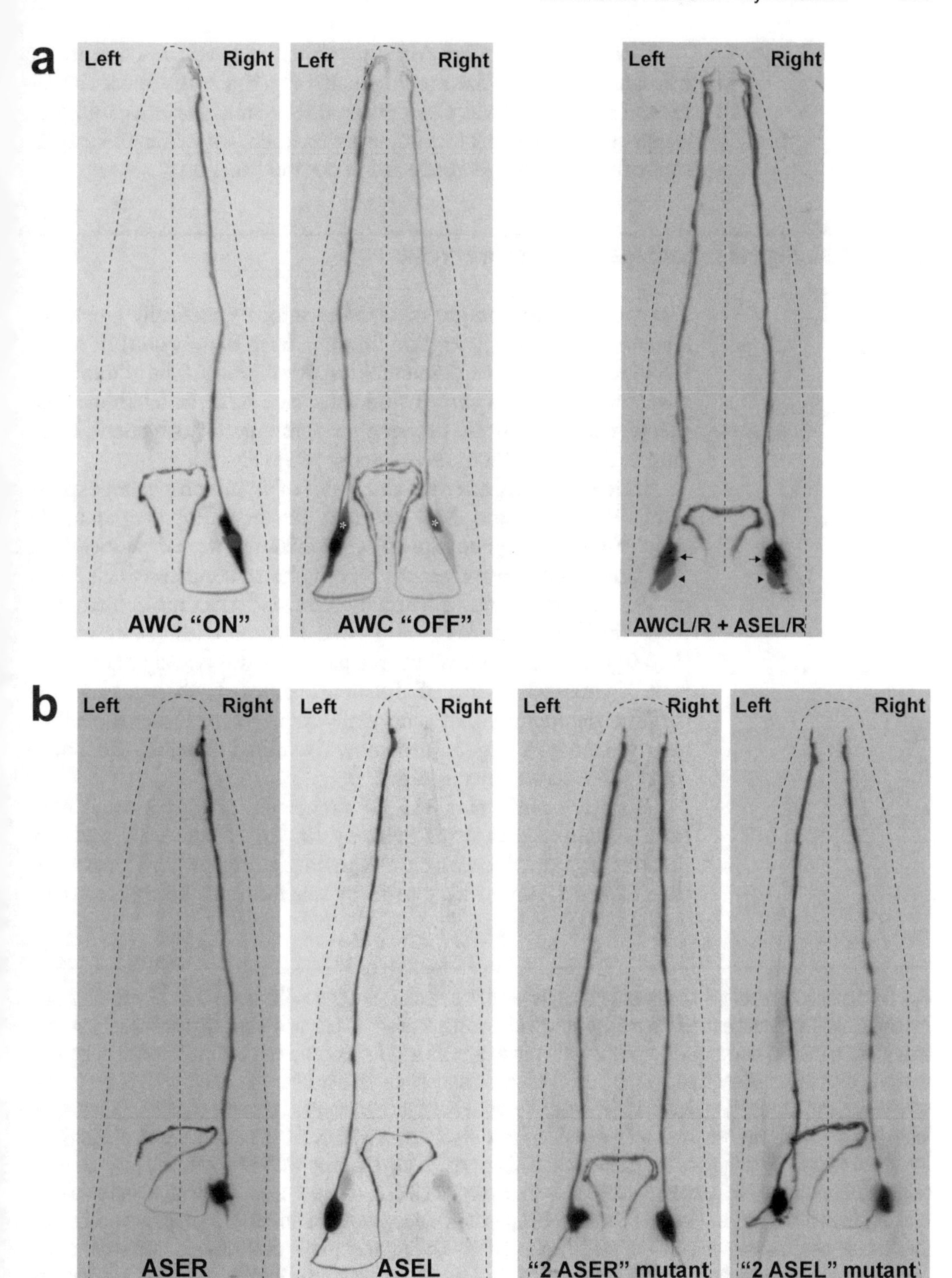

Fig. 4 Expression asymmetries in bilaterally symmetric neurons. (**a**) Representative pictures showing antisymmetric gene expression in the olfactory AWC neuron pair. The GPCR *str-2* is only expressed in AWC ON while the GPCR *srsx-3* is only expressed in AWC OFF. The *asterisk* marks bilaterally symmetric expression of *srsx-3* in another neuron pair, AWB. The *right panel* shows expression of the homeodomain transcription factor *ceh-36*,

The discovery of left/right asymmetrically-expressed reporter transgenes permits the employment of perhaps the most fundamental strength of the *C. elegans* model system: its amenability to classic genetic screens for mutants that affect a specific phenotypic trait of the animals, as discussed in the next section.

5 Methodological Advantage #4: Genetic Screens

The "classic" advantages of *C. elegans* as a genetically amenable system are its short generation time and its hermaphroditism which facilitates genetic experimental analysis [33]. The "modern" advantages of *C. elegans* include the ease with which molecular lesions can be identified through a combination of genetic mapping and whole genome sequencing [34] (Fig. 5).

Indeed, shortly after the discovery of asymmetric gene expressionin the AWC and ASE neurons, the first mutant animals in which asymmetric gene expression is disrupted were described [29, 35]. Ensuing extensive genetic screens for mutants in which AWC or ASE asymmetry was disrupted revealed distinct molecular pathways that instruct AWC and ASE asymmetries (reviewed in [9, 10]). The disruption of these pathways was usually manifested by a "symmetrization" of the neurons, i.e. both neurons now adopted the identity that is normally displayed by either the left or the right neuron (e.g. animals now contained either 2 ASEL neurons or 2 ASER neurons; Fig. 4b).

Both the AWC and ASE asymmetry-inducing pathways have been extensively reviewed recently [9, 10]. The AWC pathway involves gap junctions, calcium signaling, and a protein kinase cascade. The ASE asymmetry pathway was found to be characterized

Fig. 5 (continued) methanesulphonate (EMS), is used to induce mutations in the germ cells of wild-type hermaphrodites. The mutagenized worms are grown for two generations to produce homozygous mutants. Worms that show a mutant phenotype can be transferred to a plate and its progeny can be inspected to determine whether the mutant phenotype breeds true. Once a mutant of interest has been identified, WGS is the fastest way to pinpoint the causing mutation. However, because mutagenized strains contain a significant mutational load, it is often still necessary to map mutations to a chromosomal interval to elucidate which of the WGS-identified sequence variants is the phenotype-causing one. In the one-step WGS-SNP strategy [34] a mutant retrieved from a genetic screen is crossed with a polymorphic *C. elegans* strain (Hawaiian), individual F2 progeny from this cross is selected for the mutant phenotype, the progeny of these F2 animals are pooled and then whole-genome-sequenced. The density of polymorphic SNP markers is decreased in the region of the phenotype-causing sequence variant and therefore enables its identification in the WGS data. The *red diamond* indicates the mutation of interest. The sequence "pileup," generated by processing the WGS sequencing data, is a representation of the number of reads that are mapped to a specific position in the genome. The relative number of Bristol vs. Hawaiian nucleotides is a reflection of the relative distribution of recombinants in the pool. The use of the cloud-based pipeline, CloudMap, can greatly simplify the analysis of mutant genome sequences generated by WGS [55]

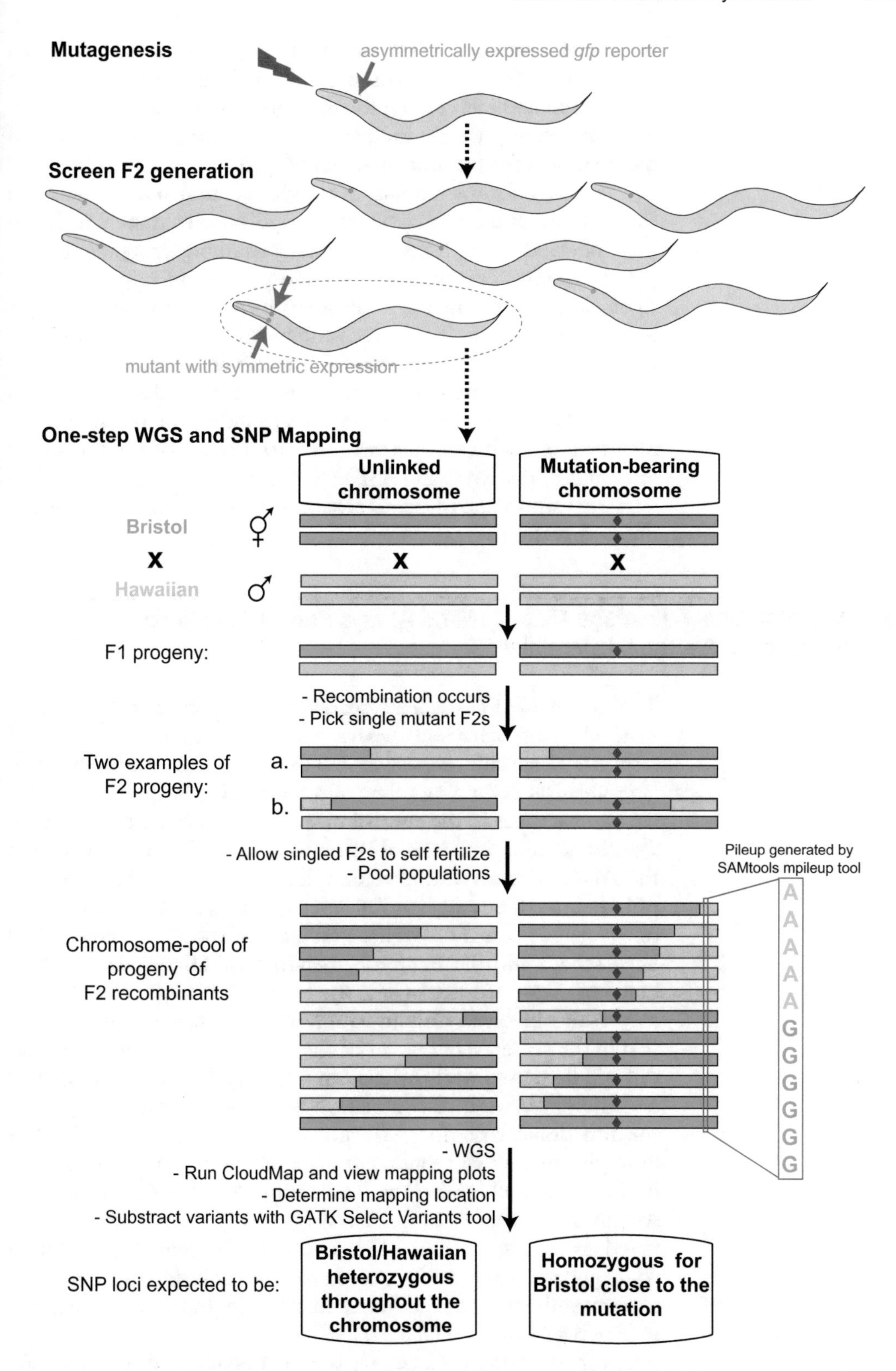

Fig. 5 Forward genetic screens and one-step WGS and SNP mapping to identify the causing mutation. Visible, recessive mutations can be identified using a simple F2 screen. In these screens, a mutagen, such as ethyl

by several layers of 3′UTR-mediated control of transcription factors. One of the 3′UTR-based mechanisms involves *lsy-6*, which constituted upon its cloning the first miRNA with a function in the nervous system, namely the control of chemosensory laterality of the ASE neurons (as described below) [36].

The isolation of mutants through chemical mutagenesis is a classical tool of the *C. elegans* trade. Traditionally, mutant isolation was followed by tedious and time-consuming mapping of the genetic lesion. The advent of second-generation sequencing has significantly short-cut the identification of the causing molecular lesion (Fig. 5). Whole genome sequencing has been employed to clone a number of mutants that affect ASE asymmetry [37–40].

The identification of the first mutants that affect AWC and ASE asymmetries preceded the understanding of the functional relevance of such asymmetries. The description of the functional relevance of the AWC and ASE asymmetries relied on another fundamental advantage of the *C. elegans* system, which we describe in the next section.

6 Methodological Advantage #5: Behavioral Assays Provide Functional Relevance to Neuronal Lateralities

The early days of *C. elegans* research have seen the establishment of a number of behavioral assays, which quantified the animal's response to specific sensory inputs [41–43]. The establishment of laser ablation technology then allowed the implication of individual neuron types in the control of specific behaviors. For example, Cori Bargmann used laser ablation to assign olfactory functions to the AWC neurons and gustatory functions to the ASE neurons [44, 45]. In these early studies, laser operated animals were placed onto agarose plates on which gradients of specific chemicals had been set up and the migratory behavior of animals within these gradients was measured (Fig. 6). More sophisticated chemosensory assays involved discrimination assays in which it was shown that in the presence of one set of chemicals, several other chemicals can still be sensed while others can not (Fig. 6) [46]. It is this easy and straightforward discrimination behavior that was found to be the functional basis for ASE and AWC laterality [46, 47]. For example, animals can sense and discriminate sodium ions (sensed by ASEL) and chloride ions (sensed by ASER) [46]. In contrast, sodium and lithium, both sensed by ASEL, cannot be discriminated from one another [48]. If the ASEL neuron is, through some genetic manipulation, made to sense chloride ions, the neuron can still sense both sodium and chloride, but can not discriminate between the two (Fig. 6) [46].

Not surprisingly, *C. elegans* is exceptionally well suited for the optogenetic manipulation of neuronal activity [49]. Even though

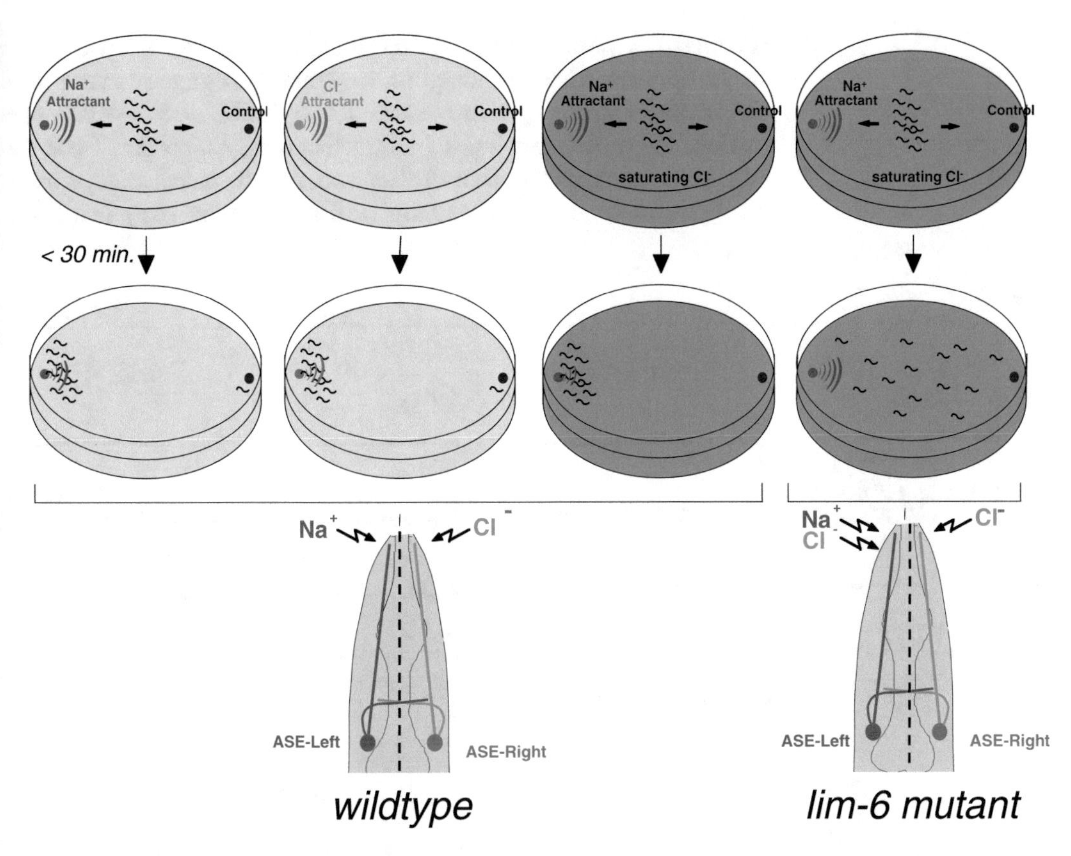

Fig. 6 Chemotaxis discrimination assays reveal lateralization of the *C. elegans* gustatory system. Worms placed on an agar plate with a specific gustatory attractant on one side (e.g. sodium, in *red*, paired up with acetate; or chloride, in *blue*, paired up with ammonium) will migrate toward the chemoattractant. Since sodium is primarily sensed by ASEL, not by ASER, and chloride is primarily sensed by ASER, wild-type animals will also be able to migrate toward sodium (*red*) in the background of saturating concentrations of chloride (entire plate shown in *blue*), i.e. animals are able to discriminate between sodium and chloride. However, if animals are genetically manipulated (*lim-6* mutants) to also be able to sense chloride by ASEL, the animals will not be able to discriminate sodium from chloride and hence not be able to migrate to sodium in the presence of chloride. Discrimination assays schematized here were performed by [46]

not explicitly done yet, the localized activation or inhibition of neurons on only one side of the animal, followed by behavioral assays, is a potential route to discover novel neuronal lateralities.

7 Methodological Advantage #6: Lateralized Activity Can Be Visualized on a Single Neuron Level

Lateralized brain function has been imaged extensively in humans through functional magnetic resonance imaging, yet with very limited cellular resolution [50]. Lateralized activity of ASE neurons has been recorded on the single neuron level using calcium-sensitive optogenetic tools [48, 51]. These imaging studies have revealed fascinating insights into functional lateralization of sensory responses. Calcium imaging, together with genetic manipulations, unambiguously demonstrated that ASEL and ASER sense distinct chemosensory cues [48]. Intriguingly, Schafer, Lockery and colleagues found that ASEL and ASER sense and process information in a fundamentally distinct manner [51] (Fig. 7).

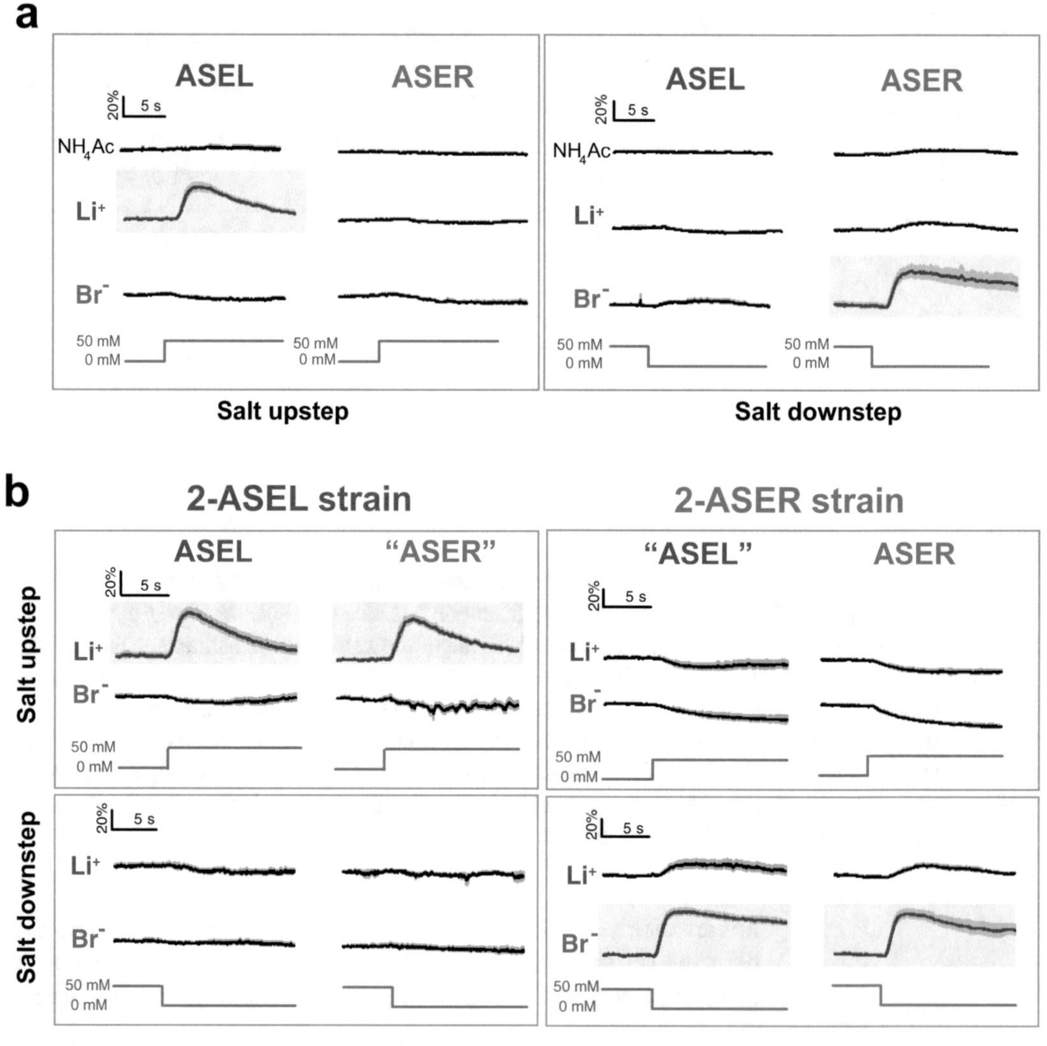

Fig. 7 Visualizing left/right asymmetric neuronal activity. Dynamic changes in Ca^{2+} were recorded in the ASEL and ASER neurons using yellow cameleon 2.12 (*YC2.12*), a Ca^{2+} indicator protein, using strain XL76 *lin-15(n765); ntIs13[flp-6::YC2.12, lin-15(+)]*. For details see [48], from which these calcium traces and parts of this figure legend were reproduced with permission. (**a**) Calcium response of ASEL and ASER to concentration steps of various salts. Average calcium transients in response to 50 mM upsteps or downsteps of indicated salts. The *gray bands* represent SEM. Traces represent the average percentage change from baseline over time of the fluorescence emission ratio of the ratiometric calcium sensor cameleon. (**b**) Average calcium transients of left and right ASE neurons in *lsy-6(ot71)* (2-ASER) and *otIs204[ceh-36::lsy-6]* (2-ASEL) animals, in response to 50 mM upsteps and downsteps of indicated salts

ASEL-sensed cues are sensed only upon increases in the concentration of the cue, and this activation resulted in ASEL promoting runs of the animal toward the chemical [51]. In contrast, ASER-sensed cues evoke a calcium response only upon decreases in their concentration and this response triggers reversal behavior, thereby pushing an animal away from unfavorable (i.e. decreasing attractants) conditions (Fig. 7) [51]. Imaging neuronal activities on a more unbiased, nervous system-wide level will surely reveal more asymmetries in nervous system function, which brings us to the concluding section, the future perspectives.

8 Future Perspectives

The vast majority of the 118 neuron classes of *C. elegans* are made up of bilaterally symmetric neuron pairs (including some cases composed of radially symmetric 4 or 6 neurons). Two of these bilateral neurons pairs show specific functional lateralities. The key unanswered question is how many other bilateral neuron pairs are functionally lateralized? A recent study revealed gene expression asymmetries in some interneuron classes [52], but the functional relevance of these asymmetries is unknown. Importantly, this observation suggests that asymmetries apparently are not restricted to sensory neurons, but also exist in downstream neurons. Lateralization of neuronal function has the potential to diversify functional properties of the very small nervous system of *C. elegans* and we therefore hypothesize that lateralities will be pervasive.

There are two strategies that we can envision to unravel new lateralities, both making use of the specific advantages of the *C. elegans* system: To discover lateralities specifically in sensory neurons, a more comprehensive analysis of predicted sensory receptor proteins in the genome (>1200 GPCRs) may reveal more lateralized neurons. Past expression pattern analyses usually have not systematically examined whether a gene is expressed symmetrically or not (mostly for technical reasons) and hence it is not surprising that no more asymmetric expression patterns have been analyzed. Once more asymmetric expression patterns have been found, it will be intriguing to examine whether these asymmetries are established via the same mechanisms with which AWC or ASE asymmetries are established or whether new left/right patterning mechanisms remain to be discovered. Such discovery will be enabled through genetic screens for mutants in which novel asymmetric gene expression patterns are disrupted.

The other strategy to discover novel lateralities should make use of advances in the ability to visualize neuronal activity on a nervous system-wide level, which has been recently reported in *C. elegans* [53]. However, activity signals cannot yet be reliably assigned to specific neuron types. Once this has been achieved, for example with the use of specific fluorescent landmarks that facilitate neuron identification, it will be possible to examine whether neuronal circuits engaged under specific conditions show lateralized activities.

Acknowledgements

Left/right asymmetry research in the Hobert laboratory has been funded by the National Institutes of Health and the Howard Hughes Medical Institute.

References

1. Rogers LJ, Vallortigara G, Andrew RJ (2013) Divided brains: the biology and behavior of brain asymmetries. Cambridge University Press, Cambridge
2. Davidson RJ, Hugdahl K (eds) (1994) Brain asymmetry. MIT Press, Cambridge, MA
3. Hugdahl K, Davidson RJ (eds) (2003) The asymmetrical brain. MIT Press, Cambridge, MA
4. Concha ML, Bianco IH, Wilson SW (2012) Encoding asymmetry within neural circuits. Nat Rev Neurosci 13:832–843
5. Sun T, Walsh CA (2006) Molecular approaches to brain asymmetry and handedness. Nat Rev Neurosci 7:655–662
6. Hobert O, Johnston RJ Jr, Chang S (2002) Left-right asymmetry in the nervous system: the Caenorhabditis elegans model. Nat Rev Neurosci 3:629–640
7. Frasnelli E, Vallortigara G, Rogers LJ (2012) Left-right asymmetries of behaviour and nervous system in invertebrates. Neurosci Biobehav Rev 36:1273–1291
8. White JG, Southgate E, Thomson JN, Brenner S (1986) The structure of the nervous system of the nematode *Caenorhabditis elegans*. Philos Trans R Soc Lond B Biol Sci 314:1–340
9. Hobert O (2014) Development of left/right asymmetry in the Caenorhabditis elegans nervous system: from zygote to postmitotic neuron. Genesis 52:528–543
10. Hsieh YW, Alqadah A, Chuang CF (2014) Asymmetric neural development in the Caenorhabditis elegans olfactory system. Genesis 52:544–554
11. Hall DH, Altun Z (2007) C. elegans atlas. Cold Spring Harbor Laboratory Press, New York
12. McIntire SL, Jorgensen E, Kaplan J, Horvitz HR (1993) The GABAergic nervous system of Caenorhabditis elegans. Nature 364:337–341
13. Turek M, Lewandrowski I, Bringmann H (2013) An AP2 transcription factor is required for a sleep-active neuron to induce sleep-like quiescence in C. elegans. Curr Biol 23:2215–2223
14. Van Buskirk C, Sternberg PW (2007) Epidermal growth factor signaling induces behavioral quiescence in Caenorhabditis elegans. Nat Neurosci 10:1300–1307
15. Sulston JE, Horvitz HR (1977) Post-embryonic cell lineages of the nematode, Caenorhabditis elegans. Dev Biol 56:110–156
16. Sulston JE, Schierenberg E, White JG, Thomson JN (1983) The embryonic cell lineage of the nematode Caenorhabditis elegans. Dev Biol 100:64–119
17. Greenwald IS, Sternberg PW, Horvitz HR (1983) The lin-12 locus specifies cell fates in Caenorhabditis elegans. Cell 34:435–444
18. Lambie EJ, Kimble J (1991) Two homologous regulatory genes, lin-12 and glp-1, have overlapping functions. Development 112:231–240
19. Hutter H, Schnabel R (1995) Establishment of left-right asymmetry in the Caenorhabditis elegans embryo: a multistep process involving a series of inductive events. Development 121:3417–3424
20. Moskowitz IP, Rothman JH (1996) lin-12 and glp-1 are required zygotically for early embryonic cellular interactions and are regulated by maternal GLP-1 signaling in Caenorhabditis elegans. Development 122:4105–4117
21. Priess JR (2005) Notch signaling in the C. elegans embryo. In: WormBook, C.e.R. Community (ed). WormBook. doi:10.1895/wormbook.1.4.1, http://www.wormbook.org
22. Hermann GJ, Leung B, Priess JR (2000) Left-right asymmetry in C. elegans intestine organ-

ogenesis involves a LIN-12/Notch signaling pathway. Development 127:3429–3440
23. Lin R, Hill RJ, Priess JR (1998) POP-1 and anterior-posterior fate decisions in C. elegans embryos. Cell 92:229–239
24. Hutter H, Schnabel R (1994) glp-1 and inductions establishing embryonic axes in C. elegans. Development 120:2051–2064
25. Cochella L, Hobert O (2012) Embryonic priming of a miRNA locus predetermines postmitotic neuronal left/right asymmetry in C. elegans. Cell 151:1229–1242
26. Chalfie M, Tu Y, Euskirchen G, Ward WW, Prasher DC (1994) Green fluorescent protein as a marker for gene expression. Science 263:802–805
27. Troemel ER, Chou JH, Dwyer ND, Colbert HA, Bargmann CI (1995) Divergent seven transmembrane receptors are candidate chemosensory receptors in C. elegans. Cell 83:207–218
28. Yu S, Avery L, Baude E, Garbers DL (1997) Guanylyl cyclase expression in specific sensory neurons: a new family of chemosensory receptors. Proc Natl Acad Sci U S A 94:3384–3387
29. Troemel ER, Sagasti A, Bargmann CI (1999) Lateral signaling mediated by axon contact and calcium entry regulates asymmetric odorant receptor expression in C. elegans. Cell 99:387–398
30. Lesch BJ, Bargmann CI (2010) The homeodomain protein hmbx-1 maintains asymmetric gene expression in adult C. elegans olfactory neurons. Genes Dev 24:1802–1815
31. Ortiz CO, Etchberger JF, Posy SL, Frokjaer-Jensen C, Lockery S, Honig B, Hobert O (2006) Searching for neuronal left/right asymmetry: genomewide analysis of nematode receptor-type guanylyl cyclases. Genetics 173:131–149
32. Johnston RJ Jr, Chang S, Etchberger JF, Ortiz CO, Hobert O (2005) MicroRNAs acting in a double-negative feedback loop to control a neuronal cell fate decision. Proc Natl Acad Sci U S A 102:12449–12454
33. Brenner S (1974) The genetics of Caenorhabditis elegans. Genetics 77:71–94
34. Doitsidou M, Poole RJ, Sarin S, Bigelow H, Hobert O (2010) C. elegans mutant identification with a one-step whole-genome-sequencing and SNP mapping strategy. PLoS ONE 5:e15435
35. Hobert O, Tessmar K, Ruvkun G (1999) The Caenorhabditis elegans lim-6 LIM homeobox gene regulates neurite outgrowth and function of particular GABAergic neurons. Development 126:1547–1562
36. Johnston RJ, Hobert O (2003) A microRNA controlling left/right neuronal asymmetry in Caenorhabditis elegans. Nature 426:845–849
37. Sarin S, Prabhu S, O'Meara MM, Pe'er I, Hobert O (2008) Caenorhabditis elegans mutant allele identification by whole-genome sequencing. Nat Methods 5:865–867
38. Zhang F, O'Meara MM, Hobert O (2011) A left/right asymmetric neuronal differentiation program is controlled by the Caenorhabditis elegans lsy-27 zinc-finger transcription factor. Genetics 188:753–759
39. Flowers EB, Poole RJ, Tursun B, Bashllari E, Pe'er I, Hobert O (2010) The Groucho ortholog UNC-37interacts with the short Groucho-like protein LSY-22 to control developmental decisions in C. elegans. Development 137:1799–1805
40. Sarin S, Bertrand V, Bigelow H, Boyanov A, Doitsidou M, Poole RJ, Narula S, Hobert O (2010) Analysis of multiple ethyl methanesulfonate-mutagenized caenorhabditis elegans strains by whole-genome sequencing. Genetics 185:417–430
41. Dusenbery DB (1974) Analysis of chemotaxis in the nematode Caenorhabditis elegans by countercurrent separation. J Exp Zool 188:41–47
42. Ward S (1973) Chemotaxis by the nematode Caenorhabditis elegans: identification of attractants and analysis of the response by use of mutants. Proc Natl Acad Sci U S A 70:817–821
43. Hedgecock EM, Russell RL (1975) Normal and mutant thermotaxis in the nematode Caenorhabditis elegans. Proc Natl Acad Sci U S A 72:4061–4065
44. Bargmann CI, Hartwieg E, Horvitz HR (1993) Odorant-selective genes and neurons mediate olfaction in C. elegans. Cell 74:515–527
45. Bargmann CI, Horvitz HR (1991) Chemosensory neurons with overlapping functions direct chemotaxis to multiple chemicals in C. elegans. Neuron 7:729–742
46. Pierce-Shimomura JT, Faumont S, Gaston MR, Pearson BJ, Lockery SR (2001) The homeobox gene lim-6 is required for distinct chemosensory representations in C. elegans. Nature 410:694–698
47. Wes PD, Bargmann CI (2001) C. elegans odour discrimination requires asymmetric diversity in olfactory neurons. Nature 410:698–701
48. Ortiz CO, Faumont S, Takayama J, Ahmed HK, Goldsmith AD, Pocock R, McCormick KE, Kunimoto H, Iino Y, Lockery S et al

(2009) Lateralized gustatory behavior of C. elegans is controlled by specific receptor-type guanylyl cyclases. Curr Biol 19: 996–1004

49. Glock C, Nagpal J, Gottschalk A (2015) Microbial rhodopsin optogenetic tools: application for analyses of synaptic transmission and of neuronal network activity in behavior. Methods Mol Biol 1327:87–103
50. Toga AW, Thompson PM (2003) Mapping brain asymmetry. Nat Rev Neurosci 4:37–48
51. Suzuki H, Thiele TR, Faumont S, Ezcurra M, Lockery SR, Schafer WR (2008) Functional asymmetry in Caenorhabditis elegans taste neurons and its computational role in chemotaxis. Nature 454:114–117
52. Bertrand V, Bisso P, Poole RJ, Hobert O (2011) Notch-dependent induction of left/right asymmetry in C. elegans interneurons and motoneurons. Curr Biol 21:1225–1231
53. Prevedel R, Yoon YG, Hoffmann M, Pak N, Wetzstein G, Kato S, Schrodel T, Raskar R, Zimmer M, Boyden ES et al (2014) Simultaneous whole-animal 3D imaging of neuronal activity using light-field microscopy. Nat Methods 11:727–730
54. Boulin T, Etchberger JF, Hobert O (2006). Reporter gene fusions. WormBook, pp 1–23
55. Minevich G, Park DS, Blankenberg D, Poole RJ, Hobert O (2012) CloudMap: a cloud-based pipeline for analysis of mutant genome sequences. Genetics 192:1249–1269

Part V

Development of Lateralization

Chapter 19

Manipulation of Strength of Cerebral Lateralization via Embryonic Light Stimulation in Birds

Cinzia Chiandetti

Abstract

Birds represent a particularly suitable model for cognitive neurosciences to study the prenatal factors at play in the establishment and modulation of brain lateralization. In pre-hatching stages, genetic factors determine the asymmetrical position of the embryo that in turn allows asymmetric environmental stimulation. In such tight gene–environment dialogue, genes promote the direction of asymmetries and environmental experience modulates the strength, or presence, of anatomical asymmetries and related cognitive specializations. Embryos of birds are easily accessible at all stages from fertilization to hatching due to development in eggs, and specific environmental effects (e.g., light-exposure) have been observed at the level of visual pathways and of related behaviors mediated by visual analysis.

Thanks to the nearly complete decussation of optic nerve fibers at the chiasma, the noninvasive temporary occlusion of one eye allows direct post-hatching investigation of single contralateral hemispheric processing to understand the effects of asymmetric prenatal manipulation. This chapter outlines the two main visual pathways of birds, their developmental timeline and the effects of light stimulation on the establishment and modulation of lateral biases. The detailed method of applying embryonic photostimulation in the domestic chick is provided and the relevance of the influence of environmental light exposure on the development of cerebral asymmetries in other species is discussed.

Key words Birds, Chicken, Pigeon, Brain, Lateralization, Asymmetry, Light, Photostimulation, Environment, Visual pathways

1 Introduction

It is widely acknowledged that experience moulds the development of brain and behavior throughout an animal's life, and this holds true especially during particular stages of development, known as critical periods [1]. In particular time windows, specific experience is crucial [2, 3] and is sometimes a determinant even for normal growth [4, 5]. Experience is relevant also during the prenatal phase with both detrimental [6] and beneficial effects [7] on brain and cognitive development. Furthermore, various forms of

Lesley J. Rogers and Giorgio Vallortigara (eds.), *Lateralized Brain Functions: Methods in Human and Non-Human Species*, Neuromethods, vol. 122, DOI 10.1007/978-1-4939-6725-4_19, © Springer Science+Business Media LLC 2017

prenatal experience can specifically modulate the expression of cerebral lateralization: these include stress hormones [8], androgens [9; see also Chap. 20] and odors [10].

In birds, whose embryonic development occurs in eggs, a major form of prenatal experience is exposure to environmental light. A critical effect of embryonic light stimulation was first described in the domestic chick (*Gallus gallus domesticus*) on the asymmetry of overt behavior mediated by visual input [11].

Broody hens, in natural environments, usually get off their eggs from time to time to fulfil their physiological needs during the incubation period, which lasts for 21 days (for a comparable behavior in pigeons see [12]). They leave the nest only for short breaks and return before the eggs cool down too much. In these periods, the light entering through the eggshell reaches the embryo.

When the embryo prepares for hatching (embryonic day E19–E21), it moves the beak toward the air sac at the top of the egg. In turn, the right eye faces outward and lies close to the shell, while the left remains shielded by the rest of the body (Fig. 1). This postural asymmetry is due to genetic factors (expression of components of the Nodal cascade [13]) and hence it is relatively invariant across individuals, with nearly all embryos (99%) turned so that the environmental light affects almost always the right eye only [14]. As a consequence of this asymmetrical input, the optic projections ascending from the left side of the thalamus (and receiving input from the right eye) to the forebrain have a strong structural asymmetry leading to lateralized responses to visual stimuli [15, 16].

Fig. 1 The chick's asymmetric position within the egg. By opening a window in the eggshell few hours before hatching, the chick is visible in its tilted pose. It folds up leaning the head over the body, with the right eye close to the eggshell and the left eye shielded by the body. The internal translucent membranes and the eyelids can be passed through by light, which reaches almost only the right eye

1.1 Basics on the Birds' Visual System and Its Developmental Stages

Birds' visual system comprises two main visual pathways. One is the thalamofugal pathway that corresponds to the mammalian geniculo-cortical pathway and projects from the retina to the geniculatus lateralis pars dorsalis (GLd) and then to the visual Wulst. Another is the tectofugal pathway that corresponds to the mammalian extrageniculate pathway and projects from the retina to the tectum opticum (TO) and then to the nucleus rotundus (RT) before arriving in the forebrain (Entopallium) [17]; more details in Chap. 3.

The developmental time course of the two visual pathways is different in different species of birds. Chickens are precocial species, meaning that they are early maturing birds. Soon after hatching, chicks are independent from parental care and, despite their need of the hen's warmth initially post-hatching, they can immediately explore and forage autonomously. Pigeons (*Columba livia*), by contrast, hatch in an immature state with their eyes closed and needing parents to feed and care for them for almost one further month within the nest.

In turn, strong differences are at play in the developmental time course of the two visual pathways in precocial and altricial birds. In chicks, the optic vesicles are already forming during the early stages of embryonic development [18], the optic lobes appear in their complete form by day E14–E15 and the optic nerves are myelinated by day E16. On days E17–E18 the eyes open and the retinal functionality starts, with recordings at day E19 comparable to that of hatchlings. By this time, pupillary constrictions and eyelid movements can be observed in response to light [19]. The formation of the tectofugal pathway is nearly already completed at around day E18 when the retina starts to respond to stimulation, and in this same time window the thalamofugal projections are rapidly developing [20].

In pigeons, by contrast, the tectofugal pathway is slowly developing at the day of hatching. When, 8 days later, the pigeon opens its eyes, the thalamofugal projections finally become functional [21].

The thalamofugal and the tectofugal visual pathways decussate nearly completely at the level of the optic chiasma [22]. Such remarkable crossing together with the presence of only small central commissures renders the chicken a kind of natural split-brain subject. In this condition, the visual input from each eye separately reaches the opposite hemisphere and the noninvasive application of a removable eye-patch (Fig. 2) reveals the processing of the hemisphere contralateral to the eye in use [23], despite the anterior commissure interconnecting the thalamofugal system (but see also Chap. 3 for more discussion of this).

In pigeons, by contrast, the interhemispheric exchange involves a larger proportion of visual areas [24].

1.2 Light Stimulation Structural Effects

In domestic chicks, the asymmetric light exposure of the embryo promotes asymmetric changes in the connectivity of the thalamofugal system fed by the stimulated right eye. An increased number

Fig. 2 Monocular noninvasive occlusion. Chick's selective occlusion of one eye is operated by means of a removable eye-patch and serves to test the functionality of the hemisphere contralateral to the eye in use. A small rectangular piece of paper-tape can be folded up in a cone-like shape and gently applied on one eye

of fibers projects from the left GLd to the right visual Wulst, approximately 60% more in comparison to the opposite projections from the right side of the thalamus to the contralateral forebrain [25–27]. The asymmetric pattern of thalamofugal connections is more pronounced in males than in females [28]. The tectofugal pathway of the chick, instead, remains unaffected by light [25] (Fig. 3).

The reverse holds true for pigeons. The thalamofugal pathway shows no effects of stimulation (and it is not lateralized either in young or in adult pigeons [29]), whereas in the tectofugal pathway there is an increase in cell size of the TO [30], in fibers projecting from the right TO to the contralateral RT [31] and other TO connections [32]; see also Chap. 9.

When the entire incubation process is carried out in complete darkness, the establishment of such asymmetries is prevented in both chicks [33] and pigeons [31].

The thalamofugal asymmetry in chicks is manifest from the second day post-hatching and it is transient, i.e. observable only during the first 3 weeks after hatching [33], whereas the tectofugal asymmetry in pigeons is permanent [34] and promoted also by top-down connections [35]. However, the possibility exists that the transient asymmetry in the thalamofugal pathway of the chick influences asymmetric neuronal expression at higher-level regions that could be enduring throughout adulthood [21, 36].

By experimentally overturning the head in order to expose the left eye to the light and covering with a patch the right eye (only

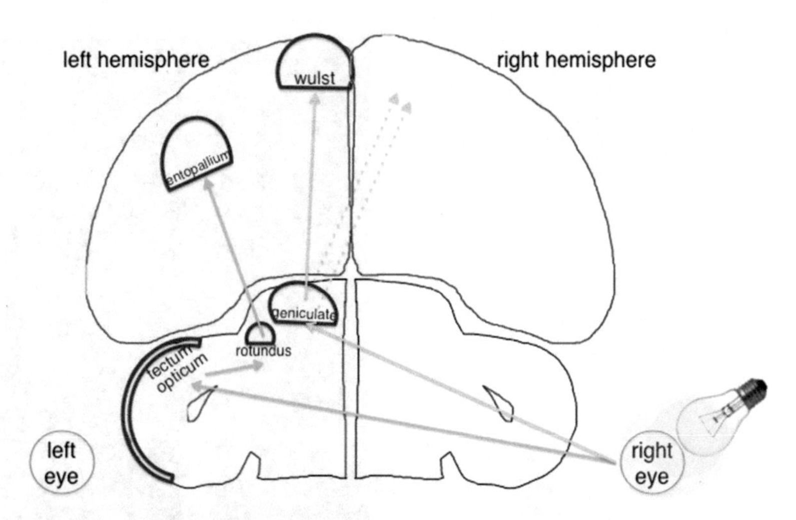

Fig. 3 The chick's visual pathways and embryonic light stimulation effect. On a horizontal section of the chick brain the main visual stations are shown in white. The tectofugal (*green lines*) and thalamofugal (*orange lines*) pathways are schematically drawn only for the right eye/left hemisphere. *Dashed orange lines* projecting to the right hemisphere represent the increased number of connections to the right hemisphere following selective light stimulation of the right eye

during pre-hatching experience), the asymmetry results in an opposite pattern [33, 37]. The reversal of the chick's head position within the egg can be operated on days E19–E20, when the beak has penetrated the air sac and the embryo has commenced breathing air. After removal of the shell at the air sac end of the egg and the membranes of the air sac overlying the embryo, it is possible to gently ease the embryo's head out to the egg leaving the body inside the egg. At this point, an eye-patch can be applied with a chinstrap and the egg with the embryo can be laid on its side within the incubator until hatching.

An alteration in the direction of asymmetries in pigeons can follow a different experimental manipulation, i.e. if one eye is selectively deprived of visual experience after hatching by an eye-patch. The delayed development of visual pathways in pigeons leaves open further sensitive periods for plastic changes to exert their effects [38, 39]. In contrast, post-hatching monocular occlusion of the stimulated right eye has no effect on the direction of lateralization in chicks: the light-dependent asymmetry determined before hatching remains unaltered [16].

1.3 Light Stimulation Functional Effects

In a classical experiment, the pebble floor task, chicks were tested for their ability to find grains of food that are intermingled with similar pebbles stuck to the floor. Grains were scattered among pebbles adherred to the ground in a 3:5 grains:pebbles ratio and number of pecks at pebbles in three blocks of 20 pecks each were

Fig. 4 The pebble floor task. Grains are alternated to pebbles on the floor and the chick's pecking behavior is observed and scored through windows placed on the bottom of the cage. In three consecutive frames, the peck at one element is shown

scored by observing the chick's behavior through openings placed at the bottom of the testing cage (Fig. 4). The performance was impaired following glutamate-treatment of left Wulst selectively in chicks hatched after light stimulation, whereas dark-incubated chicks had no asymmetry [15, 16, 40]. If the opposite eye was stimulated by light (by exposing the left eye and occluding the right), such reversal of natural conditions reverses the direction of lateralization of performance: glutamate-treatment of right Wulst impaired chicks' performance (see also Chap. 8). If light reached both eyes equally, the bilaterally stimulated birds showed no lateralization, in a similar fashion to dark-incubated birds [41].

When the pebble floor task is administered simultaneously with the presentation of a predator (a silhouette of a flying raptor of 11 cm, moving at a speed of 12 cm/s on a screen placed on the ceiling of the apparatus), chicks hatched after exposure to light outperform chicks hatched from eggs maintained for the whole incubation process in complete darkness [42]. Light-exposed (lateralized) birds can manage the two tasks at the same time successfully, whereas dark-incubated (less asymmetric) birds suffer interference between the expertise of the two hemispheres. When the chicks are tested monocularly in such a dual task, only birds hatched after exposure to light use preferentially their left hemisphere to discriminate the grain from pebbles and the right eye for vigilance, coping well with the task requirements. Symmetric unstimulated birds show no asymmetry and significantly poorer performance [43]. Also, interhemispheric transmission of information is more effectual in strongly lateralized chicks only [44].

Despite the anatomical difference in the pattern of lateralization, pigeons show the same functional division of labor between the two sides of the brain, with the right eye/left hemisphere superior in avoiding pebbles in light-stimulated birds, no functional asymmetry detectable in dark-incubated birds, and a less efficient interhemispheric exchange in dark-incubated than in light-exposed pigeons [45].

Exposure to light also affects functions governed by the unstimulated eye/hemisphere in chicks. For instance, attack responses are higher in light-stimulated chicks when they use their left eye, while dark chicks show comparable levels of attack independently of the eye used [15]. In a comparable vein, during a visuospatial task in which the chick had to peck at target elements regularly placed on an array in front of it, the right hemispheric response to novelty was modulated by the asymmetric embryonic light stimulation [46]. Specifically, in a first test, chicks faced grains of food for 3 min: dark chicks showed no bias in attending the elements whereas light chicks explored more the left hemispace of the array. Here, the effect of asymmetric stimulation is manifest in the preferential attention allocated by the right hemisphere to the left hemispace. In a second test, chicks faced grains of food intermingled with pebbles (differently from the pebble floor task, here the chick is confined and each element is regularly spaced on the array, Fig. 5): both dark and light chicks attended significantly more to the elements located to the left, likely because the right hemisphere responded to novel elements independently from light modulation. In a third test, chicks were first accustomed to learn how to discriminate grains from pebbles and then were presented with the array of grains and pebbles: in both dark and light chicks any bias disappeared. This result likely shows the recruitment of the left

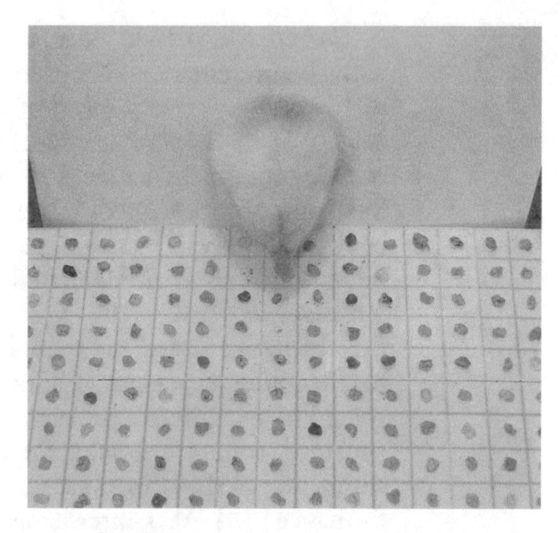

Fig. 5 The regular array of grains and pebbles. The chick protrudes its head from a circular window and pecks freely at grains intermingled with pebbles. The elements are regularly spaced, 1 cm apart from each other

hemisphere (fed by the right eye) for fine discrimination only after elimination of distraction by novel elements, something that the right hemisphere (left eye) is prone to.

1.4 Selective Effects of Light Stimulation on Visual Sensitivity

The greater mobility of the right side than of the left side of the hatching chick supports the rotation and the removal of the top of the eggshell [47]. Such motor biased experience may be related to postnatal behavioral asymmetries since, by compromising the completion of hatching movements on E20 with a cut on the eggshell, basic motor lateralization is subjected to a significant decrease [48]. However, a detailed kinematic analysis of footsteps of groups of chicks hatched in darkness or under a specific light regime revealed no difference in performance [49]. Chicks were trained to walk without interruption in the direction of the experimenter; a camera placed beneath a transparent floor recorded the foot placement in four walk trials. The absence of difference in locomotor skills indicates that asymmetries are related to visual input processing rather than simple motor asymmetries. Indeed, a systematic analysis of motor turning bias after light stimulation of both eyes, right eye or left eye only, demonstrated a specific effect of unilateral experience to the right eye in the development of the bias [50]. Consistently, no differences in the trajectory of the path followed to reach a target were apparent depending on incubation condition (dark incubation vs. light stimulation), but rather all animals retained a slight but significant leftward bias [51] when tested both within the first and the second week post-hatching. Only when an obstacle was interposed between the starting position and the target, a difference in the side of the trajectory emerged between conditions, with dark-incubated chicks that continued to run with a left bias, hence avoiding the obstacle with a detour toward the left, while light-stimulated chicks showed no systematic side preference [51]. In this case, sustained attention to the target carried out with the right eye/left hemisphere can explain the bias rather than motor asymmetries due to hatching experience.

In a comparable fashion, the postural control needed for ground scratching seems to be linked to asymmetric hemispheric control of visual inputs rather than only to motor asymmetries [52, 53].

When dark-incubated and light-stimulated animals are tested in higher-level cognitive tasks, no differences in learning have ever been reported following light or dark treatment. For instance, chicks trained to discriminate objects on the basis of left-right arrangements reach the learning criterion in a comparable number of trials and errors, independently of the prenatal manipulation: performance depends specifically on the capability of using left-right cues to discriminate between objects after light treatment [54] and in social contexts, whereas dark-reared chicks show an advantage in competing for access to food [55]. Likewise, pigeons trained to perform a color discrimination task showed comparable learning curves and discrimination accuracy irrespective of light-experience or dark-rearing, but light stimulation affected the inter-hemispheric access to the stored visual information [45, 56]. Pigeons learnt to discriminate two pairs of colors separately with the two eyes, hence the pairs of colors were different for each hemisphere. At test, the capability of each brain half of accessing to the information stored by the other half was assessed by presenting the animals with both the trained and the transfer color pairs. While there was a normal advantage of the right hemisphere in performing the task, light stimulation modulated the interhemispheric transfer by enhancing the success of the left hemisphere.

1.5 Outside the Avian World and Further Critical Periods

Photic application to embryos prior to hatching is a specific form of environmental event that models the cerebral asymmetries. Its relevance for neuroscientific studies lies especially in the fact that it can be easily controlled and administered in avian species. Light stimulation during prenatal stages, however, affects also lateralization of brain structures and processing of other species.

One notable instance is the development of lateralization in zebrafish (*Danio rerio*), asymmetry that follows normal light/dark cycles during the first week post-fertilization. If light fails to reach the embryos in two sensitive periods within this limited time window, the correct development of lateralization is either compromised or prevented [57–59]. In zebrafish, light does not act on the maturing visual system as in birds, but rather it may affect another photosensitive structure that is genetically asymmetrical, the parapineal gland and its connections to the habenular nucleus [59, 60, see also Chap. 17]; dark maintenance is responsible for late differentiation of habenular neurons [61]. Despite following a different anatomical route, the functional cerebral specialization of zebrafish is comparable to that of birds with primitive avoidance and wariness governed by the right hemisphere and a complementary left hemisphere specialization for the control of routine behaviors of feeding and analysis in familiar contexts [62]. Thus, similarly to that which has been discussed for precocial and altricial avian species, behavioral asymmetries in the zebrafish may be generated by different asymmetric neural systems.

A different (early) effective time window for light stimulation in chicks has been recently described [51]. Just-laid fertilized eggs were exposed to light from the arrival in the lab (Monday, late afternoon) and for nearly 50 h (light was switched off on Wednesday, early morning). By that time, the optic vesicles start forming but no retinal photoreceptors are in place, whereas other photosensitive cells may be already responding to light as shown in fish [63, 64]. The performance of early-stimulated chicks was compared to that of dark-incubated and late-stimulated chicks (during the canonical window of the last 3 days before hatching). Two tasks were used to assess lateralization. In a first task, chicks were confined and had to sample grains of food in the homogeneous array placed in front of them as previously described [46]. Dark chicks selected elements located both on the left and the right hemispaces, whereas late and early chicks attended preferentially to the left side of the array. In a second task, chicks were freely moving and had to reach a target located nearly 170 cm ahead from the starting point. At test, when an obstacle was placed on the way to the target, dark chicks detoured it systematically on the left while late and early chicks detoured it on the left or on the right indifferently. The results of both experiments show that there can be a similarity in the way light modulates behavioral asymmetries despite being applied in different time windows and hence likely affecting different pathways in the establishment of cerebral lateralization. However, only further research of monocular performance after light treatment following different incubation periods and complemented with histological analysis will reveal whether the precocial light stimulation is acting through the modulation of a different pathway, likely a genetic one comparable to that at play in fish, to affect brain lateralization. Also, it is still not known whether a brief early exposure is comparable to the more prolonged light stimulation during the last 3 days prior to hatching [65, 66] (but see Sect. 3.3 for discussion on length of light stimulation period) and the same amount of stimulation applied in the two critical time windows should be tested. Further systematic studies are needed to unveil all these fundamental aspects.

Light influences also the development of lateralization in a species of live-bearing fish (*Girardinus falcatus*) after prenatal exposure of gravid females [67]. In this specific case, light acts by aligning the direction of the lateral bias at the population level rather than, as observed in chicks, in modulating its strength. One possible explanation of this difference refers to the fact that light can exert its influence by modifying the mother's physiology rather than by directly affecting the fry. Whether or not light reaches the embryos is a question that applies also to mammalian fetuses within the uterus. The possibility that asymmetric light influences the establishment or the modulation of later bias in mammals is still under scrutiny. Jacques and colleagues [68] estimated the amount of light passing through the womb of rats and guinea pigs in about 2% of incident biologically relevant wavelengths and recently in mice it has been shown that prenatal light stimulation regulates the vasculature and the development of retinal neurons [69]. According to a simulation [70] the fetus can get a small quantity of light that depends on thickness of maternal abdomen and seasonal anisotropic variations. These seasonal variations have proven to be correlated with incidence of left-handedness in males born in the interval between November and January [71] showing that complex genetic-environmental-hormonal interactions are at play in the initiation of cerebral left-right differences [72]. Asymmetric fetal position has been associated to the development of lateralization also for humans. Two thirds of the human fetuses face the outer environment with the right side of the head [73], a position that is comparable to that taken by birds. It remains to be clarified whether prenatal light stimulation may comparably affect, though very likely to a lesser extent, the development of lateral biases in humans.

2 Materials

A list of elemental materials and accessories is provided and described.

2.1 Rooms

The environment in which incubators and hatchers are placed for dark incubation of eggs should preferably be a completely dark, light isolated, room. A separate room should be selected for placing incubators or hatchers for light stimulation of eggs, in order to avoid any unwanted stimulation of dark-incubated eggs.

2.2 Incubators, Hatchers, and Accessories

To incubate the embryos, insulated chambers equipped with trays providing eggs with serial automatic turning should be preferred. The automatic turning substitutes hens' behavior during brooding and prevents both unequal heat distribution and embryos' adherence to the shell walls.

The internal temperature and humidity of the incubator have to be constantly monitored either through regular thermometer and wet-bulb hygrometer or through newer digital LCD control displays connected to automatically systems that read and regulate heating and humidification inside the incubator (for incubating temperature, see Sect. 3.2).

Humidifier basins can be added on the bottom of the incubator and can be partially or completely covered to adjust the humidity level in the internal chamber (for incubating humidity level, see Sect. 3.2).

Hatching can be set either inside the same incubator machine if the machine is equipped also with hatching baskets or in a separate and dedicated machine. The use of separate but complementary machines may allow an easier separation of the experimental groups, for instance keeping dark-incubated eggs within the same incubator by transferring the eggs in the hatching basket, and separating only the light-treated eggs in the hatcher which will be predesignated for light hatching in a dedicated room, possibly (for hatching temperature and humidity level, see Sect. 3.2).

2.3 Lights and Other Accessories

The light source should be placed outside the transparent ceiling of the incubator or hatcher and the quantity and nature of sources depend exclusively on the size and egg capacity of the machine (for the use of different light sources or different positioning, see Sects. 4.1 and 4.2).

The use of a light meter is also suggested. Depending for instance on the size of the incubator and the distance of the tray from the ceiling, the light intensity at the level of the eggs should be checked at least the first time the protocol is adopted (for light intensity, see Sect. 3.3).

A programmable light switch timer can easily control the switching on and off of the light source with exact timing.

3 Methods

A description of how to use all the materials and accessories listed in Sect. 2 is detailed.

3.1 Collection and Transportation of Fertilized Eggs

To exert a complete and rigorous control over incubation conditions, fertile, freshly laid eggs have to be placed as soon as possible in opaque cooler bags for transportation to the laboratory to avoid exposure to light, large temperature changes (especially during winter and during summer), air drafts, and excessive heat or humidity. Also, careful transportation of the bags helps to avoid mechanical stress to the delicate initial stages of embryonic development (see also Sect. 4.6).

3.2 Incubation Protocol

As soon as the eggs arrive in the laboratory, it is preferable that they are marked with the date of arrival written with a pencil on the eggshell. This procedure, accompanied by placement on separate trays, will avoid mixing up eggs from different batches. Eggs have to be placed within the incubators minimizing any further delay.

The marking and moving of the eggs from the cooler bags to the incubators should occur in the dark room, in dark conditions (see also Sect. 4.3 for helpful suggestions on operations that are carried out in such a dark environment).

Extra eggs for each batch can be bought in order to have spare eggs that can be used to check both the developmental stage at arrival [18, 47] and the date of arrival whenever the operators are no longer sure of the division of batches within the incubator in the trays. In both cases, eggs will then be eliminated from the main experiment but chicks can still be used in other ongoing experiments in which no strict control over incubation conditions is required. The embryonic staging can be operated via candling, via the opening of a small window in the eggshell (that can be subsequently closed with a tape without compromising the hatching of the bird) or via complete dissection of the embryo. The exact staging supports a more accurate definition of the proper time to start with light stimulation.

From E0 to E18, the ideal incubating temperature is set at 37.7 °C with humidity at 47%. Values too high or too low can compromise the embryo's development not only by respectively accelerating or slowing down the growth but also resulting in malformations and weakness that can eventually provoke death before or immediately after hatching.

Correct ventilation within the incubator chamber is important both to maintain constant temperature and humidity and to provide proper air exchanges from the internal and the external environment for the oxygenation of the embryos and the chicks once hatched.

Around E19 the embryo enters the hatching phase: it pecks at the eggshell and the hatching process is completed within the next 1–2 days (E20–E21). By this time, eggs need to be placed on proper hatching baskets or dedicated hatchers. Independently from using either the incubator or the specifically designed hatcher machine, temperature for hatching can be kept constant at 37.7 °C but humidity has to be increased up to 65%. Extra humidifier basins can be added on the bottom of the incubator, or covers to regular basins may be removed (but always keep a thin net on the top of the basin to prevent chicks from falling into it) to adjust the humidity level in the internal chamber accordingly.

By E19, eggs randomly assigned to separate dark and light regimes enter either the dark-dedicated incubator or the light-prepared incubator/hatcher and remain separate until the hatching process is completed (see Table 1).

3.3 Light Intensity and Regimes

In controlled laboratory conditions, it has been shown that even a brief (i.e., 2 h) light stimulation applied at any time during the final days of incubation (sensitive period E19–E21) is sufficient to determine long lasting effects on lateralized behavioral responses

Table 1
Stages of the light stimulation protocol

	Collect	Transport	Mark	Incubate	Hatch	Rear
	E0			**E0–E18**	**E19–E21**	**P1–PN**
Dark	Collection of freshly laid eggs	Transportation in dark and isolated cooler bag	Marking of the eggs upon arrival in the lab	Incubation at 37.7 °C 47 % humidity Complete darkness	Hatching at 37.7 °C Humidity increase of 15 % Complete darkness	12 h:12 h light:dark Cycle Testing after P2
Light					100–400 lx 2 h minimum Homogeneous lighting Control on temperature/humidity parameters	
In the text	Section 3.1 Note 4.6	Section 3.2 Notes 4.2 and 4.5			Section 3.3 Notes 4.1–4.7	Section 3.4 Note 4.4

The main steps to conduct the embryonic light stimulation are summarized with the indication of the main important details. Direct cross-references to specific sections and notes fully described in the main text are provided in the *bottom row*

[11]. However, more extended photic stimulation seems desirable to stabilize the lateralization [15].

Light-dependent asymmetries show wavelength a-specificity as demonstrated in systematic investigation of both chicks [74] and zebrafish [75], in line with tetrachromatic color vision in these species [76, 77] and with the idea that the broad spectrum of light wavelengths stimulates all types of photoreceptors.

Intensities of around 100 lux are sufficient [74]. In recent works [44, 46, 51, 54], light stimulation was applied with intensity of 250 lux and continuously during the last 3 days, or intensity of 400 lux [42, 43] (see Sect. 4.4 for effects related to higher intensities and more extended period of administration of light).

3.4 Postnatal Rearing Environment

After the hatching process has been completed and all of the birds' feathers have dried (post-hatching day P1), they can be placed within regular home-cages with water and food available and reared and tested according to the chosen behavioral protocol. The rearing of the chick in isolation, in pairs, or little social groups of 4–5 individuals, depends on the protocol adopted for behavioral assessment. During their maintenance in the laboratory, the room temperature should be set at around 30–32 °C for the first days and illumination provided should follow a 12/12-h light/dark cycle and should consider the amount of lux at the bird level (see Sect. 4.4).

4 Notes

4.1 Light Positioning

Incubators may be provided with a dim light source placed on the rear wall of the internal chamber, just behind the fan. The built-in light bulb has more the function of a courtesy light and instructions accompanying incubators indicate to switch it off when the machine is operating and switch it on only for egg inspection because it can alter internal temperature and humidity parameters. Also, the dim light (with around 20 lux) is not placed symmetrically within the incubator to provide the suitable light stimulation for an efficacious manipulation.

The best positioning of light source is at the center, outside a transparent ceiling (made of plexiglass) of the incubator so that light will irradiate all eggs placed in the hatching basket.

Almost all incubators are manufactured with an opaque ceiling that cannot be easily modified. An electricity cable with a light bulb holder on the end that will be placed inside the machine can be easily inserted from the ventilation hole. The insertion of the cable through the ventilation hole will be minimally invasive, as it will not compromise its function (i.e., the air exchange). The use of compact fluorescent lamps (CFL) will help to keep the adequate temperature (and for discussion on types of light see Sect. 4.3).

However, note that with this method only the upper tray will receive adequate photic stimulation and trays located underneath will still receive some light but to a significant, and not comparable, less extent.

Some incubators and hatchers are supplied with a complete transparent frontal opening. This feature allows an alternative approach that, however, may be less easy to implement to provide adequate photic stimulation. The stimulation may still be effective, but some cautions should be taken:

(a) the light will have to have a certain angle and point toward the rear wall in order to illuminate the frontal part of the tray/basket;
(b) the number of eggs should be limited, depending on specific tray/basket sizes, in order to allow the most favorable illumination possible of all the eggs placed in the foremost part of the tray;
(c) the eggs will need to be swapped in position, mixing them from central to lateral places because more lateral units will get less light unless two or three light bulbs can be placed at an equal distance one from the other but complying, importantly, with the total amount of suggested lux (see Sects. 3.3 and 1.4).

Note that this alternative is neither the chosen method nor the optimal solution.

4.2 Alternative Light

Incandescent bulbs have been recently replaced by compact fluorescent lamps (CFL) since a directive from European Union has phased out the use of incandescent light bulbs. Although never tested systematically, but in consideration of the results described in Sect. 3.3, both CFL and light-emitting diode lamps (LED) can presumably be adopted without altering the light stimulation effects but efficiently overcoming the problems described in Sect. 4.1 (see also Sects. 4.3 and 4.4). The LED positioned on the ceiling will provide eggs with equal light irradiation.

4.3 Humidity and Temperature

Temperature and humidity are delicate parameters and their incorrect setting is almost always associated with uneven hatching. Room temperature, ventilation, and distance from heating sources are crucial factors even if considered only secondary. For instance, constant and regular room temperature (about 23 °C), incubators kept at least 50 cm apart from the walls and far from heaters, incubator air holes free from curtains or other type of occluders are among the things to pay attention to.

4.4 Undesired Nonvisual Light-Related Effects

In natural environments, light regulates physiological functions depending on circadian rhythms. Light pollution following increased urban life is affecting normal biological rhythms in avian species, whose reproductive physiology can be significantly altered [78]. In captive reared chicks, when light is administered at high intensity and during the entire incubation process it can alter hatching time by about 2 days [79] and the development of inter-limb coordination [80, 81]. Such acceleration in hatching time is more noticeable with incandescent than fluorescent light, showing that type of light, duration, and age of exposure are all critical factors [82]. Embryonic monochromaticLED lighting at 15 lux does not affect hatching time or hatchability in broiler chickens, nor weight at birth [83] despite the fact that it exerts effects on post-hatching muscle growth [83] that is altered especially after prolonged rearing under specific polychromatic components [84].

4.5 Absolute Darkness

The most challenging step in the whole procedure is to obtain complete darkness for dark incubation. If the incubator has lights to heat the internal chamber, they must be replaced by a heating element.

As pointed out in Sect. 3.1, the ideal solution would be a dark room. Indeed, since incubator air holes cannot be occluded, the only way to avoid any accidental light stimulation is to isolate the incubator from outer sources of light. To this aim, windows should be preferably sealed and the same caution should be applied to doors; alarm lights should be covered with dark isolating tape, with the same procedure used to cover lights built-in in the incubator (mini LCD displays, etc.). A dark thick curtain may well divide the incubators from the rest of the room so that light coming from outer sources will not affect the conditions (but see Sect. 4.3: curtains too close to the incubators or for each single incubator should be avoided because they impede the correct aeration).

In these conditions, the next problem may be management of eggs and birds within the incubator. An infrared LED hat may be an optimal device in support to all procedures that have to be carefully carried out in complete darkness.

4.6 Freshly Laid Eggs

One of the most important steps is that of collecting eggs immediately after laying and to incubate them in laboratory for the duration of incubation process, until hatching. The collection of eggs at later stages of incubation should be avoided because standard procedures in commercial aviaries may expose eggs to light in other moments during the embryonic development and the protocol may be compromised. For instance, in commercial aviaries there might be the need to candle the eggs (usually at least three times in the 21 days of incubation). Candling has to be avoided at any stage.

Just fertilized eggs should be handled with particular care in transporting and positioning them in order to avoid negative outcomes.

4.7 Absence/Presence of Visual Experience

For some experimental purposes, researchers may need to hatch birds with light exposure in the last days before hatching but, after hatching, prevent them from having visual experience of the outer physical and/or social environment at birth and before testing.

One possibility is that of turning off the lights when the peeps start. If the procedure is highly standardized, a timer can be used. However, with this strategy, the problem may be that birds hatching at slightly different times will receive different amounts of stimulation in the last stages of incubation and this may affect the result increasing the heterogeneity of measurements.

An alternative solution is that of adding a set of small boxes whose opaque partitions will effectively prevent experience other than that known by the researchers (i.e., the characteristics of the box itself). In order to hatch birds in separate small boxes, the hatching basket should be divided considering an adequate space for both the egg and the baby bird. In this condition, light has to come necessarily from the ceiling of the incubator in order to illuminate to a comparable extent all the boxes.

The new partitions may reduce the internal airflow causing the incubator to suffer for humidity/temperature maintenance.

Within the boxes, researchers interested in providing the birds with specific visual experience (for instance, the features of an imprinting object) should increase the size of the box itself, so as to host also the artificial companion.

References

1. Bateson P, Gluckman P (2011) Plasticity, robustness, development and evolution. Cambridge University Press, Cambridge
2. Lorenz KZ (1937) The companion in the bird's world. Auk 54:245–273. doi:10.2307/4078077
3. Hubel DH, Wiesel TN (1970) The period of susceptibility to the physiological effects of unilateral eye closure in kittens. J Physiol 206:419–436. doi:10.1113/jphysiol.1970.sp009022
4. Zeanah CH, Egger HL, Smyke AT et al (2009) Institutional rearing and psychiatric disorders in Romanian preschool children. Am J Psychiatry 166:777–785. doi:10.1176/appi.ajp.2009.08091438
5. Kalcher-Sommersguter E, Preuschoft S, Franz-Schaider C et al (2015) Early maternal loss affects social integration of chimpanzees throughout their lifetime. Sci Rep 5:16439. doi:10.1038/srep16439
6. Charil A, Laplante DP, Vaillancourt C, King S (2010) Prenatal stress and brain development. Brain Res Rev 65:56–79. doi:10.1016/j.brainresrev.2010.06.002
7. Chan KP (2014) Prenatal meditation influences infant behaviors. Infant Behav Dev 37:556–561. doi:10.1016/j.infbeh.2014.06.011
8. Fride E, Weinstock M (1988) Prenatal stress increases anxiety related behavior and alters cerebral lateralization of dopamine activity. Life Sci 42:1059–1065
9. Schaafsma SM, Riedstra BJ, Pfannkuche KA et al (2009) Epigenesis of behavioural lateralization in humans and other animals. Philos Trans R Soc B Biol Sci 364:915–927. doi:10.1098/rstb.2008.0244
10. Jozet-alves C, Hébert M (2013) Embryonic exposure to predator odour modulates visual lateralization in cuttlefish Embryonic exposure to predator odour modulates visual lateralization in cuttlefish. Proc R Soc B Biol Sci 280:20122575
11. Rogers LJ (1982) Light experience and asymmetry of brain function in chickens. Nature 297:223–225. doi:10.1038/297223a0

12. Buschmann J-UF, Manns M, Güntürkün O (2006) "Let There be Light!" pigeon eggs are regularly exposed to light during breeding. Behav Processes 73:62–67. doi:10.1016/j.beproc.2006.03.012
13. Levin M, Johnson RL, Stern CD et al (1995) A molecular pathway determining left-right asymmetry in chick embryogenesis. Cell 82:803–814. doi:10.1016/0092-8674(95)90477-8
14. Kovach JK (1970) Development and mechanisms of behavior in the chick embryo during the last five days of incubation. J Comp Physiol Psychol 73:392–406. doi:10.1037/h0030196
15. Rogers LJ (1990) Light input and the reversal of functional lateralization in the chicken brain. Behav Brain Res 38:211–221. doi:10.1016/0166-4328(90)90176-F
16. Rogers LJ (1997) Early experiential effects on laterality: research on chicks has relevance to other species. Laterality 2:199–219. doi:10.1080/135765097397440
17. Deng C, Rogers LJ (1998) Bilaterally projecting neurons in the two visual pathways of chicks. Brain Res 794:281–290. doi:10.1016/S0006-8993(98)00237-6
18. Hamburger V, Hamilton H (1951) A series of normal stages in the development of the chick embryo. J Morphol 88:49–92. doi:10.1002/jmor.1050880104
19. Gottlieb G (1968) Prenatal behavior of birds. Q Rev Biol 43:148–174. doi:10.1086/405726
20. Rogers L (1995) The development of brain and behaviour in the chicken. CAB International, Wallingford
21. Rogers LJ, Andrew RJ (2002) Comparative vertebrate lateralization. Cambridge University Press, Cambridge
22. Ehrlich D, Mark R (1984) Topography of primary visual centres in the brain of the chick, Gallus gallus. J Comp Neurol 223:611–625. doi:10.1002/cne.902230411
23. Andrew RJ (1991) Neural and behavioural plasticity: the use of the domestic chick as a model. Oxford Science, Oxford
24. Letzner S, Simon A, Onur G (2016) Connectivity and neurochemistry of the commissura anterior of the pigeon (Columba livia). J Comp Neurol 524:343–361. doi:10.1002/cne.23858
25. Rogers LJ, Bolden SW (1991) Light-dependent development and asymmetry of visual projections. Neurosci Lett 121:63–67
26. Rogers LJ, Deng C (1999) Light experience and lateralization of the two visual pathways in the chick. Behav Brain Res 98:277–287. doi:10.1016/S0166-4328(98)00094-1
27. Koshiba M, Kikuchi T, Yohda M, Nakamura S (2002) Inversion of the anatomical lateralization of chick thalamofugal visual pathway by light experience. Neurosci Lett 318:113–116. doi:10.1016/S0304-3940(01)02306-0
28. Rajendra S, Rogers LJ (1993) Asymmetry is present in the thalamofugal visual projections of female chicks. Exp Brain Res 92:542–544. doi:10.1007/BF00229044
29. Ströckens F, Freund N, Manns M et al (2013) Visual asymmetries and the ascending thalamofugal pathway in pigeons. Brain Struct Funct 218:1197–1209. doi:10.1007/s00429-012-0454-x
30. Güntürkün O (1997) Morphological asymmetries of the tectum opticum in the pigeon. Exp Brain Res 116:561–566. doi:10.1007/PL00005785
31. Skiba M, Diekamp B, Güntürkün O (2002) Embryonic light stimulation induces different asymmetries in visuoperceptual and visuomotor pathways of pigeons. Behav Brain Res 134:149–156. doi:10.1016/S0166-4328(01)00463-6
32. Freund N, Güntürkün O, Manns M (2008) A morphological study of the nucleus subpretectalis of the pigeon. Brain Res Bull 75:491–493. doi:10.1016/j.brainresbull.2007.10.031
33. Rogers L, Sink H (1988) Transient asymmetry in the projections of the rostral thalamus to the visual hyperstriatum of the chicken, and reversal of its direction by light exposure. Exp Brain Res 70:378–384. doi:10.1007/BF00248362
34. Güntürkün O, Hellmann B, Melsbach G, Prior H (1998) Asymmetries of representation in the visual system of pigeons. Neuroreport 9:4127–4130
35. Valencia-Alfonso C-E, Verhaal J, Güntürkün O (2009) Ascending and descending mechanisms of visual lateralization in pigeons. Philos Trans R Soc Lond B Biol Sci 364:955–963. doi:10.1098/rstb.2008.0240
36. Rogers LJ, Vallortigara G, Andrew RJ (2013) Divided brains: the biology and behaviour of brain asymmetries. Cambridge University Press, Cambridge
37. Rogers LJ (2008) Development and function of lateralization in the avian brain. Brain Res Bull 76:235–244. doi:10.1016/j.brainresbull.2008.02.001
38. Manns M, Güntürkün O (1999) Monocular deprivation alters the direction of functional and morphological asymmetries in the pigeon's (Columba livia) visual system. Behav Neurosci 113:1257–1266

39. Manns M, Güntürkün O (1999) "Natural" and artificial monocular deprivation effects on thalamic soma sizes in pigeons. Neuroreport 10:3223–3228
40. Deng C, Rogers LJ (1997) Differential contributions of the two visual pathways to functional lateralization in chicks. Behav Brain Res 87:173–182. doi:10.1016/S0166-4328(97)02276-6
41. Deng C, Rogers LJ (2000) Organization of intratelencephalic projections to the visual Wulst of the chick. Brain Res 856:152–162. doi:10.1016/S0006-8993(99)02403-8
42. Rogers LJ, Zucca P, Vallortigara G (2004) Advantages of having a lateralized brain. Proc R Soc B Biol Sci 271:S420–S422. doi:10.1098/rsbl.2004.0200
43. Dharmaretnam M, Rogers LJ (2005) Hemispheric specialization and dual processing in strongly versus weakly lateralized chicks. Behav Brain Res 162:62–70. doi:10.1016/j.bbr.2005.03.012
44. Chiandetti C, Regolin L, Rogers LJ, Vallortigara G (2005) Effects of light stimulation of embryos on the use of position-specific and object-specific cues in binocular and monocular domestic chicks (Gallus gallus). Behav Brain Res 163:10–17. doi:10.1016/j.bbr.2005.03.024
45. Manns M, Römling J (2012) The impact of asymmetrical light input on cerebral hemispheric specialization and interhemispheric cooperation. Nat Commun 3:1–5. doi:10.1038/ncomms1699
46. Chiandetti C (2011) Pseudoneglect and embryonic light stimulation in the avian brain. Behav Neurosci 125:775–782. doi:10.1037/a0024721
47. Freeman BM, Vince AM (1974) Development of the avian embryo. A behavioural and physiological study. Chapman & Hall, London
48. Casey MB, Martino CM (2000) Asymmetrical hatching behaviors influence the development of postnatal laterality in domestic chicks (Gallus gallus). Dev Psychobiol 37:13–24. doi:10.1002/1098-2302(200007)
49. Sindhurakar A, Bradley NS (2010) Kinematic analysis of overground locomotion in chicks incubated under different light conditions. Dev Psychobiol 52:802–812. doi:10.1002/dev.20476
50. Casey MB, Karpinski S (1999) The development of postnatal turning bias is influenced by prenatal visual experience in domestic chicks (Gallus gallus). Psychol Rec 49:67–74
51. Chiandetti C, Galliussi J, Andrew RJ, Vallortigara G (2013) Early-light embryonic stimulation suggests a second route, via gene activation, to cerebral lateralization in vertebrates. Sci Rep 3:2701. doi:10.1038/srep02701
52. Tommasi L, Vallortigara G (1999) Footedness in binocular and monocular chicks. Laterality 4:89–95. doi:10.1080/135765099397132
53. Dharmaretnam M, Vijitha V, Priyadharshini K et al (2002) Ground scratching and preferred leg use in domestic chicks: changes in motor control in the first two. Laterality 7:371–380
54. Chiandetti C, Vallortigara G (2009) Effects of embryonic light stimulation on the ability to discriminate left from right in the domestic chick. Behav Brain Res 198:240–246. doi:10.1016/j.bbr.2008.11.018
55. Wichman A, Freire R, Rogers LJ (2009) Light exposure during incubation and social and vigilance behaviour of domestic chicks. Laterality 14:381–394. doi:10.1080/13576500802440616
56. Letzner S, Patzke N, Verhaal J, Manns M (2014) Shaping a lateralized brain: asymmetrical light experience modulates access to visual interhemispheric information in pigeons. Sci Rep 4:4253. doi:10.1038/srep04253
57. Budaev S, Andrew RJ (2009) Patterns of early embryonic light exposure determine behavioural asymmetries in zebrafish: a habenular hypothesis. Behav Brain Res 200:91–94. doi:10.1016/j.bbr.2008.12.030
58. Budaev S, Andrew R (2009) Shyness and behavioural asymmetries in larval zebrafish (Brachydanio rerio) developed in light and dark. Behaviour 146:1037–1052. doi:10.1163/156853909X404448
59. Andrew RJ, Osorio D, Budaev S (2009) Light during embryonic development modulates patterns of lateralization strongly and similarly in both zebrafish and chick. Philos Trans R Soc Lond B Biol Sci 364:983–989. doi:10.1098/rstb.2008.0241
60. Kuan Y-S, Gamse JT, Schreiber AM, Halpern ME (2007) Selective asymmetry in a conserved forebrain to midbrain projection. J Exp Zool B Mol Dev Evol 308:669–678. doi:10.1002/jez.b.21184
61. de Borsetti NH, Dean BJ, Bain EJ et al (2011) Light and melatonin schedule neuronal differentiation in the habenular nuclei. Dev Biol 358:251–261. doi:10.1016/j.ydbio.2011.07.038
62. Andrew RJ (2009) Origins of asymmetry in the CNS. Semin Cell Dev Biol 20:485–490. doi:10.1016/j.semcdb.2008.11.001
63. Omura Y, Oguri M (1993) Early development of the pineal photoreceptors prior to the retinal

differentiation in the embryonic rainbow trout, Oncorhynchus mykiss (Teleostei). Arch Histol Cytol 56:283–291. doi:10.1679/aohc.56.283

64. Ostholm T, Briinn E, Van Veen T (1987) The pineal organ is the first differentiated light receptor in the embryonic salmon, Salmo salar L. Cell Tissue Res 249:641–646. doi:10.1007/BF00217336
65. Rogers LJ (2014) Asymmetry of brain and behavior in animals: its development, function, and human relevance. Genesis 17:1–17. doi:10.1002/dvg.22741
66. Zappia JV, Rogers LJ (1983) Light experience during development affects asymmetry of forebrain function in chickens. Dev Brain Res 11:93–106. doi:10.1016/0165-3806(83)90204-3
67. Dadda M, Bisazza A (2012) Prenatal light exposure affects development of behavioural lateralization in a livebearing fish. Behav Processes 91:115–118. doi:10.1016/j.beproc.2012.06.008
68. Jacques SL, Weaver DR, Reppert SM (1987) Penetration of light into the uterus of pregnant mammals. Photochem Photobiol 45:637–641.doi:10.1111/j.1751-1097.1987.tb07391.x
69. Rao S, Chun C, Fan J et al (2013) A direct and melanopsin-dependent fetal light response regulates mouse eye development. Nature 494:243–246. doi:10.1038/nature11823
70. Del Giudice M (2011) Alone in the dark? Modeling the conditions for visual experience in human fetuses. Dev Psychobiol 53:214–219. doi:10.1002/dev.20506
71. Tran US, Stieger S, Voracek M (2014) Latent variable analysis indicates that seasonal anisotropy accounts for the higher prevalence of left-handedness in men. Cortex 57:188–197. doi:10.1016/j.cortex.2014.04.011
72. Geschwind N, Galaburda AM (1985) Cerebral lateralization. Arch Neurol 42:634–654. doi:10.1001/archneur.1985.04060070024012
73. Previc FH (1991) A general theory concerning the prenatal origins of cerebral lateralization in humans. Psychol Rev 98:299–334. doi:10.1037/0033-295X.98.3.299
74. Rogers LJ, Krebs GA (1996) Exposure to different wavelengths of light and the development of structural and functional asymmetries in the chicken. Behav Brain Res 80:65–73. doi:10.1016/0166-4328(96)00021-6
75. Sovrano VA, Bertolucci C, Frigato E et al (2016) Influence of exposure in ovo to different light wavelengths on the lateralization of social response in zebrafish larvae. Physiol Behav 157:258–264
76. Osorio D, Vorobyev M, Jones CD (1999) Colour vision of domestic chicks. J Exp Biol 202:2951–2959
77. Fleisch VC, Neuhauss SCF (2006) Visual behavior in zebrafish. Zebrafish 3:1–11. doi:10.1089/zeb.2006.3.191
78. Dominoni D, Quetting M, Partecke J (2013) Artificial light at night advances avian reproductive physiology. Proc Biol Sci 280:20123017. doi:10.1098/rspb.2012.3017
79. Siegel PB, Isakson ST, Coleman FN, Huffman BJ (1969) Photoacceleration of development in chick embryos. Comp Biochem Physiol 28:753–758. doi:10.1016/0010-406X(69)92108-2
80. Sindhurakar A, Bradley NS (2012) Light accelerates morphogenesis and acquisition of interlimb stepping in chick embryos. PLoS One 7:1–14. doi:10.1371/journal.pone.0051348
81. Porterfield JH, Sindhurakar A, Finley JM, Bradley NS (2015) Drift during overground locomotion in newly hatched chicks varies with light exposure during embryogenesis. Dev Psychobiol 57:459–469. doi:10.1002/dev.21306
82. Lauber JK, Shutze JV (1964) Accelerated growth of embryo chicks under the influence of light. Growth 28:179–190
83. Zhang L, Zhang HJ, Wang J et al (2014) Stimulation with monochromatic green light during incubation alters satellite cell mitotic activity and gene expression in relation to embryonic and posthatch muscle growth of broiler chickens. Animal 8:86–93. doi:10.1017/S1751731113001882
84. Pan J, Yang Y, Yang B, Yu Y (2014) Artificial polychromatic light affects growth and physiology in chicks. PLoS One 9:e113595. doi:10.1371/journal.pone.0113595

Chapter 20

Investigating Effects of Steroid Hormones on Lateralization of Brain and Behavior

Tess Beking, Reint H. Geuze, and Ton G.G. Groothuis

Abstract

Steroid hormones have been proposed to influence the development of lateralization of brain and behavior. We briefly describe the available hypotheses explaining this influence. These are all based on human data. However, experimental testing is almost exclusively limited to other animal models. As a consequence, different research fields investigate the relationship between steroid hormones and lateralization, all using different techniques and study species. The aim of this chapter is to present an overview of available techniques to study this relationship with an interdisciplinary approach. To this end we describe the basics of hormone secretion and mechanisms of action for androgens, estrogens, progesterone, and corticosteroids. Next, general issues related to hormone sampling and hormone assays are discussed. We then present a critical overview of correlational and experimental methods to study the influence of prenatal and postnatal hormones on lateralization. These methods include hormone measurement in amniotic fluid, saliva, urine, feces, and blood plasma or serum of fetus, mother, and umbilical cord. We also discuss hormone-mediated maternal effects, the manipulation of hormone levels in the embryo or mother, hormone treatment in persons with Gender Dysphoria, and the 2D:4D finger length ratio as a proxy for prenatal testosterone exposure. We argue that lateralization can and should be studied at different levels of organization. Namely, structural and functional brain lateralization, perception and cognition, lateralized motor output and performance. We present tests for these different levels and argue that keeping these levels apart is important, as well as realizing that lateralization and the hormonal influence on it may be different at different levels, for different functions and different species. We conclude that the study of hormonal influences on lateralization of brain and behavior has not yet exploited the knowledge and wide array of techniques currently available, leaving an interesting research field substantially under-explored.

Key words Animals, Human, Correlational, Experimental, Prenatal, Postnatal, Testosterone, Estrogens, Hormone sampling, Organizing effects

1 Introduction

Lateralization is a fundamental aspect of the organization of brain and behavior throughout the animal kingdom, but the development of brain lateralization is not yet well understood. Brain lateralization differs in some aspects between males and females in several species [1]. This suggests that sexual differentiation is

Lesley J. Rogers and Giorgio Vallortigara (eds.), *Lateralized Brain Functions: Methods in Human and Non-Human Species*, Neuromethods, vol. 122, DOI 10.1007/978-1-4939-6725-4_20, © Springer Science+Business Media LLC 2017

responsible for the differences between males and females. Sexual differentiation, including differentiation of the brain, is influenced by steroid hormones, primarily from gonadal origin [2]. Therefore, gonadal hormones are thought to influence the development of brain lateralization as well. However, whether this is really the case remains an open question, as the underlying mechanisms are still elusive.

The influence of gonadal hormones on the development of brain lateralization is studied in different research fields, including biology, endocrinology, psychology, and neuroscience. These research fields use different concepts, methods, and study species. Moreover, the latter differ in aspects of their endocrine systems, the process of sexual differentiation, brain structures and lateralized functions. These differences are at the same time a strength as well as a weakness. The different research fields offer the strength of a more integrative explanation, in which different approaches can be used to complement each other. This may also clarify how Darwinian evolution has shaped the differences between species in relation to life history traits. However, the results from one species may not be transferred automatically to explain phenomena in other species.

The aim of this chapter is to present a comprehensive overview of the methods used to investigate the influence of steroid hormones on the development of brain lateralization, including the advantages and disadvantages of each method. To this end we start with a very brief outline of why genetic factors alone are not a sufficient explanation for the developmental process, and what other factors might be involved. In order to provide the reader with an adequate background in relevant aspects of endocrinology, we discuss some principles of the underlying endocrine systems, and we show which different explanatory models exist for potential hormonal effects. Next, we discuss relevant methodological sampling and assay issues, and we discuss many correlational and experimental methods to measure steroid hormones prenatally or postnatally. Finally, we will explain that the relationship between hormones and lateralization can be measured at different levels, and we will point out future perspectives.

1.1 The Scope for Environmental Influences

Brain lateralization is caused by an interplay between genes, other internal factors, and the environment [3]. Genes are never the only factor affecting development because they need to be expressed and transcribed in continuous interaction with other internal and external factors. Therefore, even in cases in which genes are demonstrated to affect development, other factors cannot be excluded. This can be clearly demonstrated in sexual differentiation in species with sex chromosomes, where factors other than genes, such as gonadal hormones, are indispensable for proper development. In addition, the influence of hormones opens the possibility that hormonal signals from sources other than the embryo itself, such as from the mother, siblings, or environmental pollution, affect the developmental process. Therefore, purely genetic models explaining variation in brain lateralization inevitably offer an incomplete explanation of the mechanisms underlying the developmental process and its potential plasticity.

Models of the genetic influence on brain lateralization derive primarily from psychological studies on handedness in humans. In humans, roughly 90% of the population is right-handed, and a minority of about 10% is left-handed with 1–2% being ambidextrous [4]. The classical Mendelian model [5] with a recessive allele for left-handedness was discarded, as two left-handed parents frequently produce right-handed offspring. Several other models were postulated that fit most of the heritability data for handedness in humans, including the fact that monozygotic twins can differ in handedness. These models assume an interplay of two alleles that determine handedness: one dominant allele codes for right-handedness, and another recessive allele randomly codes for left- or right-handedness [6–11]. Although the models are elegant, they are post hoc explanations since the models were made to fit the prevalence of left- and right-handedness in the population. More importantly, the models are not supported by human genome data [3, 12, 13, and see Chap. 16]. In addition, some of these models still leave substantial room for as yet unexplained environmental factors affecting lateralization. For an extensive review on this topic see [3]. More recently, parent-offspring regressions have established the heritability of handedness at around 50% [14]. These data indicate much scope for factors other than genes affecting handedness. Moreover, such studies cannot exclude parental effects, being environmental effects for the offspring, affecting the heritability estimate. Interestingly, handedness is associated with polymorphism of the androgen receptor gene [15], suggesting indeed that hormones contribute to lateralization of the brain.

Several lines of evidence suggest environmental effects on development of brain lateralization. For example, prenatal head position and asymmetrical light input (birds and fish), early postnatal head position (humans), handedness of the foster mother (chimpanzees), visual lateralization of peers (chickens), and preterm birth and pathological factors like congenital hemiplegia, have all been demonstrated to affect brain or behavioral lateralization (for a review see [3] and see Chap. 19).

There is now evidence that prenatal steroid hormones affect several aspects of brain lateralization in humans and other animals (reviewed in [1]). Below we first provide an introduction into the field of gonadal hormones, next discuss hypotheses on how they may influence lateralization, and then embark on a discussion of the relevant methodology for further studies.

2 An Introduction to Steroid Hormones

To understand how hormones might affect brain lateralization, it is necessary to know the basics of hormone secretion and mechanisms of action. This basic information can be found in several textbooks [16, 17]. The following paragraphs summarize the information that is relevant for this chapter.

There are hundreds of hormones in the animal kingdom and the endocrine system has been conserved in evolution to a great extent among vertebrates. Although vertebrate taxa differ in details of metabolic pathways, carrier proteins and receptor distributions for these hormones, the overlap among the taxa is much larger than the differences. Thus, the information in this chapter is generally applicable a wide range of species. This chapter focuses on the steroid hormones testosterone, progesterone, estrogen, and cortisol. These steroid hormones have been found to affect brain development, possibly including brain lateralization. This does not exclude other steroid hormones being important for the development of brain lateralization. Steroid hormones are produced from cholesterol and the metabolic pathway includes many steps and metabolites. Although many metabolites have been considered to be biological inactive precursors, evidence is accumulating that their assumed inactivity might be at least partially incorrect. In addition, several nonsteroid hormones are known to be important for brain development, such as thyroid hormones and neuroendocrine factors, but their potential influence on brain lateralization has, to our knowledge, hardly been studied.

The brain can be seen as the main controller of all steroid hormone production in the body, but is itself also an endocrine gland, as it is able to produce these hormones in specific brain areas. Nevertheless, the major production pathway is the following. The hypothalamus, located at the base of the brain, is the major link between the nervous system and the endocrine system, translating internal and environmental information into hormone production. The hypothalamus releases hormones that signal to the anterior pituitary. The anterior pituitary is located just below the hypothalamus and releases luteinizing hormones (LH), follicle stimulating hormones (FSH), and adrenocorticotropic hormones (ACTH), that in turn stimulate other endocrine glands in the periphery of the body, such as the gonads and the adrenal glands. The hormones of these glands influence brain and behavior, and also provide negative feedback in order to inform the brain and to control current hormone production (Fig. 1).

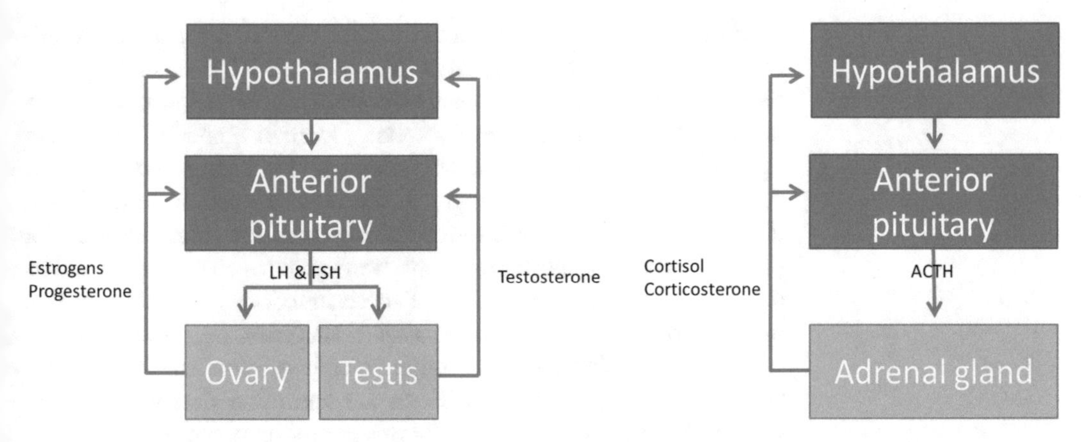

Fig. 1 Schematic representation of steroid hormone production in the gonads (*left*) and the adrenals (*right*), see text

Steroid hormones are signaling molecules that can greatly influence brain and behavior. The endocrine glands synthesize steroid hormones from cholesterol and release them into the blood. The majority of the hormones bind to a carrier protein called "globulin" and travel through the body until they arrive at target tissue. After separation from the carrier protein and binding to the receptor, the hormone enters through the cell membrane into the cell. There, the steroid hormone can be converted to other hormones and is transported into the cell nucleus to induce mRNA transcription. The mRNA is translated into proteins or enzymes that produce a physiological response. In this way, steroid hormones can affect a wide array of processes, including brain development and brain lateralization. The genomic action via these receptors is relatively slow. However, more recently, some of the steroid hormones, such as estradiol, have been found to induce action in the organisms on a very much faster time scale, by acting directly on the membrane of neurons [18]. This way, action is induced in the timeframe of minutes, instead of hours or even days.

Figure 2 depicts a simplified scheme of the metabolic pathway of the steroid hormones discussed in this paper. Progesterone, androgens, and estrogens are often called "sex hormones." Although they are not specific for one sex, the sexes produce these in different quantities. They are all synthesized from cholesterol. Progesterone is an important precursor for sex steroid hormones in vertebrates, although androgens and estrogens can also be produced via an alternative pathway.

Each hormone has its own affinity to specific receptors and induces its own specific effects. We briefly discuss here the most important ones. *Progesterone* is involved in mating behavior and pregnancy especially in mammals, affects sperm behavior, blocks receptors for the mineral corticoids and interacts with estrogens. The most important *androgens* are androstenedione, dehydroepiandrosterone

(DHEA), testosterone, and dihydrotestosterone. Androgens are primarily produced by the gonads and all act on the same androgen receptor. Their main functions are development of male secondary sex characteristics, muscle development, spermatogenesis, facilitating aggressive and sexual behavior, boldness, risk taking and alertness. Androgens also induce changes in the nervous system, both during development and in adulthood. Androgens are often crucial for proper sexual differentiation, but how and to what extent differs strongly among species. Many of these functions are, also in males, induced by the conversion of androstenedione or testosterone to estradiol, often locally in the brain. The important enzyme for this conversion is aromatase. DHEA is an androgen that is converted directly from cholesterol. Initially this hormone was overlooked, but more recently it has been demonstrated that DHEA has some biological relevance as a week androgen [19–21]. DHEA can be converted to androstenedione.

Androstenedione is produced by the gonads, but can be produced by the adrenal glands too. It is converted by the enzyme 17betaHSD to other androgens or by aromatase to estrogens in the gonads and other tissues, including the brain. Androstenedione has a low affinity to the androgen receptor and is primarily a prohormone for testosterone, dihydrotestosterone, and estrogens (see Fig. 2). *Testosterone* has a high affinity to the androgen receptor and can be converted to estrogens as well. *Dihydrotestosterone* has an even higher affinity to the androgen receptor than testosterone. Dihydrotestosterone is an androgen that cannot be converted to estradiol and can therefore be used to test whether effects of androgens are mediated by only the androgen receptor or also estrogen receptors. In fish, dihydrotestosterone is replaced by 11-keto-testosterone, which strongly affects secondary sexual characteristics.

Estrogens are secreted by the gonads but are also produced in a wide variety of tissues, including the liver and the brain. The active form 17-beta-estradiol contributes to the development and

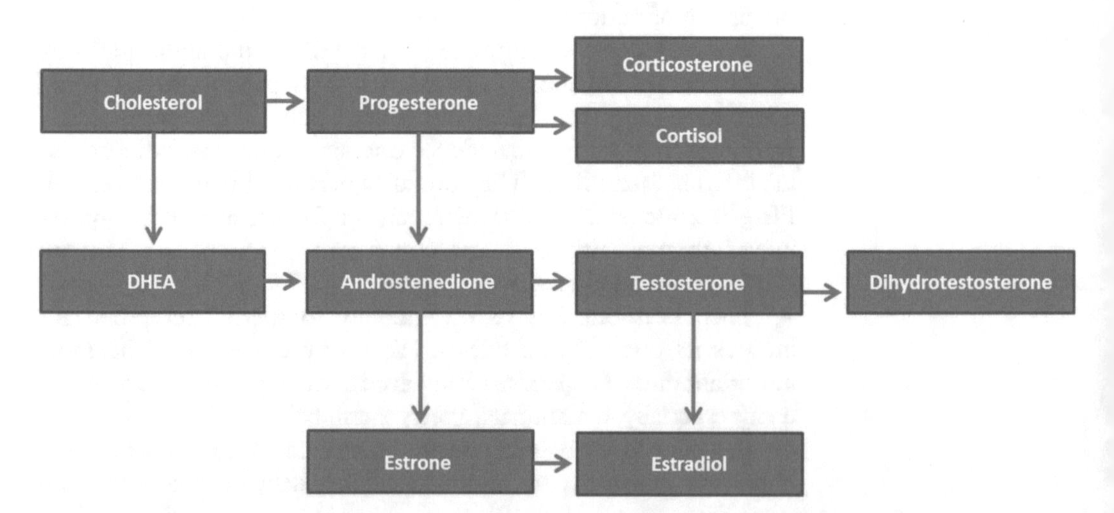

Fig. 2 Simplified scheme of the metabolic pathway of the steroid hormones discussed in this paper

maintenance of female sex characteristics and sexual behavior, but also has effects on bone condition, fat deposition, and neuroprotection. There are two different receptors for estrogens, alpha and beta, with different thresholds and distributions.

Both sexes produce biologically active androgens and estrogens, but the distribution and density of the relevant receptors differ, as well as circulating concentrations of androgens and estrogens. Males have higher levels of testosterone and dihydrotestosterone, and lower levels of estrogens and androstenedione than females. Circulating levels of these hormones can vary more rapidly than previously thought, within 10 min, depending on environmental, mostly social, factors.

The steroid hormones *cortisol and corticosterone* are glucocorticoids. Humans and primates primarily secrete cortisol, whereas birds, reptiles and some mammalian species such as rats secrete a similar hormone "corticosterone." Cortisol and corticosterone are primarily—but not exclusively—secreted in the adrenal cortex (for example, they are also produced by the immune system). Like sex steroid hormones, they are produced under the influence of hormonal signals from the hypothalamus and pituitary, on which they exert a negative feedback (see Fig. 1), and induce genomic and nongenomic actions. Cortisol and corticosterone have two main functions. The first function is their role in stressful situations. Cortisol and corticosterone are produced in only a few minutes after an environmental stressor. They affect metabolism by stimulating production of glucose (gluconeogenesis), mobilize immune-acids and breakdown fat tissue to release energy for emergency actions, including rapid decisions in the brain. The second important function is the regulation of the immune defense, for example by stimulating anti-inflammatory effects. They also play a role in development, such as lung maturation. Cortisol and corticosterone affect the brain by mediation of emotion-based learning, induction of long-term potentiation, and adaptation of structures in the hippocampus, amygdala, and frontal areas. Some studies suggest that cortisol and corticosterone are involved in sexual differentiation of the brain as well [22–24].

Historically, the effects of sex hormones on development have been classified into "organizing effects" and "activating effects." Although in 1985 Arnold and Breedlove [25] had already reported that the distinction between the two effects is not absolute, the distinction in the literature still remains. Organizing effects consist of structural changes, are permanent and generally occur before or just after birth during a sensitive period. Activating effects are additional to the organizing effects, are temporary or reversible and can occur during the whole lifespan [25, 26]. Most vertebrates have a sensitive period before or just after birth. Evidence is now accumulating that there could be other sensitive periods for organizational effects of sex hormones on the brain later in life. In humans and rodents puberty could be such a period [27].

3 Theories Regarding Steroid Hormones and Lateralization

There are three main theories on the influence of prenatal testosterone on brain lateralization. All three propose organizational effects of prenatal testosterone on brain lateralization in humans, but are at least partially applicable to other animals as well. Testosterone has been involved in these theories because it plays a major role in sexual differentiation in mammals including humans, and because of the small but consistent sex differences in human behavioral and cognitive lateralization.

The first theory is the sexual differentiation theory [28]. This theory proposes that higher prenatal testosterone levels lead to a more masculine pattern of lateralization. Human males and females differ in several aspects of lateralization, males being less strongly right- and more strongly left-handed, more strongly lateralized for spatial orientation and less for language functions, whereas the performance of these lateralized functions also differs (slightly) between the sexes (for a review see [1]).

The second theory [29] states that higher levels of prenatal exposure to testosterone delay growth of parts of the left hemisphere, resulting in compensatory growth of the homologue regions in the right hemisphere.

The third theory is the Corpus Callosum theory of Witelson and Nowakowski [30]. This theory is based on correlational evidence that prenatal exposure to testosterone, at least in males, induces pruning of the neural connections in the corpus callosum. The corpus callosum is the main connection between the hemispheres. Prenatal testosterone would reduce crosstalk between the hemispheres, and thereby increase the strength of lateralization [30].

These three theories lead to different predictions about the effect of prenatal testosterone on lateralization. The sexual differentiation theory is function specific as it is limited to those functions in which the sexes differ, whereas the Geschwind and Galaburda theory predicts a more overall effect of androgens on the strength or even the direction of lateralization. The Corpus Callosum theory specifically concerns the strength of lateralization.

Besides these proposed organizational effects, testosterone could also have activating effects on lateralization of cognitive and behavioral processes. There are as yet, however, no well-defined theories about the mechanisms underlying such activating effects. Since testosterone can be metabolized to dihydrotestosterone and estradiol, these hormones may be part of the assumed underlying mechanism, but this has never been addressed. Although a few experiments with glucocorticoids have been conducted [31, 32], no clear theory on the effect of these hormones on brain lateralization has emerged yet. Furthermore, all three theories have been both supported and challenged by correlational and experimental data (see [1]). It is not the aim of this chapter to review or discuss these data. Rather, we will discuss the proper methodologies for further testing, addressing the pros and cons of different techniques and different animal models.

4 Methodological Aspects of Hormone Sampling

4.1 Sampling Issues

A proper measurement of hormone concentrations is crucial for both correlational and experimental studies. In most cases hormone concentrations are measured in blood plasma. Mostly for convenience, hormone concentrations are also measured in saliva, urine or feces.

Plasma concentrations of hormones can fluctuate rapidly. This is both an advantage (for example when studying effects of environmental stimuli or correlations with variations in behavior) and a disadvantage (may be unreliable as a measure of basal levels). Each hormone has its specific response curves and sensitivity to environmental influences, such as time of the day, the estrous cycle in females, social factors, and stressors. It is therefore important to carefully select the timing of the samples and to avoid confounding influences. For example, the response of stress hormones (cortisol and corticosterone) to an external trigger is only 3 min, and for gonadal hormones this is about 15 min. To control for this, the time interval between approaching the individual or cage and taking the actual blood sample should always be registered and used as a covariate in the analyses.

Saliva samples show a delay in hormone concentrations of about 15 min relative to plasma samples, but have the advantage that the method of sampling is noninvasive. A major concern is that the measurement is sensitive to type and timing of food intake relative to the time of sampling. A rule of thumb is that the subject should not have eaten or drunk in the hour before sampling. Another concern is that the type of swabs used for sampling can interfere with the assay; for example, do not use cotton swabs when sampling estrogen. For reviews see [33].

Urine or fecal samples are suitable when studying long-term exposure, but this method is fraught with even more difficulties. The metabolites in the samples are more indirect measures with a poorer time solution than plasma or saliva samples, and the time course of the metabolites depends on diet and differs between species. The hormones in feces or urine are metabolized extensively to other hormones and hormone conjugates, so that extensive validation is needed. This can be done by injecting a labeled hormone into the organism, and then sampling the urine or feces over the next 24 h. Extensive determination of all possible metabolites that show up as labeled forms will then show which metabolites should be measured as a potential reflection of the target hormone, and in what time frame, the latter being dependent on the species and the kind of food ingested. Techniques to determine metabolites from labeled forms are for example high-performance liquid chromatography (HPLC) or liquid chromatography-mass spectrometry (LC-MS).

Fecal sampling has the disadvantage that it shows a much dampened dynamic over time. On the other side, it has the advantage that it is less sensitive to environmentally induced fluctuations, showing integration over a longer time-interval, such as at least a few hours. For an excellent review see [34]. Even a much longer time integration is provided by measuring hormones in hair or feathers, which has now become available for corticosterone and cortisol [35, 36].

In all methods one should realize that the measured hormone concentrations do not necessary reflect the organism's exposure. In most analyses the free and bound fraction are measured together, whereas only the free fraction is biologically active and its proportion can fluctuate rapidly. Moreover, the biological effect is determined by both the conversion to other hormones and receptor densities that all fluctuate too, and sometimes also by synergistic effects of other hormones. This becomes even more problematic when measuring hormones in the egg or amniotic fluid, as the timing of uptake and excretion of the hormone is often unclear.

4.2 Assay Issues

The hormone concentrations of blood plasma, saliva, feces, urine, hair or feathers can be determined by several techniques [36, 37]. Usually, the samples have to be extracted so that substances other than hormones do not disturb the measurement. By using different extraction methods on the same sample (such as different solutions one after the other over the same column), multiple hormones can be measured in the same sample, as the different hormones elute differently due to differences in polarity. Since such extractions are never complete, it is important to measure the extraction efficiency or recoveries by adding a known quantity of a labeled target hormone and measuring the quantity after extraction [37]. The measured concentrations should then be corrected for the loss of the hormone (% recovery) during extraction.

The classical extraction methods using columns packed with celite or commercially available columns are relatively cheap but time consuming. After extraction, the hormone concentration is determined by radio-immuno assays (RIA) or enzyme-linked immunosorbent assay (ELISA). The first method has the disadvantage that in many countries strict regulations are in place for working with radioactive material. The second method has the disadvantage that variation among different assays of the same samples can be relatively large. Both methods have the pitfall that the antibody may bind not only to the target hormone but also to other substances by cross-reactions. These cross-reactivities, or "specificities," are usually reported by the manufacturer. However, when using sample material for which the assay was not designed, such as material other than human or rodent blood plasma, this information may not be correct.

In contrast to the classical extraction methods, modern methods use liquid or gas chromatography (LC and GC) followed by mass spectrometry (MS). This allows precise determination of the many hormonal components in the sample in a rapid way, but the equipment is much more expensive. It is noteworthy that LC-MS and GC-MS generally result in much lower concentrations of the hormone than the more classical methods. This is important not only for comparing the results of different methods, labs, or meta-analyses but also for determining the physiological dosage for hormone manipulations.

Regardless of the assay used, it is important to add different known quantities of the target hormone to calibrate the procedure. One should always determine and report the intra-assay variation. In case more than one assay has been used, the inter-assay variation should be determined as well.

5 The Study of Prenatal Effects

The early influence of hormones can be studied by correlations between prenatal hormone levels and the lateralization of brain and behavior at a later stage of life, and by experimental manipulation to determine causal relationships.

5.1 Correlational Methods

5.1.1 Blood Samples from the Fetus

The most direct way to measure prenatal hormones would be to take a blood sample from the fetus. In humans this is not an option due to ethical reasons, but in birds and nonhuman mammals this has been done by dissection (e.g. [38, 39]). In a very early stage of gestation, the hormone production of the mother determines the (low) hormone levels in the fetus. This is the case in all placental animals, but also in some egg laying species in which the yolk hormones are determined by the mother. Rather soon the fetus starts producing its own hormones [40]. At that stage, the blood sample of the fetus reflects fetal hormone production, and potentially maternal hormones too. Preferably, the blood is sampled during the sensitive time that prenatal hormones affect brain development, which is different for different species. It often starts halfway through embryonic development or later and may extend during early postnatal development. Moreover, prenatal hormone production fluctuates strongly during fetal development (for birds see references in [38]; for mammals see [41]). The timing of the sensitive period differs between species, but generally coincides with the time that testosterone production peaks.

5.1.2 Amniotic Fluid

The preferred method to measure prenatal hormone levels is via amniocentesis, because direct sampling of fetal blood plasma is generally not possible due to ethical or practical constraints. During amniocentesis in mammals, a needle is inserted through the wall of the abdomen and the uterus into the amniotic sac, whereby a small amount of amniotic fluid is taken. The amniotic fluid contains waste products of the fetus, like hormones and their metabolites. In humans, an amniocentesis is only performed when there is a medical indication, for example age of the mother, genetic predisposition to specific diseases or disturbed development of the fetus. The timing of amniocentesis coincides with the sensitive period (14th–18th week) that steroid hormones affect brain differentiation [42]. In other mammals amniocentesis is most often performed for research purposes and can be repeated several times.

Hormones excreted by the fetus via fetal urine are the major contributor to the sex hormones in amniotic fluid from the second trimester onwards in humans [43, 44]. The major argument for this conclusion is that in humans sex differences in steroid hormone levels are found in the amniotic fluid but not in the maternal blood [45]. Male fetuses have higher testosterone [43, 45, 46] and androstenedione levels [45] in the amniotic fluid than female fetuses, and female fetuses have higher estradiol levels in the amniotic fluid than male fetuses [45, 47]. Moreover, few correlations between hormone levels in the amniotic fluid and maternal serum or umbilical cord have been found [45]. This is probably due to the fact that the interface between the circulation of the mother and the fetus, the placenta, is a highly active endocrine organ that produces hormones itself and converts maternal hormones to their metabolites.

5.1.3 Blood, Urine, or Saliva from the Mother During Pregnancy

Another strategy that is used to infer prenatal hormone exposure levels of the fetus is to collect urine, blood, or saliva from the mother during pregnancy. The big advantage is that sampling does not interfere with the pregnancy and can be repeated to see the hormonal fluctuations over time. However, the important downside is that, as mentioned above, the endocrine role of the placenta in mammals and the fetus' own hormonal production disturbs any correlation between maternal and fetal hormone exposure levels. Indeed, the testosterone peak of male fetuses is not reflected in the maternal blood [48].

The above should not be taken as evidence that maternal hormones do not reach the fetus. One way to investigate the amount of maternal hormone that reaches the fetus is to inject the mother or the egg yolk with labeled hormones, which are subsequently measured in the fetus. There is increasing evidence that they do reach the fetus in both placental and nonplacental animals and affect offspring development in humans and other animals (for a review see [49]), including lateralization [50]. However, maternal levels should not be used to estimate total embryonic exposure of steroid hormones once the embryo is capable of producing its own hormones. We therefore do not recommend using maternal hormone levels to estimate fetal steroid levels.

5.1.4 Umbilical Cord Blood

The umbilical cord is the connection between the developing fetus and the placenta in mammals, consisting of a vein and one or more arteries. The umbilical vein provides the fetus with fresh blood from the placenta, rich in oxygen and nutrients. The heart of the fetus pumps the blood back to the placenta via the umbilical arteries. Hormones in the fetus' circulation can be modified by the placenta, the production in the adrenals or gonads of the fetus, and their metabolism in its tissues.

One study measured testosterone levels in the umbilical artery, vein, and amniotic fluid multiple times during pregnancy in piglets [51]. Male fetuses had higher testosterone levels in the umbilical artery than females at all measurements. Testosterone levels were generally much higher in the umbilical artery than in the vein. In addition, testosterone levels in the amniotic fluid reflected the testosterone levels in the umbilical artery, but not the umbilical vein. This indicates that fetal testosterone levels are better represented in the umbilical artery than in the umbilical vein. However, in most animals, the umbilical vein is sampled as this is larger than the artery.

In this study [51], the umbilical cord was sampled during pregnancy. However, in most studies, the umbilical cord is sampled just after birth. It is important to realize that the timing of the sampling of the umbilical blood, just after birth, is not always optimal. Firstly, this may not be the sensitive period for prenatal hormones affecting brain lateralization, as this is species specific. Secondly, there is no [51] or low to moderate [45] correlation between testosterone levels in the prenatal amniotic fluid and in the umbilical blood collected at birth. Thirdly, a lot of hormones are released during labor, which could influence the hormone levels measured in the umbilical cord. It may be concluded that umbilical cord blood hormone levels are of limited value in the study of the development of lateralization.

5.1.5 2D:4D Finger Length Ratio

The 2D:4D finger length ratio is widely used as an estimate of the level of prenatal testosterone. To calculate the 2D:4D finger length ratio, the length of the right index finger is divided by the length of the right ring finger in humans. The fingers are measured from the middle of the bottom crease of the finger to the tip of the finger. The 2D:4D finger length ratio shows a small but consistent sex difference with smaller ratios in males. High finger length ratios are assumed to be negatively related to prenatal testosterone levels. The idea that the 2D:4D finger length ratio could be an estimate of prenatal testosterone comes from the finding that the Hox genes both control the growth of the finger bones and the sexual differentiation of the gonads [52]. The 2D:4D ratio is not related to adult sex hormone levels (see review [52]).

Since the formulation of this theory in 1998 by Manning and colleagues an enormous amount of research has used the 2D:4D finger length ratio as an estimate of prenatal testosterone exposure. However, the validity of the 2D:4D finger length ratio as an estimate of prenatal testosterone level is heavily debated. Hönekopp et al. [52] list eight indirect lines of evidence to support that the 2D:4D ratio may be a valid marker of prenatal testosterone exposure. However, they also make the reservation that the ratio may not be a very accurate marker, and that more longitudinal studies directly comparing prenatal testosterone with the ratio values later in life are needed. Putz et al. [53] reviewed literature on 2D:4D ratio as a predictor of the degree of expression of sexually dimorphic and other sex-hormone-mediated traits. Subsequently they tested the relationship between 57 traits claimed to be related to the 2D:4D ratio. Of all the traits, they found only a correlation between sexual orientation for both sexes and the left-hand ratio.

Only two studies have directly tested and published the relationship between prenatal testosterone levels and the 2D:4D ratio. Lutchmaya et al. [54] found a significant negative relationship between the testosterone/estradiol ratio measured in amniotic fluid and the 2D:4D finger length ratio of the right hand in 29 participants 2 years old, and this was independent of sex. However, the explained variance of the prenatal testosterone/estradiol ratio on the 2D:4D ratio was only 27%. Moreover, no relationship between prenatal testosterone or estradiol and the 2D:4D finger ratio was found, and none for the left-hand ratio of these hormones. So, this article is generally used as support for the relation between prenatal testosterone and the 2D:4D finger length ratio, but the support is actually weak. Ventura et al. [55] also reported that finger length ratio is negatively correlated with prenatal testosterone, but only in females. In a model taking into account the mother's finger length ratio and prenatal testosterone concentration, the addition of prenatal testosterone explained only about 10% of the variance.

The 2D:4D ratio has also been calculated in nonhuman primates, rodents, and birds. Some sex differences in this ratio were found in gorillas and chimpanzees [56]. In rodents [57, 58] and birds [59–61], prenatal hormone manipulations were conducted by injection in ovo to test the assumption that the digit ratios are under the influence of prenatal testosterone, but the results are not at all consistent. In conclusion, there is some evidence for a relationship between prenatal testosterone and the finger length ratio. However, the explained variance is very low and results are inconclusive and seem mostly applicable to females. Accurate measurement of the digit ratios is difficult as the finger length differences are minute. Measurement of total prenatal exposure to hormones during the sensitive period is difficult too, and has been estimated until now by a single sample within this period. Furthermore, it is likely that a considerable bias in the literature toward results supporting the relevance of digit ratios exists. Therefore we conclude that this index is not a reliable index of prenatal exposure to testosterone.

5.1.6 Intrauterine Position

In several mammalian species in which the mother produces large litter sizes, such as mice, gerbils, and pigs, hormone exposure in utero depends on the position of the embryo relative to its brothers and sisters. Males positioned in between two females develop a different phenotype than males in between brothers, and the same is true for females. This is because male embryos produce testosterone which also reaches their neighboring embryos. There is an extensive body of literature on this phenomenon, demonstrating differences in hormone exposure as well as effects on many aspects of the adult phenotype such as behavior, hormone production, reproduction (mice: [62]; Gerbils: [63]), and urogenital distance (a measurement of masculinisation). To what extend a similar phenomenon occurs in humans is as yet not completely clear. This can be studied by comparing same sex and opposite sex twins. The results suggest that in some aspects females are influenced by their brother's testosterone but more research is needed [64].

The measurement of urogenital distance provides an ideal case for the study of the relation between prenatal testosterone exposure and lateralization. Since urogenital distance is related to prenatal testosterone exposure and the intrauterine position, this simple measurement might be taken as an indirect measurement of prenatal hormone exposure and be related to lateralization. This approach has to our knowledge not yet been applied.

5.1.7 Hormone-Mediated Maternal Effects

In many animal species, ranging from insects to humans, mothers bestow their offspring with maternal steroid hormones. This occurs prenatally, either by depositing hormones in the egg, or by transferring hormones during pregnancy via the placenta. Mothers differ in the amount of hormones they pass on to their offspring, which often depends on the environment experienced by the mother during reproduction (for a review see [49]). This potentially offers the possibility of manipulating the environment and to assess the effect of these maternal hormones on lateralization of the offspring. However, since the hormone exposure of the embryo is not simply a reflection of the hormone concentrations circulating in the mother (see above and [65]), this approach has clear limitations. In humans the relationships between maternal hormones and lateralization of offspring can be studied in mothers with deviating testosterone concentrations (e.g. congenital adrenal hypoplasia (CAH) mothers, leading to an excess of testosterone) or from offspring lacking the androgen receptor. The potential problem here is that these data come from pathological cases and might be confounded by factors other than exposure to testosterone only (e.g. in de case of CAH the decrease in glucocorticoid production), or do not reflect normal variation. However, direct manipulation of embryonic exposure to these hormones is to some extent possible as will be discussed below.

5.2 Experimental Approaches: Manipulation of Hormone Levels

There is currently a wide range of methods of manipulating hormone concentrations in animals, including humans. In all cases, however, it is of crucial importance to scale the treatment against normal variation in hormone concentrations, in order to prevent levels that do not occur in a normal population. Moreover, hormone concentrations may have nonlinear effects, so that dose response curves may be necessary [66].

5.2.1 Manipulation of Hormone Levels in the Embryo

The best way to manipulate prenatal exposure to hormones is to manipulate directly hormone concentrations in the embryo. In placental animals this always affects the mother too, so the effect on the offspring may be confounded by maternal effects. Moreover, it is known that the fetus and mother communicate by means of hormones and may adjust their signals to each other [67], which makes the interpretation of the effect of the manipulation even more difficult. Therefore, a flourishing field of research makes use of egg laying species, especially birds, in which prenatal hormones are manipulated. As the embryo develops outside the mother's body in a sealed environment, hormone exposure of the embryo can be measured and manipulated without interfering with the mother. Eggs can be injected with hormones into the yolk, or even into the embryo directly in a later stage of development. This can be easily applied by cleaning the egg shell with alcohol, drilling a small hole in the shell, and injecting the hormone solution into the yolk or embryo with the help of candling the egg for visual guidance. Many studies have applied this technique in the framework of sexual differentiation, using extremely high dosages of the hormone. More recently much lower dosages within the physiological range of the species have been shown to affect many aspects of the phenotype of the chick (for a review of gonadal hormones see [68], for stress hormones [69]). Only very few studies have examined the effects of the manipulation in ovo on lateralization, so far with inconsistent results [38, 70, 71].

Recently it has been discovered that the embryo, even before the development of active gonads, is capable of metabolizing hormones, such as those received from the mother (or an injection) [37]. This warrants further research in order to establish the effective dosage to which the embryo is exposed in different phases of development.

5.2.2 Manipulation of Hormone Levels of the Mother

Another, but less preferred way (since hormone levels in the embryo do not necessarily reflect those in the mother: see above), is to manipulate circulating hormone levels in the mother during pregnancy or egg production. In many cases elevated hormone levels in the mother may elevate those in the embryo [72], as these elevated levels are often so high that maternal regulation cannot prevent the hormones from reaching the embryo. Although this may lead to other correlated maternal effects (for example, stress hormones inhibit the nutrient flow to the embryo, see [69]), such studies have been applied frequently and suggest potential flexibility and hormonal pathways for regulating brain and behavioral lateralization.

Manipulating hormones in reproductively active females can simply be done by injection or implantation techniques, or mixing the hormone through the food or water. Injection leads only to brief elevations as the hormone quickly degrades (half time value of steroids is about 15 min), unless the hormone is modified to make it more stable (for example testosterone propionate). A solution to this problem is the subcutaneous implantation of controlled release delivery systems, such as silicon tubers or mini-pumps filled with hormone (solution), for which only local anesthesia may be necessary. The dynamics of the released hormones should be checked by blood sampling. In case of cortisol or corticosterone, such implants often have a paradoxical effect in that after a short peak blood plasma concentrations decrease substantially. This is probably caused by feedback mechanisms in the animal, as a continuous high level of stress hormones is detrimental. This effect can be avoided by mini-pumps that are programmed such that the release or hormones is pulsative, or by simply mixing the hormone with the food twice a day. Finally, plasters that release hormone through the skin can also be applied.

All these techniques elevate circulating levels. Preferably, the complementary experiment in which hormone levels are lowered should also be conducted. This can be done by applying blockers of either specific enzymes that convert hormones to other components (such as aromatase converting testosterone to estradiol), or by applying receptor blockers. The problem of the first is that blocking of conversion results in higher concentrations of the hormone that would normally be converted. The problem of the latter is that the extent of their effectiveness (how many receptors are blocked for how long) is difficult to establish; this needs injections with labeled hormones and sacrificing of the embryos to estimate the amount of bound receptors.

6 The Study of Postnatal Effects

The methods discussed under Sects. 5.1.3 (blood, urine and saliva) and 5.2.2 (hormone manipulation) can also be applied to study the postnatal hormonal effects on lateralization of brain and behavior. Obviously, estimating actual hormone exposure is much easier when the subject is not confined in the mother or in the egg. Fortunately in many species the timing of the sensitive period for organizing effects of steroid hormones extends to early postnatal life [73]. In small animals the sampling of enough blood for hormone analyses without sacrificing the individual may be a problem.

6.1 Correlational Methods

For correlational research the postnatal phase offers some additional possibilities compared with studying the prenatal phase. First, additional sensitive phases for organizing effects of hormones on the brain and behavior may be detected by studying different age classes. For example, currently evidence is accumulating that organizing effects also occur during puberty in mammals. Second, plasticity in the relationships between hormones and lateralization might be detected by studying changes over time, which could also indicate reversible activating effects of the hormone. This can be analyzed by studying (changes in) direction and strength of lateralization in relation to changes in hormone levels, such as the natural occurring hormonal fluctuations during 24 h (being especially pronounced in cortisol and corticosterone, but also in gonadal hormones), during the estrous cycle in females (e.g. [74]), or the season of the year (for example, the change from nonreproductive to reproductive modes).

6.2 Experimental Methods: Manipulation of Hormone Levels

Since correlational evidence is by no means evidence for causal relationships, experimental approaches are indispensable. For this the methods discussed in Sect. 5.2.2 can also be applied to offspring postnatally. For studying short-lasting activating effects an additional method can be used, being nose sprays that can even be used on human subjects. Nose sprays have mild and short-lasting effects. This method has been applied in humans for testing hormonal effects on competitive or altruistic behavior [75, 76], but to our knowledge lateralization has not been studied with this method. In humans we cannot experimentally test longer lasting treatment effects, or organizing effects, by manipulation of sex hormones in participants without medical indication. Treatment with hormones or hormone blockers based on medical indication can be used in humans, although again confounding influence of the medical history might be in play here. A very interesting case for, and perhaps the only way for the experimental study of organizing effects of steroid hormones on lateralization in humans, is the hormonal treatment of Gender Dysphoria. Therefore we will discuss this more extensively.

6.2.1 Hormone Treatment in Persons with Gender Dysphoria

Persons with Gender Dysphoria experience a mismatch between their biological sex and their gender identity. They have the feeling they are trapped in the wrong body. In some cases, transgender individuals chose hormone treatment, sometimes together with surgery, to improve the match between their body and experienced gender.

Because substantial hormone manipulation for scientific purposes only is not allowed in humans, there is always a confounding effect of the medical condition. However, apart from their Gender Dysphoria, most transgender persons are mentally and physically fit. This is even a requirement for treatment. As the transgender clinics offer careful preparation and guidance, the participants are followed longitudinally for their treatment. Often they are enthusiastic to take part in scientific research in order to increase knowledge about their condition. The longitudinal study of the effects of cross-sex hormone treatment is therefore feasible. In addition, the age of transgenders presenting at a gender clinic differs greatly, from very young children to adults, enabling research at varying ages and treatment stages in order to study potential sensitive windows for hormonal effects on lateralization.

Two different hormone treatments are applied in the treatment of transgender individuals: puberty suppression and cross-sex hormone treatment. Puberty suppression is applied from the onset of puberty and the emergence of secondary sex characteristics and gives time for a decision to be made about further treatment or not. From about 18 years onwards, they can subsequently start cross-sex hormone treatment if they wish to do so and are found to be eligible. Adults with Gender Dysphoria may also be subject to cross-sex hormone treatment if found to be eligible. Before any treatment starts, the patients have several sessions with a psychologist and psychiatrist, they undergo numerous psychological tests and medical examinations to ensure the state of their mental and physical health, hormone levels are measured and stage of puberty is assessed by a doctor in case of young patients. This provides a wealth of interesting data for studying the direct effects of steroid hormones on lateralization of brain and behavior over time in humans. A limitation may be that it is unknown if research on persons with Gender Dysphoria translates to the general population.

6.2.2 Puberty Suppression

Normally, puberty is initiated by the hypothalamus, which releases gonadotropin-releasing hormones (GnRH). GnRH stimulates in both sexes the pituitary to release LH and FSH in order to stimulate the gonads to produce sex hormones, inducing pubertal changes in the body. Naturally, GnRH levels fluctuate during the day. In puberty suppression GnRH agonists are administered at a continuous level. After the initial stimulation this will actually block the release of LH and FSH and thereby the gonadal hormone production [77, 78]. Puberty suppression is started from an age of 12 years onwards, provided that the transgender child has experienced puberty for at least 1 year and is in Tanner's stage 2 or 3 [79, 80].

6.2.3 Cross-Sex Hormone Treatment

Cross-sex hormone treatment may follow after puberty suppression in adolescents, or it may be applied in adults who did not receive puberty suppression. In adolescents, cross-sex hormone treatment starts with the induction of puberty of the desired sex.

Feminizing puberty is induced in male-to-female transgenders by orally administering estradiol. Masculinizing puberty is induced in female-to-male transgenders by intramuscularly administering testosterone esters (a stable form of testosterone) [78]. In adolescents the dose is increased gradually over a period of 2 years. Thereafter the adolescents receive the same cross-sex hormone treatment as adult transgenders.

Adult cross-sex hormone treatment in female-to-male transgenders is relatively simple. They receive testosterone orally, via the skin (transdermal) or injection. The dose depends on the way of administration, but is within the normal range for men. Male-to-female transgenders receive estrogens orally, transdermally, or by injection. Again, the dose depends on the way of administration, but is within the normal range for women. In addition, male-to-female transgenders also receive antiandrogens and a GnRH agonist (leading to androgen suppression after some time) to suppress their endogenous sex hormones. For information on which hormone supplements and doses are used in clinical practice, see the Guidelines on the Endocrine Treatment of Transsexuals [78].

7 Different Levels to Study the Relationship Between Hormones and Lateralization in Vivo

The aim of this section is twofold. First, we would like to point out that the relation between hormone exposure and lateralization can and should be studied at different levels of brain and behavior, and that these have to be made explicit to avoid confusion. Moreover, these different levels should be studied preferably in relation to each other and in the same species. Second, we like to show the gaps in our knowledge at these different levels, and explain future perspectives in order to bring the field forward.

The different ways and levels at which hormones can affect lateralization of brain and behavior are schematically depicted in Fig. 3. We will explain the different levels of the scheme and their relationships below.

7.1 Hormones

As explained in the last paragraph of Sect. 2, steroid hormones can have organizing and activating effects. Organizing effects act among others on the anatomy of the brain, including structural lateralization. Such effects primarily happen during sensitive periods in development, for example prenatally or in puberty. As depicted in the scheme organizing effects may prepare the brain for activating effects, for example by inducing the development of the appropriate receptors for the activating effects. A strict dichotomy between organizing and activating effects is debatable. For example, brain activation *can* result in structural changes, since neuronal connections are strengthened by lasting activation. Also,

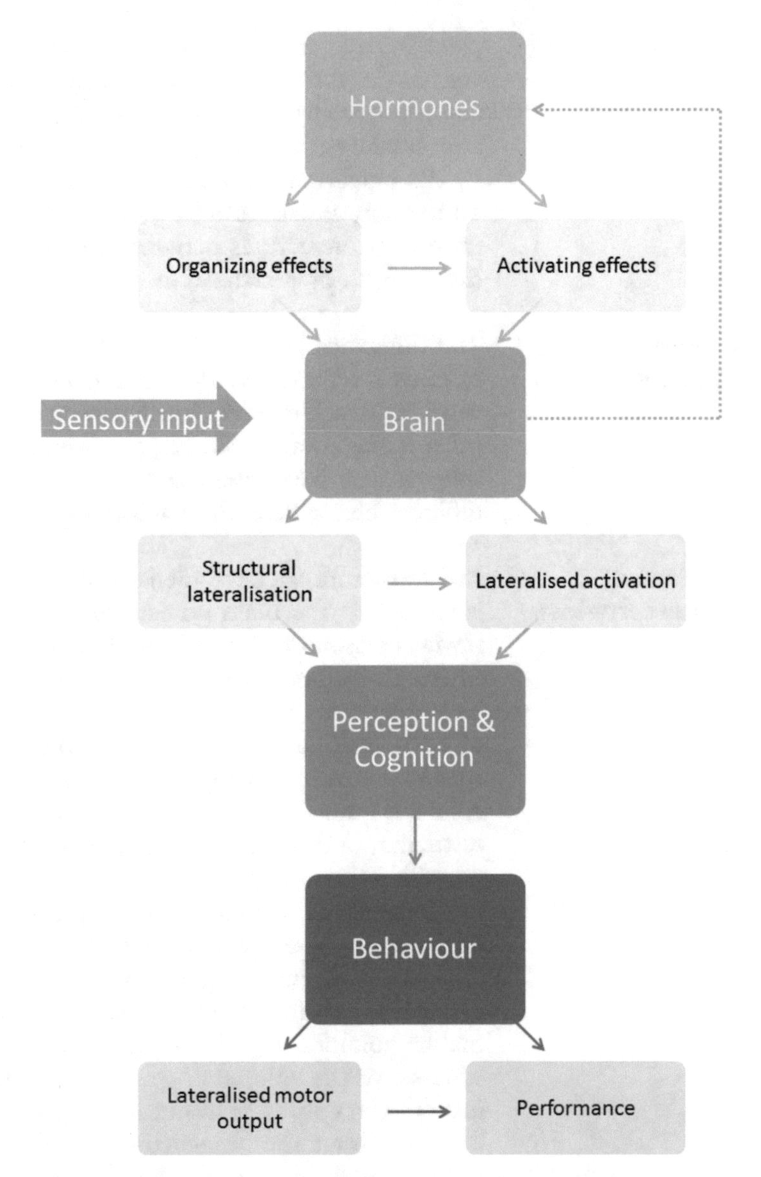

Fig. 3 Schematic and simplified illustration on how and at which levels hormones can act on the lateralization of the brain, and thereby its functioning in the form of perception, cognition, and behavior

seasonal profiles in testosterone in adult animals can lead to seasonal changes in the size and structure of adult brain areas, as well described for the lateralized song system in songbirds [81, 82]. In any case, both organizational and activating effects impact on the brain, the former affecting brain anatomy and the latter primarily affecting brain activation. By doing so, lateralization is induced if either of four requirements are fulfilled: (1) The receptor distribution for these hormones, or for downstream components in the pathway, is lateralized [83]; (2) The hormone production or conversion in the brain is lateralized; (3) The hormone acts on latently lateralized brain structures that become expressed only after hormone exposure, or that become even stronger lateralized by the performance of relevant behavior; (4) The hormone acts on the corpus callosum (if present, depending on the species), or on other connections between the hemispheres, affecting mainly the strength of lateralization.

7.2 Brain Lateralization

The difference between brain anatomy and brain activation seems obvious: it is the difference between physical brain structure and neuronal activity. However, how anatomy and activation relate to each other is not always clear. For example, anatomically identical hemispheres may have lateralized activation. Vice versa, anatomically different hemispheres do not necessarily show lateralized activation.

7.2.1 Structural Lateralization of the Brain

Brain structure can be studied in different ways. The most direct way is to look at the brain post-mortem, or even during surgery. For obvious reasons, this is easier in nonhuman animals than in humans. Alternatively, brain structure can be investigated with MRI. Other techniques are PET and DTI—techniques that can measure aspects of both brain anatomy and activation. During a scan the subject should not move, so an MRI scan in animals is possible only if the animal is trained not to move in the scanner, if the animal is tightly restricted, or if the animal is sedated. The latter option is the most prevailing one.

Magnetic Resonance Imaging (MRI) computes a 3D-image of the brain, based on information on the type and location of tissues like white matter, gray matter and cerebrospinal fluid. In lateralization research MRI can be used to compare white and gray matter volumes, the size of the surface area between the hemispheres, or the volume of specific brain areas, like the corpus callosum. The corpus callosum is especially of interest when testing theories about the influence of testosterone on the strength of brain lateralization (e.g. the sexual differentiation theory, the Callosal Theory). It is also relevant to investigate the influence of steroid hormones on lateralized development of sexually dimorphic brain areas. The bed nucleus of the stria terminalis (BNST) and the interstitial nucleus of the anterior hypothalamus (INAH, specifically the INAH-2 and INAH-3) are sexually dimorphic in humans and rodents, while hormone dependent areas controlling song are sexually dimorphic in song birds [84]. Interestingly, these brain areas are also part of lateralized networks which may be responsible for sex differences in lateralization [82]. Similarly, in the domestic chick the organization of the thalamofugal visual projections is sexually dimorphic [85].

Diffusion Tensor Imaging (DTI) may be used to study lateralization of brain networks. It is a technique to track the neural connections in the brain from MRI data. This technique is relevant in lateralization research as it can show the connections between the left and right hemisphere via the Corpus Callosum. One can also visualize and quantify the connections within hemispheres and how they differ between hemispheres.

Positron Emission Tomography (PET) may be used to study lateralized differences in receptor densities, e.g. of serotonin [86] or dopamine [87]. Therefore, PET offers a specific opportunity for studies investigating the relation between hormones and lateralization. It uses a bioactive molecule labeled with a positron-emitting radionuclide (tracer), which is injected in the blood. The tracer emits gamma-rays which are detected by the PET-scanner. The output of a PET scan is a 3D representation of the locations where the molecule binds. By labeling the steroid hormone, receptor densities in the brain can be measured with PET. In humans, the use of PET scans is mostly restricted to medical studies as a PET scan involves exposure to nuclear radiation. PET is more often used in animal research.

7.2.2 Functional Lateralization of the Brain

Functional activation of the brain is typically studied when the brain is processing information. Brain activation is commonly investigated in humans, but increasingly in animals too. Lateralized brain activation can be measured noninvasively with fMRI, fTCD, PET, and EEG. It can even be manipulated by TMS [88, 89].

Functional Magnetic Resonance Imaging (fMRI) is based on the assumption that the oxygen level in the blood is directly linked to neural activity. The oxygen level in the blood is measured with the Blood Oxygen Level Dependent (BOLD) response, which can be determined when applying a magnetic field to the brain. fMRI makes a 3D-image of the brain and shows the brain regions that are active during a specific task. It is possible to perform an fMRI-scan with other animal species than humans, if the animal is trained to lie motionless in a scanner and to look at a screen, or if the animal is sedated and receives stimuli that can still be perceived such as auditory or olfactory input. The task is then limited to processing of sensory input or measurement of resting state (see also Chap. 14 by Hopkins and Phillips).

PET can also be used to measure aspects of brain activation when radioactive labeled tracers are used that trace metabolic activity, such as oxygen or glucose [90]. The oxygen-tracer has a very short half-life, therefore the tracer is more often linked to glucose. In humans, functional PET-scans are not used for lateralization research (fMRI is the preferred technique). In animals, PET can be a preferred option to study animals that first perform a task and subsequently are scanned when sedated. The activation is measurable for a longer time than is the case with fMRI (depending on the half-life of the label).

Functional Transcranial Doppler sonography (fTCD) measures task related differences in blood supply to the left and right hemisphere, i.e. the blood flow velocity in the left and right mid-cerebral arteries during a cognitive task, relative to the blood flow velocity in each during the baseline. For example, if the left hemisphere is dominant for language, the blood flow velocity will increase more in the left than in the right hemisphere during a language task, relative to the baseline. This method is based on the assumption that blood flow velocity increases after a hemisphere becomes more active. fTCD measures both the strength and direction (left or right hemisphere) of lateralization. It is a validated, cheap, and easy method to assess and the device is portable [91, 92]. A cap with two probes is placed on the head of a participant. Both probes emit a high-pitched sound signal that reflects off the blood cells in the left and right middle cerebral artery. The middle cerebral artery supplies the major part of the cortex with blood. The faster the blood flows, the bigger the Doppler shift of the reflected signal. Movement of the probes impairs the measurement and fTCD is therefore only performed in humans.

Electroencephalography (EEG) and **Magnetoencephalography** (MEG): EEG is a technique that measures the electrical activity of the brain, via electrodes on the scalp. MEG is a related technique but uses magnetic sensors for the recording (see also Chap. 13 by Mazza and Pagano). The main advantage of EEG and MEG is that they measure brain activity with a temporal accuracy of typically 1 ms. The average activation after presentation of a series of stimuli is called the "event-related potential" (ERP) [93, 94]. These ERP's can be compared for different brain regions. With regard to brain lateralization, the ERP's can be compared between the left and the right hemisphere. Source localization software can be used to look for lateralized differences in the dynamics of activity when processing a stimulus. The change in strength of activation at specific 3D-locations is calculated, as well as the localization of the activation source over time. Thus, one can follow the sequential activation of lateral brain areas with great temporal accuracy. However, the spatial accuracy is less precise than with fMRI.

Transcranial magnetic stimulation (TMS) may result in an excitatory or an inhibitory effect on a targeted brain region by applying a conditioning stimulus sequence with a magnetic coil. The magnetic coil is placed over a specific region of the cortex. The magnetic field induces electric currents in the brain region under the coil, for example the cortex or cerebellum. Inhibition and excitation of the specific brain area depend on the characteristics of the stimulus sequence. TMS can be applied during task performance to study the involvement of these brain areas in task-related information processing. Until now TMS has rarely been used in lateralization research that evaluates hormones, but this technique has the potential to study the effects of hormones on lateralized differences in the behavioral output during a task when TMS is applied to homologous areas of the brain (see also Chap. 12 by Cataneo).

Immediate early genes (IEGs) are expressed when neurons are activated. Products of IEGs can be stained, such as the well-known c-fos protein. C-fos is rapidly and transiently expressed in many brain areas in response to a wide range of stimuli. Thus, c-fos and other IEG products can be used to detect neural activity. The procedure is as follows: an animal is subjected to a specific environmental condition or stimulus, after which it is decapitated, and the brain slices are stained. This way a detailed functional map of the brain regions activated after an environmental trigger is made and lateralized activation can be measured. For example, based on c-fos expression, lateralization of auditory input has been found [82].

The output from all of these methods (see also Chap. 10 by Patton et al.) may be used to calculate a laterality index that represents brain lateralization. In addition, lateralization can also be measured at different levels, as we will explain in the next paragraphs.

7.3 Sensory Input

Sensory input can be given to a specific hemisphere and can in this way be used to investigate brain lateralization, either by monitoring brain activity or behavioral output. For example, visual input can be presented to the left or right visual field, or auditory input to the left or right ear. The tests are based on the assumption that sensory input is differently processed in the left or right hemisphere, resulting in a different perception of the stimuli, which can be estimated with ERP, neuroimagingor behavioral output. Theoretically, tests could be designed for all sensory modalities. In humans, frequently applied tests are dichotic listening, visual half-field tasks, and—as a special case of the latter—chimeric face recognition. In animals, the left and right eye, ear, or nostril can be alternately closed off from input and the effect on behavioral output can be determined (see Chap. 3 by Rogers in this volume). In addition, visual scanning or ear movement can be registered after offering different types of stimuli.

Dichotic listening: this is a test used in humans presenting different auditory stimuli—generally brief words—to both ears simultaneously. The participant is instructed to report as many stimuli as possible. Participants typically report more stimuli presented to the right ear. Auditory information is passed to the contralateral hemisphere to be processed. As the left hemisphere is specialized in language processing in most people, the words presented to the right ear are generally processed faster. This asymmetry results in a right ear advantage, which is interpreted as evidence that language in that person is lateralized to the left hemisphere.

Visual half-field paradigms: visual information in the left visual field projects to and is initially processed in the right hemisphere and vice versa. Visual half-field paradigms make use of this knowledge. The participant has to sit exactly in front of a computer screen and stimuli are presented at the left or right side of a fixation cross in the center. Reaction time for stimuli presented in the left visual field is compared to reaction time for stimuli presented in the right visual field, and used to measure asymmetry in visual perception. Similarly, asymmetry of language processing can be measured if letters or words are used as stimuli.

Chimeric faces: this method is a special case of the visual half-field paradigm, but aims to measure visual processing at a higher level of cognition, that is the perception of facial emotion. The participant is placed in front of a computer screen. Two vertically arranged pictures are depicted in the middle of the screen. One half of each picture shows a neutral face, and the other half shows an emotional face. The two pictures are mirrored, so in one picture the emotion is shown on the left side of the face (in the left visual field) and in the other picture the emotion is shown in the right visual field. The task of the participant is to say which picture looks more emotional. The idea is that facial emotion processing is generally lateralized to the right hemisphere, and that the picture with emotion in the left visual field (which is processed by the right hemisphere) looks more emotional.

7.4 Perception and Cognition

Brain lateralization can directly influence behavior, or via perception or cognition. It is impossible to measure perception or cognition directly, so the only way is to assess it indirectly via behavior or cognitive performance, discussed below. This is important to realize, as lack of a lateralized response does not necessarily imply lack of lateralized cognition or perception.

7.5 Behavior

Behavior can be a direct result of brain lateralization, but lateralized behavior may also be a direct consequence of asymmetrical sensory input. For example, a tendency for an animal to flee to one specific direction may simply be due to asymmetry in the environment or test apparatus. In such cases care should be taken to test the animal in completely symmetrical environments, or test the animal in balanced directions. For example, testing the animal in a corridor both from south to north, and north to south, so that the right side also becomes the left side, and vice versa. Thus, when testing lateralized behavior, one should make sure the sensory input is symmetrical, or use a balanced design with respect to the symmetry of the environment.

Two types of behavioral output can be distinguished: lateralized motor output and performance. Lateralized motor output refers to asymmetry in behavior, for example, the lateralization of limb use is often quantified as the number of times the right limb is used, minus that of the left one, divided by the total number of limb movements. This is called the lateralization index. Performance refers to a measurement of quality. Performance is measured as how well an individual performs a certain task, mostly a motor task or a cognitive task, like spatial orientation or language fluency.

7.5.1 Lateralized Motor Output

In some cases the relationship between lateralized brain activation and lateralized behavior is clear. This is, for example, the case in limb movement, where lateralized activation in the motor cortex precedes movement of the limb at the contralateral side of the body. In other cases this is less clear. For example, the asymmetry in the visual pathway in the domestic chick disappears in the first weeks of development, whereas in older birds the function of the left and right eye still differ (discussed in Chap. 4 of Ref. 95). In nonhuman animals, most lateralization research is about lateralized behavior, as lateralized brain activation is difficult to assess.

One of the remarkable characteristics of motor behavior among vertebrates is that there may be preferences for left or right. Examples are the flight response and limb preference. Most studies focus on preferences in limb use, in which the direction (i.e. which limb is used most often) and the strength of the preference (i.e. the difference in frequency between the left and right limb use irrespective of the direction) is investigated. Much less frequently, and perhaps exclusively in humans, asymmetry in limb skill can also be analyzed. "Limb skill" refers to the difference in performance between the hands. For humans there are several tests to test this, such as the pegboard task, which requires moving pegs as quickly as possible from one hole to another (see Chap. 5 by Forrester in this volume). It is likely that evolution has selected asymmetries in skill rather than preference. Other lateralized motoric responses concern the direction of turning in a symmetrical environment, fleeing for a (simulated) predator, or ear movements in species where both ears can move independently, such as in horses. However, in the case of ear movement, as well as in many other orientation responses, the asymmetry in behavior represents lateralization in perception rather than lateralization in motor responses.

The lateralization index is often calculated to quantify lateralized motor output. The lateralization index describes both direction and strength of lateralization. It should be realized that this ratio requires sufficient data points per individual to be reliable. For measuring lateralization at the population level, one could in theory measure many individuals of the same population only once. However, one should realize that lateralization at the population level is different from that of the individual level. This issue relates to the difference between lateralization at the level of the individual or the population. At the population level no disparity from 50% left and right may be present, whereas within this population some individuals may be strongly lateralized to the left and others to the right. Therefore one should always measure individual lateralization. This does not mean that population lateralization is not interesting. From an evolutionary point of view a skew in population lateralization is very interesting and several hypotheses have been put forward to explain such biases [96], but this is outside the scope of this chapter.

7.5.2 Performance

The relation between brain lateralization and behavior is not always observable [97–100], for example in language. Language is generally a strongly lateralized function, yet we do not see this in the behavioral output. The same holds for visuospatial orientation. These are the two most studied lateralized cognitive functions in humans. A common mistake is to assume brain lateralization and cognitive performance are directly related. This assumption is often based on indirect evidence. An example is the following. Boys generally perform better on a visuospatial task than girls. Boys are also generally more strongly lateralized to the right hemisphere for visuospatial function. Therefore one may conclude that stronger lateralization results in better performance and that performance is a measurement of lateralization. However, the relationship between lateralization and performance is actually not clear.

8 Opportunities for Future Research

In this chapter we have presented methods to measure steroid hormones, lateralization and their relationships. Lateralization can be measured at different levels, and on every level many lateralized functions can be investigated. To this end one can select from multiple study species. From the overview it is apparent that many techniques that are developed for humans may be used in other animals as well. Animal studies offer crucial experimental manipulations that are unacceptable to be used in humans for ethical reasons, such as hormone manipulation and injection with radioactive labeled tracers. On the other hand, human studies allow for extensive analyses of complex cognition, which is more difficult to study in animals. The use of similar procedures in all species would enhance the insight in the possible evolutionary explanation for the emergence of lateralized functions.

With respect to hormones, most research until now has focused on testosterone. This has been driven by three theories that have been put forward, with predictions of organizing effects of prenatal testosterone on direction and/or strength of lateralization. There are no theories on activating effects of testosterone on brain lateralization, nor are there theories on interactions between organizing and activating effects. Steroid hormones other than testosterone have received less or no attention, and theories for their effects have not been developed yet. Moreover, the possibility that specific combinations of hormones affect lateralization has not been tested yet. Furthermore, the measurement of the differences in hormone exposure between (homologous) brain areas due to differences in receptor densities or enzyme distribution is still underdeveloped. It is also important to realize that the underlying hormonal mechanism is not necessarily the same for different lateralized functions. More accurate and complete measurements of the underlying endocrine systems, including potentially asymmetrical local conversion of hormones and receptor densities, are needed. To make it even more complex, the underlying mechanism is not necessarily the same for different species. For example, in birds, estrogens are crucial for sexual differentiation in females, whereas in mammals testosterone has this role in males. These species also differ in the presence of the corpus callosum. In addition, more attention should be given to manipulations within the physiological range of the study species. The methods to investigate all these issues are available, including research tools to measure or manipulate hormone levels that have not been used in lateralization research yet, like urogenital distance, hormonal nose-sprays, and puberty suppression and cross-sex hormone treatment in persons with Gender Dysphoria. So hopefully these issues will be investigated in the near future.

At the level of the brain, structural asymmetries can be detected, but their development is poorly understood. The function of these structural asymmetries is also not clear. Brain lateralization is not a unitary characteristic of the entire brain. There are numerous lateralized functions and the relation between them is unclear. For example, the directions and strengths of brain lateralization for language, visuospatial function, and emotion processing in humans are not at all strongly correlated [96], showing not only substantial within but also between individual variation. Moreover there is an endless amount of other functions that could possibly be lateralized, and one brain area can be part of different networks. For example, it can be part of a lateralized network, while the same area can be part of another (global) network spreading over two hemispheres. The relationship between different lateralized functions should be investigated within study species. Interestingly, human research has increasing possibilities for studying lateralization of functional brain networks, like DTI and source localization in EEG or MEG. Using these techniques—also in different species—is an opportunity for future research.

With respect to lateralized perception and cognition, we have stressed that these cannot be measured directly but have to be inferred from evoked brain responses, lateralized motor output, or performance measures. The relationship between such output measures, including behavioral measures, and brain lateralization is unlikely to be unitary and linear. For example, handedness in humans correlates only moderately with lateralized activation of the homologous motor areas in the brain. Thus, the challenge is to explore multifactorial and nonlinear relationships.

To conclude, after decades of research the relationships between steroid hormones and lateralization of brain and behavior are still unclear. This may in part be due to lack of generalizability between studies, or limitations of hormone manipulations or measurements. We especially want to highlight the need for an interdisciplinary approach, combining up to date techniques in endocrinology with those in neuroimaging, cognition, and behavioral measurements. There should also be more cross-talk between human and nonhuman studies. Focusing on one specific lateralized trait, such as limb preference or the visual system, may be of help when generalizing outcomes between different species. Comparative studies may not only generate new insights and unexpected mechanisms but also shed light on Darwinian function and evolution of lateralization, which in turn may fertilize mechanistic studies. Such comparative studies may reveal general underlying principles. We now have the knowledge and the techniques to bring the field a step further.

References

1. Pfannkuche KA, Bouma A, Groothuis TGG (2009) Does testosterone affect lateralization of brain and behaviour? A meta-analysis in humans and other animal species. Philos Trans R Soc B Biol Sci 364:929–942. doi:10.1098/rstb.2008.0282
2. Cooke B, Hegstrom CD, Villeneuve LS, Breedlove SM (1998) Sexual differentiation of the vertebrate brain: principles and mechanisms. Front Neuroendocrinol 19:323–362. doi:10.1006/frne.1998.0171
3. Schaafsma SM, Riedstra BJ, Pfannkuche KA et al (2009) Epigenesis of behavioural lateralization in humans and other animals. Philos Trans R Soc Lond B Biol Sci 364:915–927. doi:10.1098/rstb.2008.0244
4. Hardyck C, Petrinovich LF (1977) Left-handedness. Psychol Bull 84:385–404. doi:10.1037/0033-2909.84.3.385
5. Jordan HE (1911) The inheritance of left-handedness. Am Breeders Mag 19–29: 113–124
6. Annett M (1972) The distribution of manual asymmetry. Br J Psychol 63:343–358. doi:10.1111/j.2044-8295.1972.tb01282.x
7. Annett M (1985) Left, right, hand and brain: the right shift theory. Erlbaum, VA
8. Annett M (2002) Handedness and brain asymmetry: the right shift theory. Psychology Press, East Sussex
9. McManus IA (1985) Handedness, language dominance and aphasia: a genetic model. http://journals.cambridge.org/action/displayFulltext?type=1&fid=7057904&jid=PMS&volumeId=8&issueId=-1&aid=7057900&bodyId=&membershipNumber=&societyETOCSession=. Accessed 4 Jan 2016
10. McManus IA (1999) Handedness, cerebral lateralization and the evolution of language. In: Corballis MC, Lea SEG (eds) The descent of mind: psychological perspectives on homonid evolution. Oxford University Press, Oxford, pp 194–217
11. Klar AJS (1996) A single locus, RGHT, specifies preference for hand utilization in humans. Cold Spring Harb Symp Quant Biol 61:59–65. doi:10.1101/SQB.1996.061.01.009
12. Somers M, Ophoff RA, Aukes MF et al (2015) Linkage analysis in a Dutch population isolate shows no major gene for left-handedness or atypical language lateralization. J Neurosci 35:8730–8736. doi:10.1523/JNEUROSCI.3287-14.2015
13. Corballis MC (2014) Left brain, right brain: facts and fantasies. PLoS Biol 12:e1001767. doi:10.1371/journal.pbio.1001767
14. Lien Y-J, Chen WJ, Hsiao P-C, Tsuang H-C (2015) Estimation of heritability for varied indexes of handedness. Laterality 20:469–482. doi:10.1080/1357650X.2014.1000920

15. Arning L, Ocklenburg S, Schulz S et al (2015) Handedness and the X chromosome: the role of androgen receptor CAG-repeat length. Sci Rep 5:8325. doi:10.1038/srep08325
16. Nelson RJ (2005) An introduction to behavioral endocrinology, 3rd edn. Sinauer Associates, Sunderland
17. Adkins-Regan E (2005) Hormones and animal social behavior. Princeton University Press, Princeton
18. Seredynski AL, Balthazart J, Ball GF, Cornil CA (2015) Estrogen receptor β activation rapidly modulates male sexual motivation through the transactivation of metabotropic glutamate receptor 1a. J Neurosci 35:13110–13123. doi:10.1523/JNEUROSCI.2056-15.2015
19. Eberling P, Koivisto VA (1994) Physiological importance of dehydroepiandrosterone. Lancet 343:1479–1481. doi:10.1016/S0140-6736(94)92587-9
20. Nair KS, Rizza RA, O'Brien P et al (2006) DHEA in elderly women and DHEA or testosterone in elderly men. N Engl J Med 355:1647–1659. doi:10.1056/NEJMoa054629
21. Soma KK (2006) Testosterone and aggression: berthold, birds and beyond. J Neuroendocrinol 18:543–551. doi:10.1111/j.1365-2826.2006.01440.x
22. Weinstock M (2007) Gender differences in the effects of prenatal stress on brain development and behaviour. Neurochem Res 32:1730–1740. doi:10.1007/s11064-007-9339-4
23. Anderson DK, Rhees RW, Fleming DE (1985) Effects of prenatal stress on differentiation of the sexually dimorphic nucleus of the preoptic area (SDN-POA) of the rat brain. Brain Res 332:113–118. doi:10.1016/0006-8993(85)90394-4
24. Kaiser S, Kruijver FPM, Swaab DF, Sachser N (2003) Early social stress in female guinea pigs induces a masculinization of adult behavior and corresponding changes in brain and neuroendocrine function. Behav Brain Res 144:199–210. doi:10.1016/S0166-4328(03)00077-9
25. Arnold AP, Breedlove SM (1985) Organizational and activational effects of sex steroids on brain and behavior: a reanalysis. Horm Behav 19:469–498. doi:10.1016/0018-506X(85)90042-X
26. McCarthy MM, Arnold AP (2011) Reframing sexual differentiation of the brain. Nat Neurosci 14:677–683. doi:10.1038/nn.2834
27. Romeo RD (2003) Puberty: a period of both organizational and activational effects of steroid hormones on neurobehavioural development. J Neuroendocrinol 15:1185–1192. doi:10.1111/j.1365-2826.2003.01106.x
28. Hines M, Shipley C (1984) Prenatal exposure to diethylstilbestrol (DES) and the development of sexually dimorphic cognitive abilities and cerebral lateralization. Dev Psychol 20:81–94. doi:10.1037/0012-1649.20.1.81
29. Geschwind N, Galaburda AM (1985) Cerebral lateralization: biological mechanisms, associations, and pathology: I. A hypothesis and a program for research. Arch Neurol 42:428. doi:10.1001/archneur.1985.04060050026008
30. Witelson SF, Nowakowski RS (1991) Left out axons make men right: a hypothesis for the origin of handedness and functional asymmetry. Neuropsychologia 29:327–333
31. Henriksen R, Rettenbacher S, Groothuis TGG (2013) Maternal corticosterone elevation during egg formation in chickens (Gallus gallus domesticus) influences offspring traits, partly via prenatal undernutrition. Gen Comp Endocrinol 191:83–91. doi:10.1016/j.ygcen.2013.05.028
32. Freire R, van Dort S, Rogers LJ (2006) Pre- and post-hatching effects of corticosterone treatment on behavior of the domestic chick. Horm Behav 49:157–165. doi:10.1016/j.yhbeh.2005.05.015
33. Shirtcliff EA, Granger DA, Schwartz E, Curran MJ (2001) Use of salivary biomarkers in biobehavioral research: cotton-based sample collection methods can interfere with salivary immunoassay results. Psychoneuroendocrinology 26:165–173
34. Goymann W (2005) Noninvasive monitoring of hormones in bird droppings: physiological validation, sampling, extraction, sex differences, and the influence of diet on hormone metabolite levels. Ann N Y Acad Sci 1046:35–53. doi:10.1196/annals.1343.005
35. Gow R, Thomson S, Rieder M et al (2010) An assessment of cortisol analysis in hair and its clinical applications. Forensic Sci Int 196:32–37. doi:10.1016/j.forsciint.2009.12.040
36. Cook NJ (2012) Review: Minimally invasive sampling media and the measurement of corticosteroids as biomarkers of stress in animals. Can J Anim Sci 92:227–259. doi:10.4141/cjas2012-045
37. von Engelhardt N, Groothuis TGG (2005) Measuring steroid hormones in avian eggs. Ann N Y Acad Sci 1046:181–192. doi:10.1196/annals.1343.015
38. Pfannkuche KA, Gahr M, Weites IM et al (2011) Examining a pathway for hormone

mediated maternal effects—yolk testosterone affects androgen receptor expression and endogenous testosterone production in young chicks (Gallus gallus domesticus). Gen Comp Endocrinol 172:487–493. doi:10.1016/j.ygcen.2011.04.014

39. vom Saal FS (1990) Paradoxical effects of maternal stress on fetal steroids and postnatal reproductive traits in female mice from different intrauterine positions. Biol Reprod 43:751–761. doi:10.1095/biolreprod43.5.751
40. Winter JSD, Faiman C, Reyes FI (1977) Morphogenesis and malformations of the genital system. A. Liss, New York
41. Wilson CA, Davies DC (2007) The control of sexual differentiation of the reproductive system and brain. Reproduction 133:331–359. doi:10.1530/REP-06-0078
42. Knickmeyer RC, Baron-Cohen S (2006) Fetal testosterone and sex differences. Early Hum Dev 82:755–760. doi:10.1016/j.earlhumdev.2006.09.014
43. Judd HL, Robinson JD, Young PE, Jones OW (1976) Amniotic fluid testosterone levels in midpregnancy. Obstet Gynecol 48:690–692
44. Schindler AE (1982) Hormones in human amniotic fluid. Springer, Heidelberg
45. van de Beek C, Thijssen JHH, Cohen-Kettenis PT et al (2004) Relationships between sex hormones assessed in amniotic fluid, and maternal and umbilical cord serum: what is the best source of information to investigate the effects of fetal hormonal exposure? Horm Behav 46:663–669. doi:10.1016/j.yhbeh.2004.06.010
46. Finegan JA, Bartleman B, Wong PY (1989) A window for the study of prenatal sex hormone influences on postnatal development. J Genet Psychol 150:101–112. doi:10.1080/00221325.1989.9914580
47. Robinson JD, Judd HL, Young PE et al (1977) Amniotic fluid androgens and estrogens in midgestation. J Clin Endocrinol Metab 45:755–761. doi:10.1210/jcem-45-4-755
48. Sikich L, Todd RD (1988) Are the neurodevelopmental effects of gonadal hormones related to sex differences in psychiatric illnesses? Psychiatr Dev 6:277–309
49. Groothuis TGG, von Engelhardt N (2005) Investigating maternal hormones in avian eggs: measurement, manipulation, and interpretation. Ann N Y Acad Sci 1046:168–180. doi:10.1196/annals.1343.014
50. Schaafsma SM, Groothuis TGG (2012) Sex-specific effects of maternal testosterone on lateralization in a cichlid fish. Anim Behav 83:437–443. doi:10.1016/j.anbehav.2011.11.015
51. Ford JJ (1980) Serum testosterone concentrations in embryonic and fetal pigs during sexual differentiation. Biol Reprod 23:583–587. doi:10.1095/biolreprod23.3.583
52. Hönekopp J, Bartholdt L, Beier L, Liebert A (2007) Second to fourth digit length ratio (2D:4D) and adult sex hormone levels: new data and a meta-analytic review. Psychoneuroendocrinology 32:313–321
53. Putz DA, Gaulin SJC, Sporter RJ, McBurney DH (2004) Sex hormones and finger length. what does 2D:4D indicate? Evol Hum Behav 25:182–199. doi:10.1016/j.evolhumbehav.2004.03.005
54. Lutchmaya S, Baron-Cohen S, Raggatt P et al (2004) 2nd to 4th digit ratios, fetal testosterone and estradiol. Early Hum Dev 77:23–28. doi:10.1016/j.earlhumdev.2003.12.002
55. Ventura T, Gomes MC, Pita A et al (2013) Digit ratio (2D:4D) in newborns: influences of prenatal testosterone and maternal environment. Early Hum Dev 89:107–112. doi:10.1016/j.earlhumdev.2012.08.009
56. McFadden D, Bracht MS (2005) Sex differences in the relative lengths of metacarpals and metatarsals in gorillas and chimpanzees. Horm Behav 47:99–111. doi:10.1016/j.yhbeh.2004.08.013
57. Auger J, Le Denmat D, Berges R et al (2013) Environmental levels of oestrogenic and antiandrogenic compounds feminize digit ratios in male rats and their unexposed male progeny. Proc Biol Sci 280:20131532. doi:10.1098/rspb.2013.1532
58. Dean A, Sharpe RM (2013) Clinical review: Anogenital distance or digit length ratio as measures of fetal androgen exposure: relationship to male reproductive development and its disorders. J Clin Endocrinol Metab 98:2230–2238. doi:10.1210/jc.2012-4057
59. Nagy G, Blázi G, Hegyi G, Török J (2016) Side-specific effect of yolk testosterone elevation on second-to-fourth digit ratio in a wild passerine. Naturwissenschaften 103:4. doi:10.1007/s00114-015-1328-x
60. Ruuskanen S, Helle S, Ahola M et al (2011) Digit ratios have poor indicator value in a wild bird population. Behav Ecol Sociobiol 65:983–994. doi:10.1007/s00265-010-1099-5
61. Romano M, Rubolini D, Martinelli R et al (2005) Experimental manipulation of yolk testosterone affects digit length ratios in the ring-necked pheasant (Phasianus colchicus). Horm Behav 48:342–346. doi:10.1016/j.yhbeh.2005.03.007

62. Zielinski WJ, vom Saal FS, Vandenbergh JG (1992) The effect of intrauterine position on the survival, reproduction and home range size of female house mice (Mus musculus). Behav Ecol Sociobiol 30:185–191. doi:10.1007/BF00166702
63. Clark M, Galef B (1998) Effects of intrauterine position on the behavior and genital morphology of litter-bearing rodents. Dev Neuropsychol 14:197–211
64. Tapp AL, Maybery MT, Whitehouse AJO (2011) Evaluating the twin testosterone transfer hypothesis: a review of the empirical evidence. Horm Behav 60:713–722. doi:10.1016/j.yhbeh.2011.08.011
65. Okuliarova M, Groothuis TGG, Skrobánek P, Zeman M (2011) Experimental evidence for genetic heritability of maternal hormone transfer to offspring. Am Nat 177:824–834. doi:10.1086/659996
66. Costantini D, Metcalfe NB, Monaghan P (2010) Ecological processes in a hormetic framework. Ecol Lett 13:1435–1447. doi:10.1111/j.1461-0248.2010.01531.x
67. Del Giudice M (2012) Fetal programming by maternal stress: insights from a conflict perspective. Psychoneuroendocrinology 37:1614–1629. doi:10.1016/j.psyneuen.2012.05.014
68. von Engelhardt N, Groothuis TGG (2011) Hormones and reproduction of vertebrates. doi:10.1016/B978-0-12-374929-1.10004-6
69. Henriksen R, Rettenbacher S, Groothuis TGG (2011) Prenatal stress in birds: pathways, effects, function and perspectives. Neurosci Biobehav Rev 35:1484–1501. doi:10.1016/j.neubiorev.2011.04.010
70. Rogers LJ, Deng C (2005) Corticosterone treatment of the chick embryo affects light-stimulated development of the thalamofugal visual pathway. Behav Brain Res 159:63–71. doi:10.1016/j.bbr.2004.10.003
71. Schwarz IM, Rogers LJ (1992) Testosterone: a role in the development of brain asymmetry in the chick. Neurosci Lett 146:167–170. doi:10.1016/0304-3940(92)90069-J
72. Groothuis TGG, Schwabl H (2008) Hormone-mediated maternal effects in birds: mechanisms matter but what do we know of them? Philos Trans R Soc Lond B Biol Sci 363:1647–1661. doi:10.1098/rstb.2007.0007
73. Hines M (2006) Prenatal testosterone and gender-related behaviour. Eur J Endocrinol 155(Suppl):S115–S121. doi:10.1530/eje.1.02236
74. Hodgetts S, Weis S, Hausmann M (2015) Sex hormones affect language lateralisation but not cognitive control in normally cycling women. Horm Behav 74:194–200. doi:10.1016/j.yhbeh.2015.06.019
75. Fischer H, Sandblom J, Gavazzeni J et al (2005) Age-differential patterns of brain activation during perception of angry faces. Neurosci Lett 386:99–104. doi:10.1016/j.neulet.2005.06.002
76. Wright ND, Bahrami B, Johnson E et al (2012) Testosterone disrupts human collaboration by increasing egocentric choices. Proc Biol Sci 279:2275–2280. doi:10.1098/rspb.2011.2523
77. Kreukels BPC, Cohen-Kettenis PT (2011) Puberty suppression in gender identity disorder: the Amsterdam experience. Nat Rev Endocrinol 7:466–472. doi:10.1038/nrendo.2011.78
78. Hembree WC, Cohen-Kettenis P, Delemarre-van de Waal HA et al (2009) Endocrine treatment of transsexual persons: an Endocrine Society clinical practice guideline. J Clin Endocrinol Metab 94:3132–3154. doi:10.1210/jc.2009-0345
79. Marshall WA, Tanner JM (1969) Variations in pattern of pubertal changes in girls. Arch Dis Child 44:291–303
80. Marshall WA, Tanner JM (1970) Variations in the pattern of pubertal changes in boys. Arch Dis Child 45:13–23
81. Brenowitz EA (2004) Plasticity of the adult avian song control system. Ann N Y Acad Sci 1016:560–585. doi:10.1196/annals.1298.006
82. Moorman S, Gobes SMH, Kuijpers M et al (2012) Human-like brain hemispheric dominance in birdsong learning. Proc Natl Acad Sci U S A 109:12782–12787. doi:10.1073/pnas.1207207109
83. Neveu PJ, Liège S, Sarrieau A (1998) Asymmetrical distribution of hippocampal mineralocorticoid receptors depends on lateralization in mice. Neuroimmunomodulation 5:16–21
84. Gahr M (2007) Sexual differentiation of the vocal control system of birds. Adv Genet 59:67–105. doi:10.1016/S0065-2660(07)59003-6
85. Adret P, Rogers LJ (1989) Sex difference in the visual projections of young chicks: a quantitative study of the thalamofugal pathway. Brain Res 478:59–73. doi:10.1016/0006-8993(89)91477-7
86. Fink M, Wadsak W, Savli M et al (2009) Lateralization of the serotonin-1A receptor distribution in language areas revealed by PET. Neuroimage 45:598–605. doi:10.1016/j.neuroimage.2008.11.033

87. Vernaleken I, Weibrich C, Siessmeier T et al (2007) Asymmetry in dopamine D(2/3) receptors of caudate nucleus is lost with age. Neuroimage 34:870–878. doi:10.1016/j.neuroimage.2006.10.013
88. Wang H, Wang X, Wetzel W, Scheich H (2006) Rapid-rate transcranial magnetic stimulation of animal auditory cortex impairs short-term but not long-term memory formation. Eur J Neurosci 23:2176–2184. doi:10.1111/j.1460-9568.2006.04745.x
89. Rotenberg A, Muller PA, Vahabzadeh-Hagh AM et al (2010) Lateralization of forelimb motor evoked potentials by transcranial magnetic stimulation in rats. Clin Neurophysiol 121:104–108. doi:10.1016/j.clinph.2009.09.008
90. Willis MW, Ketter TA, Kimbrell TA et al (2002) Age, sex and laterality effects on cerebral glucose metabolism in healthy adults. Psychiatry Res 114:23–37
91. Stroobant N, Vingerhoets G (2000) Transcranial Doppler ultrasonography monitoring of cerebral hemodynamics during performance of cognitive tasks: a review. Neuropsychol Rev 10:213–231
92. Deppe M, Ringelstein EB, Knecht S (2004) The investigation of functional brain lateralization by transcranial Doppler sonography. Neuroimage 21:1124–1146. doi:10.1016/j.neuroimage.2003.10.016
93. Honing H, Merchant H, Háden GP et al (2012) Rhesus monkeys (Macaca mulatta) detect rhythmic groups in music, but not the beat. PLoS One 7:e51369. doi:10.1371/journal.pone.0051369
94. Rowan A, Liégeois F, Vargha-Khadem F et al (2004) Cortical lateralization during verb generation: a combined ERP and fMRI study. Neuroimage 22:665–675. doi:10.1016/j.neuroimage.2004.01.034
95. Rogers LJ, Vallortigara G, Andrew RJ (2015) Divided brains: the biology and behaviour of brain asymmetries. Cambridge University Press, Cambridge
96. Cashmore L, Uomini N, Chapelain A (2008) The evolution of handedness in humans and great apes: a review and current issues. J Anthropol Sci 86:7–35
97. Lust JM, Geuze RH, Groothuis AGG, Bouma A (2011) Functional cerebral lateralization and dual-task efficiency-testing the function of human brain lateralization using fTCD. Behav Brain Res 217:293–301. doi:10.1016/j.bbr.2010.10.029
98. Lust JM, Geuze RH, Groothuis AGG et al (2011) Driving performance during word generation—testing the function of human brain lateralization using fTCD in an ecologically relevant context. Neuropsychologia 49:2375–2383. doi:10.1016/j.neuropsychologia.2011.04.011
99. Hirnstein M, Leask S, Rose J, Hausmann M (2010) Disentangling the relationship between hemispheric asymmetry and cognitive performance. Brain Cogn 73:119–127. doi:10.1016/j.bandc.2010.04.002
100. Stroobant N, van Boxstael J, Vingerhoets G (2011) Language lateralization in children: a functional transcranial Doppler reliability study. J Neurolinguistics 24:14–24. doi:10.1016/j.jneuroling.2010.07.003

Chapter 21

Reversals of Bodies, Brains, and Behavior

Douglas J. Blackiston and Michael Levin

Abstract

Left–right asymmetries are highly prevalent throughout metazoan phyla, with bilaterally symmetrical organisms exhibiting well-conserved, consistently sided positioning and anatomy of visceral organs and central nervous system structures. Deviations from normal laterality constitute an important class of birth defects and much study has been devoted to the early mechanisms orienting the left–right axis during embryogenesis as well as lateralization of the brain. Far less understood are the potential links between laterality of the body and cognition, though recent work has begun to uncover a range of behaviors which are modified in organisms with altered left–right asymmetry. Here, we review regulatory events critical for the establishment of asymmetry and subsequent left–right patterning, using data from *Xenopus*, zebrafish, chick, *Arabidopsis*, and single cells, and discuss molecular and pharmacological reagents that disrupt these processes. We especially focus on behavioral assays which are sensitive to body laterality, presenting existing data for several model systems. Beyond classical conditioning and behavior screens, new automated machine vision platforms are powerful emerging tools to quantitatively examine the relationship between body asymmetry and lateralized and nonlateralized behaviors. This chapter serves as a primer for methods that allow the examination of cognitive and behavioral endpoints subsequent to molecular interventions in embryonic left–right asymmetry.

Key words Left–right asymmetry, Laterality, Behavior, Automated analysis, Quantification

1 Introduction

Though our universe does not distinguish left from right on a macroscopic level, many types of embryos are able to reliably distinguish left from right during development [1]. The positioning and morphology of numerous structures within the bilaterian bodyplan is consistently asymmetric, deviating from perfect left–right symmetry in the same directions in all normal individuals [2, 3]. The consistent laterality of the heart, visceral organs, and even brain is well conserved throughout the animal kingdom, and suggests many fascinating puzzles regarding the evolution and ontogenetic mechanisms of this curious developmental and neurobehavioral feature [4, 5]. In contrast to fluctuating asymmetry (where bilateral traits are imperfectly synchronized but without consistent unilateral bias)

Lesley J. Rogers and Giorgio Vallortigara (eds.), *Lateralized Brain Functions: Methods in Human and Non-Human Species*, Neuromethods, vol. 122, DOI 10.1007/978-1-4939-6725-4_21, © Springer Science+Business Media LLC 2017

[6] and externally/behaviorally imposed asymmetry [7], we focus on fixed asymmetry, in which structure and function of specific organs is consistently oriented relative to the anterior-posterior and dorso-ventral axes.

Laterality is not only of basic interest, but has considerable biomedical implications. Errors in left–right patterning include loss of asymmetry (isomerism), reversal (*situs inversus*), or randomization (heterotaxia). These are an important class of human birth defects as a primary teratology [8, 9], but many other syndromes have a laterality component to their manifestation, both in fetal development [10, 11] and in adulthood, such as the relationship between asymmetry and cancer [12–15], immune response [16, 17], and schizophrenia [18]. Beyond body anatomy, asymmetry of the brain and nervous system gives rise to lateralized behaviors, most frequently observed as handedness but influencing many aspects of an animal's function in their ecological niche [19–22].

In the last two decades, remarkable progress has been made in understanding the molecular mechanisms of left–right patterning throughout phyla [23]. The major steps (Fig. 1) involve symmetry breaking at the subcellular level [24–29], physiological signaling via small molecules [30–34], cascades of asymmetric gene expression [35–37], and asymmetric organogenesis driven by the resulting differential cell behaviors [38, 39]. Asymmetry is thus a fascinating example of multiscale integration [40–42], in which order is generated, amplified, and imposed on biological processes at many levels of organization, from molecular structures to tissues to the behavior of entire communities of individual organisms. Despite recent progress, many puzzles and open questions remain with respect to the mechanisms and consequences of asymmetry [43–46], making it essential to address the questions of asymmetry at multiple levels and in a wide variety of model systems. Here, we provide an overview of several model systems and the techniques of characterization of functional asymmetry and assays of lateralized behavior—an endpoint that enables study of the entire causal chain from genetics to CNS function.

2 Zebrafish

Among animal models used in laterality research, *Danio* spp., and related fish species, have contributed much to the field [21, 47, 48; see also Chap. 17], greatly facilitating the linking of genetics with behavioral tests. As a vertebrate model they are an attractive entry point into translational research with the goals of improving health, as well as being the focus of both ecological and evolutionary work. Large numbers of fish can be raised in aquaculture, and fertilized eggs can be microinjected with synthetic RNAs to alter gene expression in specific tissues across developmental stages.

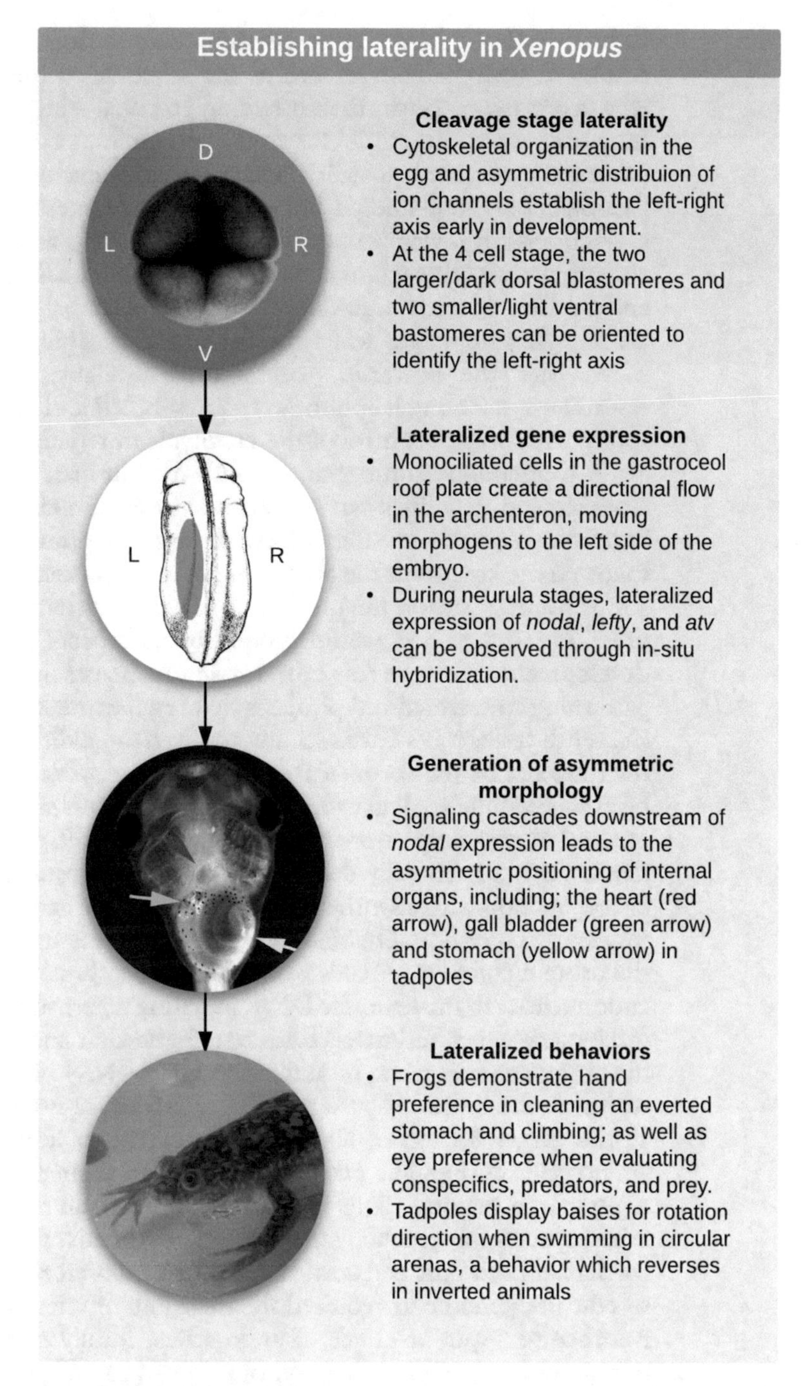

Fig. 1 Molecular, morphological, and behavioral laterality in *Xenopus laevis*. Shortly after fertilization, cytoskeletal organization initiates left–right axis specification. The resulting ion channel localization, intracellular morphogen redistribution, and chromatin acetylation serve as early markers of laterality. Following gastrulation, these early signaling events, and the later action of ciliated cells in the roof-plate, lead to asymmetric expression of *nodal* and several other laterality markers. These asymmetric cascades ensure the lateralized position of the internal organs. After hatching, tadpoles and frogs also demonstrate a number of lateralized behaviors whose relationship to internal left–right asymmetries are not yet known

Zebrafish also explore their environment; paying attention to stimuli in their habitats as well as the schedule of the individual who feeds them. Given their attention to visual stimuli, adult fish have been a favorite model for cognitive studies; examining visual processing, memory, sociality, addiction, and predator avoidance [49–58]. Less well-studied but also capable learners nevertheless are zebrafish fry, which upon hatching becoming active hunters; chasing down paramecium and brine shrimp when raised in the lab and a variety of microorganisms in the wild. The real power of this model system emerges when the behavioral capabilities are combined with the powerful genetic tools available to zebrafish researchers, enabling laboratories to cross-breed and engineer offspring with precise control of the genotypic outcome.

A tremendous number of zebrafish mutant lines are currently available and new lines can be generated through either targeted knockout methods or alternatively through random mutagenesis followed by screens for the phenotype of interest (known colloquially as fishing; pardon the pun). As such, these genetic tools position Zebrafish as a powerful model for all aspects of molecular development: from the role of transcription factors in mesodermal patterning to secreted morphogens that regulate neural induction. Zebrafish researchers have recently received an additional tool for their arsenal in the form of the CRISPR-Cas system, which has been successfully applied to *Danio* embryos as a less labor-intensive method of knocking in-or-out genes of interest [59, 60]. This system utilizes a bacterially derived CRISPR/Cas9 protein complex in conjunction with a synthetic RNA targeting the gene (or regulatory site) in question. On binding to the complementary sequence, the protein complex initiates a double stranded break by the Cas9 endonuclease in the genomic DNA; initiating an error-prone repair mechanism which ultimately leads to a frame-shift and silencing of the target gene. Further, by adding bridging RNAs which span the region of the break it is also possible to introduce novel sequence at the target locations; allowing researchers to add/substitute amino acid residues in a protein or regulatory sequence promotor regions of a genome. This technique has made an already established species for genetic work even more attractive; lowering the barriers of both time and cost, and allowing laboratories of all sizes to edit the genome as required for their line of scientific inquiry. Another very powerful set of approaches, ideally suited to the transparency of zebrafish, is optogenetics [61, 62]—the use of light to modulate the function of specific components of the CNS with incredible spatiotemporal specificity. Optogenetics is now widely used in the zebrafish model to interrogate behavioral circuits [63–67].

Given these tools, it is perhaps not surprising that a number of laterality research groups have turned to zebrafish as their model of choice; working out many of the early pathways which correctly orient the left–right axis during early development. The majority of this work has focused on a structure present in all teleost fishes during embryogenesis known as Kuppfer's vesicle, first described by its namesake in 1868 [68]. During gastrulation, cells occupying the animal pole of the embryo undergo epiboly to surround the yolk and move in a posterior direction; while also undergoing a second cell movement at one position along the equator, involuting and forming the embryonic shield. Along the leading edge of this structure approximately 25 cells maintain their position without involuting; a population referred to as the dorsal forerunner cells. At the conclusion of gastrulation, these cells occupy a position with the posterior of the tailbud, and migrate to the interior of the embryo to form Kuppfer's vesicle, a hollow region lined by monociliated cells. The rotation of these cilia coordinate in a counterclockwise motion and their combined action results in a directional fluid flow which is thought to move morphogens to lateralized positions within the embryo [69–71]. These morphogens in turn signal downstream pathways which activate genes required for lateralization of visceral and central nervous system components of the developing zebrafish.

A number of techniques that disrupt aspects of Kuppfer's vesicle result in animals with heterotaxia (randomized laterality) and *situs inversus* (reversed laterality). Genetic knockouts of genes expressed by the dorsal forerunner cells are sufficient to induce the phenotype [72], as is both laser ablation and surgical damage [71]. In addition, *left–right dynein-related 1* (*lrd1*) is expressed in the cells of Kuppfer's vesicle and knockdown of this gene by morpholino disrupts fluid flow by the cilia and again induces both heterotaxia and *situs inversus* [71]. Further, the use of a morpholino allows specific targeting of the RNA through microinjection and knockdown of the transcript in regions unrelated to Kuppfer's vesicle fail to induce any changes in laterality, restricting its action in left–right patterning to the monociliated cells only. Indeed, a number of research programs have examined the effects of disrupting genes in the formation of Kuppfer's vesicle itself, or the directional action of the monociliated cells, and all have found randomization of lateralized structures as a result [69, 73–77].

What pathways lie downstream of Kuppfer's vesicle mediated directional morphogen flow? Many transcripts show lateralized enrichment following gastrulation and three of the most well-studied include transforming growth factor-β nodal-related family members (specifically *cyclops* and *southpaw*), *lefty 2*, as well as the homeobox gene*pitx2* [78–84]. The tgf-β family members *cyclops*, *southpaw*, and *lefty 2* are first observed in the lateral plate mesoderm of the developing embryo and demonstrate a left-only expression pattern. In situ hybridization reveals overlapping staining for the transcripts and misexpression on the contralateral side induces a randomization of asymmetric organ and CNS structures with the body [79]. *Pitx2* is expressed in a similar pattern to the above transcripts and is also capable of inducing heterotaxia when misexpressed on the right side of body following gastrulation. Further, *pitx2* lies downstream of both *nodal* signaling and *lefty* as the expression of either is sufficient to drive the activation of *pitx2* in a cell-autonomous manner. Together, these proteins constitute the canonical laterality specification pathway and are both necessary and sufficient to properly orient the left–right axis in developing zebrafish.

Both the heart and gut demonstrate asymmetries in shape and position, and expression of these genes can be found in both: *cyclops*, *lefty 2*, and *pitx2* in the left heart field; *lefty 2* and *pitx2* in the left gut precursors [85]. In addition to visceral structures, two distinct asymmetries have been noted within the brain of zebrafish, both within the epithalamus. The parapineal organ begins medially oriented before migrating to a final left-biased position in the fish diencephalon, where it is thought to signal the asymmetric growth of the habenula, a central component of the forebrain-midbrain limbic system [86]. The habenula itself present in both the left and right sides of the zebrafish brain, but shows distinct differences in nuclei organization and projections between the two. In addition, the left and the right habenula send projections posteriorly, which meet at the midline at the interpeduncular nucleus (IPN) within the midbrain. This structure does not show any left–right asymmetries but is highly patterned along the dorsal ventral axis. Projections from the left-habenula make up the dorsal portion of the IPN while those of the right-habenula those of the ventral IPN [87, 88]. While the functional roles of these asymmetries are not clear, the habenula has been implicated in many aspects of cognition including anxiety, memory, sleep, stress, and attention [89–94]. Further, disruption of left-sided gene flow, as well as left-sided migration of the parapineal, both result in habenular randomization, a finding which has helped establish fish as key species for studies linking brain/body laterality and behavior.

Zebrafish (as well as other fish species) exhibit lateralized behaviors in the wild and in the laboratory, and a large number have been described in the literature [21, 58, 95–99; see also Chap. 17]. The most well-studied are the eye biases during inspection of objects in the fish's environment. When observing familiar objects including their reflection and conspecifics, zebrafish show a bias for their left eye (turning with the object to the left of their body). However, when preparing to strike at food or prey, fish instead orient to place the object on their right, using their right-eye to examine the object before biting [100, 101]. The neurological and evolutionary basis of these behaviors is not clear; they are shared among many vertebrates [102–106] but in the case of zebrafish there has been speculation the right eye possesses better depth perception and is thus more efficient at tasks requiring the animal to make precise movements in three-dimensional space. In addition, zebra-fish exhibit turning biases when entering a novel environment or when startled in the laboratory setting. Using a set of rectangular chambers connected by central openings, zebrafish will demonstrate positive phototaxis and move between arenas if the proceeding areas are illuminated. When entering these novel arenas fish may either move forward, turn left, or turn right, and in the absence of any other stimuli, demonstrate a bias for a left-sided turn upon exploring these new environments [95]. It is unclear why this bias occurs though the authors of the study [95] speculate that it may be due to the left eye creating a map of the recently exited area (as the left eye is used when viewing familiar objects [107, 108]), or it may indicate a more efficient spatial mapping function of the left eye in general. Inversely, when fish entering these environments are startled by the sudden dimming of lights they show the opposite bias; turning to the right and swimming in the direction opposite to that seen under positive phototaxis [95].

How are these behaviors altered in fish with reversals of laterality? Recent work addressed this question using a genetic line of zebrafish known as *fsi*; or *frequent situs inversus.* Fish in this line display one of two phenotypes, either completely normal laterality or complete *situs inversus.* In addition, the laterality of both the visceral organs and central nervous system display the same laterality in these animals; if the visceral organs are reversed then the asymmetry of neural structures are also reversed. This allows researchers to score animals quickly for visceral asymmetry (fish fry are transparent and can easily be assessed for laterality under a dissecting microscope) and simultaneously know the position of the parapineal organ and habenula (note: it is not known if visceral and neural asymmetries are patterned by the same mechanisms in other species; this may be a unique advantage, or disadvantage, to zebrafish depending on the study). Testing *fsi* zebrafish in the above lateralized assays yielded interesting results [95]. Two behaviors, the eye individuals use to inspect themselves in a mirror, and the eye individuals use when striking food, were reversed in animals possessing laterality inversions. This finding was one of the first revealing a direct link between laterality and cognition, and has implications for a number of open questions in the field regarding brain laterality. However, more interestingly, two behaviors did not reverse in the zebrafish *fsi* line. Inverted fish still demonstrated a left-turn preference when entering an illuminated environment as well as a right turn bias when startled by rapid light dimming [95]. How these behaviors can maintain directionality irrespective of both neural and visceral asymmetry is not clear but introduces a number of intriguing questions for us to serve as the basis for future work.

Besides explicit lateralized behaviors, do fish with reversed neural asymmetry demonstrate any other cognitive differences compared to wild-type siblings? Here the answer is also affirmative.

Zebrafish with reversed parapineal organ position spend more time inspecting a predator (a measure of boldness), spend more time in the center of an open field, and cover less distance when startled [109]. While none of these behaviors are lateralized per-se, they are still important metrics whose outcome are clearly directed by the positioning of central nervous system asymmetries and can be modulated through laterality inversions. Taken together, the large toolbox available to modulate and assay both molecular and cognitive laterality in zebrafish have made them a key species for investigations of laterality and cognition, and given modern advancements in genome editing the species will no doubt continue to be a favorite for many research groups examining this interesting question.

3 Amphibians

Amphibians have long been a favorite model system in biological research, with early records of their use in physiology dating back to the 1700s [110]. Perhaps most famous are the studies of Luigi Galvani, who discovered that by passing an electric current through the dismembered legs of frogs he could provoke them to kick, "as in life." Galvani's studies, as well as those which followed, paved the way for advances in the fields of physiology and greatly enhanced our understanding of organ systems, and in turn significantly impacted human health and medicine. More recently (over the past century) frog research has diversified and many species' have been used as tools for the study of ecology, evolutionary biology, environmental health, neurodevelopmental disorders, and animal behavior [111–113], as well as regeneration [114–117], genetics [118], cancer biology [119, 120], and immunology [121–123]. A particularly profound impact in the fields of embryology and developmental biology has been made by amphibians (particularly *Xenopus* and *Rana* species) [124].

In the laboratory *Xenopus* species are capable of producing large clutches of eggs naturally, or through subcutaneous injection of human chorionic gonadotropin. An individual female is capable of laying thousands of eggs over the course of a day and can produce these clutches on a quarterly basis. Once fertilized, either naturally or through manual application of male sperm, the resulting embryos are extremely robust tolerating temperatures between 14 and 24 °C. Further, developing embryos are also amenable to cut-and-paste grafting experiments, allowing researcher's to transplant ectoderm, mesoderm, or endoderm derived tissues between individuals to elucidate cell fate and signaling events. Similar to zebrafish, with the advent of modern molecular tools during the past 30 years frogs have become a leading model for developmental genetics, as microinjection of synthetic mRNA's into fate mapped blastomeres allows researchers to induce or inhibit the expression of specific genes across developmental stages, probing the specific pathways required for the transition from egg to organism.

Given this amenability to developmental manipulation, *Xenopus* species have become primary models for the study of left–right laterality and the early signaling events leading to asymmetric *nodal* expression [125–129]. Readouts for laterality in *Xenopus* are many: one can perform in situ hybridization for a number of asymmetrically expressed genes (such as *nodal*, *lefty*, and *pitx2*), or simply grow tadpoles to swimming stages where the asymmetric position of the heart, stomach, and gall bladder are visible through the transparent epidermis. Using the latter method, screens of molecules/pharmaceuticals which disrupt laterality can be readily accomplished [130–133]; developing embryos can simply be microinjected with constructs or raised in media containing chemicals of interest and scored at swimming stages for the position of the three lateralized visceral organs.

A number of techniques and tools exist to randomize or reverse visceral asymmetry in *Xenopus* tadpoles. *Xenopus nodal-related gene 1* is expressed asymmetrically in the left lateral plate mesoderm during neural stages of development. Knockdown of this gene on the left-side, or misexpression of the wild-type construct on the right side induces both heterotaxia and *situs inversus* in experimental animals without altering other aspects of anatomy [134]. Further, alteration of upstream and downstream signaling molecules in the *nodal* pathway, such as *atv* and *lefty*, likewise result in randomization or reversal of lateralized organ placement [42, 135]. Targeted microinjection of wild-type or dominant negative constructs of these genes can be done at early frog stages, as the left–right axis is established during the first cell cleavage (which divides the embryo into to right and left halves, though it is impossible to know which side is which until the second cleavage). When visualizing four-cell embryos, two of the blastomeres are larger and more pigmented—these give rise to ventral tissues of the embryo—while the remaining two are smaller and lighter, giving rise to dorsal tissues. Orienting the embryo with the ventral cells pointing northward and the dorsal cells southward, the two left and right cells then correspond to the left and right hand sides of the developing animal. Following this fate-map allows researchers to easily target constructs of interest to the appropriate location, giving rise to laterality disruptions [136–138].

In addition to signaling factor cascades, a number of other mechanisms can be disrupted to alter left–right asymmetry in *Xenopus* prior to *nodal* expression. One is the action of monociliated cells lining gastrocoel roof plate of early neurula-stage embryos. These cilia beat with a bias resulting in a strong leftward flow within the interior of the gastrocoel (mirroring the Kuppfer's vesicle in zebrafish), which have been proposed to move secreted determinants to the left-side of the archenteron. Molecular disruption of the cilia (either by removal or disruption of their movement) results in tadpoles with randomized laterality, although the majority of treatments used to target this ciliary motion in fact also disrupt earlier intracellular (cytoskeletal) components of asymmetry [44, 45, 125]. It has also been reported that inhibition of fluid flow within the gastrocoel (by microinjection of a viscous mixture of methylcellulose to impede movement of fluid by cilia) is sufficient to induce heterotaxia and *situs* inversus [139].

In addition to motile cilia, bioelectrical gradients resulting from ion channel and pump function are also important regulators of left–right laterality in frog embryos. Early in development (between the 1 cell and 32 cell stage), asymmetric localization of potassium channels and the two proton pumps (the H,K-ATPase and the V-ATPase) can be observed between the left and right hand sides of the embryos [33, 140–143]. This differential expression leads to changes in membrane voltage potentials on opposite halves of the developing animal, with cells on the right-hand side being more negatively polarized than those on the right. Given this voltage difference, charged morphogens (such as serotonin, which is weakly positive at physiological pH) progressively move through gap junction-mediated transcellular paths, and subsequently signal downstream to epigenetically regulate expression asymmetric genes [144]. This physiological system amplifies the initial chirality of the cytoskeleton [29, 125, 145], enabling intracellular directional asymmetries to ultimately impose positional asymmetries (with respect to the midline) of gene expression in large cell fields [146, 147]. As such, it provides a number of convenient and tractable control points, at which specific pharmacological or genetic reagents can be introduced to disrupt this sequence of events and induce heterotaxia. The pharmacological toolkit includes cytoskeletal or motor protein disruptors, potassium channel blockers, inhibitors of gap junctions, and serotonergic pathway modulators [29, 32, 33, 137, 140, 148–150]. Similarly, the heterotaxia or *situs inversus* can be induced by misexpression of mutated cytoskeletal components, mutant ion channels that regulate resting potential, dominant negative connexin proteins (on the dorsal side), constitutively open connexin proteins (across the ventral side), or mutant serotonin transporters. Since these signaling events occur extremely early in development, animals can be treated within the first 24 h of fertilization (often within the first few hours only), and then raised under normal conditions; this does not require exposure at later stages that could potentially impact additional tissues or patterning pathways. Indeed, it has proven possible to dissociate the left–right patterning roles of all of the above-mentioned components from other functions; the randomization obtained with these reagents can be very specific, giving rise to otherwise quite normal embryos with no generalized toxicity or defects that can perform well in behavioral assays.

Another convenient method for generating asymmetry defects in tadpoles is through the application of low frequency vibration, is sufficient to induce left–right randomization and reversals in frog embryos (the target is not known definitively, but is likely to involve subtle cytoskeletal disruption). Using this method, fertilized eggs are placed on a simple audio speaker, driven by a sinusoidal function generator set to a frequency of 7 or 15 Hz [151]. Vibration at these frequencies across the first 24 h of development is sufficient to induce heterotaxia in 30% or more of treated embryos, in line with other methods of left–right disruption [151, 152]. The greatest advantage of this method is the experimenter can establish precise "on-off" times for treatment, as opposed to molecular or pharmaceutical treatments, where determining definitive drug washout can be very difficult.

Disruption of *nodal*-related signaling pathways, motile cilia in the gastrocoel roof plate, ion-channel mediated bioelectricity, and early cytoskeletal organization are all established methods to randomize and reverse *Xenopus* visceral asymmetry. What about assays for cognitive and neural asymmetry? Here the research is far less established in *Xenopus.* Frog species (not *Xenopus* in particular) show hand preferences in a number of tasks including climbing and the limb used to clean an everted stomach [153]. Evolutionarily this was thought to be due to the asymmetry in gut position; an everted stomach will typically evert in the direction opposite its normal position within the body; thus the hand closest to the everted stomach may be more effective at wiping and cleaning if the frog ingested hazardous material. However, counter examples exist in which hand preference is absent or varies between tasks [105, 154, 155] suggesting the everted stomach hypotheses as an unlikely driver of behavioral laterality in amphibians. Beyond handedness, frogs also show a bias for specific eye use based on particular tasks. In Australian green tree frogs and Bufonids, individuals demonstrate a left-eye bias when engaging in antagonistic behavior with conspecifics or when evaluating potential predators, as well as a right eye bias when assessing prey items [153]. However, it is not currently known if any of these behaviors are reversed in animals possessing laterality alterations, as these species have not been reared in laboratories performing functional left–right asymmetry studies.

However, in *Xenopus* one behavioral assay has recently been developed which allows investigation of behavioral laterality. Tadpoles feed by filtering particulate matter from their environment; a process which is facilitated by natural swimming movements. When placed in circular containers, tadpoles (both individuals and groups) readily adopt a circular swimming pattern, following the edge of the arena as they feed in the water column. Observation of individuals revealed a clockwise rotation preference for wild-type animals [156]. Interestingly, populations of tadpoles with reversed visceral laterality (reversed heart, gall bladder, and stomach) revealed a counterclockwise bias for swimming direction, and tadpoles with randomized laterality showed no bias for either direction. To date, this was the first report of a lateralized behavior in *Xenopus* which tracks with organ placement. While only tested in larvae it would be interesting to determine if adults had these same biases in swimming to any number of task—including those currently used in zebrafish (bias when navigating around an obstacle, direction turned when startled, or location preferences when entering a novel environment).

Given the amount of work dedicated to molecular left–right signaling in early *Xenopus* development, why are there so few cognitive assays of lateralized behavior in this species? The answer is twofold. First, *Xenopus* tapdoles and adults are notoriously difficult to use in studies of behavior, learning, and memory. Numerous attempts were made in through the 1970s at which point one group frustratingly stated "It appears that attempts to train frogs have been rather unproductive. ...after several years of unsuccessful and unpublished studies with frogs we were inclined to agree" [157, 158]. Only recently has this line of research been reopened and with the help of automated computer software some preliminary success has been made in training *Xenopus* tadpoles using associative assays [159–161]. The recent development of several different quantitative behavioral paradigms [162–166] in this species is sure to facilitate the gathering of data that can be readily mined for directional asymmetries.

Second, and equally frustrating, is that there are currently no asymmetric brain markers in *Xenopus* or *Rana*, although asymmetries in cell behavior during neurulation have been described [167]. Thus, it is currently unknown whether alteration of visceral laterality also impacts mature neural asymmetry. A primary step towards overcoming this barrier could be cloning frog transcripts of zebrafish markers for asymmetric structures within the epithalamus including the parapineal organ and habenula. Beyond this candidate gene approach, microarrays comparing the left and right portions of the developing brain for enriched transcripts would likely yield hitherto unknown markers for cerebral asymmetry and help establish the relationship between visceral and CNS laterality in amphibian species.

4 Chick

In addition to the aquatic model organisms described above, the chick also represents a powerful model for studies of developmental and behavioral laterality. As an amniote, this model system rests closer to humans on the evolutionary tree than both teleost and amphibians. Moreover, the flat blastodisc embryonic architecture of the chick is more similar to that of humans than the popular mouse model, which gastrulates as a cylinder-like morphology [168]. Indeed, the chick model system was the first in which molecular pathways regulating organ asymmetry were characterized [23], and also contributed to the understanding of laterality disturbances in conjoined twins [169].

The chick's development provides a contrast to the epiboly based movements present in early fish and frog embryogenesis, and presents a flat, optically accessible system in which to study asymmetry. Amniotes begin development by forming a disc of epiblast tissue covering a region known as the area pellucida. Within this area, at a location that will later become the posterior of the animal, the epiblast undergoes a coordinated moment forming the primitive streak; a visible line from the posterior to anterior, dividing the left and right hand sides of the developing organism (the leading edge of this structure is known as Hensen's node, named after its discoverer Victor Hensen). It is thought that the embryonic midline is first determined with the formation of the primitive streak; however, possible very early (pre-gastrulation) mechanisms have not yet been examined due to the difficulty of working with chick embryos prior to egg-laying (by which time tens of thousands of cells are already present in the embryo). Following formation of the primitive streak, a number of transcripts are expressed in an asymmetric manner; *activin-inducible type II activating receptor* to the right of Henson's node, left-sided expression of *sonic hedgehog*, followed by *nodal* expression to the left of the primitive streak [23, 170, 171]. These gene cascades progress in a similar fashion to those in zebrafish and *Xenopus*, eventually establishing the asymmetric placement of visceral organs and CNS structures in an otherwise bilateral organism.

However, unlike fish and frog, the chick does not possess a ciliated fluid-filled organ for chiral flow [168, 172]. How do cells occupying positions to the left and right relative the primitive streak receive positional information in the absence of ciliated cells? The most recent data pointed to planar cell polarity as a mechanism for orientation within this two-dimensional plane. Immunohistochemistry against *vangl2*, a core planar cell polarity protein, revealed a polarized expression pattern with all labeled cells pointing in the direction of the primitive streak [173]. Cells to the left of the streak localized *vangl2* to their rightmost position, while those to the right of the streak localized the protein to leftward positions. Further, early disruption of *vangl2* function through electroporation of a dominant negative transcript was sufficient to randomize *sonic hedgehog* expression later in development. These novel data were significant in that such a mechanism may be possible in humans as well, as all amniotes share the formation of an embryonic disc prior to tissue specification. In addition to this, the familiar ion channels, gap junctions, and serotonergic signaling have all been implicated in chick [32, 33, 140, 149] as in frog [34, 174] and more distant species such as sea urchins [175] and protochordates [176].

In addition to shared molecular markers of laterality between chick and aquatic species, lateralized behaviors are also similar between the models. As with zebrafish, chicks also demonstrate characteristic eye biases when examining their environment. Viewing familiar objects and conspecifics favors right-eye usage after hatching; although this appears to reverse to a left-eye preference as animals age [177]. In addition, when bypassing a centrally located barrier, chicks prefer to detour around the object to the left, using their right eye in circumnavigation [178]. Covering one eye or the other during the task revealed that animals with their right eyes exposed took less time to move around the barrier than those with exposed left eyes. Chicks are also more responsive to predators when observed with the left eye (similar to zebrafish) [179; see also Chap. 3].

However, while these behaviors demonstrate lateralized biases, it is not clear how they relate to lateralization of the viscera and central nervous system as few studies exist explicitly testing this correlation [180]. It would be interesting to raise randomized or reversed animals (produced though disruption of *vangl2*, *sonic hedgehog*, or *nodal*) and determine if the above behaviors randomize in a comparable manner. Given the differences between chick and fish/frog development, similarities in the outcomes of such experiments would strengthen the idea that findings in these species would also be applicable to human, as the results would apply to both amniotes and anamniotes (and thus, potentially all vertebrates). Future multispecies comparisons to similar lateralized tasks would be greatly informative to our understanding of the evolutionary relationship between body laterality and cognition.

5 Single Cells

While studies of laterality tend to focus on whole organisms, or even vertebrates (given the link to human health), asymmetry runs deep in the evolutionary tree [181, 182]. Perhaps this idea is no better illustrated than by the finding that even single cells, grown in culture, are capable of consistently lateralized behavior, such as preferential bending to one side during neurite outgrowth or biased migration in patterned surfaces [183–188]. How can directional axes be determined in eukaryotic cells? In the case of epithelial sheets there is often a well-defined apical-basal polarity, and other tissues are known to use planar cell polarity to orient with regard to one another. But how can individual cells, unattached to neighbors, demonstrate a directional bias in the absence of external coordinates? To answer this, one must look to two landmark structures within the cell, such as the nucleus and the centrosome.

While the centrosome is duplicated during mitosis to aid in chromosome segregation, it exists as a single structure directing microtubule organization during other phases of the cell cycle. After identifying these two structures, one can draw a line through the cell which bisects both the nucleus and the centrosome. By doing so, the center of the nucleus can be considered 0 in Cartesian coordinates and the direction of the centriole the *Y* axis (or "north" alternatively). With the cell oriented in such a way, it is possible to label half of the embryo as "left" and the other half as "right" with regards to their location relative to these two structures.

Having created such a direction system, a cellular behavior has been discovered which demonstrates a left–right bias; the direction of polarization and extension in cultured neutrophil-like dHL60 cells [183]. When these cells are cultured in uniform media (no flow or gradients of chemicals or attractants) 88% showed a leftward bias of extension and cytoplasm distribution. It is important to note here that the cells were in no way directing their growth towards one another; in fact, the directions of cell growth appeared random with respect to the position under the cover slip. However, by orienting their growth to the left–right axis established by transecting the nucleus and centrosome this asymmetric polarization becomes clear.

This behavior could be abolished through disruption of centrosome or microtubule organization. Molecular knockdown of *par6*, a component of the centrosome orientation pathway, as well as application of the microtubule polymerization inhibitor nocodazole, both randomized dHL60 polarity with regard to the left–right axis. Perhaps most interesting is the role of *GSK3β* in this process, a protein known to be involved in the polarity of dHL60 cells. Overexpression of wild-type *GSK3β* resulted in a reversal of laterality, with 83% of cells polarizing to the right. Similar to the original finding, this effect could be abolished through the application of nocodazole; producing cells with randomized polarity [183]. Interestingly, nocodazole randomizes asymmetry in *Xenopus* as well [150, 151].

It is unclear why individual cells would possess an internal laterality, although there are multiple theories. A number of single-celled organisms exhibit radial asymmetry and as such an inherent left–right establishing mechanism could be a pathway upon which selection pressures could shape cellular morphology [189, 190]. Further, developing tissues, as well as communal unicellular organisms, must orient themselves in three-dimensional space and an intrinsic laterality could help coordinate these complicated movements. Finally, the centrosome is a key component of the primary cilia of eukaryotic cells, which act as sensors to the external environment in a wide variety of contexts. Disruption of cytoskeletal components affects chiral intracellular transport as well as primary cilia. The tantalizing connection between left–right asymmetry of individual cells and complex multicellular animals could thus be a sign that asymmetry is extremely ancient—a feature that predates multicellularity but was exploited during metazoan development by novel amplification and long-range coordination mechanisms.

6 Plants

While plants are less common models for studies of lateralization [147, 191–196], the kingdom has made a significant impact nonetheless. Studies of plant asymmetry typically focus on helical growth of climbing plants—honeysuckles form left-handed (clockwise) helices as they grow while flowering bindweeds form right-handed (counterclockwise) helices. As the genetic model for much of plant research, *Arabidopsis* has been the primary species to examine the molecular pathways underlying these phenotypes, and two primary pathways have been identified which contribute to helical growth. Wild-type *Arabidopsis* plants demonstrate no native helical asymmetries during growth, producing uncoiled vertical hypocotyls (the stem between the root and cotyledon of the germinating seedling) and radial cotyledons. However, dominant negative mutations to two α-tubulins, *α-tubulin 4* and *α-tubulin 6*, result in plants with left-handed helical growth and clockwise twisting of the developing cotyledons [192, 196]. There are two interesting points to make note of regarding these results. First is the fact that *Arabidopsis* shows no asymmetric growth natively, yet asymmetry can be induced by tubulin mutations; breaking the symmetry of an otherwise radially organized organism. Second and equally fascinating is that single mutations in a cytoskeletal protein, not multiple changes to gene transcription factor cascades, are sufficient to induce asymmetric growth in *Arabidopsis.* More recently, these novel findings were mirrored by work examining a core subunit of the γ-tubulin-containing complex, *GCP2*, where mutation of an interaction domain within the protein resulted in right-handed Arabidopsis hypocotyl growth and clockwise twisting of the cotyledons [197], exactly opposite to that of the α-tubulin mutations.

These findings extend far beyond the plant kingdom. Recently, researchers have examined how mutations directing Arabidopsis chirality affect laterality in animal models, introducing the α-tubulin and γ-tubulin mutants to *C. elegans* and *Xenopus laevis* during development. The result was a randomization of laterality in both species; *Xenopus* tadpoles demonstrated contralateral or bilateral expression of the marker *nodal* as well as randomized organ placement, while *C. elegans* demonstrated a mirroring of asymmetric olfactory neurons which would otherwise express different receptors on the right and left sides [29]. This study identified tubulin genes as a mechanism for orienting laterality conserved across kingdoms, further strengthening the idea of potential conservation between unicellular organisms, plants, and animals. Similarities between the auxin and serotonin pathways in plants and animals have been noted [41]; however, roles of serotonin-like signaling in plant asymmetry remain to be tested.

7 Automation

Beyond additional tools to specifically alter early left–right patterning events during embryogenesis, what other advances would significantly enhance study of the relationship between brain/body laterality and cognition? The primary roadblock to date has been the basic nature of many behavioral assays in model organisms: researchers have molecular tools to alter phenotype but few, if any, cognitive tools to examine the result on animal behavior. The barriers here are many; behavioral trials typically involve a researcher manually watching an animal perform a task in an arena, or a video recording of the animal, which is later scored by a blind observer. This process is extremely time-consuming; given a full workday without breaks, only 8–9 h of footage can be parsed by hand, limiting the ability to do rapid large-scale screens of behaviors. In addition, animals must be handled repeatedly as they are introduced and removed from arenas; adding variability to the results. Primary data generated through manual scoring is also difficult to standardize and share between laboratories, as the data is generally coded in shorthand to increase the speed of analysis. Finally, it is challenging to reevaluate trials for additional behaviors at a later date when using recorded footage (and impossible to do with live scoring). For example, if a researcher was initially interested in animal movement rate but later wanted to reevaluate all of their data for clockwise or counterclockwise rotation within the arena, the process would likely require re-analysis of all primary data.

Automation has the potential to overcome these barriers. Using a combination of motion tracking cameras and computer software, a machine vision robot can track animals in real time; recording complex behavior (such as distance and location between conspecific animals, social behavior, grooming, courtship, aggression, movement in three-dimensional space, and predator responses, among others) as well as fine movements that would be impossible to evaluate by eye. Computer software is generally free of observer bias and can run continually for days, removing the need to constantly introduce and remove animals from experimental environments. Multiple animals can be tracked simultaneously, either through the use of multiple arenas under the field of view of a single camera, or by stitching together the fields of view of multiple cameras (potentially allowing for thousands of behavior trials to be run in parallel). Further, complicated controls are readily automated through software. For example, a yoked control involves delivering a stimulus (reward or punishment) to an animal not for its own behavior, but for the behavior of its paired neighbor. Using manual training by hand, even with only two individuals, these trials are challenging to say the least. However, computers can perform these controls with minimal computing power for many animals simultaneously, greatly increasing the throughput of cognitive data. One obtained, data can rapidly be parsed for multiple behaviors at a later date. In addition, log files or spreadsheets can be stored on collaborative servers allowing researchers to mine the results for their behaviors of interest; one laboratory may be investigating animal speed and locomotion, another color preference in the arena, and a final group in collisions with barriers within the environment: all three of these could be analyzed using separate algorithms applied to the same original log file.

A number of automated systems have been developed over the past decade and are currently used with organisms ranging from invertebrate to vertebrate. These systems fall into two distinct categories (Fig. 2); off-the-shelf turn-key units which run on premade software, and custom units typically designed for specific tasks (such as 3D motion, or to deliver current or vibration as a punishment). For the off-the-shelf units, two products see frequent use; Noldus Ethovision (Wageningen, Ne.) and Viewpoint Behavior Technology (Lyon, Fr.) [198, 199]. Both are similar in operation (although at the moment Noldus Ethovision delivers more flexibility) utilizing a single camera connected to a PC capture card and operated through a visual interface. Cameras are generally placed above the animal of interest, and multiple individuals can be tracked simultaneously within the field of view of the camera. Animals can be tracked as individuals (placed in separate wells of a 96-well plate) or as a group to examine social behaviors. The researcher can also outline a testing area through the live-feed, allowing any number of environments to be monitored; circular dishes, T-mazes, Y-tubes, or other shapes that can be seen by the camera. Data acquisition and analysis is automated, providing rapid summary of animal movement rates, color preferences, choices in T-mazes, or position within the arena. Indeed, these software packages have proved extremely powerful and have been used with vertebrates, especially zebrafish, to analyze a number of behaviors including: response to predators [200–202], drug addiction [203], sleep [204, 205], associative learning [206], social learning [207], and alcohol exposure [55, 208]. It is also important to note that while these systems have historically been used with aquatic vertebrates, the motion tracking algorithms could work equally well with single cells, birds, mice, and even humans. Very flexible software platforms, such as the JAABA system [209, 210], are coming on-line with increasingly sophisticated capabilities powered by machine learning and statistical mechanics algorithms.

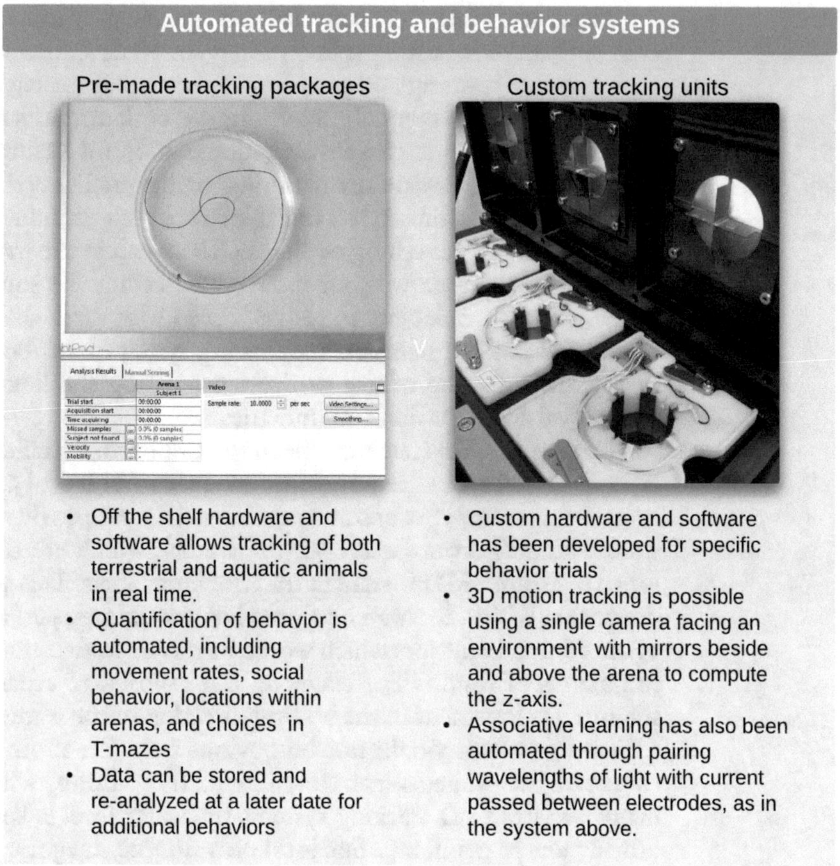

Automated tracking and behavior systems

Pre-made tracking packages

- Off the shelf hardware and software allows tracking of both terrestrial and aquatic animals in real time.
- Quantification of behavior is automated, including movement rates, social behavior, locations within arenas, and choices in T-mazes
- Data can be stored and re-analyzed at a later date for additional behaviors

Custom tracking units

- Custom hardware and software has been developed for specific behavior trials
- 3D motion tracking is possible using a single camera facing an environment with mirrors beside and above the arena to compute the z-axis.
- Associative learning has also been automated through pairing wavelengths of light with current passed between electrodes, as in the system above.

Fig. 2 Turnkey and custom analysis software for monitoring lateralized behaviors. A number of tracking and analysis hardware systems exist allowing researchers to define custom "arenas" in which behaviors including movement rates, escape responses, sociality, aggression, and fine movements can be quantified. Analysis can happen in real time or from pre-recorded video. Custom hardware and software also exists for specialized analysis that allows not only tracking, but real-time feedback (rewards and punishments) for the animals. Both types of systems allow for rapid and parallel analysis of multiple individuals, resulting in high-throughput behavioral testing of animals with altered genetics, physiology, or anatomy. The latter class of systems in addition facilitates training paradigms, which enables assays of learning and memory

While these systems have not been used extensively in laterality related research, the potential is immense. The direction animals turn when entering environments, in response to light cues, or when bypassing barriers, would not be difficult to set up using the pre-packaged software. In addition, it is possible to record direction vectors (indicating the angle at which the anterior of the animal is pointing in reference to the cameras position) that would allow researchers to automate trials examining the eye used to examine prey or familiar objects. Finally, fine movements can also be calculated, such as alternating left–right body arching while zebrafish fry and adults swim. Analysis of such fine scale motor movement may reveal yet undocumented lateralized behavior and act as new metrics for body and brain laterality studies.

However, while these software packages are powerful, there also exist limitations for their use in certain types of trials. Real-time feedback between tracking and punishment/reward in the environment is difficult to implement. In addition, the background subtraction algorithms used to track animals requires constant lighting conditions within the environment during of the trail. Further, while training multiple animals, it is difficult to deliver a stimulus (changes in light intensity, refreshing or altering chemicals in the media, etc.) to one animal/arena without indirectly affecting the surrounding animals; adding variability to results. Faced with these challenges, a number of custom behavior monitoring systems have been developed which may be of use to those examining the link between anatomical and behavioral asymmetries.

One of these systems has been developed to specifically track aquatic animals in three-dimensional space [211–214]. Using a single camera, mirrors beside and above the tank provide the software with XYZ coordinates for the animal, which are then combined to create a 3D movement track for individuals. This additional dimension within the water column has already revealed a number of interesting behaviors which would be missed in two-dimensional summaries of results. For example, fish exposed to ethanol show significant movement in the vertical direction of the water column, a direction which would not be obvious from the more standard overhead two-dimensional view [214]. In addition, when using many overhead 2D tracking systems the water level is kept intentionally low to maintain a fine focal plane during image acquisition. However, low water levels are fairly unnatural for aquatic animals in the wild and may in fact stress fish, frogs, and tadpoles; leading to variability or inconsistencies within the results. It is likely there is much more to be learned from three-dimensional tracking of animals and would be exciting to examine how laterality alterations may impact movement and behavior in 3D space.

An integrated motion tracking and feedback system has recently been developed (suitable for example for planaria, zebrafish, axolotl, and *Xenopus* tadpoles), which enables automated associative training assays using separate wavelengths of light and electric current [159–161]. As opposed to top-mounted camera units, this device uses under-mounted cameras; one for each of the individual arenas in the device. Illumination is provided from above by LEDs of different wavelengths and current can be applied to the media by six electrodes mounted around the edges of the circular arena. The power of this hardware and software is that shock can be delivered in real time to each of the experimental arenas independently in response to animal behavior; allowing for associative color learning assays to specific wavelengths of light. This particular device was used in the first study to demonstrate associative learning in *Xenopus* tadpoles [160] and has also been used to show that posteriorly located ectopic eyes confer vision to blinded animals [161]. In addition, this machine vision system was employed to explicitly test the effects of laterality inversions on tadpole cognitive ability; determining that randomization of visceral organs resulted in decreased learning compared to unaltered siblings [156].

8 Summary

Several vertebrate and invertebrate models are now routinely used to produce animals with altered laterality. A number of pharmacological, biophysical, and molecular-genetic tools exist to perturb normal asymmetry, and versatile assays are being developed to track not only the resulting neural structures but also the behavior. With continued work, even more subtle lateralized aspects of cognition will be discovered, and linked to molecular-genetic pathways for a fuller, multiscale understanding of asymmetry and its importance. Moving forward, automated devices such as those outlined above, as well as forthcoming next-generation motion tracking and real-time training systems, will be key tools for investigating the link between laterality and cognition. Such closed-loop platforms will overcome many of the barriers in the field; allowing multiple animals to be tracked simultaneously, long recording sessions leading to decreased animal handling, examination of fine motor movement possibly in three-dimensional space, rapid re-analysis of previous trials for lateralized behavior, and easy implementation of complicated controls. It is likely that a marriage of modern molecular functional techniques with robotics technology will uncover a wide variety of subtle behavioral changes related to left–right patterning and in turn give us new insight into the link between laterality and cognition.

Acknowledgements

We thank the members of the Levin lab and of the behavioral science community for many useful discussions. M.L. gratefully acknowledges an Allen Discovery Center award from The Paul G. Allen Frontiers Group, and support of the Templeton World Charity Foundation (TWCF0089/AB55) and the G. Harold and Leila Y. Mathers Charitable Foundation.

References

1. McManus C (2002) Right hand, left hand: the origins of asymmetry in brains, bodies, atoms and cultures. Weidenfeld and Nicolson, London
2. Ludwig W (1932) Rechts-Links-Problem im Tierreich und beim Menschen. Springer, Berlin
3. Neville A (1976) Animal asymmetry. Edward Arnold, London
4. Palmer AR (1996) From symmetry to asymmetry: phylogenetic patterns of asymmetry variation in animals and their evolutionary significance. Proc Natl Acad Sci U S A 93(25): 14279–14286

5. Palmer AR (2004) Symmetry breaking and the evolution of development. Science 306(5697):828–833
6. Klingenberg CP, McIntyre GS (1998) Geometric morphometrics of developmental instability: analyzing patterns of fluctuating asymmetry with Procrustes methods. Evolution 52:1363–1375
7. Govind CK (1992) Claw asymmetry in lobsters: case study in developmental neuroethology. J Neurobiol 23(10):1423–1445
8. Burn J (1991) Disturbance of morphological laterality in humans. CIBA Found Symp 162:282–296
9. Peeters H, Devriendt K (2006) Human laterality disorders. Eur J Med Genet 49(5): 349–362
10. Smith AT, Sack GH Jr, Taylor GJ (1979) Holt-Oram syndrome. J Pediatr 95(4): 538–543
11. Paulozzi LJ, Lary JM (1999) Laterality patterns in infants with external birth defects. Teratology 60(5):265–271
12. Sandson TA, Wen PY, LeMay M (1992) Reversed cerebral asymmetry in women with breast cancer. Lancet 339(8792):523–524
13. McManus IC (1992) Reversed cerebral asymmetry and breast cancer. Lancet 339(8800): 1055
14. Sotelo-Avila C, Gonzalez-Crussi F, Fowler JW (1980) Complete and incomplete forms of Beckwith-Wiedemann syndrome: their oncogenic potential. J Pediatr 96(1):47–50
15. Veltmaat JM, Ramsdell AF, Sterneck E (2013) Positional variations in mammary gland development and cancer. J Mammary Gland Biol Neoplasia 18(2):179–188
16. Neveu PJ (1993) Brain lateralization and immunomodulation. Int J Neurosci 70(1–2): 135–143
17. Neveu PJ (2002) Cerebral lateralization and the immune system. Int Rev Neurobiol 52:303–323
18. Klar AJ (1999) Genetic models for handedness, brain lateralization, schizophrenia, and manic-depression. Schizophr Res 39(3):207–218
19. Vallortigara G, Rogers LJ (2005) Survival with an asymmetrical brain: advantages and disadvantages of cerebral lateralization. Behav Brain Sci 28(4):575–589, discussion 589–633
20. McManus C (2005) Reversed bodies, reversed brains, and (some) reversed behaviors: of zebrafish and men. Dev Cell 8(6):796–797
21. Halpern ME et al (2005) Lateralization of the vertebrate brain: taking the side of model systems. J Neurosci 25(45):10351–10357
22. Frasnelli E, Vallortigara G, Rogers LJ (2012) Left-right asymmetries of behaviour and nervous system in invertebrates. Neurosci Biobehav Rev 36(4):1273–1291
23. Levin M et al (1995) A molecular pathway determining left-right asymmetry in chick embryogenesis. Cell 82(5):803–814
24. Brown NA, Wolpert L (1990) The development of handedness in left/right asymmetry. Development 109(1):1–9
25. Sauer S, Klar AJ (2012) Left-right symmetry breaking in mice by left-right dynein may occur via a biased chromatid segregation mechanism, without directly involving the Nodal gene. Front Oncol 2:166
26. Basu B, Brueckner M (2008) Cilia: multifunctional organelles at the center of vertebrate left-right asymmetry. Curr Top Dev Biol 85:151–174
27. Tee YH et al (2015) Cellular chirality arising from the self-organization of the actin cytoskeleton. Nat Cell Biol 17(4):445–457
28. Taniguchi K et al (2011) Chirality in planar cell shape contributes to left-right asymmetric epithelial morphogenesis. Science 333(6040): 339–341
29. Lobikin M et al (2012) Early, nonciliary role for microtubule proteins in left-right patterning is conserved across kingdoms. Proc Natl Acad Sci U S A 109(31):12586–12591
30. Toyoizumi R et al (1997) Adrenergic neurotransmitters and calcium ionophore-induced situs inversus viscerum in Xenopus laevis embryos. Dev Growth Differ 39(4): 505–514
31. Garic-Stankovic A et al (2008) A ryanodine receptor-dependent Cai2+ asymmetry at Hensen's node mediates avian lateral identity. Development 135(19):3271–3280
32. Fukumoto T, Blakely R, Levin M (2005) Serotonin transporter function is an early step in left-right patterning in chick and frog embryos. Dev Neurosci 27(6):349–363
33. Levin M et al (2002) Asymmetries in H+/K+-ATPase and cell membrane potentials comprise a very early step in left-right patterning. Cell 111(1):77–89
34. Raya A et al (2004) Notch activity acts as a sensor for extracellular calcium during vertebrate left-right determination. Nature 427(6970):121–128
35. Levin M (1998) Left-right asymmetry and the chick embryo. Semin Cell Dev Biol 9(1):67–76
36. Muller P et al (2012) Differential diffusivity of Nodal and Lefty underlies a reaction-diffusion patterning system. Science 336(6082):721–724

37. Kato Y (2011) The multiple roles of Notch signaling during left-right patterning. Cell Mol Life Sci 68(15):2555–2567
38. Ramasubramanian A et al (2013) On the role of intrinsic and extrinsic forces in early cardiac S-looping. Dev Dyn 242(7):801–816
39. Horne-Badovinac S, Rebagliati M, Stainier DY (2003) A cellular framework for gut-looping morphogenesis in zebrafish. Science 302(5645):662–665
40. Raya A, Belmonte JC (2006) Left-right asymmetry in the vertebrate embryo: from early information to higher-level integration. Nat Rev Genet 7(4):283–293
41. Levin M (2006) Is the early left-right axis like a plant, a kidney, or a neuron? The integration of physiological signals in embryonic asymmetry. Birth Defects Res C Embryo Today 78(3):191–223
42. Branford WW, Essner JJ, Yost HJ (2000) Regulation of gut and heart left-right asymmetry by context-dependent interactions between xenopus lefty and BMP4 signaling. Dev Biol 223(2):291–306
43. Tabin C (2005) Do we know anything about how left-right asymmetry is first established in the vertebrate embryo? J Mol Histol 36(5):317–323
44. Vandenberg LN, Levin M (2010) Far from solved: a perspective on what we know about early mechanisms of left-right asymmetry. Dev Dyn 239(12):3131–3146
45. Vandenberg LN, Levin M (2009) Perspectives and open problems in the early phases of left-right patterning. Semin Cell Dev Biol 20(4):456–463
46. Aw S, Levin M (2008) What's left in asymmetry? Dev Dyn 237(12):3453–3463
47. Ahmad N, Long S, Rebagliati M (2004) A southpaw joins the roster: the role of the zebrafish nodal-related gene southpaw in cardiac LR asymmetry. Trends Cardiovasc Med 14(2):43–49
48. Halpern ME, Liang JO, Gamse JT (2003) Leaning to the left: laterality in the zebrafish forebrain. Trends Neurosci 26(6):308–313
49. Baier H (2000) Zebrafish on the move: towards a behavior-genetic analysis of vertebrate vision. Curr Opin Neurobiol 10(4):451–455
50. Buske C, Gerlai R (2011) Shoaling develops with age in Zebrafish (Danio rerio). Prog Neuropsychopharmacol Biol Psychiatry 35(6):1409–1415
51. Darland T, Dowling JE (2001) Behavioral screening for cocaine sensitivity in mutagenized zebrafish. Proc Natl Acad Sci U S A 98(20):11691–11696
52. Engeszer RE, Ryan MJ, Parichy DM (2004) Learned social preference in zebrafish. Curr Biol 14(10):881–884
53. Fetcho JR, Liu KS (1998) Zebrafish as a model system for studying neuronal circuits and behavior. Ann N Y Acad Sci 860:333–345
54. Gerlai R (2010) High-throughput behavioral screens: the first step towards finding genes involved in vertebrate brain function using zebrafish. Molecules 15(4):2609–2622
55. Gerlai R, Lee V, Blaser R (2006) Effects of acute and chronic ethanol exposure on the behavior of adult zebrafish (Danio rerio). Pharmacol Biochem Behav 85(4):752–761
56. Goldsmith P (2001) Modelling eye diseases in zebrafish. Neuroreport 12(13):A73–A77
57. Pan Y et al (2011) Chronic alcohol exposure induced gene expression changes in the zebrafish brain. Behav Brain Res 216(1):66–76
58. Guo S (2004) Linking genes to brain, behavior and neurological diseases: what can we learn from zebrafish? Genes Brain Behav 3(2):63–74
59. Hruscha A et al (2013) Efficient CRISPR/Cas9 genome editing with low off-target effects in zebrafish. Development 140(24):4982–4987
60. Hwang WY et al (2013) Efficient genome editing in zebrafish using a CRISPR-Cas system. Nat Biotechnol 31(3):227–229
61. Bernstein JG, Garrity PA, Boyden ES (2012) Optogenetics and thermogenetics: technologies for controlling the activity of targeted cells within intact neural circuits. Curr Opin Neurobiol 22(1):61–71
62. Knopfel T et al (2010) Toward the second generation of optogenetic tools. J Neurosci 30(45):14998–15004
63. Portugues R et al (2013) Optogenetics in a transparent animal: circuit function in the larval zebrafish. Curr Opin Neurobiol 23(1):119–126
64. Simmich J, Staykov E, Scott E (2012) Zebrafish as an appealing model for optogenetic studies. Prog Brain Res 196:145–162
65. Del Bene F, Wyart C (2012) Optogenetics: a new enlightenment age for zebrafish neurobiology. Dev Neurobiol 72(3):404–414
66. Wyart C, Del Bene F (2011) Let there be light: zebrafish neurobiology and the optogenetic revolution. Rev Neurosci 22(1):121–130
67. Friedrich RW, Jacobson GA, Zhu P (2010) Circuit neuroscience in zebrafish. Curr Biol 20(8):R371–R381
68. Kupffer C (1868) Beobachtungea uber die Entwicklung der Knochenfische. Arch Mikrob Anat 4:209–272

69. Bisgrove BW et al (2005) Polaris and Polycystin-2 in dorsal forerunner cells and Kupffer's vesicle are required for specification of the zebrafish left-right axis. Dev Biol 287(2):274–288
70. Essner J et al (2002) Conserved function for embryonic nodal cilia. Nature 418:37–38
71. Essner JJ et al (2005) Kupffer's vesicle is a ciliated organ of asymmetry in the zebrafish embryo that initiates left-right development of the brain, heart and gut. Development 132(6):1247–1260
72. Amack JD, Yost HJ (2004) The T box transcription factor no tail in ciliated cells controls zebrafish left-right asymmetry. Curr Biol 14(8):685–690
73. Kramer-Zucker AG et al (2005) Cilia-driven fluid flow in the zebrafish pronephros, brain and Kupffer's vesicle is required for normal organogenesis. Development 132(8): 1907–1921
74. Amack JD, Wang X, Yost HJ (2007) Two T-box genes play independent and cooperative roles to regulate morphogenesis of ciliated Kupffer's vesicle in zebrafish. Dev Biol 310(2):196–210
75. Becker-Heck A et al (2011) The coiled-coil domain containing protein CCDC40 is essential for motile cilia function and left-right axis formation. Nat Genet 43(1): 79-U105
76. Francescatto L et al (2010) The activation of membrane targeted CaMK-II in the zebrafish Kupffer's vesicle is required for left-right asymmetry. Development 137(16):2753–2762
77. Wang GL et al (2011) The Rho kinase Rock2b establishes anteroposterior asymmetry of the ciliated Kupffer's vesicle in zebrafish. Development 138(1):45–54
78. Hamada H et al (2002) Establishment of vertebrate left-right asymmetry. Nat Rev Genet 3(2):103–113
79. Ramsdell AF, Yost HJ (1998) Molecular mechanisms of vertebrate left-right development. Trends Genet 14(11):459–465
80. Lenhart KF et al (2013) Integration of nodal and BMP signals in the heart requires FoxH1 to create left-right differences in cell migration rates that direct cardiac asymmetry. PLoS Genet 9(1):e1003109
81. Lin SY, Burdine RD (2005) Brain asymmetry: switching from left to right. Curr Biol 15(9):R343–R345
82. Concha ML et al (2000) A nodal signaling pathway regulates the laterality of neuroanatomical asymmetries in the zebrafish forebrain. Neuron 28(2):399–409
83. Bamford RN et al (2000) Loss-of-function mutations in the EGF-CFC gene CFC1 are associated with human left-right laterality defects. Nat Genet 26(3):365–369
84. Yan YT et al (1999) Conserved requirement for EGF-CFC genes in vertebrate left-right axis formation. Genes Dev 13(19):2527–2537
85. Bisgrove BW, Essner JJ, Yost HJ (2000) Multiple pathways in the midline regulate concordant brain, heart and gut left-right asymmetry. Development 127(16):3567–3579
86. Gamse JT et al (2003) The parapineal mediates left-right asymmetry in the zebrafish diencephalon. Development 130(6):1059–1068
87. Aizawa H et al (2005) Laterotopic representation of left-right information onto the dorso-ventral axis of a zebrafish midbrain target nucleus. Curr Biol 15(3):238–243
88. Gamse JT et al (2005) Directional asymmetry of the zebrafish epithalamus guides dorsoventral innervation of the midbrain target. Development 132(21):4869–4881
89. Amat J et al (2001) The role of the habenular complex in the elevation of dorsal raphe nucleus serotonin and the changes in the behavioral responses produced by uncontrollable stress. Brain Res 917(1):118–126
90. Haun F, Eckenrode TC, Murray M (1992) Habenula and thalamus cell transplants restore normal sleep behaviors disrupted by denervation of the interpeduncular nucleus. J Neurosci 12(8):3282–3290
91. Lecourtier L, Kelly PH (2005) Bilateral lesions of the habenula induce attentional disturbances in rats. Neuropsychopharmacology 30(3):484–496
92. Lecourtier L, Neijt HC, Kelly PH (2004) Habenula lesions cause impaired cognitive performance in rats: implications for schizophrenia. Eur J Neurosci 19(9):2551–2560
93. Murphy CA et al (1996) Lesion of the habenular efferent pathway produces anxiety and locomotor hyperactivity in rats: a comparison of the effects of neonatal and adult lesions. Behav Brain Res 81(1–2):43–52
94. Valjakka A et al (1998) The fasciculus retroflexus controls the integrity of REM sleep by supporting the generation of hippocampal theta rhythm and rapid eye movements in rats. Brain Res Bull 47(2):171–184
95. Barth KA et al (2005) fsi zebrafish show concordant reversal of laterality of viscera, neuroanatomy, and a subset of behavioral responses. Curr Biol 15(9):844–850
96. Miklosi A, Andrew RJ (2006) The zebrafish as a model for behavioral studies. Zebrafish 3(2):227–234

97. Sovrano VA et al (1999) Roots of brain specializations: preferential left-eye use during mirror-image inspection in six species of teleost fish. Behav Brain Res 106(1–2):175–180
98. Bisazza A, Pignatti R, Vallortigara G (1997) Laterality in detour behaviour: interspecific variation in poeciliid fish. Anim Behav 54(5):1273–1281
99. Facchin L, Bisazza A, Vallortigara G (1999) What causes lateralization of detour behavior in fish? Evidence for asymmetries in eye use. Behav Brain Res 103(2):229–234
100. Miklosi A, Andrew RJ (1999) Right eye use associated with decision to bite in zebrafish. Behav Brain Res 105(2):199–205
101. Miklosi A, Andrew RJ, Gasparini S (2001) Role of right hemifield in visual control of approach to target in zebrafish. Behav Brain Res 122(1):57–65
102. Vallortigara G et al (2001) How birds use their eyes: opposite left-right specialization for the lateral and frontal visual hemifield in the domestic chick. Curr Biol 11(1):29–33
103. Rogers LJ, Vallortigara G, Andrew RJ (2013) Divided brains: the biology and behaviour of brain asymmetries. Cambridge University Press, Cambridge
104. Gunturkun O, Kesch S (1987) Visual lateralization during feeding in pigeons. Behav Neurosci 101(3):433–435
105. Robins A et al (1998) Lateralized agonistic responses and hindlimb use in toads. Anim Behav 56:875–881
106. Yaman S et al (2003) Visual lateralization in the bottlenose dolphin (Tursiops truncatus): evidence for a population asymmetry? Behav Brain Res 142(1–2):109–114
107. Miklosi A, Andrew RJ, Savage H (1997) Behavioural lateralisation of the tetrapod type in the zebrafish (Brachydanio rerio). Physiol Behav 63(1):127–135
108. Sovrano VA (2004) Visual lateralization in response to familiar and unfamiliar stimuli in fish. Behav Brain Res 152(2):385–391
109. Dadda M et al (2010) Early differences in epithalamic left-right asymmetry influence lateralization and personality of adult zebrafish. Behav Brain Res 206(2):208–215
110. Burggren WW, Warburton S (2007) Amphibians as animal models for laboratory research in physiology. Ilar J 48(3):260–269
111. Beck CW, Slack JM (2001) An amphibian with ambition: a new role for Xenopus in the 21st century. Genome Biol 2(10):REVIEWS1029
112. Mouche I, Malesic L, Gillardeaux O (2011) FETAX assay for evaluation of developmental toxicity. Methods Mol Biol 691:257–269
113. Pratt KG, Khakhalin AS (2013) Modeling human neurodevelopmental disorders in the Xenopus tadpole: from mechanisms to therapeutic targets. Dis Model Mech 6(5): 1057–1065
114. Beck CW, Izpisua Belmonte JC, Christen B (2009) Beyond early development: Xenopus as an emerging model for the study of regenerative mechanisms. Dev Dyn 238(6): 1226–1248
115. Tseng AS, Levin M (2008) Tail regeneration in Xenopus laevis as a model for understanding tissue repair. J Dent Res 87(9):806–816
116. Gibbs KM, Chittur SV, Szaro BG (2011) Metamorphosis and the regenerative capacity of spinal cord axons in Xenopus laevis. Eur J Neurosci 33(1):9–25
117. Lee-Liu D et al (2014) Genome-wide expression profile of the response to spinal cord injury in Xenopus laevis reveals extensive differences between regenerative and non-regenerative stages. Neural Dev 9:12
118. Koide T, Hayata T, Cho KW (2005) Xenopus as a model system to study transcriptional regulatory networks. Proc Natl Acad Sci U S A 102(14):4943–4948
119. Ruben LN, Clothier RH, Balls M (2007) Cancer resistance in amphibians. Altern Lab Anim 35(5):463–470
120. Lobikin M et al (2012) Resting potential, oncogene-induced tumorigenesis, and metastasis: the bioelectric basis of cancer in vivo. Phys Biol 9(6):065002
121. Robert J, Cohen N (2011) The genus Xenopus as a multispecies model for evolutionary and comparative immunobiology of the 21st century. Dev Comp Immunol 35(9):916–923
122. Robert J, Ohta Y (2009) Comparative and developmental study of the immune system in Xenopus. Dev Dyn 238(6):1249–1270
123. Kinney KS, Cohen N (2009) Neural-immune system interactions in Xenopus. Front Biosci 14:112–129
124. Callery EM (2006) There's more than one frog in the pond: a survey of the Amphibia and their contributions to developmental biology. Semin Cell Dev Biol 17(1):80–92
125. Vandenberg LN, Lemire JM, Levin M (2013) It's never too early to get it right: a conserved role for the cytoskeleton in left-right asymmetry. Commun Integr Biol 6(6):e27155
126. Schweickert A et al (2012) Linking early determinants and cilia-driven leftward flow in left-right axis specification of Xenopus laevis: a theoretical approach. Differentiation 83(2): S67–S77

127. Blum M et al (2009) Xenopus, an ideal model system to study vertebrate left-right asymmetry. Dev Dyn 238(6):1215–1225
128. Yost HJ (1991) Development of the left-right axis in amphibians. Ciba Found Symp 162:165–176, discussion 176–181
129. Yost HJ (1990) Inhibition of proteoglycan synthesis eliminates left-right asymmetry in Xenopus laevis cardiac looping. Development 110(3):865–874
130. Adams DS, Levin M (2006) Inverse drug screens: a rapid and inexpensive method for implicating molecular targets. Genesis 44(11):530–540
131. Dush MK et al (2011) Heterotaxin: a TGF-beta signaling inhibitor identified in a multi-phenotype profiling screen in Xenopus embryos. Chem Biol 18(2):252–263
132. Wheeler GN, Liu KJ (2012) Xenopus: an ideal system for chemical genetics. Genesis 50(3):207–218
133. Wheeler GN, Brandli AW (2009) Simple vertebrate models for chemical genetics and drug discovery screens: lessons from zebrafish and Xenopus. Dev Dyn 238(6):1287–1308
134. Sampath K et al (1997) Functional differences among Xenopus nodal-related genes in left-right axis determination. Development 124(17):3293–3302
135. Cheng AM et al (2000) The lefty-related factor Xatv acts as a feedback inhibitor of nodal signaling in mesoderm induction and L-R axis development in xenopus. Development 127(5):1049–1061
136. Vandenberg LN, Levin M (2010) Consistent left-right asymmetry cannot be established by late organizers in Xenopus unless the late organizer is a conjoined twin. Development 137(7):1095–1105
137. Levin M, Mercola M (1998) Gap junctions are involved in the early generation of left-right asymmetry. Dev Biol 203(1):90–105
138. Vandenberg LN, Lemire JM, Levin M (2013) Serotonin has early, cilia-independent roles in Xenopus left-right patterning. Dis Model Mech 6(1):261–268
139. Schweickert A et al (2007) Cilia-driven leftward flow determines laterality in Xenopus. Curr Biol 17(1):60–66
140. Adams DS et al (2006) Early, H+-V-ATPase-dependent proton flux is necessary for consistent left-right patterning of non-mammalian vertebrates. Development 133(9):1657–1671
141. Aw S et al (2008) H,K-ATPase protein localization and Kir4.1 function reveal concordance of three axes during early determination of left-right asymmetry. Mech Dev 125(3–4): 353–372
142. Morokuma J, Blackiston D, Levin M (2008) KCNQ1 and KCNE1 K+ channel components are involved in early left-right patterning in Xenopus laevis embryos. Cell Physiol Biochem 21(5–6):357–372
143. Aw S et al (2010) The ATP-sensitive K(+)-channel (K(ATP)) controls early left-right patterning in Xenopus and chick embryos. Dev Biol 346:39–53
144. Carneiro K et al (2011) Histone deacetylase activity is necessary for left-right patterning during vertebrate development. BMC Dev Biol 11(1):29
145. Vandenberg LN et al (2013) Rab GTPases are required for early orientation of the left-right axis in Xenopus. Mech Dev 130:254–271
146. Aw S, Levin M (2009) Is left-right asymmetry a form of planar cell polarity? Development 136(3):355–366
147. Levin M, Palmer AR (2007) Left-right patterning from the inside out: widespread evidence for intracellular control. Bioessays 29(3):271–287
148. Bunney TD, De Boer AH, Levin M (2003) Fusicoccin signaling reveals 14-3-3 protein function as a novel step in left-right patterning during amphibian embryogenesis. Development 130(20):4847–4858
149. Fukumoto T, Kema IP, Levin M (2005) Serotonin signaling is a very early step in patterning of the left-right axis in chick and frog embryos. Curr Biol 15(9):794–803
150. Qiu D et al (2005) Localization and loss-of-function implicates ciliary proteins in early, cytoplasmic roles in left-right asymmetry. Dev Dyn 234(1):176–189
151. Vandenberg LN, Pennarola BW, Levin M (2011) Low frequency vibrations disrupt left-right patterning in the Xenopus embryo. PLoS One 6(8):e23306
152. Vandenberg LN, Stevenson C, Levin M (2012) Low frequency vibrations induce malformations in two aquatic species in a frequency-, waveform-, and direction-specific manner. PLoS One 7(12):e51473
153. Robins A, Rogers LJ (2006) Lateralized visual and motor responses in the green tree frog, *Litoria caerulea*. Anim Behav 72: 843–852
154. Bisazza A et al (1996) Right-pawedness in toads. Nature 379(6564):408
155. Bisazza A et al (1997) Pawedness and motor asymmetries in toads. Laterality 2(1):49–64
156. Blackiston DJ, Levin M (2013) Inversion of left-right asymmetry alters performance of

Xenopus tadpoles in nonlateralized cognitive tasks. Anim Behav 86(2):459–466
157. Mcgill TE (1960) Response of the leopard frog to electric shock in an escape-learning situation. J Comp Physiol Psychol 53(4): 443–445
158. Thompson PA, Boice R (1975) Attempts to train frogs—review and experiments. J Biol Psychol 17(1):3–13
159. Blackiston D et al (2010) A second-generation device for automated training and quantitative behavior analyses of molecularly-tractable model organisms. PLoS One 5(12):e14370
160. Blackiston DJ, Levin M (2012) Aversive training methods in Xenopus laevis: general principles. Cold Spring Harb Protoc 2012(5)
161. Blackiston DJ, Levin M (2013) Ectopic eyes outside the head in Xenopus tadpoles provide sensory data for light-mediated learning. J Exp Biol 216(Pt 6):1031–1040
162. James EJ et al (2015) Valproate-induced neurodevelopmental deficits in Xenopus laevis tadpoles. J Neurosci 35(7):3218–3229
163. Khakhalin AS et al (2014) Excitation and inhibition in recurrent networks mediate collision avoidance in Xenopus tadpoles. Eur J Neurosci 40(6):2948–2962
164. Spawn A, Aizenman CD (2012) Abnormal visual processing and increased seizure susceptibility result from developmental exposure to the biocide methylisothiazolinone. Neuroscience 205:194–204
165. Bell MR et al (2011) A neuroprotective role for polyamines in a Xenopus tadpole model of epilepsy. Nat Neurosci 14(4):505–512
166. Dong W et al (2009) Visual avoidance in Xenopus tadpoles is correlated with the maturation of visual responses in the optic tectum. J Neurophysiol 101(2):803–815
167. Pai VP et al (2012) Neurally derived tissues in Xenopus laevis embryos exhibit a consistent bioelectrical left-right asymmetry. Stem Cells Int 2012:353491
168. Gros J et al (2009) Cell movements at Hensen's node establish left/right asymmetric gene expression in the chick. Science 324(5929):941–944
169. Levin M et al (1996) Laterality defects in conjoined twins. Nature 384(6607):321
170. Raya A, Belmonte JCI (2004) Unveiling the establishment of left-right asymmetry in the chick embryo. Mech Dev 121(9):1043–1054
171. Monsoro-Burq A, Le Douarin NM (2001) BMP4 plays a key role in left-right patterning in chick embryos by maintaining Sonic Hedgehog asymmetry. Mol Cell 7(4): 789–799
172. Manner J (2001) Does an equivalent of the "ventral node" exist in chick embryos? A scanning electron microscopic study. Anat Embryol 203(6):481–490
173. Zhang Y, Levin M (2009) Left-right asymmetry in the chick embryo requires core planar cell polarity protein Vangl2. Genesis 47(11):719–728
174. Raya A et al (2003) Notch activity induces Nodal expression and mediates the establishment of left-right asymmetry in vertebrate embryos. Genes Dev 17(10):1213–1218
175. Hibino T et al (2006) Ion flow regulates left-right asymmetry in sea urchin development. Dev Genes Evol 216(5):265–276
176. Shimeld SM, Levin M (2006) Evidence for the regulation of left-right asymmetry in Ciona intestinalis by ion flux. Dev Dyn 235(6):1543–1553
177. Dharmaretnam M, Andrew RJ (1994) Age-specific and stimulus-specific use of right and left eyes by the domestic chick. Anim Behav 48(6):1395–1406
178. Vallortigara G, Regolin L, Pagni P (1999) Detour behaviour, imprinting and visual lateralization in the domestic chick. Cogn Brain Res 7(3):307–320
179. Rogers LJ (2000) Evolution of hemispheric specialization: advantages and disadvantages. Brain Lang 73(2):236–253
180. Rogers LJ (2008) Development and function of lateralization in the avian brain. Brain Res Bull 76(3):235–244
181. Okumura T et al (2008) The development and evolution of left-right asymmetry in invertebrates: lessons from Drosophila and snails. Dev Dyn 237(12):3497–3515
182. Speder P et al (2007) Strategies to establish left/right asymmetry in vertebrates and invertebrates. Curr Opin Genet Dev 17(4): 351–358
183. Xu J et al (2007) Polarity reveals intrinsic cell chirality. Proc Natl Acad Sci U S A 104(22): 9296–9300
184. Chen TH et al (2012) Left-right symmetry breaking in tissue morphogenesis via cytoskeletal mechanics. Circ Res 110(4):551–559
185. Tamada A et al (2010) Autonomous right-screw rotation of growth cone filopodia drives neurite turning. J Cell Biol 188(3): 429–441
186. Heacock AM, Agranoff BW (1977) Clockwise growth of neurites from retinal explants. Science 198(4312):64–66
187. Wan LQ, Vunjak-Novakovic G (2011) Micropatterning chiral morphogenesis. Commun Integr Biol 4(6):745–748

188. Wan LQ et al (2011) Micropatterned mammalian cells exhibit phenotype-specific left-right asymmetry. Proc Natl Acad Sci U S A 108(30):12295–12300
189. Aufderheide KJ, Frankel J, Williams NE (1980) Formation and positioning of surface-related structures in protozoa. Microbiol Rev 44(2):252–302
190. Geimer S, Melkonian M (2004) The ultrastructure of the Chlamydomonas reinhardtii basal apparatus: identification of an early marker of radial asymmetry inherent in the basal body. J Cell Sci 117(Pt 13): 2663–2674
191. Munoz-Nortes T et al (2014) Symmetry, asymmetry, and the cell cycle in plants: known knowns and some known unknowns. J Exp Bot 65(10):2645–2655
192. Abe T, Thitamadee S, Hashimoto T (2004) Microtubule defects and cell morphogenesis in the lefty1lefty2 tubulin mutant of Arabidopsis thaliana. Plant Cell Physiol 45(2):211–220
193. Costa MM et al (2005) Evolution of regulatory interactions controlling floral asymmetry. Development 132(22):5093–5101
194. Hashimoto T (2002) Molecular genetic analysis of left-right handedness in plants. Philos Trans R Soc Lond B Biol Sci 357(1422):799–808
195. Henley CL (2012) Possible origins of macroscopic left-right asymmetry in organisms. J Stat Phys 148(4):740–774
196. Thitamadee S, Tuchihara K, Hashimoto T (2002) Microtubule basis for left-handed helical growth in Arabidopsis. Nature 417(6885):193–196
197. Nakamura M, Hashimoto T (2009) A mutation in the Arabidopsis gamma-tubulin-containing complex causes helical growth and abnormal microtubule branching. J Cell Sci 122(Pt 13):2208–2217
198. Blaser R, Gerlai R (2006) Behavioral phenotyping in zebrafish: comparison of three behavioral quantification methods. Behav Res Methods 38(3):456–469
199. Delcourt J et al (2006) Comparing the EthoVision 2.3 system and a new computerized multitracking prototype system to measure the swimming behavior in fry fish. Behav Res Methods 38(4):704–710
200. Bass SL, Gerlai R (2008) Zebrafish (Danio rerio) responds differentially to stimulus fish: the effects of sympatric and allopatric predators and harmless fish. Behav Brain Res 186(1):107–117
201. Gerlai R, Fernandes Y, Pereira T (2009) Zebrafish (Danio rerio) responds to the animated image of a predator: towards the development of an automated aversive task. Behav Brain Res 201(2):318–324
202. Speedie N, Gerlai R (2008) Alarm substance induced behavioral responses in zebrafish (Danio rerio). Behav Brain Res 188(1): 168–177
203. Lopez Patino MA et al (2008) Gender differences in zebrafish responses to cocaine withdrawal. Physiol Behav 95(1–2):36–47
204. Prober DA et al (2006) Hypocretin/orexin overexpression induces an insomnia-like phenotype in zebrafish. J Neurosci 26(51): 13400–13410
205. Zhdanova IV et al (2008) Aging of the circadian system in zebrafish and the effects of melatonin on sleep and cognitive performance. Brain Res Bull 75(2–4):433–441
206. Sison M, Gerlai R (2010) Associative learning in zebrafish (Danio rerio) in the plus maze. Behav Brain Res 207(1):99–104
207. Al-Imari L, Gerlai R (2008) Sight of conspecifics as reward in associative learning in zebrafish (Danio rerio). Behav Brain Res 189(1):216–219
208. Gerlai R et al (2000) Drinks like a fish: zebra fish (Danio rerio) as a behavior genetic model to study alcohol effects. Pharmacol Biochem Behav 67(4):773–782
209. Berman GJ et al (2014) Mapping the stereotyped behaviour of freely moving fruit flies. J R Soc Interface 11(99)
210. Kabra M et al (2013) JAABA: interactive machine learning for automatic annotation of animal behavior. Nat Methods 10(1):64–67
211. Maaswinkel H et al (2013) Dissociating the effects of habituation, black walls, buspirone and ethanol on anxiety-like behavioral responses in shoaling zebrafish. A 3D approach to social behavior. Pharmacol Biochem Behav 108:16–27
212. Maaswinkel H, Zhu L, Weng W (2012) The immediate and the delayed effects of buspirone on zebrafish (Danio rerio) in an open field test: a 3-D approach. Behav Brain Res 234(2):365–374
213. Maaswinkel H, Zhu LQ, Weng W (2013) Using an automated 3D-tracking system to record individual and shoals of adult zebrafish. J Vis Exp (82): 50681
214. Zhu L, Weng W (2007) Catadioptric stereovision system for the real-time monitoring of 3D behavior in aquatic animals. Physiol Behav 91(1):106–119

ERRATUM

Lateralized Brain Functions: Methods in Human and Non-Human Species

Lesley J. Rogers and Giorgio Vallortigara

Lesley J. Rogers and Giorgio Vallortigara (eds.), *Lateralized Brain Functions: Methods in Human and Non-Human Species*, Neuromethods, vol. 122, DOI 10.1007/978-1-4939-6725-4, © Springer Science+Business Media LLC 2017

DOI 10.1007/978-1-4939-6725-4_22

In the original version of this book, the affiliations for Nicoletta Foschi and Gabriele Polonara, authors of Chapter 2, were published erroneously. The correct affiliations are as below:

Nicoletta Foschi: Centro Epilessia, Clinica di Neurologia, Azienda Ospedaliera-Universitaria Umberto I, Ancona, Italy

Gabriele Polonara: Dipartimento di Scienze Cliniche Specialistiche ed Odontostomatologiche, Sezione di Scienze Radiologiche, Università Politecnica delle Marche, Ancona, Italy

The updated original online version for this book can be found at http://dx.doi.org/10.1007/978-1-4939-6725-4

Lesley J. Rogers and Giorgio Vallortigara (eds.), *Lateralized Brain Functions: Methods in Human and Non-Human Species*, Neuromethods, vol. 122, DOI 10.1007/978-1-4939-6725-4_22, © Springer Science+Business Media LLC 2017

INDEX

A

Lesley J. Rogers and Giorgio Vallortigara (eds.), *Lateralized Brain Functions: Methods in Human and Non-Human Species*, Neuromethods, vol. 122, DOI 10.1007/978-1-4939-6725-4, © Springer Science+Business Media LLC 2017

B

C

D

E

F

G

H

I

M

O

P

Q

R

T

U

V

W

X

Y

Z

Printed in the [illegible] tes
By Bookmast [illegible]